T0137170

Lecture Notes in Computer Science 13625

The series Lecture Notes in Computer Science (LNCS), including its subseries Lecture Notes in Artificial Intelligence (LNAI) and Lecture Notes in Bioinformatics (LNBI), has established itself as a medium for the publication of new developments in computer science and information technology research, teaching, and education.

LNCS enjoys close cooperation with the computer science R & D community, the series counts many renowned academics among its volume editors and paper authors, and collaborates with prestigious societies. Its mission is to serve this international community by providing an invaluable service, mainly focused on the publication of conference and workshop proceedings and postproceedings. LNCS commenced publication in 1973.

Mohammad Tanveer · Sonali Agarwal ·
Seiichi Ozawa · Asif Ekbal · Adam Jatowt
Editors

Neural Information Processing

29th International Conference, ICONIP 2022
Virtual Event, November 22–26, 2022
Proceedings, Part III

 Springer

Editors
Mohammad Tanveer
Indian Institute of Technology Indore
Indore, India

Seiichi Ozawa
Kobe University
Kobe, Japan

Adam Jatowt
University of Innsbruck
Innsbruck, Austria

Sonali Agarwal 🆔
Indian Institute of Information Technology -
Allahabad
Prayagraj, India

Asif Ekbal
Indian Institute of Technology Patna
Patna, India

ISSN 0302-9743   ISSN 1611-3349 (electronic)
Lecture Notes in Computer Science
ISBN 978-3-031-30110-0   ISBN 978-3-031-30111-7 (eBook)
https://doi.org/10.1007/978-3-031-30111-7

This Springer imprint is published by the registered company Springer Nature Switzerland AG
The registered company address is: Gewerbestrasse 11, 6330 Cham, Switzerland

Preface

Welcome to the proceedings of the 29th International Conference on Neural Information Processing (ICONIP 2022) of the Asia-Pacific Neural Network Society (APNNS), held virtually from Indore, India, during November 22–26, 2022.

The mission of the Asia-Pacific Neural Network Society is to promote active interactions among researchers, scientists, and industry professionals who are working in neural networks and related fields in the Asia-Pacific region. APNNS has Governing Board Members from 13 countries/regions – Australia, China, Hong Kong, India, Japan, Malaysia, New Zealand, Singapore, South Korea, Qatar, Taiwan, Thailand, and Turkey. The society's flagship annual conference is the International Conference of Neural Information Processing (ICONIP).

The ICONIP conference aims to provide a leading international forum for researchers, scientists, and industry professionals who are working in neuroscience, neural networks, deep learning, and related fields to share their new ideas, progress, and achievements. Due to the current situation regarding the pandemic and international travel, ICONIP 2022, which was planned to be held in New Delhi, India, was organized as a fully virtual conference.

The proceedings of ICONIP 2022 consists of a multi-volume set in LNCS and CCIS, which includes 146 and 213 papers, respectively, selected from 1003 submissions reflecting the increasingly high quality of research in neural networks and related areas. The conference focused on four main areas, i.e., "Theory and Algorithms," "Cognitive Neurosciences," "Human Centered Computing," and "Applications." The conference also had special sessions in 12 niche areas, namely

1. International Workshop on Artificial Intelligence and Cyber Security (AICS)
2. Computationally Intelligent Techniques in Processing and Analysis of Neuronal Information (PANI)
3. Learning with Fewer Labels in Medical Computing (FMC)
4. Computational Intelligence for Biomedical Image Analysis (BIA)
5. Optimized AI Models with Interpretability, Security, and Uncertainty Estimation in Healthcare (OAI)
6. Advances in Deep Learning for Biometrics and Forensics (ADBF)
7. Machine Learning for Decision-Making in Healthcare: Challenges and Opportunities (MDH)
8. Reliable, Robust and Secure Machine Learning Algorithms (RRS)
9. Evolutionary Machine Learning Technologies in Healthcare (EMLH)
10. High Performance Computing Based Scalable Machine Learning Techniques for Big Data and Their Applications (HPCML)
11. Intelligent Transportation Analytics (ITA)
12. Deep Learning and Security Techniques for Secure Video Processing (DLST)

Our great appreciation goes to the Program Committee members and the reviewers who devoted their time and effort to our rigorous peer-review process. Their insightful reviews and timely feedback ensured the high quality of the papers accepted for publication.

The submitted papers in the main conference and special sessions were reviewed following the same process, and we ensured that every paper has at least two high-quality single-blind reviews. The PC Chairs discussed the reviews of every paper very meticulously before making a final decision. Finally, thank you to all the authors of papers, presenters, and participants, which made the conference a grand success. Your support and engagement made it all worthwhile.

December 2022

Mohammad Tanveer
Sonali Agarwal
Seiichi Ozawa
Asif Ekbal
Adam Jatowt

Organization

Program Committee

General Chairs

M. Tanveer — Indian Institute of Technology Indore, India
Sonali Agarwal — IIIT Allahabad, India
Seiichi Ozawa — Kobe University, Japan

Honorary Chairs

Jonathan Chan — King Mongkut's University of Technology Thonburi, Thailand
P. N. Suganthan — Nanyang Technological University, Singapore

Program Chairs

Asif Ekbal — Indian Institute of Technology Patna, India
Adam Jatowt — University of Innsbruck, Austria

Technical Chairs

Shandar Ahmad — JNU, India
Derong Liu — University of Chicago, USA

Special Session Chairs

Kai Qin — Swinburne University of Technology, Australia
Kaizhu Huang — Duke Kunshan University, China
Amit Kumar Singh — NIT Patna, India

Tutorial Chairs

Swagatam Das — ISI Kolkata, India
Partha Pratim Roy — IIT Roorkee, India

Finance Chairs

Shekhar Verma   Indian Institute of Information Technology
                                   Allahabad, India
Hayaru Shouno   University of Electro-Communications, Japan
R. B. Pachori   IIT Indore, India

Publicity Chairs

Jerry Chun-Wei Lin   Western Norway University of Applied Sciences,
                                   Norway
Chandan Gautam   A*STAR, Singapore

Publication Chairs

Deepak Ranjan Nayak   MNIT Jaipur, India
Tripti Goel   NIT Silchar, India

Sponsorship Chairs

Asoke K. Talukder   NIT Surathkal, India
Vrijendra Singh   IIIT Allahabad, India

Website Chairs

M. Arshad   IIT Indore, India
Navjot Singh   IIIT Allahabad, India

Local Arrangement Chairs

Pallavi Somvanshi   JNU, India
Yogendra Meena   University of Delhi, India
M. Javed   IIIT Allahabad, India
Vinay Kumar Gupta   IIT Indore, India
Iqbal Hasan   National Informatics Centre, Ministry of
                                   Electronics and Information Technology, India

Regional Liaison Committee

Sansanee Auephanwiriyakul   Chiang Mai University, Thailand
Nia Kurnianingsih   Politeknik Negeri Semarang, Indonesia

Md Rafiqul Islam	University of Technology Sydney, Australia
Bharat Richhariya	IISc Bangalore, India
Sanjay Kumar Sonbhadra	Shiksha 'O' Anusandhan, India
Mufti Mahmud	Nottingham Trent University, UK
Francesco Piccialli	University of Naples Federico II, Italy

Program Committee

Balamurali A. R.	IITB-Monash Research Academy, India
Ibrahim A. Hameed	Norwegian University of Science and Technology (NTNU), Norway
Fazly Salleh Abas	Multimedia University, Malaysia
Prabath Abeysekara	RMIT University, Australia
Adamu Abubakar Ibrahim	International Islamic University, Malaysia
Muhammad Abulaish	South Asian University, India
Saptakatha Adak	Philips, India
Abhijit Adhikary	King's College, London, UK
Hasin Afzal Ahmed	Gauhati University, India
Rohit Agarwal	UiT The Arctic University of Norway, Norway
A. K. Agarwal	Sharda University, India
Fenty Eka Muzayyana Agustin	UIN Syarif Hidayatullah Jakarta, Indonesia
Gulfam Ahamad	BGSB University, India
Farhad Ahamed	Kent Institute, Australia
Zishan Ahmad	Indian Institute of Technology Patna, India
Mohammad Faizal Ahmad Fauzi	Multimedia University, Malaysia
Mudasir Ahmadganaie	Indian Institute of Technology Indore, India
Hasin Afzal Ahmed	Gauhati University, India
Sangtae Ahn	Kyungpook National University, South Korea
Md. Shad Akhtar	Indraprastha Institute of Information Technology, Delhi, India
Abdulrazak Yahya Saleh Alhababi	University of Malaysia, Sarawak, Malaysia
Ahmed Alharbi	RMIT University, Australia
Irfan Ali	Aligarh Muslim University, India
Ali Anaissi	CSIRO, Australia
Ashish Anand	Indian Institute of Technology, Guwahati, India
C. Anantaram	Indraprastha Institute of Information Technology and Tata Consultancy Services Ltd., India
Nur Afny C. Andryani	Universiti Teknologi Petronas, Malaysia
Marco Anisetti	Università degli Studi di Milano, Italy
Mohd Zeeshan Ansari	Jamia Millia Islamia, India
J. Anuradha	VIT, India
Ramakrishna Appicharla	Indian Institute of Technology Patna, India

He Chen	Hebei University of Technology, China
Hongxu Chen	University of Queensland, Australia
J. Chen	Dalian University of Technology, China
Jianhui Chen	Beijing University of Technology, China
Junxin Chen	Dalian University of Technology, China
Junyi Chen	City University of Hong Kong, China
Junying Chen	South China University of Technology, China
Lisi Chen	Hong Kong Baptist University, China
Mulin Chen	Northwestern Polytechnical University, China
Xiaocong Chen	University of New South Wales, Australia
Xiaofeng Chen	Chongqing Jiaotong University, China
Zhuangbin Chen	The Chinese University of Hong Kong, China
Long Cheng	Institute of Automation, China
Qingrong Cheng	Fudan University, China
Ruting Cheng	George Washington University, USA
Girija Chetty	University of Canberra, Australia
Manoj Chinnakotla	Microsoft R&D Pvt. Ltd., India
Andrew Chiou	CQ University, Australia
Sung-Bae Cho	Yonsei University, South Korea
Kupsze Choi	The Hong Kong Polytechnic University, China
Phatthanaphong Chomphuwiset	Mahasarakham University, Thailand
Fengyu Cong	Dalian University of Technology, China
Jose Alfredo Ferreira Costa	UFRN, Brazil
Ruxandra Liana Costea	Polytechnic University of Bucharest, Romania
Raphaël Couturier	University of Franche-Comte, France
Zhenyu Cui	Peking University, China
Zhihong Cui	Shandong University, China
Juan D. Velasquez	University of Chile, Chile
Rukshima Dabare	Murdoch University, Australia
Cherifi Dalila	University of Boumerdes, Algeria
Minh-Son Dao	National Institute of Information and Communications Technology, Japan
Tedjo Darmanto	STMIK AMIK Bandung, Indonesia
Debasmit Das	IIT Roorkee, India
Dipankar Das	Jadavpur University, India
Niladri Sekhar Dash	Indian Statistical Institute, Kolkata, India
Satya Ranjan Dash	KIIT University, India
Shubhajit Datta	Indian Institute of Technology, Kharagpur, India
Alok Debnath	Trinity College Dublin, Ireland
Amir Dehsarvi	Ludwig Maximilian University of Munich, Germany
Hangyu Deng	Waseda University, Japan

Mingcong Deng	Tokyo University of Agriculture and Technology, Japan
Zhaohong Deng	Jiangnan University, China
V. Susheela Devi	Indian Institute of Science, Bangalore, India
M. M. Dhabu	VNIT Nagpur, India
Dhimas Arief Dharmawan	Universitas Indonesia, Indonesia
Khaldoon Dhou	Texas A&M University Central Texas, USA
Gihan Dias	University of Moratuwa, Sri Lanka
Nat Dilokthanakul	Vidyasirimedhi Institute of Science and Technology, Thailand
Tai Dinh	Kyoto College of Graduate Studies for Informatics, Japan
Gaurav Dixit	Indian Institute of Technology Roorkee, India
Youcef Djenouri	SINTEF Digital, Norway
Hai Dong	RMIT University, Australia
Shichao Dong	Ping An Insurance Group, China
Mohit Dua	NIT Kurukshetra, India
Yijun Duan	Kyoto University, Japan
Shiv Ram Dubey	Indian Institute of Information Technology, Allahabad, India
Piotr Duda	Institute of Computational Intelligence/Czestochowa University of Technology, Poland
Sri Harsha Dumpala	Dalhousie University and Vector Institute, Canada
Hridoy Sankar Dutta	University of Cambridge, UK
Indranil Dutta	Jadavpur University, India
Pratik Dutta	Indian Institute of Technology Patna, India
Rudresh Dwivedi	Netaji Subhas University of Technology, India
Heba El-Fiqi	UNSW Canberra, Australia
Felix Engel	Leibniz Information Centre for Science and Technology (TIB), Germany
Akshay Fajge	Indian Institute of Technology Patna, India
Yuchun Fang	Shanghai University, China
Mohd Fazil	JMI, India
Zhengyang Feng	Shanghai Jiao Tong University, China
Zunlei Feng	Zhejiang University, China
Mauajama Firdaus	University of Alberta, Canada
Devi Fitrianah	Bina Nusantara University, Indonesia
Philippe Fournierviger	Shenzhen University, China
Wai-Keung Fung	Cardiff Metropolitan University, UK
Baban Gain	Indian Institute of Technology, Patna, India
Claudio Gallicchio	University of Pisa, Italy
Yongsheng Gao	Griffith University, Australia

Yunjun Gao	Zhejiang University, China
Vicente García Díaz	University of Oviedo, Spain
Arpit Garg	University of Adelaide, Australia
Chandan Gautam	I2R, A*STAR, Singapore
Yaswanth Gavini	University of Hyderabad, India
Tom Gedeon	Australian National University, Australia
Iuliana Georgescu	University of Bucharest, Romania
Deepanway Ghosal	Indian Institute of Technology Patna, India
Arjun Ghosh	National Institute of Technology Durgapur, India
Sanjukta Ghosh	IIT (BHU) Varanasi, India
Soumitra Ghosh	Indian Institute of Technology Patna, India
Pranav Goel	Bloomberg L.P., India
Tripti Goel	National Institute of Technology Silchar, India
Kah Ong Michael Goh	Multimedia University, Malaysia
Kam Meng Goh	Tunku Abdul Rahman University of Management and Technology, Malaysia
Iqbal Gondal	RMIT University, Australia
Puneet Goyal	Indian Institute of Technology Ropar, India
Vishal Goyal	Punjabi University Patiala, India
Xiaotong Gu	University of Tasmania, Australia
Radha Krishna Guntur	VNRVJIET, India
Li Guo	University of Macau, China
Ping Guo	Beijing Normal University, China
Yu Guo	Xi'an Jiaotong University, China
Akshansh Gupta	CSIR-Central Electronics Engineering Research Institute, India
Deepak Gupta	National Library of Medicine, National Institutes of Health (NIH), USA
Deepak Gupta	NIT Arunachal Pradesh, India
Kamal Gupta	NIT Patna, India
Kapil Gupta	PDPM IIITDM, Jabalpur, India
Komal Gupta	IIT Patna, India
Christophe Guyeux	University of Franche-Comte, France
Katsuyuki Hagiwara	Mie University, Japan
Soyeon Han	University of Sydney, Australia
Palak Handa	IGDTUW, India
Rahmadya Handayanto	Universitas Islam 45 Bekasi, Indonesia
Ahteshamul Haq	Aligarh Muslim University, India
Muhammad Haris	Universitas Nusa Mandiri, Indonesia
Harith Al-Sahaf	Victoria University of Wellington, New Zealand
Md Rakibul Hasan	BRAC University, Bangladesh
Mohammed Hasanuzzaman	ADAPT Centre, Ireland

Takako Hashimoto	Chiba University of Commerce, Japan
Bipan Hazarika	Gauhati University, India
Huiguang He	Institute of Automation, Chinese Academy of Sciences, China
Wei He	University of Science and Technology Beijing, China
Xinwei He	University of Illinois Urbana-Champaign, USA
Enna Hirata	Kobe University, Japan
Akira Hirose	University of Tokyo, Japan
Katsuhiro Honda	Osaka Metropolitan University, Japan
Huy Hongnguyen	National Institute of Informatics, Japan
Wai Lam Hoo	University of Malaya, Malaysia
Shih Hsiung Lee	National Cheng Kung University, Taiwan
Jiankun Hu	UNSW@ADFA, Australia
Yanyan Hu	University of Science and Technology Beijing, China
Chaoran Huang	UNSW Sydney, Australia
He Huang	Soochow University, Taiwan
Ko-Wei Huang	National Kaohsiung University of Science and Technology, Taiwan
Shudong Huang	Sichuan University, China
Chih-Chieh Hung	National Chung Hsing University, Taiwan
Mohamed Ibn Khedher	IRT-SystemX, France
David Iclanzan	Sapientia Hungarian University of Transylvania, Romania
Cosimo Ieracitano	University "Mediterranea" of Reggio Calabria, Italy
Kazushi Ikeda	Nara Institute of Science and Technology, Japan
Hiroaki Inoue	Kobe University, Japan
Teijiro Isokawa	University of Hyogo, Japan
Kokila Jagadeesh	Indian Institute of Information Technology, Allahabad, India
Mukesh Jain	Jawaharlal Nehru University, India
Fuad Jamour	AWS, USA
Mohd. Javed	Indian Institute of Information Technology, Allahabad, India
Balasubramaniam Jayaram	Indian Institute of Technology Hyderabad, India
Jin-Tsong Jeng	National Formosa University, Taiwan
Sungmoon Jeong	Kyungpook National University Hospital, South Korea
Yizhang Jiang	Jiangnan University, China
Ferdinjoe Johnjoseph	Thai-Nichi Institute of Technology, Thailand
Alireza Jolfaei	Federation University, Australia

Ratnesh Joshi	Indian Institute of Technology Patna, India
Roshan Joymartis	Global Academy of Technology, India
Chen Junjie	IMAU, The Netherlands
Ashwini K.	Global Academy of Technology, India
Asoke K. Talukder	National Institute of Technology Karnataka - Surathkal, India
Ashad Kabir	Charles Sturt University, Australia
Narendra Kadoo	CSIR-National Chemical Laboratory, India
Seifedine Kadry	Noroff University College, Norway
M. Shamim Kaiser	Jahangirnagar University, Bangladesh
Ashraf Kamal	ACL Digital, India
Sabyasachi Kamila	Indian Institute of Technology Patna, India
Tomoyuki Kaneko	University of Tokyo, Japan
Rajkumar Kannan	Bishop Heber College, India
Hamid Karimi	Utah State University, USA
Nikola Kasabov	AUT, New Zealand
Dermot Kerr	University of Ulster, UK
Abhishek Kesarwani	NIT Rourkela, India
Shwet Ketu	Shambhunath Institute of Engineering and Technology, India
Asif Khan	Integral University, India
Tariq Khan	UNSW, Australia
Thaweesak Khongtuk	Rajamangala University of Technology Suvarnabhumi (RMUTSB), India
Abbas Khosravi	Deakin University, Australia
Thanh Tung Khuat	University of Technology Sydney, Australia
Junae Kim	DST Group, Australia
Sangwook Kim	Kobe University, Japan
Mutsumi Kimura	Ryukoku University, Japan
Uday Kiran	University of Aizu, Japan
Hisashi Koga	University of Electro-Communications, Japan
Yasuharu Koike	Tokyo Institute of Technology, Japan
Ven Jyn Kok	Universiti Kebangsaan Malaysia, Malaysia
Praveen Kolli	Pinterest Inc, USA
Sunil Kumar Kopparapu	Tata Consultancy Services Ltd., India
Fajri Koto	MBZUAI, UAE
Aneesh Krishna	Curtin University, Australia
Parameswari Krishnamurthy	University of Hyderabad, India
Malhar Kulkarni	IIT Bombay, India
Abhinav Kumar	NIT, Patna, India
Abhishek Kumar	Indian Institute of Technology Patna, India
Amit Kumar	Tarento Technologies Pvt Limited, India

Nagendra Kumar	IIT Indore, India
Pranaw Kumar	Centre for Development of Advanced Computing (CDAC) Mumbai, India
Puneet Kumar	Jawaharlal Nehru University, India
Raja Kumar	Taylor's University, Malaysia
Sachin Kumar	University of Delhi, India
Sandeep Kumar	IIT Patna, India
Sanjaya Kumar Panda	National Institute of Technology, Warangal, India
Chouhan Kumar Rath	National Institute of Technology, Durgapur, India
Sovan Kumar Sahoo	Indian Institute of Technology Patna, India
Anil Kumar Singh	IIT (BHU) Varanasi, India
Vikash Kumar Singh	VIT-AP University, India
Sanjay Kumar Sonbhadra	ITER, SoA, Odisha, India
Gitanjali Kumari	Indian Institute of Technology Patna, India
Rina Kumari	KIIT, India
Amit Kumarsingh	National Institute of Technology Patna, India
Sanjay Kumarsonbhadra	SSITM, India
Vishesh Kumar Tanwar	Missouri University of Science and Technology, USA
Bibekananda Kundu	CDAC Kolkata, India
Yoshimitsu Kuroki	Kurume National College of Technology, Japan
Susumu Kuroyanagi	Nagoya Institute of Technology, Japan
Retno Kusumaningrum	Universitas Diponegoro, Indonesia
Dwina Kuswardani	Institut Teknologi PLN, Indonesia
Stephen Kwok	Murdoch University, Australia
Hamid Laga	Murdoch University, Australia
Edmund Lai	Auckland University of Technology, New Zealand
Weng Kin Lai	Tunku Abdul Rahman University of Management & Technology (TAR UMT), Malaysia
Kittichai Lavangnananda	King Mongkut's University of Technology Thonburi (KMUTT), Thailand
Anwesha Law	Indian Statistical Institute, India
Thao Le	Deakin University, Australia
Xinyi Le	Shanghai Jiao Tong University, China
Dong-Gyu Lee	Kyungpook National University, South Korea
Eui Chul Lee	Sangmyung University, South Korea
Minho Lee	Kyungpook National University, South Korea
Shih Hsiung Lee	National Kaohsiung University of Science and Technology, Taiwan
Gurpreet Lehal	Punjabi University, India
Jiahuan Lei	Meituan-Dianping Group, China

Pui Huang Leong	Tunku Abdul Rahman University of Management and Technology, Malaysia
Chi Sing Leung	City University of Hong Kong, China
Man-Fai Leung	Anglia Ruskin University, UK
Bing-Zhao Li	Beijing Institute of Technology, China
Gang Li	Deakin University, Australia
Jiawei Li	Tsinghua University, China
Mengmeng Li	Zhengzhou University, China
Xiangtao Li	Jilin University, China
Yang Li	East China Normal University, China
Yantao Li	Chongqing University, China
Yaxin Li	Michigan State University, USA
Yiming Li	Tsinghua University, China
Yuankai Li	University of Science and Technology of China, China
Yun Li	Nanjing University of Posts and Telecommunications, China
Zhipeng Li	Tsinghua University, China
Hualou Liang	Drexel University, USA
Xiao Liang	Nankai University, China
Hao Liao	Shenzhen University, China
Alan Wee-Chung Liew	Griffith University, Australia
Chern Hong Lim	Monash University Malaysia, Malaysia
Kok Lim Yau	Universiti Tunku Abdul Rahman (UTAR), Malaysia
Chin-Teng Lin	UTS, Australia
Jerry Chun-Wei Lin	Western Norway University of Applied Sciences, Norway
Jiecong Lin	City University of Hong Kong, China
Dugang Liu	Shenzhen University, China
Feng Liu	Stevens Institute of Technology, USA
Hongtao Liu	Du Xiaoman Financial, China
Ju Liu	Shandong University, China
Linjing Liu	City University of Hong Kong, China
Weifeng Liu	China University of Petroleum (East China), China
Wenqiang Liu	Hong Kong Polytechnic University, China
Xin Liu	National Institute of Advanced Industrial Science and Technology (AIST), Japan
Yang Liu	Harbin Institute of Technology, China
Zhi-Yong Liu	Institute of Automation, Chinese Academy of Sciences, China
Zongying Liu	Dalian Maritime University, China

Jaime Lloret	Universitat Politècnica de València, Spain
Sye Loong Keoh	University of Glasgow, Singapore, Singapore
Hongtao Lu	Shanghai Jiao Tong University, China
Wenlian Lu	Fudan University, China
Xuequan Lu	Deakin University, Australia
Xiao Luo	UCLA, USA
Guozheng Ma	Shenzhen International Graduate School, Tsinghua University, China
Qianli Ma	South China University of Technology, China
Wanli Ma	University of Canberra, Australia
Muhammad Anwar Ma'sum	Universitas Indonesia, Indonesia
Michele Magno	University of Bologna, Italy
Sainik Kumar Mahata	JU, India
Shalni Mahato	Indian Institute of Information Technology (IIIT) Ranchi, India
Adnan Mahmood	Macquarie University, Australia
Mohammed Mahmoud	October University for Modern Sciences & Arts - MSA University, Egypt
Mufti Mahmud	University of Padova, Italy
Krishanu Maity	Indian Institute of Technology Patna, India
Mamta	IIT Patna, India
Aprinaldi Mantau	Kyushu Institute of Technology, Japan
Mohsen Marjani	Taylor's University, Malaysia
Sanparith Marukatat	NECTEC, Thailand
José María Luna	Universidad de Córdoba, Spain
Archana Mathur	Nitte Meenakshi Institute of Technology, India
Patrick McAllister	Ulster University, UK
Piotr Milczarski	Lodz University of Technology, Poland
Kshitij Mishra	IIT Patna, India
Pruthwik Mishra	IIIT-Hyderabad, India
Santosh Mishra	Indian Institute of Technology Patna, India
Sajib Mistry	Curtin University, Australia
Sayantan Mitra	Accenture Labs, India
Vinay Kumar Mittal	Neti International Research Center, India
Daisuke Miyamoto	University of Tokyo, Japan
Kazuteru Miyazaki	National Institution for Academic Degrees and Quality Enhancement of Higher Education, Japan
U. Mmodibbo	Modibbo Adama University Yola, Nigeria
Aditya Mogadala	Saarland University, Germany
Reem Mohamed	Mansoura University, Egypt
Muhammad Syafiq Mohd Pozi	Universiti Utara Malaysia, Malaysia

Anirban Mondal	University of Tokyo, Japan
Anupam Mondal	Jadavpur University, India
Supriyo Mondal	ZBW - Leibniz Information Centre for Economics, Germany
J. Manuel Moreno	Universitat Politècnica de Catalunya, Spain
Francisco J. Moreno-Barea	Universidad de Málaga, Spain
Sakchai Muangsrinoon	Walailak University, Thailand
Siti Anizah Muhamed	Politeknik Sultan Salahuddin Abdul Aziz Shah, Malaysia
Samrat Mukherjee	Indian Institute of Technology, Patna, India
Siddhartha Mukherjee	Samsung R&D Institute India, Bangalore, India
Dharmalingam Muthusamy	Bharathiar University, India
Abhijith Athreya Mysore Gopinath	Pennsylvania State University, USA
Harikrishnan N. B.	BITS Pilani K K Birla Goa Campus, India
Usman Naseem	University of Sydney, Australia
Deepak Nayak	Malaviya National Institute of Technology, Jaipur, India
Hamada Nayel	Benha University, Egypt
Usman Nazir	Lahore University of Management Sciences, Pakistan
Vasudevan Nedumpozhimana	TU Dublin, Ireland
Atul Negi	University of Hyderabad, India
Aneta Neumann	University of Adelaide, Australia
Hea Choon Ngo	Universiti Teknikal Malaysia Melaka, Malaysia
Dang Nguyen	University of Canberra, Australia
Duy Khuong Nguyen	FPT Software Ltd., FPT Group, Vietnam
Hoang D. Nguyen	University College Cork, Ireland
Hong Huy Nguyen	National Institute of Informatics, Japan
Tam Nguyen	Leibniz University Hannover, Germany
Thanh-Son Nguyen	Agency for Science, Technology and Research (A*STAR), Singapore
Vu-Linh Nguyen	Eindhoven University of Technology, Netherlands
Nick Nikzad	Griffith University, Australia
Boda Ning	Swinburne University of Technology, Australia
Haruhiko Nishimura	University of Hyogo, Japan
Kishorjit Nongmeikapam	Indian Institute of Information Technology (IIIT) Manipur, India
Aleksandra Nowak	Jagiellonian University, Poland
Stavros Ntalampiras	University of Milan, Italy
Anupiya Nugaliyadde	Sri Lanka Institute of Information Technology, Sri Lanka

Anto Satriyo Nugroho	Agency for Assessment & Application of Technology, Indonesia
Aparajita Ojha	PDPM IIITDM Jabalpur, India
Akeem Olowolayemo	International Islamic University Malaysia, Malaysia
Toshiaki Omori	Kobe University, Japan
Shih Yin Ooi	Multimedia University, Malaysia
Sidali Ouadfeul	Algerian Petroleum Institute, Algeria
Samir Ouchani	CESI Lineact, France
Srinivas P. Y. K. L.	IIIT Sri City, India
Neelamadhab Padhy	GIET University, India
Worapat Paireekreng	Dhurakij Pundit University, Thailand
Partha Pakray	National Institute of Technology Silchar, India
Santanu Pal	Wipro Limited, India
Bin Pan	Nankai University, China
Rrubaa Panchendrarajan	Sri Lanka Institute of Information Technology, Sri Lanka
Pankaj Pandey	Indian Institute of Technology, Gandhinagar, India
Lie Meng Pang	Southern University of Science and Technology, China
Sweta Panigrahi	National Institute of Technology Warangal, India
T. Pant	IIIT Allahabad, India
Shantipriya Parida	Idiap Research Institute, Switzerland
Hyeyoung Park	Kyungpook National University, South Korea
Md Aslam Parwez	Jamia Millia Islamia, India
Leandro Pasa	Federal University of Technology - Parana (UTFPR), Brazil
Kitsuchart Pasupa	King Mongkut's Institute of Technology Ladkrabang, Thailand
Debanjan Pathak	Kalinga Institute of Industrial Technology (KIIT), India
Vyom Pathak	University of Florida, USA
Sangameshwar Patil	TCS Research, India
Bidyut Kr. Patra	IIT (BHU) Varanasi, India
Dipanjyoti Paul	Indian Institute of Technology Patna, India
Sayanta Paul	Ola, India
Sachin Pawar	Tata Consultancy Services Ltd., India
Pornntiwa Pawara	Mahasarakham University, Thailand
Yong Peng	Hangzhou Dianzi University, China
Yusuf Perwej	Ambalika Institute of Management and Technology (AIMT), India
Olutomilayo Olayemi Petinrin	City University of Hong Kong, China
Arpan Phukan	Indian Institute of Technology Patna, India

Chiara Picardi	University of York, UK
Francesco Piccialli	University of Naples Federico II, Italy
Josephine Plested	University of New South Wales, Australia
Krishna Reddy Polepalli	IIIT Hyderabad, India
Dan Popescu	University Politehnica of Bucharest, Romania
Heru Praptono	Bank Indonesia/UI, Indonesia
Mukesh Prasad	University of Technology Sydney, Australia
Yamuna Prasad	Thompson Rivers University, Canada
Krishna Prasadmiyapuram	IIT Gandhinagar, India
Partha Pratim Sarangi	KIIT Deemed to be University, India
Emanuele Principi	Università Politecnica delle Marche, Italy
Dimeter Prodonov	Imec, Belgium
Ratchakoon Pruengkarn	College of Innovative Technology and Engineering, Dhurakij Pundit University, Thailand
Michal Ptaszynski	Kitami Institute of Technology, Japan
Narinder Singh Punn	Mayo Clinic, Arizona, USA
Abhinanda Ranjit Punnakkal	UiT The Arctic University of Norway, Norway
Zico Pratama Putra	Queen Mary University of London, UK
Zhenyue Qin	Tencent, China
Nawab Muhammad Faseeh Qureshi	SU, South Korea
Md Rafiqul	UTS, Australia
Saifur Rahaman	City University of Hong Kong, China
Shri Rai	Murdoch University, Australia
Vartika Rai	IIIT Hyderabad, India
Kiran Raja	Norwegian University of Science and Technology, Norway
Sutharshan Rajasegarar	Deakin University, Australia
Arief Ramadhan	Bina Nusantara University, Indonesia
Mallipeddi Rammohan	Kyungpook National University, South Korea
Md. Mashud Rana	Commonwealth Scientific and Industrial Research Organisation (CSIRO), Australia
Surangika Ranathunga	University of Moratuwa, Sri Lanka
Soumya Ranjan Mishra	KIIT University, India
Hemant Rathore	Birla Institute of Technology & Science, Pilani, India
Imran Razzak	UNSW, Australia
Yazhou Ren	University of Science and Technology of China, China
Motahar Reza	GITAM University Hyderabad, India
Dwiza Riana	STMIK Nusa Mandiri, Indonesia
Bharat Richhariya	BITS Pilani, India

Pattabhi R. K. Rao	AU-KBC Research Centre, India
Heejun Roh	Korea University, South Korea
Vijay Rowtula	IIIT Hyderabad, India
Aniruddha Roy	IIT Kharagpur, India
Sudipta Roy	Jio Institute, India
Narendra S. Chaudhari	Indian Institute of Technology Indore, India
Fariza Sabrina	Central Queensland University, Australia
Debanjan Sadhya	ABV-IIITM Gwalior, India
Sumit Sah	IIT Dharwad, India
Atanu Saha	Jadavpur University, India
Sajib Saha	Commonwealth Scientific and Industrial Research Organisation, Australia
Snehanshu Saha	BITS Pilani K K Birla Goa Campus, India
Tulika Saha	IIT Patna, India
Navanath Saharia	Indian Institute of Information Technology Manipur, India
Pracheta Sahoo	University of Texas at Dallas, USA
Sovan Kumar Sahoo	Indian Institute of Technology Patna, India
Tanik Saikh	L3S Research Center, Germany
Naveen Saini	Indian Institute of Information Technology Lucknow, India
Fumiaki Saitoh	Chiba Institute of Technology, Japan
Rohit Salgotra	Swansea University, UK
Michel Salomon	Univ. Bourgogne Franche-Comté, France
Yu Sang	Research Institute of Institute of Computing Technology, Exploration and Development, Liaohe Oilfield, PetroChina, China
Suyash Sangwan	Indian Institute of Technology Patna, India
Soubhagya Sankar Barpanda	VIT-AP University, India
Jose A. Santos	Ulster University, UK
Kamal Sarkar	Jadavpur University, India
Sandip Sarkar	Jadavpur University, India
Naoyuki Sato	Future University Hakodate, Japan
Eri Sato-Shimokawara	Tokyo Metropolitan University, Japan
Sunil Saumya	Indian Institute of Information Technology Dharwad, India
Gerald Schaefer	Loughborough University, UK
Rafal Scherer	Czestochowa University of Technology, Poland
Arvind Selwal	Central University of Jammu, India
Noor Akhmad Setiawan	Universitas Gadjah Mada, Indonesia
Mohammad Shahid	Aligarh Muslim University, India
Jie Shao	University of Science and Technology of China, China

Nabin Sharma	University of Technology Sydney, Australia
Raksha Sharma	IIT Bombay, India
Sourabh Sharma	Avantika University, India
Suraj Sharma	International Institute of Information Technology Bhubaneswar, India
Ravi Shekhar	Queen Mary University of London, UK
Michael Sheng	Macquarie University, Australia
Yin Sheng	Huazhong University of Science and Technology, China
Yongpan Sheng	Southwest University, China
Liu Shenglan	Dalian University of Technology, China
Tomohiro Shibata	Kyushu Institute of Technology, Japan
Iksoo Shin	University of Science & Technology, China
Mohd Fairuz Shiratuddin	Murdoch University, Australia
Hayaru Shouno	University of Electro-Communications, Japan
Sanyam Shukla	MANIT, Bhopal, India
Udom Silparcha	KMUTT, Thailand
Apoorva Singh	Indian Institute of Technology Patna, India
Divya Singh	Central University of Bihar, India
Gitanjali Singh	Indian Institute of Technology Patna, India
Gopendra Singh	Indian Institute of Technology Patna, India
K. P. Singh	IIIT Allahabad, India
Navjot Singh	IIIT Allahabad, India
Om Singh	NIT Patna, India
Pardeep Singh	Jawaharlal Nehru University, India
Rajiv Singh	Banasthali Vidyapith, India
Sandhya Singh	Indian Institute of Technology Bombay, India
Smriti Singh	IIT Bombay, India
Narinder Singhpunn	Mayo Clinic, Arizona, USA
Saaveethya Sivakumar	Curtin University, Malaysia
Ferdous Sohel	Murdoch University, Australia
Chattrakul Sombattheera	Mahasarakham University, Thailand
Lei Song	Unitec Institute of Technology, New Zealand
Linqi Song	City University of Hong Kong, China
Yuhua Song	University of Science and Technology Beijing, China
Gautam Srivastava	Brandon University, Canada
Rajeev Srivastava	Banaras Hindu University (IT-BHU), Varanasi, India
Jérémie Sublime	ISEP - Institut Supérieur d'Électronique de Paris, France
P. N. Suganthan	Nanyang Technological University, Singapore

Derwin Suhartono	Bina Nusantara University, Indonesia
Indra Adji Sulistijono	Politeknik Elektronika Negeri Surabaya (PENS), Indonesia
John Sum	National Chung Hsing University, Taiwan
Fuchun Sun	Tsinghua University, China
Ning Sun	Nankai University, China
Anindya Sundar Das	Indian Institute of Technology Patna, India
Bapi Raju Surampudi	International Institute of Information Technology Hyderabad, India
Olarik Surinta	Mahasarakham University, Thailand
Maria Susan Anggreainy	Bina Nusantara University, Indonesia
M. Syafrullah	Universitas Budi Luhur, Indonesia
Murtaza Taj	Lahore University of Management Sciences, Pakistan
Norikazu Takahashi	Okayama University, Japan
Abdelmalik Taleb-Ahmed	Polytechnic University of Hauts-de-France, France
Hakaru Tamukoh	Kyushu Institute of Technology, Japan
Choo Jun Tan	Wawasan Open University, Malaysia
Chuanqi Tan	BIT, China
Shing Chiang Tan	Multimedia University, Malaysia
Xiao Jian Tan	Tunku Abdul Rahman University of Management and Technology (TAR UMT), Malaysia
Xin Tan	East China Normal University, China
Ying Tan	Peking University, China
Gouhei Tanaka	University of Tokyo, Japan
Yang Tang	East China University of Science and Technology, China
Zhiri Tang	City University of Hong Kong, China
Tanveer Tarray	Islamic University of Science and Technology, India
Chee Siong Teh	Universiti Malaysia Sarawak (UNIMAS), Malaysia
Ya-Wen Teng	Academia Sinica, Taiwan
Gaurish Thakkar	University of Zagreb, Croatia
Medari Tham	St. Anthony's College, India
Selvarajah Thuseethan	Sabaragamuwa University of Sri Lanka, Sri Lanka
Shu Tian	University of Science and Technology Beijing, China
Massimo Tistarelli	University of Sassari, Italy
Abhisek Tiwari	IIT Patna, India
Uma Shanker Tiwary	Indian Institute of Information Technology, Allahabad, India

Alex To	University of Sydney, Australia
Stefania Tomasiello	University of Tartu, Estonia
Anh Duong Trinh	Technological University Dublin, Ireland
Enkhtur Tsogbaatar	Mongolian University of Science and Technology, Mongolia
Enmei Tu	Shanghai Jiao Tong University, China
Eiji Uchino	Yamaguchi University, Japan
Prajna Upadhyay	IIT Delhi, India
Sahand Vahidnia	University of New South Wales, Australia
Ashwini Vaidya	IIT Delhi, India
Deeksha Varshney	Indian Institute of Technology, Patna, India
Sowmini Devi Veeramachaneni	Mahindra University, India
Samudra Vijaya	Koneru Lakshmaiah Education Foundation, India
Surbhi Vijh	JSS Academy of Technical Education, Noida, India
Nhi N. Y. Vo	University of Technology Sydney, Australia
Xuan-Son Vu	Umeå University, Sweden
Anil Kumar Vuppala	IIIT Hyderabad, India
Nobuhiko Wagatsuma	Toho University, Japan
Feng Wan	University of Macau, China
Bingshu Wang	Northwestern Polytechnical University Taicang Campus, China
Dianhui Wang	La Trobe University, Australia
Ding Wang	Beijing University of Technology, China
Guanjin Wang	Murdoch University, Australia
Jiasen Wang	City University of Hong Kong, China
Lei Wang	Beihang University, China
Libo Wang	Xiamen University of Technology, China
Meng Wang	Southeast University, China
Qiu-Feng Wang	Xi'an Jiaotong-Liverpool University, China
Sheng Wang	Henan University, China
Weiqun Wang	Institute of Automation, Chinese Academy of Sciences, China
Wentao Wang	Michigan State University, USA
Yongyu Wang	Michigan Technological University, USA
Zhijin Wang	Jimei University, China
Bunthit Watanapa	KMUTT-SIT, Thailand
Yanling Wei	TU Berlin, Germany
Guanghui Wen	RMIT University, Australia
Ari Wibisono	Universitas Indonesia, Indonesia
Adi Wibowo	Diponegoro University, Indonesia
Ka-Chun Wong	City University of Hong Kong, China

Kevin Wong   Murdoch University, Australia
Raymond Wong   Universiti Malaya, Malaysia
Kuntpong Woraratpanya   King Mongkut's Institute of Technology
                              Ladkrabang (KMITL), Thailand
Marcin Woźniak   Silesian University of Technology, Poland
Chengwei Wu   Harbin Institute of Technology, China
Jing Wu   Shanghai Jiao Tong University, China
Weibin Wu   Sun Yat-sen University, China
Hongbing Xia   Beijing Normal University, China
Tao Xiang   Chongqing University, China
Qiang Xiao   Huazhong University of Science and Technology,
                              China
Guandong Xu   University of Technology Sydney, Australia
Qing Xu   Tianjin University, China
Yifan Xu   Huazhong University of Science and Technology,
                              China
Junyu Xuan   University of Technology Sydney, Australia
Hui Xue   Southeast University, China
Saumitra Yadav   IIIT-Hyderabad, India
Shekhar Yadav   Madan Mohan Malaviya University of
                              Technology, India
Sweta Yadav   University of Illinois at Chicago, USA
Tarun Yadav   Defence Research and Development
                              Organisation, India
Shankai Yan   Hainan University, China
Feidiao Yang   Microsoft, China
Gang Yang   Renmin University of China, China
Haiqin Yang   International Digital Economy Academy, China
Jianyi Yang   Shandong University, China
Jinfu Yang   BJUT, China
Minghao Yang   Institute of Automation, Chinese Academy of
                              Sciences, China
Shaofu Yang   Southeast University, China
Wachira Yangyuen   Rajamangala University of Technology Srivijaya,
                              Thailand
Xinye Yi   Guilin University of Electronic Technology, China
Hang Yu   Shanghai University, China
Wen Yu   Cinvestav, Mexico
Wenxin Yu   Southwest University of Science and Technology,
                              China
Zhaoyuan Yu   Nanjing Normal University, China
Ye Yuan   Xi'an Jiaotong University, China
Xiaodong Yue   Shanghai University, China

Aizan Zafar	Indian Institute of Technology Patna, India
Jichuan Zeng	Bytedance, China
Jie Zhang	Newcastle University, UK
Shixiong Zhang	Xidian University, China
Tianlin Zhang	University of Manchester, UK
Mingbo Zhao	Donghua University, China
Shenglin Zhao	Zhejiang University, China
Guoqiang Zhong	Ocean University of China, China
Jinghui Zhong	South China University of Technology, China
Bo Zhou	Southwest University, China
Yucheng Zhou	University of Technology Sydney, Australia
Dengya Zhu	Curtin University, Australia
Xuanying Zhu	ANU, Australia
Hua Zuo	University of Technology Sydney, Australia

Additional Reviewers

Acharya, Rajul
Afrin, Mahbuba
Alsuhaibani, Abdullah
Amarnath
Appicharla, Ramakrishna
Arora, Ridhi
Azar, Joseph
Bai, Weiwei
Bao, Xiwen
Barawi, Mohamad Hardyman
Bhat, Mohammad Idrees Bhat
Cai, Taotao
Cao, Feiqi
Chakraborty, Bodhi
Chang, Yu-Cheng
Chen
Chen, Jianpeng
Chen, Yong
Chhipa, Priyank
Cho, Joshua
Chongyang, Chen
Cuenat, Stéphane
Dang, Lili
Das Chakladar, Debashis
Das, Kishalay
Dey, Monalisa

Doborjeh, Maryam
Dong, Zhuben
Dutta, Subhabrata
Dybala, Pawel
El Achkar, Charbel
Feng, Zhengyang
Galkowski, Tomasz
Garg, Arpit
Ghobakhlou, Akbar
Ghosh, Soumitra
Guo, Hui
Gupta, Ankur
Gupta, Deepak
Gupta, Megha
Han, Yanyang
Han, Yiyan
Hang, Bin
Harshit
He, Silu
Hua, Ning
Huang, Meng
Huang, Rongting
Huang, Xiuyu
Hussain, Zawar
Imran, Javed
Islam, Md Rafiqul

Jain, Samir
Jia, Mei
Jiang, Jincen
Jiang, Xiao
Jiangyu, Wang
Jiaxin, Lou
Jiaxu, Hou
Jinzhou, Bao
Ju, Wei
Kasyap, Harsh
Katai, Zoltan
Keserwani, Prateek
Khan, Asif
Khan, Muhammad Fawad Akbar
Khari, Manju
Kheiri, Kiana
Kirk, Nathan
Kiyani, Arslan
Kolya, Anup Kumar
Krdzavac, Nenad
Kumar, Lov
Kumar, Mukesh
Kumar, Puneet
Kumar, Rahul
Kumar, Sunil
Lan, Meng
Lavangnananda, Kittichai
Li, Qian
Li, Xiaoou
Li, Xin
Li, Xinjia
Liang, Mengnan
Liang, Shuai
Liquan, Li
Liu, Boyang
Liu, Chang
Liu, Feng
Liu, Linjing
Liu, Xinglan
Liu, Xinling
Liu, Zhe
Lotey, Taveena
Ma, Bing
Ma, Zeyu
Madanian, Samaneh

Mahata, Sainik Kumar
Mahmud, Md. Redowan
Man, Jingtao
Meena, Kunj Bihari
Mishra, Pragnyaban
Mistry, Sajib
Modibbo, Umar Muhammad
Na, Na
Nag Choudhury, Somenath
Nampalle, Kishore
Nandi, Palash
Neupane, Dhiraj
Nigam, Nitika
Nigam, Swati
Ning, Jianbo
Oumer, Jehad
Pandey, Abhineet Kumar
Pandey, Sandeep
Paramita, Adi Suryaputra
Paul, Apurba
Petinrin, Olutomilayo Olayemi
Phan Trong, Dat
Pradana, Muhamad Hilmil Muchtar Aditya
Pundhir, Anshul
Rahman, Sheikh Shah Mohammad Motiur
Rai, Sawan
Rajesh, Bulla
Rajput, Amitesh Singh
Rao, Raghunandan K. R.
Rathore, Santosh Singh
Ray, Payel
Roy, Satyaki
Saini, Nikhil
Saki, Mahdi
Salimath, Nagesh
Sang, Haiwei
Shao, Jian
Sharma, Anshul
Sharma, Shivam
Shi, Jichen
Shi, Jun
Shi, Kaize
Shi, Li
Singh, Nagendra Pratap
Singh, Pritpal

Singh, Rituraj
Singh, Shrey
Singh, Tribhuvan
Song, Meilun
Song, Yuhua
Soni, Bharat
Stommel, Martin
Su, Yanchi
Sun, Xiaoxuan
Suryodiningrat, Satrio Pradono
Swarnkar, Mayank
Tammewar, Aniruddha
Tan, Xiaosu
Tanoni, Giulia
Tanwar, Vishesh
Tao, Yuwen
To, Alex
Tran, Khuong
Varshney, Ayush
Vo, Anh-Khoa
Vuppala, Anil
Wang, Hui
Wang, Kai
Wang, Rui
Wang, Xia
Wang, Yansong

Wang, Yuan
Wang, Yunhe
Watanapa, Saowaluk
Wenqian, Fan
Xia, Hongbing
Xie, Weidun
Xiong, Wenxin
Xu, Zhehao
Xu, Zhikun
Yan, Bosheng
Yang, Haoran
Yang, Jie
Yang, Xin
Yansui, Song
Yu, Cunzhe
Yu, Zhuohan
Zandavi, Seid Miad
Zeng, Longbin
Zhang, Jane
Zhang, Ruolan
Zhang, Ziqi
Zhao, Chen
Zhou, Xinxin
Zhou, Zihang
Zhu, Liao
Zhu, Linghui

Contents – Part III

Applications

A Comparative Analysis of Loss Functions for Handling Foreground-Background Imbalance in Image Segmentation

Ali Braytee[1,2(✉)], Ali Anaissi[1], and Mohamad Naji[2]

[1] School of Computer Science, The University of Sydney, Camperdown, Australia
ali.braytee@uts.edu.au
[2] School of Computer Science, University of Technology Sydney, Ultimo, Australia

Abstract. Foreground-Background (F-B) imbalance problem has emerged as a fundamental challenge to building accurate image segmentation models in computer vision. F-B imbalance problem occurs due to a disproportionate ratio of observations of foreground and background samples. It degrades the model performance of image segmentation applications. Several loss functions are proposed to improve the classification performance of models on the traditional imbalanced data. But there is no sufficient information on whether these loss functions can improve the accuracy performance of the image segmentation models in the presence of F-B imbalance problem. In this paper, we perform a comparative analysis between four loss functions, namely Focal loss, Dice loss, Tversky and Mixed Focal loss to handle the F-B imbalance problem on datasets from various domains. We embed each loss function in the deep learning model using U-Net architecture and experimentally perform the evaluation in terms of accuracy, precision, F1-score and recall metrics. Then, we present a comparative discussion to the researchers to identify the appropriate loss function that achieves better detection accuracy in different domains.

Keywords: Image segmentation · Foreground-Background (F-B) imbalance problem · U-Net · Loss functions

1 Introduction

In computer vision, object detection creates a bounding box for each class in the image. However, this approach only provides information about the bounding box coordinates and does not consider the specific shape of the object. Image segmentation is an extension of object detection in which the researchers mark an object's presence through pixelwise masks generated for each object in the image. Semantic segmentation is dividing an image into regions in which the pixels have similar properties in colour, texture, or belonging to the same object [3, 4]. By segmenting into different segments for each pixel class, this approach is more granular than object detection, allowing us to determine the shape of the objects in the image.

© The Author(s), under exclusive license to Springer Nature Switzerland AG 2023
M. Tanveer et al. (Eds.): ICONIP 2022, LNCS 13625, pp. 3–13, 2023.
https://doi.org/10.1007/978-3-031-30111-7_1

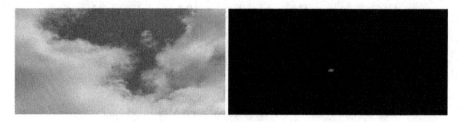

Fig. 1. An example of highly foreground-background class imbalance

Recently, the foreground-background (F-B) imbalance problem has attracted aware-ness for researchers in computer vision [5]. In the F-B imbalance problem, the over-represented and under-represented classes are background and foreground classes, respectively. Thus, it occurs when foreground objects are a smaller proportion of the image than background objects. For example, in Fig. 1, the drone is segmented as fore-ground class and marked in red while the sky as background class is in black. The F-B class imbalance indicates that the drone occupies only a few pixels in the whole image, while the sky takes up almost all proportion. This results in performance degradation in the image segmentation methods such as the popular U-Net deep learning model.

U-Net is a state-of-art semantic pixel-wise segmentation based on a Convolutional Neural Network (CNN) [6]. It transforms image pixels into pixel categories by utilising low-resolution features and classifying the different regions of a colour input image. It is commonly used in medical image analysis for lesion segmentation, anatomical segmentation, and classification. F-B imbalance problem poses a challenge for the U-Net image segmentation model, and it leads to performance reduction [14]. Several loss functions have been proposed to handle the traditional class imbalance problem in deep learning. They efficiently handle the class imbalance by weighing the predicted class and true output (ground truth).

However, practitioners and researchers lack the confidence in which loss functions are more effective in handling F-B imbalance problems in real-world datasets such as medical imaging and remote sensing. Therefore, it is necessary to evaluate various loss functions in the U-Net deep learning model to handle the F-B imbalance problem. We select four loss functions from three algorithm categories that are used in the traditional class imbalance problem namely distribution-based Focal loss, distribution-based Dice and Tversky loss, and compound Mixed Focal loss function. We evaluate the perfor-mance for each loss function in U-Net deep learning with F-B class imbalanced data. In summary, our main contributions are:

- to compare several U-Net deep learning models using different loss functions to handle the F-B imbalance problem.
- to evaluate and compare the efficiency of these loss functions in terms of accuracy, precision, F1-score and recall metrics.
- to help the researchers identify and enhance loss function in image segmentation applications.

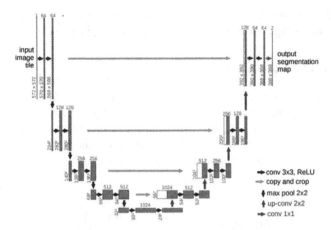

Fig. 2. U-Net architecture (example for 32 × 32 pixels in the lowest resolution)

2 Background

2.1 U-Net Architecture

U-Net is a deep learning model used to segment images accurately. The U-Net architecture contains two paths as shown in Fig. 2, one is the encoder, which can capture the features of the images. Another path is called the decoder, which is used to enable localization using transposed convolutions. The following figure is the U-Net architecture using examples for 32 × 32 pixels [9].

2.2 Loss Functions Handling Imbalanced Datasets in Deep Learning

Several loss functions have been proposed to handle the class imbalance problems in deep learning models, including Cross-entropy, Focal loss, Dice loss, Tversky loss, and others.

Cross-entropy measures the difference between two probability distributions for a given random variable or set of events. Cross-entropy loss focuses on minimizing pixel-wise error, as a result, it has poor performance on the segmentation of smaller objects. Focal loss is a variant of the binary cross-entropy that can address the problem brought by the class imbalanced problem by using the normal cross-entropy. Focal loss down weighting the contribution of simple examples and enables the learning progress to learn from the more complex examples based on the classification error [8]. Dice loss is based on cross-entropy loss. It is the most used metric for calculating segmentation accuracy. However, the shortcoming of Dice loss is it ignores the background rigid, which makes the loss of information. Based on Dice loss, Tversky loss is proposed to handle the class imbalance. But, it can optimize the output imbalance by adjusting the weights assigned to false positives and false negatives. By assigning a higher weight to false negatives, the recall rate is improved, and a better balance of precision and recall is achieved [10]. Another loss function called Focal Tversky loss modifies Tversky loss by down-weighting the easily classified regions and adding more weight to the difficult regions. It achieves high performance in both precision and recall [1].

2.3 Evaluated Loss Functions

Focal Loss. Focal loss has been proposed to focus more on difficult samples during training by reducing the weight of easy classification [8]. It is based on standard Cross-Entropy Loss as shown in Eq. 1. The Cross-Entropy Loss is biased toward the majority of samples in model training, which leads to a small loss function during training, but the recognition accuracy of the minority samples is low.

$$CE(p, y) = \begin{array}{ll} -\log(p) & \text{if } y = 1 \\ -\log(1 - p) & \text{otherwise} \end{array} \qquad (1)$$

where y represents the ground value and p is the probability when the model predicts the label $y = 1$. To improve this problem, two adjustment factors α and γ need to be added, and thus the following formula of focal loss is formed as follows

$$L_{fl} = \begin{array}{ll} -\alpha(1 - y|)^{\gamma} \log(y|), & y = 1 \\ -(1 - \alpha)y|^{\gamma} \log(1 - y|), & y = 0 \end{array} \qquad (2)$$

The α is a weight factor, which can be set to control the weight of majority samples and minority samples. Reducing the weight of the majority of samples can be achieved by using smaller values for α, $(1 - y|)\gamma$ and $y|\gamma$, which can make the model focus more on the difficult samples during training by reducing the weight of the easy samples [8].

The larger the y the more important the influence of the modulation factor.

Dice Loss. Dice Loss is often used to address medical image segmentation tasks since it can solve the imbalance problem between foreground and background [12]. Dice Loss is derived from the Dice Coefficient (DSC), which has been acknowledged as the famous evaluation metric for image segmentation. Unlike Focal loss, DSC only considers the segmentation class and not the background class. Dice similarity coefficient (DSC) is defined as

$$DSC = \frac{2TP}{2TP + FP + FN} = \frac{2|P \cap G|}{|P| + |G|}$$
$$\frac{\sum_{i=1}^{N} p_{0i} g_{0i}}{\sum_{i=1}^{N} p_{0i} g_{0i} + \frac{1}{2} \sum_{i=1}^{N} p_{0i} g_{1i} \frac{1}{2} \sum_{i=1}^{N} p_{1i} g_{0i}} \qquad (3)$$

where TP is a true positive, FP is false positive, and FN is a false negative. P is the set of predicted binary labels, G is the set of ground truth labels, 0 represents the background class, and 1 represents the foreground class. The p_{0i} represents the probability of a pixel i belonging to the foreground class and g_{0i} is the ground truth training label.

Tversky Loss. Similar to DSC, the Tversky index is defined as follows

$$TI = \frac{2TP}{2TP + \alpha FP + \beta FN} = \frac{|P \cap G|}{|P \cap G| + \alpha|P - G| + \beta|G - P|}$$

$$= \frac{\sum_{i=1}^{N} p_{0i} g_{0i}}{\sum_{i=1}^{N} p_{0i} g_{0i} + \alpha \sum_{i=1}^{N} p_{0i} g_{1i} + \beta \sum_{i=1}^{N} p_{1i} g_{0i}} \tag{4}$$

where $|P - G|$ is FP and $|G - P|$ is FN , α and β control the trade-off between FP and FN respectively. TI equals to DSC when α and β equals 0.5. In our experiments, we set $\alpha = 0.3$ and $\beta = 0.7$ then Tversky Loss (LT) is defined as follows

$$L_T = (1 - TI) \tag{5}$$

The resulting segmentation using the Dice loss function usually has a high precision rate but a low recall rate. By adding tunable parameters α and β, a higher weight can be assigned to FN, so the recall rate can be improved. Then, a better balance of precision and recall can be achieved [7].

Mixed Focal Loss. Mixed Focal Loss function is proposed by Yeung (2021) [13] that involves further modifications of Dice-based and Cross entropy-based loss functions with adjustable parameters to solve output imbalance, also with focal parameters to handle input imbalance. Firstly, the Modified Cross entropy loss $(LmCE)$ is defined as follows

$$L_{mCE} = -\frac{1}{N} \sum_{i=1}^{N} (\beta(t_i - log(p_i)) + (1 - \beta)[(1 - t_i) \ln(1 - p_i)] \tag{6}$$

where β controls the relative weights assigned to false positive and negative samples. Also, the modified Dice loss (L_{md}) is derived from the standard dice loss function and defined as follows

$$L_{md} = (1 - mDSC) \tag{7}$$

$mDSC$ is the Modified Dice Similarity Coefficient which is defined as follows

$$mDSC = \frac{\sum_{i=1}^{N} p_{0i} g_{0i}}{\sum_{i=1}^{N} p_{0i} g_{0i} + \delta \sum_{i=1}^{N} p_{0i} g_{1i} + (1 - \delta) \sum_{i=1}^{N} p_{1i} g_{0i}} \tag{8}$$

where δ controls the relative contribution of false positive and false negative predictions to the loss, by adding focal parameters γ, Dice and Cross-Entropy losses can better handle the class imbalance inputs. Therefore, separate focal parameters were added to both the modified Cross Entropy Loss $(LmCE)$ and modified Dice Loss (LmD), then we can find Modified Focal Loss (LmF) and Modified Focal Dice Loss $(LmFD)$ in Eq. 9 and 10 respectively.

$$L_{mF} = -\alpha(1 - p_t)^{\gamma} L_{mCE} \tag{9}$$

$$L_{mFD} = (1 - mDSC)^{\frac{1}{r}} \qquad (10)$$

Mixed Focal Loss (LMF) is defined as the weighted sum of Modified Focal Loss (LmF) and Modified Focal Dice Loss ($LmFD$) as follows

$$L_{MF} = \lambda L_{mF} + (1 + \lambda)L_{mFD} \qquad (11)$$

where $\lambda \in [0, 1]$ and λ can adjust the related weighting of the two modified loss functions. Mixed Focal Loss function is a compound Loss function involving modification on the Dice-based and cross-entropy based loss functions. With further modification on the loss functions, Dice-based and cross-entropy based loss functions with new tunable parameters can better handle the output imbalance, and new focal parameters can better handle the input imbalance [13].

3   Datasets

Our experiments are performed on four datasets namely Ships in Google Earth, Breast Ultrasound, Drone images, and Crack Segmentation, in which the images in these datasets have foreground-background imbalance problems.

Table 1. Imbalance ratio for each dataset

Dataset	Foreground	Background
Ships in Google Earth dataset	8.3% (Ships)	91.7%
Breast Ultrasound Image dataset	5.2% (Breast Tumor)	94.8%
Drone Image Dataset	0.4% (Drones)	99.6%
Crack Segmentation Dataset	7.6% (Cracks)	92.4%

3.1   Ships in Google Earth

Ships in Google Earth[1] is a satellite image dataset containing 794 Google earth images and are partitioned into training and testing sets. Each image has at least one ship, which is considered as the foreground, and the background in each picture is the sea. The ships in each picture take a small portion of the whole page, and it can be considered a foreground-background imbalanced image.

---

[1] https://www.kaggle.com/tomluther/ships-in-google-earth.

3.2 Breast Ultrasound Images

Breast Ultrasound images dataset [2] is a collection of images among women in different age groups, ranging from 25 to 75 years old. It contains 780 images with an average size of 500*500 pixels, and this dataset was divided into three classes: normal, begin, and malignant. All images are stored in PNG format. For the images in the group begin and malignant, the tumour is the foreground, and other tissues are considered as the background. For the images in the group normal, there is no tumour images. This dataset is foreground-background imbalanced since the tumour in each picture takes a small portion.

3.3 Drone Images

The Drone images dataset[2] contains pictures of drones produced by the MetaVision team. It contains 4449 images of drones and the corresponding masks. The pictures of drones were taken in different shooting positions and angles. Also, the flying heights of drones were different in each picture. The foreground in each image is the drone, and the background is the sky.

3.4 Crack Segmentation

The Crack Segmentation dataset [11] collects the pictures of line crackers on the wall, and some of the pictures do not have any crackers. It contains around 11,200 images merged from 12 available crack segmentation datasets, and the image size is 448*448. The foreground is the line crackers in each image, and the background is the wall.

4 Experiments

4.1 Imbalance Ratio Description

The four datasets are all highly imbalanced imaging datasets. We calculate the imbalance ratio for each dataset to assess the degree of the imbalance. The imbalance ratio is defined as the percentage of foreground and background pixels/vowels as shown in Table 1.

4.2 Data Annotation

The Ships in Google Earth dataset only include the original image, and it does not originally contain the mask (or ground truth). The deep learning training input needs both original image and ground truth. It is necessary to build its relative ground truth and apply it to build our deep learning models. Specifically, we manually annotated a total of 556 images by using both Photoshop and Matlab software. Firstly, we used Photoshop to annotate the photos by using the Lasso tool to select the ship in the picture. Secondly, we convert the ship part to white colour as the foreground class while the sea part as the background class in black colour, and then we save it as the BMP file format. Finally,

---

[2] https://www.kaggle.com/metavision/accurate-drone-shapessegmentation.

we use MATLAB to convert the BMP file format to a PNG. BMP is both uncompressed and lossless, whereas PNG is compressed but lossless. In this way, the PNG file size is smaller than BMP, which is convenient to reduce the burden of memory consumption of later network training. For the other three data sets, data annotation is not required because the mask is already included in the original dataset. Figures 3a and 3b show the samples of the original and annotated images respectively.

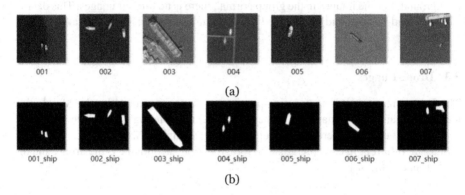

Fig. 3. The example of the original images (a) and their masks after data annotation (b).

4.3   Data Preprocessing

Datasets contain different image sizes, so it is necessary to resize all image input before training the model. According to the specification of U-Net architecture, the original images with corresponding labels have been resized to 256 × 256. The reduced size of the training set can reduce the burden on the computer without affecting the accuracy of recognition. Then rotate the tensor of shape from [width, height, channels] to [channels, height, width], which is best suitable for the U-Net training process. To verify the correctness of the model and prevent overfitting, we divide the resized image and mask into training and validation sets.

4.4   Model Training and Evaluation

In the model training step, to be consistent, we define four loss functions and use the same parameter values. We set $\alpha = 0.5$ and $\gamma = 2$ for Focal loss. In contrast, we set the trade-off of Tversky loss with hyperparameters $\alpha = 0.7$ and $\beta = 0.3$. For the Mixed Focal loss, we use $\delta = 0.4$, $\alpha = 0.25$, $\beta = 0.5$, and $\gamma = 2$. We conduct the experiments with a simple train-test sequence on all four loss functions for 400 epochs with batch size = 8. Further, we use $\beta = 1$ for the F1-score. We also define our four performance metrics by comparing the prediction and target results, including accuracy, precision, recall, and F1-score. Then we use a pre-trained Resnet-34 (imageNet) as the encoder, which is a 34-layer convolutional neural network and has been regarded as a state-of the-art image classification model. It takes residuals from each layer and uses them in the subsequent connected layers. The decoder is based on the Fastai U-Net learner, which automatically constructs the decoder side in transforming the ResNet-34 with cross-connections.

The results are presented through TensorBoard with loss value and four different evaluation metrics, namely Accuracy, Precision, Recall and F1-score. The predicted image segmentations are also described in order to compare with the ground truth. The specific evaluation and analysis will be given in the group report later. In the model evaluation step, we used Tensorboard to evaluate our results. The Tensorboard is a tool for providing measurements and visualization.

5 Results and Discussion

In this section, we describe the results of Focal loss, Dice loss, Tversky loss, and Mixed Focal loss for the four different datasets. The results are shown in Table 2.

For the breast ultrasound segmentation, Mixed Focal loss, Tversky loss and Dice loss have precision and recall scores reaching above 0.9 after merely 150 epochs. On the contrary, the highest-scoring loss function peaked at around 0.8 for precision and recall scores after training for the full 400 epochs. Noticeably, Focal loss exhibits the worst performance in all metrics for all datasets, especially in the drone segmentation dataset, where Focal loss struggled to reach a precision score of 0.05. Comparing the compound Mixed Focal loss with Focal loss, Mixed Focal loss outperformed Focal loss across all metrics by a long shot. The Mixed Focal loss is associated with the highest precision score for both ships and crack segmentation and a similarly strong performance in F1-score. Mixed Focal loss was associated with the highest F1-score for both ships and crack segmentation despite slightly lower recall scores. In contrast, higher recall scores are obtained by Tversky loss, including the best recall score among all loss functions for ship segmentation, and the best recall scores among all the metrics across breast ultrasound, drone and crack segmentations, although decreased precision scores balance out the latter. On the other hand, Mixed Focal loss is seen to have a better recall score than the precision score for the breast ultrasound segmentation, and the highest precision score for this dataset is associated with Dice loss instead. Despite high precision scores, Dice loss is observed to have poor recall scores, with around a 3% performance decrease for the ships and drone segmentation and a 14% drop for the crack segmentation. Dice loss achieves the best scores from drone segmentation, with a 3-4% lead over the trailing Mixed Focal loss across all metrics.

As an apparent implication from the results, Focal loss is the worst loss function among the four loss functions, especially for highly imbalanced problems. However, the comparison might be biased because Focal loss is a distribution-based loss, yet the other three loss functions are all region-based. Region-based losses are designed to reduce the difference between the predicted and the actual segmentation or increase the overlap between these two. On the other hand, Focal loss reduces the loss attributed to well-classified instances while adapting the Cross-Entropy loss to deal with extreme class imbalances. The value of γ in Focal Loss is based on empirical evidence, combined with the fact that in practice, larger γ seems always to produce better results. It is unclear the methodology to compare these metrics with other losses. Moreover, the highly contrasting results in drone segmentation scores suggest that Focal loss has difficulty reconciling accuracy with recall, as several recent papers have also pointed out.

Classic loss functions tend to prioritise one of the component measures over the other, resulting in output imbalance.

In our experiments, the cross entropy-based Dice loss exhibited greater affinity to accuracy scores, and the Tversky loss scored higher in the recall. We also confirm that the compound Mixed Focal loss is associated with the best recall-precision balance due to a more balanced recall-precision trade-off with the combination of Focal, Dice and Tversky losses. Such characteristics may also contribute to its resilience to the class imbalance in satellite, medical and engineering-related problems. However, in severely imbalanced cases, the conventional Dice loss attained better performance than the innovative Mixed Focal loss function, reminding us that compound loss is not a one- size-fits-all solution.

Table 2. Results on ships, breast ultrasound, Drone, and crack segmentation datasets with imbalance ratio 8.3%, 5.2%, 0.4%, and 7.6% respectively. Numbers in boldface denote the highest values for each metric.

Dataset	Loss function	Focal loss	Dice loss	Tversky loss	Mixed focal loss
Ships in Google earth	Accuracy	0.993	0.997	0.997	**0.997**
	Precision	0.776	0.946	0.935	**0.954**
	Recall	0.891	0.913	**0.934**	0.926
	F1-score	0.829	0.928	0.934	**0.939**
Breast ultrasound	Accuracy	0.992	0.993	**0.996**	0.995
	Precision	0.942	**0.973**	0.970	0.961
	Recall	0.952	0.937	0.974	**0.976**
	F1-score	0.947	0.953	**0.972**	0.968
Drone	Accuracy	0.997	**0.999**	**0.999**	**0.999**
	Precision	0.054	**0.798**	0.695	0.770
	Recall	0.688	**0.821**	0.732	0.797
	F1-score	0.100	**0.807**	0.704	0.780
Crack segmentation	Accuracy	0.986	0.988	0.989	**0.992**
	Precision	0.774	0.883	0.799	**0.887**
	Recall	0.789	0.740	**0.892**	0.859
	F1-score	0.781	0.802	0.842	**0.872**

6   Conclusion

We present a comparative analysis between four loss functions to identify the most appropriate loss function to handle the foreground-background imbalance problem. Four highly imbalanced datasets from different domains are evaluated. The most suitable loss function for each dataset is selected by comparing various metrics such as accuracy,

F1-score, recall, and precision results. The difference in model performance among the evaluated loss functions highlights the importance of loss function selection in image segmentation tasks with imbalanced categories. Therefore, this study presents a fair comparison to evaluate the efficacy of the loss function to handle the F-B imbalance problem in image segmentation. The results showed that the loss functions: Tversky loss and mixed focal loss achieved the best performance compared to the other loss functions in most of the datasets.

References

1. Abraham, N., Khan, N.M.: A novel focal tversky loss function with improved attention u-net for lesion segmentation. In: 2019 IEEE 16th International Symposium on Biomedical Imaging (ISBI 2019), pp. 683–687. IEEE (2019)
2. Al-Dhabyani, W., Gomaa, M., Khaled, H., Fahmy, A.: Dataset of breast ultrasound images. Data Brief **28**, 104863 (2020)
3. Braytee, A., Liu, W., Kennedy, P.: A cost-sensitive learning strategy for feature extraction from imbalanced data. In: Hirose, A., Ozawa, S., Doya, K., Ikeda, K., Lee, M., Liu, D. (eds.) ICONIP 2016. LNCS, vol. 9949, pp. 78–86. Springer, Cham (2016). https://doi.org/10.1007/978-3-319-46675-0_9
4. Bressan, P.O., et al.: Semantic segmentation with labeling uncertainty and class imbalance. arXiv preprint arXiv:2102.04566 (2021)
5. Chen, J., Wu, Q., Liu, D., Xu, T.: Foreground-background imbalance problem in deep object detectors: a review. In: 2020 IEEE Conference on Multimedia Information Processing and Retrieval (MIPR), pp. 285–290. IEEE (2020)
6. Du, G., Cao, X., Liang, J., Chen, X., Zhan, Y.: Medical image segmentation based on u-net: a review. J. Imaging Sci. Technol. **64**(2), 20508–20511 (2020)
7. Hashemi, S.R., Salehi, S.S.M., Erdogmus, D., Prabhu, S.P., Warfield, S.K., Gholipour, A.: Tversky as a loss function for highly unbalanced image segmentation using 3d fully convolutional deep networks. arXiv preprint arXiv:1803.11078 (2018)
8. Lin, T.Y., Goyal, P., Girshick, R., He, K., Doll'ar, P.: Focal loss for dense object detection. In: Proceedings of the IEEE International Conference on Computer Vision, pp. 2980–2988 (2017)
9. Ronneberger, O., Fischer, P., Brox, T.: U-net: Convolutional networks for biomedical image segmentation. In: Navab, N., Hornegger, J., Wells, W.M., Frangi, A.F. (eds.) MICCAI 2015. LNCS, vol. 9351, pp. 234–241. Springer, Cham (2015). https://doi.org/10.1007/978-3-319-24574-4_28
10. Salehi, S.S.M., Erdogmus, D., Gholipour, A.: Tversky loss function for image segmentation using 3d fully convolutional deep networks. In: Wang, Q., Shi, Y., Suk, H.-I., Suzuki, K. (eds.) MLMI 2017. LNCS, vol. 10541, pp. 379–387. Springer, Cham (2017). https://doi.org/10.1007/978-3-319-67389-9_44
11. Shi, Y., Cui, L., Qi, Z., Meng, F., Chen, Z.: Automatic road crack detection using random structured forests. IEEE Trans. Intell. Transp. Syst. **17**(12), 3434–3445 (2016)
12. Wang, L., Wang, C., Sun, Z., Chen, S.: An improved dice loss for pneumothorax segmentation by mining the information of negative areas. IEEE Access **8**, 167939–167949 (2020)
13. Yeung, M., Sala, E., Sch"onlieb, C.B., Rundo, L.: A mixed focal loss function for handling class imbalanced medical image segmentation. arXiv preprint arXiv:2102.04525 (2021)
14. Yudistira, N., Kavitha, M., Itabashi, T., Iwane, A.H., Kurita, T.: Prediction of sequential organelles localization under imbalance using a balanced deep u-net. Sci. Rep. **10**(1), 1–11 (2020)

Electron Microscopy Image Registration with Transformers

Fuyu Feng[1(✉)], Tianzhu Zhang[1,2], Rui Sun[1], Jianfeng He[1], Zhiwei Xiong[1,2], and Feng Wu[1,2]

[1] School of Information Science and Technology,
University of Science and Technology of China, Hefei, China
ffy0701@mail.ustc.edu.cn
[2] Institute of Artificial Intelligence, Hefei Comprehensive National Science Center, Hefei, China

Abstract. We propose a new encoder-decoder framework based on transformer for electron microscopy (EM) image registration. Current image registration methods mostly require long running time and complicated parameter tuning or cannot model the correlation of pixels between adjacent images. In this paper, we propose an encoder-decoder framework based on transformer to consider the relationship of serial images. In the transformer encoder, we model the long-range dependencies of multiple reference images and focus on their relevant regions by self-attention mechanism. Then in the transformer decoder, we predict the feature of the undeformed source image by introducing a prediction query to interact with the features from the encoder. To our best knowledge, our method is the first to apply transformers to EM image registration tasks, which needs no extra parameter tuning and can produce more accurate deformation fields. Evaluated on two datasets, Cremi and FIB25, our method outperforms state-of-the-art methods with more precise registration results and competitive speed.

Keywords: Image registration · Electron microscopy images · Transformers

1 Introduction

Electron microscopy (EM) images play an important role in biomedical science, they provide scientists with the opportunity to explore brains by showing the structure of neurons at nanoscale and generating a comprehensive map of brain connections [8,12,26]. The current methods to acquire large-scale EM image data usually require cutting and imaging from a complete piece of tissue. In this process, the three-dimensional integrity of the tissue is destroyed, and mechanical cutting may introduce nonlinear deformation and artifacts to the sliced samples. Therefore, the recovery of these deformed sliced images is essential for subsequent analysis such as neuron segmentation [8] and reconstruction of 3D volumes [4].

© The Author(s), under exclusive license to Springer Nature Switzerland AG 2023
M. Tanveer et al. (Eds.): ICONIP 2022, LNCS 13625, pp. 14–25, 2023.
https://doi.org/10.1007/978-3-031-30111-7_2

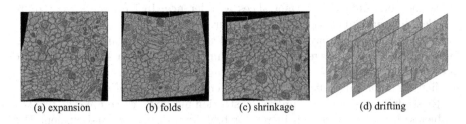

(a) expansion   (b) folds   (c) shrinkage   (d) drifting

Fig. 1. Examples of various tissue deformation in EM images including (a) expansion, (b) folds, (c) shrinkage, (d) drifting. Therefore, EM image registration is a challenging task.

Image registration is a classic computer vision task that focuses on aligning the structure of two images and warping one (source) image to a second (reference) image with the estimated deformation fields [9,10,19,24]. Compared with other types of medical images, EM images have more complex tissues and structures. There are also relatively similar content and texture as well as structural differences between adjacent EM images, bringing more challenges to EM image registration. Moreover, as shown in Fig. 1, there exists extensive tissue deformation such as shrinkage, folds, expansion and drifting of long-range image sequences, simply aligning one image to another by maximizing pixel-wise similarity is not able to recover the image structure very well.

Recently, there have been some traditional methods trying to solve the above challenges with hand-crafted features such as SIFT [15] and SURF [5]. BUnwarpJ [2] calculates a cubic B-splines spatial transformation to warp source images. The elastic alignment method [17] applies a global elastic constraint to images and utilizes block matching to minimize artificial deformation. However, all these methods are time-consuming because they are based on iterative optimizations. Moreover, they usually need complicated manual parameter tuning, which is undesirable in practical applications. With the rapid development of deep learning in the field of computer vision, some deep learning methods have been proposed for EM image registration. SsEMnet [25] is the first method to propose an end-to-end deep learning architecture for EM image registration. It takes a pair of images, a reference image and a source image, as input and utilizes spatial transformer networks [11] to align them. DualNet [18] is the first work to divide the transformation between images into a linear part and a nonlinear part. They use convolutional neural networks to estimate an affine transformation and a dense deformation field to warp source images. However, these two methods only take one image as reference, which may result in the accumulation of error and inaccurate registration results due to insufficient reference information. To address this issue, Zhou et al. [27] attempt to utilize multiple images with 3D convolutional neural networks to predict the deformation field. However, they just simply feed the source image together with multiple reference images into the neural networks, but fail to make good use of the relationship between consecutive reference images and model global information. Moreover, most of the above methods only take former slices as reference. In fact, there exist quite detailed differences between former and current slices. Compared with the current slice, former slices may not provide sufficient information and complete clues for recovering source images. Therefore, simply taking former slices as reference may result in the inaccuracy of the deformation field. It is more desirable to predict the deformation field with the current slice instead of former slices.

By studying previous EM image registration methods, we summarize two key problems that need to be considered for building an accurate and efficient EM image registration model. (1) The first is how to fuse the information from multiple reference images effectively. Current methods usually estimate the deformation field with two images. They fail to fuse and make good use of the information from multiple reference images. Previous slices of reference images contain the appearance characteristics of the tissue in different sections and the latent moving tendency of cells, enabling the model to predict the displacement of pixels more accurately. Therefore, modeling the long-range dependencies of multiple reference images is essential for estimating the deformation field of the source image. (2) The second one is how to predict the feature of undeformed source images. Current methods mostly take former slices as reference. However, there are a lot of differences in detail between former and current slices. It is more reasonable to estimate the deformation field with the undeformed source image. Therefore, it is desirable to consider how to predict the feature of undeformed source images for estimating more accurate deformation fields. Moreover, the deformation of slices may be different, making it difficult to directly predict the current slice with the information of former slices. Therefore, given different inputs, it is essential for our model to dynamically exploit different regions of reference images to predict the feature of undeformed source images.

Transformers were first proposed to model the relationship of words in natural language processing (NLP) [21]. Inspired by the success of transformers in NLP, more and more works apply transformers to computer vision tasks or combine CNN frameworks with attention mechanism [7,14,22,23]. In this paper, we propose an encoder-decoder framework based on transformers [23] to learn the relationship of reference images, which is able to predict the feature of the undeformed source image with multiple reference images for registration. In specific, we follow DualNet [18] to divide the deformation between images into a linear part and a nonlinear part. After we linearly warp the source image with the linear net, the source image and multiple reference images are fed into a weight-shared convolutional neural network to extract features. Then the features are fed into the transformer to predict the feature of the undeformed source image. Our transformer has an encoder-decoder architecture. **In the encoder**, we consider the long-range dependencies of multiple reference images and focus on the relevant regions between the feature of multiple reference images by self-attention mechanism. In this way, the challenges mentioned above can be overcome by aggregating valuable features for better representation. **In the decoder**, we predict the feature of the undeformed source image with multiple reference images. To be specific, we introduce a prediction query to do cross-attention with the

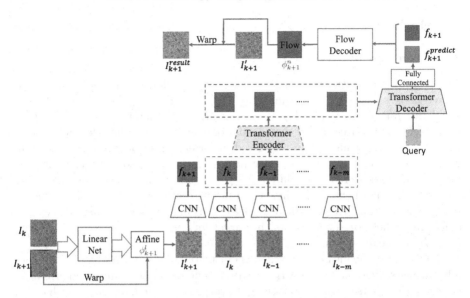

Fig. 2. The framework of our method. First, we use a linear net to estimate an affine deformation field for coarse registration. Then we utilize CNN to extract features and feed them into the transformer to predict the feature of the undeformed source image. Finally, we use a flow decoder to predict the dense deformation flow.

features from the encoder, which enables our model to focus on more meaningful regions for dynamically predicting the feature of source images. We observe that the output of the decoder adapts to the changes of reference images. Therefore, given different reference images as input, the query is able to predict the feature of undeformed source images dynamically. In this way, our method can make full use of multiple reference images and generate more accurate deformation fields by jointly considering the relationship between several adjacent images.

In summary, our main contributions are as follows:

- We model the long-range dependencies of multiple reference images with the proposed transformer encoder. It enables the network to effectively fuse the information from multiple reference images and focus more on their relevant regions.
- We introduce a prediction query in the transformer decoder, which is able to interact with the output features from the encoder and dynamically predict the feature of undeformed source images for estimating more accurate deformation fields.
- To our best knowledge, our method is the first to apply transformers to EM image registration tasks. Evaluated on Cremi and FIB25 [20] datasets, experimental results show that our method outperforms state-of-the-art methods with competitive accuracy and speed.

2 Method

2.1 Overview of Method

The task of image registration is to estimate a deformation field that warps source images to its reference image(s). Let $I = \{I_{k-m}, ..., I_{k-1}, I_k, I_{k+1}\}$ be a serial EM image set, where $\{I_{k-m}, ..., I_{k-1}, I_k\}$ are reference images and I_{k+1} is the deformed source image that needs to be registered. The reference images $\{I_{k-m}, ..., I_{k-1}, I_k\}$ have been registered in former iterations. In each iteration, we feed $m + 1$ registered former reference images and the source image into the model. After registered, the source image will be put into the reference image set for registering next source images in later iterations. The goal of our model is to estimate a deformation field ϕ_{k+1} for warping the source image I_{k+1}. The deformation field ϕ_{k+1} represents the target location of each pixel in I_{k+1}, and $\phi_{k+1}(I_{k+1})$ means warping I_{k+1} with ϕ_{k+1}.

The overall pipeline of our method is illustrated in Fig. 2. Firstly, we construct a linear net which takes I_k and I_{k+1} as input and estimates a linear deformation field ϕ_{k+1}^l for coarse registration, then we warp the source image I_{k+1} with ϕ_{k+1}^l, the linearly deformed source image is noted as:

$$I_{k+1}' = \phi_{k+1}^l(I_{k+1}) \tag{1}$$

With the linear deformed source image I_{k+1}', we then feed $\{I_{k-m}, ..., I_{k-1}, I_k, I_{k+1}'\}$ into a convolutional neural network to extract features for the transformer:

$$f_{k-m}, ... f_k, f_{k+1} = CNN(I_{k-m}, ..., I_k, I_{k+1}') \tag{2}$$

where $f_i \in \mathbf{R}^{H_f \times W_f \times C}$, C is the number of channels, H_f and W_f are the height and width of the feature map respectively. Then we feed the features of reference images into the transformer to predict the feature of the undeformed source image:

$$f_{k+1}^{predict} = Transformer(f_{k-m}, ..., f_k) \tag{3}$$

The $f_{k+1}^{predict}$ represents the predicted feature of the undeformed source image, and then we feed $f_{k+1}^{predict}$ and f_{k+1} into a flow decoder to predict a dense deformation field ϕ_{k+1}^n and finally warp I_{k+1}' with ϕ_{k+1}^n:

$$I_{k+1}^{result} = \phi_{k+1}^n(I_{k+1}') \tag{4}$$

Therefore, our final registration result can be obtained in two steps:

$$I_{k+1}^{result} = \phi_{k+1}(I_{k+1}) = \phi_{k+1}^n(\phi_{k+1}^l(I_{k+1})) \tag{5}$$

2.2 Linear Net

Since the real deformation of source images may be very drastic, simply predicting a dense deformation field to warp source images will result in an unsmooth deformation field. Therefore, we utilize a linear net to estimate an affine transformation to coarsely register the source image as shown in Eq. 1. The linear net includes a sequence of convolutional layers and outputs an affine transformation with six degrees of freedom. We utilize the affine transformation to achieve linear transformations such as translation, scaling, and rotation of images.

2.3 Transformer

In order to address the above challenges, we utilize the transformer to model multiple reference images. The detailed structure is shown in Fig. 3, which is mainly composed of an encoder and a decoder. We first flatten the features of reference images into $\bar{f} \in \mathbf{R}^{H_f W_f \times C}$. Then we concatenate these flattened features as $\bar{F} \in \mathbf{R}^{(m+1) \times H_f W_f \times C}$ and feed them into the transformer encoder. The core component of the transformer encoder is the multi-head self-attention layer, which is able to gather information from multiple reference images with attention mechanism. The multi-head self-attention procedure is defined as:

$$\bar{F}' = Norm(\bar{F}) \tag{6}$$

$$Q_i = \bar{F}' W_i^q, K_i = \bar{F}' W_i^k, V_i = \bar{F}' W_i^v, \tag{7}$$

$$Attention(Q_i, K_i, V_i) = softmax(\frac{Q_i K_i^T}{\sqrt{d_k}})V_i \tag{8}$$

$$F_i = Attention(Q_i; K_i; V_i) \tag{9}$$

where Norm is layer normalization [3], i is the index of heads number, $\sqrt{d_k}$ is a scaling factor, N_{head} is the number of heads and $W_i^q, W_i^k, W_i^v \in \mathbf{R}^{C \times \frac{C}{N_{head}}}$ are three linear layers applied to the input sequence to learn queries, keys and values. Then the features from several heads interacting with each other in the self-attention procedure are concatenated and passed to a linear layer by:

$$\hat{F} = Concat(F_1, F_2, ..., F_{N_{head}})W_d \in \mathbf{R}^{(m+1) \times H_f W_f \times C} \tag{10}$$

Then, as shown in Fig. 3, the output features are passed to the feed forward layer followed by layer normalization [3] in a residual manner for N times.

The transformer decoder is mostly the same as the encoder, the task of the decoder is to predict the feature of the undeformed source image with output features from the encoder. In the decoder, we propose a prediction query $Q \in \mathbf{R}^{H_f W_f \times C}$ whose learnable parameters are randomly initialized. It can be regarded as the prototype of the output predicted feature, which is able to focus on more meaningful regions for dynamically predicting the feature of the undeformed source image. In the decoder, the prediction query is first passed to a self-attention layer and then interacts with the features from the encoder in a cross-attention layer where the keys and the values are learned from the output features of the encoder and the queries are learned from the prediction query after self-attention. Then the output feature of the decoder is fed into a fully connected layer to produce the predicted feature of the undeformed source image

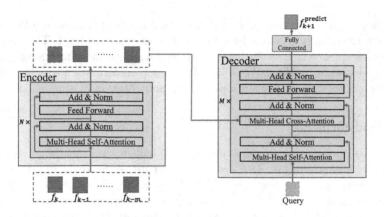

Fig. 3. The detailed architecture of the transformer. Our transformer is composed of an encoder and a decoder, which takes the features of reference images as input and outputs the predicted feature of the undeformed source image.

$f_{k+1}^{predict}$. Finally, $f_{k+1}^{predict}$ and f_{k+1} are concatenated and fed into a flow decoder to produce a dense deformation field ϕ_{k+1}^n. We notice that the output of the decoder can adapt to the change of reference images. Therefore, given different reference images, our model is able to dynamically predict the feature of undeformed source images and a more accurate deformation field with respect to the inputs.

2.4   Loss Function

The loss function consists of four parts. The first loss is the image intensity loss, which forces the linear deformed image and the final registered image to be similar with the nearest reference image:

$$L_I = \parallel I_{k+1}^{'} - I_k \parallel_2 + \parallel I_{k+1}^{result} - I_k \parallel_2 \tag{11}$$

Secondly, in order to promise $f_{k+1}^{predict}$ to be similar with the feature of the real undeformed source image, we feed the final registered source image I_{k+1}^{result} into the CNN of Eq. 2 to extract the feature map of I_{k+1}^{result} and compute the difference with $f_{k+1}^{predict}$:

$$L_F = \parallel f_{k+1}^{predict} - CNN(I_{k+1}^{result}) \parallel_2 \tag{12}$$

Then we follow [18] to add a constraint to avoid the linear transformation being too impractical by forcing every value of the affine transformation to be near zero:

$$L_A = \parallel \phi_{k+1}^l \parallel_1 \tag{13}$$

The last term of the loss function is the smoothness constraint of the dense deformation field to avoid the existence of discontinuous regions and randomness in the deformation field:

$$L_S = \parallel \nabla \phi_{k+1}^n \parallel_2 \tag{14}$$

where $\nabla\phi_{k+1}^{n}$ is the gradient of the nonlinear deformation field. Therefore, the total loss function is:

$$L = \lambda_1 L_I + \lambda_2 L_F + \lambda_3 L_A + \lambda_4 L_S \tag{15}$$

where λ_i is the regularization parameter.

3 Experiments and Results

3.1 Implementation Details

The training images are from Cremi dataset. We select 1,125 images at 1250 × 1250 resolution for training. In order to be consistent with the settings of [25,27], we apply random deformation on each image of the training set, testing set and their corresponding labels to mimic practical image deformation with a thin plate spline transformation [6] defined by random vectors on random positions, where the random vectors are sampled from a normal distribution with zero mean value and the random positions are uniformly distributed in the image. In the transformer encoder, we additionally embed learnable position encoding into the input features by element-wise addition. The flow decoder is composed of multiple convolutional layers and upsample layers followed by ReLU activation function except for the last layer which is activated by *tanh* function. Our model is implemented in PyTorch framework [16] and trained for 500 epochs using the Adam [13] optimizer with a batch size of 2 on a single NVIDIA V100 GPU. The learning rate is set to 0.001 and halved twice at the 20th epoch. The weights in the total loss are set as $\lambda_1 = 5$, $\lambda_2 = 0.2$, $\lambda_3 = 1$, $\lambda_4 = 0.1$. We take three reference images into the transformer by setting $m = 2$.

3.2 Evaluation on Cremi Testing Set

The Cremi dataset is a benchmark of MICCAI 2017 Challenge [1] which consists of 5 um^3 volumes of serial section EM images of the adult fly brain. It has been manually registered and segmented. To evaluate the performance of our method, we register the manually deformed image labels with the output deformation field from our model. Then we select the 50 largest neurons and calculate the average Dice coefficient by evaluating the overlap of neurons between the registered labels and the ground truth labels.

The quantitative result is shown in Table 1. Our method achieves the highest accuracy among the compared methods. As for the speed, our model is faster than most of the compared methods but slower than ssEMnet [25] and Dual-Net [18] because they use simple 2D convolution in their methods. However, our method achieves 4.6% and 1.7% higher Dice compared with ssEMnet [25] and DualNet [18]. Figure 4 shows a qualitative comparison between other methods. Our method achieves the most accurate registration result among all the compared methods. BUnwarpJ [2] and elastic method [17] fail to recover the source image when large deformation exists. The deep learning methods SsEMnet [25],

Table 1. Quantitative result and average running time on Cremi test set.

Method	Dice	Time(s)
bUnwarpJ [2]	0.647	2.43
elastic [17]	0.766	1.58
ssEMnet [25]	0.787	0.762
Fast 3D-Net [27]	0.793	1.72
DualNet [18]	0.816	0.126
Ours	**0.833**	**0.978**

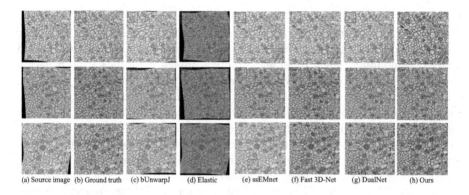

(a) Source image (b) Ground truth (c) bUnwarpJ (d) Elastic (e) ssEMnet (f) Fast 3D-Net (g) DualNet (h) Ours

Fig. 4. Registration results on Cremi test set where (a)–(h) columns represent source images, ground truth images, bUnwarpJ [2], elastic method [17], ssEMnet [25], Fast 3D-Net [27], DualNet [18] and our method respectively.

DualNet [18] and Fast 3D-Net [27] can remove most of the deformation, but there still exist artifacts and distortions at the boundaries of the images and complex areas. On the contrary, our method can achieve more accurate registration performance with competitive speed.

3.3 Evaluation on FIB25 Testing Set

FIB25 [20] is a dataset from the brain of drosophila. To test the generalization of our method, we evaluate our method on FIB25 dataset after fine-tuning on it for extra 50 epochs. The input images of our method are cropped to 512×512 resolution. Like Cremi testing set, we also randomly deform the ground images to generate data. Since FIB25 dataset does not contain neuron segmentation labels, we only report qualitative results in Fig. 5. Compared with other methods, our method can still achieve better registration quality, which proves the generalization and robustness of our method.

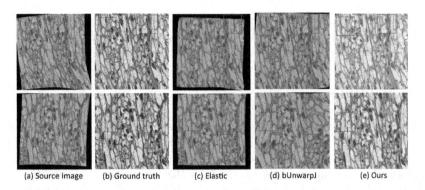

(a) Source image   (b) Ground truth   (c) Elastic   (d) bUnwarpJ   (e) Ours

Fig. 5. Registration results on FIB25 dataset. Our method is fine-tuned for extra 50 epochs. (a)–(e) columns represent source images, ground truth images, elastic method [17], bUnwarpJ [2] and our method respectively.

3.4   Ablation Study

To study the effectiveness of our proposed transformer and the number of reference images, we conduct a detailed ablation study to evaluate different settings of our method on the Cremi testing set. In the baseline, we remove the proposed transformer and directly feed the feature map of the linear deformed source image f_{k+1} and the feature map of the nearest reference image f_k into the flow decoder. Then, in order to study the influence of the numbers of reference images, we report the comparison experiments of different numbers of reference images in our proposed method. As shown in Table 2, our proposed transformer can achieve 1.6% higher Dice when only one reference image is available, which proves the effectiveness of the transformer. As the number of reference images increases, the network is able to take more adequate information as reference. Therefore, our method can benefit from more details of multiple reference images and produce higher Dice. However, we observe that Dice tends to gradually saturate when we take more than three images as reference. The reason may be that the contents of the reference images change dramatically as the distance

Table 2. Ablation study of the effectiveness of our proposed transformer and the number of reference images on Cremi testing set.

Method	#Reference Image	Dice	Time(s)
Baseline	1	0.791	0.265
Baseline+transformer	1	0.807	0.638
Baseline+transformer	2	0.824	0.797
Baseline+transformer(ours)	3	0.833	0.978
Baseline+transformer	4	0.835	1.169
Baseline+transformer	5	0.836	1.412

increases. As a result, taking more images as reference is not able to provide more helpful information but costs more running time. Therefore, we take three images as reference considering the trade-off between effectiveness and speed.

4   Conclusions

In this paper, we propose an unsupervised registration method for EM images with transformer. In our method, we utilize the transformer to model the long-range dependencies of multiple reference images and predict the feature of the undeformed source image. Experiments show that our method achieves the most accurate and robust performance compared with state-of-the-art methods.

References

1. Miccai 2017 challenge website. https://cremi.org/
2. Arganda-Carreras, I., Sorzano, C.O.S., Marabini, R., Carazo, J.M., Ortiz-de-Solorzano, C., Kybic, J.: Consistent and elastic registration of histological sections using vector-spline regularization. In: Beichel, R.R., Sonka, M. (eds.) CVAMIA 2006. LNCS, vol. 4241, pp. 85–95. Springer, Heidelberg (2006). https://doi.org/10.1007/11889762_8
3. Ba, J.L., Kiros, J.R., Hinton, G.E.: Layer normalization. arXiv preprint arXiv:1607.06450 (2016)
4. Bailoni, A., Pape, C., Wolf, S., Beier, T., Kreshuk, A., Hamprecht, F.A.: A generalized framework for agglomerative clustering of signed graphs applied to instance segmentation. arXiv preprint arXiv:1906.11713 (2019)
5. Bay, H., Tuytelaars, T., Van Gool, L.: SURF: speeded up robust features. In: Leonardis, A., Bischof, H., Pinz, A. (eds.) ECCV 2006. LNCS, vol. 3951, pp. 404–417. Springer, Heidelberg (2006). https://doi.org/10.1007/11744023_32
6. Bookstein, F.L.: Principal warps: thin-plate splines and the decomposition of deformations. IEEE Trans. Pattern Anal. Mach. Intell. 11(6), 567–585 (1989)
7. Carion, N., Massa, F., Synnaeve, G., Usunier, N., Kirillov, A., Zagoruyko, S.: End-to-end object detection with transformers. In: Vedaldi, A., Bischof, H., Brox, T., Frahm, J.-M. (eds.) ECCV 2020. LNCS, vol. 12346, pp. 213–229. Springer, Cham (2020). https://doi.org/10.1007/978-3-030-58452-8_13
8. Funke, J., et al.: Large scale image segmentation with structured loss based deep learning for connectome reconstruction. IEEE Trans. Pattern Anal. Mach. Intell. 41(7), 1669–1680 (2018)
9. Hu, B., Zhou, S., Xiong, Z., Wu, F.: Self-recursive contextual network for unsupervised 3d medical image registration. In: Liu, M., Yan, P., Lian, C., Cao, X. (eds.) MLMI 2020. LNCS, vol. 12436, pp. 60–69. Springer, Cham (2020). https://doi.org/10.1007/978-3-030-59861-7_7
10. Hu, X., Kang, M., Huang, W., Scott, M.R., Wiest, R., Reyes, M.: Dual-stream pyramid registration network. In: Shen, D., et al. (eds.) MICCAI 2019. LNCS, vol. 11765, pp. 382–390. Springer, Cham (2019). https://doi.org/10.1007/978-3-030-32245-8_43
11. Jaderberg, M., Simonyan, K., Zisserman, A., et al.: Spatial transformer networks. In: Advances in Neural Information Processing Systems, vol. 28 (2015)

12. Januszewski, M., et al.: High-precision automated reconstruction of neurons with flood-filling networks. Nat. Methods **15**(8), 605–610 (2018)
13. Kingma, D.P., Ba, J.: Adam: a method for stochastic optimization. arXiv preprint arXiv:1412.6980 (2014)
14. Liu, Z., et al.: Swin transformer: hierarchical vision transformer using shifted windows. In: Proceedings of the IEEE/CVF International Conference on Computer Vision, pp. 10012–10022 (2021)
15. Lowe, D.G.: Distinctive image features from scale-invariant keypoints. Int. J. Comput. Vis. **60**(2), 91–110 (2004)
16. Paszke, A., et al.: Automatic differentiation in pytorch (2017)
17. Saalfeld, S., Fetter, R., Cardona, A., Tomancak, P.: Elastic volume reconstruction from series of ultra-thin microscopy sections. Nat. Methods **9**(7), 717–720 (2012)
18. Shu, C., Xin, T., Zhou, F., Chen, X., Han, H.: Dual networks for high-precision and high-speed registration of brain electron microscopy images. Brain Sci. **10**(2), 86 (2020)
19. Sokooti, H., de Vos, B., Berendsen, F., Lelieveldt, B.P.F., Išgum, I., Staring, M.: Nonrigid image registration using multi-scale 3D convolutional neural networks. In: Descoteaux, M., Maier-Hein, L., Franz, A., Jannin, P., Collins, D.L., Duchesne, S. (eds.) MICCAI 2017. LNCS, vol. 10433, pp. 232–239. Springer, Cham (2017). https://doi.org/10.1007/978-3-319-66182-7_27
20. Takemura, S.Y., et al.: Synaptic circuits and their variations within different columns in the visual system of drosophila. Proc. Nat. Acad. Sci. **112**(44), 13711–13716 (2015)
21. Vaswani, A., et al.: Attention is all you need. In: Advances in Neural Information Processing Systems, vol. 30 (2017)
22. Wang, W., et al.: Pyramid vision transformer: a versatile backbone for dense prediction without convolutions. In: Proceedings of the IEEE/CVF International Conference on Computer Vision, pp. 568–578 (2021)
23. Wang, X., Zhang, S., Qing, Z., Shao, Y., Zuo, Z., Gao, C., Sang, N.: OadTR: online action detection with transformers. arXiv preprint arXiv:2106.11149 (2021)
24. Wu, G., Kim, M., Wang, Q., Munsell, B.C., Shen, D.: Scalable high-performance image registration framework by unsupervised deep feature representations learning. IEEE Trans. Biomed. Eng. **63**(7), 1505–1516 (2015)
25. Yoo, I., Hildebrand, D.G.C., Tobin, W.F., Lee, W.-C.A., Jeong, W.-K.: ssEMnet: serial-section electron microscopy image registration using a spatial transformer network with learned features. In: Cardoso, M.J., et al. (eds.) DLMIA/ML-CDS -2017. LNCS, vol. 10553, pp. 249–257. Springer, Cham (2017). https://doi.org/10.1007/978-3-319-67558-9_29
26. Zheng, Z., et al.: A complete electron microscopy volume of the brain of adult drosophila melanogaster. Cell **174**(3), 730-743.e22 (2018)
27. Zhou, S., et al.: Fast and accurate electron microscopy image registration with 3D convolution. In: Shen, D., et al. (eds.) MICCAI 2019. LNCS, vol. 11764, pp. 478–486. Springer, Cham (2019). https://doi.org/10.1007/978-3-030-32239-7_53

Deps-SAN: Neural Machine Translation with Dependency-Scaled Self-Attention Network

Ru Peng[1], Nankai Lin[2], Yi Fang[2], Shengyi Jiang[3], Tianyong Hao[4], Boyu Chen[5], and Junbo Zhao[1(✉)]

[1] College of Computer Science and Technology, Zhejiang University, Hangzhou, China
j.zhao@zju.edu.cn
[2] School of Information, Guangdong University of Technology, Guangzhou, China
fangyi@gdut.edu.cn
[3] School of Information Science and Technology,
Guangdong University of Foreign Studies, Guangzhou, China
[4] School of Computer Science, South China Normal University, Guangzhou, China
haoty@m.scnu.edu.cn
[5] Institute of Health Informatics, University College London, London, UK
boyu.chen.19@ucl.ac.uk

Abstract. Syntax knowledge contributes its powerful strength in Neural machine translation (NMT) tasks. Early NMT works supposed that syntax details can be automatically learned from numerous texts via attention networks. However, succeeding researches pointed out that limited by the uncontrolled nature of attention computation, the NMT model requires an external syntax to capture the deep syntactic awareness. Although existing syntax-aware NMT methods have born great fruits in combining syntax, the additional workloads they introduced render the model heavy and slow. Particularly, these efforts scarcely involve the Transformer-based NMT and modify its core self-attention network (SAN). To this end, we propose a parameter-free, **Dep**endency-scaled **S**elf-**A**ttention **N**etwork (Deps-SAN) for syntax-aware Transformer-based NMT. A quantified matrix of dependency closeness between tokens is constructed to impose explicit syntactic constraints into the SAN for learning syntactic details and dispelling the dispersion of attention distributions. Two knowledge sparsing techniques are further integrated to avoid the model overfitting the dependency noises introduced by the external parser. Experiments and analyses on IWSLT14 German-to-English and WMT16 German-to-English benchmark NMT tasks verify the effectiveness of our approach.

Keywords: Neural machine translation · Syntax knowledge · Transformer · Self-attention network

1 Introduction

Syntax knowledge occupies a pivotal position in learning the context of translation, i.e., the deep semantic modelling of the sentence. Consequently,

M. Tanveer et al. (Eds.): ICONIP 2022, LNCS 13625, pp. 26–37, 2023.
https://doi.org/10.1007/978-3-031-30111-7_3

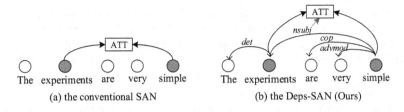

Fig. 1. Illustration of different attention network learning the correspondence between "experiments" and "simple".

incorporating syntax knowledge has attracted massive attention from the neural machine translation (NMT) community. Early works such as [7,8,38,40] assumed that the NMT model can draw out syntax knowledge from large-scale bilingual texts via attention networks [1,15,32]. However, were the facts really like that? Shi et al. [30] and Li et al. [14] found that the NMT network failed to capture sufficient internal syntax details of a sentence. The rationale behind the above failure is that the NMT model requires a parse-task-specific training paradigm to recover the hidden syntax in sentences. Going further, Yang et al. [36] revealed that this limitation of syntactic distillation arose from a deficiency in the attention mechanism. That is, the self-attention network (SAN) is solely controlled by two trainable parameter matrices when modeling the correspondences of query and key vectors:

$$QK^T = (HW_Q)(HW_K)^T = H(W_Q W_K^T)H^T, \tag{1}$$

where $\{Q, K\}$ are the query and key vectors, and H is the hidden representation of input token. $\{W_Q, W_K\}$ are the trainable parameter matrices. From this viewpoint of mathematical theory, in the absence of external syntactic constraints, the attention network is trained for syntactic parsing in a weakly supervised manner. To give an intuitive example in Fig. 1(a), the SAN individually computes the correspondence between "experiments" and "simple" without considering their syntactic dependency. Naturally, we expect that exploiting dependency syntax can guide the attention network to further improve the NMT performance, as done by Fig. 1(b).

Such being the case, let us revisit the existing research on syntax knowledge in NMT. Recent promising syntax-aware MMT methods mostly focused on RNN-based architecture, which roughly fall into four categories: a) extra tree-RNN/CNN modules [4,10,35]; b) linearizing dependency tree [16,18,27]; c) intermediate syntactic representations [5,39]; d) multi-task with dependence parsing [11,34]; e) neural syntactic distance [17,21,24]; e) syntax-aware data augmentation [9]. Although the great efforts these works have made, the additional syntactic workloads (i.e. modules, linearization, representations, etc) render the NMT model heavy and slow. Moreover, the referred approaches scarcely explored to the more advanced Transformer-based NMT architecture[1] [32], also do not modify the core SAN component by syntax information.

In response, we innovatively propose a parameter-free, **Dep**endency-scaled **S**elf-**A**ttention **N**etwork (Deps-SAN) for syntax-aware Transformer-based NMT. Motivated by the uncontrolled attention weight in Eq. 1, an idea that comes into our mind, we can inject the closeness of dependencies between tokens into the training of SANs to learn syntactic details. We enable this by proposing a dependency-scaled matrix employing the dependency distance of words derived from the parsed dependency tree. From our experiments, we find that this quantified matrix has notably imposed strong syntactic constraints as well as dispelled the dispersion of attention distribution. Besides, to tackle the problem of overfitting the dependency noises for our approach, we further partially modify the tuned attention weight by incorporating random sampling sparsing (RS-Sparsing) or k-value window sparsing (Wink-Sparsing).

Overall, our contributions are summarized as follows:

I. To guide the attention distribution of unconstrained SANs, we contribute a Deps-SAN for Syntax-aware NMT, which introduces quantitative syntax dependencies for the computation of traditional SANs.
II. Two knowledge sparsing methods are proposed to prevent Deps-SAN from overfitting dependency noises, which sheds some light on the anti-overfitting for SAN-based methods.
III. Extensive experiments and analyses on two benchmark NMT datasets demonstrate the effectiveness of our system.

2   Approach

In this section, we describe the methodological details of our approach. The overall framework of our Deps-SAN is depicted in Fig. 3. Specifically, applying our method to Transformer-based NMT is supposed to go through three steps: 1) deriving dependency-scaled matrix; 2) constructing Deps-SAN; and 3) adding knowledge sparsing.

2.1   Dependency-Scaled Matrix

To construct Deps-SAN, we first derive a dependency-scaled matrix to measure the closeness of syntactic dependencies, as shown in Fig. 2. Given the input sentence X with length I, in line with previous works [6,25,33], we extract the dependency tree T by an external syntax parser. Next, we compute the word-level dependency distance based on the parsed dependency tree. Specifically, we defined the dependency distance as the length of the path traversed from a word to another word on the tree, and the distance between two directly-connected

---

[1] Transformer, as the new generation of NMT baseline, abandons the recurrence and convolutions, and solely relies on SANs to achieve the incredible progress.

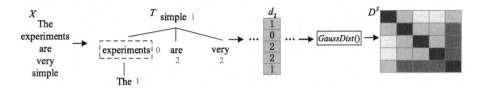

Fig. 2. The derivation pipeline of the dependency-scaled matrix.

words is assigned as 1. For example, as displayed in Fig. 2, the dependency distance of "experiments" itself is 0, the dependency distance of "simple" and "The" is 1, and the dependency distance of "are" and "very" is 2. In this way, the nearer the dependency distance between a word with the word "experiments", the closer the syntactic dependence between the word and the word "experiments". Then we traverse each word via the original word order and simultaneously count the dependency distance between the current traversed word and other words (including itself). After that, we combine all dependency distance sequences d_i and derive a dependency-scaled matrix $D^s \in \mathbb{R}^{I \times I}$ by a Gaussian distribution. Each row of the matrix denotes the closeness of syntactic dependence of each word to other words. As thus, when encoding the word "experiments", the SAN can discriminatively focus on words that are more correlated with their dependence with the help of the dependency-scaled matrix.

2.2 Dependency-Scaled Self-Attention Network

In this part, we will comprehensively introduce the overall architecture of Deps-SAN (i.e. Fig. 3) and how to apply it to Transformer-based NMT. For the source sentence X, the source annotation sequence H was initialized by the sum of the word embeddings E_x and the position encoding [12]. Further, the dependency-scaled matrix D^s and the source annotation sequence H are both fed as the input of the N attention heads in the Deps-SAN. Analogous to the Transformer[2] [32], the query, key and value vectors of l-th encoder layer is calculated as follow:

$$\{Q^l, K^l, V^l\} = \{H^l W_Q^l, H^l W_K^l, H^l W_V^l\}, \tag{2}$$

where $Q^l, K^l, V^l \in \mathbb{R}^{I \times d_k}$ are three vectors that are linearly projected from the sentence annotation H^l of the l-th encoder layer. Here, $d_k = d_{model}/H$ and H^0 is the initial source annotation sequence. Next, we compute the dot product between each query and all keys, and divide by $\sqrt{d_k}$ to obtain the alignment score S^l, which indicates how much attention should be placed on other words when annotating the current word. Following upon the alignment score, we explicitly impose syntactic constrains on this score by point-wise weighting the dependency-scaled matrix D^s, and forcing the model to focus on the syntactic dependencies among words.

---

[2] To distinctly illustrate our model, please refer to the original paper for details of Transformer-based NMT.

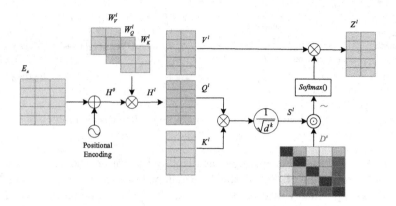

Fig. 3. The overall architecture of our Deps-SAN.

$$S^l = \frac{Q^l K^{l^T}}{\sqrt{d^k}} \tag{3}$$

$$\widetilde{S}^l = S^l \odot D^s \tag{4}$$

$$D_{ij}^s = GaussDist\,(d_{ij}) = \frac{1}{\sqrt{2\pi\sigma^2}} exp\left(-\frac{(d_{ij})^2}{2\sigma^2}\right), (i,j) = 1,...,I, \tag{5}$$

where $\widetilde{S}_i^l$ is the i-th row of $\widetilde{S}^l \in \mathbb{R}^{I \times I}$, which represents the scaled alignment score based on the dependency distribution of the i-th word x_i. σ is the Gaussian variance set with respect to a empirical value, usually a tiny number. Namely, we quantify the dependency closeness among words that are used for re-weighting alignment scores as the Gaussian probability density value. Here, d_{ij} is the dependency distance of the word pair x_i and x_j. Following by this, $GaussDist\,(d_{ij})$ as the $(i,j)^{th}$ entry of the D_{ij}^s, which is actually a Gaussian distribution with the variance σ^2 and the input d_{ij}.

$$Z^l = softmax(\widetilde{S}^l)V^l \tag{6}$$

$$O^l = Concat(Z_1^l, \ldots, Z_N^l)W_O^l. \tag{7}$$

After that, the scaled attention weights are normalized by the softmax function to represent the correspondences between words. So far, we have achieved that the closer the syntactic dependency to the current encoding word, the higher the attention weight can be assigned. The attention weight is multiplied by the value vector to produce the output representation of a single attention head. Finally, the concatenations of all output representations are passed into a linear projection layer to generate context vectors for subsequent decoding. In summary, when encoding sentences, the proposed Deps-SAN guides the model to attend to each word in varying scales according to the dependence closeness with the encoding word.

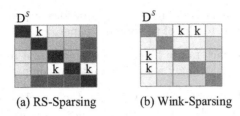

(a) RS-Sparsing   (b) Wink-Sparsing

Fig. 4. The illustration of two knowledge sparsing techniques.

2.3   Knowledge Sparsing

In this section, we narrate the motivation for introducing knowledge sparsing and elaborate on two proposed knowledge sparsing techniques in Fig. 4. Since the bilingual corpus does not equip with the parsed result, we are obliged to extract it from an external syntax parser. Unfortunately, the parsing accuracy of the syntax parser declines significantly as the distance between the head and dependent increases [19]. To alleviate this obstacle, two knowledge sparsing techniques are provided to confront over-fitting the syntax noise.

RS-Sparsing. In this practice, we randomly set each element of D^s to k with probability q, as illustrated in Fig. 4(a). We dilute partial syntactic information following the essence of dropout [31].

Wink-Sparsing. In this variant, we explicitly concentrate on position whose dependency distance is only within k and masking out other positions. This hard clipping mechanism tends to preserving partial syntactic knowledge instead of discarding them, which avert drastic information crash.

3   Experiment

3.1   Setup

Datasets. We evaluated the proposed approach on the widely-used IWSLT14 German-to-English (DE-EN) and WMT14 German-to-English translation tasks. For the low-resource IWSLT14 DE-EN translation task, the training set contained 160K sentence pairs[3] Following the standard instructions (Cettolo. et al. [3]), we sampled 5% of the training data for validation and combined multiple test sets IWSLT14.TED.{dev2010, dev2012, tst2010, tst1011, tst2012} for testing[4] For the large-scale WMT14 DE-EN translation task, the training set consisted of 4.5M sentence pairs. We selected the newstest2015 and newstest2016 as validation sets and test sets respectively. We used the raw word rather than the sub-word unit segmented by byte pair encoding algorithm [28] for better

---

[3] https://wit3.fbk.eu/archive/2014-01/texts/de/en/de-en.tgz.
[4] https://github.com/pytorch/fairseq/blob/master/examples/translation/prepare-iwslt14.sh.

dependency parsing. That is the reason our Transformer performance is slightly inferior to the reported results of the original paper.

Baseline Systems. We train a Transformer model as a robust baseline, and reproduce the following strongly related variants for comparison:

- **PASCAL** [2]: The parent-scaled SAN enable the model to focus on the dependency parent of each token in the encoding phase. An auxiliary regularization technique *parent ignoring* is formulated to avert over-fitting noisy dependencies.
- **Localness** [37]: The localness-aware SAN casts localness modeling as a learnable Gaussian bias to incorporate into SAN. In this paper, three alternatives of *fixed, layer-specific* and *query-specific* localness are proposed according to the local region that need to be paid attention.
- **Context** [36]: The context-aware SAN aim to leverage the internal states as context vectors to improve the SAN effect. It is divided into *Global, Deep* and *Deep-Global* three types, according to the diversity of embedded internal state and contextualization.
- **RPE** [29]: Shaw et al. extend the SAN by efficiently considering representations of the relative positions, or distance between sequence elements.

Settings. We extract the dependency parsing results from Stanford parser [20]. The source and target vocabularies size are both limited to 60K. All sentences are limited to 80 words and without aggressive hyphen splitting. Case-sensitive 4-gram BLEU [23] is used as the evaluation metric and paired bootstrap sampling [13] is applied for statistical significance test[5]. We train 80K and 200K steps on the low-resource ISWLT14 dataset and large-scale WMT14 dataset respectively to guarantee that the model has reached convergence. The hyperparameters k and q for knowledge sparsing are assigned to 6 and 0.1 respectively, which are sufficient to dilute the noisy dependencies of the sentence. We employ the beam search with a beam size of 5 and length penalty $\alpha = 0.6$ for inference. Other configurations keep the same as the paper reported by Vaswani et al. [32]. All NMT models are trained on an NVIDIA TITAN RTX using *Fairseq* toolkit [22].

3.2   Main Results

Table 1 shows the translation results of all NMT systems on the IWSLT14 DE-EN and WMT16 DE-EN translation tasks. Comparing all baselines, we draw the following major conclusions:

First, our model *+Deps-SAN+RS-Sparsing* accomplishes the best performance on both translation tasks without additional computation and parameters cost. These considerable gains are attributed to the dependency-scaled constraints of Deps-SAN and the anti-syntax-noise ability of RS-Sparsing.

---

[5] https://github.com/moses-smt/mosesdecoder/blob/master/scripts/analysis/bootstrap-hypothesis-difference-significance.pl

Table 1. Translation results of different NMT systems on IWSLT14 DE-EN and WMT16 DE-EN tasks. ‡/† indicate that the significance of our models is significantly better than that of the Transformer ($p < 0.01/0.05$). "#Speed" and "#Param" denote the training speed (seconds/each 100 batches) and the size of model parameters, respectively. We highlight the best results in bold for both tasks.

Systems	IWSLT14 DE-EN			WMT16 DE-EN	
	BLEU	#Speed	#Param	BLEU	#Param
Reproduced NMT systems of the existing work					
Transformer	29.88	23 s	74.85M	27.48	105.57M
+PASCAL	29.75	23 s	74.85M	27.55	105.57M
+PASCAL+parent ignoring	29.95	24 s	74.85M	27.73	105.57M
+Fixed Localness	28.72	24 s	74.87M	27.21	105.59M
+Layer-Spec. Localness	29.89	25 s	74.88M	27.23	105.60M
+Query-Spec. Localness	29.63	25 s	74.87M	27.46	105.59M
+Global Context	29.92	26 s	74.90M	27.45	105.62M
+Deep Context	30.03	26 s	74.98M	27.63	105.70M
+Deep-Global Context	30.07	26 s	75.03M	27.55	105.75M
+RPE	30.34	25 s	74.88M	27.71	105.60M
Our NMT systems					
+Deps-SAN	30.44‡	29 s	74.85M	27.94†	105.57M
+Deps-SAN+RS-Sparsing	**30.75‡**	29 s	74.85M	**28.21†**	105.57M
+Deps-SAN+Wink-Sparsing	30.64‡	29 s	74.85M	28.13†	105.57M

Second, all kinds of localness-aware SANs are inferior to the context-aware SANs. This fact reflects that the context vector directly enhances the sentence semantics compared with implicit localization. In our implementation, PASCAL does not derive practical improvement from focusing on the dependency parent of the word. This probably is related to the word closest to each word is not its dependency parent but its multiple dependency children. Injecting relative position representation into SAN improves translation performance to a certain extent.

Finally, towards anti-overfitting noisy dependencies, both +*parent ignoring* and our knowledge sparsing techniques (i.e. +*RS-Sparsing* and +*Wink-Sparsing*) bring a fundamental improvement over both translation tasks.

4 Analysis

4.1 Ablation Studies

In this section, we conduct two ablation experiments on the IWSLT14 DE-EN translation task. First, we investigate the effect of any combination of encoder layers equipped with Deps-SAN. Then, we consider to find the best variance value of Gaussian distribution via a grid search. To control the interference of variables, all ablation experiments are implemented on the encoder side solely.

Deps-SAN Layer. As shown in Table 2(a), the performance of the encoder layers plugged with Deps-SAN go up with the increase of layers from bottom to

Table 2. Validation and test BLEU score for ablation experiments on the IWSLT14 DE-EN translation task.

(a) Combinations of encoder layers

Layer	Valid	Test	Δ_v	Δ_t
[1-3]	31.42	30.44	-	-
[1-1]	30.71	29.83	-0.71	-0.63
[1-2]	30.82	30.00	-0.6	-0.44
[1-4]	30.82	29.89	-0.6	-0.55
[4-6]	30.41	29.46	-1.01	-0.98
[1-6]	29.20	28.79	-2.42	-1.65

(b) Grid search of Gaussian variance

Variance	Valid	Test	Δ_v	Δ_t
1	31.42	30.44	-	-
2	30.77	29.84	-0.65	-0.6
4	30.64	29.82	-0.78	-0.62
8	30.60	30.03	-0.82	-0.41
16	30.86	30.08	-0.56	-0.36
32	30.46	29.64	-0.96	-0.8

top, until the 3th-layer. Our model benefits more from replacing the lower three layers than that of the higher three layers. This interesting finding points to the identical conclusion as recent studies [26]— different layers tend to capture different features. Further, we infer that the lower layers of the SAN are incline to focus on syntactic information among words, while the higher layers prefer to concentrate on the semantic information of sentence level. Based on this impressive insight, we fix the Deps-SAN into the lower three layers to maximize performance.

Gaussian Variance. From Table 2(b), we observe $\sigma = 1$ as the optimal Gaussian variance. The best results with a variance of 1 benefit from strong supervision of syntactic dependencies by setting a minimum scale. Through the bell-shaped curve and non-zero properties, the Gaussian distribution allows the model to focus on the centre of distribution and ensured the integrity of information simultaneously.

4.2   Performance over Sentence Length

Following Bahdanau et al. [1], on the IWSLT14 DE-EN translation task, we divided entire test sentences into 9 disjoint groups according to their lengths[6]. Figure 5 presents the performance of varying NMT systems over groups of different sentence lengths. We discover that the performance of the proposed method maintains superiority over different sentence lengths. This success symbolizes our method is available on both long and short sentence translation. We are firmly sure this advancement is attributable to explicit syntactic guidance from the Deps-SAN. The sub-optimal RPE model in the main results is also the second-best in terms of the effect on different sentence lengths.

4.3   Case Study

We exhibit the translation results of different NMT models for a test example sentence in Table 3. For this interrogative sentence, the translation mistakes of

---

[6] These groups along with its sentence number are listed below: ([0–10],1657), ([10–20],2637), ([20–30],1381), ([30–40],614), ([40–50],252), ([50–60],122), ([60–70],43), ([70-80],27) and ([80-],17)

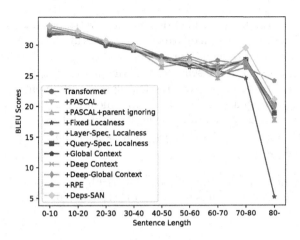

Fig. 5. Performance of different NMT models on test sentences of different lengths.

Table 3. Translation examples generated by different models. Parts of mistake, correct, and under-translation are highlighted in red, blue and underline respectively.

Source sentence	wie wäre es, länger bei guter gesundheit zu leben ?
Reference sentence	how about living longer with good health ?
Transformer	what would it be like to live on good health longer ?
+PASCAL	what would it be to live longer with good health ?
+PASCAL+parent ignoring	how about living in good health longer ?
+Fixed Localness	what would it be like to live longer on good health ?
+Layer-Spec. Localness	what would it be like to live longer on good health ?
+Query-Spec. Localness	what would it be like to live longer on good health ?
+Global Context	how would it be to live longer with good health ?
+Deep Context	how about living longer ?
+Deep-Global Context	how about living longer for good health ?
+RPE	what would it be like to live in good health longer ?
+Deps-SAN (ours)	**how about living longer with good health ?**

other models were mainly divided into three categories: sentence pattern errors, misuse of prepositions, and under-translation. Our model relies on explicit constraints of syntactic dependencies to capture the correct sentence pattern. The attained accurate translation without any omissions substantiates the reliability of our approach once again.

5 Conclusion

This research sheds new light on elevating the Transformer-based syntax-aware NMT. Specifically, this paper proposes a dependency-scaled self-attention network embedded with the quantified dependency distribution to learn syntactic details and dispel the dispersion of attention distributions. Two knowledge sparsing techniques are used to avoid our model overfitting the external dependency noises. We also investigate the sensitivity of the proposed approach to hyperparameters and their performance over the translation of different sentence lengths and a test example sentence. Substantial experimental results and analyses show that our system yields considerable benefits. In the future, we plan to extend this solution to SAN-based NLP tasks, such as grammar error correction and syntactic parsing tasks.

Acknowledgment. This work was supported in part by the National Natural Science Foundation of China under Grants 62071131, 61771149 and 61772146.

References

1. Bahdanau, D., Cho, K., Bengio, Y.: Neural machine translation by jointly learning to align and translate. In: ICLR (2015)
2. Bugliarello, E., Okazaki, N.: Enhancing machine translation with dependency-aware self-attention. In: ACL, pp. 1618–1627 (2020)
3. Cettolo, M., Niehues, J., Stüker, S., Bentivogli, L., Federico, M.: Report on the 11th iwslt evaluation campaign, IWSLT 2014. In: IWSLT, vol. 57 (2014)
4. Chen, H., Huang, S., Chiang, D., Chen, J.: Improved neural machine translation with a syntax-aware encoder and decoder. In: ACL, pp. 1936–1945 (2017)
5. Chen, K., et al.: Neural machine translation with source dependency representation. In: EMNLP, pp. 2846–2852 (2017)
6. Chen, K., Wang, R., Utiyama, M., Sumita, E., Zhao, T.: Syntax-directed attention for neural machine translation. In: AAAI, pp. 4792–4799 (2018)
7. Chen, K., Zhao, T., Yang, M., Liu, L.: Translation prediction with source dependency-based context representation. In: AAAI, vol. 31 (2017)
8. Choi, H., Cho, K., Bengio, Y.: Context-dependent word representation for neural machine translation. Comput. Speech Lang. **45**, 149–160 (2017)
9. Duan, S., Zhao, H., Zhang, D., Wang, R.: Syntax-aware data augmentation for neural machine translation. arXiv preprint arXiv:2004.14200 (2020)
10. Eriguchi, A., Hashimoto, K., Tsuruoka, Y.: Tree-to-sequence attentional neural machine translation. In: ACL, pp. 823–833 (2016)
11. Eriguchi, A., Tsuruoka, Y., Cho, K.: Learning to parse and translate improves neural machine translation. In: ACL, pp. 72–78 (2017)
12. Gehring, J., Auli, M., Grangier, D., Dauphin, Y.N.: A convolutional encoder model for neural machine translation. arXiv preprint arXiv:1611.02344 (2016)
13. Koehn, P.: Statistical significance tests for machine translation evaluation. In: EMNLP, pp. 388–395 (2004)
14. Li, J., Xiong, D., Tu, Z., Zhu, M., Zhang, M., Zhou, G.: Modeling source syntax for neural machine translation. In: ACL, pp. 688–697 (2017)
15. Luong, T., Pham, H., Manning, C.D.: Effective approaches to attention-based neural machine translation. In: EMNLP, pp. 1412–1421 (2015)
16. Ma, C., Liu, L., Tamura, A., Zhao, T., Sumita, E.: Deterministic attention for sequence-to-sequence constituent parsing. In: AAAI 2017, pp. 3237–3243 (2017)
17. Ma, C., Tamura, A., Utiyama, M., Sumita, E., Zhao, T.: Improving neural machine translation with neural syntactic distance. In: NAACL, pp. 2032–2037 (2019)

18. Ma, C., Tamura, A., Utiyama, M., Zhao, T., Sumita, E.: Forest-based neural machine translation. In: ACL, pp. 1253–1263 (2018)
19. McDonald, R., Nivre, J.: Analyzing and integrating dependency parsers. Comput. Linguist. **37**(1), 197–230 (2011)
20. Nivre, J., et al.: Universal dependencies v1: a multilingual treebank collection. In: LREC, pp. 1659–1666 (2016)
21. Omote, Y., Tamura, A., Ninomiya, T.: Dependency-based relative positional encoding for transformer NMT. In: RANLP, pp. 854–861 (2019)
22. Ott, M., et al.: FAIRSEQ: a fast, extensible toolkit for sequence modeling. In: NAACL, pp. 48–53 (2019)
23. Papineni, K., Roukos, S., Ward, T., Zhu, W.J.: Bleu: a method for automatic evaluation of machine translation. In: ACL, pp. 311–318 (2002)
24. Peng, R., Chen, Z., Hao, T., Fang, Y.: Neural machine translation with attention based on a new syntactic branch distance. In: Huang, S., Knight, K. (eds.) CCMT 2019. CCIS, vol. 1104, pp. 47–57. Springer, Singapore (2019). https://doi.org/10.1007/978-981-15-1721-1_5
25. Peng, R., Hao, T., Fang, Y.: Syntax-aware neural machine translation directed by syntactic dependency degree. Neural Comput. Appl. **33**(23), 16609–16625 (2021). https://doi.org/10.1007/s00521-021-06256-4
26. Raganato, A., Tiedemann, J.: An analysis of encoder representations in transformer-based machine translation. In: EMNLP Workshop, pp. 287–297 (2018)
27. Sennrich, R., Haddow, B.: Linguistic input features improve neural machine translation. In: WMT, pp. 83–91 (2016)
28. Sennrich, R., Haddow, B., Birch, A.: Neural machine translation of rare words with subword units. In: ACL. pp. 1715–1725 (2016)
29. Shaw, P., Uszkoreit, J., Vaswani, A.: Self-attention with relative position representations. In: NAACL, pp. 464–468 (2018)
30. Shi, X., Padhi, I., Knight, K.: Does string-based neural MT learn source syntax? In: EMNLP, pp. 1526–1534 (2016)
31. Srivastava, N., Hinton, G., Krizhevsky, A., Sutskever, I., Salakhutdinov, R.: Dropout: a simple way to prevent neural networks from overfitting. JMLR. **15**(1), 1929–1958 (2014)
32. Vaswani, A., et al.: Attention is all you need. In: Neurips, pp. 5998–6008 (2017)
33. Wang, W., Knight, K., Marcu, D.: Binarizing syntax trees to improve syntax-based machine translation accuracy. In: EMNLP-CoNLL, pp. 746–754 (2007)
34. Wu, S., Zhang, D., Zhang, Z., Yang, N., Li, M., Zhou, M.: Dependency-to-dependency neural machine translation. IEEE-ACM Trans. Audio Speech Lang. Process. **26**(11), 2132–2141 (2018)
35. Wu, S., Zhou, M., Zhang, D.: Improved neural machine translation with source syntax. In: IJCAI, pp. 4179–4185 (2017)
36. Yang, B., Li, J., Wong, D.F., Chao, L.S., Wang, X., Tu, Z.: Context-aware self-attention networks. In: AAAI, pp. 387–394 (2019)
37. Yang, B., Tu, Z., Wong, D.F., Meng, F., Chao, L.S., Zhang, T.: Modeling localness for self-attention networks. In: EMNLP, pp. 4449–4458 (2018)
38. Zhang, B., Xiong, D., Su, J., Duan, H.: A context-aware recurrent encoder for neural machine translation. TASLP. **25**(12), 2424–2432 (2017)
39. Zhang, M., Li, Z., Fu, G., Zhang, M.: Syntax-enhanced neural machine translation with syntax-aware word representations. In: NAACL, pp. 1151–1161 (2019)
40. Zhou, H., Tu, Z., Huang, S., Liu, X., Li, H., Chen, J.: Chunk-based bi-scale decoder for neural machine translation. In: ACL, pp. 580–586 (2017)

A Measurement-Based Quantum-Like Language Model for Text Matching

Wantong Zhang, Guobing Gan, Hui Gao, Peng Zhang$^{(\boxtimes)}$, Wenjie Hui, and Zipeng Fan

College of Intelligence and Computing, Tianjin University, Tianjin, China
{wt_zhang123,pzhang}@tju.edu.cn

Abstract. Quantum language models have attracted extensive attention in natural language processing tasks. Nevertheless, in this field measurement operators are mostly generated by randomly initialized parameter matrices, which cannot well explain the role of measurement operators in quantum theory. In this paper, we propose a Measurement-Based Quantum-like Language Model (MBQLM). Specifically, each word is considered a fundamental event in quantum probability space, which is a quantum state represented by a density matrix. We take the word density matrix in one sentence as a set of measurement operators to measure another sentence, which is consistent with the definition of measurement operators in quantum theory and has a specific semantic interpretation. The measured sentence state matrix representation effectively interact the feature information between sentences. To evaluate the performance of our model, we conduct experiments on WikiQA and TREC-QA datasets. The results show that our model achieves better performance than all benchmarks.

Keywords: Quantum Language Model · Quantum Measurement · Text Matching · Neural Networks

1 Introduction

Quantum language models have drawn increasing attention in Natural Language Processing tasks due to their transparency and interpretability [1,2]. Some studies have shown that there exist quantum-like phenomena in language cognition and language understanding [3,4]. A sentence can be viewed as a physical system composed of multiple correlated words. Therefore, the uncertainty of quantum probability can better represent the uncertainty of sentence semantics than classical probability [5]. In text matching tasks, the information interaction between sentences plays an important role in the final matching effect. Quantum theory is good at dealing with the complex interaction of multiple systems, so modeling text matching tasks using quantum language models is effective.

Sordoni, Nie and Bengio [6] for the first time proposed a Quantum Language Model (QLM) in Information Retrieval [7]. They used quantum theory to extend

© The Author(s), under exclusive license to Springer Nature Switzerland AG 2023
M. Tanveer et al. (Eds.): ICONIP 2022, LNCS 13625, pp. 38–47, 2023.
https://doi.org/10.1007/978-3-031-30111-7_4

statistical language model, which encodes the probability uncertainty of words using density matrices. Zhang et al. [8] creatively extended the quantum language model to the neural network architecture and proposed an end-to-end Neural Network based Quantum-like Language Model (NNQLM). The NNQLM used word embedding vectors as the analogy of state vectors, which is applied to build question and answer density matrices. Performing matrix multiplication on density matrices of two sentences obtained a joint matrix representation. To further explore the interpretability of neural networks, Li et al. [2] built an interpretable Complex-valued Network for Matching (CNM). The CNM encoded each word using a complex-valued vector, whose length represents the relative weight of the word, while the direction is considered as a superposition state.

Despite the exciting progress in quantum-inspired language models, there are still two challenges. First, the CNM uses a learnable parameter matrix as a measurement operator to extract features, which cannot be well explained in quantum theory and does not reflect the information interaction between sentences. Secondly, in the above models, the word density matrix obtained directly from the word embedding vector ignores semantic association between words.

To address the above two challenges, we come up with a Measurement-Based Quantum-like Language Model (MBQLM), which utilizes the word density matrix as a measurement operator to reconstruct the sentence representation and is more consistent with quantum theory. Our model use a Bidirectional Gated Recurrent Unit (BiGRU) [9] encoder to encode word vectors. Each word representation can better incorporates sentence context information, which makes up the defect that the word density matrix ignores semantic association between words. To evaluate the performance of our model, we conduct experiments on WikiQA and TREC-QA datasets. The empirical results demonstrate that our model achieves competitive performance compared with all benchmark quantum language models.

To summarize, the overall contributions of our work are as follows:

1. We propose a Measurement-Based Quantum-like Language Model, which utilizes word density matrix as measurement operator and has stronger interpretability.
2. We apply Bidirectional Gated Recurrent Unit (BiGRU) to encode word vectors, which strengthens the semantic association between the word density matrix.
3. We conduct extensive experiments on Wiki-QA and TREC-QA datasets. Our model outperforms state-of-the-art quantum language models on all datasets.

2   Quantum Preliminaries

In this section, we briefly introduce some basic concepts of quantum probability.

2.1   Quantum State

Because of the uncertainty of quantum states, we often describe them by quantum probability [10]. Generally, we represent the quantum probability space in the Hilbert space $\mathbb{H}^n$, but in order to be realistic, we limit the problem to the real space $\mathbb{R}^n$ [6]. In quantum system, we usually use the Dirac notation to represent a quantum state, and the Dirac notation is also called bra-ket notation. We denote a pure state $\vec{u} \in \mathbb{R}^n$ as a ket $|u\rangle$, and $|u\rangle$ is a column vector. The conjugate transpose of $|u\rangle$ is a bra $\langle u|$, which is a row vector. Any quantum pure state can be written as a linear superposition of a complete orthonormal basis $\{|e_i\rangle\}_{i=1}^n$:

$$|u\rangle = \sum_{i=1}^n \alpha_i |e_i\rangle \tag{1}$$

where the probability amplitudes $\{\alpha_i\}$ satisfy $\sum_{i=1}^n |\alpha_i|^2 = 1$.

A generalization of the conventional finite probability distributions in quantum probability theory is called density matrices. A density matrix ρ can be defined as a mixture over dyads:

$$\rho = \sum_i p_i |u_i\rangle\langle u_i| \tag{2}$$

where $\{u_i\}_{i=1}^n$ are pure states and $p_i \geq 0$ is the corresponding probability.

2.2   Quantum Measurement

As we all know, each particle has a number of different states [11], and it collapses into one of them when observed. Due to the uncertainty of the quantum system, we introduce quantum measurement to observe the internal situation of the quantum system. Then we define the measurement operator as M_m, which satisfies the completeness equation as:

$$\sum_m M_m^\dagger M_m = I \tag{3}$$

where $M_m^\dagger$ is the conjugate transposition of M_m.

Measurement operator affects the state space of the measured system, and index m represents the possible measurement result in experiment. If current state of quantum system is $|u\rangle$, the occurrence probability of the result m is:

$$p(m) = \langle u|M_m^\dagger M_m|u\rangle \tag{4}$$

The state of system after measurement is:

$$\frac{M_m|u\rangle}{\sqrt{\langle u|M_m^\dagger M_m|u\rangle}} \tag{5}$$

Define $E_m = M_m^\dagger M_m$, E_m is a semi-definite operator that satisfies Eq. 3 and Eq. 4, so the operator set E_m is sufficient to determine the probability of different measurement results. The complete combination E_m is called a positive operator-valued measure (**POVM**).

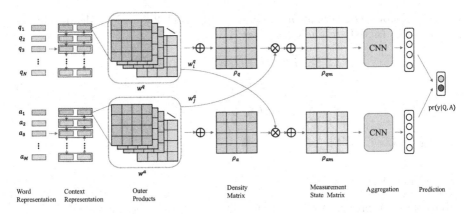

Fig. 1. Architecture for Measurement-Based Quantum-Like Language Models (MBQLM).

3 Methods

In this section, we describe our Measurement-based Quantum-like Language Model (MBQLM). The overall architecture is shown in Fig. 1. In answer selection Question Answering task, given a pair of sentences Q and A, the MBQLM model predicts the probability distribution $Pr(y \mid Q, A)$ through the flowing layers.

3.1 Word Representation

First, we choose pre-trained word embedding (GloVe) to represent each word in question and answer with a d-dimensional vector. We denote two input sentences as $Q = \{q_1, q_2, \cdots, q_N\}$ and $A = \{a_1, a_2, \cdots, a_M\}$, where N and M are the number of words in the question and answer sentences respectively. Finally, we obtain the question $Q \in \mathbb{R}^{d \times N}$ and answer $A \in \mathbb{R}^{d \times M}$.

3.2 Context Representation

QLM and NNQLM directly adopt the form of converting word vectors into density matrices, without considering the semantic connections between words. To build word context information, we utilize a bidirectional Gated Recurrent Unit (BiGRU) to encode question Q as follows:

$$\overrightarrow{h}_i^q = \overrightarrow{GRU}(\overrightarrow{h}_{i-1}^q, q_i)$$
$$\overleftarrow{h}_i^q = \overleftarrow{GRU}(\overleftarrow{h}_{i+1}^q, q_i) \qquad (6)$$
$$h_i^q = [\overrightarrow{h}_i^q; \overleftarrow{h}_i^q]$$

where $[;]$ denotes concatenation operation and $i \in \{1, 2, \cdots, N\}$. Meanwhile, we use the same BiGRU to encode answer A:

$$\overrightarrow{h}_j^a = \overrightarrow{GRU}(\overrightarrow{h}_{j-1}^a, a_j)$$
$$\overleftarrow{h}_j^a = \overleftarrow{GRU}(\overleftarrow{h}_{j+1}^a, a_j) \tag{7}$$
$$h_j^a = [\overrightarrow{h}_j^a; \overleftarrow{h}_j^a]$$

where $j \in \{1, 2, \cdots, M\}$.

3.3   Outer Products

After context representation layer encoding, we obtain the question $Q \in \mathbb{R}^{l \times N}$ and answer $A \in \mathbb{R}^{l \times M}$, where l is the dimension of word-embedding. Next, we introduce quantum theory into our MBQLM model. the ith word in a sentence can be represented as a state vector in semantic space:

$$c_i = \alpha_1 e_1 + \alpha_2 e_2 + \cdots + \alpha_n e_n \tag{8}$$

where α_i is its coefficient of the basis vector e_i.

Then, we map each word in the question (answer) sentence to a quantum event and construct a density matrix to represent each word in the question (answer) sentence through the outer product operation. Each density matrix is a pure state, which can be expressed as:

$$w_i^q = (c_i^q)^T c_i^q \tag{9}$$

$$w_j^a = (c_j^a)^T c_j^a \tag{10}$$

where the w_i is the density matrix representation of the word c_i in the sentence.

3.4   Density Matrix

Sentence representation based on density matrix, that is, add the density matrix of all the words in the sentence, which can be expressed as a mixed state density matrix:

$$\rho_q = \sum_i^N p_i w_i^q \tag{11}$$

$$\rho_a = \sum_j^M p_j w_j^a \tag{12}$$

where ρ_q is a question sentence, ρ_a is an answer sentence, they are all represented by the density matrix, and p_i is the corresponding probability of the ith word in a given sentence.

3.5   Measurement State Matrix

This is the core layer within our model. The goal of this layer is to achieve the feature interaction between question and answer sentences. We use the density matrix of each word in the answer (question) sentence as a measurement operator to measure the question (answer) sentence, and then get the state of the question sentence collapsed by the measurement operator. The state of the question (answer) sentence after being measured is expressed as:

$$\rho_{qm} = \sum_{i}^{M} p_i \frac{w_i^a \rho_q (w_i^a)^\dagger}{tr(w_i^a (w_i^a)^\dagger \rho_q)} \tag{13}$$

$$\rho_{am} = \sum_{j}^{N} p_j \frac{w_j^q \rho_a (w_j^q)^\dagger}{tr(w_j^q (w_j^q)^\dagger \rho_a)} \tag{14}$$

3.6   Aggregation

In this layer, we first use a 2D convolutional neural network to extract local feature information from the measurement state matrix ρ_{qm} and ρ_{am}. Then, the two vector sequences output by the convolutional neural network are aggregated into a fixed-length matching vector.

$$r_q = CNN(\rho_{qm})$$
$$r_a = CNN(\rho_{am}) \tag{15}$$
$$r_{qa} = [r_q; r_a]$$

where [;] denotes concatenation of each vector r_q and r_a.

3.7   Prediction

The goal of this layer is to predict the probability distribution $Pr(y \mid Q, A)$. We use a two-layer feed-forward neural network and ReLU activation function to calculate the degree of matching between the question and the answer.

4   Experiment

4.1   Datasets and Evaluation Metrics

We evaluate our model on WIKIQA and TREC-QA datasets. Relevant statistics of the two datasets are shown in Table 1. TREC-QA is the standard marker dataset used in the QA Track (8–13) of the Text REtrieval Conference (TREC) [13]. WIKIQA is a German open domain question answering system that uses Wikipedia as a knowledge base to answer natural language questions [14]. We use two evaluation indicators, Mean Average Precision (MAP) and Mean Reciprocal Rank (MRR) to evaluate the model.

Table 1. Statistics of TREC-QA and WikiQA datasets.

Dataset	Split	#Question	#Pairs	%PosRate
TrecQA	TRAIN	1229	53417	12.0
	DEV	65	1117	18.4
	TEST	68	1442	17.2
WikiQA	TRAIN	873	8672	12.0
	DEV	126	1130	12.4
	TEST	243	2351	12.5

Table 2. Results on TREC-QA and WIKIQA

Model	TREC-QA		Model	WIKIQA	
	MAP	MRR		MAP	MRR
LSTM-CNN-attn	0.7279	0.8322	CNN-Cnt	0.6520	0.6652
aNMM	0.7495	0.8109	LSTM-attn	0.6639	0.6828
PWIM	0.7588	0.8219	AP-BILSTM	0.6705	0.6842
MP-CNN	0.7770	0.8360	QA-BILSTM	0.6557	0.6695
PairwiseRank	0.7620	0.8540	PairwiseRank	0.6930	0.7100
QLM	0.6780	0.7260	QLM	0.5120	0.5150
NNQLM-I	0.6791	0.7529	NNQLM-I	0.5462	0.5574
NNQLM-II	0.7589	0.8254	NNQLM-II	0.6496	0.6594
QMWF-LM	0.7520	0.8140	QMWF-LM	0.6950	0.7100
CNM	0.7701	0.8591	CNM	0.6748	0.6864
QLM-EE	0.7713	0.8542	QLM-EE	0.6956	0.7003
MBQLM	**0.7787**	**0.8739**	MBQLM	**0.7017**	**0.7187**

4.2 Experiment Settings and Training

The initialization method of word embeddings in the word representation utilizes the 300-dimensional GloVe word vectors pretrained from the 840B Common Crawl corpus [20]. Meanwhile, the word embeddings for the out-of-vocabulary words were initialized randomly. In context representation layer, we set the hidden size as 100 for the BiGRU layer. For the aggregation layer, we apply batchNorm2d to normalize the output of the 2D convolutional neural network. ReLU activation function is used between two layers of feed-forward neural network. To reduce the occurrence of model overfitting, we use dropout in each layer of the model. When the model is trained, we utilize the ADAM optimizer to update the model parameters.

Table 3. Ablation study on two datasets.

Model	TREC-QA		WIKIQA	
	MAP	MRR	MAP	MRR
MBQLM-no-bigru	0.7553	0.8452	0.6714	0.6853
MBQLM-joint	0.7610	0.8460	0.6650	0.6770
MBQLM	**0.7787**	**0.8739**	**0.7017**	**0.7187**

4.3 Experimental Results

In order to intuitively evaluate the performance of MBQLM over TREC-QA and WIKIQA datasets, we compare our model with the neural network language models [15–19] and the quantum language models respectively. Table 2 shows the experiment results on TREC-QA and WIKIQA datasets. The experimental results show that the language modeling based on quantum theory is effective in QA tasks, and can greatly surpass the language modeling method based on machine learning under the framework of neural network, which can be further popularized.

We can also see that the performance of MBQLM is better than all benchmark quantum-inspired language models on both datasets. We can clearly notice that MBQLM improves the NNQLM-II [8] by 2.60% MAP and 5.87% MRR on TREC-QA and by 8.02% MAP and 8.99% MRR on WIKIQA, which proves that quantum measurement can significantly enhance the information interaction between question and answer sentences. Compared with QMWF-LM [12], MBQLM improves by 3.55% MAP and 7.35% MRR on TREC-QA. In addition, compared to CNM [2] and QLM-EE [21] using complex value to construct word embedding, our model achieves more efficient information interaction and has better performance.

4.4 Ablation Test

We conduct an ablation analysis to investigate the influence of each component on our proposed model. The ablation studies are divided into two groups to investigate the respective effects of the Bidirectional Gated Recurrent Unit (BiGRU) and the quantum measurement. MBQLM-no-bigru replaces joint matrix with the interactive method of quantum measurement and removes BiGRU encoder. MBQLM-joint applies the word vector encoded by Bidirectional Gated Recurrent Unit (BiGRU) as the quantum state representation of the word, and then get the joint matrix representation of the question and answer sentences. The test results in Table 3 demonstrate that each component plays a crucial role in the MBQLM model.

5   Conclusions and Future Work

In this paper, we propose a Measurement-based Quantum-like Language Model (MBQLM). The model utilizes the word matrix in the answer (question) sentence as a measurement operator to measure the question (answer) sentence, and then obtains the state of the question (answer) sentence collapsed after being measured. We show how to use quantum measurement to enhance the information interaction between question and answer, and BiGRU encoder is introduced to obtain sentence context information. We conduct experiments on WikiQA and TREC-QA datasets to evaluate the performance of our model. The empirical results demonstrate that our model achieves competitive performance compared with all benchmark quantum language models.

A future job is to consider using quantum measurement to do other tasks, such as short text pair matching tasks, and other ways to enhance the information interaction between question and answer pairs can also be explored. On the other hand, we believe that word order is also an important factor affecting the results. It is also worthy of further exploration to study the word order information from the perspective of quantum theory.

Acknowledgement. This work is supported in part by the Natural Science Foundation of China (grant No. 62276188 and No. 61876129), TJU-Wenge joint laboratory funding, Tianjin Research Innovation ' Students (grant No. 2021YJSB167), and Mind-Spore (https://www.mindspore.cn/) [22].

References

1. Basile, I., Tamburini, F.: Towards quantum language models. In: Proceedings of the 2017 Conference on Empirical Methods in Natural Language Processing, pp. 1840–1849 (2017)
2. Li, Q., Wang, B., Melucci, M.: CNM: an interpretable complex-valued network for matching. arXiv preprint arXiv:1904.05298 (2019)
3. Bruza, P., Kitto, K., McEvoy, D.: Entangling words and meaning. In: Quantum Interaction: Proceedings of the Second Quantum Interaction Symposium (QI-2008), pp. 118–124. College Publications (2008)
4. Aerts, D., Sozzo, S.: Quantum entanglement in concept combinations. Int. J. Theory Phys. **53**(10), 3587–3603 (2014)
5. Wang, B., Zhang, P., Li, J., Song, D., Hou, Y., Shang, Z.: Exploration of quantum interference in document relevance judgement discrepancy. Entropy **18**(4), 144 (2016)
6. Sordoni, A., Nie, J.Y., Bengio, Y.: Modeling term dependencies with quantum language models for IR. In: Proceedings of the 36th International ACM SIGIR Conference on Research and Development in Information Retrieval, pp. 653–662 (2013)
7. Blacoe, W., Kashefi, E., Lapata, M.: A quantum-theoretic approach to distributional semantics. In: Proceedings of the 2013 Conference of the North American Chapter of the Association for Computational Linguistics: Human Language Technologies, pp. 847–857 (2013)

8. Zhang, P., Niu, J., Su, Z., Wang, B., Ma, L., Song, D.: End-to-end quantum-like language models with application to question answering. In: Proceedings of the AAAI Conference on Artificial Intelligence, vol. 32 (2018)

9. Cho, K., et al.: Learning phrase representations using rnn encoder-decoder for statistical machine translation. arXiv preprint arXiv:1406.1078 (2014)

10. Von Neumann, J.: Mathematical Foundations of Quantum Mechanics. Princeton University Press, Princeton (2018)

11. Nielsen, M.A., Chuang, I.: Quantum Computation and Quantum Information (2002)

12. Zhang, P., Su, Z., Zhang, L., Wang, B., Song, D.: A quantum many-body wave function inspired language modeling approach. In: Proceedings of the 27th ACM International Conference on Information and Knowledge Management, pp. 1303–1312 (2018)

13. Wang, M., Smith, N.A., Mitamura, T.: What is the jeopardy model? a quasi-synchronous grammar for QA. In: Proceedings of the 2007 Joint Conference on Empirical Methods in Natural Language Processing and Computational Natural Language Learning (EMNLP-CoNLL), pp. 22–32 (2007)

14. Yang, Y., Yih, W.T., Meek, C.: WikiQA: a challenge dataset for open-domain question answering. In: Proceedings of the 2015 Conference on Empirical Methods in Natural Language Processing, pp. 2013–2018 (2015)

15. Tang, D., Qin, B., Liu, T.: Document modeling with gated recurrent neural network for sentiment classification. In: Proceedings of the 2015 Conference on Empirical Methods in Natural Language Processing, pp. 1422–1432 (2015)

16. Yang, L., Ai, Q., Guo, J., Croft, W.B.: aNMM: ranking short answer texts with attention-based neural matching model. In: Proceedings of the 25th ACM International on Conference on Information and Knowledge Management, pp. 287–296 (2016)

17. He, H., Lin, J.: Pairwise word interaction modeling with deep neural networks for semantic similarity measurement. In: Proceedings of the 2016 Conference of the North American Chapter of the Association for Computational Linguistics: Human Language Technologies, pp. 937–948 (2016)

18. Rao, J., He, H., Lin, J.: Noise-contrastive estimation for answer selection with deep neural networks. In: Proceedings of the 25th ACM International on Conference on Information and Knowledge Management, pp. 1913–1916 (2016)

19. dos Santos, C.N., Tan, M., Xiang, B., Zhou, B.: Attentive pooling networks. CoRR abs/1602.03609 (2016). http://arxiv.org/abs/1602.03609

20. Pennington, J., Socher, R., Manning, C.D.: Glove: global vectors for word representation. In: Proceedings of the 2014 Conference on Empirical Methods in Natural Language Processing (EMNLP), pp. 1532–1543 (2014)

21. Chen, Y., Pan, Y., Dong, D.: Quantum language model with entanglement embedding for question answering. arXiv preprint arXiv:2008.09943 (2020)

22. Mindspore (2020). https://www.mindspore.cn/

Virtual Try-on via Matching Relation
with Landmark

Hao Wu[1], Xingxing Yao[2], Baodi Liu[1], Xiaoping Lu[2(✉)], and Weifeng Liu[1(✉)]

[1] College of Control Science and Engineering, China University of Petroleum
(East China), Qingdao, China
liuwf@upc.edu.cn
[2] COSMOPlat Institute of Industrial Intelligence, Qingdao, China
luxiaoping@haier.com

Abstract. Virtual try-on based on image synthesis aims to combine
the customer's photo with in-shop clothes to acquire a try-on image.
The key to generating a realistic try-on image is whether the in-shop
clothes is spatially aligned with the customer's body. Prior methods
usually directly adopt the spatial transformation network to complete
the clothing deformation, but they cannot generate high-quality try-on
images when facing the customer's posture changes or complex clothes
pattern. To address it, we propose a virtual try-on network based on
landmark constraint (LCVTON) in this work. Specifically, we notice the
corresponding relationship between the clothes feature points and the
customer body feature points, making the clothes match more closely
with the customer's body. The matching of the feature points enables us
to introduce the landmark constraint into the spatial transformer network for naturally and smoothly warping clothes. Moreover, we construct
a refinement network and introduce the landmark constraint into it to
preserve the texture details of clothes. We conducted experiments on the
try-on dataset and compared our method with existing methods. Both
qualitative and quantitative results demonstrate the superiority of our
method compared to existing state-of-the-art method.

Keywords: Virtual try-on · Adversarial network generation ·
Landmark detection · Semantic segmentation

1 Introduction

Since the COVID-19 pandemic has blocked the traditional offline clothing industry, plenty of clothing companies have turned their attention to e-commerce platforms. One of the critical offline experiences missed by the online is the changing

This work was supported in part by the National Natural Science Foundation of China
under Grant 61671480, in part by the Major Scientific and Technological Projects
of CNPC under Grant ZD2019-183-008, in part by the Natural Science Foundation
of Shandong Province under Grant ZR2019MF073, and in part by the Fundamental Research Funds for the Central Universities, China University of Petroleum (East
China) under Grant 20CX05001A.

M. Tanveer et al. (Eds.): ICONIP 2022, LNCS 13625, pp. 48–60, 2023.
https://doi.org/10.1007/978-3-031-30111-7_5

room where a clothes item can be tried-on. Under such circumstances, virtual try-on (VTON) has developed rapidly, which requires the creation of a virtual try-on environment close to reality. To achieve it, recently, many virtual try-on methods have emerged [1–8].

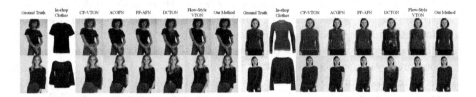

Fig. 1. Try-on results of our method and previous methods, CP-VTON [1], ACGPN [2], PF-AFN [3], DCTON [4] and Flow-Style-VTON [5]. In order to display our results intuitively, the customer's original clothes (ground-truth) are displayed here to compare the superiority of our method.

The existing methods such as PF-AFN [4] and the state-of-the-art method Flow-Style-VTON [5] are parsing-free models without the customer's human parsing for virtual try-on. They usually adopt a structure based on appearance flow to match the correspondence between the customer and the in-shop clothes. Although the parsing-free models simplify the try-on process, they could cause a lack of edge coverage and preserve few texture features. Besides, methods such as CP-VTON [1], ACGPN [2] and DCTON [4], which rely on customer parsing, usually build clothes warping module to complete the matching of customer and clothes. They adopted Thin Plate Spline (TPS) [9] or Spatial Transformation Network (STN) [10] warp the in-shop clothes. However, according to the previous analysis [2,3], directly implementing the warping network does not guarantee accurate transformations for challenging clothes with large posture changes. The comparison of the above methods is shown in Fig. 1.

To address the above-mentioned challenges, we propose a method named LCVTON, which consists of four modules. The first is the Pre-processing Module, which generates reasonable representations. Moreover, we detect landmark from the customer's photo and the in-shop clothes image. The second is the Semantic Generation Module, we learn a semantic indicator that gradually generates the customer's body mask to try on the in-shop clothes naturally. The third is the Clothes Warping Module, which warps the clothes to match the customer's body. It consists of three stages: the first stage synthesizes the clothes mask; the second stage combines the second-order difference constraint with the landmark constraint based on the Spatial Transformation Network (STN) [10]; the third stage is the refinement network, it is also constrained by the landmark, which generates the target clothes based on better warping clothes. The fourth module is the Image Fusion Module, which gathers the customer's body mask, the refined clothes image and the non-clothes regions image, adaptively determines the different body parts in the composite image, and finally generates the try-on image. With the above method, we produce the reality-photo successfully.

The main contributions of our work are summarized as follows:

(1) We propose a new virtual try-on network, LCVTON, which solves the problem of semantic synthesis of high-quality clothes generation and authentic try-on images.
(2) We combine the landmark constraints with second-order difference constraints in the clothes warping module, which will cause better warping clothes. Moreover, we construct a refinement network and introduce the landmark constraint into it to preserve the texture details of clothes, further promoting subsequent virtual try-on.
(3) Experiments prove that the proposed method is superior in qualitative and quantitative terms on VITON dataset compared to existing state-of-the-art method.

2   Related Work

2.1   Generative Adversarial Networks

Generative Adversarial Networks (GAN) [11] has become a significant field in image synthesis and image manipulation. Due to the powerful advantages of GAN, it drives the development of plenty of tasks such as computer vision [12–14], semantic segmentation [15–17] and image processing [18–20].

Pix2PixHD [17] proposes a coarse-to-refinement generator, a multi-scalar discriminator, and an improved loss to synthesize a realistic 2048×1024 image. StyleGAN [21] proposed an alternative generator structure for GAN, in which the generator structure starts with a learned input and adjusts the characteristics of the image in each convolution layer using different learned potential variables.

2.2   Virtual Try-On

Image-based virtual try-on aims to generate try-on images with photorealism. VITON [6] proposed the clothing agnostic human representation and the strategy which converts the clothes from coarse to refinement. CP-VTON [1] proposed a new learnable sheet spline transformation network (TPS) [9] to describe clothes warping, and a try-on module(TOM) to reduce the appearance of artifacts. However, simply applying TPS transformation cannot guarantee precise transformation, which will also deteriorate the afterward try-on synthesis. As a result, ACGPN [2] introduced second-order difference constraints to eliminate apparent warping, which can resist severe deformation. The DCTON [4] adopted a self-supervised manner for training following cyclic consistent learning to produce highly-realistic try-on images. Apart from the model with the human-parser, WUTON [8] guides the student network by training a network of teachers based on body analysis, which with parser-free. PF-AFN [3] formulates the knowledge extraction of matching problems to extract the appearance flow between the human body image and the garment image, which is beneficial for discovering the dense correspondence and generating high-quality images. Flow-Style-VTON [5] proposed a StyleGAN-based architecture is adopted for appearance flow estimation. However, previous methods also contain artifacts, as shown in Fig. 1.

3 Our Approach

The LCVTON is mainly composed of four modules, and its architecture is shown in Fig. 2. First, the Pre-processing Module mainly generates the in-shop clothes landmark $\mathcal{L}_{IM}$ and customer's clothes landmark $\mathcal{L}_{CM}$. In addition generates the customer's representations: the keypoint of human body posture $\mathcal{M}_P$, the body-fused mask $\mathcal{M}^F$ and the in-shop clothes $\mathcal{T}_C$. Second, the Semantic Generate Module predicts semantic segmentation image via the customer's representations, yielding the synthetic body mask $\mathcal{M}_\omega^S$. Third, the Clothes Warping Module unites $\mathcal{M}_\omega^S$ and $\mathcal{M}_P$ to predict the mask of warped clothes $\mathcal{M}_C^S$. The warped stage and the refinement stage are designed to warp in-shop clothes according to $\mathcal{M}_C^S$, where we could obtain the preliminary warping clothes $\mathcal{T}_C^W$ and the refined clothes $\mathcal{T}_C^R$. We introduce the landmark constraint to STN and refinement network to optimize the generation of refined clothes. Finally, the Image Fusion Module gathers the information from the processes mentioned above to determines the location information of each region of the customer's and finally synthesizes the try-on images.

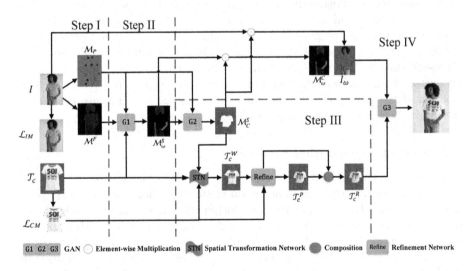

Fig. 2. The architecture above where there are Pre-processing Module (Step I), Semantic Generation Module (Step II), Clothes Warping Module (Step III) and Image Fusion Module (Step IV).

3.1   Pre-processing Module

The Pre-processing Module consists of two stages. Stage one is for landmark detection. Stage two is for the source image $\mathcal{I}$, preprocessed into two representations: person representation and semantic representation.

Fashion Landmark. We explicit the landmark constraint to generate realistic and exquisite clothes when generating warped clothes. We detect the landmark via HRNet [22] for both the in-shop clothes $\mathcal{T}_C$ and the customer's clothes, respectively denoted as $\mathcal{L}_{IM}$ and $\mathcal{L}_{CM}$. We adopt [23] detect different clothes categories and obtain clothing bounding box. The landmark is shown in Fig. 3.

Fig. 3. Landmark example for top clothes. Landmark is detected on clothes of different colors and in different poses at the top of image.

Customer Representation. Customer representation is composed of person representation and semantic representation. For person representation, we adopted the body shape map from the clothing-agnostic person representation obtained by [24], a 1-channel feature map of a blurred binary mask. For semantic representation, we adopt [24] for human parsing, which assigns semantic labels for customer photo. Semantic representation is obtained by removing the composition of the parsed clothes and the customer's arms to indicate the rest of the original customer's body characteristics.

3.2   Semantic Generation Module

The Semantic Generation Module generates semantic segmentation, which accurately indicates the semantic information of the customer's original body parts and the layout information of in-shop clothes and guides the synthesis of subsequent try-on images. As shown in Fig. 2, given the in-shop clothes $\mathcal{T}_C$, the body-fused mask $\mathcal{M}^F$ and the keypoint of customer posture $\mathcal{M}_P$, build a triplet of elements $[\mathcal{T}_C, \mathcal{M}^F, \mathcal{M}_P]$. We train U-Net [25] as generator to generate body mask $\mathcal{M}_\omega^S$, while we adopted a discriminator given in pix2pixHD [17] as discriminator to distinguish the generated mask. The CGAN loss consists of the adversarial loss $\mathcal{L}_{adv}$ and the pixel-wise cross-entropy loss $\mathcal{L}_{ce}$ [26], which can be formulated as

$$\mathcal{L}_{adv} = E_{x,y}[\log(\mathcal{D}(x,y))] + E_{x,z}[\log(1 - \mathcal{D}(x, \mathcal{G}(x,z)))], \tag{1}$$

$$\mathcal{L}_{ce} = -\sum_{i=1}^{n} S_i(i) \log S_t(i), \tag{2}$$

where x represents input, y represents ground-truth, and z represents noise. In Eq. 2, S_i is the output semantic segmentation, S_t is the ground-truth, and n is the number of pixels. Consequently, the loss for the Semantic Generation Module can be formulated as $\mathcal{L}_{SGM}$:

$$\mathcal{L}_{SGM} = \lambda_1 \mathcal{L}_{adv} + \lambda_2 \mathcal{L}_{ce}, \tag{3}$$

where λ_1 and λ_2 are the weight coefficients that distribute the loss terms.

3.3   Clothes Warping Module

The Clothes Warping Module aims to synthesize warped clothes to match customer's posture and preserve the original character of in-shop clothes. It contains three stages: warped clothes mask generation, clothes warping and clothes refinement. We introduce the second-order difference constraint and the landmark constraint on the clothes warping network to realize geometric matching and character retention. And we also construct a refinement network, which introduces the landmark constraint. As shown in Fig. 4, the transformation without landmark constraint shows obvious distortion in shape compared with the transformation with landmark constraint.

Warped Clothes Mask Generation. This stage aims to generate the warped clothes mask $\mathcal{M}_C^S$. Given $\mathcal{M}_\omega^S$ and $\mathcal{M}_P$ as input, we train Conditional GAN, which adopts U-Net as a generator and the discriminator given in pix2pixHD as a discriminator. Refer to Eq. 1 and Eq. 2, the CGAN loss can be formulated as

$$\mathcal{L}_{cm} = \lambda_3 \mathcal{L}_{adv} + \lambda_4 \mathcal{L}_{ce}, \tag{4}$$

where λ_3 and λ_4 are the weight coefficients that distribute the loss terms.

Clothes Warping. This stage aims to warp in-shop clothes. Given $\mathcal{T}_C$ and $\mathcal{M}_C^S$ as input, we adopt STN to match the in-shop clothes and the warped clothes. We introduce the second-order difference constraint and the landmark constraint to this stage. To supervise the training of the clothes warping model, we apply $\mathcal{L}_{warping}$ on it, which can be formulated as

$$\mathcal{L}_{warping} = \lambda_{lcw} \mathcal{L}_{lcwarping} + \lambda_{scw} \mathcal{L}_{scwarping} + \lambda_{cw} \mathcal{L}_{cwarping}, \tag{5}$$

where $\mathcal{L}_{scwarping}$ is originated from the description of the second-order difference constraint in ACGPN [2]. $\mathcal{L}_{lcwarping}$ is designed for the landmark constraint. $\mathcal{L}_{cwarping}$ is designed for the difference between the warped clothes $\mathcal{T}_C^W$ and the groud-truth clothes $\mathcal{T}_{CT}$. λ_{lcw}, λ_{scw}, λ_{cw} are used to balance each item. $\mathcal{L}_{lcwarping}$, $\mathcal{L}_{cwarping}$ can be formulated as

$$\mathcal{L}_{lcwarping} = \|W(\mathcal{L}_{IM}) - \mathcal{L}_{CM}\|_1, \tag{6}$$

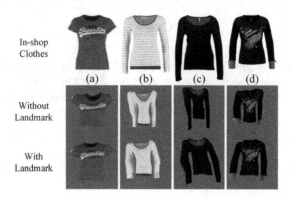

Fig. 4. Comparison of warping results with and without the landmark constraint. (a) the difference lies in the texture and characteristics of clothing; (b) the difference lies in the collar; (c)(d) the difference lies in the sleeves

$$\mathcal{L}_{cwarping} = \|W(\mathcal{T}_C) - \mathcal{T}_{CT}\|_1, \tag{7}$$

where $W(\ )$ denotes the STN transformation and $\|\ \|_1$ represents L_1 loss.

Refinement. This stage aims to synthesize the refined clothes with more details (refer to CP-VTON [1]). Given warped clothes $\mathcal{T}_C^W$ as input, we train U-Net to generate rendered clothes $\mathcal{T}_C^P$ and a composition mask $\mathcal{M}$. The loss for refinement $\mathcal{L}_{refinement}$ can be formulated as

$$\mathcal{L}_{refinement} = \lambda_{cr}\|R(W(\mathcal{T}_C)) - \mathcal{T}_{CT}\|_1 + \lambda_{lcr}\|R(W(\mathcal{L}_{IM})) - \mathcal{L}_{CM}\|_1 + \lambda_m\|\mathbf{1} - M\|_1, \tag{8}$$

where $R(\ )$ denotes the refinement network and λ_{cr}, λ_{lcr}, λ_m are the weighting parameters for each loss term.

Then utilize rendered clothes $\mathcal{T}_C^P$ and a composition mask $\mathcal{M}$ to finally combine the two clothing images as the refined clothing image $\mathcal{T}_C^R$ by

$$\mathcal{T}_C^R = (1 - \mathcal{M}) \odot \mathcal{T}_C^W + \mathcal{M} \odot \mathcal{T}_C^P, \tag{9}$$

where $\odot$ denotes element-wise multiplication.

Perceptual Loss. We utilize the perceptual loss [27] to improve the correspondence accuracy during the clothes warping stage and the refinement stage. The perceptual loss $\mathcal{L}_{vgg}$ can be formulated as

$$\mathcal{L}_{vgg} = \lambda_{v1} \sum_{i=1} \|\phi_i\left((W(\mathcal{T}_C)) - \phi_i\left(\mathcal{T}_{CT}\right)\right\|_1 + \lambda_{v2} \sum_{i=1} \|\phi_i\left(R(W(\mathcal{T}_C))\right) - \phi_i\left(\mathcal{T}_{CT}\right)\|_1, \tag{10}$$

where ϕ_i denotes the feature of the i-th layer in VGG19 [28] and λ_{v1}, λ_{v2} are the weighting parameters for each loss term.

Consequently, refer to Eq. 4, Eq. 5, Eq. 8 and Eq. 10, the Clothes Warping Module total loss can be formulated as

$$\mathcal{L}_{CWM} = \mathcal{L}_{cm} + \mathcal{L}_{warping} + \mathcal{L}_{refinement} + \mathcal{L}_{vgg}. \tag{11}$$

3.4 Image Fusion Module

The Image Fusion Module integrates the information from the mention-above module, adaptively determines the different body parts in the composite image, finally generates the try-on image. To avoid the interference from the original body area to the new clothing area in the try-on image, given the mask of warped clothes $\mathcal{M}_C^S$ and the customer photo $\mathcal{I}$ as input, generate the customer image without clothes $\mathcal{I}_w$, which can be formulated as

$$\mathcal{I}_w = \mathcal{I}' \odot \left(1 - \mathcal{M}_C^S\right), \tag{12}$$

where $\mathcal{I}'$ is the customer photo $\mathcal{I}$ subtracting clothing region.

Besides, given the mask of warped clothes $\mathcal{M}_C^S$ and the body mask $\mathcal{M}_w^S$ as input, generate the composited body mask $\mathcal{M}_w^C$, which can be formulated as

$$\mathcal{M}_w^C = \mathcal{M}_w^S \odot \left(1 - \mathcal{M}_C^S\right). \tag{13}$$

To acquire the final try-on image, we combine the $\mathcal{M}_C^S$, I_w and $\mathcal{T}_C^R$ as the input. Conditional GAN can handle the inputs, we also train U-Net as a generator and the discriminator given in pix2pixHD as a discriminator to yield the photo-realistic try-on image. Refer to Eq. 1 and Eq. 2, the Image Fusion Module total loss $\mathcal{L}_{IMF}$ can be formulated as

$$\mathcal{L}_{IMF} = \lambda_5 \mathcal{L}_{adv} + \lambda_6 \mathcal{L}_{ce}, \tag{14}$$

where λ_5 and λ_6 are the weight coefficients that distribute the loss terms.

4 Experiments

4.1 Datasets

We conducted experiments on the VITON dataset [6]. It contains approximately 19k image pairs, each consisting of a front-view woman image and a top clothing image. There are 16,253 available pairs for training and testing, which are further divided into a set of 14,221 pairs for training and a set of 2,032 pairs for testing.

4.2 Training and Testing

Model as shown in Fig. 2, the modules are separately trained. For landmark detection, we train the model [23] on deepfashion2 dataset [29], then predict landmark on VITON dataset. Finally, We obtain the two-dimensional coordinates of each landmark. The results are shown in Fig. 3. For Semantic Generation

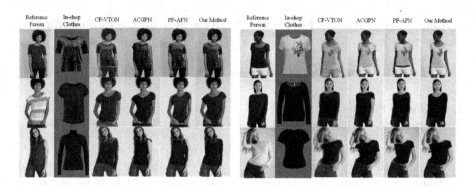

Fig. 5. Visual comparison of four virtual try-on methods. For the 1st row, our method generates photo-realistic try-on results, preserving the clothing texture and characteristics. For the 2nd row, we show our results by short-sleeve to short-sleeve substitutions and long-sleeve to long-sleeve substitutions. For the 3rd row, we show our results by short-sleeve to long-sleeve substitutions and long-sleeve to short-sleeve substitutions.

Module, we set $\lambda_1 = \lambda_2 = 1$. For Clothes Warping Module, we set $\lambda_3 = \lambda_4 = 1$, $\lambda_{lcw} = 0.4$, $\lambda_{scw} = 0.1$, $\lambda_{cw} = 1$, $\lambda_{cr} = \lambda_{lcr} = 0.2$, $\lambda_m = 1$, $\lambda_{v1} = 10$, $\lambda_{v2} = 20$. For Image Fusion Module, we set $\lambda_5 = 10$, $\lambda_6 = 10$. The learning rate is initialized to 0.0002. For the Adam optimizer, we set the parameter $\beta_1 = 0.5$ and $\beta_2 = 0.999$. We implemented the codes by using PyTorch and run them on an NVIDIA 3090 GPU.

The testing procedure follows the training phase as illustrated in Fig. 2.

4.3 Qualitative Evaluations

We conduct comprehensive experiments on popular virtual try-on method, including CP-VTON [1], ACGPN [2] and PF-AFN [3]. All methods are tested on the VITON dataset [6]. Figure 5 shows the comparison results.

In the first row, we aim to show these methods' preservation capability of clothing characteristic and texture. The landmark constraint can guide the model to pay more attention to the details of the clothes. The results show that our method effectively preserves texture and characteristics compared to the existing methods.

In the second and last rows, we aim to indicate that these methods maintain the non-clothes regions when facing short-sleeve to short-sleeve substitutions, long-sleeve to long-sleeve substitutions, short-sleeve to long-sleeve substitutions and long-sleeve to short-sleeve substitutions. We conclude from these examples that the existing methods are ineffective in preserving the customer's non-clothes regions (e.g. arm and sleeves). CP-VTON produces blurry images and cannot match the customer's body. ACGPN's problems usually arise in the retention of the arm regions. As for PF-AFN, although it is satisfactory in preserving the characteristics and texture of clothing, it cannot match the customers' shoulders naturally. Our method is able to generate high-quality try-on image. Landmark constraint can guide the model to distinguish clothes regions and customer body regions, and generate non-clothes regions accurately after try-on. The results show that our method plays a significant role in retaining the customer's body information and optimizing the clothes' textures and characteristics.

Table 1. Comparison results on VITON datasets. **Our method⋆** denotes our method without landmark.

Method	SSIM [31]	IS [30]
CP-VTON [1]	0.724	2.65 ± 0.16
ACGPN [2]	0.860	2.83 ± 0.08
PF-AFN [3]	0.868	2.91 ± 0.10
DCTON [4]	0.870	2.92 ± 0.13
Flow-Style-VTON [5]	**0.873**	2.89 ± 0.12
Our method⋆	0.832	2.80 ± 0.11
Our method	**0.870**	**2.94 ± 0.10**

4.4 Quantitative Evaluations

We use the Structural SIMilarity (SSIM) [31] to measure the similarity between the generated try-on image and the ground-truth. Also, we use the Inception Score (IS) [30] to evaluate the effectiveness of generative adversarial networks and the visual authenticity of generated images. The higher the score of both indicators, the higher the quality of the results. To make the fair comparison, we set the same configurations on different methods.

Table 1 shows our method results on SSIM scores and IS scores compared to CP-VTON [1], ACGPN [2], PF-AFN [3] and DCTON [4]. The SSIM results indicate that our method surpasses CP-VTON, ACGPN and PF-AFN by 0.146, 0.025, 0.002, and our method is equaled with DCTON's score. In the IS, our method surpasses these methods by 0.29, 0.11, 0.03 and 0.02. We also compared our method with the method without landmark, the results surpass 0.038 in SSIM and 0.14 in IS. Note the Flow-Style-VTON method [5], the SSIM score surpasses our method 0.003, but the IS score is below our method 0.05. The results demonstrate the superiority of our method.

4.5 Ablation Study

Ablation study aims to validate the effectiveness of the landmark in our method. Here, we indicate the difference in our method with or without landmark when generating try-on image.

The results in Fig. 6 show that the landmark contributes to better visual results. With the landmark constraint, the customer's arm and clothes sleeve is naturally stratified. Otherwise, the customer's arms are shown thick and the sleeve is ignored.

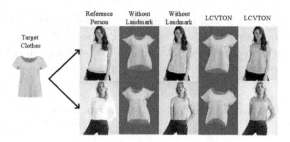

Fig. 6. Comparison of results with and without the landmark constraint.

Experiment is also conducted to verify the effectiveness of the landmark in facing the complex texture of the clothes. We tried on clothes with intricate texture. As shown in Fig. 7, in maintaining the texture of the clothes, the method with the landmark has a better visual effect. We also compare the quantized scores of the two methods in the Table 1. The results indicate that the landmark constraint could improve the quality of the try-on image.

Fig. 7. Comparison of results with and without the landmark constraint in facing the complex texture of the clothes.

5   Conclusion

In this paper, we propose a new virtual try-on network, LCVTON, which solves the problem of semantic synthesis of high-quality clothes generation and authentic try-on images. For the generation of refined clothes, we combine the second-order difference constraint with the landmark constraint in the clothes warping module, which will cause better warping clothes. Moreover, we construct a refinement network and introduce the landmark constraint into it to preserve the texture details of clothes, further promoting subsequent virtual try-on. Experiment results prove that the proposed method is superior in qualitative and quantitative terms on the VITON benchmark. Our method achieves favorable performance compared to state-of-the-art virtual try-on approaches.

References

1. Wang, B., Zheng, H., Liang, X., Chen, Y., Lin, L., Yang, M.: toward characteristic-preserving image-based virtual try-on network. In: Proceedings of the European Conference on Computer Vision (ECCV), pp. 589–604 (2018)
2. Yang, H., Zhang, R., Guo, X., Liu, W., Zuo, W., Luo, P.: Towards photo-realistic virtual try-on by adaptively generating-preserving image content. In: Proceedings of the IEEE/CVF Conference on Computer Vision and Pattern Recognition, pp. 7850–59 (2020)
3. Ge, Y., Song, Y., Zhang, R., Ge, C., Liu, W., Luo, P.: Parser-free virtual try-on via distilling appearance flows. In: Proceedings of the IEEE/CVF Conference on Computer Vision and Pattern Recognition, pp. 8485–93 (2021)
4. Ge, C., Song, Y., Ge, Y., Yang, H., Liu, W., Luo, P.: Disentangled cycle consistency for highly-realistic virtual try-on. In: Proceedings of the IEEE/CVF Conference on Computer Vision and Pattern Recognition, pp. 16928–16937 (2021)
5. He, S., Song, Y.Z., Xiang, T.: Style-based global appearance flow for virtual try-on. In: Proceedings of the IEEE/CVF Conference on Computer Vision and Pattern Recognition, pp. 3470–3479 (2022)
6. Han, X., Wu, Z., Wu, Z., Yu, R., Davis, L.S.: VITON: an image-based virtual try-on network. In: Proceedings of the IEEE Conference on Computer Vision and Pattern Recognition, pp. 7543–7552 (2018)
7. Song, D., Li, T., Mao, Z., Liu, A.-A.: SP-VITON: shape-preserving image-based virtual try-on network. Multimedia Tools Appl. **79**(45), 33757–33769 (2020). https://doi.org/10.1007/s11042-019-08363-w
8. Issenhuth, T., Mary, J., Calauzènes, C.: Do not mask what you do not need to mask: a parser-free virtual try-on. In: Vedaldi, A., Bischof, H., Brox, T., Frahm, J.-M. (eds.) ECCV 2020. LNCS, vol. 12365, pp. 619–635. Springer, Cham (2020). https://doi.org/10.1007/978-3-030-58565-5_37
9. Duchon, J.: Splines minimizing rotation-invariant semi-norms in Sobolev spaces. In: Schempp, W., Zeller, K. (eds.) Constructive Theory of Functions of Several Variables. LNM, vol. 571, pp. 85–100. Springer, Heidelberg (1977). https://doi.org/10.1007/BFb0086566
10. Jaderberg, M., Simonyan, K., Zisserman, A., et al.: Spatial transformer networks. In: Advances in Neural Information Processing Systems, vol. 28 (2015)
11. Goodfellow, I., et al.: Generative adversarial nets. In: Advances in Neural Information Processing Systems, vol. 27 (2014)
12. Karnewar, A., Wang, O.: MSG-GAN: multi-scale gradients for generative adversarial networks. In: Proceedings of the IEEE/CVF Conference on Computer Vision and Pattern Recognition, pp. 7799–7808 (2020)
13. Hussein, S.A., Tirer, T., Giryes, R.: Image-adaptive GAN based reconstruction. In: Proceedings of the AAAI Conference on Artificial Intelligence, vol. 34, pp. 3121–3129 (2020)
14. Karras, T., Aila, T., Laine, S., Lehtinen, J.: Progressive growing of GANs for improved quality, stability, and variation. In: International Conference on Learning Representations (2018). https://openreview.net/forum?id=Hk99zCeAb
15. Luc, P., Couprie, C., Chintala, S., Verbeek, J.: Semantic segmentation using adversarial networks. In: NIPS Workshop on Adversarial Training. Barcelona, Spain (2016). https://hal.inria.fr/hal-01398049
16. Isola, P., Zhu, J.Y., Zhou, T., Efros, A.A.: Image-to-image translation with conditional adversarial networks. In: Proceedings of the IEEE Conference on Computer Vision and Pattern Recognition, pp. 1125–1134 (2017)

17. Wang, T.C., Liu, M.Y., Zhu, J.Y., Tao, A., Kautz, J., Catanzaro, B.: High-resolution image synthesis and semantic manipulation with conditional GANs. In: Proceedings of the IEEE Conference on Computer Vision and Pattern Recognition, pp. 8798–8807 (2018)

18. Karras, T., Laine, S., Aittala, M., Hellsten, J., Lehtinen, J., Aila, T.: Analyzing and improving the image quality of StyleGAN. In: Proceedings of the IEEE/CVF Conference on Computer Vision and Pattern Recognition, pp. 8110–19 (2020)

19. Shaham, T.R., Dekel, T., Michaeli, T.: Singan: learning a generative model from a single natural image. In: Proceedings of the IEEE/CVF International Conference on Computer Vision, pp. 4570–80 (2019)

20. Abdal, R., Zhu, P., Mitra, N.J., Wonka, P.: Styleflow: attribute-conditioned exploration of styleGAN-generated images using conditional continuous normalizing flows. ACM Trans. Graph. (TOG) **40**(3), 1–21 (2021)

21. Karras, T., Laine, S., Aila, T.: A style-based generator architecture for generative adversarial networks. In: Proceedings of the IEEE/CVF Conference on Computer Vision and Pattern Recognition, pp. 4401–4410 (2019)

22. Sun, K., Xiao, B., Liu, D., Wang, J.: Deep high-resolution representation learning for human pose estimation. In: Proceedings of the IEEE/CVF Conference on Computer Vision and Pattern Recognition, pp. 5693–5703 (2019)

23. Lin, T.H.: Aggregation and Finetuning for Clothes Landmark Detection (2020)

24. Gong, K., Liang, X., Zhang, D., Shen, X., Lin, L.: Look into person: self-supervised structure-sensitive learning and a new benchmark for human parsing. In: Proceedings of the IEEE Conference on Computer Vision and Pattern Recognition, pp. 932–940 (2017)

25. Ronneberger, O., Fischer, P., Brox, T.: U-Net: convolutional networks for biomedical image segmentation. In: Navab, N., Hornegger, J., Wells, W.M., Frangi, A.F. (eds.) MICCAI 2015. LNCS, vol. 9351, pp. 234–241. Springer, Cham (2015). https://doi.org/10.1007/978-3-319-24574-4_28

26. Goodfellow, I., Bengio, Y., Courville, A.: Deep Learning. MIT press, Cambridge (2016)

27. Simonyan, K., Zisserman, A.: Very deep convolutional networks for large-scale image recognition. ArXiv Preprint ArXiv:1409.1556 (2014)

28. Johnson, J., Alahi, A., Fei-Fei, L.: Perceptual losses for real-time style transfer and super-resolution. In: Leibe, B., Matas, J., Sebe, N., Welling, M. (eds.) ECCV 2016. LNCS, vol. 9906, pp. 694–711. Springer, Cham (2016). https://doi.org/10.1007/978-3-319-46475-6_43

29. Ge, Y., Zhang, R., Wang, X., Tang, X., Luo, P.: Deepfashion2: a versatile benchmark for detection, pose estimation, segmentation and re-identification of clothing images. In: Proceedings of the IEEE/CVF Conference on Computer Vision and Pattern Recognition, pp. 5337–5345 (2019)

30. Salimans, T., Goodfellow, I., Zaremba, W., Cheung, V., Radford, A., Chen, X.: Improved techniques for training GANs. In: Advances in Neural Information Processing Systems, vol. 29 (2016)

31. Wang, Z., Bovik, A.C., Sheikh, H.R., Simoncelli, E.P.: Image quality assessment: from error visibility to structural similarity. IEEE Trans. Image Process. **13**(4), 600–6112 (2004)

WINMLP: Quantum & Involution Inspire False Positive Reduction in Lung Nodule Detection

Zepeng Zhang[1], Fenghui Liu[2], Lin Qi[1], and Yun Tie[1(✉)]

[1] School of Electrical and Information Engineering, Zhengzhou University, Zhengzhou, China
ieytie@zzu.edu.cn
[2] Department of Respiratory and Sleep Medicine, The First Affiliated Hospital of Zhengzhou University, Zhengzhou, China

Abstract. Improving the accuracy of early diagnosis is the key to prolong the survival of lung cancer. Lung Nodule Detection algorithms based on Deep Learning have made significant contributions to improving the accuracy. However, it remains a challenge to reduce the False Positive rate while maintaining high sensitivity. In this paper, we propose a novel MLP-based False Positive Reduction network, Wave-Involution MLP. We design a progressive multi-scale fusion block based on the novel operator Involution to fuse global features preferably. Moreover, inspired by quantum theory, we design a CT-WaveMLP feature extraction backbone, which transforms CT images into wave functions and enhances feature extraction capability. We performed experiments on LUNAV2 dataset, and the results show that our network achieves the average CPM of 0.861, which has a better performance compared with mainstream methods.

Keywords: Involution · MLP · 3D CT · Wave function · False Positive Reduction · Lung Nodule Detection

1 Introduction

Lung cancer is the leading cause of death worldwide [18]. It's well documented screening with chest Computed Tomography (CT) can significantly improve the early diagnosis rate of lung cancer [6]. Early lung cancers often present as pulmonary nodules [23], which are often round in the chest CT slices and vary in size, usually 3–30 mm [15]. Therefore, the main imaging screening method for early lung cancer is to identify malignant pulmonary nodules in chest CT.

With the increasing application of high-resolution CT devices, it's a time-consuming and challenging task for radiologists to make accurate diagnosis based on hundreds of CT slices [12]. In recent years, several Lung Nodule Detection (LND) algorithms have emerged and provided beneficial aids for radiologists to make decisions. These algorithms usually pursue high sensitivity, but their relatively high False Positive (FP) rate can't be ignored. Pulmonary nodules' size,

M. Tanveer et al. (Eds.): ICONIP 2022, LNCS 13625, pp. 61–72, 2023.
https://doi.org/10.1007/978-3-031-30111-7_6

shape, and intensity of CT values are highly variable. What's worse, some normal tissues such as blood vessels and intrapulmonary lymph nodes are difficult to distinguish from pulmonary nodules. So it's still a challenge to make LND algorithms have both high sensitivity and low FP rate. Some researchers have focused on the False Positive Reduction (FPR) task in LND [4,5,16,19,22,24].

Motivated by mainstream FPR algorithms and latest trends in general-purpose Computer Vision's (CV) research, we propose a novel MultiLayer Perceptron (MLP)-based FPR network: Wave-Involution MLP (WINMLP), which is inspired by the excellent work of Kim et al. [5]. We use multi-scale 3D CT slices as input and design a Progressive Multi-Scale Fusion (PMSF) block based on the novel operator Involution [10]. We design a CT-WaveMLP backbone based on VisionMLP, using quantum theory to transform CT images into wave functions and thus improve the feature extraction capability while maintaining strong robustness. Overall, the major contributions are shown below.

1. We are the first to introduce VisionMLP into the FPR task.
2. We use the novel operator Involution to design a PMSF block, and it can fuse global features preferably.
3. We propose a new network to the FPR task called Wave-Involution MLP (WINMLP), which achieves better performance than existing excellent FPR works as demonstrated by experiments on LUNAV2 dataset.

2 Related Work

2.1 Lung Nodule Detection and False Positive Reduction

The early researchers created the LND algorithm based on traditional image processing or Machine Learning (ML) theories, such as Linear Discriminant Analysis (LDA) [13], Support Vector Machine (SVM) [2], and Artificial Neural Network (ANN) [1]. These early algorithms are groundbreaking, but the sensitivity and the FP rate can't satisfy clinical needs. With Convolutional Neural Network (CNN)-based Deep Learning (DL) algorithms dramatically improving the CV algorithms' performance, several CNN-based LND algorithms have achieved higher sensitivity, but their FP rates are still unsatisfactory. There have been several FPR algorithms made inspiring progress. Setio et al. [16] extracted 2D CT slices from 9 different angles, which improved the FPR performance but had to be trained independently in different views. Dou et al.'s method [4] was similar to Setio's, they extracted features through a multi-stage 3D CNN. What's more, their weights were designed manually, resulting in poor robustness. Xie et al. [22] used boosting strategy based on 2D CNN in FPR and another similar method proposed by Zuo et al. [24] merged a multi-branch 3D CNN. However, both Xie et al.'s and Zuo et al's method was complex and difficult to train. Kim et al. [5]. Improved performance by proposing a progressive feature fusion block and using multi-scale 3D CT slices as input. Based on Kim's work, Sun et al. [19] introduced Attention Mechanism into the FPR task.

2.2 Long-Range Dependency and Backbone

The ability of capturing long-range dependency (hereinafter, *ability*) directly affects the global features extraction. CNN's small receptive field results in its weak *ability*. The diagnosis of pulmonary nodules is often based on both local and global features, which indicates the *ability* is essential for FPR tasks. Recently, Vision Transformer (ViT) has significantly improved the *ability*, but its complex structure and huge resource usage can't be ignored. Li et al. [10] proposed Involution, a new operator whose operation principle is opposite to that of Convolution and achieves excellent *ability* with a simple structure.

Backbone is the core component of DL-based CV algorithms and is typically used for feature extraction. Since the CNN-based backbone designed by LeCun et al. [8]. Has gained attention due to the huge performance improvement, it has dominated the mainstream for a long time. The ViT-based backbone has become a new trend. Transformer is based on Self-Attentive (SA) Mechanism and has powerful *ability*. Alexey et al. [3] introduced it into CV algorithm and proposed ViT. Since then, the ViT-based backbone has been highly pursued by researchers and achieved SOTA performance in various CV tasks.

Gradually, some researchers concern about its complex structure and huge calculation. Li et al. [10] have pointed out SA may not be essential for ViT's performance improvement, and SA mechanism seems to have a large amount of redundant information. The latest work has refocused on traditional MLP. Researchers followed the form of input in ViT and replaced the complex structures such as SA blocks with MLP to design VisionMLP-based backbone. It achieves excellent performance comparable to ViT's with a fewer computation and a simpler structure. Benefiting from the simple structure, the VisionMLP-based backbone has fewer inductive bias than the CNN-based backbone, resulting in better robustness and greater adaptability to different tasks. Several recent VisionMLP works have shown their powerful potential: Tolstikhim et al. [21] used two MLP blocks to extract spatial and channel features. Tang et al. [20] introduced quantum theory into CV research by treating each Token as a wave function with amplitude and phase, refreshing VisionMLP's performance.

3 Method

This chapter will discuss our WINMLP. Figure 1 shows its framework includes 4 parts: Data preprocessing, Involution Progressive Multi-Scale Fusion (PMSF) block, CT-WaveMLP backbone and Classification block.

3.1 Data Preprocessing

The design of this part is inspired by Kim et al. [5] and the 3 scale settings follows Dou et al. [4] As shown in the left side of Fig. 1(a), based on the candidate nodule's central coordinates, we extract the pixels of its surrounding $40 \times 40 \times 26$, $30 \times 30 \times 10$, and $20 \times 20 \times 6$ (Height $\times$ Width $\times$ Slices) from the

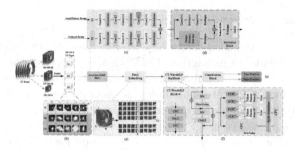

Fig. 1. The framework of the proposed WINMLP FPR algorithm. (a)WINMLP's overall structure. (b)Multi-scale input (c)Involution PMSF block. The term "Con" is Concatenation. (d)Involution block. (e)Patch embedding. (f)CT-WaveMLP block.

original CT. Since uniform data size helps training,we use the nearest neighbor interpolation algorithm to resize into 20 × 20 × 6. Please note while the sizes are the same, the different resolutions lead to changeable visual fields, as shown in Fig. 1(b). The min-max normalization is applied to data in the range of [−1000, 400] Hounsfield Units (HU) to accelerate training. Data augmentation for True Positive (TP) samples is necessary due to serious Data-Imbalance. We rotate them by 90°, 180° and 270° along the transverse plane and shift them by 0, 1, 2 pixel(s) along the x, y, z axis respectively. Finally, the number of TP samples in the training set increased 108 times. After the above preprocessing, three types of input data $S1, S2, S3$ are generated as shown in Fig. 1 *3D Input*.

3.2 Involution Progressive Multi-Scale Fusion (PMSF) Block

We design a PMSF block based on Involution [10] to better fuse features. As shown in Fig. 1(b), the data stream $S1 - S2 - S3$ is defined as an amplificatory stream, in which the visual field is gradually focused on local features. Conversely, the data stream $S3 - S2 - S1$ is defined as a reduced stream, in which the visual field is gradually expanded to global features. The changeable visual fields simulate the diagnostic process of radiologists who attach importance to both local and global features, so this block can extract and fuse features in different visual fields. The block's structure is shown in Fig. 1(c) and the fusion is finally completed by summation. Please note 1×1 Convolution is only for Channel Transformation.

Before introducing Involution, let's first review Convolution used in existing works. Let $X \in \mathbb{R}^{H*W*C_i}$ be the input, where H, W, C_i denote the Height, Width and input Channels. Macroscopically, the Convolution kernel is stated as $F \in \mathbb{R}^{C_o*C_i*K*K}$, where C_o represents the output Channels and K represents the kernel size. It can be further written as $F_{c,l} \in \mathbb{R}^{K*K}, c \in [0, C_o), l \in [0, C_i)$ for a single channel. Its formula is shown in Eq. 1:

$$Y_{i,j,n} = \sum_{c=0}^{C_i} \sum_{(u,v)\in\Delta M} F_{n,c,u+\lfloor K/2 \rfloor,v+\lfloor K/2 \rfloor} X_{i+u,j+v,c}, n \in [0, C_o) \qquad (1)$$

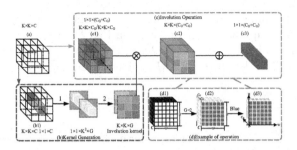

Fig. 2. Involution schematic diagram(G=2). (a)Extracted Data. (b)Kernel generation process. b1: Data needed to generate the kernel(red cube). Arrow1: Transformations shown in Eq. 5. Arrow2: Reshape. (c)Involution operation process. c1: Data after grouping. c2: Output after multiplying with kernel. c3: Final output. (d) An example of Involution's operation. d1: Original input. d2: Grouped data are shown in blue and green. d3: Select the blue group for detailed instruction. (Color figure online)

(ΔM:Center pixel's neighborhood, as shown below.$\times$:Cartesian product)

$$\Delta M = \{-\lfloor K/2 \rfloor, \ldots, \lfloor K/2 \rfloor\} \times \{-\lfloor K/2 \rfloor, \ldots, \lfloor K/2 \rfloor\} \qquad (2)$$

According to the above analysis, we have two findings. The first one is that different locations in a feature map share the same Convolution kernel. Although it reduces parameters, it leads to the lack of specificity perception of different spatial features. The second one is that the Convolution kernel's dimension is $C_o * C_i * K * K$. Since the number of channels in DL practice is usually up to hundreds. In order to avoid Parameter Explosion, the kernel size K must be compressed and 3×3 Convolution is commonly used. The above two findings lead to poor *ability* for Convolution.

Involution can realize the long-range & self-adaptive relationship modeling for nodules and its principle is shown in Fig. 2. To enhance efficiency, the input needs to be divided into G groups in the channel dimension. The same location of all channels in a group shares the kernel. Let $X \in \mathbb{R}^{H*W*C}$ be the input and C_G be the channels contained in a group which can be calculate by $C_G = C \div G$.Involution kernel is $I \in \mathbb{R}^{H*W*K*K*G}$, where K is the kernel size. A pixel located at (i,j) in a feature map is $X_{g,l,(i,j)} \in \mathbb{R}^{1*1}, g \in [0,G), l \in [0,C_G)$ and the Involution kernel belonging to it is $I_{g,(i,j)} \in \mathbb{R}^{K*K}$. Figure 2(c) shows the operation process of Involution and Fig. 2(d) shows a specific example with $X \in \mathbb{R}^{6*6*6}, G = 2, K = 3$.

In Fig. 2(d3), the kernel is $I_0 \in \mathbb{R}^{6*6*3*3}$. For simplicity of description, we only consider performing the operation on the red region with coordinates $(1,2)$ in $W - H$ plane($C_G = 0$,purple area is origin), the kernel of red region is $I_{0,(1,2)} \in \mathbb{R}^{3*3}$, and kernel size is the same as that of the gray region (containing red area). Then the output $Y_{0,0,(1,2)} \in \mathbb{R}^{1*1}$ as shown below. ($\Delta M_0 = \{-1,0,1\} \times \{-1,0,1\}$)

$$Y_{0,0,(1,2)} = \sum_{(u,v) \in \Delta M_0} I_{0,(1,2),u+1,v+1} X_{0,0,(1+u,2+v)} \qquad (3)$$

The kernel for the same position in another two channels is also $I_{0,(1,2)}$. For other regions of green group and the green group in Fig. 2(d2), the operation is similarly except for the kernel differences. Thus, the general formula for Involution is shown below: (The definition of ΔM is given in Eq. 2)

$$Y_{g,(i,j)} = \sum_{(u,v)\in\Delta M} I_{g,(i,j),u+\lfloor K/2\rfloor,v+\lfloor K/2\rfloor} X_{g,(i+u,j+v)} \qquad (4)$$

In Eq. 4, $Y_{g,(i,j)} \in \mathbb{R}^{1*1*C_G}$. Repeat Eq. 4 for each position and then concatenate outputs along channel dimension, we can get the final result $Y \in \mathbb{R}^{H*W*C}$.

The generation process of Involution kernel is shown in Fig. 2(b). Please note the kernels are generated without grouping. Figure 2(a) denotes the center pixel and its neighborhood extracted from original CT slices. We select $X_{\psi_{(i,j)}} \in \mathbb{R}^{1*1*C}$ from the same position as the center pixel's in all channels (red cube in Fig. 2(b1)) to generate the kernel, which can ensure spatial specificity. Set the kernel as $I_{\psi_{(i,j)}} \in \mathbb{R}^{K*K*G}$ and its generation formula is:

$$I_{\psi_{(i,j)}} = \omega_1 \sigma \left(\omega_0 X_{\psi_{(i,j)}}\right) \qquad (5)$$

Equation 5 consists of two linear transformations and one nonlinear transformation: ω_0 compresses the data into $X_{\psi_{(i,j)}} \in \mathbb{R}^{1*1*(C/r)}$ (r: compression ratio). $\sigma(*)$ denotes the nonlinear transformation consisting of Batch Normalization(BN) and ReLU. ω_1 generates the final kernel. The two linear transformations are executed by 1×1 Convolution and detailed Involution's structure is shown in Fig. 1(d).

In this section, we have illustrated the theory of PMSF block based on Involution. The kernel's generation considers the context between channels which enhances 3D fusion ability. Moreover, the spatially specific kernel realizes self-adaptive modeling for nodules. Due to $H * W * G$ is much smaller than $C_o * C_i$, the kernel size can be set larger and thus the *ability* can be improved.

3.3   CT-WaveMLP Backbone

In Sect. 2.2, we have recalled the design concepts of mainstream backbones. This section will discuss our CT-WaveMLP backbone based on VisionMLP.

MLP consists of FC layers and nonlinear activation functions. Since FC layers are unable to obtain information between local regions, features can't be fully extracted if original images are used as input. VisionMLP introduces image splitting in ViT to solve the limitation of FC layers, as shown in Fig. 1(e). Set the input as $X \in \mathbb{R}^{H*W*C_i}$, and the split tokens as $X_T \in \mathbb{R}^{S*C_o}$, where $S = (H * W)/p^2$. Existing VisionMLP-based backbones usually contain two components: Channel-FC(CFC) and Token-FC(TFC), CFC is used to extract channel-domain features and TFC is used to extract spatial-domain features. However, in order to reduce calculation, existing methods usually share the parameters of TFC resulting in ignoring differences of distinct feature maps. Inspired by Tang et al. [20], we design our backbone by stacking CT-WaveMLP block 4 times.

The structure of CT-WaveMLP block is shown in Fig. 1(f), which is inspired by quantum mechanics. Quantum mechanics describes the laws of particles' motion and *Wave-Particle Duality* is a basic property. Waves usually contain two properties: Amplitude and Phase, and wave functions (e.g., de Broglie wave) are more complete description of particles' motion. Motivated by the above quantum theory, we think converting CT images into wave functions may also extract features more completely. We take X_T as input and one of the tokens is $X_{T,j} \in \mathbb{R}^{1*C}, j \in [0, S)$, $\widetilde{X}_{T,j}$ is the Wave form of it. It's formulated as below:

$$\widetilde{X}_{T,j} = X_{m(T,j)} \odot e^{i\theta_j} \ (\odot : Hadamard \ product) \tag{6}$$

In Eq. 6, $X_{m(T,j)}$ denotes Real-Value feature (Amplitude), $X_{m(T,j)} = |X_{T,j}|$. $e^{i\theta_j}$ is a periodic function, where i is an Imaginary Unit satisfying $i^2 = -1$, θ_j denotes Phase, expressing position. Phase θ_j has a significant effect on the fused results, which can be proved by Maths: according to Euclidean Geometry and Cosine Theorem, take two tokens $\widetilde{X}_1, \widetilde{X}_2$ as an example, and their Phase Difference is $\varphi = \theta_2 - \theta_1$. The fused token is $\widetilde{X}_f$, its Amplitude $|X_f|$ and Phase θ_f are calculated as follows: ($atan2(y, x)$: two-argument arctangent function)

$$|X_f| = \sqrt{|X_1|^2 + |X_2|^2 + 2 \odot |X_1| \odot |X_2| \odot \cos \varphi} \tag{7}$$

$$\theta_f = \theta_1 + atan2(|X_2| \odot \sin \varphi, |X_1| + |X_2| \odot \sin \varphi) \tag{8}$$

Both Amplitude and Phase are essential to convert CT images. When $X_{T,j,k} < 0, |X_{T,j,k}|e^{i\theta_{j,k}} = X_{T,j,k}e^{i(\theta_{j,k}+\pi)}$, where $X_{T,j,k}, \theta_{j,k}$ represent the $k - th$ element in $X_{T,j}$ and θ_j. It shows Absolute Operation can be absorbed into the phase term, so Absolute Operation can be omitted. Set the original Token $X_{T,j}$ as input, its Amplitude $X_{m(T,j)}$ and Phase θ_j are generated as follows:

$$X_{m(T,j)} = \sigma(\Omega_A X_{T,j}), \ \theta_j = \sigma(\Omega_P X_{T,j}) \tag{9}$$

where Ω_A, Ω_P are linear transformations used to generate Amplitude and Phase respectively, which are realized by 2 individual CFC layers, as ACFC and PCFC shown in Fig. 1(f). $\sigma(*)$ denotes the nonlinear transformation consists of BN and ReLU, which can enhance the network's robustness.

The complex form of $X_{T,j}$'s wave function can be obtained by Eq. 6 and Eq. 9. It can be transformed by Euler's formula for ease of operation:

$$\widetilde{X}_{T,j} = X_{m(T,j)} \odot \cos \theta_j + i(X_{m(T,j)} \odot \sin \theta_j) \tag{10}$$

The output $\widetilde{Y}_j$ of Quantum State can be obtained by fusing wave functions of different Tokens. It must be transformed into Observable to extract its features and classify nodules. Following usual Quantum Measurement methods [14], we use 2 distinct Token-Fusion blocks to fuse Real and Imaginary parts respectively, and then obtain Observable Y_j by summing them:

$$Y_j = \xi_r(X_{m(T,j)} \odot \cos \theta_j) + \xi_i(X_{m(T,j)} \odot \sin \theta_j) \tag{11}$$

where ξ_r, ξ_i denote Token-Fusion for Real and Imaginary parts respectively. We realize Token-Fusion with Depthwise Convolution (DC).

Our CT-WaveMLP block consists of Wave Fusion, ChannelMLP(CMLP) and BN. Wave Fusion is the core block which converts CT images into wave functions and completes fusion. We operate along W,H separately to better adapt to 2D images, as $Wave\text{-}W, Wave\text{-}H$ shown in Fig. 1(f). Besides, an extra CFC layer is added to preserve maximal info of original input. The 3 paths are fused by Split-Attention and then are fed to the final CFC layer for enhancing feature expression. CMLP consists of 2 CFC layers and GeLU to extract features.

3.4   Classification Block

Classification block is used to output the final pulmonary nodules' classification results, which consists of a Global Average Pooling(GAP) layer [11] and a FC layer. The use of GAP layer can reduce parameters, which is difficult to overfit and enhances robustness. Since there are only 2 classes of nodules in FPR task: TP and FP, the FC layer is designed to contain only 1 neuron which outputs the probability of a nodule being TP after normalization by Sigmoid function.

4   Experiments and Results

4.1   Datasets and Experimental Settings

We perform the experiments on LUng Nodule Analysis 2016 (LUNA16) [17] dataset and it provides two versions: LUNAV1 and LUNAV2, as shown in Table 1(a). As LUNAV2 dataset is a newer version which contains more candidate nodules, we select it. Table 1(a) clearly shows the dataset has a serious Data-Imbalance, so we perform Data Augmentation on the TP nodules used for training as described in Sect. 3.1. We divide the dataset into 5 subsets and use 5-fold cross-validation to evaluate WINMLP, as shown in Table 1(b).

We utilize DropPath [7] regularization and Gradient Harmonizing Mechanism (GHM) [9] loss function to enhance robustness. For optimization, we use Adam with a mini-batch size of 128 and a learning rate of 0.003. We perform training and experiments on a NVIDIA RTX3060 GPU with Ubuntu20.04, writing the code based on Python3.7, TensorFlow2.5 and CUDA11.2. We use Free Response Operating Characteristic(FROC) and Competition Performance Metric(CPM) to evaluate performance. To obtain them, *Sensitivity* is essential, as defined in Eq. 12: (*TP*: Nodules predicted correctly. *FN*: Nodules predicted as non-nodules.)

$$Sensitivity = \frac{TP}{TP + FN} \tag{12}$$

Following the official evaluation method, the FROC curve can be calculated over the range of $[0.125, 8]$ FPs per scan and the CPM score is the average of *Sensitivity* at 7 predefined FPs per scan: $1/8, 1/4, 1/2, 1, 2, 4, 8$.

Table 1. Dataset.

(a) Two versions of dataset in LUNA16 for the FPR challenge.

Dataset Version	Nodules(TP)	Non-nodules(FP)	Total
LUNAV1	1351	549,714	551,065
LUNAV2	1557	753,418	754,975

(b) Details of dataset(Format:Training Set/Testing Set)

Item	Run No.1	Run No.2	Run No.3	Run No.4	Run No.5
Patients(Scans)	710/178	710/178	709/179	711/177	710/178
Nodules(TP)	1262/295	1250/307	1207/350	1270/287	1285/272
Augmentation	136,296/-	135,000/-	130,356/-	137,160/-	138,780/-
Non-nodules(FP)	609,124/ 144,294	607,765/ 145,653	598,678/ 154,740	606,760/ 146,658	604,569/ 148,849

4.2 Experimental Results

We compare WINMLP with several existing FPR methods [4,5,16,19,22,24] on LUNAV2 dataset. The experimental results are shown in Table 2 and Fig. 3(a). Table 2 indicates our WINMLP achieves a better CPM score. Xie's [22] method has a higher sensitivity in low FP rate(0.125 FP/scan) due to the boosting strategy based on several CNNs, but our WINMLP still achieves a very close (0.27%) sensitivity with a single-branch structure. Zuo's [24] method is like Xie's, which adopts a multi-branch 3D CNN strategy. Both Setio's [16] and Dou's [4] method need to be trained several times under different views. Compared with them, both Involution and Vision MLP are based 2D CNN and FC layers, so our hardware requirements are significantly reduced. Meanwhile, we achieve better performance by training only once. Both Sun's [19] method and WINMLP are based on Kim's [5] idea. Our CPM score improvement is 16.98% compared to Kim's(Base) [5], and 12.84% compared to Sun's [19].

Table 2. Comparison of different methods.

Method	Average number of FP per scan							
	0.125	0.25	0.5	1	2	4	8	CPM
Setio et al. [16]	0.692	**0.771**	0.809	0.863	0.895	0.914	0.923	0.838
Dou et al. [4]	0.677	0.737	0.815	0.848	0.879	0.907	0.922	0.826
Xie et al. [22]	**0.734**	0.744	0.763	0.796	0.824	0.832	0.834	0.790
Zuo et al. [24]	0.630	0.753	0.819	0.869	0.903	0.915	0.920	0.830
Base [5]	0.459	0.563	0.706	0.758	0.856	0.902	0.911	0.736
Sun et al. [19]	0.467	0.602	0.730	0.812	0.877	0.920	0.931	0.763
WINMLP(Ours)	0.732	0.769	**0.853**	**0.885**	**0.916**	**0.926**	**0.945**	**0.861**

Involution PMSF block (Section 3.2) contains two major Hyper-Parameters: channels contained in a single group C_G and kernel size K. Table 3(a) shows their effect and their FROC curves are in Fig. 3(b),(c). In the upper part of Table 3(a), although it is clearly shown that $C_G = 1$ give the best performance, we still believe it's not the best choice: its performance improvement is quite small ($\sim$0.58%), but it nearly doubles the parameters. So $C_G = 2$ is the best choice to balance performance and efficiency. The result for $C_G = 4$ proves that a large C_G may cause network to ignore the specificity information between different channels. The results for different K are showed in the lower part of Table 3(a), and the best choice is $K = 5$. In fact, both local and global features are essential for the FPR task. Too small $K(K = 3)$ is not good for extracting global features and too large $K(K = 7)$ is not good for extracting local features, so $K = 5$ is more balanced and can achieve the best performance.

Our WINMLP contains two main components: Involution PMSF(Inv) block (Sect. 3.2) and CT-WaveMLP (MLP) backbone (Sect. 3.3). To verify the effectiveness of them, we take CNN PMSF and CNN backbone in Baseline method [5] as a comparison. The results show in Table 3(b) & Fig. 3(d), and they confirm that both Inv block and MLP backbone help to improve the FPR performance, and the combination of them performs better.

Table 3. Ablation study

(a) Different C_G and K in Involution. When adjusting one hyper-parameter, the other hyper-parameter uses the best setting denoted by *.

Channels C_G	Average number of FP per scan							
	0.125	0.25	0.5	1	2	4	8	CPM
1	**0.733**	**0.786**	**0.862**	**0.887**	**0.919**	**0.927**	**0.948**	**0.866**
2*	0.732	0.769	0.853	0.885	0.916	0.926	0.945	0.861
4	0.725	0.758	0.815	0.847	0.886	0.915	0.932	0.840

Kernel size K	Average number of FP per scan							
	0.125	0.25	0.5	1	2	4	8	CPM
3	0.689	0.736	0.754	0.842	0.867	0.909	0.918	0.816
5*	**0.732**	**0.769**	**0.853**	**0.885**	**0.916**	**0.926**	**0.945**	**0.861**
7	0.725	0.761	0.796	0.862	0.887	0.917	0.935	0.840

(b) Effects of Involution(Inv) PMSF block and CT-WaveMLP(MLP) backbone. CNN PMSF and CNN backbone in Baseline method [5] as comparison.

No.	PMSF block		Backbone		Average number of FP per scan							
	Inv	CNN	MLP	CNN	0.125	0.25	0.5	1	2	4	8	CPM
1		✔		✔	0.459	0.563	0.706	0.758	0.856	0.902	0.911	0.736
2	✔			✔	0.625	0.678	0.718	0.763	0.859	0.906	0.914	0.780
3		✔	✔		0.604	0.657	0.715	0.859	0.874	0.913	0.929	0.793
4	✔		✔		**0.732**	**0.769**	**0.853**	**0.885**	**0.916**	**0.926**	**0.945**	**0.861**

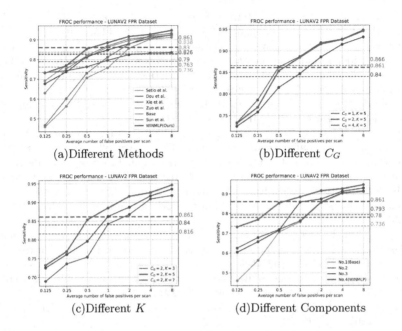

Fig. 3. FROC Curves. Dotted Line represents CPM Score.

5 Conclusion

In this work, we propose a novel FPR network: WINMLP. We introduce Involution and design Involution PMSF block for better fusion of multi-scale data. We also design CT-WaveMLP backbone based on VisionMLP, which improves the feature extraction ability of pulmonary nodules. The experimental results on LUNAV2 dataset prove the effectiveness of the proposed components, and our WINMLP achieves better performance than mainstream methods. In future work, we will design a novel LND model and try to embed WINMLP thus realizing an end-to-end LND model with both high sensitivity and low FP rate.

References

1. Cascio, D., Magro, R., Fauci, F., Iacomi, M., Raso, G.: Automatic detection of lung nodules in CT datasets based on stable 3D mass-spring models. Comput. Biol. Med. **42**(11), 1098–1109 (2012)
2. Choi, W.J., Choi, T.S.: Automated pulmonary nodule detection based on three-dimensional shape-based feature descriptor. Comput. Methods Programs Biomed. **113**(1), 37–54 (2014)
3. Dosovitskiy, A., Beyer, L., Kolesnikov, A., Weissenborn, D., et al.: An image is worth 16 × 16 words: transformers for image recognition at scale. In: 2020 International Conference on Learning Representations (ICLR) (2020)
4. Dou, Q., Chen, H., Yu, L., Qin, J., Heng, P.A.: Multilevel contextual 3-D CNNs for false positive reduction in pulmonary nodule detection. IEEE Trans. Biomed. Eng. **64**(7), 1558–1567 (2016)

5. Kim, B.C., Yoon, J.S., et al.: Multi-scale gradual integration CNN for false positive reduction in pulmonary nodule detection. Neural Netw. **115**, 1–10 (2019)
6. Lancaster, H.L., et al.: Low-dose computed tomography lung cancer screening: clinical evidence and implementation research. J. Intern. Med. **292**, 68–80 (2022)
7. Larsson, G., Maire, M., Shakhnarovich, G.: Fractalnet: ultra-deep neural networks without residuals. arXiv preprint arXiv:1605.07648 (2016)
8. LeCun, Y., Bottou, L., Bengio, Y., Haffner, P.: Gradient-based learning applied to document recognition. Proc. IEEE **86**(11), 2278–2324 (1998)
9. Li, B., Liu, Y., Wang, X.: Gradient harmonized single-stage detector. In: Proceedings of the AAAI conference on artificial intelligence, vol. 33, pp. 8577–8584 (2019)
10. Li, D., Hu, J., Wang, C., Li, X., et al.: Involution: inverting the inherence of convolution for visual recognition. In: 2021 IEEE/CVF Conference on Computer Vision and Pattern Recognition (CVPR), pp. 12316–12325. IEEE (2021)
11. Lin, M., Chen, Q., Yan, S.: Network in network. arXiv preprint arXiv:1312.4400 (2013)
12. Liu, D., Liu, F., Tie, Y., Qi, L., Wang, F.: Res-trans networks for lung nodule classification. Int. J. Comput. Assist. Radiol. Surg. **17**, 1–10 (2022). https://doi.org/10.1007/s11548-022-02576-5
13. Messay, T., et al.: A new computationally efficient cad system for pulmonary nodule detection in CT imagery. Med. Image Anal. **14**(3), 390–406 (2010)
14. Morgan, P.: Classical and quantum measurement theory. arXiv preprint arXiv: 2201.04667 (2022)
15. Murchison, J.T., Ritchie, G., Senyszak, D., et al.: Validation of a deep learning computer aided system for CT based lung nodule detection, classification, and growth rate estimation in a routine clinical population. PLoS ONE **17**(5), e0266799 (2022)
16. Setio, A.A.A., Ciompi, F., Litjens, G., Gerke, P., Jacobs, C., et al.: Pulmonary nodule detection in CT images: false positive reduction using multi-view convolutional networks. IEEE Trans. Med. Imaging **35**(5), 1160–1169 (2016)
17. Setio, A.A.A., Traverso, A., et al.: Validation, comparison, and combination of algorithms for automatic detection of pulmonary nodules in computed tomography images: the luna16 challenge. Med. Image Anal. **42**, 1–13 (2017)
18. Sharma, R.: Mapping of global, regional and national incidence, mortality and mortality-to-incidence ratio of lung cancer in 2020 and 2050. Int. J. Clin. Oncol. **37**, 1–11 (2022)
19. Sun, L., et al.: Attention-embedded complementary-stream CNN for false positive reduction in pulmonary nodule detection. Comput. Biol. Med. **133**, 104357 (2021)
20. Tang, Y., et al.: An image patch is a wave: phase-aware vision MLP. In: 2022 IEEE/CVF Conference on Computer Vision and Pattern Recognition (CVPR), pp. 10935–10944 (2022)
21. Tolstikhin, I.O., Houlsby, N., Kolesnikov, A., et al.: MLP-mixer: an all-MLP architecture for vision. In: Advances in Neural Information Processing Systems, vol. 34 (2021)
22. Xie, H., Yang, et al.: Automated pulmonary nodule detection in CT images using deep convolutional neural networks. Pattern Recogn. **85**, 109–119 (2019)
23. Zhang, H., Peng, Y., Guo, Y.: Pulmonary nodules detection based on multi-scale attention networks. Sci. Rep. **12**(1), 1–14 (2022)
24. Zuo, W., Zhou, F., He, Y.: An embedded multi-branch 3D convolution neural network for false positive reduction in lung nodule detection. J. Digit. Imaging **33**(4), 846–857 (2020)

Incorporating Generation Method and Discourse Structure to Event Coreference Resolution

Congcheng Huang, Sheng Xu, Longwang He, Peifeng Li[(⊠)], and Qiaoming Zhu

School of Computer Science and Technology, Soochow University, Jiangsu, China
{cchuangnlp,sxu,20205227115}@stu.suda.edu.cn, {pfli,qmzhu}@suda.edu.cn

Abstract. Event coreference resolution is an important task of information extraction. Previous work often focused on modeling the sentence structure, ignoring the structure between paragraphs which is also important to event coreference resolution. Moreover, almost all previous work modeled event coreference resolution as a classification task. In this paper, we introduce macro discourse structure to help event coreference resolution through a Relational Graph Convolutional Network (R-GCN), which can take advantage of structure and relations between paragraphs. Moreover, we are the first to introduce an encoder-decoder style generation model to further boost event coreference resolution task. The experimental results on the English KBP2016 and KBP2017 datasets show that our model CGECR (Classification and Generation models for Event Coreference Resolution) outperforms the SOTA baselines.

Keywords: Event Coreference Resolution · Discourse Structure · Generation model

1 Introduction

Event coreference resolution is an important task in the field of information extraction, which is beneficial to many downstream tasks, such as question answering [1] and text summarization [2]. This task focuses on identifying whether different event mentions[1] in a document refer to the same event in the real world and linking coreferential event mentions to an event chain.

Take a document in Fig. 1 as an example, there are seven event mentions (color texts) in this document. The event mentions marked with the same color are refer to the same event in the real world. For example, the event mentions with the triggers "proposed" and "proposals" both refer to the same real-world event of "White House proposed gun control regulations" and these event mentions are coreferential.

The relation or structure between paragraphs is an important part of linguistic information, which can effectively help event coreference resolution. According to statistic analysis, the more relevant the paragraphs, the greater the probability of the occurrence of coreferential event mentions. In the other word,

---

[1] An event mention refers to a phrase or sentence within which an event is described.

M. Tanveer et al. (Eds.): ICONIP 2022, LNCS 13625, pp. 73–84, 2023.
https://doi.org/10.1007/978-3-031-30111-7_7

(1) U.S. Attorney General urges Congress to pass gun control legislation

(2) WASHINGTON, Jan. 18 (Xinhua) -- U.S. Attorney General Eric Holder on Friday urged Congress to move quickly to pass a package of gun control measures proposed by the White House earlier this week. Speaking at the U.S. Conference of Mayors in the capital, Holder called for immediate congressional action, to make "significant strides in reducing the violence that too often fills our headlines and afflicts our communities."

... Omit six paragraphs ...

(9) Obama on Wednesday unveiled a sweeping package of gun violence reduction proposals, a month after the Sandy Hook Elementary School mass shooting that killed 26 people including 20 schoolchildren. The measures include 23 executive actions which the president signed at the scene.

... ...

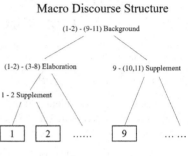

Fig. 1. An example of event coreference resolution and a macro discourse tree of a document.

we can find more coreferential event mentions in relevant paragraphs, even if they are far away from each other in the text, through modeling structure between paragraphs. As shown in Fig. 1, although the coreferential event mention pairs appear in the paragraphs 2 and 9 are far away from each other, they are connected by the *Background* relation in a discourse structure tree. A model can find these coreferential event mention pair more easily through this relation.

However, previous work has not taken advantage of this information. In this paper, we introduce macro discourse structure [3] to boost our model, which focuses on the relations and structures between paragraphs in a document and reveals the topic and the overall structure of an document from a higher level. Specially, we apply a Relational Graph Convolutional Network(R-GCN) [4] to model the macro discourse structure. Consistent with previous works [5], we also introduce entity coreference information and text semantic similarity information to R-GCN to help event coreference resolution.

Most of the existing methods [5–8] of event coreference resolution are based on encoder-classification architecture. Nowadays, the generation models in encoder-decoder style have been proved to be effective in many NLP tasks such as machine translation [9] and event extraction [10]. In this paper, we are the first to introduce a generation model to event coreference resolution as an auxiliary task. Specially, we formulate a special generation template for event coreference resolution, which makes the generation model have the ability of event extraction and event coreference resolution at the same time after training.

Due to the limitation of text encoding length by the popular encoders (e.g., BERT), existing studies [5–7] usually cut a long document into segments, and then take each segment as the input. This will lead to the macro discourse structure and generation model can not be fully utilized. To solve this issue, we use Longformer [11] and LongT5 [12] as encoders in our model, which allow for scaling both input length and model scale at the same time.

In summary, we propose a novel event coreference resolution model CGECR (Classification and Generation models for Event Coreference Resolution) in this paper. First, we introduce the macro discourse structure to mine the relations and structures between paragraphs, and then apply a generation model to CGECR as an auxiliary model. Moreover, to better use of macro discourse structure and generation model, we apply Longformer and LongT5 to encode the whole documents. Experimental results on the KBP2016 and KBP2017 corpora show that our proposed model outperforms the strong baselines.

2   Related Work

The studies on event coreference resolution can be divided into within-document and cross-document task [13]. This paper focuses on within-document event coreference resolution, which is critical to the further cross-document task.

Early studies are mostly based on traditional machine learning architecture. For example, Chen et al. [14] use four basic event attributes (e.g., subtype and factuality) of events as event mention features to help events coreference resolution. Liu et al. [15] extract abundant linguistic features and use SVM to solve event coreference.

Currently, researchers begin to apply neural model to event coreference resolution. Choubey et al. [8] propose several document level neural-network models to automatically construct news content structures. Huang et al. [16] propose a method of transfer learning to obtain argument compatibility information. Lu and Ng [6] prove that SpanBERT works well on event coreference resolution, and propose a joint model of event coreference resolution and entity coreference resolution. Tran et al. [5] introduce GCN to event coreference resolution and define some rules to constrain the consistency of event mentions. However, different from R-GCN, GCN does not have the ability to distinguish the types of edges in the graphs, and they ignore the advantages of the relation and structure between paragraphs to event coreference resolution.

All the above methods are based on the encoding-classification architecture. Nowadays, the encoder-decoder style generation architecture has been used in many NLP tasks such as machine translation [9] and event extraction [10]. Hence, we introduce the generation model to event coreference resolution as an auxiliary model.

3   Method

Our event coreference resolution model CGECR (Classification and Generation models for Event Coreference Resolution) consists of three modules: the event detection module, the classification module and the generation module of event coreference resolution, as shown in Fig. 2.

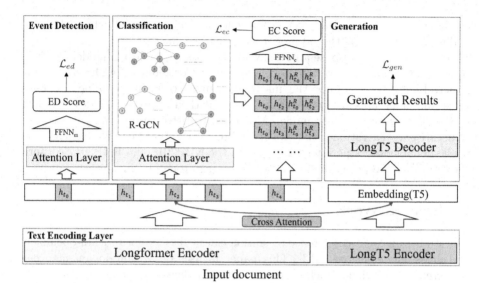

Fig. 2. The framework of our model CGECR.

We first input the whole document into the text encoding layer to obtain the text representation, and detect event mentions via the event detection module. And then we use R-GCN to update the representation and each pair of event mentions will be determined whether they are coreferential via the event coreference classification module. Meanwhile, as an auxiliary module, the generation module will generate predictions according to the formulated templates. Finally, we jointly train these modules simultaneously.

3.1   Text Encoding Layer

Because most of the existing work [6,7] use encoders with length limitation (e.g., BERT) to encode texts, those long documents need to be cut into short segments to meet this limitation, which will lead to the incompletion of discourse structure and semantics. To address this issue, we use Longformer [11] and LongT5 [12] to encode the whole document. Longformer and LongT5 are both built for encoding long texts, which can capture both local and global context information at the same time.

Longformer is more suitable for classification tasks, while LongT5 is better for generation tasks. Hence, we use different encoder in different modules to make better use of them. For a document $D = \{w_1, ..., w_i, ..., w_n\}$ where w_i is the i-th token in the document, we pass it into Longformer and LongT5 to encode each token as shown in Eq. (1) and Eq. (2).

$$\{\mathbf{h}_1^{LF}, ..., \mathbf{h}_n^{LF}\} = \text{Longformer}(\{w_1, ..., w_n\}) \tag{1}$$

$$\{\mathbf{h}_1^{T5}, ..., \mathbf{h}_n^{T5}\} = \text{LongT5Encoder}(\{w_1, ..., w_n\}) \tag{2}$$

To enhance the interaction between the two encoders and the three modules, we use cross attention mechanism to fuse the results of the two encoders as follows.

$$\{\mathbf{h}_1, ..., \mathbf{h}_n\} = \text{CrossAttention}(\{\mathbf{h}_1^{LF}, ..., \mathbf{h}_n^{LF}\}, \{\mathbf{h}_1^{T5}, ..., \mathbf{h}_n^{T5}\}) \qquad (3)$$

As shown in Fig. 2, the event detection and the classification module share the result of cross attention mechanism while the generation module uses the result of LongT5.

3.2 Event Detection Module

Trigger word is the core of event mention and each event mention is triggered by a trigger word. Consistent with existing work, we simplify event detection to trigger detection and use the sequence annotation framework. Specifically, after getting the text representation from the text encoding layer, we construct a neural network scorer s_m for each word and judge its possibility of being an trigger as follows.

$$s_m(w_i) = \text{sigmoid}(\mathbf{W}_m \text{FFNN}_m(\mathbf{h}_i)) \qquad (4)$$

where FFNN_m is a one-layer feed-forward neural network, $\mathbf{W}_m$ is a parameter matrix that can be trained. We input the representation $\mathbf{h}_i$ into the feed-forward neural network, and then pass the result to a sigmoid activation function. The token w_i will be considered as the trigger of an event, if $s_m(w_i)$ is greater than 0.5.

In KBP corpus, most triggers are composed of one word and only a few are composed of multiple words. Therefore, we define a sliding window with the size w. If the scores of multiple words in the sliding window are all greater than 0.5, we will combine these words and consider them as one trigger word. To get the trigger word representation, we use the attention mechanism as follows.

$$\mathbf{h}_{t_k} = \text{AttentionLayer}(\mathbf{h}_p, ..., \mathbf{h}_q) \qquad (5)$$

where p, q represents the starting position and ending position of sliding window. Finally, we use $\mathbf{h}_{t_k}$ indicates the k-th event mention (trigger word) in this document.

Following the existing studies, we directly use the representation of trigger word as that of event mention. In this paper, we use binary cross-entropy loss to train the event detection module as follows.

$$\mathcal{L}_{ed} = \sum_{i=0}^{n} [y_i \log(s_m(i)) + (1 - y_i)\log(1 - s_m(i))] \qquad (6)$$

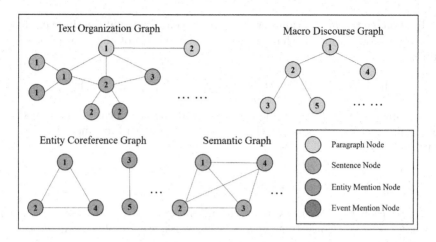

Fig. 3. The construction of relational graphs.

3.3 Classification Module

The classification module for event coreference resolution is the main module in our model CGECR. To engage more information when classifying event mention pairs, we introduce relational graphs to this module. The representations will be updated through R-GCN and have richer information. As shown in Fig. 3, the nodes and graphs for the classification module are constructed as follows.

Nodes. In each document, there are four elements that play important roles in judging event coreference: paragraphs, sentences, entity mentions and event mentions. To capture the structure and relation information between them, we set four types of nodes in our graph: paragraph nodes, sentence nodes, entity mention nodes and event mention nodes. The representation of each node is obtained by the attention mechanism shown in Eq. (5).

Text Organization Graph. An input document usually contains multiple paragraphs, sentences and event/entity mentions. We need to understand where such nodes located in the document and how they relate to each other. Therefore, we construct a text organization graph for every document, in which undirected edges exist if two nodes have inclusive or adjacency relation.

Macro Discourse Graph. Coreferential event mention pairs often appear in related paragraphs, even if they are far away from each other. Therefore, capturing the relation and structure between paragraphs and constructing a graph for them can effectively help the model identify coreferential event mention pairs far apart. In this paper, we use the method proposed by [3] to build a macro discourse tree for each document. Macro discourse tree takes paragraphs as the elementary discourse units (EDUs), and it can reveal the structure of paragraphs from a high level and is conducive to clarify their organization. If two paragraph nodes are connected in the macro discourse tree, they will be connected by an undirected edge in the macro discourse graph.

Entity Coreference Graph. The arguments of the coreferential event mention pairs are always coreferential too. Hence, coreferential entity mentions often appear around the coreferential event mention pairs. To take advantage of entity coreference information, we build the entity coreference graph. We utilize AllenNLP[2] to extract the entity coreference chains. The entity mention nodes on the same entity coreference chain are connected with each other.

Semantic Graph. Intuitively, if two sentences describe the same event, their semantic similarity will be also relatively high. To capture the semantic information between sentences, we build the semantic graph. Edges exist between all sentence nodes in a document, and the value of edges are that of their semantic similarity. We use sentence-BERT [17] to get sentence embedding and calculate semantic similarity by cosine similarity function.

Given the graphs mentioned above, we use R-GCN [4] to update representation for all the nodes as follows.

$$\mathbf{h}_{t_i}^R = \mathrm{R-GCN}(\mathbf{h}_{t_i}) \qquad (7)$$

For the event mentions obtained by the event detection module (or annotated in the training set when training), let $\mathbf{h}_{t_i}^R$ be the representation of the i-th event mention updated by R-GCN. Then we pair all of them and classify weather each event mention pair is coreferential. We concat each pair of event mention representation as follows.

$$\mathbf{g}_{ij} = \mathrm{concat}(\mathbf{h}_{t_i}; \mathbf{h}_{t_j}; \mathbf{h}_{t_i}^R; \mathbf{h}_{t_j}^R; \mathbf{h}_{t_i}^R \circ \mathbf{h}_{t_j}^R; \mathbf{h}_{t_i}^R - \mathbf{h}_{t_j}^R) \qquad (8)$$

where $\circ$ and $-$ denotes element-wise multiplication and subtraction, which enable the model to capture similarities and differences between two event mentions. Then, we input the event mention pair representation $\mathbf{g}_{ij}$ into the scorer as follows.

$$s_c(i, j) = \mathrm{sigmoid}(\mathbf{W}_c \mathrm{FFNN}_c(\mathbf{g}_{ij})) \qquad (9)$$

where FFNN_c is a two layer feed-forward neural network, $\mathbf{W}_c$ is a parameter matrix that can be trained. We also use the binary cross-entropy loss to train the classification module as follows.

$$\mathcal{L}_{ec} = \sum_{i=0}^{n} \sum_{j=i+1}^{n} [y_{ij}\mathrm{log}(s_c(i,j)) + (1 - y_{ij})\mathrm{log}(1 - s_c(i,j))] \qquad (10)$$

3.4   Generation Module

Inspired by the success of the generation model in many NLP tasks, we introduce the generation module to help event coreference resolution. Because of the need for long document encoding/decoding, we adopt the pre-trained model LongT5 [12] as our transformer-based encoder-decoder model.

---

[2] https://allenai.org/allennlp/software/allennlp-library.

To make better use of the generation model in the task of event coreference resolution, we formulate a structure normal sentence as the target sentence of LongT5. Specifically, the form of the target sentence is as follows.

The triggers of the 1st event are <pad> *trigger1, trigger2* ... <pad>.
The triggers of the 2nd event are <pad> *trigger3, trigger4* ... <pad>.
... ...

Between the two <pad> in each sentence are the trigger words of the coreferential event mentions. That is, each sentence contains a trigger cluster. There are two purposes for setting the target sentence in this way. Firstly, we use the generation model to help trigger detection and strengthen the model's attention to trigger words in the document. Second, we hope that the generation model can help our classification model cluster event trigger words to discover coreferential event mentions.

As shown in Fig. 2, after getting the embedding by LongT5-encoder, we decode it with LongT5-decoder. Then we calculate the cross entropy loss between the results and the target sentences. Finally, let $\mathcal{L}_{gen}$ represent the loss of generation module.

$$\mathcal{L}_{gen} = \text{Loss}(results, targets) \tag{11}$$

3.5   Joint Learning

In this paper, we apply joint learning framework to train the model. We add up the loss function of the three modules: event detection, classification and generation as follows. We tune the parameters of the three modules at the same time with the back propagation algorithm.

$$\mathcal{L}(\Theta) = \mathcal{L}_{ed} + \mathcal{L}_{ec} + \mathcal{L}_{gen} \tag{12}$$

4   Experimentation

4.1   Experimental Settings

KBP and ECB+ are the main datasets to evaluate event coreference resolution. We evaluate our model on KBP datasets, because our model is built for within-document event coreference resolution while ECB+ is used to evaluate cross-document event coreference resolution. Following Tran et al. [5] and Choubey et al. [8], we use KBP 2015 as the training set, and use official KBP 2016 and KBP 2017 for evaluation. KBP 2015 contains 648 documents, of which 583 documents are selected for training, and the remaining 65 documents are for parameter adjustment; KBP 2016 and KBP 2017 contain 169 and 167 documents respectively.

We use the official evaluation tools provided in KBP 2017 Evaluation to evaluate the performance of our model. Specifically, it includes four basic metrics: MUC [18], B^3 [19], $CEAF_e$ [20] and BLANC [21]. Consistent with previous studies, we also report the metrics AVG-F which is the unweighted average of the above four metrics. Though event coreference resolution is the main task of our model, we also report the result of event detection module, and the metrics is the macro value.

The window size w in the event detection module is set to 2, and we set 2 layers for R-GCN in the classification module. When joint learning, we use Adam as the optimizer to train all parameters, and set the learning rate to $1e-5$ and the dropout to 0.2.

4.2 Experimental Results

To evaluate the performance of our model CGECR, we compare it with two SOTA models as follows.

1) Lu2021 [6]: It is based on the SpanBERT model and jointly learned entity coreference resolution and event coreference resolution.
2) Tran2021 [5]: It introduces GCN to event coreference resolution and define some rules to constrain the consistency of event mentions.

Table 1. Result comparison of event coreference resolution on KBP2016 and KBP2017.

KBP 2016				
Model	MUC	B^3	$CEAF_e$	BLANC	AVG-F
Lu2021	38.9	52.6	51.9	35.0	44.6
Tran2021	38.4	52.8	52.3	35.7	44.8
CGECR(ours)	**40.4**	**56.8**	**56.5**	**37.8**	**47.9**
KBP 2017				
Model	MUC	B^3	$CEAF_e$	BLANC	AVG-F
Lu2021	42.8	53.7	51.5	36.4	46.1
Tran2021	40.7	51.9	52.8	34.8	45.1
CGECR(ours)	**44.2**	**56.0**	**55.9**	**37.3**	**48.4**

Table 1 shows the results of event coreference resolution of our model CGECR and two benchmarks on the KBP2016 and KBP2017 datasets. We can find that our model CGECR outperforms Lu2021 and improves the average metric AVG-F by 3.3 and 2.3 on the KBP2016 and KBP2017 datasets, respectively. Compared with Tran2021, CGECR gains the improvement of AVG-F by 3.1 and 3.3, respectively. These results verify the effectiveness of our CGEDR on event coreference resolution.

According to the definitions of metrics, MUC and BLANC prefer the number of recalled coreferential pairs when computing the results, while B^3 and $CEAF_e$ prefer the precision of event chains. Our model is higher than the above two benchmarks in all metrics. It is proved that our model is better than the existing models in both recall and precision, benefit from our relational graphs and generation module.

Table 2 shows the results of the event detection module[3]. As shown in the second row of Table 2, compared with our model CGDCR, the F1 value of the simplified model without the generation model (w/o Generation) is reduced by 0.91

---

[3] Lu et al. [6] and Tran et al. [5] did not reported the perfromance of event detection.

and 0.65, respectively. It proves that our generation module can make CGECR better capture event information, and then improve the performance of event detection task.

Table 2. Results of event detection on KBP2016 and KBP2017.

Model	KBP 2016	KBP 2017
CGECR	**66.72**	**66.41**
w/o Generation	65.81	65.76

4.3 Ablation Analysis

To further analyze the contribution of each part of our model to the performance improvement, We conduct some ablation experiments, as shown in Table 3.

Table 3. Ablation results on test sets.

KBP 2016						
Model	MUC	B^3	$CEAF_e$	BLANC	AVG-F	Δ
CGECR	**40.4**	**56.8**	**56.5**	**37.8**	**47.9**	
w/o Macro Discourse	38.7	55.7	55.5	36.9	46.8	−1.1
w/o Entity Coreference	39.3	55.9	56.2	37.2	47.2	−0.7
w/o Semantic	39.2	55.6	55.9	36.7	46.9	−1.0
w/o All Graphs	38.4	54.7	53.7	36.6	45.9	−2.0
w/o Generation	39.1	55.6	55.2	36.0	46.4	−1.5
KBP 2017						
Model	MUC	B^3	$CEAF_e$	BLANC	AVG-F	Δ
CGECR	**44.2**	**56.0**	**55.9**	**37.3**	**48.4**	
w/o Macro Discourse	42.6	55.3	54.7	36.7	47.3	−1.1
w/o Entity Coreference	43.5	55.4	55.3	37.1	47.8	−0.6
w/o Semantic	43.1	54.9	55.3	36.3	47.5	−0.9
w/o All Graphs	43.0	53.8	52.8	36.1	46.5	−1.9
w/o Generation	42.0	55.7	54.0	35.8	46.8	−1.6

Effect of Relational Graphs. As shown in Table 3, after removing all graphs, the AVG-F value of the model decreases by 2.0 and 1.9 on KBP 2016 and KBP 2017, respectively, and the value of other metrics also has declined. Specifically, the second to fourth rows in Table 3 indicate the results of removing the macro discourse graph, entity coreference graph and semantic graph, respectively. We can see that these three graphs both have a positive effect on event coreference resolution. In particular, the macro discourse graph improve our model greatly, which proves the importance of the relations and structure between paragraphs for event coreference resolution.

Effect of Generation Module. The experimental results show that after deleting the generation module, the AVG-F value decreases by 1.5 and 1.6 on KBP

2016 and KBP 2017 datasets, respectively. Take the following generation result as an example:

The triggers of the 1st event are <pad> Retire, retirement, retire, retiring, go <pad>. ...

The generation result shows that our generation module has a certain ability of trigger detection and trigger clustering, and can identify the event of "retirement". Meanwhile, the experimental results in Table 2 and Table 3 prove that the generation module can improve the performance of our model significantly.

5   Conclusion

In this paper, we first introduce the macro discourse structure to mine the relations and structures between paragraphs, and then apply a generation model to event coreference resolution as an auxiliary model. Experimental results on the KBP2016 and KBP2017 corpora show that our proposed model outperforms the strong baselines. In the future, we will focus on how to use more other NLP tasks (e.g., question answering) to boost event coreference resolution.

Acknowledgments. The authors would like to thank the three anonymous reviewers for their comments on this paper. This research was supported by the National Natural Science Foundation of China (Nos. 61836007, 62276177 and 62006167.), and Project Funded by the Priority Academic Program Development of Jiangsu Higher Education Institutions (PAPD).

References

1. Weissenborn, D., Wiese, G., Seiffe, L.: Making neural QA as simple as possible but not simple. In: Proceedings of the CoNLL 2017, pp. 271–280 (2017)
2. Huang, Y.J., Kurohashi, S.: Extractive summarization considering discourse and coreference relations based on heterogeneous graph. In: Proceedings of the 16th Conference of the European Chapter of the Association for Computational Linguistics(EACL), pp. 3046–3052 (2021)
3. Jiang, F., Fan, Y., Chu, X., Li, P., Zhu, Q., Kong, F.: Hierarchical macro discourse parsing based on topic segmentation. In: Proceedings of the Conference on Artificial Intelligence (AAAI), pp. 13152–13160 (2021)
4. Schtkrull, M., Kipf, T.N., Bloem, P., Berg, R.V.D., Titov, I., Welling, M.: Modeling relational data with graph convolutional networks. In: Proceedings of the European Semantic Web Conference, pp. 593–607 (2018)
5. Tran, H.M., Phung, D., Nguyen, T.H.: Exploiting document structures and cluster consistencies for event coreference resolution. In: Proceedings of the 59th Annual Meeting of the Association for Computational Linguistics and the 11th International Joint Conference on Natural Language Processing, pp. 4840–4850 (2021)
6. Lu, J., Ng, V.: Span-based event coreference resolution. In: Proceedings of the AAAI Conference on Artificial Intelligence, vol. 35, pp. 13489–13497 (2021)
7. Lu, Y., Lin, H., Tang, J., Han, X., Sun, L.: End-to-end neural event coreference resolution. Artif. Intell. **303**, 103632 (2022)

8. Choubey, P.K., Huang, R.: Improving event coreference resolution by modeling correlations between event coreference chains and document topic structures. In: Proceedings of the 56th Annual Meeting of the Association for Computational Linguistics, pp. 485–495 (2018)

9. Kalchbrenner, N., Blunsom, P.: Recurrent continuous translation models. In: Proceedings of the 2013 Conference on Empirical Methods in Natural Language Processing, pp. 1700–1709 (2013)

10. Lu, Y., et al.: Text2event: controllable sequence-to-structure generation for end-to-end event extraction. In: Proceedings of the 59th Annual Meeting of the Association for Computational Linguistics and the 11th International Joint Conference on Natural Language Processing, pp. 2795–2806 (2021)

11. Beltagy, I., Peters, M.E., Cohan, A.: Longformer: the long-document transformer. arXiv preprint arXiv:2004.05150 (2020)

12. Guo, M., et al.: Longt5: efficient text-to-text transformer for long sequences. arXiv preprint arXiv:2112.07916 (2021)

13. Barhom, S., Shwartz, V., Eirew, A., Bugert, M., Reimers, N., Dagan, I.: Revisiting joint modeling of cross-document entity and event coreference resolution. arXiv preprint arXiv:1906.01753 (2019)

14. Chen, Z., Ji, H., Haralick, R.M.: A pairwise event coreference model, feature impact and evaluation for event coreference resolution. In: Proceedings of the Workshop on Events in Emerging Text Types, pp. 17–22 (2009)

15. Liu, Z., Araki, J., Hovy, E.H., Mitamura, T.: Supervised within-document event coreference using information propagation. In: Proceedings of LREC, pp. 4539–4544 (2014)

16. Huang, Y.J., Lu, J., Kurohashi, S., Ng, V.: Improving event coreference resolution by learning argument compatibility from unlabeled data. In: Proceedings of the 2019 Conference of the North American Chapter of the Association for Computational Linguistics: Human Language Technologies, pp. 785–795 (2019)

17. Reimers, N., Gurevych, I.: Sentence-Bert: sentence embeddings using siamese bert-networks. In: Proceedings of the 2019 Conference on Empirical Methods in Natural Language Processing and the 9th International Joint Conference on Natural Language Processing (EMNLP-IJCNLP), pp. 3982–3992 (2019)

18. Vilain, M., Burger, J.D., Aberdeen, J., Connolly, D., Hirschman, L.: A modeltheoretic coreference scoring scheme. In: Proceedings of Sixth Message Understanding Conference (MUC-6) (1995)

19. Bagga, A., Baldwin, B.: Algorithms for scoring coreference chains. In: The first International Conference on Language Resources and Evaluation Workshop on Linguistics Coreference, vol. 1, pp. 563–566 (1998)

20. Luo, X.: On coreference resolution performance metrics. In: Proceedings of Human Language Technology Conference and Conference on Empirical Methods in Natural Language Processing, pp. 25–32 (2005)

21. Recasens, M., Hovy, E.: Blanc: implementing the rand index for coreference evaluation. Nat. Lang. Eng. **17**(4), 485–510 (2011)

CCN: Pavement Crack Detection with Context Contrasted Net

Yihuan Zhu$^{(\boxtimes)}$ [ID], Sheng Zhang [ID], and Chengfeng Ruan [ID]

Department of Computer Science and Technology, Nanjing University,
Nanjing, China
zhuyihuan521@126.com

Abstract. Different from general object detection, there are three special characteristics in pavement crack detection: the tiny foreground proportion, the weak semanticity, and the local-whole similarity. To overcome the above challenges, we first construct Context Contrasted Network (CCN) to capture multi-scales features. By contrasting contextual features with local features, CCN contains the ability of local-whole discriminating. Then, we add an attention module in CCN to force the network to pay more attention to the foreground with small proportion instead of the background with large proportion, which improves the sensitivity of the model for pavement cracks. We achieve state-of-the-art results on two pavement crack datasets (i.e. CRACK500 and GRDDC), which demonstrates the effectiveness of our proposed method. Our source code will be opened later.

Keywords: Pavement Crack Detection · Object Detection · Context Contrasted Net

1 Introduction

Pavement crack detection is a sub-task of general object detection which is necessary to traffic safety. Since the remarkable success reached by deep neural networks [1] in the field of computer vision, many CNNs-based general object detection algorithms [2–6] have achieved state-of-the-art performance. Some works [7–9] have applied general object detection models to the pavement crack detection task directly but achieved limited performance. This is because the pavement crack has several particularities different from generic object detection targets, we conclude that as follows: Firstly, compared with the generic object detection targets, the foreground proportion of the pavement crack is much smaller. Secondly, cracks do not have a fixed pattern and have a large variation in size, which leads to lacking strong semantics; Finally, there is a significant visual similarity between the local and the whole of a crack region, which is hard to distinguish.

It can be divided into two ways that the existing approaches improve general object detection models to suit the pavement crack detection task, namely fusing high-level features with low-level features [10,12,13] and learning the context information by a large receptive field [11,14]. However, the above methods

© The Author(s), under exclusive license to Springer Nature Switzerland AG 2023
M. Tanveer et al. (Eds.): ICONIP 2022, LNCS 13625, pp. 85–96, 2023.
https://doi.org/10.1007/978-3-031-30111-7_8

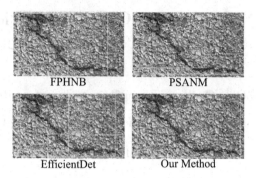

FPHNB   PSANM

EfficientDet   Our Method

Fig. 1. Detecting results of FPHNB [10], PSANM [11], EfficientDet [6], and our method. Due to the lack of local-whole discrimination, the detector treats a crack as multi-cracks thus outputting multi-boxes to one crack.

just ignore the issue that even if the receptive field is expanded, there is still a relatively small receptive field that will cover the local part of the crack, but the network is not specifically designed to distinguish whether the feature corresponds to the local or to the whole part of the crack, which leads to the performance degradation as shown in Fig. 1.

For making the network obtain the local-whole discrimination of the crack, we utilize such a prior: If a region still contains the foreground in its neighborhood, then this region is likely to be a local of the crack; On the contrary, if a region's neighborhood does not contain any foreground, then this region is likely to be the whole of the crack, as shown in Fig. 2. We can see that the region between the red dotted box and the red solid box still contains cracks marked by red circles. While the region between the yellow dotted box and yellow solid box does not contain any crack. Therefore, it should not only consider the different sizes of the receptive field coverage but also the feature differences of its neighborhoods when learning the contextual information of pavement cracks.

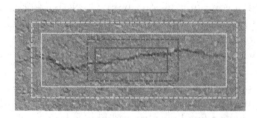

Fig. 2. The prior we adopted in this paper. The red solid box indicates a local part of crack and the red dotted box indicates its neighborhood. The yellow solid box indicates the whole crack and the the yellow dotted box indicates its neighborhood. We can see there is no crack between the yellow dotted box and solid box, which is different between the red boxes. (Color figure online)

In this paper, we propose Context Contrasted Net(CCN), which utilizes the common convolution and the dilated convolution to extract features of a region at the same time and get the neighborhood information by contrasting features extracted by two types of convolution. Our CCN can not only cover receptive fields in different scales but also obtain the context contrasted features from the neighborhood. After that, by adding the CBAM module [15] in CCN, the network is more sensitive to the crack foreground, which leads to a more outstanding detecting performance. In summary, the contributions of our work are as follows:

- We propose Context Contrasted Network (CCN) in the detection pipeline. To our best knowledge, this is the first work that adopt the neighborhood feature as the contextual information in pavement crack detection.
- The proposed CCN makes the network obtain the local-whole discrimination of pavement cracks. In addition, we add CBAM modules in CCN to enhance the sensitivity of our network to the foreground of cracks.
- We carry out experiments on two pavement detection datasets. Our method achieves state-of-the-art results on both datasets, which demonstrates the effectiveness of the context contrasted features we extracted.

2   Related Works

2.1   Object Detection

According to the processing flow, CNN-based object detection methods can be divided into two parts: two-stage object detectors (e.g. Faster RCNN [2] and its variants [16,17]) and one-stage object detectors (e.g. YOLO [3], RetinaNet [4], FCOS [5], EffcientDet [6]). Two-stage object detectors first select the regions that may contain objects as candidate regions, then conduct category determination and bounding box regression on these candidate regions. One-stage object detectors treat object detection as a regression process and first assign positive and negative samples to anchor boxes or anchor points, then conduct category and bounding box regression on the positive samples. Overall, two-stage detectors have better task adaptability and one-stage detectors have higher computational efficiency.

2.2   Pavement Crack Detection

The existing pavement crack detectors are modifications of general object detectors to suit the pavement crack detection task. FPHBN [10] uses Feature Pyramid Network (FPN [18]) to obtain multi-scale features by fusing high-level features with low-level features and to detect cracks on feature maps with different receptive fields; However, the scaling flexibility brought by FPN is limited and can not meet the demand of pavement crack detection. PSAMN [11] uses dilated convolution to expand the receptive field to capture the context information and adds the Bottleneck Attention Module (BAM) [19] to enhance the foreground

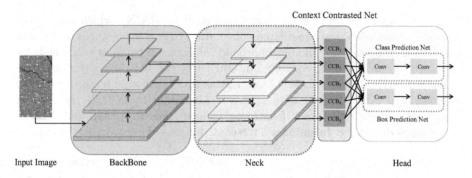

Fig. 3. The network structure of our proposed method. We first use a backbone to extract features from input images and send them to a neck. Then we utilize our proposed CCN to deal with features out from each layer of the neck. After the process, we use detection heads with same structure to detect objects after each layer of CCN.

sensitivity of the detector. However, the range of the captured context information is limited by adopting only doubled dilation rate. HFAN [14] adds an MDC module to the top layer of FPN, by utilizing dilated convolution with different rates to expand the receptive field, thus improving the feature extraction performance of the network for cracks in different scales.

3   Method

3.1   Network Overview

The network structure of the proposed method is shown in Fig. 3, which contains four parts: the backbone, the neck, our proposed CCN, and the detection head. The backbone is used as a feature extractor that extracts features from the input image, and the commonly used backbone includes VGG [20], ResNet [21], EfficientNet [22], etc.

The neck is used to enhance or fuse the features extracted from the backbone, and the commonly used neck includes FPN [18] and its variants [6,23–25].

CCN is our proposed Context Contrasted Network, which can make the detector obtain the local-whole discrimination of cracks by contrasting contextual features with local features. CCN is made up of several Context Contrasted Blocks (CCB), and CCB is cascaded by numbers of Context Contrasted Units (CCU) in a residual structure. The details of CCU and CCB are in Sects. 3.2 and 3.3, respectively.

The head is a common module for localization and classification, such as the same used in [5]; In our method, the backbone, the neck, and the head are all replaceable, and the backbone used in this paper is EfficientNet0 [22], the neck is BiFPN [6].

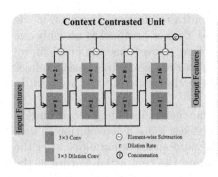
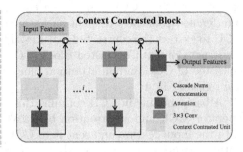

Fig. 4. Left: The structure of Context Contrasted Unit. **Right**: The structure of Context Contrasted Block.

3.2 Context Contrasted Unit

As mentioned above, for pavement crack detection, the contextual information refers not only to the large receptive field coverage, but also to compare the features under different receptive fields with that in their neighborhood. Thus, we build Context Contrasted Unit (CCU) to extract the context contrasted features effectively. The architecture of CCU is shown in the left of Fig. 4.

The main operations in CCU are 3×3 convolution and 3×3 dilated convolution with different dilated rates. We set 3×3 convolution as the local convolution and 3×3 dilated convolution as the contextual convolution. We first deal with input features of CCU by the two different types of convolutions at the same time. To extract context contrasted features in various scales, we use four dilated convolutions with different dilated rates for extracting contextual features of different ranges. Then, we can obtain context contrasted features by doing an element-wise subtraction. After concatenating that subtraction results in a parallel way, output features of CCU can be represented as context contrasted features we expected. The mathematical description of CCU is:

$$CCU = f_{context}(F, \Theta_{context}) - f_{local}(F, \Theta_{local}) \tag{1}$$

where F indicates input features, f_{local} is the convolution with the kernel size 3×3; $f_{context}$ is the 3×3 dilated convolution with a dilation rate x; Θ_{local} and $\Theta_{context}$ denote the parameters respectively; $-$ is the element-wise subtraction; CCU indicates the context contrasted features.

To ensure that the awareness range of CCU can reach the complete region of the object and learn the context contrasted features at different scales, the dilation rate x is set to 2, 4, 8 and 16 in this paper.

3.3 Context Contrasted Block

Since pavement cracks involve a larger scale span compared to general object detection targets, a diverse context contrasted ranges need to be considered.

At the same time, CCU can extract context contrasted features but contain the interference from the pavement background. So it is not practical to use only one CCU to extract the contrasted features. Therefore, to overcome the problem, we construct Context Contrasted Block (CCB) by cascading multiple basic modules consisted of CCU and an attention block. The architecture of CCB is shown in the right of Fig. 4.

A basic module in CCB contains three components, including a convolution operator, CCU and an attention block. We first use a 3×3 convolution to deal with the input feature and sent it to CCU. We add an attention block after the output feature from CCU, so that we can obtain the contrasted feature of the crack foreground. As shown in Fig. 3, inputs of different CCBs are from different layers of FPN. Thus we should consider the difference in receptive field of input feature maps.

In this paper, we cascade numbers of CCB basic modules using a residual structure to adjust the variation in receptive field of input features. Small receptive field corresponds to a small region of a image, for which we should set a large contrasting range to find the object bounding. Large receptive field corresponds to a large region, for which we should set a small contrasting range to reduce the background interference. For this reason, we add CCB with large cascaded number to the bottom layer of FPN, because the receptive field of the layer is small; on the contrary, we add CCB with small cascaded number to the top layer of FPN. The mathematical description of CCB is defined as follows:

$$CCB_i = Attention(CCU_1 \Delta \cdots \Delta CCU_i) \qquad (2)$$

where $Attention(\cdot)$ denotes the attention module, as described in Sect. 3.4; Δ indicates the cascade operation, i is the number of cascades of CCUs, which is set to 1, 2, 3 and 4 in this paper.

3.4 Attention Module

The pavement crack target is long but slim, so its foreground coverage is still tiny even though it occupies a large image region. In order to force our CCN to pay more attention to the foreground area and reduce the interference from the pavement background, we not only use Convolutional Block Attention Module (CBAM) [15] in a basic module of CCB but also deal with concatenated features of different CCB basic modules through it.

4 Experiments

4.1 Settings

Datasets. Two public pavement crack detection datasets are used in our experiments, including CRACK500 [10] and GRDDC [26]. CRACK500 is a crack detection dataset collected from different pavements. It contains 2975 training images, 500 validation images, and 1124 testing images with bounding box ground truth labels. GRDDC is a large-scale dataset, collected from different countries, for pavement defect detection which contains 7481 training images, 2111 validation images, and 4382 testing images with annotations.

Implementations. We implement our proposed methods using the PyTorch v1.2.0 on two NVIDIA 1080T GPUs (11G memory). We utilize EfficientNet0 [22] as the backbone, and the parameters of the backbone are initialized from the model pre-trained on ImageNet [27]. We adopt BiFPN [6] as the neck of our network. Each model is trained using SGD [28] optimizer with momentum 0.9 and weight decay 4e−5. Learning rate is linearly increased from 0 to 0.16 in the first training epoch and then annealed down using cosine decay rule [29]. Synchronized batch norm [30] is added after every convolution with batch norm decay 0.99 and epsilon 1e−3.

Table 1. Comparison with state-of-the-art object detection methods and pavement crack detection methods, and experimental results are reported on CRACK500 and GRDDC.

Methods	CRACK500						GRDDC					
	AP_{50}	AP_{75}	AP_s	AP_m	AP_l	mAP	AP_{50}	AP_{75}	AP_s	AP_m	AP_l	mAP
Faster RCNN [2]	79.3	48.8	31.5	46.3	52.6	45.8	47.3	13.2	11.2	15.4	20.3	19.9
RetinaNet [4]	72.8	35.9	48.7	37.4	46.8	38.7	45.9	11.9	7.0	14.7	21.2	19.2
FCOS [5]	73.2	37.3	50.6	38.9	47.6	39.5	45.0	11.6	7.3	13.9	21.0	18.7
EfficientDet [6]	82.6	33.9	64.8	41.9	44.5	40.5	46.7	10.5	8.0	14.2	21.5	18.5
DSIC [23]	78.3	34.5	52.3	40.6	47.2	39.8	46.1	10.2	7.6	14.8	20.7	18.8
FPHBN [10]	83.7	55.1	66.2	48.5	58.3	53.9	52.0	11.8	8.9	16.5	24.4	20.9
PSAMN [11]	81.1	56.9	73.2	49.5	57.8	51.3	51.9	12.4	8.3	16.3	24.3	20.7
HFAN [14]	80.5	49.7	65.1	50.0	58.6	50.6	51.3	11.7	8.5	16.0	23.9	20.4
Ours	**86.7**	**65.3**	**79.3**	**53.1**	**65.9**	**57.9**	**54.2**	**14.3**	**10.9**	**18.3**	**26.8**	**23.0**

4.2 Compared Experiments

For demonstrating the detection performance of the proposed method, quantitative experiments and qualitative experiments are conducted to compare with the recent state-of-the-art methods on two datasets. The comparison detectors include one-stage object detectors (FCOS [5] and RetinaNet [4], EfficientDet [6], and DSIC [23]), two-stage object detector Faster RCNN and recent pavement crack detection methods [10,11,14]. One-stage detectors and two-stage detector are experimented on MMDetection [31]. Since there is no open source code in [10,11,14], we reproduce these works as described in their papers.

Quantitative experiment results are shown in Table 1. Experiment results show that our method achieves the state-of-the-art performance on both two datasets. Comparing with the baseline detector EfficientDet [6], our proposed method improves detecting accuracy by 17.9% on dataset CRACK500 and 4.5% on dataset GRDDC.

For further demonstrating the ability of local-whole discrimination, we conduct qualitative experiments on different detectors in Fig. 5. We can see that our proposed method can detect pavement cracks completely and exactly. However, experimental results of other detectors show a similar phenomenon that they may treat a crack as multi-cracks so output multi-boxes to one target. That is because our proposed CCN makes the network obtain the local-whole discrimination, which is an advantage that others lack.

Table 2. Analysis of detecting performances using different components of our method. Experimental results show that the committed step for CCN to work is to make use of the context contrasted features, instead of expanding the receptive field simply (CCN_+ and CCN_{D-only}). Because neither of these two ways can allow CCN have the ability of local-whole discrimination.

Model	mAP
$Baseline$	40.5
$Baseline + CCN$	43.6
$Baseline + Attention$	42.8
$Baseline + CCN_+ + Attention$	48.6
$Baseline + CCN_{D-only} + Attention$	52.4
$Baseline + CCN + Attention(\textbf{Ours})$	**57.9**

4.3   Ablation Study

Firstly, we conduct ablation experiments with the baseline to investigate the contributions of different parts of our network, and experimental results are shown in Table 2. We use mAP (%) as the evaluation metric on CRACK500 dataset.

"$Baseline$" indicates that the backbone is EffcientNet0, the neck is BiFPN and the head is shown in Fig. 3; "$Baseline + CCN$" indicates that the CNN without CBAM is added to "$Baseline$"; "$Baseline + Attention$" indicates that the CBAM module is inserted between the neck and the head of "$Baseline$"; "$Baseline + CCN + Attention$" is our proposed network.

For proving that the context contrasting is the primary cause for our CCN working instead of expanding the receptive field, we also compare the experimental results with "$Baseline + CCN_{D-only} + Attention$" and "$Baseline + CCN_+ + Attention$".

CCN_{D-only} and CCN_+ indicate that CCUs in CCN are CCU_{D-only} and CCU_+ respectively. CCU_{D-only} indicates that we only extract features with a large receptive field by dilated convolution; CCU_+ indicates that we combine local features with contextual features by doing an element-wise addition. CCU_{D-only} and CCU_+ are defined as follows:

$$CCU_{D-only} = f_{context}(F, \Theta_{context}) \tag{3}$$

$$CCU_+ = f_{context}(F, \Theta_{context}) + f_{local}(F, \Theta_{local}) \tag{4}$$

where mathematical symbols are the same as that in the Formula 1.

FPHBN   PSAMN   EffcientDet   Ours

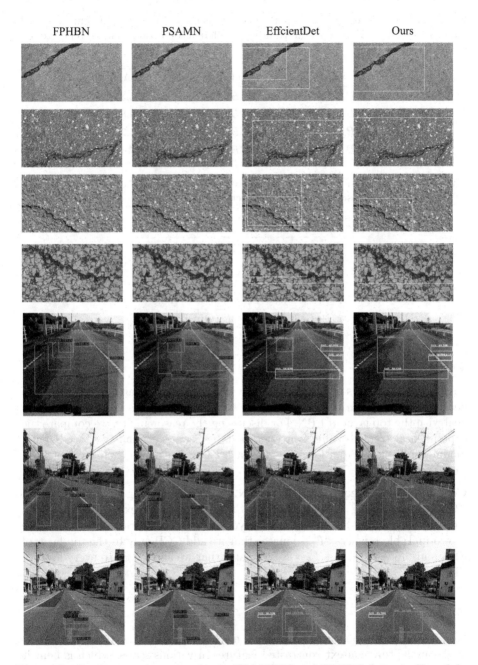

Fig. 5. Qualitative experiment results on CRACK500 and GRDDC datasets. From left to right: Results in each line are FPHBN [10], PSAMN [11], EfficientDet [6], Ground Truth, and our method, respectively. From top to bottom: First four lines are detection results on CRACK500 and others are detection results on GRDDC. The experiment results show that our method can reduce repeated detection of the same object effi-ciently.

Table 3. Analysis of detecting performances using CCBs with different cascaded numbers in CCN. Experimental results of different detection models reported on CRACK500.

CCBs with different cascaded numbers	mAP
$CCB_4, CCB_4, CCB_4, CCB_4, CCB_4$	52.3
$CCB_1, CCB_1, CCB_1, CCB_1, CCB_1$	54.8
$CCB_4, CCB_3, CCB_2, CCB_1, CCB_1$	47.7
$CCB_1, CCB_2, CCB_3, CCB_4, CCB_4$(**Ours**)	**57.9**

Table 4. Analysis of detecting performances using CCBs added to different layers of FPN. Experimental results of different detection models reported on CRACK500.

CCB(s) added to	mAP
F_1	44.6
$F_1 + F_2$	47.7
$F_1 + F_2 + F_3$	49.1
$F_1 + F_2 + F_3 + F_4$	53.3
$F_1 + F_2 + F_3 + F_4 + F_5$(**Ours**)	**57.9**

4.4 Discussion

Cascaded numbers in CCB. We use CCB with large cascaded numbers to deal with the bottom layer of FPN and CCB with small cascaded numbers to deal with the top layer of FPN. For discussing the reasonability, we compare the experimental results of different conditions in cascaded number of CCB basic module in Table 3. From left to right, each CCB deals with each layer of FPN from top to bottom.

Contributions of each CCB in CCN. For demonstrating that our proposed CCN is contributing to every layer in FPN, we compare experimental results when adding CCB(s) after different layers of FPN. Experimental results are shown in Table 4, where "F_1" donates that we add CCB only to the top layer of FPN and "$F_1 + F_2 + F_3 + F_4 + F_5$" donates that we add CCBs to every layer of FPN.

5 Conclusion

We propose a pavement crack detection method in this paper. CCN in the network can capture context contrasted features in various scales, which is benefit for pavement crack boundary detecting. Meanwhile, we also utilize the attention module in our method to enhance the sensitivity of the network to the crack foreground and reduce the background interference. By comparing with some excellent detectors, our method achieves the state-of-the-art experimental results on CRACK500 and GRDDC datasets, which demonstrates the superiority and efficientiveness.

References

1. LeCun, Y., Bengio, Y., Hinton, G.: Deep learning. Nature **521**(7533), 436–444 (2015)
2. Ren, S., He, K., Girshick, R., Sun, J.: Faster R-CNN: towards real-time object detection with region proposal networks. In: Neural Information Processing Systems, vol. 28, pp. 91–99 (2015)
3. Redmon, J., Divvala, S., Girshick, R., Farhadi, A.: You only look once: Unified, real-time object detection. In: CVPR, pp. 779–788 (2016)
4. Lin, T. Y., Goyal, P., Girshick, R., He, K., Dollár, P. : Focal loss for dense object detection. In: ICCV, pp. 2980–2988 (2017)
5. Tian, Z., Shen, C., Chen, H., He, T. : FCOS: fully convolutional one-stage object detection. In: ICCV, pp. 9626–9635 (2019)
6. Tan, M., Pang, R., Le, Q.V.: EfficientDet: scalable and efficient object detection. In: CVPR, pp. 10781–10790 (2020)
7. Nguyen, N.T.H., Le, T.H., Perry, S., Nguyen, T.T.: Pavement crack detection using convolutional neural network. In: SoICT, pp. 251–256 (2018)
8. Gou, C., Peng, B., Li, T., Gao, Z.: Pavement crack detection based on the improved faster-RCNN. In: ISKE, pp. 962–967 (2019)
9. Yusof, N.A.M., et al.: Automated asphalt pavement crack detection and classification using deep convolution neural network. In: ICCSCE, pp. 215–220 (2019)
10. Yang, F., Zhang, L., Yu, S., Prokhorov, D., Mei, X., Ling, H.: Feature pyramid and hierarchical boosting network for pavement crack detection. IEEE Trans. Intell. Transp. Syst. **21**(4), 1525–1535 (2019)
11. Xiang, X., Zhang, Y., El Saddik, A.: Pavement crack detection network based on pyramid structure and attention mechanism. IET Image Process. **14**(8), 1580–1586 (2020)
12. Wang, J., Liu, F., Yang, W., Xu, G., Tao, Z.: Pavement crack detection using attention U-Net with multiple sources. In: Peng, Y., et al. (eds.) PRCV 2020. LNCS, vol. 12306, pp. 664–672. Springer, Cham (2020). https://doi.org/10.1007/978-3-030-60639-8_55
13. Zhang, K., Zhang, Y., Cheng, H.D.: Crack-GAN: pavement crack detection using partially accurate ground truths based on generative adversarial learning. IEEE Trans. Intell. Transp. Syst. **22**(2), 1306–1319 (2020)
14. Cheng, W., Zhou, Y.: Automatic pavement crack detection based on hierarchical feature augmentation. In: ICAIIS, pp. 1–7 (2021)
15. Woo, S., Park, J., Lee, J.-Y., Kweon, I.S.: CBAM: convolutional block attention module. In: Ferrari, V., Hebert, M., Sminchisescu, C., Weiss, Y. (eds.) ECCV 2018. LNCS, vol. 11211, pp. 3–19. Springer, Cham (2018). https://doi.org/10.1007/978-3-030-01234-2_1
16. He, K., Gkioxari, G., Dollár, P., Girshick, R.: Mask R-CNN. In: ICCV, pp. 2961–2969 (2017)
17. Pang, J., Chen, K., Shi, J., Feng, H., Ouyang, W., Lin, D.: Libra R-CNN: towards balanced learning for object detection. In: CVPR, pp. 821–830 (2019)
18. Lin, T.Y., Dollár, P., Girshick, R.B., He, K., Hariharan, B., Belongie, S.J.: Feature pyramid networks for object detection. In: CVPR, pp. 936–944 (2017)
19. Park, J., Woo, S., Lee, J.Y., Kweon, I.S.: Bam: bottleneck attention module. arXiv preprint arXiv:1807.06514 (2018)
20. Simonyan, K., Zisserman, A.: Very deep convolutional networks for large-scale image recognition. arXiv preprint arXiv:1409.1556 (2014)

21. He, K., Zhang, X., Ren, S., Sun, J.: Deep residual learning for image recognition. In: CVPR, pp. 770–778 (2016)
22. Tan, M., Le, Q.: EfficientNet: rethinking model scaling for convolutional neural networks. In: ICML, pp. 6105–6114 (2019)
23. Li, Z., Liu, Y., Li, B., Hu, W., Miao, Y., Zhang, H.: DSIC: dynamic sample-individualized connector for multi-scale object detection. In: ICME (2021)
24. Liu, S., Qi, L., Qin, H., Shi, J., Jia, J.: Path aggregation network for instance segmentation. In: CVPR, pp. 8759–8768 (2018)
25. Ghiasi, G., Lin, T.Y., Le, Q.V.: NAS-FPN: learning scalable feature pyramid architecture for object detection. In: CVPR, pp. 7036–7045 (2019)
26. Arya, D., Maeda, H., Ghosh, S.K., et al.: Global road damage detection: state-of-the-art solutions. In: IEEE BigData, pp. 5533–5539 (2020)
27. Deng, J., Dong, W., Socher, R., Li, L.J., Li, K., Fei-Fei, L.: ImageNet: a large-scale hierarchical image database. In: CVPR, pp. 248–255 (2009)
28. Bottou, L.: Stochastic gradient descent tricks. In: Montavon, G., Orr, G.B., Müller, K.-R. (eds.) Neural Networks: Tricks of the Trade. LNCS, vol. 7700, pp. 421–436. Springer, Heidelberg (2012). https://doi.org/10.1007/978-3-642-35289-8_25
29. Loshchilov, I., Hutter, F.: SGDR: stochastic gradient descent with warm restarts. arXiv preprint arXiv:1608.03983 (2016)
30. Peng, C., et al.: MegDet: a large mini-batch object detector. In: CVPR, pp. 6181–6189 (2018)
31. Chen, K., Wang, J., Pang, J., et al.: MMDetection: open MMLab detection toolbox and benchmark. CoRR (2019)

Spatial and Temporal Guidance for Semi-supervised Video Object Segmentation

Guoqiang Li[1], Shengrong Gong[1,2(✉)], Shan Zhong[2], and Lifan Zhou[2]

[1] Soochow University, Soochow, China
gqlicong@stu.suda.edu.cn
[2] Changshu Institute of Technology, Soochow, China
{shrgong,sunshine620,zhoulifan}@cslg.edu.cn

Abstract. Semi-supervised video object segmentation aims to segment the object in the video when only the annotated mask of the first frame is given. Recently, memory-based methods have attracted increasing attention with significant performance improvements. However, these methods employ pixel-level matching according to the similarity without considering the trajectory and the feature of the object, which may result in mismatching between the object and non-object region in complex scenarios. To relieve this problem, we propose spatial and temporal guidance for semi-supervised video object segmentation. The proposed method takes into account the consistency of the object in spatiotemporal domain and employs global matching to conduct pixel-level matching. Moreover, we design the spatial guidance module (SGM) to track the trajectory of the object. And we design the temporal guidance module (TGM) to focus on long-term object-level feature from the first frame. The proposed spatial and temporal guidance effectively alleviates mismatching and makes the model more robust and efficient. Experiments on YouTube-VOS and DAVIS benchmarks show that our method outperforms previous state-of-the-art methods with a fast inference speed.

Keywords: Video object segmentation · Spatial and temporal guidance · Memory network

1 Introduction

Video object segmentation (VOS) aims to segment the moving object appearing in the video, which has wide application in the fields of autonomous driving and video analysis. This paper focuses on semi-supervised VOS when only the annotated mask of the first frame is given during inference. The segmentation process is illustrated in Fig. 1. Although the object mask in the first frame is provided, semi-supervised VOS is still a difficult task due to appearance changes, object deformation, occlusion and similar objects confusion, etc.

Earlier VOS methods [2,17] fine-tuned their models by the first frame. These methods have a poor accuracy in segmenting similar objects due to treating each

M. Tanveer et al. (Eds.): ICONIP 2022, LNCS 13625, pp. 97–109, 2023.
https://doi.org/10.1007/978-3-031-30111-7_9

Fig. 1. An example of semi-supervised VOS. The first image and mask (red border) are the reference frame, and the other segmentation (purple border) are the outputs of our method. (Color figure online)

frame seperately. Propagation-based methods [18,26,31] attempted to build the coherence between frames. However, these methods are subjective to mask drift caused by the disappearing of the object. Matching-based methods tried to find the correspondence between frames by performing feature matching. Among them, STM [19] and its variants [5,6,10,24,30] explored the pixel-level matching between all past frames and the current frame, so they can gain significant performance improvements. However, the pixel-level matching between the fixed object feature of the first frame and the subsequent frames will inevitably bring accumulated error because the object appearance may change heavily during inference. If the spatial-irrelative object-level clues in the first frame can be utilized, the accumulative error can be rectified and the object consistency can be guaranteed. Therefore, we attempt to capture object-level information from the first frame. To adapt to the change of the object appearance, the object-level clues of the first frame are adaptively extracted and then smoothly transferred to the subsequent frames in the video. Moreover, we argue that the object in the video is smoothly moving, so the object position in the previous frame can provide an important instruction to guide the latter segmentation. If the position information and the relationship between the object and non-object can be utilized, the error accumulation can be further rectified. Therefore, we attempt to capture the rich contextual information through integrating object position information and establishing the connection between the object and non-object.

In this paper, we present spatial and temporal guidance with memory-based network for semi-supervised video object segmentation (STGVOS). Three submodules are designed: the global matching module (GMM), the spatial guidance module (SGM) and the temporal guidance module (TGM). GMM performs global matching to find correspondence between pixels. To better leverage spatial information, SGM combines the position information of the previous frame and aggregates the contextual information of the current frame. To better leverage the object feature, TGM can adaptively adjust object-level clues over time for every frame according to the first frame. Experimental results on YouTube-VOS and DAVIS benchmarks indicate that the proposed STGVOS outperforms previous state-of-the-art methods with a fast inference speed.

2   Related Work

Detection-Based Methods. Earlier VOS methods [2,17] learned a detector from the first frame so that they can recognize the object appearance, and then segmented the remain video frames. These methods do not use temporal information, so it is difficult for them to distinguish the objects with similar appearances. Furthermore, the online fine-tuning in the first frame is time-consuming.

Propagation-Based Methods. Propagation-based methods mainly used temporal relationship between frames for segmentation. MaskTrack [20] and RGMP [18] employed optical flow or the previous mask to construct temporal propagation. S2S [31] and RVOS [26] used recurrent neural network to build spatiotemporal coherence in videos. Compared with detection-based methods, propagation-based methods have a large improvement in performance. However, they still perform poorly in cases of partial or complete occlusion in the propagation.

Matching-Based Methods. Matching-based methods applied pixel-level matching between the current frame and template frames. FEEL [27] and CFBI [34,35] employed global matching on the first frame and local matching on the previous frame. These methods only use the first frame and the previous frame to match the current frame. Although occlusion and drifting can be relieved, they often fail when the appearance of the object changes heavily during inference.

Recently, more and more work attempted to introduce memory network to matching-based methods. STM [19] stored all previous frames in the memory network and then performed pixel-level matching. Based on STM, KMN [24] employed a 2D Gaussian kernel to reduce the degree of non-localization. RMNet [30] built an efficient region memory network using optical flow. MiVOS [5] utilized a lightweight top-k operation to filter the noise in the memory network. LCM [10] learned position and target consistency. STCN [6] established image-to-image correspondences instead of encoding features for each object. Matching-based methods are benefited from memory network, but the isolate treatment of the historical frames makes the segment unstable due to ignoring the utilization of both object spatial information and long-term object-level feature.

3   Method

3.1   Overview

The overview of our method is shown in Fig. 2. We adopt a typical encoder-decoder architecture. The key encoder is designed to extract the key feature for each image, while the value encoder using the image and the object mask as input is applied to obtain value features for all objects appeared in historical frames. Three sub-modules are designed: the global matching module (GMM), the spatial guidance module (SGM) and the temporal guidance module (TGM). GMM performs pixel-level matching by estimating the similarities between the key features of historical frames and the key feature of the current frame. To obtain richer spatial-specific feature, SGM exploits object's moving trajectory and contextual information. TGM can capture the spatial-irrelative object clues derived from the first frame and then update for segmenting the subsequent frames. In the following sections, we will further present a detailed description.

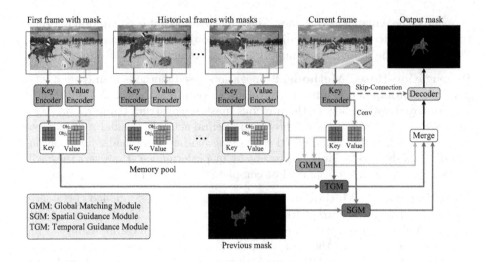

Fig. 2. Overview of STGVOS. we adopt a typical encoder-decoder architecture. Three sub-modules are designed: the global matching module (GMM), the spatial guidance module (SGM) and the temporal guidance module (TGM).

3.2   Global Matching Module

The architecture of the global matching module (GMM) is illustrated in Fig. 2 and Fig. 3(a). Similar with STCN [6], the current frame is used as the query, and historical frames with their masks pay a role of memory. ResNet50 [8] is adopted here as the key encoder to generate the key feature for each frame, while ResNet18 [8] is constructed as the value encoder to get the value feature for any arbitrary object appeared in historical frames. The key feature and the value feature with a size of 1/16 of the original image through a 3×3 convolutional layer following stage-4 (res4). For the tth frame, the outputs from key encoder and value encoder are defined as $k_t \in \mathbb{R}^{H \times W \times C_k}$ and $v_t \in \mathbb{R}^{H \times W \times C_v}$, where H is the height, W is the width, C is the channel dimension. We set C_k to be 64 and C_v to be 512 following STCN [6].

Given T historical frames as memory frames, their features are stored in memory pool and then concatenated along the temporal dimension, which are defined as $K^M = \{k_t^M, t \in [1, T]\}$, $k_t^M \in \mathbb{R}^{H \times W \times C_k}$, and $V^M = \{v_t^M, t \in [1, T]\}$, $v_t^M \in \mathbb{R}^{H \times W \times C_v}$. For a query image, we can get $K^Q \in \mathbb{R}^{H \times W \times C_k}$ by feeding it into the key encoder. $r \in \mathbb{R}^{H \times W \times C_v}$ is obtained by calculating the similarity of each pixel pair between the query key and the memory key and then using the similarity to weight the memory value:

$$r_{(j)} = \sum_i \frac{exp(L_2(K_{(i)}^M, K_{(j)}^Q))}{\sum_i exp(L_2(K_{(i)}^M, K_{(j)}^Q))} V_{(i)}^M, \tag{1}$$

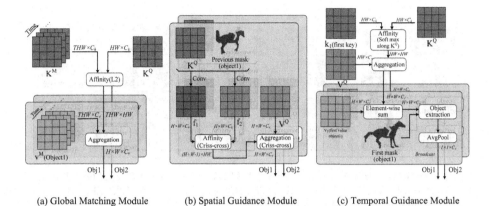

(a) Global Matching Module   (b) Spatial Guidance Module   (c) Temporal Guidance Module

Fig. 3. The detailed implementation of our proposed three sub-modules.

where $i \in [1, THW]$ and $j \in [1, HW]$ indicate the positions of K^M and K^Q, respectively. So $K_{(i)} \in \mathbb{R}^{1 \times 1 \times C_k}$ indicates the feature vector at the ith position. $L_2(:,:)$ indicate the negative squared Euclidean distance of two vectors.

The main advantage of the global matching module is that it can match to the object even if the object is occluded or disappeared. However, this module only performs pixel-level matching without considering the relationship between the object and the other regions, which may lead the object to be matched to non-object regions when facing a complex scene. Therefore we propose the spatial guidance module (SGM) and the temporal guidance module (TGM) to solve the above problem in the following sections.

3.3   Spatial Guidance Module

In most scenes, the object moves smoothly in the video. This means that the object's position of the previous frame may provide an estimated position for the current frame. In addition, rich contextual information can help establish the connection between the object and non-object, so the object feature can be better captured. Based on the above analysis, we designed the spatial guidance module (SGM) to track the trajectory of the object and then extract rich contextual information. The architecture is shown in Fig. 3(b).

Inspired by Criss-Cross Attention [11], we built the spatial guidance module. For a query image, we can get $V^Q \in \mathbb{R}^{H \times W \times C_v}$ via a 3×3 convolutional layer from the stage-4 (res4) of the key encoder. In order to track the trajectory of the object, we use the previous mask with size of $H \times W \times 1$. K^Q and the previous mask are concatenated along the channel dimension. The module first applies two different convolutional layers with 1×1 filters on new features to generate two feature maps $f_1 \in \mathbb{R}^{H \times W \times C_k}$ and $f_2 \in \mathbb{R}^{H \times W \times C_k}$. Then the module calculates affinity operation between f_1 and f_2. Finally, the contextual information is collected by an aggregation operation. We depict the process as follows:

Fig. 4. Qualitative results of our proposed STGVOS on the DAVIS-2017 validation set. We re-implement STCN [6], MiVOS [5], KMN [24] and STM [19]. All methods do not use simulated data in the training process.

$$V_{(j)}^S = \sum_i \frac{exp(f_{1(i)} \otimes f_{2(j)})}{\sum_i exp(f_{1(i)} \otimes f_{2(j)})} V_{(i)}^Q, \tag{2}$$

where $j \in [1, HW]$ indicates the position in spatial dimension of f_2. For each position j, we can obtain vectors $\{f_{1(i)}, i \in [1, H + W - 1]\}$, $f_{1(i)} \in \mathbb{R}^{1 \times 1 \times C_k}$, by extracting feature vectors in the same row or same column with position j. $V_{(j)}^S \in \mathbb{R}^{1 \times 1 \times C_v}$ aggregates horizontal and vertical contextual information at position j. $\otimes$ represents the matrix inner production.

The spatial guidance module first use the previous mask to improve the response of the possible object area. Then it will further aggregate contextual information. This efficient way provides richer spatial-detail feature without more computation.

3.4 Temporal Guidance Module

The true mask, which is believed to contain object-level feature, is solely provided for the first frame during inference. Unfortunately, such object-level feature is weakened as the video length increases. We designed the temporal guidance module (TGM) to remain the object-level information in the process of passing from the first frame to the subsequent frames. The architecture is shown in Fig. 3(c).

The key feature, the value feature and the mask for the first frame can be denote as k_1, v_1 and M^F, respectively. For the current frame, we have K^Q and V^Q. We first design a reverse matching mechanism to match object feature according to the current frame by the following formulation:

$$v_{(j)}^{obj} = \sum_i \frac{exp(L_2(K_{(i)}^Q, k_{1(j)}))}{\sum_i exp(L_2(K_{(i)}^Q, k_{1(j)}))} V_{(i)}^Q, \tag{3}$$

where $v^{obj} \in \mathbb{R}^{H \times W \times C_v}$. Note that this process is different from GMM although we both use K^Q. In GMM, we use K^Q as the query. But we reverse its role in here

Fig. 5. Qualitative results of our proposed STGVOS on the DAVIS-2017 testdev set and YouTube-VOS validation set. Our method performs well in some complex scenes, such as similar objects, object deformation and occlusions.

(k_1 as the query). The advantage of this setting is that the object information can be adaptively adjusted according to the current frame. Then v_1 and v^{obj} will be fused, and the object-level information is extracted through M^F. Finally, we use average pooling on new feature followed by two fully-connected layers and Sigmoid function as in the design of SENet [9]. We depict the process as follows:

$$V^T = AvgPool((v_1 + v^{obj}) \circ M^F), \qquad (4)$$

where $\circ$ represents Hadamard product. $V^T \in \mathbb{R}^{1 \times 1 \times C_v}$ merges with features of different modules by using broadcast:

$$v^{total} = \lambda_1 V^S + \lambda_2 V^T + V^Q, \qquad (5)$$

where $v^{total} \in \mathbb{R}^{H \times W \times C_v}$. $\lambda 1$ and $\lambda 2$ are learnable parameters which are initialized to zero. With the training of the network, $\lambda 1$ and $\lambda 2$ will be set to suitable values. v^{total} concatenated with r from GMM as the final feature to be decoded. We employ the decoder which consists of a residual block and two stacks of refinement modules. Soft aggregation [19] is used when facing a multi-object scene.

4 Experiments

In this section, we describe the implementation details of our model and evaluate our model on two popular benchmarks, YouTube-VOS [32] and DAVIS [21,22]. We take region similarity J, contour accuracy F and their average $J\&F$ as the evaluation measures for DAVIS. While in YouTube-VOS, we report J and F for both seen and unseen categories, and the averaged score avg.

4.1 Implementation Details

Following [5,6,10,19,24,30], our model is trained in two phases. First, we **pre-train** the model using simulated videos generated from image datasets ECSSD [25], FSS1000 [12], HRSOD [36], BIG [4], DUTS [29]. A simulated video containing three frames is generated using image transformations. Second, we **main-train** our pre-trained model through VOS datasets [22,32]. Each training sample contains three temporally ordered frames sampled from the training video.

We initialize the network with ImageNet pretrained parameters. The model are trained with two 16 GB Tesla P100 GPUs. The Adam optimizer is used with $\beta_1 = 0.9$ and $\beta_2 = 0.999$, a base learning rate of 10^{-5}. Bootstrapped cross entropy is used following [5,6]. We use the batch sizes of 16 during pre-training and 8 during main-training. The pre-training and main-training have lasted for about 3 days (300K iterations) and 2 days (200K iterations), respectively.

Table 1. The quantitative evaluation on DAVIS-2017 validation set and YouTube-VOS set. S: using simulated data in the training process. *: the results that we reproduce in the same environment.

(a) DAVIS-2017 validation set

Pub.	Methods	S	J&F	J	F
CVPR18	OSMN [33]		54.8	52.5	57.1
CVPR19	RVOS [26]		50.3	48.0	52.6
CVPR20	SAT [3]		72.3	68.6	76.0
ACCV18	PReM [16]		77.8	73.9	81.7
CVPR19	FEEL [27]		71.5	69.1	74.0
CVPR20	TVOS [37]		72.3	69.9	74.7
ECCV20	LWL [1]		81.6	79.1	84.1
ECCV20	CFBI [34]		81.9	79.1	84.6
PAMI21	CFBI+ [35]		82.9	80.1	85.7
	KMN* [24]		80.7	78.1	83.2
	MiVOS* [5]		81.2	78.6	83.7
CVPR21	SST [7]		82.5	79.9	85.1
	STCN [6]		82.5	79.3	85.7
	Ours		**83.8**	**80.5**	**87.0**
CVPR18	RGMP [18]	✓	66.7	64.8	68.6
ECCV20	GC [13]	✓	71.4	69.3	73.5
NIPS20	AFBURR [14]	✓	74.6	73.0	76.1
CVPR21	SwiftNet [28]	✓	81.1	78.3	83.9
ICCV19	STM [19]	✓	81.8	79.2	84.3
ECCV20	GraphM [15]	✓	82.8	80.2	85.2
ECCV20	KMN [24]	✓	82.8	80.0	85.6
CVPR21	MiVOS [5]	✓	83.3	80.6	85.9
CVPR21	RMNet[30]	✓	83.5	81.0	86.0
CVPR21	LCM[10]	✓	83.5	80.5	86.5
NIPS21	STCN[6]	✓	85.4	82.2	88.6
	Ours	✓	**85.6**	**82.4**	**88.8**

(b) YouTube-VOS set

Methods	avg	Seen		Unseen	
		J	F	J	F
Validation 2018 set					
FRTM [23]	72.1	72.3	76.2	65.9	74.1
GC [13]	73.2	72.6	68.9	75.6	75.7
SwiftNet [28]	77.8	77.8	81.8	72.3	79.5
STM [19]	79.4	79.7	84.2	72.8	80.9
AFBURR [14]	79.6	78.8	83.1	74.1	82.6
GraphM [15]	80.2	80.7	85.1	74.0	80.9
MiVOS [5]	80.4	80.0	84.6	74.8	82.4
CFBI [34]	81.4	81.1	85.8	75.3	83.4
KMN [24]	81.4	81.4	85.6	75.3	83.3
RMNet [30]	81.5	82.1	85.7	75.7	82.4
LWL [1]	81.5	80.4	84.9	76.4	84.4
SST [7]	81.7	81.2	-	76.0	-
CFBI+ [35]	82.0	81.2	86.0	76.2	84.6
LCM [10]	82.0	82.2	86.7	75.7	83.4
STCN [6]	83.0	81.9	86.5	**77.9**	85.7
Ours	**83.5**	**83.1**	**87.9**	77.1	**85.9**
Validation 2019 set					
MiVOS [5]	80.3	79.3	83.7	75.3	82.8
CFBI [34]	81.0	80.6	85.1	75.2	83.0
SST [7]	81.8	80.9	-	76.6	-
CFBI+ [35]	82.6	81.7	86.2	77.1	85.2
STCN [6]	82.7	81.1	85.4	**78.2**	85.9
Ours	**83.6**	**82.8**	**87.3**	77.8	**86.4**

4.2 Compare with the State-of-the-Art Methods

DAVIS is one of the most commonly-used benchmarks for VOS. In detail, DAVIS-2016 contains a small number of single-object videos. As a multi-object

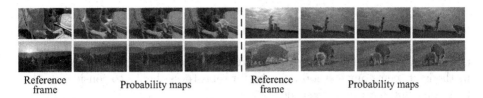

Reference frame   Probability maps   Reference frame   Probability maps

Fig. 6. Probability maps for different objects sampled at important moments.

Table 2. The quantitative evaluation on DAVIS-2017 test-dev set and DAVIS-2016 validation set. The time is measured on a Tesla P100 GPU without I/O time. ‡: recorded on our device.

(a) DAVIS-2017 test-dev set

Methods	$J\&F$	J	F
OSMN [33]	41.3	37.3	44.9
RGMP [18]	52.9	51.3	54.4
FEEL [27]	57.8	55.2	60.5
PReM [16]	71.6	67.5	75.7
STM [19]	72.2	69.3	75.2
RMNet [30]	75.0	71.9	78.1
CFBI [34]	74.8	71.1	78.5
STCN [6]	76.5	73.1	80.0
Ours	**77.3**	**73.8**	**80.8**

(b) DAVIS-2016 validation set

Methods	Time(s)	$J\&F$	J	F
FEEL [27]	0.286	82.2	81.7	88.1
STM [19]	0.112	89.3	88.7	89.9
CFBI [34]	0.160	89.4	88.3	90.5
KMN [24]	0.120	90.5	89.5	91.5
CFBI+ [35]	0.170	89.9	88.7	91.1
RMNet [30]	0.084	88.8	88.9	88.7
LCM [10]	0.118	90.7	89.9	91.4
STCN ‡ [6]	0.071	**91.6**	**90.8**	**92.5**
Ours	0.074	91.5	90.6	92.3

Table 3. Ablation study on the DAVIS-2017 validation set

(a) Effects of sub-module

G M M	S G M	T G M	$J\&F$	J	F
✓			82.7	79.5	85.8
✓	✓		83.4	80.1	86.6
✓	✓	✓	**83.8**	**80.5**	**87.0**

(b) Training strategy

Pre	Main	$J\&F$	J	F
✓		74.5	72.0	76.9
	✓	83.8	80.5	87.0
✓	✓	**85.6**	**82.4**	**88.8**

(c) Number of objects

Number of Objects	$J\&F$	J	F
1	95.6	94.1	97.1
2	84.1	80.1	88.0
3	79.8	75.9	83.6

extension of DAVIS 2016, DAVIS-2017 contains 60 videos for training, 30 videos for validating and 30 videos for testing.

As shown in Table 1(a), STGVOS achieves **83.8%** on the DAVIS-2017 validation set, which surpasses the previous state-of-the-art method, STCN (82.5%). After using simulated data, our method achieves a better accuracy and outperforms the latest methods. As shown in Table 2(a), STGVOS is also better than

STCN (**77.3%** *vs.* 76.5%) on the DAVIS-2017 testdev set and achieves state-of-the-art performance. As shown in Table 2(b), STGVOS achieves a competitive level on the DAVIS-2016 validation set with a fast inference speed. The qualitative results are shown in Fig. 4 and Fig. 5. Compare with the latest memory-based methods, it seems our method has a better robustness thanks to considering the trajectory and feature of the object.

YouTube-VOS is currently the largest dataset for VOS which consists of 4453 video sequences annotated with multiple objects. In detail, the training set contains 3471 videos with 65 categories. The validation set contains 507 videos and additional 26 unseen categories.

The results on YouTube-VOS is shown in Table 1(b). Without using any bells and whistles, like post-processing and fine-tuning, STGVOS obtains a final score of **83.5%** in Validation 2018 and **83.6%** in Validation 2019, both of which outperforms previous methods. The qualitative results are shown in Fig. 5, which demonstrates that STGVOS can accurately recognize the object appearance by spatial and temporal guidance, even in some complex scenarios.

4.3   Ablation Study

Network Sub-module. As shown in Table 3(a), we analyzed the effectiveness of each sub-module. All experiments do not employ pre-training to save time. The results indicate that the spatial guidance module (SGM) and the temporal guidance module (TGM) can further boost performance.

Training Strategy. As shown in Table 3(b), we analyzed the impact of our training strategy. When only the main-training is used, the result achieves 83.8% in terms of $J\&F$. We further improve our performance to 85.6% by combining pre-training and main-training.

The effect of Number of Objects. As shown in Table 3(c), we found that the number of objects is inversely proportional to the accuracy. In multi-object scenario, it is more complex since the objects may be similar or occluded. Owing to the consideration of object-level feature and trajectory information, our method performs well in muti-object scenarios. Figure 6 shows probability maps of our method for different objects in complex scenarios.

5   Conclusion

In this paper, we propose spatial and temporal guidance with memory-based network for semi-supervised video object segmentation (STGVOS). Compared to other memory-based methods, we apply the global matching module to find correspondence between pixels, design the spatial guidance module to focus on the trajectory and contextual information of the object and introduce the temporal guidance module to integrate long-term object-level clues. Extensive experiments on YouTube-VOS and DAVIS benchmarks indicate that STGVOS achieves state-of-the-art performance with a fast inference speed.

Acknowledgements. This work was supported by the National Natural Science Foundation of China (61972059, 61773272, 62102347), China Postdoctoral Science Foundation (2021M69236), Key Laboratory of Symbolic Computation and Knowledge Engineering of Ministry of Education, Jilin University (93K172017K18), Natural Science Foundation of Jiangsu Province under Grant (BK20191474, BK20191475, BK20161268), Qinglan Project of Jiangsu Province (No. 2020).

References

1. Bhat, G., et al.: Learning what to learn for video object segmentation. In: Vedaldi, A., Bischof, H., Brox, T., Frahm, J.-M. (eds.) ECCV 2020. LNCS, vol. 12347, pp. 777–794. Springer, Cham (2020). https://doi.org/10.1007/978-3-030-58536-5_46

2. Caelles, S., Maninis, K.K., Pont-Tuset, J., Leal-Taixé, L., Cremers, D., Van Gool, L.: One-shot video object segmentation. In: CVPR, pp. 221–230 (2017)

3. Chen, X., Li, Z., Yuan, Y., Yu, G., Shen, J., Qi, D.: State-aware tracker for real-time video object segmentation. In: CVPR, pp. 9384–9393 (2020)

4. Cheng, H.K., Chung, J., Tai, Y.W., Tang, C.K.: CascadePSP: toward class-agnostic and very high-resolution segmentation via global and local refinement. In: CVPR, pp. 8890–8899 (2020)

5. Cheng, H.K., Tai, Y.W., Tang, C.K.: Modular interactive video object segmentation: Interaction-to-mask, propagation and difference-aware fusion. In: CVPR, pp. 5559–5568 (2021)

6. Cheng, H.K., Tai, Y.W., Tang, C.K.: Rethinking space-time networks with improved memory coverage for efficient video object segmentation. In: NIPS (2021)

7. Duke, B., Ahmed, A., Wolf, C., et al.: SSTVOS: sparse spatiotemporal transformers for video object segmentation. In: CVPR, pp. 5912–5921 (2021)

8. He, K., Zhang, X., Ren, S., Sun, J.: Deep residual learning for image recognition. In: CVPR, pp. 770–778 (2016)

9. Hu, J., Shen, L., Sun, G.: Squeeze-and-excitation networks. In: CVPR, pp. 7132–7141 (2018)

10. Hu, L., Zhang, P., Zhang, B., et al.: Learning position and target consistency for memory-based video object segmentation. In: CVPR, pp. 4144–4154 (2021)

11. Huang, Z., Wang, X., Huang, L., Huang, C., Wei, Y., Liu, W.: CCNet: Criss-Cross attention for semantic segmentation. In: ICCV, pp. 603–612 (2019)

12. Li, X., Wei, T., Chen, Y.P., Tai, Y.W., Tang, C.K.: FSS-1000: A 1000-class dataset for few-shot segmentation. In: CVPR, pp. 2869–2878 (2020)

13. Li, Yu., Shen, Z., Shan, Y.: Fast video object segmentation using the global context module. In: Vedaldi, A., Bischof, H., Brox, T., Frahm, J.-M. (eds.) ECCV 2020. LNCS, vol. 12355, pp. 735–750. Springer, Cham (2020). https://doi.org/10.1007/978-3-030-58607-2_43

14. Liang, Y., Li, X., Jafari, N., Chen, J.: Video object segmentation with adaptive feature bank and uncertain-region refinement. In: NIPS, vol. 33, pp. 3430–3441 (2020)

15. Lu, X., Wang, W., Danelljan, M., Zhou, T., Shen, J., Van Gool, L.: Video object segmentation with episodic graph memory networks. In: Vedaldi, A., Bischof, H., Brox, T., Frahm, J.-M. (eds.) ECCV 2020. LNCS, vol. 12348, pp. 661–679. Springer, Cham (2020). https://doi.org/10.1007/978-3-030-58580-8_39

16. Luiten, J., Voigtlaender, P., Leibe, B.: PReMVOS: proposal-generation, refinement and merging for video object segmentation. In: Jawahar, C.V., Li, H., Mori, G., Schindler, K. (eds.) ACCV 2018. LNCS, vol. 11364, pp. 565–580. Springer, Cham (2019). https://doi.org/10.1007/978-3-030-20870-7_35
17. Maninis, K.K.: Video object segmentation without temporal information. TPAMI **41**(6), 1515–1530 (2018)
18. Oh, S.W., Lee, J.Y., Sunkavalli, K., Kim, S.J.: Fast video object segmentation by reference-guided mask propagation. In: CVPR, pp. 7376–7385 (2018)
19. Oh, S.W., Lee, J.Y., Xu, N., Kim, S.J.: Video object segmentation using space-time memory networks. In: ICCV, pp. 9226–9235 (2019)
20. Perazzi, F., Khoreva, A., Benenson, R., Schiele, B., Sorkine-Hornung, A.: Learning video object segmentation from static images. In: CVPR, pp. 2663–2672 (2017)
21. Perazzi, F., Pont-Tuset, J., McWilliams, B., Van Gool, L., Gross, M., Sorkine-Hornung, A.: A benchmark dataset and evaluation methodology for video object segmentation. In: CVPR, pp. 724–732 (2016)
22. Pont-Tuset, J., Perazzi, F., Caelles, S., Arbeláez, P., Sorkine-Hornung, A., Van Gool, L.: The 2017 DAVIS challenge on video object segmentation. arXiv preprint arXiv:1704.00675 (2017)
23. Robinson, A., Lawin, F.J., Danelljan, M., et al.: Learning fast and robust target models for video object segmentation. In: CVPR, pp. 7406–7415 (2020)
24. Seong, H., Hyun, J., Kim, E.: Kernelized memory network for video object segmentation. In: Vedaldi, A., Bischof, H., Brox, T., Frahm, J.-M. (eds.) ECCV 2020. LNCS, vol. 12367, pp. 629–645. Springer, Cham (2020). https://doi.org/10.1007/978-3-030-58542-6_38
25. Shi, J., Yan, Q., Xu, L., Jia, J.: Hierarchical image saliency detection on extended CSSD. TPAMI **38**(4), 717–729 (2015)
26. Ventura, C., Bellver, M., Girbau, A., Salvador, A., Marques, F., Giro-i Nieto, X.: RVOS: end-to-end recurrent network for video object segmentation. In: CVPR, pp. 5277–5286 (2019)
27. Voigtlaender, P., Chai, Y., Schroff, F., Adam, H., Leibe, B., Chen, L.C.: FEELVOS: fast end-to-end embedding learning for video object segmentation. In: CVPR, pp. 9481–9490 (2019)
28. Wang, H., Jiang, X., Ren, H., Hu, Y., Bai, S.: SwiftNet: real-time video object segmentation. In: CVPR, pp. 1296–1305 (2021)
29. Wang, L., Lu, H., Wang, Y., Feng, M., Wang, D., et al.: Learning to detect salient objects with image-level supervision. In: CVPR, pp. 136–145 (2017)
30. Xie, H., Yao, H., Zhou, S., Zhang, S., Sun, W.: Efficient regional memory network for video object segmentation. In: CVPR, pp. 1286–1295 (2021)
31. Xu, N., et al.: YouTube-VOS: sequence-to-sequence video object segmentation. In: ECCV, pp. 585–601 (2018)
32. Xu, N., Yang, L., Fan, Y., Yue, D., Liang, Y., et al.: YouTube-VOS: a large-scale video object segmentation benchmark. In: ECCV, pp. 585–601 (2018)
33. Yang, L., Wang, Y., Xiong, X., Yang, J., Katsaggelos, A.K.: Efficient video object segmentation via network modulation. In: CVPR, pp. 6499–6507 (2018)
34. Yang, Z., Wei, Y., Yang, Y.: Collaborative video object segmentation by foreground-background integration. In: Vedaldi, A., Bischof, H., Brox, T., Frahm, J.-M. (eds.) ECCV 2020. LNCS, vol. 12350, pp. 332–348. Springer, Cham (2020). https://doi.org/10.1007/978-3-030-58558-7_20
35. Yang, Z., Wei, Y., Yang, Y.: Collaborative video object segmentation by multi-scale foreground-background integration. TPAMI **49**, 4701–4712 (2021)

36. Zeng, Y., Zhang, P., Zhang, J., Lin, Z., Lu, H.: Towards high-resolution salient object detection. In: ICCV, pp. 7234–7243 (2019)
37. Zhang, Y., Wu, Z., Peng, H., Lin, S.: A transductive approach for video object segmentation. In: CVPR, pp. 6949–6958 (2020)

A Hybrid Framework Based on Classifier Calibration for Imbalanced Aerial Scene Recognition

Yihong Zhuang[1], Changxing Jing[1], Senlin Cai[1], Lexing Huang[1], Yue Huang[1,2], Xiaotong Tu[1,2], and Xinghao Ding[1,2(✉)]

[1] School of Informatics, Xiamen University, Xiamen, China
dxh@xmu.edu.cn
[2] Institute of Artificial Intelligent, Xiamen University, Xiamen, China

Abstract. Aerial scene images are often imbalanced, where the most common classes as majorities and a few significant classes as minorities. We observe that the majority classes not only dominate the classification optimization but also generate deviations that affect the classifier weight matrices. In this work, we propose a hybrid framework based on classifier calibration, which mitigate the effect of the class imbalance problem in aerial scene recognition. In particular, the framework progressively incorporates feature representation and classifier learning branches, while building a memory bank of learned representations for approximating deviations derived from imbalanced data. We calibrate the classifier by excluding the deviations in the prediction of the testing stage. Extensive experiments are evaluated on class imbalanced aerial scene image datasets, which show the advantages of the proposed hybrid framework with classifier calibration outperforming state-of-the-art aerial scene recognition methods.

Keywords: Remote sensing · Scene recognition · Class imbalanced · Contrastive learning · Classifier calibration

1 Introduction

Aerial scene recognition is an essential aspect of remote sensing image analysis. It plays a critical role in practical remote sensing applications [17], such as natural disaster detection [1], urban planning [8], and management of agricultural sources [27]. With the development of satellite technology, an increasing number of aerial images are being acquired, and the realization of accurate and efficient aerial image scene recognition is an immediate demand for current remote sensing image analysis applications. As the quantity of aerial scene images increases, the

The study is supported partly by the National Natural Science Foundation of China under Grants 61971369, 52105126, 82172033, U19B2031, Science and Technology Key Project of Fujian Province(No. 2019HZ020009).

M. Tanveer et al. (Eds.): ICONIP 2022, LNCS 13625, pp. 110–121, 2023.
https://doi.org/10.1007/978-3-031-30111-7_10

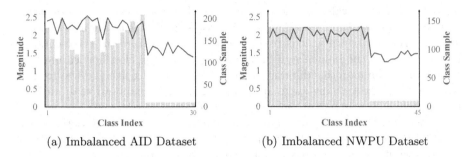

(a) Imbalanced AID Dataset   (b) Imbalanced NWPU Dataset

Fig. 1. The magnitudes of classifier weights for each class after training on two imbalanced AID and NWPU datasets. The class index represents the index of classes with training samples in descending order. The class sample indicates the number of samples, shown as the histogram. The magnitude indicates the L_2 normalization for each class of the classifier weight matrix, shown as the line graph. We show the magnitude of model weights of classes is proportional to their sample sizes, from which the classifier weights suffer from imbalanced training samples.

data distribution becomes imbalanced, including the most common scene classes as majorities (such as farmland, forest, etc.) and a few significant scene classes as minorities (such as airport, gas station, etc.). In an aerial scene recognition, the imbalanced distribution typically leads to label bias, which makes the decision boundary of the classifier biased to the majority classes.

For the problem of class imbalance in aerial scene classification, most of the previous work has been studied in the data balancing or compensating for the loss of minority classes, including re-sampling [22], re-weighting [15], which will cause under/over-fitting to the majority/minority classes. Recent studies suggest that less biased representations can be obtained by decoupling representation learning and classifier learning [18], thus effectively alleviating the impact of imbalanced data distribution. Although the two-stage learning scheme is intuitive and experimentally effective, we argue that such independent learning may not be optimal for representation learning with specific classification tasks.

In this paper, we observe that the magnitudes of classifier weight matrices learned are affected by imbalanced data distributions. As shown in Fig. 1, the magnitudes of the classifier weights for the minorities are smaller than that of the majorities. Since the majorities dominate the gradients in optimization on imbalanced data [24], we further notice that such deviations will reduce the classifier weights of the minorities in the training stage. Thus, we mitigate the adverse effect of the class imbalance problem by introducing a classifier calibration strategy to substitute ineffective re-balancing strategies.

Accordingly, we present an effective hybrid framework based on classifier calibration for imbalanced aerial scene recognition (Hybrid-CAL) that (1) progressively joints learn representations and classifier in model training, (2) creates moving average representation as deviation value retained within the memory bank, to approximate the majority direction, and (3) calibrates the classifier by removing deviation value in the testing stage. The framework adopts a parallel network architecture, including feature representation learning and classifier learning. As it is advantageous to learn class-related features in representation learning. Due to the influence of imbalanced data on the classifier learning, we approximate the majority direction by estimating moving average feature representation in the training stage and applying classifier calibration to mitigate the impact of class imbalance in aerial scene recognition testing.

Experiments on two manually processed class imbalanced aerial scene datasets demonstrate that the proposed framework outperforms alternatives. The contributions of this work can be summarized as follows:

- We observe that the majority classes not only dominate the classification in optimization but also result in deviation that affects the weight matrix of the classifier.
- We propose a hybrid framework with classifier calibration that progressively incorporates representation and classifier learning without re-balancing strategies, and approximate the deviation magnitude by estimating the moving average feature representation. We calibrate the classifier by excluding the deviation in the prediction of the testing stage.
- We evaluate Hybrid-CAL on two aerial scene benchmarks with various degrees of class imbalance, and our method outperforms alternatives on imbalanced aerial scene recognition.

2   Related Work

2.1   Imbalanced Aerial Image Recognition

In recent years, many deep learning methods have achieved good performance in aerial image recognition. In practical application, aerial images are usually imbalanced. Most of the existing researches focus on the poor recognition accuracy of the model in the minority classes to solve the problem of imbalanced aerial image recognition. The mainly several methods are as follows.

Re-balancing Strategies. Most of the current approaches to solve the imbalanced problem are based on re-balancing strategies, including data re-sampling or loss re-weighting. The data re-sampling strategy under-samples the majority classes [7,11] and over-samples the minority classes [3,10], but some studies [24] have shown that this strategy can lead to over-fitting the minority classes and miss partial information for the majority classes. The idea of re-weighting is to reduce the weight of the majority classes and increase the weight of the minority classes in the loss function [15,26]. Still, this strategy can distort the original data distribution and damage representation learning. Overall, re-balance strategies are limited in improving the recognition performance of the minority classes.

Decoupling Learning. Decoupling learning is a novel approach to solve the imbalanced image recognition problem. To address the effects of imbalanced distribution on each stage of the model training, [2,6,18] indicates that two-stage training can have unexpected effects, including the representation learning stage and the classifier learning stage. In addition, random sampling strategy is adopted in the representation learning stage and re-balancing strategy is adopted in the classifier learning stage.

2.2 Contrastive Learning

In recent years, contrastive learning [4] has achieved great success in the field of unsupervised representation learning [12]. By comparing the similarity between different augmented images of the unified image, the features from the same class are clustered. Its purpose is to find the most appropriate feature embedding for each sample, which is clusters of features belonging to the same class are pulled together in embedding space, while simultaneously pushing apart clusters of features from different classes. Supervised contrast learning [19] is the extension of contrastive learning by introducing labels, and constructs more appropriate positive samples and negative samples by combining label information. The article [19] achieves convincing performance by combining supervised contrastive learning with decoupling learning, applying random sampling in supervised contrastive learning at the first stage, and applying a re-balancing strategy to train the classifier in the second stage. However, this serial architecture decouples the learning stage into two stages, which makes the feature extractor and the classifier incompatible [25].

3   Method

In this section, we first elaborate on the hybrid framework that progressively combines representation and classifier learning while aggregating representations learned into a memory bank. Then, we detail the classifier calibration for aerial scene recognition.

3.1   A Hybrid Framework for Imbalanced Aerial Scene Recognition

As shown in Fig. 2, The proposed hybrid framework consists of two branches: the feature representation learning (FRL) branch, which aims to learn the representation of imbalanced aerial images, and the classifier learning (CL) branch, which is optimized by cross-entropy loss without re-balancing strategies. In the FRL branch, we adopt the recently proposed supervised contrastive learning [19], which maps samples to representations, pulls the representation with the same class, and pushes those from different classes. In the CL branch, the classifier can benefit from the representations learned from the FRL branch and then optimize by cross-entropy loss. Concurrently, the hybrid framework results in an extensive memory bank of representations, where a moving average of the representation over all samples describe the direction toward the majorities. In the testing stage, we exclude the deviation in testing to eliminate the wrong prediction, such as misclassifying a minority sample for the majority.

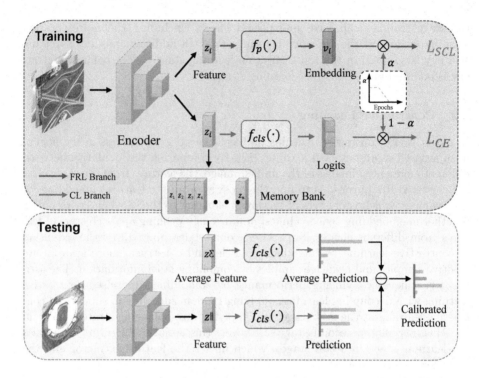

Fig. 2. Overview of the proposed framework Hybrid-CAL. The structure consists of a hybrid framework with: (1) the feature representation learning (FRL) branch, which learns discriminative representations from the imbalanced data distribution; (2) the classifier learning (CL) branch applies the representations to predict the classification logits; (3) the cumulative learning controls the weights between these two branches. The memory bank is designed to collect the learned representations, wherein the moving average of representations is described the deviations during training. At test time, output prediction and average prediction are subtracted by classifier calibration.

For a set of D sample-label pairs $\{(x_i, y_i)\}_{i=1}^{D}$ where x_i denote a training sample, $y_i \in \{1, 2, \cdots, C\}$ is the corresponding label, C is the number of classes. The hybrid framework shared the backbone between the two branches and learns to map a sample $x_i \in \mathcal{R}^d$ to a feature representation $z_i \in \mathcal{R}^g$, where $d \gg g$. In the FRL branch, the representation z_i is normalized into the projection network $f_p(\cdot)$ as a feature embedding v_i for supervised contrastive learning. Meanwhile, the representation z_i is used to train the classifier with a single linear layer $f_{cls}(\cdot)$ that projects the representation z_i to predict the classification logits.

We follow cumulative learning to bridge and shift focus between the two proposed branches by controlling the weight α between supervised contrastive learning and classification learning. It is designed first to learn the FRL branch for t_r epochs and then gradually shift the attention to the CL branch as the number of training epochs increases. Concretely, we denote t_{max} as the number of total training epochs and t as the current epoch. The weight α is defined as

$$\alpha = \begin{cases} 1, & t \leq t_r \\ -\dfrac{t - t_r}{t_{max} - t_r} + 1, & t > t_r \end{cases} \tag{1}$$

The final loss function for the hybrid framework Hybrid-CAL can be:

$$L = \alpha \cdot L_{SCL} + (1 - \alpha) \cdot L_{CE}. \tag{2}$$

Feature Representation Learning (FRL) Branch. The FRL branch adopts supervised contrastive learning [19]. We define sample-label pairs $S = \{(x_i, y_i)\}_{i=1}^{2D}$ with data augmentation in which each sample x_i is augmented into two views, y_i is the corresponding label. Each sample x_i acts as an anchor while other samples from the same class as positives $\{x_i^+\} = \{x_j | y_i = y_j, i \neq j\}$ and the rest from other classes as negatives. For the set of randomly sampled example pairs mapped to representations $z_i, i \in [1, 2, \cdots, D]$ and then projected by projector $f_p(\cdot)$ as normalized embeddings $v_i, i \in [1, 2, \cdots, D]$, the objective is to cluster samples from the same classes and separate them from negatives. The objective of SCL loss as follows:

$$L_{SCL} = \sum_{i=1}^{2D} \ell_{SCL}(v_i) \tag{3}$$

$$\ell_{SCL}(v_i) = -\frac{1}{|\{v_i^+\}|} \sum_{v_j \in \{v_i^+\}} \log \frac{\exp(v_i \cdot v_j / \tau)}{\sum_{v_a, a \neq i} \exp(v_i \cdot v_a / \tau)}, \tag{4}$$

where v_i denotes the normalized projection of sample x_i, and $\{v_i^+\}$ denotes the same class of projection of anchor v_i, τ is a scalar temperature parameter.

Classifier Learning (CL) Branch. We followed article [24] and normalized the classifier weights to alleviate the label bias in classifier learning. In particular, the classifier is a linear layer with softmax loss to optimize the normalized feature representations $\tilde{z}_i = z_i / ||z_i||_2$ to predict the classification logits. Given a feature vector z_i, the definition of the softmax loss is

$$L_{CE} = -\frac{1}{D} \sum_{i=1}^{D} \log \frac{\exp(w_i^T \tilde{z}_i / ||w_i|| + \epsilon)}{\sum_{j=1}^{C} \exp(w_j^T \tilde{z}_i / ||w_j|| + \epsilon)}, \tag{5}$$

where D is the number of training samples, C is the number of classes, w_j is the weight corresponding to the j-th class, ϵ is a small positive value to prevent dividing zero.

3.2   Classifier Calibration

Memory Bank. In the training stage, we build an extensive memory bank of learned representations $z_i \in \mathcal{Z}_D$ for each sample $x_i \in D$ to approximate the biased direction. Notably, the memory bank is simply used to store representation vectors; it does not participate in the training and is not constrained by the loss function. For all representations $z_i \in \mathcal{Z_D}$, the memory bank is then defined as

$$\mathcal{M} = \bigcup_{z_i \in \mathcal{Z_D}} z_i. \tag{6}$$

The moving average of the representation obtained from the memory bank $\mathcal{M}$ can be described as the deviation:

$$\hat{z} = \frac{1}{|\mathcal{M}|} \sum_{z_i \in \mathcal{M}} \sum_{t=0}^{t_{max}} \mu^{(t_{max}-t)} z_i^t, \tag{7}$$

where μ is weight decay, t is the number of current training epoch.

Classifier Calibration. In classifier testing stage, the test sample x_k is fed into the encoder, and the feature z_k is obtained and normalized as $\tilde{z}_k = z_k/||z_k||_2$. With the representation memory bank $\mathcal{M}$, we estimate deviations derived from the impact of imbalanced data in training as $\hat{z}$. We exclude the deviation when predicting the classification logits to eliminate the biased prediction of the classifier with the following:

$$P(y|z_k) = softmax \left(\frac{w_k^T \tilde{z}_k}{||w_k^T|| + \epsilon} - \beta \frac{\cos(\tilde{z}_k, \hat{z}) w_k^T \hat{z}}{||w_k^T|| + \epsilon} \right). \tag{8}$$

In the case of classifier weight normalization, the first term indicates the logit output by the classifier, which are subject to label bias. The second term is implemented as logits to represent the bias effect, wherein the cosine similarity measures for the representation $\tilde{z}_k$ and the deviation $\hat{z}$, which are applied to determine the exact value for calibration. β is a weight coefficient correlated with the dataset's degree of imbalance. The value of β increases proportionally with the imbalance of the dataset.

4   Experiments

In this section, we introduce the two class imbalanced aerial image classification datasets. Then we present the implementation details of our experiments.Next, we compare the our proposed method with existing class imbalanced aerial image classification methods. Finally, the ablation experiments are conducted to discuss some important component of our Hybird-CAL.

4.1   Dataset

In this work, the Aerial Image Dataset (AID) [28] and the NWPU-RESISC45 Dataset (NWPU) [5] are selected to evaluate the proposed model. The AID dataset consists of 10,000 remote sensing images with 30 aerial scenes, from which each scene has a range of 220 to 420 aerial images. The NWPU-RESISC45 dataset includes 31,500 images, covering 45 scene classes with 700 images in each category. All samples in the dataset are RGB color images. Here, we generate imbalanced datasets based on the two datasets mentioned above following the agreement mentioned in [9,16], which divide all classes into the majority and minority classes.

Class Imbalanced AID(CI-AID). For the AID dataset, the majority comprises 20 classes, and the minority contains the remaining 10 classes. We randomly select 50% of the samples in each majority class, and for each minority class, only a few examples are chosen (e.g., $s = 5$, or $s = 10$). Then, the balanced test set is constructed by selecting 100 samples from the remaining samples of each class.

Class Imbalanced NEWPU-RESISC45(CI-NWPU). For the NWPU-RESISC45 dataset, the majority and minority classes are set as the first 30 classes and the last 15 classes, respectively. The training dataset is constructed as follows: for each majority class, randomly sample 140 images and for each minority class, only choose a few samples (e.g., $s = 5$, or $s = 10$). The number of images in each category as testing data is 400.

4.2   Implementation Details

For all datasets, we use ResNet-50 [13] as the backbone to extract image representation. For FRL Branch we use a 2-layers MLP as projection network, the temperature τ is set to 0.1 in SCL loss function. For CL branch, the momentum μ and β is set to 0.9 and 3.0 in the classifier. Relu is used as the activation function of the model. We follow the data augmentation strategies proposed in [16]: random crop and resize with size 224×224 [20,23], random rotation, random color distort [14,23], random flip.

Hybrid-CAL is implemented using an initial learning rate of 0.5 with batch size of 64. All the training process is used the cosine annealing decay strategy [21], when the momentum and weight decay of SGD are 0.9 and 1×10^{-4}. As shown in formula 1, the t_r is set to 300 and t_{max} is set to 900.

4.3   Main Result

We compare our proposed Hybrid-CAL method with other state-of-the-art methods for addressing the class imbalanced of aerial images classification, including traditional training without any strategy [29], training with re-sampling and re-weighting methods [15,22], meta metric learning [9] and two-stage fine-tuning strategies [16]. Different from the RF-MML [9], this method may introduce additional information by using a pre-trained model with ImageNet dataset. SCL-CB and SCL-CSCE [16] is a serial method that trains the representation learning first and then trains the classifier learning using re-balancing strategies. Unlike them, our proposed approach is based on the parallel structure of pain ResNet-50 and does not require any re-balancing strategies.

Table 1. Top-1 accuracy rates(%) on the CI-AID and CI-NWPU.

Datasets	Methods	Majority	Minority	Overall	Majority	Minority	Overall
		$s > 100$	$s = 10$		$s > 100$	$s = 5$	
CI-AID	Plain [29]	94.75	48.60	79.37	94.85	37.00	75.57
	Re-Weighting [15]	93.35	56.70	81.13	91.75	47.80	77.10
	Re-Sampling [22]	93.50	50.60	79.20	93.10	40.30	75.50
	RF-MML-Proto [9]	92.15	63.20	82.50	90.80	54.50	78.67
	RF-MML-SVM [9]	92.15	62.70	82.33	91.80	53.09	78.99
	SCL-CSCE [16]	94.10	69.90	86.03	94.10	55.70	81.29
	SCL-CB [16]	94.00	70.50	86.16	93.65	57.90	81.73
	Hybrid-CAL (ours)	94.20	**73.00**	**87.13**	92.65	**61.80**	**82.37**
CI-NWPU	Plain [29]	89.46	50.75	76.56	90.96	30.38	70.77
	Re-Weighting [15]	89.07	55.62	77.92	86.41	39.10	70.64
	Re-Sampling [22]	90.04	50.73	76.94	88.67	32.52	69.95
	RF-MML-Proto [9]	87.52	61.49	78.73	86.13	50.24	74.11
	RF-MML-SVM [9]	88.31	63.86	80.80	87.87	50.99	75.51
	SCL-CSCE [16]	88.66	63.21	80.18	89.56	51.00	76.71
	SCL-CB [16]	88.67	64.15	80.50	89.07	53.28	77.14
	Hybrid-CAL (ours)	90.64	**65.17**	**82.15**	88.88	**54.95**	**77.57**

Experimental Results on CI-AID and CI-NWPU. Table 1 reports the accuracy rates of various methods int the CI-AID and CI-NWPU. Compared to other methods, Our method achieve the best overall accuracy, while outperforming other methods on minority classes that require attention. Among these compared methods, Plain denotes the baseline which directly uses cross-entropy loss to train the model. As expected, the results of plain method are the worst, which reflects that the original cross-entropy loss is affected by the imbalance dataset. Although the overall results of the two-stage SCL method are close to our method, in minority classes, our method has a great improvement compared with other methods, especially in $s = 5$. The most critical reason is that our method pays more attention to the compatibility of classifier and feature representation, so that a few classes can be greatly improved without decreasing the overall accuracy.

4.4   Ablation Studies

In this section, we conduct two ablation studies to highlight some important properties of our method.

Ablation Studies on the Classifier Calibration. In order to explore the performance of the classifier calibration in Hybrid-CAL, we adopted other methods to train the classifier branches, including removing the classifier calibration and using the ordinary single-layer classifier. The results are shown in Table 2,

Table 2. Ablation studies of classifier calibration on the CI-AID.

Methods	Majority $s > 100$	Minority $s = 10$	Overall –	Majority $s > 100$	Minority $s = 5$	Overall –
1-Layer classifier	92.70	70.30	85.23	93.15	53.60	79.96
w/o Classifier Calibration	92.65	69.40	84.90	93.70	53.50	80.30
Hybrid-CAL (ours)	94.20	**73.00**	**87.13**	92.65	**61.80**	**82.37**

classifier calibration plays an important role in improving the performance of minority classes. The effect of removing the classifier calibration is basically the same as that of the single-layer classifier, in fact they are equivalent.

Different Re-balancing Strategy for the Classifier Branch. For better understanding our classifier branch, we conduct experiments on different re-balance strategy including re-sampling, re-weighting and random sampling. A lot of work has shown that in the representation learning stage, the re-balancing strategy can change the original distribution and impair the feature representation. However, most decoupled learning methods use original distribution to train feature representation and re-balance strategy to train classifier. Therefore, we further explore the re-balancing strategy for classifier branch.As it can be seen from Table 3, the random sampling strategy yielded relatively good results because it can obtain more accurate majority direction using the original distribution and remove this effect during testing.The results of re-sampling and re-weighting strategies do not improve significantly in the case of more time and calculation cost.

Table 3. Ablation studies of different re-balance strategy for the classifier branch on the CI-NWPU.

Methods	Majority $s = 140$	Minority $s = 10$	Overall –	Majority $s = 140$	Minority $s = 5$	Overall –
Re-Sampling [22]	88.92	62.77	90.2	91.08	48.13	76.77
Re-Weighting [15]	90.68	64.23	81.86	88.53	54.72	77.26
Random Sampling	90.64	**65.17**	**82.15**	88.88	**54.95**	**77.57**

5   Conclusion

In this paper, based on the observation of the imbalanced weight of the classifier, we notice that imbalanced data will produce classifier bias, which will eventually affect feature representations. We propose a hybrid framework based on classifier calibration for imbalanced aerial scene recognition, which consists of a feature representation learning branch and a classifier learning branch. We use supervised contrastive learning in the feature representation learning branch to learn the classification-related representations. In classifier learning, we created a memory bank to describe the direction of majorities and calibrated the offsets during the test stage. The two branches adopt a progressive joint learning strategy. Extensive experiments on two imbalanced aerial scene datasets showed that our method not only achieves the most advanced recognition performance but also does not require any re-sampling strategy. As moving forward, we will continue to explore more feature representation and classifiers.

References

1. Bello, O.M., Aina, Y.A.: Satellite remote sensing as a tool in disaster management and sustainable development: towards a synergistic approach. Procedia Soc. Behav. Sci. **120**, 365–373 (2014)
2. Cao, K., Wei, C., Gaidon, A., Arechiga, N., Ma, T.: Learning imbalanced datasets with label-distribution-aware margin loss. In: Advances in Neural Information Processing Systems, vol. 32 (2019)
3. Chawla, N.V., Bowyer, K.W., Hall, L.O., Kegelmeyer, W.P.: SMOTE: synthetic minority over-sampling technique. J. Artif. Intell. Res. **16**, 321–357 (2002)
4. Chen, T., Kornblith, S., Norouzi, M., Hinton, G.: A simple framework for contrastive learning of visual representations. In: International Conference on Machine Learning, pp. 1597–1607. PMLR (2020)
5. Cheng, G., Han, J., Lu, X.: Remote sensing image scene classification: benchmark and state of the art. Proc. IEEE **105**(10), 1865–1883 (2017)
6. Cui, Y., Song, Y., Sun, C., Howard, A., Belongie, S.: Large scale fine-grained categorization and domain-specific transfer learning. In: Proceedings of the IEEE Conference on Computer Vision and Pattern Recognition, pp. 4109–4118 (2018)
7. Drummond, C., Holte, R.: Class imbalance and cost sensitivity: why under-sampling beats oversampling. In: ICML-KDD 2003 Workshop: Learning from Imbalanced Datasets, vol. 3 (2003)
8. Fauvel, M., Chanussot, J., Benediktsson, J.A.: Decision fusion for the classification of urban remote sensing images. IEEE Trans. Geosci. Remote Sens. **44**(10), 2828–2838 (2006)
9. Guan, J., Liu, J., Sun, J., Feng, P., Shuai, T., Wang, W.: Meta metric learning for highly imbalanced aerial scene classification. In: ICASSP 2020, IEEE International Conference on Acoustics, Speech and Signal Processing (ICASSP), pp. 4047–4051. IEEE (2020)
10. Han, H., Wang, W.-Y., Mao, B.-H.: Borderline-SMOTE: a new over-sampling method in imbalanced data sets learning. In: Huang, D.-S., Zhang, X.-P., Huang, G.-B. (eds.) ICIC 2005. LNCS, vol. 3644, pp. 878–887. Springer, Heidelberg (2005). https://doi.org/10.1007/11538059_91
11. He, H., Garcia, E.A.: Learning from imbalanced data. IEEE Trans. Knowl. Data Eng. **21**(9), 1263–1284 (2009)
12. He, K., Fan, H., Wu, Y., Xie, S., Girshick, R.: Momentum contrast for unsupervised visual representation learning. In: Proceedings of the IEEE/CVF Conference on Computer Vision and Pattern Recognition, pp. 9729–9738 (2020)
13. He, K., Zhang, X., Ren, S., Sun, J.: Deep residual learning for image recognition. In: Proceedings of the IEEE Conference on Computer Vision and Pattern Recognition, pp. 770–778 (2016)

14. Howard, A.G.: Some improvements on deep convolutional neural network based image classification. arXiv preprint arXiv:1312.5402 (2013)
15. Huang, C., Li, Y., Loy, C.C., Tang, X.: Learning deep representation for imbalanced classification. In: Proceedings of the IEEE Conference on Computer Vision and Pattern Recognition, pp. 5375–5384 (2016)
16. Huang, L., et al.: A two stage contrastive learning framework for imbalanced aerial scene recognition. In: ICASSP 2022, IEEE International Conference on Acoustics, Speech and Signal Processing (ICASSP), pp. 3518–3522. IEEE (2022)
17. Huang, N., Yang, Y., Liu, J., Gu, X., Cai, H.: Single-image super-resolution for remote sensing data using deep residual-learning neural network. In: Liu, D., Xie, S., Li, Y., Zhao, D., El-Alfy, E.S. (eds.) Neural Information Processing, ICONIP 2017. Lecture Notes in Computer Science(), vol. 10635, pp. 622–630. Springer, Cham (2017). https://doi.org/10.1007/978-3-319-70096-0_64
18. Kang, B., Xie, S., Rohrbach, M., Yan, Z., Gordo, A., Feng, J., Kalantidis, Y.: Decoupling representation and classifier for long-tailed recognition. arXiv preprint arXiv:1910.09217 (2019)
19. Khosla, P., et al.: Supervised contrastive learning. In: Advances in Neural Information Processing Systems, vol. 33, pp. 18661–18673 (2020)
20. Krizhevsky, A., Sutskever, I., Hinton, G.E.: ImageNet classification with deep convolutional neural networks. In: Advances in Neural Information Processing System, vol. 25, pp. 1097–1105 (2012)
21. Loshchilov, I., Hutter, F.: SGDR: stochastic gradient descent with warm restarts. arXiv preprint arXiv:1608.03983 (2016)
22. Shen, L., Lin, Z., Huang, Q.: Relay backpropagation for effective learning of deep convolutional neural networks. In: Leibe, B., Matas, J., Sebe, N., Welling, M. (eds.) ECCV 2016. LNCS, vol. 9911, pp. 467–482. Springer, Cham (2016). https://doi.org/10.1007/978-3-319-46478-7_29
23. Szegedy, C., et al.: Going deeper with convolutions. In: Proceedings of the IEEE Conference on Computer Vision and Pattern Recognition, pp. 1–9 (2015)
24. Tang, K., Huang, J., Zhang, H.: Long-tailed classification by keeping the good and removing the bad momentum causal effect. In: Advances in Neural Information Processing Systems, vol. 33, pp. 1513–1524 (2020)
25. Wang, P., Han, K., Wei, X.S., Zhang, L., Wang, L.: Contrastive learning based hybrid networks for long-tailed image classification. In: Proceedings of the IEEE/CVF Conference on Computer Vision and Pattern Recognition, pp. 943–952 (2021)
26. Wang, Y.X., Ramanan, D., Hebert, M.: Learning to model the tail. In: Advances in Neural Information Processing Systems, vol. 30 (2017)
27. Weiss, M., Jacob, F., Duveiller, G.: Remote sensing for agricultural applications: a meta-review. Remote Sens. Environ. **236**, 111402 (2020)
28. Xia, G.S., et al.: AID: a benchmark data set for performance evaluation of aerial scene classification. IEEE Trans. Geosci. Remote Sens. **55**(7), 3965–3981 (2017)
29. Zagoruyko, S., Komodakis, N.: Wide residual networks. arXiv preprint arXiv:1605.07146 (2016)

Enhancing BERT for Short Text Classification with Latent Information

Ailing Tang[1,2,3], Yufan Hu[1,2,3], and Rong Yan[1,2,3(✉)]

[1] College of Computer Science, Inner Mongolia University, Hohhot 010021, China
`csyanr@imu.edu.cn`
[2] Inner Mongolia Key Laboratory of Mongolian, Information Processing Technology,
Hohhot 010021, China
[3] National & Local Joint Engineering Research Center of Intelligent Information
Processing Technology for Mongolian, Hohhot 010021, China

Abstract. With explosive growth of short text, short text categorization has been attracted increasing attention. How to alleviate the sparsity of short texts is a research hotspot, and takes a enormous challenge for classical text categorization technique. In this paper, we focus on short text expansion based on multi-granularity and explore to construct an EBLI (Enhancing BERT with Latent Information) model by combining BERT and latent information for addressing short text classification task. Additionally, we establish a memory bank to store the whole document topic information that assists in the joint training of deep semantic features and topic features. Experimental results with five widely datasets show that our proposed model achieves better performance of short text classification as well as promote the generalization ability and strong competition ability for the classifier.

Keywords: Short Text Classification · BERT Model · Topic Model · Text Expansion · Memory Bank

1 Introduction

Text classification is one of the core tasks in natural language processing (NLP) and has been used in many real-world applications such as opinion mining [1], sentiment analysis [4], and news classification [21]. Different from the standard text classification, short text classification has to face with a series of difficulties and problems, such as sparsity, shortness, lack of contextual information and semantic inadequacy, which brings an enormous challenge for traditional text classification methods. In recent years, many scholars have put forward some ingenious strategies to solve these problems in short texts. Some way try to solve the sparsity and shortness of short texts by using internal, external resources and deep learning methods to expand text [3,10,11]. Li et al. [10] identified the concept of the text based on Wikipedia as a knowledge base and attempted to add the corresponding information into short text. Nevertheless, these methods strongly depend on the quality of external resources, and these

M. Tanveer et al. (Eds.): ICONIP 2022, LNCS 13625, pp. 122–132, 2023.
https://doi.org/10.1007/978-3-031-30111-7_11

resources are really scarce in fact. Another way explores to construct the classification model for short text. Recently, combining pre-training models [6,9,14,19] are popular for addressing short text classification task. However, these models are lack of generalization ability. Then, Peinelt et al. [16] proposed a novel topic-based architecture to enhance performance of pre-training model.

Inspired by their work, in this paper, we focus on enhancing the classification effect of single label short text by improving text expansion techniques. Firstly, we explore to carry out short text expansion based on multi-granularity to make up for the sparsity of short texts. Based on this, we propose a significantly more lightweight model named EBLI (Enhancing BERT with Latent Information). It can increase the interpretability of the model by latent topic information and improve the efficiency of semantic extraction for short texts. Furthermore, we adopt a memory bank mechanism to achieve the joint training of these features. In summary, EBLI model is able to leverage the advantages of both pre-training model and topic model, not only it can extract the dependencies between words from the input word sequence, but also capture the global semantic information of multiple documents by using the underlying topic. Experimental results show that our framework based on expanded text outperforms the state-of-the-art baselines on five public datasets.

2   Related Work

Our work is most relevant to two bodies of research efforts: text expansion and pre-training model. In this section, we describe a brief related work about these two respects for the specific scenario of short text classification task.

Text expansion has always been fundamental research in short text classification tasks. It is common that using topic model [15] to extract the additional information from short text as expanded words. For instance, Gao et al. [7] proposed a regularized model based on conditional random fields and expanded the content by extracting appropriate words from topic model. These methods can effectively alleviate the impact of sparsity and shortness on the classification effect. However, they ignored the correlation between the words, which makes the topic model unable to clearly express the semantic information of short text. In addition, the emerging keywords extraction techniques [2,17] are also favored, which rests on the most relevant words from single documents based on 'word' granularity. Sharma et al. [17] proposed a novel self-supervised approach using contextual and semantic features to extract the keywords. However, they had to face an awkward situation of these information merely reflected the semantic information from 'word' granularity, and unable to consider multi-granularity information.

Besides above text expansion techniques, some researches tried to improve pre-training models [9,14] for short text classification, which are typically trained on large-scale corpora unrelated to a specific NLP task. And they are convenient to fine-tune for specific NLP tasks. Compared to other known pre-training models, BERT model [6] captured deep semantic representation and achieved prominent performance on lots of NLP tasks [5,16,18]. Nevertheless, sparsity, shortness, lack of contextual information and semantic inadequacy of short texts are still challenges for BERT model, which will limit the ability of this model during classifying procedure. In the work of [13], it indicated that integrating structured information in the knowledge base into the pre-training model would improving the representation ability. However, their models still needed to mine related information from external knowledge and they did not take ambiguity of the noise of external information. Therefore, how to extract more reliable knowledge and explore appropriate combination method for short text classification is still a dominant problem.

3   Methodology

In this section, we introduce the overall architecture based on text expansion and classification model in detail. Obviously, the improvement of text extension focuses on the construction of extended word set based on multi-granularity resulting in short text found strong related words. Despite of this, single text expanding is not always useful for short text expressions of the flexibility and diversity. Thus, EBLI model is absorbed to accommodate more complex short texts.

3.1   Text Expansion Scheme

The goal of this scheme is to expand strong relevant words from the internal resources. Taking the Biomedical dataset as an example, Fig. 1 shows the text expansion method based on keywords and topic model.

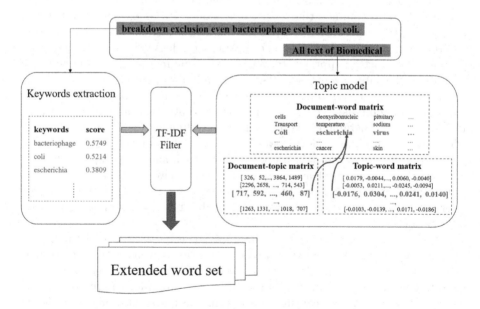

Fig. 1. Text expansion based on multi-granularity information.

Short text is generally of shortness and sparsity which makes it difficult to model. Accordingly, we use the topic model [15] and keyBERT model [17] to retrieve multi-granularity information and obtain an extended candidate word set Y. According to the statistics, the part of speech (POS) of the feature words are generally nouns, verbs and adjectives in short texts. Therefore, in order to ensure the quality of the extended words, we calculate the Term Frequency Inverse Documentation Frequency [8] (TF-IDF) values of Y. And then the minimum value of TF-IDF is regarded as the threshold according to the POS. To be specific, if the calculated TF-IDF value is no less than the threshold, the sample word will be appended to expanded word set Y. Otherwise, the sample word will delete. With high-quality extended word set, we can quickly classify data into a certain category as shown in Algorithm 1. The text expansion technique can be divided into three steps, including construct candidate word set, calculate threshold and determine expanded word set in Algorithm 1.

Algorithm 1. Text expansion

Input: $D=\{d_1,\cdots,d_N\}$
Output: extended word set Y
1: #(Step 1)construct candidate word set
2: **for** $i = 1$ **to** N **do**
3:      $EK=keyBERT(d_i)$
4:      Append EK to $ekList[i]$
5: $etList=LDA(D)$
6: $Y=(etList; ekList)$
7: #(Step 2) calculate threshold
8: Compute w in $(Y_1,\cdots,Y_N)\in Y$ TF-IDF value
9: $threshold=min(w.tfidf)$ while $w.pos$ in (n,v,adj)
10: #(Step 3) determine expanded word set
11: **for** w in $(Y_1,\cdots,Y_N)\in Y$ **do**
12:      **if** $w.tfidf¡threshold$ **then**
13:          Remove w from Y

3.2 EBLI Model

Figure 2 shows the architecture of our proposed EBLI model based on expanded text. In our model, we encode two aspects original text $(X_1,\cdots,X_n)$ and the extended word set $(Y_1,\cdots,Y_m)$ with BERT model. Afterwards, the expanded set is mapped into input embedding layer, which outputs by adding token, segment and position embedding. Then feeding input embedding into bi-directional transformer. We use the final hidden state $[CLS]$ token from Eq.(1-2) as the semantic representation of BERT model. Indeed, $C_x \in R^{b \times k}$ and $C_y \in R^{b \times k}$ vectors are from original text and extended word set respectively, which are corresponding to the CLS after a linear classification layer, where b and k denote the batch-size and category of BERT model respectively.

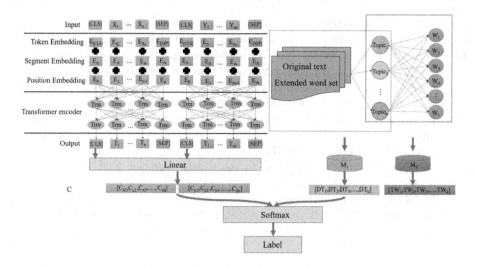

Fig. 2. Architecture of EBLI with DT.

$$G = LayerNorm(T^{i-1} + MultiAttention(T^{i-1})) \qquad (1)$$

$$T^i = LayerNorm(G + FeedForward(G)) \qquad (2)$$

where, T^i denotes the output of the i-th layer, it can learn and store the semantic relationships and syntactic structure information of document d_i. LayerNorm is layer normalization, MultiAttention is a multiple attention mechanism and FeedForward is a feedforward network with RELU. Notably, it is assumed that the dataset $D = \{d_1, \cdots, d_N\}$ contains N documents, we represent each document as $d_i = [DT, TW]$, which are given by the following Eq. (3). The document-topic distribution $DT \in R^{n \times k}$ represents the overall topic of the document, while the topic-word distribution $TW \in R^{k \times v}$ of word w_i essentially captures how topical the word is in itself. Among them, α and β are the hyperparameters of topic model, $n_i^{(k)}$ and $n_k^{(w)}$ represent the number of times that d_i is sampled as topic k and word w is sampled as topic k.

$$DT_{i,k} = \frac{n_i^{(k)} + \alpha}{\sum_{k=1}^{k} \left(n_i^{(k)} + \alpha \right)}, TW_{k,w} = \frac{n_k^{(w)} + \beta}{\sum_{v=1}^{v} \left(n_k^{(w)} + \beta \right)} \qquad (3)$$

For the sake of subsequent calculations, we compress the TW distribution to a fixed-dimension feature representation. However, the topic model uses the full-batch datasets for training, which is intractable for BERT model due to the memory limitation. Inspired by techniques in [12] which decouples training batch size from the total number of nodes in the graph, we maintain two memory banks M_1 and M_2 that track document-topic features and topic-word features for all documents. During each iteration, we sample a mini-batch b from two memory banks. Correspondingly, the M_1 and M_2 banks are injected into BERT model.

As shown in Eq. 4, we adopt full connection layer and use softmax as activation function to calculate a category probability distribution $P_{DT} \in R^{b \times k}$. Similarly, we can get $P_{TW} \in R^{b \times k}$ according to this function. In this way, BERT model stores the semantic relationships among words and we integrate deep semantic features with topic features resulting in words found in close proximity to one another semantically disambiguated.

$$p\left(L_k \mid C_{Xi}, C_{Yi}, DT_i\right) = \frac{\exp\left(C_{Xi} + C_{Yi} + DT_i\right)}{\sum_{k=1}^{k} \exp\left(C_{Xk} + C_{Yk} + DT_k\right)} \tag{4}$$

4 Experiments

Five popular short text classification datasets (Biomedical[1], Movie Review (MR)[2], Tweet[3], SearchSnippets[4] and StackOverflow[5]) with multiple domains are applied here to run the experiments. To evaluate the robustness and effectiveness of our proposed EBLI model, we compare EBLI model with five baselines.

4.1 Dataset

In the experiment, we randomly divide each dataset into train set, test set and the validation set according to the proportion of 7:1.5:1.5. After preprocessing, we summarize the detailed information of each dataset in Table 1, where category_num denotes the category number of each dataset, and len_avg and len_max denote the average length and maximum length of each document respectively. Moreover, vocabulary and doc_num denote the size of the vocabulary and the number of documents in each dataset respectively.

Table 1. Statistic information of the datasets.

Dataset	category_num	len_avg	len_max	vocabulary	doc_num
Biomedical	20	7.44	28	4,498	19,448
MR	2	11.17	27	4,081	10,662
SearchSnippets	8	14.40	37	5,547	12,295
Tweet	3	17.15	26	5,574	20,735
StackOverflow	20	5.03	17	2,638	16,407

---

[1] https://github.com/rashadulrakib/short-text-clustering-enhancement/tree/master/data/biomedical/.

[2] http://disi.unitn.it/moschitti/corpora.html.

[3] https://github.com/haroonshakeel/multisenti/.

[4] http://jwebpro.sourceforge.net/data-web-snippets.tar.gz.

[5] https://github.com/jacoxu/StackOverflow/.

4.2 Baselines

For comparison, five baselines are used to illustrate the effectiveness of EBLI model, including BERT[6], ERNIE[7], Roberta[8], Albert[9] and SHINE[10] with heterogeneous information networks in our experiment [20]. For EBLI model, we not only use BERT model to represent over the text documents, but also yield more interpretable results due to the involvement of topic modeling.

4.3 Experiment Settings

In the experiment, we set hyperparameters $\alpha = 50/k$, $\beta = 0.01$, and the number of topics k is set according to the specific dataset. In particular, the k is also equal to the number of categories in short texts. During encoding, we use a batch size of 16, the maximum sequence length is set to 128 for all datasets. For EBLI model, the training epochs are set to 3. We set the learning rate = 5e−5 when updating BERT model. It is worth mentioning that the hidden size of Albert model is set to 312 and ERNIE model with a learning rate of 2e−5. We train our model for a dropout of 0.1 and optimize cross entropy loss using Adam[11] optimizer. Once the performance of model has not improved after more than 1,000 batches, we will stop train model early.

4.4 Results and Discussion

Table 2 presents the comparisons of our model with baselines, where line 6 (EBLI+TW) and line 7 (EBLI+DT) denote the different combination latent information in EBLI model from the word level and the collection level respectively.

Table 2. Results of different models on all datasets.

	Biomedical	MR	SearchSnippets	Tweet	StackOverflow
ERNIE	60.62	71.11	90.95	61.76	81.91
Albert	63.08	69.11	88.29	63.66	79.27
Roberta	61.71	74.73	94.15	64.29	83.21
SHINE	65.25	77.50	94.37	63.27	81.00
BERT	62.62	72.61	92.14	64.91	81.54
EBLI+TW	**79.02**	79.55	93.98	67.28	83.90
EBLI+DT	78.09	**83.68**	**95.61**	**68.00**	**84.11**

---

[6] https://huggingface.co/bert-base-uncased/.
[7] https://huggingface.co/nghuyong/ernie-2.0-en/.
[8] https://huggingface.co/roberta-base/.
[9] https://huggingface.co/albert-base-v1/.
[10] https://github.com/tata1661/shine-emnlp21/.
[11] https://arxiv.org/pdf/1412.6980.pdf.

As shown in Table 2, it shows that our proposed EBLI model is remarkable and stable no matter how the domain information changed. EBLI model can acquire more robust performance than other models by capturing interpretable information both on word and collection point of view. To be specific, the accuracy of EBLI model reaches 78.09 in EBLI+DT from the document level and 79.02 in EBLI+TW from the word level, but BERT model reaches 62.62 in Biomedical. We analysis that it is due to the text characters of Biomedical with lots of biomedical terms. However, for BERT model, the performance of classification is heavily depend on the integrating degree of data to the model. Apparently, it is difficult for BERT model to learn specifical words, but topic model serves as a simple and efficient way, which can efficient to extract strong interpretable information and alleviate this awkwardness. Simultaneously, the same scene is happened on MR. Besides, we find that a slight improvement for StackOverflow. Its average sequences length ranges from a few to more than 10 and its vocabulary is not enough to capture rich topic information even after expanding. The result validates the effectiveness and importance of the proposed joint optimization framework in leveraging the strengths of both BERT and topic model to complement each other.

Furthermore, in order to measure the impact of text expansion methods on the performance of EBLI model, we exhibit *Precision (P)*, *Recall (R)*, and *F*1 metrics comparisons on MR in Table 3.

Table 3. Results of EBLI on MR with/without expanding.

	Original text			Expanded text		
	$P(\%)$	$R(\%)$	$F1(\%)$	$P(\%)$	$R(\%)$	$F1(\%)$
Positive	73.71	62.47	67.63	**74.01**	**72.40**	**73.19**
Negative	65.52	**76.20**	70.45	**71.18**	72.83	**71.99**

From Table 3, it is obvious find that EBLI model achieve better on MR when using extension scheme. It further illuminates that our extension scheme improves the amount of information contained in the short text itself and text expansion based on multi-granularity provides rich extension features, which makes our models with joint training have more available contextual information, and further extract the deep semantic information of short texts.

Figure 3 further shows the comparisons on all datasets. As shown in Fig. 3, it further illustrates that the effectiveness of our expending strategy for the extended text has longer length and stronger semantic correlation.

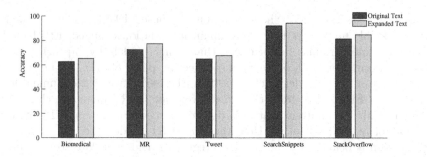

Fig. 3. Comparisons on all datasets with and without expansion.

Next, we exhibit the classification performance of EBLI model with different representations of topic information in Table 4. Compared with traditional BERT model, it is obvious that the EBLI model combined different topic representations (DW and DT+TW) can significantly reduces the lack of contextual information and semantic inadequacy. Especially for Biomedical, the EBLI model achieves the best effect in EBLI+DT+TW. We analysis that it is because topic representation from the multi-angle can better integrating the background constraints. Meanwhile, it is further verified that launching a smaller and simpler model to BERT model will capture richer information for short text classification.

Table 4. Classification performance of EBLI model with different representations.

	Biomedical	MR	SearchSnippets	Tweet	StackOverflow
BERT	62.62	72.61	92.14	64.91	81.54
EBLI+DW	78.54	81.99	**94.36**	**67.64**	**83.90**
EBLI+DT+TW	**79.84**	**82.55**	92.36	66.10	83.66

Furthermore, we attempt to add different topic representations for our model, and obtain $F1$ score under 20 categories of Biomedical as shown in Fig. 4, due to the large number of categories, we use the number to represent the categories to show the experimental results. Results from experiments demonstrate that $F1$ score of BERT model is worse than EBLI model on different categories. The reason is that the 20 categories of Biomedical have a strong correlation, and EBLI model can efficiently boost their representative ability using lexicon information. On the whole, based on the comparison results of the above experiments, the effectiveness of EBLI model with latent information in solving sparsity, shortness,lack of semantic and contextual information has also been fully illustrated.

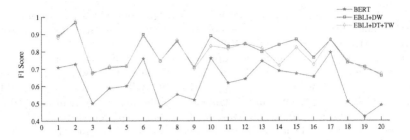

Fig. 4. The $F1$ score of models with different topic represents on Biomedical.

5   Conclusion and Future Work

In this paper, we propose a flexible framework named EBLI for enhancing BERT with latent information for addressing short text classification task. What makes it reasonable is that semantic knowledge and topic knowledge can provide enough diversity for short texts. Experimental results indicate that the effectiveness and generalization of our proposed approach. In future work, we try to explore how to directly reduce the loss of hierarchical semantics and preserve rich and complex semantic information of the document in training procedure in order to better apply it to the multi-classification task for short text.

Acknowledgment. This research is supported by the National Natural Science Foundation of China (Grant No. 61866029).

References

1. Bakshi, R.K., Kaur, N., Kaur, R., Kaur, G.: Opinion mining and sentiment analysis. In: Proceedings the 3rd International Conference on Computing for Sustainable Global Development (INDIACom), pp. 452–455. IEEE (2016)
2. Campos, R., Mangaravite, V., Pasquali, A., Nunes, C., Jatowt, A.: YAKE! keyword extraction from single documents using multiple local features. Inf. Sci. **509**, 257–289 (2020)
3. Chen, C., Ren, J.: An improved PLDA model for short text. In: Frasincar, F., Ittoo, A., Nguyen, L.M., Métais, E. (eds.) NLDB 2017. LNCS, vol. 10260, pp. 58–70. Springer, Cham (2017). https://doi.org/10.1007/978-3-319-59569-6_7
4. Chen, J., Hu, Y., Liu, J., Xiao, Y., Jiang, H.: Deep short text classification with knowledge powered attention. In: Proceedings of the AAAI Conference on Artificial Intelligence, pp. 6252–6259. AAAI, Honolulu, Hawaii, USA (2019)
5. Chi, S., Huang, L., Qiu, X.: Utilizing BERT for aspect-based sentiment analysis via constructing auxiliary sentence. In: Proceedings of the Conference of the North American Chapter of the Association for Computational Linguistics: Human Language Technologies, pp. 380–385. Minneapolis, MN, USA (2019)
6. Devlin, J., Chang, M.W., Lee, K., Toutanova, K.: BERT: pre-training of deep bidirectional transformers for language understanding. In: Proceedings of the Conference of the North American Chapter of the Association for Computational Linguistics: Human Language Technologies (NAACL-HL), pp. 4171–4186 (2019)

7. Gao, W., Peng, M., Wang, H., Zhang, Y., Xie, Q., Tian, G.: Incorporating word embeddings into topic modeling of short text. Knowl. Inf. Syst. **61**(2), 1123–1145 (2018). https://doi.org/10.1007/s10115-018-1314-7

8. Guo, A., Yang, T.: Research and improvement of feature words weight based on TFIDF algorithm. In: 2016 IEEE Information Technology. Networking, Electronic and Automation Control Conference, pp. 415–419. IEEE, Chongqing, China (2016)

9. Lan, Z., Chen, M., Goodman, S., Gimpel, K., Sharma, P., Soricut, R.: Albert: A lite BERT for self-supervised learning of language representations. In: International Conference on Learning Representations, pp. 26–30. OpenReview.net, Addis Ababa, Ethiopia (2020)

10. Li, J., Cai, Y., Cai, Z., Leung, H., Yang, K.: Wikipedia based short text classification method. In: Bao, Z., Trajcevski, G., Chang, L., Hua, W. (eds.) DASFAA 2017. LNCS, vol. 10179, pp. 275–286. Springer, Cham (2017). https://doi.org/10.1007/978-3-319-55705-2_22

11. Li, Y.: Short text classification improved by feature space extension. In: IOP Conference Series: Materials Science and Engineering, vol. 533, p. 012046 (2019)

12. Lin, Y., et al.: BERTGCN: Transductive text classification by combining GNN and BERT. In: Findings of the Association for Computational Linguistics: ACL-IJCNLP 2021, pp. 1456–1462. Association for Computational Linguistics, Bangkok, Thailand (2021)

13. Liu, W., et al.: K-BERT: enabling language representation with knowledge graph. In: Proceedings of the AAAI Conference on Artificial Intelligence, vol. 34, pp. 2901–2908 (2020)

14. Liu, Y., et al.: Roberta a robustly optimized BERT pretraining approach. CoRR abs/1907.11692 (2019)

15. Blei, D.M., Ng, A.Y., Jordan, M.I.: Latent dirichlet allocation. J. Mach. Learn. Res. **3**, 993–1022 (2003)

16. Peinelt, N., Nguyen, D., Liakata, M.: tBERT: topic models and BERT joining forces for semantic similarity detection. In: Proceedings of the 58th Annual Meeting of the Association for Computational Linguistics, pp. 7047–7055 (2020)

17. Sharma, P., Li, Y.: Self-supervised contextual keyword and keyphrase retrieval with self-labelling. arXiv e-prints (2019)

18. Sun, C., Qiu, X., Xu, Y., Huang, X.: How to fine-tune BERT for text classification? In: Sun, M., Huang, X., Ji, H., Liu, Z., Liu, Y. (eds.) CCL 2019. LNCS (LNAI), vol. 11856, pp. 194–206. Springer, Cham (2019). https://doi.org/10.1007/978-3-030-32381-3_16

19. Sun, Y., et al.: Ernie 2.0: a continual pre-training framework for language understanding. In: Proceedings of the AAAI Conference on Artificial Intelligence, pp. 8968–8975. AAAI, New York, NY, USA (2020)

20. Wang, Y., Wang, S., Yao, Q., Dou, D.: Hierarchical heterogeneous graph representation learning for short text classification. In: Proceedings of the 2021 Conference on Empirical Methods in Natural Language Processing, pp. 3091–3101. Association for Computational Linguistics, Cana, Dominican Republic (2021)

21. Yao, L., Mao, C., Luo, Y.: Graph convolutional networks for text classification. In: Proceedings of the AAAI Conference on Artificial Intelligence, pp. 7370–7377. AAAI, Honolulu, Hawaii, USA (2019)

Unsupervised Anomaly Segmentation for Brain Lesions Using Dual Semantic-Manifold Reconstruction

Zhiyuan Ding[1], Qi Dong[1], Haote Xu[1], Chenxin Li[1], Xinghao Ding[1,2(✉)], and Yue Huang[1,2]

[1] School of Informatics, Xiamen University, Xiamen, China
[2] Institue of Artificial Intelligent, Xiamen University, Xiamen, China
`dxh@xmu.edu.cn`

Abstract. Unsupervised anomaly segmentation (UAS) is promising in many computer vision applications, *e.g.*, the analysis of brain MRI, thanks to the advantage of detecting the anomalies (lesions) by only using the normal samples (healthy anatomies) in the training phase. Existing methods utilize the reconstruction process to model the normative distribution but inevitably lead to the impairment of localization information, which is critical for the pixel-level detection task. In this paper, we address this challenge by formulating a semantic layout of the healthy anatomy as the reconstruction manifold, which naturally forces the embedding to explicitly encode more semantic features as well as facilitates the preservation of spatial information during the reconstruction. Based on this special autoencoder framework of Semantic-Manifold Reconstruction (SMR), we further apply two consistency regularizations not only on the semantic layout but also the image appearance. In this way a Dual Semantic-Manifold Reconstruction (DSMR) is trained and then used to detect the anomalies accurately. Experiments reveal that the proposed DSMR approach exceeds the state-of-the-art performance on the benchmark datasets of BraTS and ISLES.

Keywords: Unsupervised anomaly segmentation · Brain lesion localization · Autoencoder

1 Introduction

The automatic analysis of brain MRI, *e.g.*, brain lesion segmentation, is essential in clinical practice and catches much attention in the community [16,18,23]. Although deep learning methods have achieved progress in this field, they are usually quite dependent on the massive well-labeled medical imaging datasets, which are very expensive to acquire. In comparison, unsupervised anomaly segmentation (UAS), which gets rid of the laborious annotations and even the

The study is supported partly by the National Natural Science Foundation of China under Grants 82172033, U19B2031, 52105126, 82272071, 62271430, and 61971369.

M. Tanveer et al. (Eds.): ICONIP 2022, LNCS 13625, pp. 133–144, 2023.
https://doi.org/10.1007/978-3-031-30111-7_12

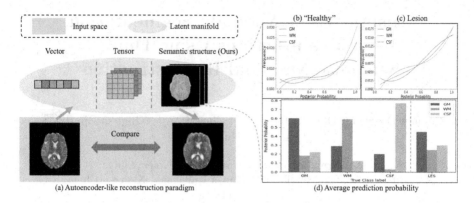

Fig. 1. Motivation of the proposed work. The semantic layout of anatomical tissues (gray matter (GM), white matter (WM) and cord spinal fluid (CSF) in brain MRI.) not only brings a less loss of spatial information but also encodes rich semantic information that shows somehow discriminative power for "healthy" and lesion distributions.

images for specific lesions and diseases, shows its potential in clinical practice. In other words, UAS can detect the unseen diseases and lesions using only the "healthy" training images, which is closer to the pattern in which a radiologist reads an imaging scan.

Till now, considerable effort has been devoted to UAS, especially in medical imaging analysis. Traditional approaches [9,10,15,27] are based on statistical modeling, content-based retrieval, clustering or outlier-detection. Recently, deep learning methods [5–8,20,22,25,29–31] have emerged and showed their superiority. The common principle behind them is to model the distribution of normative "healthy" anatomy via learning to compress and reconstruct it, *i.e.*, an Autoencoder (AE) [11]. Figure 1(a) illustrates this process, where the samples in the input images space are first mapped to the latent manifold, and then back to the input space. In this way, the anomalies can be detected from the imperfect reconstructions, as they are usually the outliers from the normative distribution learned in the training phase.

Nonetheless, current methods heavily involve the reconstruction of the anatomy, which requires accurate localization information. However, during the process of reconstruction in UAS, the localization-oriented detail information is fragile and tends to be devastated, which leads to the inaccurate reconstruction and anomaly detection. To address this problem, some pioneering works [5–7,20] introduce the special designs, to enhance the features of this localization representation. For instance, Baur *et al.* [7] focused on the loss of localization information during the dimensionality reduction of AE, and proposes to introduce Laplacian Pyramid to alleviate it. Besides, a study [6] proposes to utilize a spatial AE which utilizes the feature tensors as the latent manifold rather than the dense vector in previous works, which is claimed to preserve more information on spatial details. Despite the progress, these designs fail to fully exploit some domain knowledge in brain MRI. Instead, given the goal of promoting the preservation of localization information, our motivation is to exploit the semantic segmentation maps, which naturally contain rich spatial details as well as explicitly encode the semantic representation of brain tissues.

Specially, we draw the spirit from a different but related field, namely the semantic segmentation, and propose to exploit the semantic feature maps as the reconstruction manifold for UAS. Figure 1 illustrates our insight. Compared with the dense or tensor manifold, the semantic layout of anatomy tissue is expected to preserve enough spatial information otherwise it can not meet the pixel-wise segmentation task. Meanwhile, it explicitly encodes rich semantic information, which may be valuable for the other high-level task, *i.e.*, UAS. For example, as shown in Fig. 1(b) and Fig. 1(c), the histograms of the prediction probability of its true class show different patterns for "healthy" and lesion pixels. The average prediction distribution for different categories of tissues as well as lesions also reflects a similar situation about its discriminative power between "healthy" and "diseased", as illustrated in Fig. 1(d). Compared with the former three classes of brain tissues, the distribution of average prediction probability in the lesion category shows its discriminability. This observation motivates this paper to forge an autoencoder-like reconstruction paradigm in virtue of this semantic layout as the intermediate manifold, as shown in Fig. 1.

In this paper, we propose a Semantic-Manifold Reconstruction (SMR) framework for UAS, in which the specially designed semantic manifold promotes not only the preservation of detailed location information but also the discrimination of "healthy" and outlier distribution. Furthermore, we introduce a regularization of semantic consistency on the output of SMR, which constrains the origin input images as well as the recovered ones to emphasize the consistency between their semantic layout. By this combination, the proposed framework constitutes a dual reconstruction process, in which the two reconstruction paradigms of "Im→Se→Im"[1] and "Se→Im→Se" complete each other and collaboratively facilitate the modeling of normative distribution as well as the segmentation of the anomalous structures. We summarize the contribution of our dual semantic-manifold reconstruction (DSMR) framework as follows:

- We investigate a key challenge in exploiting UAS in brain MRI, *i.e.*, the loss of location information during the reconstruction process, which is essential for the pixel-wise detection task.
- We propose a semantic-manifold reconstruction (SMR) framework, which ensures the discriminability between normative/abnormal distribution as well as facilitates the locality of pixel-wise anomalies.
- We further construct a dual semantic-manifold reconstruction (DSMR) framework by applying the consistency constraints not only on the image space but also the semantic embedding.
- Experiments on the UAS benchmarks of brain MRI demonstrate the superiority of the proposed DSMR over prior state-of-the-art methods.

---

[1] where Im denotes image and Se denotes semantic layout.

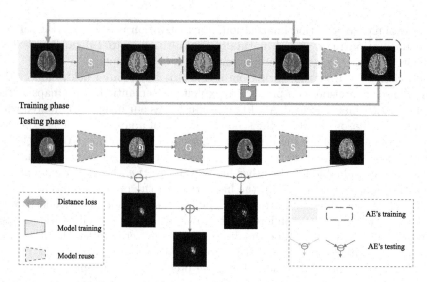

Fig. 2. The pipeline of the proposed dual semantic-manifold reconstruction (DSMR) framework. It consists of two dual autoencoders (AEs), whose manifolds are based on a semantic domain and image domain, respectively. In the training phase, the segmentor S learns to map a healthy image to the semantic layout of anatomy tissues, while the generator G learns to map it back to the image domain, associated with an adversarial discriminator D. For the inference, given a query image, we utilize the dual AE to calculate the ensembled residual maps for the anomaly score.

2   Methodology

Figure 2 depicts the flowchart of the proposed method. In this section, we first introduce the basic paradigm of unsupervised anomaly segmentation (UAS) and then discuss how to introduce the semantic layout as the manifold. Based on this intermediate embedding, we design a novel reconstruction process, by which accurate pixel-wise anomaly detection can be facilitated.

2.1   Preliminary of UAS

For the task of unsupervised anomaly segmentation (UAS), the reconstruction paradigm aims to train an autoencoder (AE) by minimizing the empirical reconstruction risk on the "healthy" scans x:

$$min \ \mathbb{E}_x \ \ell_1[f_{Dec}(f_{Enc}(x)), x] \tag{1}$$

where f_{Enc} and f_{Dec} denotes the network function of encoder and decoder, respectively. The image scan $x \in \mathbb{R}^{H \times W}$, and H, W denotes the height and width of the MR scans. ℓ_1 is the $L1$ distance metric. During training, AE models the normative "healthy" distribution of x, and can reconstruct the images that lie within this distribution well. Then, given a query image x^* which may contain anomalies like lesions, the anomalous regions can be detected by AE from the reconstruction residual scores:

$$r = \ell_1[f_{Dec}(f_{Enc}(x^*)), x^*] \tag{2}$$

where $r \in \mathbb{R}^{H \times W}$ is a map of pixel-wise anomaly scores, in which the pixel with a higher r means more likelihood to be anomalous. Next, we will show how the proposed DSMR meets the goal of autoencoder in Eq. (1) in essence.

2.2 Intermediate Manifold of Semantic Layout

Typically, the manifold in a regular AE is a dense vector, while we propose to utilize the high-resolution semantic space of healthy tissues as the intermediate manifold, which constitutes a special semantic-manifold autoencoder. Our motivation mainly comes from two folds. First, this high-resolution manifold is expected to retain more localization information to cope with the semantic segmentation task, which is also critical for the UAS task. Second, the encoded semantic structures of tissue anatomy show something discriminative about the "healthy" and lesion pixels, as shown in the right part of Fig. 1.

In our context, brain MRI, we use common types of anatomical tissues, namely gray matter (GM), white matter (WM), and cord spinal fluid (CSF). In this way, the manifold space can be depicted as $z \in \mathbb{R}^{H \times W \times C}$, where C denotes the number of tissue classes in the introduced semantic layout. In other words, the manifold representation at every position is in essence the probability vector of dimension C. There are two common approaches for this probability vector, the hard-label like (i.e., one-hot distribution) and soft-label like. Inspired by the development of knowledge distillation [12] and label smoothing [19,26], we utilize the soft-label fashion for the semantic layout, as the soft distribution tends to encode more potential semantic structures.

2.3 Semantic-Manifold Reconstruction

From Image to Semantic Domain. We first need to encode the "healthy" data x from image domain I to the latent code z from semantic layout domain Se. We utilize a segmentor S of anatomical tissues to meet it. Concretely, a three-class semantic segmentation task is introduced, namely for GM, WM, and CSF in brain MRI, as:

$$\mathcal{L}_{Im \rightarrow Se}(S) = \mathbb{E}_x \, \ell_{ce}(S(x), z) \tag{3}$$

where $z \in \mathbb{R}^{H \times W \times C}$ is the prior (ground-truth) probabilistic distribution that we desire the latent manifold to obey.

From Semantic to Image Domain. Then it is required to map the semantic encoding z back to the image domain x. We implement it via a generator G in the image translation module of pixel-to-pixel [13,17]. Specially, a generator G and a discriminator D are trained by an adversarial loss:

$$\mathcal{L}_{GAN}(G, D) = \mathbb{E}_x[log\, D(x)] + \mathbb{E}_z[1 - log\, D(G(z))] \tag{4}$$

where G learns to translate the semantic maps to the image-like counterpart while D aims to discriminate between the synthesized images and the real ones. Besides, a ℓ_1 loss is introduced to visually keep the image consistency for the synthesis:

$$\mathcal{L}_{IC}(G) = \mathbb{E}_{(x,z)}\, \ell_1(G(z), x) \tag{5}$$

where IC denotes the image-wise consistency regularization. Next, by combining Eq. (4) and Eq. (5), the translation module can be optimized in a minimax two-player game, as:

$$\min_G \max_D \mathcal{L}_{GAN}(G, D) + \lambda_I \mathcal{L}_{IC}(G) \tag{6}$$

where λ_{IC} controls the trade-off parameter.

Regularization of Semantic Consistency. In this step, we further reuse the trained segmentor S to acquire the semantic layout of the synthesized images $G(z)$ and regularize the consistency between the semantic output of $G(z)$ and input image x. For the probabilistic space, usually a ℓ_2 loss is used:

$$\mathcal{L}_{SC}(G) = \mathbb{E}_{(x,z)}\, \ell_2(S[G(z)], z) \tag{7}$$

where SC denotes semantic-wise consistency regularization. The motivation is twofold. First, it has been investigated that the performance of semantic segmentation can reveal the quality of semantic synthesis [13,21] (*i.e.*, synthesizing images from semantic layout). So this regularization term can also improve the synthesis process from semantics to image domain. Second, it is discussed in [8] that sometimes the similarity structure among the training data can not be well preserved to be consistent in the latent manifold. Thus, applying this consistency term facilitates the discriminability in the latent semantic space.

In this regard, combining Eq. (4), Eq. (5), Eq. (7), the overall loss for the translation module from semantic to image domain can be expressed by:

$$\mathcal{L}_{Se \to Im}(G, D) = \mathcal{L}_{GAN}(G, D) + \lambda_{IC}\mathcal{L}_{IC}(G) + \lambda_{SC}\mathcal{L}_{SC}(G) \tag{8}$$

where λ_{IC} and λ_{SC} are the trade-off parameters. Based on the above objective, we can formulate the optimization of the translation module in a minimax two-player game:

$$\min_G \max_D \mathcal{L}_{Se \to Im}(G, D) \tag{9}$$

2.4   Dual Reconstruction for Detecting Anomalies

In this section, we reveal how the goal of AE-based reconstruction in Eq. (1) is achieved in our training framework. As shown in Fig. 2, two autoencoders are in essence trained, namely semantic-manifold autoencoder (SMAE) and image-manifold autoencoder (IMAE), in virtue of the objective in Eq. (3) and Eq. (9). Concretely, in SMAE, the input space is image domain, and the encoder and decoder correspond to the segmentor S and generator G, *i.e.*, $f_{Enc} \Leftrightarrow S$ and $f_{Dec} \Leftrightarrow G$. While in IMAE, the input is semantic layout, with the encoder as generator G and decoder as the reused segmentor S, *i.e.*, $f_{Enc} \Leftrightarrow G$ and $f_{Dec} \Leftrightarrow S$. In this regard, the two AEs model the normative "healthy" distribution but from different perspectives, namely "$Im \rightarrow Se \rightarrow Im$" and "$Se \rightarrow Im \rightarrow Se$", which are bridged by the generator G. On the other hand, they are complementary to each other since the results in one AE can be utilized in the other one.

Then in the testing phase, given a query image x^*, we combine the introduced two AEs, *i.e.*, SMAE and IMAE. In this regard, the anomaly score of our DSMR framework can be calculated by:

$$r = \lambda_{IC}\, \ell_1((G \circ S)(x^*), x^*) + \lambda_{SC}\, \ell_1((S \circ G \circ S)(x^*), S(x^*)) \qquad (10)$$

where $r \in \mathbb{R}^{H \times W}$ reflects the anomaly probability in every pixel and $\circ$ represents the composition operator.

3   Experiments and Results

3.1   Dataset

We evaluated our method using three public datasets: Human Connectome Project (HCP) [28], Multimodal Brain Tumor Image Segmentation (BRATS) 2019 [4,16] and Ischemic Stroke Lesion Segmentation (ISLES) 2015 [3]. A version called S500 of HCP [28], which is composed of 526 healthy subjects, has been released and is used as the training set in this paper. The BRATS has 259 GBM (i.e., glioblastoma) and 76 LGG (i.e., lower-grade glioma) volumes, and the ISLES has 28 acute ischemic stroke volumes, which are different in lesion type. We utilize simply the T2 modality among them. Considering there are no available labels of anatomical tissues in the training set HCP, we use a well-known open source Matlab kit for functional imaging data, called SPM[2] [2], in which the tissue probability maps of input brain MRI can be acquired by using the average template based on many subjects' data. Concretely, we use SPM to obtain the probability maps of gray matter (GM), white matter (WM), cerebrospinal fluid (CSF) in HCP, and regard them as the ground-truth of tissue segmentation, which are utilized to train the generator model from semantic to image domain. The labels for training the segmentor of anatomical tissues are in the one-hot fashion.

3.2   Implementation Details

All the methods for comparison as well as ours are implemented based on the public code[3] of UAS [1]. The input images are down-sampled to the size of

---

[2] https://www.fil.ion.ucl.ac.uk/spm/software/download/.

[3] https://github.com/StefanDenn3r/Unsupervised-Anomaly-Detection-Brain-MRI.

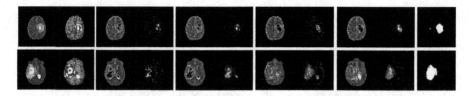

Fig. 3. Ablation studies with visualizing two cases on BraTS. 1st block: input slice and anatomical tissue prediction; 2th~5th block: reconstruction and anomaly score of the variants (detailed in Table 1) of our overall method; last block: lesion ground-truth.

Table 1. Ablation study of our DSMR on BraTS.

Method	SM w/ DSL	SM w/ CSL	DSMR w/o dual testing	DSMR (full)
AUPRC	0.278(±0.015)	0.303(±0.008)	0.365(±0.006)	**0.401**(±0.011)
[DICE]	0.402(±0.007)	0.425(±0.005)	0.447(±0.003)	**0.459**(±0.005)

128×128 and normalized to the range of $[0, 1]$. The experiments are implemented on TensorFlow and an NVIDIA TITAN XP GPU. For the basic component in our method, we use U-Net [24] as the segmentor S. For the synthesis module from semantic to image domain, we use the architecture of generator and discriminator following the setting in [14]. λ_{IC} in Eq. (8) is set as 10 to achieve better visual quality, and λ_{SC} is set as 0.5 such that the loss terms are in a similar scale. We use Adam optimizer with learning rate as $2e-4$ and β as (0.5,0.999). We train the segmentor for 100 epochs with batch size of 128, and the synthesis module for 30 epochs with batch size of 64. All results are reported on the average of three random runs.

3.3   Ablation Study

We conduct comprehensive ablation studies to verify the effect of each design in the proposed work, as shown in Table 1 and Fig. 3. Regarding the results in Table 1, two common evaluation metrics for UAS [1,5,7] are adopted. Among them, Area Under the Precision-Recall curve (AUPRC) evaluates the comprehensive performance of a classifier under heavy class imbalance and [DICE] refers to the generally best achievable DICE score, which reveals the theoretical upper bound for segmentation performance and is determined via a greedy search.

We first investigate the designs of semantic manifold. As shown in Table 1, SM w/ DSL or w/ CSL denotes introducing semantic manifold of the discrete (*i.e.*, one-hot) or continuous (*i.e.*, probabilistic) semantic layout, respectively. We can see that a continuous semantic layout performs better, which is intuitive since the continuous probabilistic distribution has a more powerful capacity. DSMR w/o dual testing denotes only using the image-manifold reconstruction to calculate the anomaly score. It appears that when equipping the baseline model with the proposed components, the performance results go up accordingly, showing the effectiveness and complementarity of our framework.

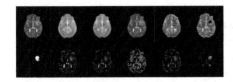 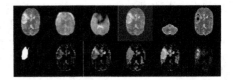

Fig. 4. Visual results on BraTS (left block) and ISLES (right block). In each block, 1st col.: input image and lesion ground truth; 2nd~6th col.: reconstructions and residual maps from AE [6], VAE [30], GMVAE [29], f-AnoGAN [25] and our proposed method.

Table 2. Quantitative results on two benchmark datasets.

Databases	Metrics	Methods				
		AE [6]	VAE [30]	GMVAE [29]	f-AnoGAN [25]	DSMR (Ours)
BraTS	AUPRC	0.170	0.174	0.209	0.237	**0.401**
		(±0.007)	(±0.009)	(±0.014)	(±0.019)	(±0.011)
	[DICE]	0.302	0.318	0.317	0.361	**0.459**
		(±0.004)	(±0.004)	(±0.006)	(±0.010)	(±0.005)
ISLES	AUPRC	0.120	0.141	0.128	0.115	**0.173**
		(±0.005)	(±0.012)	(±0.015)	(±0.013)	(±0.009)
	[DICE]	0.242	0.287	0.282	0.247	**0.298**
		(±0.002)	(±0.006)	(±0.009)	(±0.007)	(±0.005)

3.4   Comparison with Existing Methods

We compare our approach with the existing state-of-the-art methods including AE [6], VAE [30], GMVAE [29] and f-AnoGAN [25]. Table 2 and Fig. 4 present the results. Results in the table show that the proposed DSMR generates more obvious contrast information between the lesions and healthy tissues, which facilitates the discriminability of the lesion detection.

The reasons are mainly twofold. First, the healthy anatomy is recovered successfully. The proposed approach has high reconstruction fidelity in the regions of healthy tissues, where AE and VAE can generate only blurry reconstructions. f-AnoGAN can reconstruct more obviously in those healthy regions, but the visual similarity of normal tissues is still poor. This also detects some normal tissues highlighted in the image, such as WM and CSF, which leads to false positive results. It also confirms the concerns about the loss of localization information in the compression to a low-resolution manifold. Second, most anomalous regions fail to be recovered by our method. In contrast, some methods like GMVAE, generate nearly the same images as origin input, which ruins the contrast relationship between the anomaly and healthy anatomy.

The qualitative results on ISLES are shown in the right block of Fig. 4. It is noteworthy that the lesions in ISLES are more diverse in quantity, pixel intensity, and shape (*e.g.*, sometimes very small), which dramatically increases the difficulty of this task and thus causes the significant performance degradation.

As shown in Table 2, the above challenges make the obvious performance degradation compared to that on BraTS. However, the proposed approach can still achieve the best performance. For the examples shown in Fig. 4, other methods generate massive false positives, but our proposed method avoids most of them and still retains the majority of the effect of anomaly segmentation, which indicates the strong modeling ability and generalization of our method.

4   Conclusion

In this paper, we focus on the issue of Unsupervised Anomaly Segmentation (UAS), which is promising in brain MRI and medical imaging analysis. The challenge is the loss of spatial location information during the reconstruction process of UAS, which is essential for the pixel-wise detection task. To tackle it, we introduce the semantic layout of healthy anatomy in the process of modeling normative distribution. It preserves more localization information as well as the semantic encoding. We further present a consistency regularization in the semantic layout and forge a dual semantic-manifold reconstruction framework for UAS, which can be used collaboratively for detecting lesions. Experimental results on benchmarks show that our approach achieves significantly superior performance compared to several prior methods and segments the anomalies more accurately.

References

1. Baur, C., Denner, S., Wiestler, B., Navab, N., Albarqouni, S.: Autoencoders for unsupervised anomaly segmentation in brain MR images: a comparative study - science direct. Med. Image Anal. **69**, 101952 (2021)
2. Ashburner, J., Friston, K.J.: Computing average shaped tissue probability templates. Neuroimage **45**(2), 333–341 (2009)
3. Maier, O., et al.: ISLES 2015 - a public evaluation benchmark for ischemic stroke lesion segmentation from multispectral MRI. Med. Image Anal. **35**, 250–269 (2017)
4. Bakas, S., et al.: Advancing the cancer genome atlas glioma MRI collections with expert segmentation labels and radiomic features. Sci. Data **4**, 170117 (2017)
5. Baur, C., Graf, R., Wiestler, B., Albarqouni, S., Navab, N.: SteGANomaly: inhibiting CycleGAN steganography for unsupervised anomaly detection in brain MRI. In: Martel, A.L., et al. (eds.) MICCAI 2020. LNCS, vol. 12262, pp. 718–727. Springer, Cham (2020). https://doi.org/10.1007/978-3-030-59713-9_69
6. Baur, C., Wiestler, B., Albarqouni, S., Navab, N.: Deep autoencoding models for unsupervised anomaly segmentation in brain MR images. In: Crimi, A., Bakas, S., Kuijf, H., Keyvan, F., Reyes, M., van Walsum, T. (eds.) BrainLes 2018. LNCS, vol. 11383, pp. 161–169. Springer, Cham (2019). https://doi.org/10.1007/978-3-030-11723-8_16
7. Baur, C., Wiestler, B., Albarqouni, S., Navab, N.: Scale-space autoencoders for unsupervised anomaly segmentation in brain MRI. In: Martel, A.L., et al. (eds.) MICCAI 2020. LNCS, vol. 12264, pp. 552–561. Springer, Cham (2020). https://doi.org/10.1007/978-3-030-59719-1_54

8. Chen, X., Konukoglu, E.: Unsupervised detection of lesions in brain MRI using constrained adversarial auto-encoders. arXiv preprint arXiv:1806.04972 (2018)

9. Eskin, E., Arnold, A., Prerau, M., Portnoy, L., Stolfo, S.: A geometric framework for unsupervised anomaly detection. In: Barbará, D., Jajodia, S. (eds.) Applications of Data Mining in Computer Security. Advances in Information Security, vol. 6, pp. 77–101. Springer, Boston (2002). https://doi.org/10.1007/978-1-4615-0953-0_4

10. Guthrie, D., Guthrie, L., Allison, B., Wilks, Y.: Unsupervised anomaly detection. In: International Joint Conferences on Artificial Intelligence (IJCAI), pp. 1624–1628 (2007)

11. Hinton, G.E., Salakhutdinov, R.R.: Reducing the dimensionality of data with neural networks. Science **313**(5786), 504–507 (2006)

12. Hinton, G., Vinyals, O., Dean, J.: Distilling the knowledge in a neural network. arXiv preprint arXiv:1503.02531 (2015)

13. Isola, P., Zhu, J.Y., Zhou, T., Efros, A.A.: Image-to-image translation with conditional adversarial networks. In: Proceedings of the IEEE Conference on Computer Vision and Pattern Recognition (CVPR), pp. 1125–1134 (2017)

14. Johnson, J., Alahi, A., Fei-Fei, L.: Perceptual losses for real-time style transfer and super-resolution. In: Leibe, B., Matas, J., Sebe, N., Welling, M. (eds.) ECCV 2016. LNCS, vol. 9906, pp. 694–711. Springer, Cham (2016). https://doi.org/10.1007/978-3-319-46475-6_43

15. Leung, K., Leckie, C.: Unsupervised anomaly detection in network intrusion detection using clusters. In: Proceedings of the Twenty-Eighth Australasian Conference on Computer Science, pp. 333–342 (2005)

16. Menze, B.H., et al.: The multimodal brain tumor image segmentation benchmark (BRATS). IEEE Trans. Med. Imaging **34**(10), 1993–2024 (2014)

17. Mirza, M., Osindero, S.: Conditional generative adversarial nets. arXiv preprint arXiv:1411.1784 (2014)

18. Mohan, G., Subashini, M.M.: MRI based medical image analysis: survey on brain tumor grade classification. Biomed. Sign. Process. Control **39**, 139–161 (2018)

19. Müller, R., Kornblith, S., Hinton, G.E.: When does label smoothing help?. In: Advances in Neural Information Processing Systems (NeurIPS), vol. 32 (2019)

20. Nguyen, B., Feldman, A., Bethapudi, S., Jennings, A., Willcocks, C.G.: Unsupervised region-based anomaly detection in brain MRI with adversarial image inpainting. In: 2021 IEEE 18th International Symposium on Biomedical Imaging (ISBI), pp. 1127–1131. IEEE (2021)

21. Park, T., Liu, M.Y., Wang, T.C., Zhu, J.Y.: Semantic image synthesis with spatially-adaptive normalization. In: Proceedings of the IEEE Conference on Computer Vision and Pattern Recognition (CVPR), pp. 2337–2346 (2019)

22. Pawlowski, N., et al.: Unsupervised lesion detection in brain CT using Bayesian convolutional autoencoders. In: Medical Imaging with Deep Learning (2018)

23. Rezaei, M., Yang, H., Meinel, C.: Deep neural network with l2-norm unit for brain lesions detection. In: Liu, D., Xie, S., Li, Y., Zhao, D., El-Alfy, E.S. (eds.) ICONIP 2017. Lecture Notes in Computer Science(), vol. 10637, pp. 798–807. Springer, Cham (2017). https://doi.org/10.1007/978-3-319-70093-9_85

24. Ronneberger, O., Fischer, P., Brox, T.: U-Net: convolutional networks for biomedical image segmentation. In: Navab, N., Hornegger, J., Wells, W.M., Frangi, A.F. (eds.) MICCAI 2015. LNCS, vol. 9351, pp. 234–241. Springer, Cham (2015). https://doi.org/10.1007/978-3-319-24574-4_28

25. Schlegl, T., Seeböck, P., Waldstein, S.M., Langs, G., Schmidt-Erfurth, U.: f-AnoGAN: fast unsupervised anomaly detection with generative adversarial networks. Med. Image Anal. **54**, 30–44 (2019)

26. Shen, Z., Liu, Z., Xu, D., Chen, Z., Cheng, K.T., Savvides, M.: Is label smoothing truly incompatible with knowledge distillation: An empirical study. In: Proceedings of the International Conference of Learning Representation (ICLR) (2020)

27. Taboada-Crispi, A., Sahli, H., Hernandez-Pacheco, D., Falcon-Ruiz, A.: Anomaly detection in medical image analysis. In: Handbook of Research on Advanced Techniques in Diagnostic Imaging and Biomedical Applications, pp. 426–446 (2009)

28. Van Essen, D.C., et al.: The WU-Minn human connectome project: an overview. Neuroimage **80**, 62–79 (2013)

29. You, S., Tezcan, K.C., Chen, X., Konukoglu, E.: Unsupervised lesion detection via image restoration with a normative prior. In: International Conference on Medical Imaging with Deep Learning (ICDCI), pp. 540–556 (2019)

30. Zimmerer, D., Isensee, F., Petersen, J., Kohl, S., Maier-Hein, K.: Unsupervised anomaly localization using variational auto-encoders. In: International Conference on Medical Image Computing and Computer-Assisted Intervention (MICCAI), pp. 289–297 (2019)

31. Zimmerer, D., Kohl, S.A., Petersen, J., Isensee, F., Maier-Hein, K.H.: Context-encoding variational autoencoder for unsupervised anomaly detection. arXiv preprint arXiv:1812.05941 (2018)

Transformer Based High-Frequency Predictive Model for Visual-Haptic Feedback of Virtual Surgery Navigation

Jianyong Huang, Jianhui Zhao[(✉)], Zhekang Qiu, and Zhiyong Yuan

School of Computer Science, Wuhan University, Wuhan, China
{huangjianyong,jianhuizhao,qiuzekang,zhiyongyuan}@whu.edu.cn

Abstract. In virtual reality surgery training, magnetic levitation instruments have gained popularity due to the advantages of non-mechanical friction and low inertia. However, it is difficult to obtain high accuracy, frequency, and robust navigation stability, and this will not capture the subtle changes in the user's actions resulting in a much weaker sense of immersion. To tackle this issue, previous works have used inconvenient motion tracking sensors for navigation. Nevertheless, these techniques did not consider the navigation effects caused by the environmental limitations of the sensors. In this work, we propose a Transformer-based high-frequency prediction model (HPformer) to predict the direction and position data by designing an incremental module to learn the increment of navigation information in an accumulative manner. Also, to reduce the position prediction value error, we propose an initialization module related to uniform acceleration. By building a testbed, experimental results show that our method can obtain accurate navigation (the mean absolute error is less than 0.026) and increase the navigation frequency 200 Hz.

Keywords: Transformer · Time series forecasting · Virtual surgery · Deep learning

1 Introduction

Nowadays, the latest developments in virtual reality (VR) and augmented reality (AR) have facilitated the development of related applications, such as surgery [5,15], aerospace technology, teaching systems, and video entertainment [20].

In virtual reality surgery, haptics is an important part of the user's immersive interactive experience. It is common to use force feedback to train surgeons [11]. Virtual surgery is aimed at surgeons practicing surgery and letting them experience the feeling of real operation, which can give surgeons more practice time and significantly improve the success rate of surgical training [5,15]. Examples include virtual training for cardiac interventions and minimally invasive procedures, etc.

Common magnetic levitation haptic devices use visual sensors to capture the user's movements through marker points [7]. However, two issues limit the

M. Tanveer et al. (Eds.): ICONIP 2022, LNCS 13625, pp. 145–157, 2023.
https://doi.org/10.1007/978-3-031-30111-7_13

immersive use, including the low frequency (20 Hz) of navigation data output and the inability to capture positioning data due to occlusion and fast move.

Research on multi-sensor fusion navigation catches on as autonomous driving and path-planning require navigation accuracy. Because the vision sensor has the advantage of high precision positioning, the Inertial Measurement Unit (IMU) has the advantage of high-frequency sampling, being lightweight, and having no environmental limitations. Many research improves navigation accuracy by fusing navigation using filtering [12] and optimization [14] methods, they also suffer from low-frequency output and decimeter-level accuracy problems due to limitations in improving vehicle navigation accuracy.

We are inspired by the above visual-inertial navigation and the rapid development of attention networks in recent years. We embed the IMU in a magnetic levitation device and propose the HPformer to treat visual-inertial navigation as a regression problem.

To reduce the bias caused by high-frequency noise, the position data of IMU is usually obtained by a fast Fourier transform in the frequency domain. But it can not be eliminated and still accumulates with time [14]. And vision sensors can only output low-frequency position information.

To solve this problem, we propose a network named HPformer, which can improve the frequency and accuracy of navigation data. According to the characteristics of the predictive network, it can also help the camera to capture the position data in the case of rapid movement and occlusion.

In summary, the crux of improving immersion is to increase the navigation frequency and prevent the instability problem of losing navigation data when moving fast or occlusion. The main contributions are as follows:

- To improve user immersion in magnetic levitation haptic devices, we propose the HPformer model using IMU fusion with vision sensors. Utilizing the advantages of both sensors, HPformer applies a regression manner to improve the problem of vision sensors failing to capture marker points.
- We propose an incremental module and an initialization module in the deep network to learn and optimize the incremental information of direction and position in navigation.
- The experimental results show that our HPformer model can ensure high accuracy while outputting direction and position information at high-frequency and can effectively alleviate the problems of occlusion and fast movement.

The rest of the paper is organized as follows. In Sect. 2, the related work of the paper is presented. Section 3 describes the system architecture. Section 4 presents the details of HPformer. In Sect. 5, we give the experimental results. Finally in Sect. 6 the conclusion is given.

2 Related Work

2.1 Magnetic Haptic Devices

The main applications of magnetic haptic devices are touch screen or magnetically levitated haptic devices [3,4,13]. It is common to provide contactless feedback by designing a stylus with a magnetic tip in a touch screen or maglev haptic interface. In terms of providing haptic feedback force, Tong et al. [17,18] proposed a coil configuration method to control virtual haptic feedback force, which improve the accuracy of sensing the tissue stiffness of the virtual model.

However, the sampling speed of the position navigation tracking system and the noise of the position information in the magnetically levitated haptic device limit its operational effectiveness [10]. To provide real-time and accurate navigation information in magnetically levitated haptic devices, Berkelman et al. [4,13] used an optical motion tracker (Northern Digital Inc.) containing six degrees of freedom. However, when tilted at a large angle, the infrared light-emitting diodes will obscure each other, losing positioning information. To solve this issue, Tong et al. [17] fixed two-colored marker dots in the middle of the magnetic stylus and tracked the marker dots of the stylus by binocular vision sensors to locate (40 Hz) the interaction behavior made by the user.

The above methods also create low immersion and masking problems due to low-frequency sampling. To achieve a distinguishable human-eye navigation frequency, we constructed a neural network to predict high-frequency navigation data by regression form.

2.2 Fusion Navigation and Time Series Forecasting

To solve the problem of target loss due to occlusion and fast-moving, Teixeira et al. [16] used a deep learning approach to the navigation problem of robotic underwater tasks. Likewise, Clark et al. [6] first used deep learning applied to vision-inertial fusion networks, which can skip the process of synchronizing vision sensors and over time obtain accurate navigation. They also demonstrated that deep learning is competitive with traditional methods.

In time series forecasting (TSF), various methods have been well developed. LSTNet [9] proposes a temporal prediction framework combining CNN and RNN to extract short-term local dependency patterns and discover long-term patterns among variables. TCN [2] applies causal convolution to model temporal causality.

Recently, Transformer [19] has relied on an attention mechanism to learn the global relationship, which can capture long-range dependencies and interactions. Reformer [8] uses locality-sensitive hashing to depress complexity for very long sequences. Informer [22] extends the Transformer by proposing a KL-divergence based ProbSparse attention.

However, all the above methods can only handle low-frequency information. To solve the above problem, we propose a high-frequency prediction network based on Transformer. By designing lightweight encoders combined with our proposed incremental model, it can extract short-term patterns among variables and output high-frequency navigation information.

3   System Overview

The system we simulated is shown in Fig. 1. The magnetic haptic interaction system comprises a binocular camera, IMU, a magnetic stylus with marking

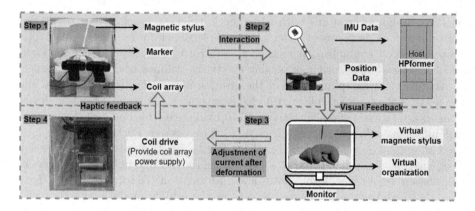

Fig. 1. An overall process for our maglev haptic system testbed.

points, a virtual surgery program, an artificial intelligence program, a coil drive, a haptic feedback interface, and so on. Two sensors send navigation data to the host via LAN. The detailed steps are shown below.

Step 1: The user controls the device with a magnetic stylus.

Step 2: The binocular vision sensor and the IMU at the top of the magnetic stylus acquire acceleration, angular velocity, direction, and position information at this time. Our host computer receives the above navigation information and Inputs it into our proposed HPfomer method, which then estimates the direction and position of the magnetic stylus in a high-frequency manner.

Step 3: The output navigation information is calculated as to whether there is a collision in the virtual organization module. If a collision occurs, a deformation is calculated to provide visual feedback in the display and trigger step 4 simultaneously.

Step 4: The current of the coil array is adjusted to give a user haptic feedback force through the magnetic stylus, which is the same as the force of the virtual magnetic stylus. Then go back to step 1 and repeat the whole process.

4 Method

We propose a new Transformer-based model called HPformer and the general process of the method is illustrated in Fig. 2. The collected direction and position data are preprocessed and input to our model based on the frequency ratio of IMU to predicted data. The model's output is the total increment of position or direction for a period of time and N sub-increments of the total increment. Unlike machine translation and speech recognition, our tasks do not have standard translations. We only have encoders that reference the standard Transformer. Depending on the requirements of our task, the features extracted by the encoder are fed into our proposed incremental module for the high-frequency prediction of navigation data.

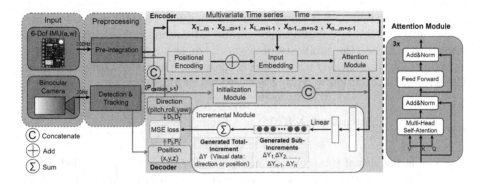

Fig. 2. The overview of our proposed HPformer.

4.1   Encoder

In HPformer, the input of the network is a 1-dimensional token embedding. We flatten the IMU data of dimension $T \times P$ to one-dimensional data of dimension $(T \times P)$, where T is the frequency ratio of inertial data to direction or position information, and P is the acceleration a and angular velocity w collected by the accelerometer and gyroscope in the IMU.

Before entering the encoder, a learnable position embedding is also prepared. We view the IMU data of the same time period as the same position encoding. Specifically, we define different position embeddings for each input data to the encoder. Concerning the encoding function for the position information, we apply the default position encoding since other functions do not improve this work much. This sine function can handle longer sequences of information. The equation for the position encoding is as follows:

$$PE_{(pos,2i)} = \sin\left(pos/10000^{2i/d_{\text{model}}}\right) \tag{1}$$

$$PE_{(pos,2i+1)} = \cos\left(pos/10000^{2i/d_{\text{model}}}\right) \tag{2}$$

where the equation shows that the input sequence gets a position encoding calculated by the sine and cosine functions with different periods as the input sequence, where the wavelengths are of geometric order from 2π to $10000 \cdot 2\pi$. Moreover, for any distance k within the sequence, $PE_{\text{pos}+k}$ can be obtained from the linear function of PE_{pos}. The purpose of adding the position encoding to the input sequence is to allow the multi-headed attention module to learn the position information of the sequence.

After the encoder receives the one-dimensional inertial data and the prepared position embedding, the next step is to perform a one-dimensional convolution of the inertial data to match the position encoding and then sum the position encoding. HPformer consists of three stacked attention modules which have two main sub-modules.

The first sub-modules is a multi-headed self-attentive layer followed by normalization and residual connection, where self-attentive means self-learning. The concept of the attention function is to map a query and a set of key-value pairs to the output and set multiple inner product attention modules with different weights to calculate the similarity between the query and the keys. And finally, learn different information by normalizing the weights with the softmax function. The formula of attention weighting is expressed as follows:

$$\text{Attention } (\boldsymbol{Q}, \boldsymbol{K}, \boldsymbol{V}) = \text{softmax} \left(\frac{\boldsymbol{Q}\boldsymbol{K}^T}{\sqrt{d_k}} \right) \boldsymbol{V} \qquad (3)$$

where d_k is the size of the attention head, a scale factor of $\frac{1}{\sqrt{d_k}}$ is added to the inner product attention in order to prevent the input from falling into a region of very small gradient due to the softmax function when the value is very large.

The second sub-module is a feedforward network followed by normalization and residual connectivity.

4.2   Direction Estimation and Incremental Module

Because of the different boosting frequencies for direction and position data, we construct two different decoders to predict the direction and the position data. In direction estimation network, we only apply incremental modules in the decoder (the red line in Fig. 2).

After received from the encoder, the features enter the incremental module. Two convolution operations are performed in the incremental module. The first convolution layer is downsampling and extracting features. And the second convolution layer is being the P-dimensional convolution, where P is a P-fold increase in the input data frequency. Each convolutional layer is followed by a Relu activation function. The next step is a fully connected layer to obtain the N-dimensional output, the high-frequency sub-incremental information. Finally, the sub-increments are summed to obtain the total increment ΔY of the direction estimate, according to $\Delta Y = \sum_{n=1}^{N} \Delta Y_n$.

To be concrete, the input is $T \times (a + w)$, where a and w are the six-degree-of-freedom acceleration and angular rate obtained from IMUs. After passing through HPformer will get $N + 1$ outputs, increments (ΔY_1, ΔY_2, ..., ΔY_N) and the total increment ΔY to them, here ΔY is the increment between the directional data at the current time I and the directional data at time step $I + 5$, according to:

$$\boldsymbol{Direction}_{I+5} = \boldsymbol{Direction}_I + \boldsymbol{\Delta Y} \qquad (4)$$

4.3   Position Estimation and Initialization Module

Different from directional estimation, position information is more complex than direction information (Unlike the position, the surgical tools are mostly oriented towards the organ below). Therefore, it is impossible to match the accuracy of the

---

Algorithm 1: Initialization module

Input: Position Information $Distance$, Total Time T_n
Output: The uniform acceleration $Uacc$ of the $Distance$ in time
1 Initialized to an all-zero matrix M_{ij}, where i is equal to the i axes of the position information, j is equal to the j time acceleration to be calculated;
2 **if** T_n *is equal to one* **then**
3 $\quad$ return $Distance$
4 **for** *Receive the position information of N axes* **do**
5 $\quad$ Create an acceleration a equal to the sum of two times the distance divided by the square of the time T_n.
6 $\quad$ **for** Tn **do**
7 $\quad\quad$ Return $Uacc$ with matrix size M_{ij}, so that it is equal a multiplied by the square of the time Tn and divided by two.

---

position prediction with the direction prediction by using only the incremental module in the decoder (The experiments are described in Sect. 5.3).

To solve the above problems, we add an initialization module shown in Algorithm 1. And then concatenate the position data and IMU data from the previous time step to the decoder (the blue line in Fig. 2). It is worth noting that the position and the IMU data have different frequencies. Thus, we use the same position data in the IMU data of the previous time step and set the position data to zero in the current time step.

We calculate the uniform acceleration using the known distance difference and time based on the kinematic theory. Then concatenate the output of the initialization module with the features extracted from the encoder and input to the incremental module outputting the predicted position information.

5 Experiments

5.1 Data Acquisition

We simulated the surgeon touching the heart with a scalpel on the test platform. The steps are as follows: the user moves the magnetic stylus in a fixed position on the platform. Two sensors get the acceleration, angular velocity, directional information, and position information when the magnetic stylus is moving.

According to the above steps, 40,000 sets of data are collected. Specifically, the position information was obtained from the binocular vision sensor with a sampling frequency 20 Hz. The acceleration angular velocity information was obtained from two IMUs with a sampling frequency 200 Hz and the direction angle data with a frequency 100 Hz. We split all datasets into training, validation and test set by the ratio of 8:1:1. All experiments are repeated three times, implemented in PyTorch and conducted on a single NVIDIA 3090 RTX 24GB GPUs. The code is written in python 3.8, using CUDA version 11.1.

5.2   Evaluation Metrics

In this paper, we use mean absolute error (MAE) and root mean square error (RMSE) to evaluate the predicted results of direction and position. The definitions of the two metrics are as follows:

$$MAE = \frac{1}{n} \sum_{i=1}^{n} |y_i - \hat{y}_i| \tag{5}$$

$$RMSE = \sqrt{\frac{1}{n} \sum_{i=1}^{n} (y_i - \hat{y}_i)^2} \tag{6}$$

where y_i is the actual value and $\hat{y}_i$ is the predicted value. The evaluation metric closer to zero represents better predictive performance of the model. MAE can reflect the actual situation of the prediction value error of the navigation data. RMSE represents the sample standard deviation of the difference between the predicted and observed values.

5.3   Experimental Results and Discussion

Direction-Based High-Frequency Forecasting. We perform experiments on IMU inertial data and directional data. To get the high-frequency direction information from the network output (the sampling frequency of the direction data is increased 100 Hz to 200 Hz), the data is preprocessed as the real value for the network training. Specifically, an incremental data set is created by subtracting the directional value of time period I from the directional value of time period $I + 5$. And then fitting the total increment obtained by accumulating the ten increments outputted by Hpformer to that true total increment.

To verify the robustness of our proposed network, we used the directional data of time period I plus the 2, 4, 6, and 8 increments obtained by HPformer to compare with the directional data of time periods $I + 1$, $I + 2$, $I + 3$, and $I + 4$. For example, $D_{I+1} = D_I + \sum_{n=1}^{2} \Delta Y_n$, $D_{I+2} = D_I + \sum_{n=1}^{4} \Delta Y_n$, where D denotes direction. The robustness of HPformer is effectively shown in Fig. 3. It can observe that our incremental module can capture the feature of incremental information. The accuracy of the direction data (60, 62–65, 67–70 time periods) without the true value fit is about the same compared to the time period with the true value fit (61, 66, 71 time periods).

As shown in Fig. 3, We compare the prediction performance of HPformer with other state-of-the-art TSF methods in the same experiment environment and direction navigation dataset. For a fair comparison, all methods are used as the backbone and then input to the incremental module. It is obvious that HPformer can predict subtle directions better than the state-of-the-art TSF network. Their MAE and RMSE comparisons are shown in Table 1. The MAE between our predicted direction angle and the true direction angle (pitch, roll, and yaw) reaches 0.0253 mm.

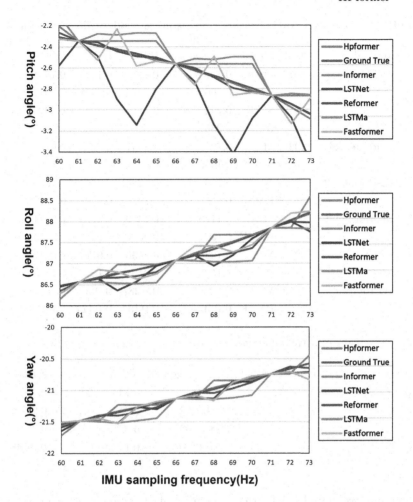

Fig. 3. Comparison of HPformer with other time series forecasting methods in direction estimation. From left to right are directional angles (pitch, roll, and yaw) for IMU sampling periods 60 to 73 (x-axis), where the y-axis denotes the angle (°).

Position-Based High-Frequency Forecasting. Experimented using IMU inertial data and position data. In order to enable the network output to high-frequency position information (so that the sampling frequency of position data 20 Hz to 200 Hz), the incremental data set of position data in time period N and time period $N + 1$ is created. Then the 10 increments output by HPformer is summed to get the total predicted increments and fitted to the real increments, that is $Position_{I+1} = Position_I + \Delta Y$.

Since the position data is more confusing than the direction data, highly accurate prediction information cannot be obtained from HPformer. To solve this issue, we proposed two modules. Table 2 shows the experiment result, where $Position_{(t-1)}$ refers to the position information of the previous time period.

Table 1. Comparison with other advanced forecasting methods in direction estimation.

Methods	Year	Metric	Average loss	Pitch	Roll	Yaw
LSTMa [1]	2015	MAE	0.1585	0.1790	0.1703	0.1261
		RMSE	0.2602	0.2874	0.2826	0.2017
TCN [2]	2018	MAE	0.1453	0.1418	0.1691	0.1249
		RMSE	0.2361	0.2204	0.2808	0.1997
Reformer [8]	2020	MAE	0.1239	0.1541	0.1272	0.0903
		RMSE	0.1913	0.2290	0.1966	0.1365
Fastformer [21]	2021	MAE	0.1166	0.1287	0.1180	0.1030
		RMSE	0.1585	0.1727	0.1623	0.1386
Informer [22]	2021	MAE	0.1128	0.1359	0.1125	0.0901
		RMSE	0.1902	0.2156	0.2038	0.1434
LSTNet [9]	2018	MAE	0.1091	0.1590	0.0772	0.0910
		RMSE	0.1522	0.2101	0.1046	0.1201
HPformer	2022	MAE	**0.0253**	**0.0288**	**0.0308**	**0.0164**
		RMSE	**0.0398**	**0.0461**	**0.0477**	**0.0256**

Table 2. Ablation study for position prediction networks.

HPformer	Init Module	Position$_{(t-1)}$	Average loss	X-axis	Y-axis	Z-axis
√	×	×	1.6041	0.7402	3.2729	0.7992
√	×	√	0.0262	0.0147	**0.0305**	0.0333
√	√	×	0.0260	0.0213	0.0335	0.0231
√	√	√	**0.0216**	**0.0129**	0.0374	**0.0146**

Without our two modules, HPformer does not learn position information features efficiently. By adding any of the above modules, HPformer can better fit the position information and approach the accuracy of direction estimation. This verifies the necessity of our proposed two modules.

Experimental results show that the MAE is 1.6041 mm mm when only using HPformer. Adding either of the proposed methods can improve the accuracy of the position information. HPformer finally adopted two modules, reducing the MAE loss to 0.0216 mm.

5.4   Ablation Study

Key Parameters. To get the best performance of the HPformer network, the key parameters should be set to appropriate values. In this paper, we try various of encoder layers and attention heads. The experiment results are shown in Table 3. For the number of encoder layers N, we tried the cases of $N = 3, 6$. For the number of attention heads h, we performed the cases with $h = 4, 8, 12$. We

Table 3. Performance of different hyperparameters.

N	h	d_{model}	$dropout$	Pitch	Roll	Yaw	X-axis	Y-axis	Z-axis
3	4	32	0.1	**0.0238**	**0.0289**	**0.0292**	0.0129	**0.0374**	0.0146
3	4	128	0.1	0.0310	0.0329	0.0172	0.0095	0.0728	**0.0100**
6	4	32	0.1	0.0300	0.0303	0.0154	0.0135	0.0740	0.0108
6	8	128	0.1	0.0350	0.0352	0.0209	0.0119	0.0858	0.0122
6	12	32	0.1	0.0302	0.0326	0.0166	**0.0092**	0.0657	0.0122

Table 4. Ablation study of the initialization module.

Methods	Average MAE loss	X-axis	Y-axis	Z-axis
Random speed	0.0581	0.0447	0.0730	0.0565
Uniform speed	0.0271	0.0261	**0.0334**	**0.0218**
Uniform acceleration	**0.0257**	**0.0137**	0.0379	0.0254

can observe that the best four of the six results occur in the result of $N = 3$ and $h = 4$. To balance performance and computational complexity, HPformer finally adopts $N = 3$ and $h = 4$ in the direction and position prediction.

Initialization Module. The operator uses speed variations such as uniform speed, uniform acceleration, or random speed in the magnetic stylus. Then we constructed initialization modules based on these velocity variations and compared their accuracy. As shown in Table 4, the initialization module associated with uniform acceleration can better capture changes in position. The acceleration features can be incorporated into our network to help us better predict subtle changes in position information.

6   Conclusion

This paper proposes a high-frequency prediction network named HPformer for improving user immersion in virtual devices. The navigation frequency of the visual-inertial sensor is increased based on the premise of accuracy and stability of navigation. HPformer achieved this by modeling the relationship between IMU and navigation data and transmitting it to our proposed incremental module. We construct two independent prediction networks to forecast the direction and position data. Additionally, proposing an initialization module based on the dynamics further improves position estimation accuracy. Intuitively, our model can input direction and position data with a lower sampling frequency to achieve high-frequency navigation 200 Hz. By building a magnetic levitation haptic testbed, we demonstrated the effectiveness of HPformer in high-frequency prediction (direction and position) by comparing it with the state-of-the-art TSF method.

Acknowledgment. The work was supported by National Natural Science Foundation of China (62073248) and Translational Medicine and Interdisciplinary Research Joint Fund of Zhongnan Hospital of Wuhan University (ZNJC201926).

References

1. Bahdanau, D., Cho, K., Bengio, Y.: Neural machine translation by jointly learning to align and translate. In: ICLR (2015)
2. Bai, S., Kolter, J.Z., Koltun, V.: An empirical evaluation of generic convolutional and recurrent networks for sequence modeling. arXiv preprint arXiv:1803.01271 (2018)
3. Berkelman, P., Bozlee, S., Miyasaka, M.: Interactive rigid-body dynamics and deformable surface simulations with co-located maglev haptic and 3d graphic display (2013)
4. Berkelman, P., Miyasaka, M., Anderson, J.: Co-located 3d graphic and haptic display using electromagnetic levitation. In: 2012 IEEE Haptics Symposium (HAPTICS), pp. 77–81. IEEE (2012)
5. Boulanger, P., Wu, G., Bischof, W., Yang, X.: Hapto-audio-visual environments for collaborative training of ophthalmic surgery over optical network. In: 2006 IEEE International Workshop on Haptic Audio Visual Environments and their Applications (HAVE 2006), pp. 21–26. IEEE (2006)
6. Clark, R., Wang, S., Wen, H., Markham, A., Trigoni, N.: Vinet: visual-inertial odometry as a sequence-to-sequence learning problem. In: Proceedings of the AAAI Conference on Artificial Intelligence, vol. 31 (2017)
7. Hamza-Lup, F.G., Bogdan, C.M., Popovici, D.M., Costea, O.D.: A survey of visuo-haptic simulation in surgical training. arXiv preprint arXiv:1903.03272 (2019)
8. Kitaev, N., Kaiser, L., Levskaya, A.: Reformer: the efficient transformer. In: ICLR (2020). https://openreview.net/forum?id=rkgNKkHtvB
9. Lai, G., Chang, W.C., Yang, Y., Liu, H.: Modeling long-and short-term temporal patterns with deep neural networks. In: SIGIR (2018)
10. Miyasaka, M., Berkelman, P.: Magnetic levitation with unlimited omnidirectional rotation range. Mechatronics **24**(3), 252–264 (2014)
11. Moody, L., Baber, C., Arvanitis, T.N.: The role of haptic feedback in the training and assessment of surgeons using a virtual environment. In: proceedings of Eurohaptics, pp. 170–173 (2001)
12. Mourikis, A.I., Roumeliotis, S.I.: A multi-state constraint kalman filter for vision-aided inertial navigation. In: Proceedings 2007 IEEE International Conference on Robotics and Automation, pp. 3565–3572. IEEE (2007)
13. Pedram, S.A., Klatzky, R.L., Berkelman, P.: Torque contribution to haptic rendering of virtual textures. IEEE Trans. Haptics **10**(4), 567–579 (2017)
14. Qin, T., Li, P., Shen, S.: Vins-mono: a robust and versatile monocular visual-inertial state estimator. IEEE Trans. Rob. **34**(4), 1004–1020 (2018)
15. Satava, R.M., Jones, S.B.: Current and future applications of virtual reality for medicine. In: Proceedings of the IEEE, vol. 86, no. 3, pp. 484–489 (1998)
16. Teixeira, B., Silva, H., Matos, A., Silva, E.: Deep learning approaches assessment for underwater scene understanding and egomotion estimation. In: OCEANS 2019 MTS/IEEE SEATTLE, pp. 1–9. IEEE (2019)
17. Tong, Q., Yuan, Z., Liao, X., Zheng, M., Yuan, T., Zhao, J.: Magnetic levitation haptic augmentation for virtual tissue stiffness perception. IEEE Trans. Visual Comput. Graph. **24**(12), 3123–3136 (2017)

18. Tong, Q., Yuan, Z., Zheng, M., Zhu, W., Zhang, G., Liao, X.: A novel magnetic levitation haptic device for augmentation of tissue stiffness perception. In: Proceedings of the 22nd ACM Conference on Virtual Reality Software and Technology, pp. 143–152 (2016)
19. Vaswani, A., et al.: Attention is all you need. In: Advances in Neural Information Processing Systems, pp. 5998–6008 (2017)
20. Viitanen, M., Vanne, J., Hämäläinen, T.D., Kulmala, A.: Low latency edge rendering scheme for interactive 360 degree virtual reality gaming. In: 2018 IEEE 38th International Conference on Distributed Computing Systems (ICDCS), pp. 1557–1560. IEEE (2018)
21. Wu, C., Wu, F., Qi, T., Huang, Y., Xie, X.: Fastformer: additive attention can be all you need. arXiv preprint arXiv:2108.09084 (2021)
22. Zhou, H., et al.: Informer: beyond efficient transformer for long sequence time-series forecasting. In: Proceedings of AAAI (2021)

Hierarchical Multimodal Attention Network Based on Semantically Textual Guidance for Video Captioning

Caihua Liu[1,2], Xiaoyi Ma[1,2], Xinyu He[1,2], and Tao Xu[1,2(✉)]

[1] College of Computer Science and Technology, Civil Aviation University of China,
Tianjin, China
{chliu,2021051010,2019052045,txu}@cauc.edu.cn
[2] Key Laboratory of Smart Airport Theory and System, CAAC, 2898 Jinbei Road,
Dongli, Tianjin 300300, China

Abstract. Utilizing multiple modal information to understand video semantics is quite natural when humans watch a video and describe its contents with natural language. In this paper, a hierarchical multimodal attention network that promotes the information interactions of visual-textual and visual-visual is proposed for video captioning, which is composed of two attention modules to learn multimodal visual representations in a hierarchical manner. Specifically, visual-textual attention modules are designed for achieving the alignment of the semantic textual guidance and global-local visual representations, thereby leading to a comprehensive understanding of the video-language correspondence. Moreover, the joint modeling of diverse visual representations is learned by the visual-visual attention modules, which can generate compact and powerful video representations to the caption model. Extensive experiments on two public benchmark datasets demonstrate that our approach is pretty competitive with the state-of-the-art methods.

Keywords: Video captioning · Hierarchical multimodal attention · Semantic textual guidance · Diverse visual representations

1 Introduction

Video captioning is a typical cross-domain task that involves research in both computer vision and natural language processing, which plays an important role in various practical applications, such as video retrieval, assisting visually impaired people and human-robot interaction [7,19]. It is necessary not only to understand the main content of a video intuitively but also to automatically generate meaningful sentences.

This work was supported by the Scientific Research Project of Tianjin Educational Committee under Grant 2021KJ037 and Fundamental Research Funds for the Central Universities under Grant 3122021052.

Video is naturally multimodal, and diverse modalities such as appearance, motion, object features differently contribute to the syntax components. Different from image captioning which mainly uses the unimodal attention mechanism and describes a static image with little visual content, video captioning is more challenging due to the complicated temporal structure and the multiple modal features that need to be carefully exploited. Early advancements of video captioning methods can be mainly attributed to the exploration of the attention mechanism [10, 19], which exploits the temporal structure of videos to dynamically weight the visual representations.

However, these methods mainly focus on global-temporal features and neglect local-spatial region features, lacking fine-grained visual modalities to generate detailed captions. Recently, STG-KD [9] and ORG-TRL [21] utilize GCNs to model object interactions on spatial and temporal relations. But they only utilize the visual representations and leave the textual representations unexploited, causing the semantic gap between video representations and linguistic captions.

In this work, we propose a hierarchical multimodal attention network based on semantically textual guidance to tackle the issues mentioned above. Our model aims to learn three genres of video representations at two hierarchical levels. We term this method as HMAN, which contains two-hierarchy multimodal attention modules as shown in Fig. 1: (I) Visual-Textual Attention Level, which learns the semantical alignment between textual representations and global-local visual representations. (II) Visual-Visual Attention Level, which fuses three genres of visual-semantic representations to generate context-related visual embedding. At last, we use the Long Short-Term Memory (LSTM) as our language model to decode the visual representations to captions.

The main contributions of this paper are summarized as follows:

1) We propose a novel hierarchical multimodal attention network (HMAN) to learn multimodal video representations by two hierarchical attention levels: Visual-Textual Attention Level and Visual-Visual Attention Level. We mine the peculiarities and complementarities among multimodal visual representations at two levels to boost the video captioning performance.
2) Semantic and Contextual text guidance is introduced to align the different visual modalities like appearance, motion, and object feature separately, narrowing the gap between video representations and linguistic captions.
3) Our method significantly outperforms the state-of-the-art methods on the MSVD dataset and is quite competitive on the MSR-VTT dataset.

2   Related Work

Current approaches for video captioning mainly adopt the encoder-decoder architecture and most of them focus on designing complex encoders to learn better video representations, which could be categorized into two types of work: (I)Utilizing multiple visual features. (II) Utilizing multiple modal features.

Utilizing Multiple Visual Features. Diverse visual features play an important role in generating accurate captions, which are briefly divided into global-temporal representations and local-spatial representations. For global representations, Yao et al. [19] design a soft attention mechanism to dynamically weight the frame representations and model the global temporal structure of videos. Pan et al. [10] exploit video temporal structure in a longer range by reducing the length of input information flow. For local representations, a large number of researchers [6,9,21] explore the local attention mechanism to fully utilize the semantic object features and explicitly model the object interactions. Chen et al. [5] introduce spatial attention effectively to video frames under the guidance of motion information. Aafaq et al. [1] develop a visual encoding method, which utilizes object attributes to enhance the semantics of visual representations. However, the above approaches focus on utilizing either the global temporal sequence features or the local spatial region features respectively, without considering the joint modeling of these different visual modalities.

Utilizing Multiple Modal Features. As captioning involves both visual and textual modalities, learning the relationship between visual features and textual features can promote a comprehensive understanding of video contents. You et al. [20] use visual attributes to guide the image captioning based on semantic attention. Pei et al. [12] propose a multimodal memory network to enhance the relevant dependency between visual and textual modalities. Ryu et al. [14] use a semantic grouping network to capture the most discriminating word phrases. However, they model visual-semantic relevance simply considering the visual representations as unimodality, which may inter-pollute the different visual modalities to generate incorrect captions.

Motivated by these observations, we construct a hierarchical multimodal attention network to model the relevance between semantic context and diverse video representations separately and then learn the generated visual-semantic representations complementarily. In this way, the language model is able to learn the better visual groundings of high-level semantic context.

3   Method

The framework of the proposed HMAN approach is illustrated in Fig. 1, which can be divided into three stages. At the visual encoding stage, the encoder extracts coarse-grained global representations and fine-grained local representations. And the text encoder encodes the words of the partially decoded caption to generate textual representations for high-level semantics. At the hierarchical multimodal attention stage, global semantic attention (GSA) and local semantic attention (LSA) are firstly introduced to complete the alignment between the visual representations and the semantic context on global and local perspectives. Then these aligned visual-semantic representations are incorporated by the semantic fusion attention to produce the context-related visual results. At the decoding stage, we decode the semantic visual results into words.

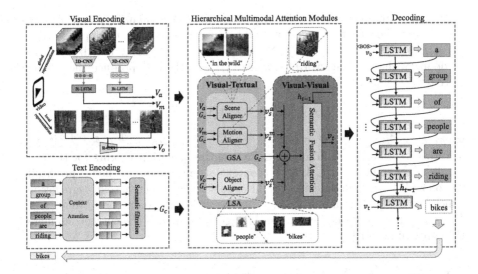

Fig. 1. The framework of our hierarchical multimodal attention network.

3.1   Encoder

Visual Encoder. Given a sequence of video frames with length T, we firstly extract global temporal representations including appearance representations $V_a = \{v_t^a\}_{t=1}^T$ and motion representations $V_m = \{v_t^m\}_{t=1}^T$ from 2D-CNN and 3D-CNN, respectively. Then we feed these two kinds of representations to Bi-LSTM separately, which can process sequence data. In addition, local semantic features play an indispensable role in generating correct descriptions, i.e. frequent objects in video usually serve as the cornerstone of captions. To mine the complementarities among global and local information, object representations $V_o = \{v_t^o\}_{t=1}^T$ are extracted by R-CNNs on each frame, where v_t^o ($v_t^o = \{v_t^i\}_{i=1}^N$) is a sequence of object features in the t-th frame and N denotes the number of region features in each frame. Thus, we construct three genres of visual modalities like appearance, motion, and object features at the visual encoding stage, which contribute differently to the generation of words.

Text Encoder. Potential contextual relationships in textual representations play an important role in understanding specific video content. As the example of "a group of people are riding bikes in the wild", when we need to come up with the object of an action like "bikes", we always want to seek the related video contents of motion or object summarized by the contextual text representations "people" and "riding". Thus, we apply a text encoder to learn the semantically textual context, subsequently guiding the visual-semantic alignment.

As the self-attention has a great faculty for modeling the intra-dependencies in a sentence, we construct a Context Attention to effectively learn the attentive latent semantics, and then a semantic filter is adopted for masking the redundant components and selecting discriminative semantics within the processed text. For time t, we first build a word representation matrix $W_t = [E(y_1)E(y_2)...E(y_{t-1})]^T \in \mathbb{R}^{(t-1) \times d_w}$ ($E[\cdot]$ denotes embedding matrix), which is made up of the previous word sequence $\{y_i\}_{i=1}^{t-1}$. Then we produce the semantically textual matrix U_t by:

$$U_t, A_t = \mathbf{CA}(W_t) \tag{1}$$

where $U_t = [u_{1,t}...u_{t-1,t}]^T \in \mathbb{R}^{(t-1) \times d_w}$, $A_t = [\alpha_{1,t}...\alpha_{t-1,t}]^T \in \mathbb{R}^{(t-1) \times (t-1)}$ is a word attention matrix and $\alpha_{j,t} \in \mathbb{R}^{t-1}$ is the attention weights for the $t-1$ words to construct $u_{j,t}$, $\mathbf{CA}$ is the context attention we designed. Specifically, we follow [14] and build a semantic filter to select distinctive text aggregations, which leaves the discriminative ones and discards redundant ones. We get $Gc = \{g_c^t\}_{t=1}^n$ to represent our text semantic aggregations, n is the current number of the captioning words.

3.2   Hierarchical Multimodal Attention

Visual-Textual Attention Module. This module is designed to learn the semantic alignment between visual and textual representations. As we know, the sentence usually consists of diverse visually-related syntax components, which might find corresponding visual cues in video contents. In this module, we design two components to model global and local visual-textual alignment respectively. As shown in Fig. 1, appearance features often represent the scene information, and motion features represent the object motion information, which may correlate to the linguistic counterpart "in the wild" and "riding" separately in this captioning example. To this end, a novel global semantic attention(GSA) is proposed to achieve the alignment of two coarse-grained temporal representations. In detail, we take the guidance context G_c as the query to guide the corresponding generation of the global visual semantic aggregations $v_s^a = \{\hat{v}_t^a\}_{t=1}^n$ and $v_s^m = \{\hat{v}_t^m\}_{t=1}^n$, respectively. For the t-th guidance context g_t^c, we compute text-related scene semantic aggregations $\hat{v}_t^a$ by dynamically summarizing temporal appearance features:

$$\hat{v}_t^a = \sum_{i=1}^T \alpha_{t,i} v_i^a \tag{2}$$

$$\alpha_{t,i} = \exp(e_{t,i}) / \sum_{j=1}^T \exp(e_{t,j}) \tag{3}$$

$$e_{t,i} = w_a^T \tanh(W_a^T g_t^c + U_a^T v_i^a + b_a) \tag{4}$$

where $\alpha_{t,i}$ is the weight of i-th appearance features v_i^a regarding t-th guidance context; W_a, w_a, U_a and b_a are learnable parameters. $\hat{v}_t^m$ is computed in a similar way, so we only take $\hat{v}_t^a$ as an example here.

Furthermore, local-spatial region features contain meaningful object information which usually serves as the cornerstone of a video caption. Therefore, our local semantic attention (LSA) tends to model latent object visual aggregations by fine-grained alignment. Finally, $\hat{v}_t^o$ is obtained via attentively summarizing total object proposals $\{v_i^o\}_{i=1}^{T \times N}$ in all frames:

$$\hat{v}_t^o = \sum_{i=1}^{T \times N} \beta_{t,i} v_i^o \tag{5}$$

$$\beta_{t,i} = \exp(e_{t,i}) / \sum_{j=1}^{T \times N} \exp(e_{t,j}) \tag{6}$$

$$e_{t,i} = w_o^T \tanh(W_o^T g_t^c + U_o^T v_i^o + b_o) \tag{7}$$

where $\beta_{t,i}$ is the weight of i-th object features v_i^o regarding t-th guidance context; W_o, w_o, U_o and b_o are learnable parameters.

Visual-Visual Attention Module. In this module, we design a Semantic Fusion Attention to jointly model the relevance among the global-local multimodal visual semantic aggregations that bridge visual content and the entire linguistic caption. Since most captions comprise the information of the scene, objects, and object motion, our Visual-Visual Attention Module takes the three text-related visual aggregations as input. To enhance the visual fusion, we also concatenate the t-th guidance context g_c^t with the three visual aggregations.

$$\hat{v}_t^f = [\hat{v}_t^a; \hat{v}_t^m; \hat{v}_t^o; g_c^t] \tag{8}$$

where $\hat{v}_t^f$ is the enhanced visual fusion and $[\cdot; \cdot]$ denotes concatenation. For time t, the final visual representations v_t are obtained by summarizing the fused visual representations $\hat{v}_t^f$ according to the history hidden state h_{t-1}^{lang} via:

$$v_t = \sum_{i=1}^{n} \gamma_{t,i} \hat{v}_{t,i}^f \tag{9}$$

$$\gamma_{t,i} = \exp(e_{t,i}) / \sum_{j=1}^{n} \exp(e_{t,j}) \tag{10}$$

$$e_{t,i} = w^T \tanh(W^T h_{t-1}^{lang} + U^T \hat{v}_{t,i}^f + b) \tag{11}$$

where $\gamma_{t,i}$ is the weight of i-th fused features $\hat{v}_{t,i}^f$ at the t-th decoding step; W, w, U and b are learnable parameters.

3.3 Decoder

We employ an LSTM network as our language generation decoder to produce accurate captions for each timestep.

$$h_t^{lang} = \textbf{deLSTM}([v_t; E(y_{t-1})], h_{t-1}^{lang}) \tag{12}$$

where $E(y_{t-1}) \in \mathbb{R}^{d_w}$ is the word embedding of y_{t-1} and d_w is the embedding size. Further, the probability distribution P_t is computed by a softmax operation:

$$P_t = softmax(S^T[v_t; h_t^{lang}; E(y_{t-1})]) \tag{13}$$

where S denotes the learnable parameters.

Loss Function. Given a video with the ground-truth words $\{y_t^*\}$, where $t \in [1, N]$ and N is the caption length. We compute the Cross-Entropy loss to optimize our captioning model:

$$\mathcal{L}_{ce} = -\sum_{t=1}^{N} log(P_t\theta(y_t^*)^T) \tag{14}$$

where θ is the one-hot encoding of words.

4   Experiments

4.1   Datasets and Experimental Details

We evaluate our proposed HMAN model on two public datasets: MSVD [4] and MSR-VTT [18], with four evaluation metrics: BLEU-4 [11], METEOR [2], CIDEr [17] and ROUGE-L [8]. MSVD is a widely used video captioning dataset composed of 1970 YouTube short video clips with an average video length of 10.2 s. Each video clip is annotated with around 40 English captions. Similar to existing methods [6,9,21,22], we take 1200 video clips for training, 100 clips for validation, and 670 clips for testing. MSR-VTT is a large-scale dataset that consists of 10k video clips with an average video length of 14.8 s from 20 categories, and each video clip is annotated with around 20 English expressions. For a fair comparison, we follow the standard splits that 6,513 clips for training, 497 clips for validation and 2,990 clips for testing.

For feature extraction, we first uniformly sample 26 equally-spaced frames for each video, then we extract appearance features using InceptionResNetV2 [15] and motion features using the I3D [3] respectively. In addition, we adopt Faster-RCNN [13] to extract 36 region features for each frame out of the 26 sampled frames. For corpus preprocessing, we first remove punctuations and convert all words into lowercase. Then captions with less than 26 words are padded with zero vectors and captions with more than 26 words are truncated. Each word is embedded to a randomly initialized 300D embedding vector and jointly learned with the whole model. The size of hidden states for language LSTM is set to 512, and a dropout layer with ratio 0.5 is applied after each LSTM cell. We adopt the Adam optimizer with an initial learning rate of 1e-4 to optimize our model. The training batch size is set to 16 for both datasets, and the training epochs is set to 20. During inference, we use beam search with size 5 to generate captions.

Table 1. Comparisons with the state-of-the-art methods on the MSVD dataset in terms of BLEU-4, METEOR, ROUGE-L and CIDEr scores.

Models	Year	BLEU-4	METEOR	ROUGE-L	CIDEr
PickNet [7]	2018	52.3	33.3	69.6	76.5
MARN [12]	2019	48.6	35.1	71.9	92.2
STG-KD [9]	2020	52.2	36.9	73.9	93.0
SAAT [22]	2020	46.5	33.5	69.4	81.0
ORG-TRL [21]	2020	54.3	36.4	73.9	95.2
SGN [14]	2021	52.8	35.5	72.9	94.3
RMN [16]	2021	54.6	36.5	73.4	94.4
MGRMP [6]	2021	55.8	36.9	74.5	98.5
HMAN(ours)	2022	**57.7**	**38.3**	**74.6**	**102.5**

4.2 Quantitative Evaluations

We compare our method with current state-of-the-art methods on MSVD and MSR-VTT datasets. The results illustrate that our method achieves the best performance on MSVD in terms of all four metrics(See Table 1). In particular, our HMAN achieves 38.3 of METEOR (with +1.4 improvement) and 102.5 of CIDEr (with +4.0 improvement). Note that CIDEr is more consistent with human judgment than other metrics [17]. The large margin improvement under CIDEr demonstrates that our proposed method can generate more accurate descriptions than the compared methods.

Our proposed model has great improvement over traditional encoder-decoder based models including PickNet [7] and MARN [12], which means the fine-grained object information plays a vital role in video captioning task. Note that STG-KD [9] and ORG-TRL [21] focus on designing more complex encoders to learn better video representations, and ignore the textual semantic information. SAAT [22] and RMN [16] associate the Part-of-Speech(POS) tags with visual information to generate local words. SGN [14] utilizes the partially decoded caption to group video frames with discriminating word phrases but leaves the multimodal visual representations unexploited.

However, our model outperforms them by a significant margin ($+7.3 \sim 26.0$ of CIDEr on MSVD) according to the visual-textual and visual-visual hierarchical joint modeling, which indicates the advantage of mining the peculiarities and complementarities among multimodal visual representations for video captioning.

Table 2 shows the comparison results on the MSR-VTT dataset. As we can see, our method achieves competitive performance compared with the state-of-the-art models. However, the scores of our method out of four metrics are not all the best as on MSVD, this is probably because MSR-VTT has a larger scale in terms of both length of captions and number of videos, which makes it difficult to simply explore the relation of different modal representations for effective

Table 2. Comparisons with the state-of-the-art methods on the MSR-VTT dataset in terms of BLEU-4, METEOR, ROUGE-L and CIDEr scores. The Red- and the Blue-colored numbers denote the best and the second best results across all approaches, respectively.

Models	Year	BLEU-4	METEOR	ROUGE-L	CIDEr
PickNet [7]	2018	41.3	27.7	59.8	44.1
MARN [12]	2019	40.4	28.1	60.7	47.1
STG-KD [9]	2020	40.5	28.3	60.9	47.1
SAAT [22]	2020	40.5	28.2	60.9	49.1
ORG-TRL [21]	2020	43.6	28.8	62.1	50.9
SGN [14]	2021	40.8	28.3	60.8	49.5
RMN [16]	2021	42.5	28.4	61.6	49.6
MGRMP [6]	2021	41.7	28.9	62.1	51.4
HMAN(ours)	2022	43.3	29.1	61.6	51.0

video captioning. Note that, ORG-TRL [21] achieves higher scores under BLEU-4 and ROUGE-L on MSR-VTT, and these are probably caused by the reason that they apply a teacher-recommended learning (TRL) method to guide the caption model to learn abundant linguistic knowledge. MGRMP [6] has better performance under ROUGE-L and CIDEr which might be due to its ability to capture the relations between fine-grained region features across frames.

It is noteworthy that our HMAN is simple but more effective than most above state-of-the-art methods. Although we merely focus on modeling the mutual semantic concepts between visual and textual representations and leave the sophisticated interactions among objects unexploited, our method still acquires quite competitive results and achieves almost state-of-the-art performance.

4.3   Ablation Study

To verify the effectiveness of our proposed HMAN method, we conduct ablation studies on the MSVD dataset as shown in Table 3. Baseline(#0): the model only utilizes global-modal representations including appearance and motion representations to generate captions directly. Ours w/Gc(#1): the model only employs global semantic attention(GSA). It takes the guidance context G_c as the query to guide the generation of the global visual semantic aggregations. Ours w/Gc, Obj(#2): the model regards the concatenation of appearance, motion, and object representations as a simple visual modality and aligns it with guidance context G_c that ignores the hierarchical structure. Full model(#3): our proposed complete hierarchical multimodal attention model.

Comparing the results of #0 and #1 variant, we observe a slight performance increase, which demonstrates the benefits of modeling the alignment for visual-semantic information. The #2 variant adds the fine-grained local region

Table 3. Ablation study on the MSVD dataset in terms of BLEU-4, METEOR, ROUGEL and CIDEr scores.

#	Models	BLEU-4	METEOR	ROUGE-L	CIDEr
0	baseline	53.9	36.3	73.1	92.3
1	ours w/G_c	55.7	37.1	74.0	92.9
2	ours w/G_c, Obj	55.6	37.6	74.0	96.9
3	full model	**57.7**	**38.3**	**74.6**	**102.5**

representations and has a significant improvement, which validates the effectiveness of learning multiple visual representations. We also further investigate the specific design of our hierarchical attention structure. Comparing the results of #2 and #3 variant, a noticeable improvement demonstrates that the hierarchical joint modeling of visual-textual and visual-visual improves video captioning performance, which is mainly due to the effectiveness of separate visual-semantic alignment and united visual-visual enhancement successively.

4.4 Qualitative Results

GT: a group of elephants are walking
Baseline: a group of people are walking
Ours: a group of elephants are walking

GT: a man is doing motorcycle tricks
Baseline: a man is riding a motorcycle
Ours: a man is doing tricks on a motorcycle

GT: a woman is mixing ingredients
Baseline: a man cooking his kichen
Ours: a woman is mixing ingredients
in a bowl

GT: a train is moving on track
Baseline: a train is running
Ours: a train is moving on a track

Fig. 2. Visualization examples for qualitative comparisons between HMAN and the baseline model (better viewed in color).

We show some qualitative examples in Fig. 2. As can be seen, the captions generated by our method contain more accurate and richer content than the baseline model(#0 variant in Table 3). For instance, the example at the top-left and bottom-right shows our method can identify the correct objects and motions which belong to different visual modalities. Similarly, in the examples at the top-right, the core video content "doing tricks" produced by our model is more detailed rather than the rough caption "riding a motorcycle". Note that the baseline model depicts the video clip at the bottom-left in Fig. 2 in a wrong manner, and this is mainly because it's difficult for the baseline model to deal with complex scene switching problems. In contrast, our method depicts the video with an accurate and rich caption "a woman is mixing ingredients in a bowl" which demonstrates that our model can select the key scene and recognize the principal activity to a certain extent.

Overall, these visualization examples demonstrate our proposed HMAN is capable of generating more comprehensive, more accurate, and richer captions. Meanwhile, the above observations validate the introduced hierarchical multimodal attention network can contribute to the quality of descriptions.

5 Conclusion

In this paper, we proposed a hierarchical multimodal attention network based on semantically textual guidance for video captioning, which promotes the information interactions of visual-textual and visual-visual. By implementing the alignment from the semantically textual representations to visual representations of different modalities: scene, motion, and object, diverse visual semantic aggregations can be obtained for understanding video effectively. Furthermore, semantic fusion attention is adopted to incorporate visual aggregations and generate context-related visual results. Extensive experiments demonstrate that our proposed HMAN achieves competitive results with the state-of-the-art models on MSVD and MSR-VTT datasets.

References

1. Aafaq, N., Akhtar, N., Liu, W., Gilani, S.Z., Mian, A.: Spatio-temporal dynamics and semantic attribute enriched visual encoding for video captioning. In: Proceedings of the IEEE/CVF Conference on Computer Vision and Pattern Recognition, pp. 12487–12496 (2019)
2. Banerjee, S., Lavie, A.: Meteor: an automatic metric for MT evaluation with improved correlation with human judgments. In: Proceedings of the ACL workshop on intrinsic and extrinsic evaluation measures for machine translation and/or summarization, pp. 65–72 (2005)
3. Carreira, J., Zisserman, A.: Quo vadis, action recognition? a new model and the kinetics dataset. In: proceedings of the IEEE Conference on Computer Vision and Pattern Recognition, pp. 6299–6308 (2017)
4. Chen, D., Dolan, W.B.: Collecting highly parallel data for paraphrase evaluation. In: Proceedings of the 49th annual meeting of the association for computational linguistics: human language technologies, pp. 190–200 (2011)
5. Chen, S., Jiang, Y.G.: Motion guided spatial attention for video captioning. In: Proceedings of the AAAI Conference on Artificial Intelligence, pp. 8191–8198 (2019)

6. Chen, S., Jiang, Y.G.: Motion guided region message passing for video captioning. In: Proceedings of the IEEE/CVF International Conference on Computer Vision, pp. 1543–1552 (2021)
7. Chen, Y., Wang, S., Zhang, W., Huang, Q.: Less is more: picking informative frames for video captioning. In: Proceedings of the European conference on computer vision (ECCV), pp. 358–373 (2018)
8. Lin, C.Y.: Rouge: a package for automatic evaluation of summaries. In: Text summarization branches out, pp. 74–81 (2004)
9. Pan, B., et al.: Spatio-temporal graph for video captioning with knowledge distillation. In: Proceedings of the IEEE/CVF Conference on Computer Vision and Pattern Recognition, pp. 10870–10879 (2020)
10. Pan, P., Xu, Z., Yang, Y., Wu, F., Zhuang, Y.: Hierarchical recurrent neural encoder for video representation with application to captioning. In: Proceedings of the IEEE Conference on Computer Vision and Pattern Recognition, pp. 1029–1038 (2016)
11. Papineni, K., Roukos, S., Ward, T., Zhu, W.J.: Bleu: a method for automatic evaluation of machine translation. In: Proceedings of the 40th Annual Meeting of the Association for Computational Linguistics, pp. 311–318 (2002)
12. Pei, W., Zhang, J., Wang, X., Ke, L., Shen, X., Tai, Y.W.: Memory-attended recurrent network for video captioning. In: Proceedings of the IEEE/CVF Conference on Computer Vision and Pattern Recognition, pp. 8347–8356 (2019)
13. Ren, S., He, K., Girshick, R.B., Sun, J.: Faster R-CNN: towards real-time object detection with region proposal networks. IEEE Trans. Pattern Anal. Mach. Intell. 39(6), 1137–1149 (2017)
14. Ryu, H., Kang, S., Kang, H., Yoo, C.D.: Semantic grouping network for video captioning. In: Proceedings of the AAAI Conference on Artificial Intelligence, pp. 2514–2522 (2021)
15. Szegedy, C., Ioffe, S., Vanhoucke, V., Alemi, A.A.: Inception-v4, inception-resnet and the impact of residual connections on learning. In: 31st AAAI Conference on Artificial Intelligence (2017)
16. Tan, G., Liu, D., Wang, M., Zha, Z.J.: Learning to discretely compose reasoning module networks for video captioning. In: Proceedings of the 29th International Joint Conference on Artificial Intelligence, pp. 745–752 (2020)
17. Vedantam, R., Lawrence Zitnick, C., Parikh, D.: Cider: consensus-based image description evaluation. In: Proceedings of the IEEE Conference on Computer Vision and Pattern Recognition, pp. 4566–4575 (2015)
18. Xu, J., Mei, T., Yao, T., Rui, Y.: Msr-vtt: a large video description dataset for bridging video and language. In: Proceedings of the IEEE Conference on Computer Vision and Pattern Recognition, pp. 5288–5296 (2016)
19. Yao, L., et al.: Describing videos by exploiting temporal structure. In: Proceedings of the IEEE International Conference on Computer Vision, pp. 4507–4515 (2015)
20. You, Q., Jin, H., Wang, Z., Fang, C., Luo, J.: Image captioning with semantic attention. In: Proceedings of the IEEE Conference on Computer Vision and Pattern Recognition, pp. 4651–4659 (2016)
21. Zhang, Z., et al.: Object relational graph with teacher-recommended learning for video captioning. In: Proceedings of the IEEE/CVF Conference on Computer Vision and Pattern Recognition, pp. 13278–13288 (2020)
22. Zheng, Q., Wang, C., Tao, D.: Syntax-aware action targeting for video captioning. In: Proceedings of the IEEE/CVF Conference on Computer Vision and Pattern Recognition, pp. 13096–13105 (2020)

Autism Spectrum Disorder Classification of Facial Images Using Xception Model and Transfer Learning with Image Augmentation

Trapti Shrivastava$^{(\boxtimes)}$ ⓘ, Vrijendra Singh ⓘ, and Anupam Agrawal ⓘ

Department of Information Technology, Indian Institute of Information Technology, Allahabad 211015, India
shri.taps02@gmail.com, {vrij,anupam}@iiita.ac.in

Abstract. Autism spectrum disorder (ASD) is a neurodevelopmental disorder in which the neurology of an autistic person is severely hampered. A child with autism has difficulty answering his name, avoids eye contact, and cannot express his emotions. Early diagnosis can assist children with ASD enhance their intellectual abilities while reducing autistic symptoms. In computer vision, determining developmental disorder problems from facial image data is a significant but largely unexplored challenge. This paper proposed a method to classify autistic and non-autistic facial images using model 1 (Xception) and model 2 (Augmentation + Xception). Among Model 1 and Model 2, Model 2 achieved higher accuracy of 98% and a minimum loss of 0.08.

Keywords: Autism Spectrum Disorder · Facial Image · Image Augmentation · Transfer learning · Xception Model

1 Introduction

Autism is primarily a hereditary disorder that causes difficulties with social skills, repetitive behaviors, speech, and non-verbal communication; however, some environmental variables play a role. According to a study conducted by Dr. Narendra Arora [1] in 2018 based on different regions of India, one in every 100 people under the age of ten has autism, and one out of every eight children has at least one neurodevelopmental problem. A child as young as 18 months old might be diagnosed with ASD. Although ASD is considered a hereditary disorder, it is primarily diagnosed based on behavioral characteristics.

Perceptions of parents and pediatricians are essential in the early detection of ASD. If children with ASD are not addressed early on, they will have low academic performance, hyperactivity, and violence. Early diagnosis and treatment are essential steps in normalizing a child's psycho-social attitude. There is no

Financially Supported by IIIT Allahabad.

M. Tanveer et al. (Eds.): ICONIP 2022, LNCS 13625, pp. 170–181, 2023.
https://doi.org/10.1007/978-3-031-30111-7_15

medical test that can determine whether someone has autism. There is no medical test that can determine whether someone has autism. Due to the enormous number of treatment centers and therapists available in India, more than 30 lakh children require therapists regularly, which is not practicable due to the lack of therapists. As a result, parental therapy becomes needed. In this paper, we have classified facial image data as having 3014 images of Autistic and non-autistic children. In our proposed work, in model 1, we used the Xception with ImageNet weights, and in model 2, we used augmentation with Xception with ImageNet weights and then retrained both models. The remaining sections of this paper are written as follows: The literature survey is concluded in Sect. 2. The data-set and proposed methodology is described in Sect. 3. The experimental results are discussed in Sect. 4. The research report comes to a conclusion with Sect. 5.

2   Literature Review

Various methods are currently used to diagnose and classify ASD according to child behavior, such as questionnaires [13–16], eye gaze [22–24], functional magnetic resonance imaging (fMRI) [17–21, 34], stereotyped behavior [25, 26] and multiple modalities in a single study [27–29]. Most of these studies aim to identify and classify ASD and Typical developed children with different modalities of data. Some research has also been conducted that focuses on the classification of Autism using brain imaging techniques.

Shukla et al. [2] worked on Deep Convolution Neural Networks (DCNN) for feature extraction and Support Vector Machine (SVM) for the classification of different spectral disorders on their own collected facial image data-set and achieve the accuracy of 98.80%. Li et al. [3] introduced a combined facial expression and action unit model to classify facial image data using CNN. They combined the two publicly available data sets AffectNet and EmotionNet and achieved the F1 score of 76%. Rabbi et al. [4] worked on autism detection using various AI algorithms such as MLP, RF, GBM, AB and CNN on the Kaggle facial image data set. They achieved 92.31% highest accuracy with the CNN model. Arumugam et al. [5] used the CNN-based model to predict ASD in the Kaggle facial image dataset. They used an 80:20 ratio of training and test split to train their model and achieve 91% with a maximum loss of 0.53. A. Lu and M. Perkowsk [7] proposed an ASD screening solution using two Kaggle facial image data set and East Asian Dataset collected by them. They used the VGG16 based transfer learning model and get 95% using the Kaggle dataset as a training and the East Asian dataset as a test dataset. Khosla et al. [8] worked on classification of facial image dataset using ImageNet weight and re-trained mobileNet, InceptionV3, and InceptionResNetV2. They claimed that from all three models, MobileNet has the highest accuracy of 87%. Li et al. [9] introduce multi-modal the framework, which consists of facial model recognition and speech model recognition. They used their own collected data, having 20-hour therapy video recordings, and worked on spectrograms to differentiate between non-negative and positive videos with an accuracy of 72.40%. Akter T. et al. [10] introduce

MobileNet-V1 with k-mean clustering to classify facial image of the ASD and non-ASD Kaggle dataset and achieve an accuracy of 92.10%. They claimed that the proposed model has 1.67% higher accuracy than normal MobileNet-V1 model. Ahmed et al. [11] performed experiments using deep learning models such as flask framework, MobileNet, Xception and InceptionV3 on the Kaggle facial image dataset. They achieved the highest accuracy of 95% with the MobileNet model. Similarly, Alsaade et al. [12] worked on the same deep learning model with different weights, and the Xception model achieved 91% accuracy, higher than VGG19 (80%) and NASNETMobile (78%). Kalaiselvi et al. [35] used various CNN based transfer learning models such as VGG16, ResNet50, InceptionV3, and NASNetLarge. They claimed that NASNetlarge gives highest accuracy of 87.05% with loss of 0.372 by using 88,999,818 trainable parameters.

3   Material and Methods

Autistic children have different facial features, and the distance between various facial landmarks point varies among typically developed children. The study by Praveen and Lakshmi briefs that autistic children have an abnormal facial characteristic ratio, such as a comprehensive set of eyes, a broad upper face, and a shorter middle region [30]. Aldridge, K. et al. [33] conducted a medical study to compare the facial features of autistic and non-autistic people. They used the euclidean distance matrix analysis (EDMA) concept to distinguish between the facial features of autistic and non-autistic people and concluded that autistic people have a broader upper face (including the forehead area), a shorter middle face region (eyes, cheeks, and nose area), and a wider mouth and phitrum area as compared to non-autistic people.

3.1   Dataset Description

In this study, we have used Kaggle dataset [31] to analyze the facial image of autistic and non-autistic children. The dataset consisted of a total of 6028 facial images and was equally divided into two classes, i.e. Autistic and Non-Autistic in Fig. 1 shows some images from Autistic and Non-autistic children images. Most photos in this dataset were obtained from websites, and social media sites. Due to images collected from different sources, Autistic and Non-autistic images have different sizes. To make them the same size, we resize all the images and remove duplicates. To make all the facial image alignment, image size, and obstacle removal, such as background noise, we write a Python script for data preprocessing. After removing all non-facial images, resizing images into $300 \times 300 \times 3$ and facial alignment, we get a total of 3014 facial images divided into autistic and non-autistic classes. The further data set is divided into three parts, namely Training-set, Testing-set, and Validation-set in the ratio of (88:10:2) having an equal number of facial images of both the classes. Table 1 shows the distribution of the split dataset into all three parts. Consequently, the primary Kaggle facial dataset was analyzed and validated.

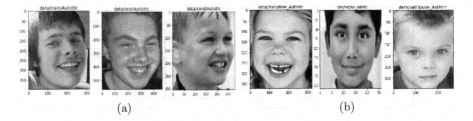

(a)   (b)

Fig. 1. (a) Autistic children images (b) Non-Autistic children images [31]

Table 1. Dataset distribution

Dataset Split	Autistic	Non-Autistic
train	1327	1327
test	140	140
validation	40	40

3.2 Methods

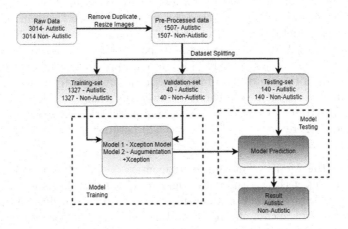

Fig. 2. Work Flow

The proposed approach aimed to classify autism images by evaluating children's facial patterns using basic Xception and Xception with augmentation transfer-learning-based. Figure 2 shows the Schematic workflow of our proposed transfer-learning-based framework. The two proposed methods: Model 1- Xception model and model 2- augmentation+Xception model and analyze the trained models in terms of accuracy, loss, and confusion matrix.

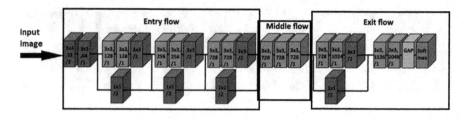

Fig. 3. Xception Model architecture. [32]

Model 1: In model 1 we are using basic Xception model. There are three segments in the Xception model: *entry, middle, and exit*, as illustrated in Fig. 3. In the Xception model, every segment has feedback connections. In each convolution layer, maximum pooling was applied. Deep learning concept of linear stack used over convolution layers with depth-wise separation. The original input image is $299 \times 299 \times 3$ but in our model, we passed the input image size of $300 \times 300 \times 3$. The data passing started from the entry flow, with $19 \times 19 \times 728$ feature maps being output and worked as input for the middle flow segment. Before traveling through the exit flow, the data passed from the input flow to the middle flow. This loop continued eight times and maintained the feature map size. The exit flow's output has 2048 features for a standard-sized input picture. The fully connected layer and the logistic regressor are two optional layers in this model. The Xception model has a total of 21,132,714 parameters, of which 58,800 are non-trainable and 21,073,834 are trainable parameters. As shown in Fig. 4 (a), the input size image is $300 \times 300 \times 3$ because all preprocessed facial images are the same size. In this model, we introduce an input layer to resonate with the Xception model's input layer. There are two truediv layers. Model 1 has used MobileNet pre-trained weights, and Xception model training will be set as false. In this model, there are a total of 20,863,529 parameters, with 2,049 trainable and 20,861,480 non-trainable parameters.

Model 2: In this Model we introduce augmentation concept as a sequential layer as shown in Fig. 4 (b). In the data augmentation layer, we used random flips with factor $= 0.1$. To apply random flip, we use horizontal flip, which creates mirror images. In the case of a horizontal flip, the pixel at position (p, q) in the newly generated image will be at coordinate $(image\_width - p - 1, q)$. Horizontal flip with random factor $= 0.1$ creates augmented output rotating images by a random amount with the horizontal flip in the range $[-10\% \times 1pi, 10\% \times 1pi]$ for each input image. Figure 5 shows some resultant augmented images generated by this layer for a particular image. In this model, we have used MobileNet pre-trained weights as well and set Xception model training as true. There are a total of 20,863,529 parameters, with 20,809,001 trainable and 2049 non-trainable parameters.

Model: "model"

Layer (type)	Output Shape	Param #
input_2 (InputLayer)	[(None, 300, 300, 3)]	0
tf.math.truediv (TFOpLambda)	(None, 300, 300, 3)	0
tf.math.subtract (TFOpLambd a)	(None, 300, 300, 3)	0
xception (Functional)	(None, 10, 10, 2048)	20861480
global_average_pooling2d (G lobalAveragePooling2D)	(None, 2048)	0
dropout (Dropout)	(None, 2048)	0
dense (Dense)	(None, 1)	2049

Total params: 20,863,529
Trainable params: 2,049
Non-trainable params: 20,861,480

(a)

Model: "model"

Layer (type)	Output Shape	Param #
input_2 (InputLayer)	[(None, 300, 300, 3)]	0
data_augmentation (Sequenti al)	(None, 300, 300, 3)	0
tf.math.truediv (TFOpLambda)	(None, 300, 300, 3)	0
tf.math.subtract (TFOpLambd a)	(None, 300, 300, 3)	0
xception (Functional)	(None, 10, 10, 2048)	20861480
global_average_pooling2d (G lobalAveragePooling2D)	(None, 2048)	0
dropout (Dropout)	(None, 2048)	0
dense (Dense)	(None, 1)	2049

Total params: 20,863,529
Trainable params: 20,809,001
Non-trainable params: 54,528

(b)

Fig. 4. (a) Model 1 architecture (b) Model 2 architecture images [31]

Fig. 5. Example of generated augmented images.

4   Results and Discussion

We trained both the models with different epoch values, for example, 30, 50, 60 and 100. After 60, we get fewer fluctuations in performance parameters of models with batch size equal to 8. We used accuracy and loss performance parameters to analyze the model's performance.

4.1   Accuracy

For binary labels, Binary Accuracy determines the percentage of anticipated values (yPred) that match actual values (yTrue). Because the label is binary, the Autistic label represents 0, the non-autistic label represents 1, and yPred is equal to the probability value of the predictions. The expression of calculating the binary accuracy is as following:

$$accuracy = sum(yPred == yTrue)/len(yPred) \tag{1}$$

Based on the calculated value of accuracy, analyzed in each epoch, specific rules are used:

1. If the possibility exceeds the threshold, a value of 1 is assigned. Otherwise, a value of 0 is assigned.
2. It's recognized as accurate if the assigned value is the same as the actual value.

Figure 6 (a)–(b) shows the training and validation accuracy analysis in 100 epochs for model 1 and model 2 respectively. We can analyze the validation and training accuracy of model 1 has 0.15% at max peek of epoch and validation and training accuracy of model 2 has 0.015% at the maximum peek of the epoch. After applying augmentation with the Xception model difference between validation and training accuracy is decreased.

4.2   Loss

We have used the binary cross-entropy function to calculate training and validation loss. When we fit the model to classify the images, it will assign a probability of being autistic. It should provide high scores for wrong predictions and low scores for correct predictions. The expression of calculating the binary cross entropy loss function is as following:

$$L_p(q) = -\frac{1}{N} \sum_{i=1}^{N} x^i \cdot log(p(x_i)) + (1 - x_i) \cdot log(1 - p(x_i)) \tag{2}$$

where x is the label of classes, i.e., 1 for autistic and 0 for non-autistic, and p(x) is the probability of input being autistic for all N images. Figure 6 (c) shows the training and validation loss analysis for model 1 in 100 epochs. Figure 6 (d) shows the training and validation loss analysis for model 2 in 100 epochs.

4.3   Confusion Matrix

The Confusion matrix is used to find the actual and predicted value of classes. The confusion matrix of model 1 depicts that 129 autistic images predicted as autistic, 135 non-autistic images predicted as non-autistic as shown in Fig. 7 (a). The confusion matrix of model 2 depicts that 137 autistic images predicted as autistic, 138 non-autistic images predicted as non-autistic as shown in Fig. 7 (b). In model 1 at epoch = 86 it gives max validation accuracy = 0.9375, training

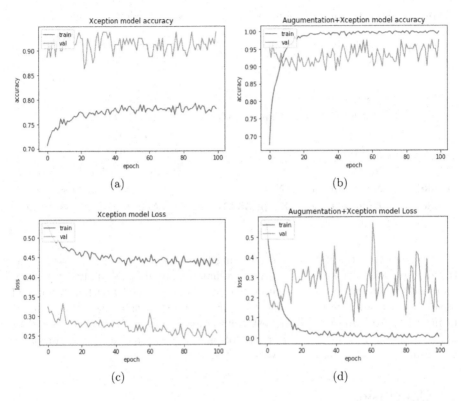

Fig. 6. Results analysis of both Models: (a) and (b) Accuracy analysis of Model 1 and Model 2. (c) and (d) Loss analysis of Model 1 and Model 2 over training and validation set

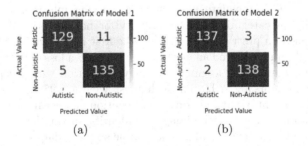

Fig. 7. Confusion Matrix (a) Model 1 (b) Model 2

accuracy = 0.7814 with training loss = 0.43 and validation loss = 0.25. After combining the augmentation and Xception model in model 2 achieved training accuracy = 0.99, validation accuracy = 0.975 with training loss = 0.0042 and validation loss = 0.1506 at 100 epoch.

Table 2. Result Comparison of proposed methods with previously published work

References	Methodologies	Performance Matrix
[4]	CNN	Acc = 92.31%
[5]	CNN	Acc = 91%, Loss = 0.53
[8]	MobileNet	Acc = 87%
[10]	MobileNet-V1+K-Mean clustering	Acc = 92.10%
[11]	MobileNet	Acc = 95%
[12]	Xception	Acc = 91%
[35]	NASNetLarge	Acc = 87.05%, Loss = 0.372
Model 1	**Xception+ImageNet**	**Acc = 94 %, Loss = 0.24**
Model 2	**Augmentation+Xception+ImageNet**	**Acc = 98 %, Loss = 0.08**

After training the models and checking the accuracy (Acc) and loss for the training and validation set, we further evaluate our model and analyze the accuracy and loss on the test set. Model 1 gives an accuracy of 0.94 with a loss of 0.24 in the test set, and Model 2 provides accuracy of 0.98 with a loss of 0.08. Hence in model 2, we achieved the highest accuracy and minimum loss compared to model 1. Table 2 shows the best result of our proposed methods and compares the result with other published work on the same dataset.

5   Conclusion

This study proposed a transfer learning-based classification method for autism spectrum disorder using an image augmentation approach. The Xception model architecture provides adequate models for extracting information from facial images, which can categorize faces as autistic or non-autistic. We implement two models based on pretrained Xception. The first model was re-trained using ImageNet weight and achieved an accuracy of 94% and a loss of 0.24. The second model was re-trained using augmentation layer addition with Xception model oven ImageNet weights and achieved the highest accuracy, 98%, with a minimum loss of 0.08. As a result, an augmentation + Xception model can be used to create a web application that uses facial images to identify autism. This model can be improved in the future by expanding the sample size and data set obtained from psychologists' diagnoses of autistic children of various ages and can be used in various other medical domain problem.

Acknowledgment. The authors are grateful to the Ministry of Education, Govt. of India and Indian Institute of Information Technology, Allahabad for providing all resources and financial support to complete this work.

References

1. Arora, N.K., et al.: Neurodevelopmental disorders in children aged 2–9 years: population-based burden estimates across five regions in India. PLoS Med. **15**(7), e1002615 (2018)
2. Shukla, P., Gupta, T., Saini, A., Singh, P., Balasubramanian, R.: A deep learning frame-work for recognizing developmental disorders. In: 2017 IEEE Winter Conference on Applications of Computer Vision (WACV), pp. 705–714. IEEE (2017)
3. Li, B., et al.: A facial affect analysis system for autism spectrum disorder. In: 2019 IEEE International Conference on Image Processing (ICIP), pp. 4549–4553. IEEE (2019)
4. Rabbi, M.F., Hasan, S.M., Champa, A.I., Zaman, M.A.: A convolutional neural network model for early-stage detection of autism spectrum disorder. In: 2021 International Conference on Information and Communication Technology for Sustainable Development (ICICT4SD), pp. 110–114. IEEE (2021)
5. Arumugam, S.R., Karuppasamy, S.G., Gowr, S., Manoj, O., Kalaivani, K.: A deep convolutional neural network based detection system for autism spectrum disorder in facial images. In: 2021 5th International Conference on I-SMAC (IoT in Social, Mobile, Analytics and Cloud)(I-SMAC), pp. 1255–1259. IEEE (2021)
6. Mujeeb Rahman, K.K., Subashini, M.M.: Identification of autism in children using static facial features and deep neural networks. Brain Sci. **12**(1), 94 (2022)
7. Lu, A., Perkowski, M.: Deep learning approach for screening autism spectrum disorder in children with facial images and analysis of ethnoracial factors in model development and application. Brain Sci. **11**(11), 1446 (2021)
8. Khosla, Y., Ramachandra, P., Chaitra, N.: Detection of autistic individuals using facial images and deep learning. In: 2021 IEEE International Conference on Computation System and Information Technology for Sustainable Solutions (CSITSS), pp. 1–5. IEEE (2021)
9. Li, J., Bhat, A., Barmaki, R.: A two-stage multi-modal affect analysis framework for children with autism spectrum disorder. arXiv preprint arXiv:2106.09199 (2021)
10. Akter, T., et al.: Improved transfer-learning-based facial recognition framework to detect autistic children at an early stage. Brain Sci. **11**(6), 734 (2021)
11. Ahmed, Z.A.T., et al.: Facial features detection system to identify children with autism spectrum disorder: deep learning models. Comput. Math. Methods Med. **2022** (2022)
12. Alsaade, F.W., Alzahrani, M.S.: Classification and detection of autism spectrum disorder based on deep learning algorithms. Comput. Intell. Neurosci. **2022** (2022)
13. Akter, T., et al.: Machine learning-based models for early stage detection of autism spectrum disorders. IEEE Access **7**, 166509–166527 (2019)
14. Kumar, C.J., Das, P.R.: The diagnosis of ASD using multiple machine learning techniques. Int. J. Dev. Disabil. **68**(6), 1–11 (2021)
15. Mohan, P., Paramasivam, I.: Feature reduction using SVM-RFE technique to detect autism spectrum disorder. Evol. Intell. **14**(2), 989–997 (2020). https://doi.org/10.1007/s12065-020-00498-2
16. Mujeeb Rahman, K.K., Monica Subashini, M.: A deep neural network-based model for screening autism spectrum disorder using the quantitative checklist for autism in toddlers (QCHAT). J. Autism Dev. Disord. **52**, 1–15 (2021). https://doi.org/10.1007/s10803-021-05141-2
17. Eslami, T., Saeed, F.: Auto-ASD-network: a technique based on deep learning and support vector machines for diagnosing autism spectrum disorder using fMRI

data. In: Proceedings of the 10th ACM International Conference on Bioinformatics, Computational Biology and Health Informatics, pp. 646–651 (2019)

18. Li, H., Parikh, N.A., He, L.: A novel transfer learning approach to enhance deep neural network classification of brain functional connectomes. Front. Neurosci. **12**, 491 (2018)

19. Crimi, A., Dodero, L., Murino, V., Sona, D.: Case-control discrimination through effective brain connectivity. In: 2017 IEEE 14th International Symposium on Biomedical Imaging (ISBI 2017), pp. 970–973 (2017)

20. Zheng, W., et al.: Multi-feature based network revealing the structural abnormalities in autism spectrum disorder. IEEE Trans. Affect. Comput. **12**(3), 732–742 (2019)

21. Almuqhim, F., Saeed, F.: ASD-SAENet: a sparse autoencoder, and deep-neural network model for detecting autism spectrum disorder (ASD) using fMRI data. Frontiers Comput. Neurosci. **15**, 27 (2021)

22. Chen, S., Zhao, Q.: Attention-based autism spectrum disorder screening with privileged modality. In: Proceedings of the IEEE/CVF International Conference on Computer Vision, pp. 1181–1190 (2019)

23. Kollias, K.F., Syriopoulou-Delli, C.K., Sarigiannidis, P., Fragulis, G.F.: The contribution of machine learning and eye-tracking technology in autism spectrum disorder research: a review study. In: 2021 10th International Conference on Modern Circuits and Systems Technologies (MOCAST), pp. 1–4 (2021)

24. Chong, E., et al.: Detecting gaze towards eyes in natural social interactions and its use in child assessment. In: Proceedings of the ACM on Interactive, Mobile, Wearable and Ubiquitous Technologies, vol. 1, no. 3, pp. 1–20. ACM New York, NY, USA (2017)

25. Hashemi, J., et al.: A scalable app for measuring autism risk behaviors in young children: a technical validity and feasibility study. In: Proceedings of the 5th EAI International Conference on Wireless Mobile Communication and Healthcare, pp. 23–27 (2015)

26. Bovery, M., et al.: A scalable off-the-shelf framework for measuring patterns of attention in young children and its application in autism spectrum disorder. IEEE Trans. Affect. Comput. **12**(3), 722–731 (2019)

27. Wang, Z., Xu, K., Liu, H.: Screening early children with autism spectrum disorder via expressing needs with index finger pointing. In: Proceedings of the 13th International Conference on Distributed Smart Cameras, pp. 1–6 (2019)

28. Mehmood, F., Ayaz, Y., Ali, S., Amadeu, R.D.C., Sadia, H.: Dominance in visual space of ASD children using multi-robot joint attention integrated distributed imitation system. IEEE Access **7**, 168815–168827 (2019)

29. Marinoiu, E., Zanfir, M., Olaru, V., Sminchisescu, C.: 3d human sensing, action and emotion recognition in robot assisted therapy of children with autism. In: Proceedings of the IEEE Conference on Computer Vision and Pattern Recognition, pp. 2158–2167 (2018)

30. Lakshmi Praveena, T., Muthu Lakshmi, N.V.: A methodology for detecting ASD from facial images efficiently using artificial neural networks. In: Jyothi, S., Mamatha, D.M., Satapathy, S.C., Raju, K.S., Favorskaya, M.N. (eds.) CBE 2019. LAIS, vol. 15, pp. 365–373. Springer, Cham (2020). https://doi.org/10.1007/978-3-030-46939-9_31

31. Senol, C.: Autism image data (2020). Accepted 02 June 2022. https://www.kaggle.com/datasets/cihan063/autism-image-data

32. Chollet, F.: Xception: deep learning with depth wise separable convolutions. In: Proceedings of the IEEE Conference on Computer Vision and Pattern Recognition, pp. 1251–1258 (2017)
33. Aldridge, K., et al.: Facial phenotypes in subgroups of prepubertal boys with autism spectrum disorders are correlated with clinical phenotypes. Mol. Autism **2**(1), 1–12 (2011)
34. Kumar, S., Sharma, S.: A hybrid deep model with concatenating framework of convolutional neural networks for identification of autism spectrum disorder. In: Enabling Technology for Neurodevelopmental Disorders, pp. 230–239 (2022)
35. Kalaiselvi, A., Nagarathinam, S., Paul, T.D., Alagumeenaakshi, M.: Detection of autism spectrum disorder using transfer learning. Turkish J. Physiotherapy Rehabil. **32**(2), 926–933 (2021)

A Comprehensive Vision-Based Model for Commercial Truck Driver Fatigue Detection

Zhen Gao[1], Xiaowen Chen[1], Jingning Xu[1], Rongjie Yu[2], and Jiaqi Zong[1(✉)]

[1] School of Software Engineering, Tongji University, Shanghai 201804, China
zjq1010@tongji.edu.cn
[2] Key Laboratory of Road and Traffic Engineering of the Ministry of Education, Tongji University, Shanghai 201804, China

Abstract. Fatigue driving is a primary reason for traffic accidents for commercial truck drivers. Using new technology to detect fatigue in advance is very important to improve road safety. Most of the existing research on fatigue detection is based on the facial key-points of drivers, which are prone to inaccurate detection and thus affect the accuracy of fatigue judgment in complex driving environments. This paper first proposes an end-to-end fatigue detection model based on the CNN-LSTM network structure. The model does not detect facial key-points directly but extracts the driver's fatigue-related high-level features in full video image frames through a CNN network. Then, the LSTM network is used to calculate the fatigue probability of the driver on the time-series fatigue features of the multi-frame images. To alleviate the impact of complex driving environments, an attention mechanism for CNN is introduced to help the network to focus on the region of interest. To further eliminate background disturbance, this paper proposes a comprehensive fatigue detection method (CFDM) combined with facial recognition. The results of the ablation experiments show that the attention mechanism and face recognition can effectively improve the accuracy of the fatigue detection model, increasing the model's AUC by 6.7% and 3.8%, respectively. And even when the face is occluded or the driver wears a mask, the comprehensive model we propose can still correctly detect the fatigue state, and the AUC of the model reaches 0.92, while the existing fatigue detection algorithms based on facial key-points will basically fail.

Keywords: Fatigue detection · CNN · LSTM · Attention mechanism

1 Introduction

In the last few decades, the world has seen a boom in infrastructure development, with almost all cities connected by road, enabling the transfer of goods from one place to another. Road transport, more economical and convenient than air transport, is rapidly developing, with the number of commercial trucks and

M. Tanveer et al. (Eds.): ICONIP 2022, LNCS 13625, pp. 182–193, 2023.
https://doi.org/10.1007/978-3-031-30111-7_16

truck drivers growing quickly. However, commercial drivers mostly drive for long periods during the transportation of goods, which can easily lead to fatigue and even traffic accidents. According to the National Highway Traffic Safety Administration [1], more than 100,000 crashes were caused by fatigue driving, and in the United States alone, about 1,500 people were killed each year [2]. Of these, 54% of fatigue driving accidents are caused by commercial truck drivers. The results of a survey conducted by the Chinese authorities on a sample of commercial truck drivers show that 84% of freight vehicle drivers have an average daily driving time of more than 8 h, with 40% of them having more than 12 h. 64% of freight vehicles have only one driver, which has a huge potential for traffic accidents. Drivers driving in a fatigued state are very dangerous. Therefore, it is essential to detect and classify the fatigue state of commercial truck drivers along the way and make corresponding warnings when fatigue is detected to protect transportation safety.

At present, there are three main methods of fatigue detection research, namely, to study the driver's physiological signals by wearing electronic sampling equipment, to study the driver's behavior by observing the vehicle state, and to study the driver's fatigue-related features through image recognition. Among them, the third method stands out due to its advantages of non-intrusiveness and anti-interference ability. The study by Charlotte et al. [3] showed that among the three fatigue assessment methods, the one based on fatigue-related features performs best.

Facial feature-based methods usually identify the driver's behaviors such as yawning, closing eyes, and nodding during driving and calculate the percentage of eye closure per unit time (PERCLOS) [4] or the degree of mouth closure [5] to assist in judging the driver's fatigue state. Jo et al. [6] used methods such as AdaBoost to extract the driver's head posture and eye state to determine whether the driver fatigued. Alioua et al. [5] adopted a support vector machine (SVM) based on face features extraction and a new method for mouth detection using Circular Hough Transform (CHT) to judge the driver's fatigue by analyzing whether the driver yawns or not. In recent years, with the development of deep learning, image recognition technology has become more and more mature, and facial key-point recognition, CNN network, and DBN network have also been applied to facial feature extraction and fatigue assessment. Dwivedi [7] et al. used CNN to extract various latent fatigue-related features and capture the interaction between complex non-linear features, and then judge whether the driver is fatigued using a softmax layer. Han et al. [8] collected 36 drivers' PERCLOS in a simulated driving environment and established an LSTM model to detect fatigue. They obtained high accuracy and found it to be positively correlated with driving time and fatigue level.

However, despite the advanced fatigue-related features extraction technique, it is still difficult to accurately judge the driver's fatigue state, especially when processing video data [9]. At the same time, it should be noted that most of the existing research is carried out in a simulated experimental environment. Even if it is applied to a real driving environment, the illumination brightness, camera position, and angle will be carefully designed, so that the driver's face can be clearly seen in collected videos, and then the extraction of fatigue-related features is relatively easy. However, to detect the fatigue state of a real driver in a real driving environment, the video quality is often disturbed by factors such as external lighting conditions, image background, camera position, and angle, etc., causing the video to have problems such as color disturbance, loss of clarity, too large or too small faces, and incomplete faces or faces covered by masks. As a result, the model detection accuracy decreases, which can be very common in commercial truck driving.

To solve these problems and make our model applicable to commercial truck driving in the real world, we propose an end-to-end Comprehensive Fatigue Detection Method (CFDM) combined with facial recognition. Major contributions of this study can be summarized as (1) designing a data-driven fatigue extraction method based on an end-to-end model with a CNN-LSTM network structure instead of the traditional rule-based method; (2) introducing an attention mechanism in CNN layers and utilizing MTCNN to extract the driver's face image to eliminate the environmental disturbance; and (3) proposing a CFDM framework to combine the advantages of the two improvements and to deal with face detection failure in complex environments.

The experiment results show that even in complex driving environments, the CFDM can still effectively detect fatigue states, and its performance is greatly improved compared with existing methods.

The rest of the paper is organized as follows. The preliminaries are described in Sect. 2. Section 3 introduces our research ideas and the structure of the three proposed models. A detailed description of the dataset and experiments and the performance evaluation results are given in Sect. 4. Conclusions are presented in Sect. 5.

2   Preliminaries

Our proposed end-to-end model mainly uses CNN and LSTM to build the network and MTCNN to recognize the faces of commercial truck drivers. To evaluate the model, we use some indices such as AUC, accuracy, precision, recall, and F1-Score. This section mainly introduces the design of network structure and the formula for calculating indices.

2.1   Convolutional Neural Network

Convolutional Neural Network (CNN) is a feedforward artificial neural network widely used in image recognition, including classic network structures such as AlexNet, VGG16, GoogleNet, and ResNet. It can help us identify, localize and extract features of target images quickly and efficiently. In this paper, we input the driver's face image to CNN, that is, a frame of size $3 \times 256 \times 256$, so that the CNN outputs an array of length 128 as the extracted fatigue-related features. The activation function of the network uses ReLU, and the pooling layer uses the maximum function.

2.2 Long Short-Term Memory

Long Short-Term Memory (LSTM) [10], an evolutionary version of Recurrent Neural Network (RNN), can carry out long and short-term memory and effectively solve the problem of gradient disappearance and gradient explosion in long sequence training. LSTM introduces a gate mechanism to control the flow and loss of features, including forget gate, input gate, and output gate, usually using the sigmoid function as the activation function. If we use x_t as the input and y_t as the output at time t, and use C_t as a parameter of long-term memory that changes slowly with time while h_t as a parameter of short-term memory that changes rapidly with time (C_t and h_t will be passed to the next operation). Here we list the formula of LSTM at each stage.

- Forget stage: Selectively forget the input C^{t-1} from the last node, controlled by the forget gate f_t.

$$f_t = \sigma \left(W_f \cdot [h_{t-1}, x_t] + b_f \right) \tag{1}$$

- Selective memory stage: Selectively memorise the input x_t and h_{t-1}, controlled by the input gate i_t.

$$i_t = \sigma \left(W_i \cdot [h_{t-1}, x_t] + b_i \right) \tag{2}$$

- Output stage: The predicted value y_t comes from the linear change of h_t at the output of the implicit node, controlled by the output gate o_t.

$$\widetilde{C_t} = tanh \left(W_C \cdot [h_{t-1}, x_t] + b_C \right) \tag{3}$$

$$C_t = f_t * C_{t-1} + i_t * \widetilde{C_t} \tag{4}$$

$$o_t = \sigma \left(W_o [h_{t-1}, x_t] \right) + b_o \tag{5}$$

$$h_t = o_t * tanh \left(C_t \right) \tag{6}$$

In this paper, after each frame of image is processed by CNN in the previous section, the corresponding fatigue-related features are obtained. We convert these features into time series data according to the time sequence and put them into the LSTM, which can finally indicate whether the driver is in a state of fatigue.

2.3 Multi-Task Cascaded Convolutional Network

Multi-Task Cascaded Convolutional Network (MTCNN) [11] is a multi-task cascaded CNN model that combines face detection and alignment simultaneously with high accuracy and wide application. MTCNN mainly consists of three tasks, namely face classification, bounding box regression, and face landmark localization. In the face classification section, the learning objective of MTCNN is a 2-class classification problem. And for each example x_i in the training process, its cross-entropy loss is:

$$L_i^{det} = -(y_i^{det}log(p_i) + (1 - y_i^{det})(1 - log(p_i))) \tag{7}$$

p_i is the probability output by the network, indicating whether the sample x_i is a human's face, and the notation $y_i^{det} \in \{0,1\}$ is the ground truth label. In the bounding box regression section, we use Euclidean Distance as the loss function for the distance metric:

$$L_i^{box} = \left|\left|\hat{y}_i^{box} - y_i^{box}\right|\right|_2^2 \tag{8}$$

Face landmark localization is also regarded as a regression problem, and the loss function is as follows.

$$L_i^{landmark} = \left|\left|\hat{y}_i^{landmark} - y_i^{landmark}\right|\right|_2^2 \tag{9}$$

In this paper, we use MTCNN to detect and localize the driver's face position for facial feature extraction.

2.4   Performance Indices

In this paper, we use accuracy, recall, precision, F1-score to evaluate the quality of the model. These indices are derived from the confusion matrix and the classification threshold of the model is calculated by Youden Index [12], which is formulated as follows.

$$Y = S_e(n) + S_p(n) - 1 \tag{10}$$

n is the set of all points in the ROC curve. The ROC curve is drawn with the true positive rate (TPR) as the y-axis and the false positive rate (FPR) as the x-axis. Calculate the sensitivity (S_e) and specificity (S_p) of each point in the ROC curve and calculate the corresponding Youden index of each point, and select the threshold corresponding to the largest Youden index as the classification threshold of the model.

In addition, AUC is also used as an evaluation metric for our fatigue detection model. AUC (the Area Under the ROC Curve) is the area enclosed by the ROC Curve and the coordinate axis. The value of AUC is generally between 0.5 and 1, indicating the authenticity of the detection method. The closer it is to 1, the better the performance of the model.

3   Proposed Approach

In this section, the methodology of the CFDM is covered. First, an end-to-end model based on the CNN-LSTM network structure is built for fatigue detection. To eliminate the environmental disturbance, the attention mechanism CBAM [13] is introduced into the CNN layers of the model, and MTCNN is used to extract the driver's face image as the region of interest. Since sometimes MTCNN fails in complex driving environments, Kalman Filter is used for supplementary smoothing in the case of face recognition failure in a single frame. If failures happen on multiple continuous frames, the fatigue state will be inferred using the end-to-end model with whole images as input instead of detected face regions.

3.1   End-to-End Model Based on CNN and LSTM

Most of the existing fatigue detection methods localize the driver's facial landmarks, especially the key points of the driver's eyes and mouth, and then estimate the driver's fatigue state according to certain rules. These methods rely
heavily on the accuracy of key-point extraction and thus tend to fail when the
face is partially occluded or the driver is in vigorous motion. Therefore, We
propose an end-to-end CNN-LSTM network to extract richer features, such as
frowns, facial expressions, head or body movement, etc. The model extracts the
fatigue-related features from a single-frame image through the CNN network and
then calculates the probability that the driver is in a fatigued state through the
LSTM network, based on the time-series features from multiple-frame images. If
the probability exceeds a certain threshold, the driver is identified as fatigued.
This method avoids extracting facial key-points and instead extracts more latent
features from the network intelligently.

The detailed structure of the end-to-end model is shown in Fig. 1. The demonstrated input images here are from the public dataset YawDD [14].

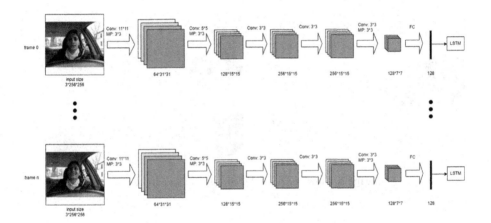

Fig. 1. The structure of the basic end-to-end model based on CNN and LSTM.

3.2   Enhancement with CBAM

To mitigate the impact of complex driving environments, such as different interior decorations or driver's clothing, on fatigue detection, we introduce an attention mechanism, CBAM, into CNN. This approach effectively makes the model
pay more attention to the changes in the driver's face area, thereby improving
the performance of the model. We name this the enhanced CNN-LSTM model.

3.3   Comprehensive Model with Face Detection

Despite introducing CBAM into the model, some samples still suffer from environmental disturbances. To further address this issue, we propose using MTCNN to detect facial regions in advance, and then feed the face regions instead of the whole image as input to the enhanced CNN-LSTM model. The new model can be called facial CNN-LSTM Model. Since the background is removed, it is not necessary to add CBAM to all CNN layers, and adding it in the first and second CNN layers of the facial CNN-LSTM model is enough. For the instability of MTCNN, which may lead to the failure of individual frame in face detection, Kalman Filter is used to supplement and smooth the detected frames.

However, this model still has defects. In a few cases, MTCNN fails in the continuous frames due to lighting, improper camera positions, incomplete faces and other complex driving environments. To solve this problem, we integrate enhanced end-to-end CNN-LSTM in 3.2 and facial CNN-LSTM in 3.3. When a large number of frames in the video fail in face detection, the enhanced end-to-end CNN-LSTM model replaces the facial CNN-LSTM model for fatigue detection, which is called a comprehensive model shown in Fig. 2.

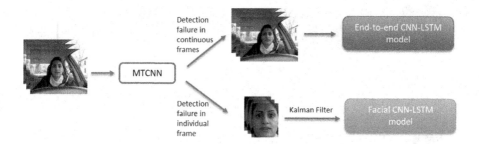

Fig. 2. Flow chart of the comprehensive model

4   Experiments

4.1   Data Preparation

In this paper, the data came from the on-board vision systems of a large delivery company. The data covered a large number of drivers, with a total of 720. The distribution of drivers' age is shown in Fig. 3. Among them, 90% of drivers were between 35–50 years old, with an average age of 42.2. The data also covered different driving environments, including different lighting conditions during day and night, and varying degrees of shaking or shifting of cameras depending on different road types. For example, country roads are bumpier than highways, and the camera shakes more frequently.

A total of 2721 driver videos were collected, including 1739 non-fatigue videos and 982 fatigue videos. The videos' length was 10 s and the resolution was

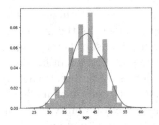

Fig. 3. Driver age distribution histogram.

352×288. The drivers were manually identified as non-fatigue or fatigue by 3 experts. In the experiments, the ratio of the number of videos in training set, validation set and testing set was 6:1:3. The data also contained many complex environmental problems, such as the improper camera installation, which resulted in only half of the driver's face being recorded in the camera, or the driver's face being obscured due to wearing hats, sunglasses and masks, etc. All these issues can greatly affect the accuracy of fatigue detection.

The public dataset YawDD [14] was also used in the experiment to compare with existing methods. We used a total of 311 videos from the forward facing vehicle camera for training and validation. The sample ratio between the training set and the validation set was 2:1. There are 47 male drivers and 43 female drivers in the data and some of them wear near-sighted glasses or sunglasses.

4.2 Performance Measurement

In this section, we compare the performance of the above mentioned three models, all trained on the same training set and tested on the same test set. In the experiments, the models were implemented with PyTorch, and were trained and tested on a workstation equipped with NVIDIA Tesla K40c GPU and Intel Core I7 processor.

The performance indices of the model on test set are shown in Table 1. Each index is calculated according to the confusion matrix, and the optimal classification threshold is provided by the Youden index [12].

Table 1. Performance indices for the proposed models.

Model	Accuracy	Recall	Precision	F1	AUC
Basic end-to-end CNN-LSTM	84.8%	57.1%	76.2%	0.65	0.83
Enhanced end-to-end CNN-LSTM	80.6%	**88.1%**	68.4%	0.77	0.89
Comprehensive Model	**83.5%**	70.3%	**85.0%**	**0.77**	**0.92**

Figure 4 shows the ROC curves of the three models. As can be seen, the comprehensive model outperforms the other two models. As shown in Table 1, the AUC of the enhanced model is 6.7% higher than that of basic end-to-end CNN-LSTM, indicating that the focus of the model has been improved after the introduction of the attention mechanism, thereby improving the fatigue detection ability of the model. After the introduction of facial recognition, the AUC of the model improves to 0.92, which is 3.8% higher than that of the enhanced end-to-end model, indicating that facial recognition effectively avoids environmental interference, and the fatigue detection ability of the model is further improved. For a 10 s video, the comprehensive model only takes about 0.27 s to complete an inference, which proves that our model has good real-time performance for fatigue detection.

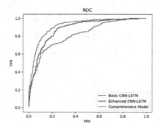

Fig. 4. ROC curves of the three models.

4.3   Improvement by CBAM

In this paper, in order to eliminate the influence of background on driver fatigue detection, we introduce the attention mechanism CBAM in CNN layers, which will make feature extraction in CNNs focus more on region of interests. To explore whether the model captures fatigue-related features as expected, we visually interpret the network using the Grad-CAM [15] method.

Grad-CAM uses gradient information flowing into the last convolutional layer of CNN to assign important values to each neuron for specific attention decisions. We can identify the focus of the CNN network during feature extraction based on the heat map generated by Grad-CAM. In the heat map, the redder the color of the region, the higher the CNN pays attention to the region. On the contrary, if the color of the region is close to blue, it indicates that CNN ignores this region.

As shown in Fig. 5, in the model without CBAM, the focus is shifted to the surrounding environment when extracting features. However, in the model with the CBAM, the focus of feature extraction is preferred around the driver's face region, which makes the feature extraction of the model more accurate.

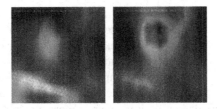

Fig. 5. Visualized by Grad-CAM, the comparison of the focus area of the model before (left) and after (right) CBAM was introduced .

4.4 Comparison with Other Methods

In this section, we compare the comprehensive model proposed in this paper with the state-of-the-art algorithms. Most of the existing fatigue detection algorithms are based on the extractions of facial key-points. The extracted features can be divided into two categories: (1) statistical features such as PERCLOS calculated from all key-points of eyes. [16]; (2) temporal sequence features based on key-points of each frame. [17]. Usually the facial key-points attraction algorithm such as Dlib is used to identify 68 or more key-points from the driver's face, and then the statistical or temporal feature is extracted and calculated based on the coordinates of the key-points. The first method is to use the coordinates of the eyes and mouth to judge the opening and closing of the driver's eyes and mouth at that time, and then evaluate the fatigue according to the rules. For example, if the eyes are closed for 3 s continuously, the driver is considered to be fatigued. And fatigue states can also be judged according to PERCLOS. In the second method, LSTM is believed to have excellent temporal information processing ability and can better process video data. So it is more reasonable to input the temporal key-points related features extracted from each single frame into LSTM for fatigue detection. The accuracy of the second algorithm is usually higher than that of the first type.

Trained and tested on the public dataset YawDD, the AUC of our proposed model reached 0.93, and the accuracy reached 92.23% which is better than the accuracy 88% of the existing study [17] with key-points extraction method.

In the complex natural commercial truck driving environment, the performance of the proposed comprehensive model is far superior to the existing key-points detecion-based algorithms, with the best AUC of 0.92. We refer to the Dlib algorithm used in the article [16], detect 68 facial key-points, and calculate 5 statistical features, namely PERCLOS, the number of yawns, the number of mouth openings, the longest eyes closing time, and the number of nods in 10 s. Then we use Random Forest to build a fatigue detection model on these features, with an AUC of only 0.67. At the same time, we refer to the method used in the literature [17], calculating left eye closure rate, right eye closure rate and mouth closure rate on each frame based on the key-points detected by Dlib. We convert them into time-series data and feed them into a bidirectional LSTM network for fatigue detection, the model's AUC reaches 0.71, which is better than the statistical features method, but far worse than our method.

The algorithms based on key-points extraction basically failed without seeing the driver's eyes or mouth, but our model can still continue to run and focus on valid regions, such as the driver's mouth and elbow movements to accurately capture the yawning related features when the drivers' eyes can not be observed because of wearing sunglasses or improper camera position. Furthermore, our model can also smartly focus on the driver's eye area when the driver's mouth cannot be seen from the camera. As shown in Fig. 6, all of these can explain why our model is much superior than the existing key-points methods.

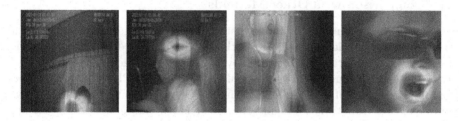

Fig. 6. The focus area of the proposed model.

5   Conclusion

In this paper, we proposed a comprehensive vision-based fatigue detection method CFDM to infer commercial truck drivers' fatigue state. Firstly, we designed an end-to-end model based on a CNN-LSTM network structure, which is proved to significantly outperform key-point based methods. Then attention mechanism was introduced to obtain enhanced performance by alleviating the impact of background in complex driving environments. To further eliminate the environmental disturbance, MTCNN was used to detect facial contours and the extracted face region was directly fed into the network to infer the driver's fatigue state. And finally, a CFDM framework was proposed to combine the advantages of the two improvements and to deal with face detection failure in complex environments. Experiments showed the proposed CFDM can still detect fatigue state correctly even in complex driving environments with much better performance compared to existing facial key-points based fatigue detection methods.

The performance of the proposed model could be further improved by data augmentation. Since the robustness of the model is very important when it is actually put into use, the adversarial training would be adopted to enhance the robustness of the model in the future.

Acknowledgements. This work was supported by the National Natural Science Foundation of China (No. 71771174) and Natural Science Foundation of Shanghai (No. 23ZR1465300 & No. 21ZR1465100).

References

1. National Highway Traffic Safety Administration. Drowsy Driving 2015 (Crash-Stats Brief Statistical Summary. Report No. DOT HS 812 446). Washington, DC (2017)
2. Alkinani, M.H., Khan, W.Z., Arshad, Q.: Detecting human driver inattentive and aggressive driving behavior using deep learning: recent advances, requirements and open challenges. IEEE Access **8**, 105008–105030 (2020)
3. de Naurois, C.J., Bourdin, C., Stratulat, A., et al.: Detection and prediction of driver drowsiness using artificial neural network models. Accid. Anal. Prev. **126**, 95–104 (2019)
4. Dinges, D.F., Grace, R.: Perclos: a valid psychophysiological measure of alertness as assessed by psychomotor vigilance. US Department of Transportation, Federal Highway Administration, Publication Number FHWA-MCRT-98-006 (1998)
5. Alioua, N., Amine, A., Rziza, M.: Driver's fatigue detection based on yawning extraction. Int. J. Veh. Technol. (2014)
6. Jo, J., Lee, S.J., Kim, J., et al.: Vision-based method for detecting driver drowsiness and distraction in driver monitoring system. Opt. Eng. **50**(12), 127202 (2011)
7. Dwivedi, K., Biswaranjan, K., Sethi, A.: Drowsy driver detection using representation learning. In: 2014 IEEE International Advance Computing Conference (IACC), pp. 995–999. IEEE (2014)
8. Han, H., Li, K., Li, Y.: Monitoring driving in a monotonous environment: classification and recognition of driving fatigue based on long short-term memory network. J. Adv. Transp. (2022)
9. Zhao, Y., Xie, K., Zou, Z., et al.: Intelligent recognition of fatigue and sleepiness based on InceptionV3-LSTM via multi-feature fusion. IEEE Access **8**, 144205–144217 (2020)
10. Hochreiter, S., Schmidhuber, J.: Long short-term memory. Neural Comput. **9**(8), 1735–1780 (1997)
11. Zhang, K., Zhang, Z., Li, Z., et al.: Joint face detection and alignment using multitask cascaded convolutional networks. IEEE Signal Process. Lett. **23**(10), 1499–1503 (2016)
12. Lan, Y., Zhou, D., Zhang, H., et al.: Development of early warning models. In: Early Warning for Infectious Disease Outbreak, pp. 35–74. Academic Press (2017)
13. Woo, S., Park, J., Lee, J.Y., et al.: Cbam: convolutional block attention module. In: Proceedings of the European Conference on Computer Vision (ECCV), pp. 3–19 (2018)
14. Abtahi, S., Omidyeganeh, M., Shirmohammadi, S., et al.: YawDD: a yawning detection dataset. In: Proceedings of the 5th ACM Multimedia Systems Conference, pp. 24–28 (2014)
15. Selvaraju, R.R., Cogswell, M., Das, A., et al.: Grad-cam: visual explanations from deep networks via gradient-based localization. In: Proceedings of the IEEE International Conference on Computer Vision, pp. 618–626 (2017)
16. Tibrewal, M., Srivastava, A., Kayalvizhi, R.: A deep learning approach to detect driver drowsiness. Int. J. Eng. Res. Technol. **10**, 183–189 (2021)
17. Chen, L., Xin, G., Liu, Y., et al.: Driver fatigue detection based on facial key points and LSTM. Secur. Commun. Netw. (2021)

Automatic Identification of Class Level Refactoring Using Abstract Syntax Tree and Embedding Technique

Rasmita Panigrahi[1]($\boxtimes$), Sanjay K. Kuanar[1], and Lov Kumar[2]

[1] School of Engineering and Technology, Department of CSE, GIET University, Gunupur 765022, Odisha, India
rasmita@giet.edu
[2] Department of Computer Engineering, National Institute of Technology, Kurukshetra 136119, Haryana, India
lovkumar@hyderabad.bits-pilani.ac.in

Abstract. Software refactoring helps improve the quality of the software and increases the execution speed. However, in large and complex systems, it is difficult for software developers to identify the exact code segments that need to be refactored. Many researchers predicted the refactoring model based on source code metrics. The metric value assumption and the tool used to extract metrics values differ from company to company. Our objective is to develop an automated refactoring model based on the code directly. The predictive power of refactoring model depends on the input features computed from the code description. In this work, we have implemented three different word embeddings intending to extract numeric vectors for the refactoring prediction model. Additionally, we have employed three data sampling techniques to solve the data imbalance issue and improve the effectiveness of the proposed model. Furthermore, we have instigated thirteen machine learning classifiers to access the predictive capability of word embedding techniques. Moreover, the Friedman test has been conducted to find the best technique based on rank of all the techniques. The experimental result on four data sets shows that our proposed model with TF-IDF word embedding and BLSMOTE data sampling, RF ensemble learner, improves its predictive capability.

Keywords: Machine Learning · SOA · Anti-patterns · Web Services

1 Introduction

Software quality is a multifaceted characteristic of every program that depends on the design, complexity, and a multitude of other factors. It is also inevitable that, given sufficient time, modifications and additions to the code base may erode its design integrity. Refactoring is the process of enhancing the internal

GIET University.

design of software without modifying its behavior. It can be accomplished using a variety of refactoring actions, such as renaming attributes, relocating classes, and dividing packages, among others. To restore the negative consequences of continuous development without starting from scratch every time the design deviates too far from its origins, refactoring code is required. Thus, the system's architecture can be modified without affecting its behavior or functionality. Source code metric is taken by several researchers [1,7] into machine learning framework for refactoring prediction and build metrics-based refactoring model. Author [2] has build metrics based as well as text based refactoring model based on commit message to identify the refactoring type. The author [] suggested a method (LANLAN) to classify Q&A forum postings into assistance requests and problem reports using word embeddings and machine learning. To derive sentence embeddings from individual words, author [3] calculated a weighted average of the word embeddings using the TF-IDF(term frequency-inverse document frequency scores of individual words as weight factors. The author [4] delves into the design and implementation of a system that initially employs some preprocessing techniques, as well as TF-IDF bag of words and word embeddings, which are classic Natural Language Processing techniques. These methods can then be utilized as inputs for multiple classifiers and Long Short Term Memory Neural Networks.

2 Related Work

This section describes the related work on refactoring by different source codes as an input for the machine learning framework. Refactoring is increasingly being used to improve the internal structure of code without changing its external behavior [5]. Previously many researchers [6–8] have worked on source code metrics as features for refactoring prediction. Their models are completely based on assumptions. Several authors have used different tools for retrieving source code metrics from the real data like ckjm [9], source meter [7,10], and Prest [11]. The tools retrieving the source code metrics may vary from company to company. So it is essential to develop an automated generalized model to give an acceptable result in industrial applications. Past research says refactoring prediction is one of the classification problems, which is solved by various data mining approaches to extract features from the textual description present in the abstract syntax tree (AST) code. In this study, we will not compute source code metrics rather, we have generated AST from each project's code. Then we collected text from the generated AST through checkstyle 9.0.1 tool. There are other tools also available for generating AST called as PSIMIner [12]. Further, we have applied tokenization to convert the text to the word level. We have used NLTK library, which has a tool called Word tokenizer that is used to break up text into words. Text is also cleaned, which means that all of the punctuation marks, special symbols, numbers, and extra spaces are taken out. Text is transformed into vector representation to feed into the neural network, which is known as word embedding. Specifically, word embedding is used for natural language processing (NLP). Word embedding is another common way to show the vocabulary of a document. It has been used in a number of recent studies [13] for defect prediction. It can figure out what a word means in the context of a document. Authors [13] have used the Word embedding technique for finding the defect severity value from the defect report, whereas the refactoring model with word embedding is unexplored yet. **Research Contributions:** In context to existing research, the work presented in this paper has several innovative contributions. As far as we know, ours is the first analysis of refactoring prediction on four projects using word embedding. As our data set is not uniformly distributed, we have applied SMOTE, BLSMOTE, and SVMSMOTE to each data set to make it balance. Finally, several classifiers, including eight machine learning classifiers (DT, SVML, SVMP, SVMR, GNB, MNB, BNB, LOGR), DL with five hidden layers (DL1, DL2, DL3, DL4, DL5), and Ensemble learners (BAG-KNN, BAG-DT, BAG-NB, RF, EXTR, ADBST, GRABST, MVE) using five-fold cross-validation approach have been applied to develop the refactoring prediction model. The performance of the proposed model is presented through AUC, accuracy parameter, and statistical description with a box-plot diagram. Based on our previous analysis, the followings are the issues to resolve in this paper:

RQ1: Is there any possibility to predict refactoring through word Embedding?
Previously several researchers have worked on refactoring prediction through software metrics. So our aim is to develop an automated refactoring model based on text vector, which is to be computed by word embedding techniques. We have applied three-word embedding techniques for retrieving n-dimensional text vectors to predict refactoring.

RQ2: Which data sampling technique achieves better performance to handle data imbalance issue?
Author Kadar et al. [14] validated the data set present in the tera-PROMISE repository, which is not evenly distributed. The data set is unbalanced, which complicates a model for candidate prediction for refactoring. We addressed the class imbalance issue through three data sampling techniques SMOTE, BLSMOTE, and SVMSMOTE. The primary aim is to find the best sampling technique for refactoring prediction through a descriptive statistical method and box-plot diagram.

RQ3: Which classifier performs best among all the machine learning classifiers, including ensemble and deep learning?
A large-scale study of how Machine Learning (ML)-based techniques and object-oriented source code metrics for refactoring prediction is already explored by different authors. Our aim is to use combine the machine learning classifiers, ensemble classifiers, and deep learning classifiers and then finally to choose the best classifier for refactoring prediction through word embedding.

3 Research Methodology

This section consists of a data set description and multiple phases included in our proposed model.

3.1 Data Set

The data set used in this study is publicly available at tera-PROMISE repository and used for different purposes like effort estimation, faults, refactoring, etc. In this study, We have utilized four open-source projects such as Antlr4, Junit, Mct, and Titan, from tera-PROMISE repository to validate our proposed model. This data set is manually verified by the author kadar et al. [14]. Table 1 contains four projects with refactored and non-refactored classes. Table 1 consists of the name of projects (project), number of classes(NOC), number of classes non-refactored (NONRC), number of refactored classes (NORC), refactored classes percentage (RC), non-refactored classes percentage (NRC). Table 1 shows that data set is not evenly distributed because of the %RC values 5.64%, 1.37%, 0.74%, 1.12% .

Table 1. Data set Description

project	NOC	NORC	NONRC	%RC	%NRC
Antlr4	408	23	385	5.64	94.36
Junit	655	9	646	1.37	98.63
Mct	2028	15	2013	0.74	99.26
Titan	1158	13	1145	1.12	98.88

3.2 Proposed Methodology

This part explains the proposed methodology used in this experiment to predict refactoring code using different variants of word-embedding based on AST. Figure 1 shows a refactoring prediction model proposed to predict refactoring. From Fig. 1, it is clearly observed that it consists of multiple steps, which are described below:-

- STEP 1: Initially, we have generated AST for each project described in Table 1 through checkstyle 9.0.1 tool. Checkstyle tool is used After the generation of AST, the tokenization process is applied to remove the irrelevant comments and punctuation marks. Then we have applied three-word embedding techniques (TFIDF, CBOW and SKG) to retrieve the text features as vectors in the form of n-dimensional numbers.
- After examining the data set considered, we found that there is a imbalance nature of data sets. So we have implemented three data sampling techniques such as SMOTE, BLSMOTE, and SVMSMOTE to handle the problem of data imbalance.

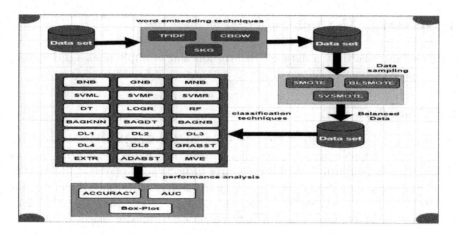

Fig. 1. The proposed model for refactoring prediction through word embedding

- After completion of data sets preprocessing, we have implemented three verities of classification techniques including four frequently used classifiers (Three variants of Naive Bayes, Decision tree, SVM with three different kernels and logistic regression), eight ensemble learning classifiers (Bagging with KNN, DT, NB and Random forest, Extract Tree, Adaboost, Gradiant boost and MVE) and one deep learning classifier with five hidden layers (DL1, DL2, DL3, DL4, DL5) for training the model for refactoring prediction. These models are validated using a 5-fold cross-validation approach.
- Finally, the performance of the model is measured through various parameter metrics such as AUC, accuracy and box-plot diagram. We have also implemented Friedman test, hypothesis testing and wilcoxon rank sum test to test the significant affect of applying different techniques. Through out our work we have used some abbreviations shown in Table 2.

Table 2. Acronyms and its corresponding name

Acronym	Corresponding Name	Acronym	Corresponding Name
TF-IDF	Term Frequency Inverse Document Frequency	DL1	Deep learning with 1 layer
CBOW	Continuous Bag of Words	DL2	Deep learning with 2 layers
SKG	Skip-gram	DL3	Deep learning with 3 layers
SMOTE	Synthetic Minority Over-sampling Technique	DL4	Deep learning with 4 layers
BLSMOTE	border line SMOTE	DL5	Deep learning with 5 layers
GNB	Gaussian Naive Bayes	RF	random forest
MNB	Multinomial Naive Bayes	EXTR	Extra tree
BNB	Bernoulli Naive Bayes	ADBST	Adaboost
LOGR	Logistic Regression	GRABST	Gadient Boost
SVML	Support vector machine with Linear kernel	SVMSMOTE	support vector machine SMOTE
SVMP	SVM with Polynomial kernel	BAG-DT	Bagging with Decision Tree
SVMR	SVM with radial basis Function kernel	BAG-KNN	Bagging with K- nearest Neighbour
BAG-NB	Bagging with Naive Bayes		

4   Result Analysis

Individual words are represented as real-valued vectors in a predetermined vector space in a process known as a word embedding. The method is frequently referred to as deep learning since each word is assigned to a single vector, and the vector values are learned in a manner like a neural network. This section shows the result of different word Embedding techniques used for refactoring prediction at class level. Table 3 shows the performance of the refactoring model on project1 in the form of accuracy and AUC. The performance of the proposed models using TF-IDF and CBOW word embedding achieves better performance as 0.81 on original data by using deep learning. The information in Table 3 also suggested that the refactoring prediction models trained on sampled data using different variants of SMOTE have a better prediction ability than the original data. These SMOTE bases models achieved AUC of 0.99 using ensemble learners such as RF, EXTR, ADBST, and MVE. Finally, we also observed from Table 3 that the models trained using the ensemble learning approach also have similar performance to DL approach.

Table 3. AUC for Proj-1

	DL1	DL2	DL3	DL4	DL5	BAG-KNN	BAG-DT	BAG-NB	RF	EXTR	ADBST	GRABST	MVE
OD													
TFIDF	0.8	**0.81**	0.79	0.77	0.78	0.79	0.76	0.65	0.76	0.74	0.74	0.59	0.73
CBOW	**0.81**	0.79	0.8	0.8	0.77	0.73	0.64	0.67	0.75	0.71	0.6	0.51	0.73
SKG	0.79	0.77	0.78	0.78	0.76	0.73	0.6	0.68	0.68	0.77	0.63	0.54	0.73
SMOTE													
TFIDF	0.98	0.98	0.98	0.98	0.98	0.98	0.99	0.92	**0.99**	**0.99**	**0.99**	0.98	**0.99**
CBOW	0.98	0.97	0.98	0.97	0.98	0.96	0.98	0.75	0.98	0.97	0.94	0.95	0.97
SKG	0.98	0.98	0.99	0.98	0.98	0.97	0.98	0.91	0.98	0.98	0.96	0.96	0.98
BLSMOTE													
TFIDF	0.98	0.98	0.98	0.98	0.98	0.98	**0.99**	0.93	**0.99**	0.98	0.98	0.98	**0.99**
CBOW	0.98	0.98	0.98	0.98	0.97	0.97	0.98	0.87	0.98	0.98	0.95	0.96	0.97
SKG	0.98	0.98	0.98	**0.99**	0.98	0.98	**0.99**	0.94	**0.99**	0.98	0.97	0.97	**0.99**
SVMSMOTE													
TFIDF	0.97	0.97	0.97	0.97	0.97	**0.98**	**0.98**	0.92	**0.98**	0.97	0.97	0.96	**0.98**
CBOW	**0.98**	**0.98**	**0.98**	0.97	**0.98**	0.96	0.97	0.86	0.97	0.96	0.92	0.93	0.96
SKG	**0.98**	0.97	0.97	**0.98**	**0.98**	0.97	**0.98**	0.92	0.97	0.96	0.97	0.97	0.97

5   Comparative Analysis

This section shows the comparison of the predictive capability of the refactoring model developed using classification techniques, data sampling techniques, and word embedding techniques in the form of statistical description and box-plot diagram. The performance of the model presents min, max, Q1, Q3, and median values.

5.1   Comparative Study of Word Embedding Techniques

In this paper, we have employed three-word embedding techniques to extract the features from AST of each project, such as TF-IDF, CBOW, and SKG. These embeddings are applied to 4 data sets and prediction capability of these techniques is computed with 3 data sampling techniques and 13 classifiers. Therefore, a combination of 4 datasets *(1 OD and 3 SMOTE) * 13 classifier* 5-fold cross-validation = 1040 models are evaluated in order to compare the predictive capacity of various word-embedding techniques and determine the optimal word-embedding strategy for refactoring prediction. The descriptive statistics of performance parameters are represented using Fig. 2 and Table 4. The data present in Fig. 2 and Table 4 confirms that the models trained by taking extracted vector using TF-IDF and SKG have a better ability of refactoring prediction as compared to CBOW. These models achieve a 0.84 mean AUC value. After finding the best embedding based on descriptive statistics, we also have applied hypothesis testing using the Friedman test to find the significant impact of using a specific technique. This test applied to test the null hypothesis that "the capability of refactoring prediction models will not significantly be improved after using different embedding techniques". The result of the Friedman test in terms of mean-rank is present in Table 4. The low value of the mean-rank shows that the models trained on TF-IDF vectors have significant improvement in performance.

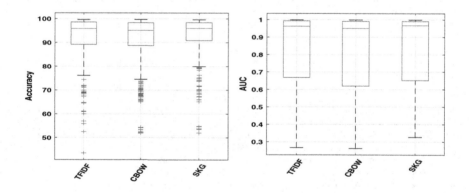

Fig. 2. The word embedding techniques for refactoring prediction

5.2   Comparative Study of Data Sampling Techniques

The considered data sets used for validation refactoring models are quite unbalanced, which may result in poor performance when attempting to predict the minority class. In this study, the SMOTE method was employed to convert unbalanced data to balanced data by oversampling minority groups. In addition to SMOTE, two additional sampling methods (BLSMOTE and SVSMOTE) are

Table 4. Statistical descriptive performance of word Embedding techniques in SRM

	Accuracy							AUC						
	Min	Q1	Median	Mean	Q3	Max	Rank	Min	Q1	Median	Mean	Q3	Max	Rank
TFIDF	43.54	89.20	95.79	**91.58**	98.58	99.58	1.8557	0.27	0.67	0.96	**0.84**	0.99	1.00	1.8006
CBOW	51.95	88.73	95.06	**90.52**	98.30	99.61	2.3229	0.26	0.62	0.95	**0.82**	0.99	1.00	2.3482
SKG	52.06	90.87	95.97	**91.85**	98.35	99.58	1.8214	0.33	0.65	0.97	**0.84**	0.99	1.00	1.8512

utilized to balance the original data. These techniques are applied to vectors
extracted using three approaches and compared using 13 classifiers. Therefore, a
combination of 4 data sets* 3 word-embedding techniques * 13 classifiers* 5-fold
cross-validation = 780 models are evaluated to compare the predictive capacity of various sampling techniques for refactoring prediction. The descriptive
statistics of performance parameters of the models trained on sampled data and
original data are represented using Fig. 3 and Table 5. The information in Fig. 3
and Table 5 confirm that the models trained on sampled data using SMOTE
have much better prediction ability than original data. These trained models on
SVMSMOTE achieve an average AUC of 0.95. After finding the best sampling
techniques, we also applied hypothesis testing like in the previous section to find
the significant impact of using a sampling technique. Here, the null hypothesis
is that "the capability of refactoring prediction models will not significantly be
improved after training on sampled data". The result of the Friedman test in
terms of mean-rank is present in Table 5. The low mean rank value shows that
the models trained on sampled data significantly improve performance. Table 5
also shows the models trained on sampled data using BLSMOTE have a better
capability of prediction as compared to other techniques.

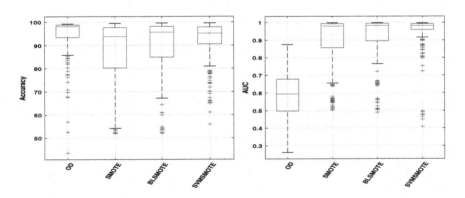

Fig. 3. The box-plot diagram of SMOTE technique for refactoring prediction through
word embedding

Table 5. Statistical descriptive performance of data sampling techniques in SRM

	Accuracy							AUC						
	Min	Q1	Median	Mean	Q3	Max	Rank	Min	Q1	Median	Mean	Q3	Max	Rank
OD	43.54	93.51	98.19	**94.78**	98.88	99.26	2.058	0.26	0.50	0.59	**0.59**	0.68	0.88	3.853
SMOTE	51.95	80.27	93.84	**88.34**	97.67	99.50	3.008	0.50	0.86	0.98	**0.89**	0.99	1.00	2.194
BLSMOTE	52.06	84.87	95.65	**90.01**	98.16	99.61	2.230	0.49	0.90	0.98	**0.91**	0.99	1.00	1.665
SVMSMOTE	55.90	90.57	95.18	**92.12**	97.90	99.58	2.704	0.41	0.96	0.98	**0.95**	0.99	1.00	2.288

5.3   Comparative Study of Machine Learning Techniques

In this work, we have used five variants of deep learning approaches (DL1, DL2, DL3, DL4, DL5) and eight ensemble learners (BAG-KNN, BAG-DT, BAG-NB, RF, EXTR, ADBST, GRABST, MVE) to train models for refactoring prediction. These techniques are applied to vector sampled data. Therefore, a combination of 4 datasets* 3 word-embedding techniques * 4 datasets *(1 OD and 3 SMOTE)* 5-fold cross-validation = 240 models are evaluated to compare the predictive capacity of various classifier techniques for refactoring prediction. The descriptive statistics of the models trained using different classifiers are represented using Figs. 5 and Table 6.

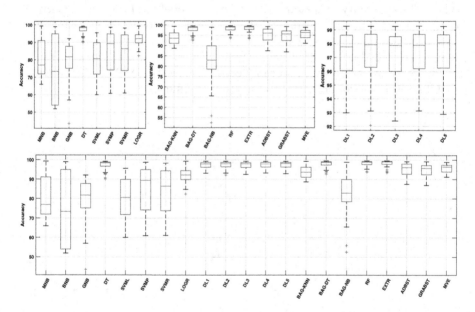

Fig. 4. The box-plot diagram of Frequently used classifiers performance in the form of Accuracy

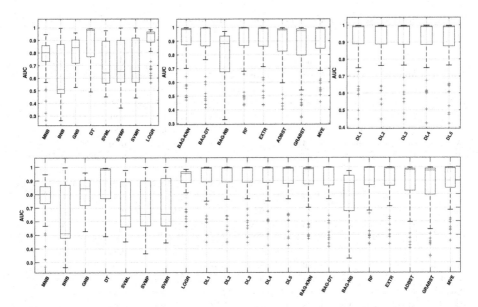

Fig. 5. The box-plot diagram of Frequently used classifiers performance in the form of AUC

The information in Figs. 5 and Table 6 confirm that the models trained using DL1 have much better prediction ability. The trained models using DL1 achieve an average AUC of 0.91. In this paper, we have also compared the performance of these thirteen classifiers with the most frequently used classifiers like DT, SVML, SVMP, SVMR, GNB, MNB, BNB, LOGR. The information in Figs. 4 and 5 and Table 6 confirm models trained using deep-learning and ensemble learning have a better ability of prediction as compared to frequently used classifiers. Finally, we have applied the Friedman test to find best classifier. Here, the null hypothesis is that "the capability of refactoring prediction models will not significantly be improved after changing classification techniques". The low mean rank value shows that the models trained using random forest classifiers significantly improve performance. The high value of the mean rank shows that the company must use ensemble learning of deep-leaning in place of the most frequently used classifiers for refactoring prediction.

Table 6. Statistical Descriptive Performance: Machine Learning Techniques: AUC and Accuracy

	Accuracy							AUC						
	Min	Q1	Median	Mean	Q3	Max	Rank	Min	Q1	Median	Mean	Q3	Max	Rank
MNB	65.98	71.95	76.94	**81.24**	91.01	99.26	15.02	0.27	0.73	0.80	**0.76**	0.86	0.95	16.73
BNB	51.95	54.01	73.38	**75.11**	94.79	98.96	17.81	0.26	0.48	0.51	**0.60**	0.87	1.00	18.17
GNB	43.54	75.18	81.76	**79.97**	87.65	92.03	17.83	0.53	0.72	0.84	**0.81**	0.90	0.96	15.27
DT	89.95	96.57	98.39	**97.12**	98.79	99.16	8.16	0.49	0.77	0.98	**0.87**	0.99	0.99	12.85
SVML	60.00	71.84	80.65	**80.45**	89.78	95.47	18.73	0.45	0.56	0.64	**0.70**	0.89	0.98	16.21
SVMP	60.91	74.22	89.31	**84.85**	94.81	98.66	15.16	0.36	0.57	0.65	**0.70**	0.90	1.00	14.08
SVMR	61.04	73.83	86.40	**84.01**	94.26	98.18	16.80	0.44	0.57	0.65	**0.71**	0.92	1.00	12.77
LOGR	82.34	89.73	92.13	**92.42**	94.36	99.26	12.03	0.56	0.89	0.95	**0.89**	0.96	0.98	11.98
DL1	92.99	96.04	97.77	**97.14**	98.62	99.26	7.57	0.45	0.89	0.99	**0.91**	0.99	1.00	6.17
DL2	92.08	96.30	97.93	**97.32**	98.69	99.26	6.30	0.45	0.89	0.99	**0.90**	0.99	1.00	6.40
DL3	92.40	96.01	97.88	**97.23**	98.53	99.26	7.24	0.44	0.89	0.99	**0.90**	1.00	1.00	6.79
DL4	93.14	96.21	97.90	**97.35**	98.69	99.28	6.57	0.42	0.89	0.99	**0.90**	0.99	1.00	6.92
DL5	92.89	96.26	98.08	**97.39**	98.66	99.26	6.28	0.42	0.88	0.99	**0.90**	0.99	1.00	7.94
BAG-KNN	88.70	91.07	93.56	**93.76**	96.12	99.26	11.52	0.47	0.87	0.98	**0.89**	0.99	1.00	9.92
BAG-DT	92.65	97.28	98.78	**97.84**	99.09	99.50	5.03	0.42	0.87	0.99	**0.88**	1.00	1.00	5.78
BAG-NB	52.53	78.59	82.91	**82.60**	89.93	98.88	16.61	0.33	0.68	0.88	**0.80**	0.94	0.97	16.46
RF	93.63	97.59	98.96	**98.15**	99.27	99.61	3.96	0.43	0.87	0.99	**0.89**	1.00	1.00	5.64
EXTR	93.38	97.79	99.08	**98.32**	99.38	99.60	3.30	0.43	0.86	0.99	**0.89**	1.00	1.00	8.90
ADBST	87.48	92.70	96.02	**95.15**	97.97	99.21	10.84	0.40	0.83	0.98	**0.87**	0.99	1.00	12.17
GRABST	86.95	92.48	95.65	**94.54**	97.15	99.21	12.22	0.34	0.80	0.97	**0.86**	0.99	1.00	13.06
MVE	91.17	93.84	96.36	**95.61**	97.33	98.87	12.00	0.46	0.85	0.99	**0.89**	1.00	1.00	6.81

6   Conclusion

Softwares pervade all aspects of our existence, and it is predictable to refactor the software earlier during the software development life cycle. In this paper, we have employed three-word embedding techniques to extract the features from AST, three sampling techniques to balance the unbalanced nature of the data set, and thirteen classifiers to train refactoring prediction models. These models are compared and validated using 5-fold cross-validation with AUC and Accuracy parameters. The focused insights of this research are:

- The models developed using vectors computed with TF-IDF have significant improvement in performance as compared to CBOW and SKG.
- The use of sampling techniques to balance data sets will help to improve the performance of our prediction model. The models trained on sampled data using BLSMOTE have a better capability of prediction as compared to other techniques.
- The use of ensemble learning or deep learning for model training will help to improve prediction models' performance. The models trained using random forest classifiers have a better ability of prediction as compared to other classifiers.

References

1. Aniche, M., Maziero, E., Durelli, R., Durelli, V.: The effectiveness of supervised machine learning algorithms in predicting software refactoring. IEEE Trans. Softw. Eng. **48**(4), 1432–1450 (2020)
2. Sagar, P.S., AlOmar, E.A., Mkaouer, M.W., Ouni, A., Newman, C.D.: Comparing commit messages and source code metrics for the prediction refactoring activities. Algorithms **14**(10), 289 (2021)
3. Ni, A., et al.: Soar: a synthesis approach for data science api refactoring. In: 2021 IEEE/ACM 43rd International Conference on Software Engineering (ICSE), pp. 112–124. IEEE (2021)
4. Bompotas, A., et al.: Implementation and repeatability aspects combined with refactoring for a reviews manager system. In: WEBIST, pp. 607–615 (2021)
5. Fowler, M.: Refactoring: Improving the Design of Existing Code, 2nd edn. Addison-Wesley Professional, Boston (2018)
6. Alenezi, M., Akour, M., Al Qasem, O.: Harnessing deep learning algorithms to predict software refactoring. Telkomnika **18**(6), 2977–2982 (2020)
7. Patnaik, A., Panigrahi, R., Padhy, N.: Prediction of accuracy on open source java projects using class level refactoring. In: 2020 International Conference on Computer Science, Engineering and Applications (ICCSEA), pp. 1–6. IEEE (2020)
8. Kumar, L., Satapathy, S.M., Murthy,L.B.: Method level refactoring prediction on five open source java projects using machine learning techniques. In: Proceedings of the 12th Innovations on Software Engineering Conference (Formerly Known as India Software Engineering Conference), pp. 1–10 (2019)
9. Jureczko, M., Spinellis, D.: Using object-oriented design metrics to predict software defects. Models and Methods of System Dependability. Oficyna Wydawnicza Politechniki Wrocławskiej, pp. 69–81 (2010)
10. Kumar, L., Sripada, S.K., Sureka, A., Rath, S.K.: Effective fault prediction model developed using least square support vector machine (lSSVM). J. Syst. Softw. **137**, 686–712 (2018)
11. Kocaguneli, E., Tosun, A., Bener, A.B., Turhan, B., Caglayan, B.: Prest: an intelligent software metrics extraction, analysis and defect prediction tool. In: SEKE, pp. 637–642 (2009)
12. Spirin, E., Bogomolov, E., Kovalenko, V., Bryksin, T.: Psiminer: a tool for mining rich abstract syntax trees from code. In: 2021 IEEE/ACM 18th International Conference on Mining Software Repositories (MSR), pp. 13–17. IEEE (2021)
13. Kumar, L., Kumar, M., Murthy, L.B., Misra, S., Kocher, V., Padmanabhuni, S.: An empirical study on application of word embedding techniques for kumar2021empiricalprediction of software defect severity level. In: 2021 16th Conference on Computer Science and Intelligence Systems (FedCSIS), pp. 477–484. IEEE (2021)
14. Kádár, I., Hegedus, P., Ferenc, R., Gyimóthy, T.: A code refactoring dataset and its assessment regarding software maintainability. In: 2016 IEEE 23rd International Conference on Software Analysis, Evolution, and Reengineering (SANER), vol. 1, pp. 599–603. IEEE (2016)

Universal Distributional Decision-Based Black-Box Adversarial Attack with Reinforcement Learning

Yiran Huang[(⊠)], Yexu Zhou, Michael Hefenbrock, Till Riedel, Likun Fang, and Michael Beigl

Karlsruhe Institute of Technology, Karlsruhe, Germany
{yhuang,zhou,hefenbrock,riedel,fang,beigl}@teco.edu

Abstract. The vulnerability of the high-performance machine learning models implies a security risk in applications with real-world consequences. Research on adversarial attacks is beneficial in guiding the development of machine learning models on the one hand and finding targeted defenses on the other. However, most of the adversarial attacks today leverage the gradient or logit information from the models to generate adversarial perturbation. Works in the more realistic domain: decision-based attacks, which generate adversarial perturbation solely based on observing the output label of the targeted model, are still relatively rare and mostly use gradient-estimation strategies. In this work, we propose a pixel-wise decision-based attack algorithm that finds a distribution of adversarial perturbation through a reinforcement learning algorithm. We call this method Decision-based Black-box Attack with Reinforcement learning (DBAR). Experiments show that the proposed approach outperforms state-of-the-art decision-based attacks with a higher attack success rate and greater transferability.

Keywords: Adversarial attack · Decision attack · Reinforcement Learning

1 Introduction

Many high-performing machine learning algorithms used in computer vision, speech recognition and other areas are susceptible to minimal changes of their inputs [1]. Despite their good performance, the vulnerability of machine learning models has raised widespread concerns. Small perturbation on road signs can have a serious impact on automated driving. These actions, which modify the benign input by imperceptible perturbations and thus manipulate the machine learning model to suit the attacker's interests, are called adversarial attacks.

Most adversarial attacks used to construct adversarial perturbation rely either on gradient information (white-box attack) or logit output (score-based attack) of the model. While these approaches help to study the limitations of current machine learning algorithms [15], they do not reflect the level of information a real attacker would have access to in most scenarios. In contrast, decision-based attacks consider limited access to the targeted model, allowing only the label information output by the model to be used. Such limited access is far more common in the real-world scenarios making its study more practical.

M. Tanveer et al. (Eds.): ICONIP 2022, LNCS 13625, pp. 206–215, 2023.
https://doi.org/10.1007/978-3-031-30111-7_18

Most of decision-based attacks start with an adversarial example with large perturbation. Then, adversarial examples with smaller perturbations are gradually found by sample-based gradient estimation. Different attacks exploit the samples in different ways, and therefore the efficiency of the algorithms varies. Such gradient-estimation-based approaches, however, require a large number of queries to the targeted model, which affects the efficiency of the algorithm and makes it impossible to perform real-time attack. In addition, the perturbations generated by gradient-based approach are too specific to the particular targeted model and benign example, and therefore lack transferability. To address these shortcomings, we propose a novel pixel-wise decision-based attack approach, called Decision-based Black-box Attack with Reinforcement learning (DBAR), which is guided by rewards instead of gradients. We therefore phrase the search for adversarial perturbations as an reinforcement learning task. Depending on whether the learned agent is targeting a single or multiple benign examples, two different attacks are designed, i.e., context-free attack and context-aware attack.

Our contributions can be summarized as: (1) Context-free DBAR achieves state-of-the-art performance, and perturbations sampled from the discovered distribution are more transferable then those generated by the other decision based attacks. In addition, the context-free DBAR is an universal attack which can also attack time-series data and super pixel of image data. (2) The context-aware attack achieves an effective attack without any queries on the targeted model after training, which is not possible for most existing decision-based attacks. (3) The algorithm generates a distribution which can be used to sample multiple different attacks.

2   Related Work

The definition of decision-based attack was first proposed in [1]. It starts with an example in the target category and optimizes the attack with random selection and validation. This method is simple and effective; however, it is inefficient because the information from the sampled examples is not fully utilized e.g., information from the 'worse' samples. Several methods attempt to bridge this gap. For example, [2] biases the sampling process by combining low-frequency noise with gradients from surrogate models. However, its performance depends on the transferablity between the surrogate model and the target model. Similarly, transfer-based attacks [15] also rely on carefully chosen surrogate models. However they obtain an attack on the original model by attacking the surrogate model. Opt attack [5] transforms the adversarial attack problem into a continuous real-valued optimization problem, i.e., the direction and distance to the decision boundary. This optimization problem can be solved by any zeroth-order optimization algorithm. However, distance calculation and gradient estimation in large dimensions will consume a large number of queries, which reduces the efficiency of the algorithm. Evo attack [8] applies evolutionary algorithms to generate adversarial perturbations and employs some techniques to reduce the dimensionality of the search space. It uses a custom variant in normal distribution and update the variant with (1+1)-CMA-ES. However, the variance is sign-independent and the sampling is therefore unstable. Rays [4] uses the dichotomous method to find perturbations. Although it has achieved good results on many datasets, the effectiveness of the algorithm is difficult to prove as results depend strongly on the test set. HSJ [3] estimate gradient in different way and achieve a decision-based attack. However, gradient estimation is time-consuming and, at the same time, reduces the transferability of the generated adversarial perturbations. In this paper, we try to solve the problem without estimating the gradient.

3 Methodology

We model the decision-based black-box adversarial attack problem as finding the adversarial distribution p_Θ with parameters Θ of an m-class deep classification model $\mathcal{M} : \mathbb{R}^d \to [m]$ that accepts an input $x \in [0, 1]^d$ and outputs $y \in [m] = \{1, \cdots, m\}$. The objective function can be described as

$$\min_{\Theta} \left(\lambda \mathop{\mathbb{E}}_{\eta \sim p_\Theta} \|\eta\|_\infty - \mathop{\mathcal{P}}_{\eta \sim p_\Theta} (\mathcal{M}(x + \eta) \neq \mathcal{M}(x)) \right), \tag{1}$$

where η is the adversarial perturbation sampled from the distribution p_Θ, $\|\cdot\|_\infty$ denotes the l_∞ norm and $\mathcal{P}$ evaluates a probability. The objective function consists of two components, the expected l_∞ norm of the perturbations sampled from the distribution and the attack success rate of the perturbations sampled from the distribution. λ is a parameter that trade-off between the expectation and the success rate. The goal of the problem definition is to find a distribution such that the center of the distribution found is as close as possible to the benign example x while the adversarial perturbations sampled from the distribution maintain a high attack success rate.

To solve (1) through a reinforcement learning algorithm, depending on whether single or multiple benign examples are considered, we design two different environments: a context-free environment and a context-aware environment, which correspond to context-free attack and context-aware attack. Both environments share the same setting except the transition model. The state and action space are both set to $\mathbb{R}^d$. The perturbation distribution p_Θ to optimized is regarded as the agent. Each adversarial perturbation η sampled from the distribution is an action and each benign example x is a state. Since the action is continuous, we model the agent with normal distribution $p_\Theta(\eta \mid x) = \mathcal{N}(\eta \mid \mu_\Theta(x), \mathrm{diag}(\Sigma_\Theta(x)))$. A trajectory τ consists of fixed number of decision step. In each decision step, a perturbation (action) is sample from the distribution (agent) and send to the environment to get the reward and next state. To achieve the optimization goal, we define the reward function as

$$r(x, \eta) = \frac{2 \cdot \mathbb{1}_{\{\mathcal{M}(x+\eta) \neq \mathcal{M}(x)\}} - 1}{\|\eta\|_\infty}, \tag{2}$$

where $\mathbb{1}$ is the indicator function to identify whether adversarial perturbations mislead the classifier $\mathcal{M}$. When the attack is successful, $2 \cdot \left(\mathbb{1}_{\{\mathcal{M}(x+\eta) \neq \mathcal{M}(x)\}} - 1 \right) = 1$, the algorithm can try to increases the reward by shrinking the perturbation. On the other hand, if the attack fails, the algorithm may try to find a successful attack by increasing the perturbation[1]. The expectation and probability in (1) are approximated by Monte Carlo estimation using sampled trajectories.

---

[1] Equation 2 can be modified to a target attack by setting the condition of indicator function to $\mathcal{M}(x + \eta) = \mathrm{target}$.

In both environment settings, the next state is selected independent of the current state and action. In the context-free environment, only one state, the benign example, exists, while in the context-aware environment, the next state is selected by random sampling.

The objective function can be written as:

$$J(\Theta) = \int p_\Theta(\tau) \left(\sum_{t=0}^{T} r(x_t, \eta_t) \right) d\tau,$$

$$p_\Theta(\tau) \approx \prod_{t=0}^{T-1} p_\Theta(\eta_t | x_t). \tag{3}$$

To learn Θ, we need to calculate the gradient of the objective function (3). In the black-box adversarial attack, each reward in one trajectory is treated equally and does not depend on the actions in the other time step of the same trajectory. Therefore, when calculating the gradient, at time step t, terms that do not depend on the action η_t can be omitted. Using the log derivation trick and Monte Carlo sampling [19], the gradient of the objective function can be expressed as

$$\nabla J(\Theta) \approx \sum_{i=0}^{I} \left[\left(\nabla_\Theta \log \left(\prod_{t=0}^{T-1} p_\Theta(\eta_{t,i} | x_{t,i}) \right) \right) \left(\sum_{t=0}^{T} r(x_{t,i}, \eta_{t,i}) \right) \right]$$

$$\approx \sum_{i=0}^{I} \left[\sum_{t=0}^{T-1} r(x_{t,i}, \eta_{t,i}) \nabla_\Theta \log p_\Theta(\eta_{t,i} | x_{t,i}) \right]$$

$$= \sum_{i=0}^{I \cdot T} r(x_i, \eta_i) \nabla_\Theta \log p_\Theta(\eta_i | x_i)$$

So far, this gradient is valid only for the samples generated by p_Θ. We apply the importance sampling technique so that old trajectories can be reused. In addition, we limit the update step size as suggested in [18], since importance sampling only works when the update size is small. Together with the stable training trick mentioned in [14], the gradient can be expressed as

$$\nabla J(\Theta) = \sum_{i=0}^{M} (\nabla_\Theta \min(w_i(\Theta), \mathrm{clip}(w_i(\Theta), 1 - \epsilon, 1 + \epsilon))) \cdot (r(x_i, \eta_i) - V(x_i))$$

$$\text{with} \quad w_i(\Theta) = \frac{p_\Theta(\eta_i | x_i)}{p_{\Theta_{old}}(\eta_i | x_i)},$$

where $p_{\Theta_{old}}$ is the distribution that generates the training samples and p_Θ is the distribution, that is frequently updated. The parameter ϵ limits the update size and $V(x_i)$ is the expected reward given to a benign example x_i.

Algorithm 1 summarizes the process of generating adversarial distribution, where Actor: $\mathbb{R}^d \rightarrow \mathbb{R}^d \times \mathbb{R}^d$ and Critic: $\mathbb{R}^d \rightarrow \mathbb{R}$ are two neural networks with the same ResNet architecture except for the output layer.

Algorithm 1. Generating adversarial distribution through DBAR

Input: x_0 (benign example), N (number of iterations), M (number of samples in one iteration), K (number of training), L (size of minibatch), $init\_mean$, $init\_std$, ϵ

Output: Actor

1: Initialization: Initialize Actor($\cdot$) with $init\_mean$ and $init\_std$, Critic($\cdot$), $x \leftarrow x_0$
2: **for** $i \leftarrow 1$ to N **do**
3:     $B \leftarrow [\,]$
4:     **for** $j \leftarrow 1$ to M **do**
5:         $\mu, \mathbf{I} \cdot \sigma^2 \leftarrow$ Actor(x)
6:         $r, p, x' \leftarrow$ sample action from $\mathcal{N}(\mu, \mathbf{I} \cdot \sigma^2)$, calculate its log-probability and applied it to the environment to get reward and the next benign example
7:         $B \leftarrow B \cup \{\mathrm{r} : r, \mathrm{lp} : lp, \mathrm{x} : x\}, x \leftarrow x'$
8:     **end for**
9:     **for** $k \leftarrow 1$ to K **do**
10:        $\{B_1, \cdots B_{\lfloor M/L \rfloor}\} \leftarrow$ generate mini-batch from B
11:        **for** $b \leftarrow \{B_1, \cdots B_{\lfloor M/L \rfloor}\}$ **do**
12:           $\mu', \mathbf{I} \cdot (\sigma')^2 \leftarrow$ Actor($b[\mathrm{x}]$)
13:           $lp' \leftarrow$ compute log-probability of $b[\mathrm{x}]$ in $\mathcal{N}(\mu', \mathbf{I} \cdot (\sigma')^2)$
14:           $v \leftarrow$ Critic($b[\mathrm{x}]$)
15:           $a \leftarrow b[\mathrm{r}] - v$
16:           $w \leftarrow \exp(lp' - b[\mathrm{lp}])$
17:           loss_actor $\leftarrow \min(w, \mathrm{clip}(w, 1 - \epsilon, 1 + \epsilon)) \cdot a$
18:           loss_critic $\leftarrow$ MSE($b[\mathrm{r}], v$)
19:           update Actor with the gradient of loss_actor
20:           update Critic with the gradient of loss_critic
21:        **end for**
22:     **end for**
23: **end for**
24: **return** Actor

---

4   Experiments

In this section, we perform experiments to investigate the following questions: (i) How does the context-free DBAR algorithm perform on image datasets compared to state-of-the-art decision-based attack methods? (ii) Can context-free DBAR be applied to time-series datasets? (iii) Are the perturbations discovered by the context-free DBAR algorithm more transferable than those discovered by state-of-the-art decision-based attack methods? (iv) Can context-aware DBAR perform real-time attacks after training? (v) How do the hyper-parameters affect the performance of context-free DBAR?

4.1   Experiment Setting

Baselines and Hyper-Parameters: To evaluate DBAR, we compare it with the following decision-based attacks: (i) the state-of-the-art Decision-based black-box Boundary attack [1] and the HopSkipJump Attack [3] for image datasets. (ii) the Universal White-box attack FGSM [21] and BIM [13] for time series datasets.

All attacks are implemented by the python package Foolbox [17]. We use the default hyper-parameter settings for all attacks with a fixed random seed. We limit the maximum number of queries for all the attacks to 20000. For the DBAR algorithm, the number of training epoch K is set to $K = 10$, with a mini-batch size of $L = 10$ and $\epsilon = 0.02$. Additionally $init\_mean = 0$ and $init\_std = 0.5$.

Datasets and Models: We carried out attacks over the following image datasets with varied dimensions and dataset sizes: CIFAR10 [12], CIFAR100 [12], STL10 [6], Caltech101 [10]. The pixel values of all images are normalized to $[0, 1]$. We also attack models with different structures such as ResNet20 [11] with 272474 parameters and VGG11 [20] with 9756426 parameters. Both models were obtained from Pytorch [16]. In addition, we carried out attacks over the publicly available time series UCR archiv dataset [7] and attack the time series ResNet-ts model as defined by [9].

4.2 Adversarial Examples for Image and Time Series Data

We perform non-target attacks on all the image datasets mentioned above and summarise the attack success rate (ASR) of each methods in Table 1. Concretely, all attacks are applied to 1000 correctly classified test examples from each dataset. If an adversarial example can mislead the classification model and is in a 0.04 (10/255) l_∞ neighborhood of the benign example, we denote this attack as a success. From the result, we see that, context-free DBAR achieves better results on most of the datasets, except for the STL10 dataset on the VGG11 model. Struggling with finding the first success attack is probably the main reason for the failure of the attack. This happens when the $init\_std$ is set too small for the given data set. The influence of the hyper-parameters on the attack performance an run-time is analyzed in experiment 4.4.

Table 1. Attack success rate (ASR) of three different decision-based attack methods: Boudary Attack (BA), HopSkipJump Attack (HSJ) and the proposed context-free DBAR, against model ResNet20 and VGG11 on four different image datasets with varies sizes.

		Model Accuracy	BA	HSJ	DBAR			Model Accuracy	BA	HSJ	DBAR
Cifar10	ResNet20	0.91	**1.00**	**1.00**	**1.00**	STL10	ResNet20	0.83	**1.00**	**1.00**	**1.00**
	VGG11	0.90	0.71	0.78	**0.84**		VGG11	0.84	0.59	**0.90**	0.78
Cifar100	ResNet20	0.66	0.95	0.95	**0.96**	CalTech101	ResNet20	0.79	0.93	**1.00**	**1.00**
	VGG11	0.61	0.63	0.80	**0.84**		VGG11	0.68	0.51	0.80	**0.82**

DBAR is universal in the sense that it is able to attack any example in form of $\mathcal{R}^d$. To prove this, we compare the performance of the proposed context-free DBAR against ResNet-ts model on the UCR open source time series datasets with the two popular white-box attacks FGSM and BIM. When the size of the perturbation found by an attack is smaller than 0.1, as proposed in [9], and the adversarial example can mislead the classification, we regard the attack as success. The parameter settings of this experiment are different from other experiments because the time series data are not normalized to $[0, 1]$. Furthermore, note that DBAR, as opposed to FGSM and BIM is a black-box attack. We remove the step limit and set the initial standard deviation to 80 to avoid struggling with finding the first successful attack.

Table 2. Attack success rate (ASR) of three different adversarial attack methods: FGSM [21], BIM [13] and the proposed context-free DBAR, against ResNet-ts on ten different time series datasets with varies size and dimensions.

	Model Accuracy	FGSM	BIM	DBAR		Model Accuracy	FGSM	BIM	DBAR
50words	0.73	0.77	0.88	**0.91**	DistalAge	0.80	0.78	**0.79**	0.64
Adiac	0.83	0.96	0.98	**0.99**	FaceAll	0.86	0.10	0.15	**0.20**
Beef	0.77	0.74	**0.87**	**0.87**	FaceUCR	0.95	0.17	**0.20**	0.19
Car	0.93	0.76	**0.92**	0.80	ElectricDevices	0.74	0.34	0.58	**1.00**
Diatom	0.30	0.00	0.00	**0.41**	ItalyPowerDemand	0.96	0.04	0.04	**0.20**

The results can be seen in Table 2. Although compared to the white-box algorithms, the context-free DBAR achieves better results on six datasets, ties on one dataset, and worse results on three datasets. As for the image datasets, the ASR score on the time series data is independent of the accuracy of the target model. And probably because of the different principles of generating attacks, DBAR performs well on some datasets where BIM performs very poorly, e.g., Diatom, ItalyPowerDemand. At the same time, the opposite situation also exists, see DistalAge. It is important to note that although FGSM has the worst results, it is attacking in real time, while both BIM and the proposed context-free DBAR require multiple iterations. However, a similar real-time attack can be achieved by context-aware DBAR. This is demonstrated in Sect. 4.3.

4.3   Transferability of the Perturbation Distribution and Real-Time Attack

The high transferability of the perturbation allows attacks generated against one platform to be applied to the other. To demonstrate the transferability of the perturbation found by the proposed method, we run the experiment in Sect. 4.2 on the Cifar datasets again and apply the attacks generated against the ResNet20 model to the VGG11 model and vice versa. The ASR score is given in Table 3. The performance of the proposed method is significantly better than that of the other methods. In particular, the perturbations generated against the VGG11 model on Cifar100 dataset have a near-average ASR score against the ResNet20 model. Besides, we can see that perturbations generated against simple model (with fewer parameters, ResNet20) are difficult to perform success attack against the more complex VGG11 model.

Table 3. Attack success rate (ASR) of attacks generated for VGG11 and applied to ResNet20 and vice versa. Targeted model denotes the model used to generate the perturbation (attack).

	Targeted model	Boundary Attack	HSJ	DBAR		Targeted model	Boundary Attack	HSJ	DBAR
Cifar10	ResNet20	0.04	0.04	**0.13**	Cifar100	ResNet20	0.14	0.14	**0.24**
	VGG11	0.10	0.10	**0.36**		VGG11	0.14	0.09	**0.45**

To evaluate the capabilities for real time attacks, we apply context-aware DBAR against ResNet20 on the Cifar10 and Cifar100. The results can be seen in Fig. 1. We

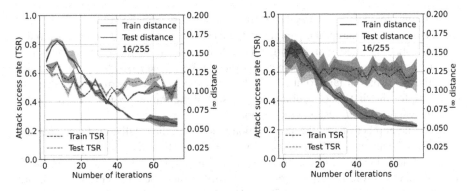

Fig. 1. Performance of context-aware DBAR against ResNet20 on Ciar10 (left) and Cifar100 (right) datasets. The picture on the left shows the performance of Cifar10 and the one on the right shows the performance of Cifar100.

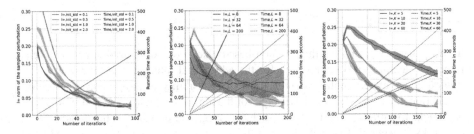

Fig. 2. Performance comparison of context-free DBAR with different hyper-parameter settings. The l_∞ norm is shown as a solid line in the plot, while the run time is shown as a dashed line.

can find that after 60 iterations, the adversarial perturbation obtained by sampling are able to implement effective attacks that have a success rate of about 50% on Cifar10 and 60% on Cifar100. We observe higher l_∞ norms for the perturbations at the beginning, which is most likely caused by the different update directions between the different benign examples. Ideally, the algorithm should reduce the size of the perturbation while increasing the success rate of the training. However, the attack success rate decreases as the training progresses, although the decrease is not significant. The algorithm improves the rewards obtained by reducing the perturbation size. This problem can be mitigated by increasing the contribution of the success attack in the reward function, see (2).

4.4 Impact of the Parameter Selection

In this experiment, we analyse the effect of following hyper-parameters on the performance of context-free DBAR regarding $init\_std$, number of training epoch in each iteration K and the size of mini batch L. The baseline hyper-parameters used in the experiments are set as follows as $init\_std = 0.5$, $L = 64$, $K = 30$. In each trial we modify one of the above parameters and summarise the result in Fig. 2. Evidently, the parameters have a great impact on the convergence speed, stability and runtime of DBAR.

The leftmost plot in Fig. 2 shows the effect of the standard deviation of the initial distribution $init\_std$. The larger the parameter, the larger the perturbation sampled from the initial distribution. In general, the larger the initial perturbation, the higher the probability that it will successfully attack a benign example. When the $init\_std$ is too small and the initial perturbation fails to attack the benign example, the algorithm will struggle in looking for the first successful attack. The size of the training batches L affects the stability, where lower L results in less stability.

Since the number of samples per iteration is constant, the smaller the batch size, the more times the agent is updated and the longer each iteration takes.

In addition, due to the importance sampling, the log-probability of actions needs to be recalculated before updating agent, which further increases the runtime. These are reflected in the plot in the center of Fig. 2. The rightmost plot shows that larger number of training K in one iteration converges faster. However, since there are more updates per iteration, it also takes longer to run.

5   Conclusion

In this paper, we formulate the decision-based black-box adversarial attack as a reinforcement learning task and search for the adversarial attack based on reward criterion. We have experimentally demonstrated the feasibility of the proposed algorithm. In addition, have shown some advantages of the algorithm such as the ability to generate attacks for different kinds of data such as images and time series, the transferability of attacks between different models and the ability for real-time attack. Besides, there is still room for further exploration in the proposed approach through using different reward functions or investigating the usage bounds by attacking very small perturbation or high resolution images.

Acknowledgements. This work was partially funded by the Ministry of The Ministry of Science, Research and the Arts Baden-Wuerttemberg as part of the SDSC-BW and by the German Ministry for Research as well as by Education as part of SDI-C (Grant 01IS19030A)

References

1. Brendel, W., Rauber, J., Bethge, M.: Decision-based adversarial attacks: reliable attacks against black-box machine learning models. arXiv preprint arXiv:1712.04248 (2017)
2. Brunner, T., Diehl, F., Le, M.T., Knoll, A.: Guessing smart: biased sampling for efficient black-box adversarial attacks. In: Proceedings of the IEEE/CVF International Conference on Computer Vision, pp. 4958–4966 (2019)
3. Chen, J., Jordan, M.I., Wainwright, M.J.: Hopskipjumpattack: a query-efficient decision-based attack. In: 2020 IEEE Symposium on Security and Privacy (sp), pp. 1277–1294. IEEE (2020)
4. Chen, J., Gu, Q.: Rays: a ray searching method for hard-label adversarial attack. In: Proceedings of the 26th ACM SIGKDD International Conference on Knowledge Discovery & Data Mining, pp. 1739–1747 (2020)
5. Cheng, M., Le, T., Chen, P.Y., Yi, J., Zhang, H., Hsieh, C.J.: Query-efficient hard-label black-box attack: an optimization-based approach. arXiv preprint arXiv:1807.04457 (2018)

6. Coates, A., Ng, A., Lee, H.: An analysis of single-layer networks in unsupervised feature learning. In: Proceedings of the 14th International Conference on Artificial Intelligence and Statistics, pp. 215–223. JMLR Workshop and Conference Proceedings (2011)
7. Dau, H.A., et al.: The UCR time series archive. IEEE/CAA J. Automatica Sin. **6**(6), 1293–1305 (2019)
8. Dong, Y., et al.: Efficient decision-based black-box adversarial attacks on face recognition. In: Proceedings of the IEEE/CVF Conference on Computer Vision and Pattern Recognition, pp. 7714–7722 (2019)
9. Fawaz, H.I., Forestier, G., Weber, J., Idoumghar, L., Muller, P.A.: Adversarial attacks on deep neural networks for time series classification. In: 2019 International Joint Conference on Neural Networks (IJCNN), pp. 1–8. IEEE (2019)
10. Griffin, G., Holub, A., Perona, P.: Caltech-256 object category dataset (2007)
11. He, K., Zhang, X., Ren, S., Sun, J.: Deep residual learning for image recognition. In: Proceedings of the IEEE Conference on Computer Vision and Pattern Recognition, pp. 770–778 (2016)
12. Krizhevsky, A., Hinton, G., et al.: Learning multiple layers of features from tiny images (2009)
13. Kurakin, A., Goodfellow, I., Bengio, S.: Adversarial machine learning at scale. arXiv preprint arXiv:1611.01236 (2016)
14. Mnih, V., et al.: Asynchronous methods for deep reinforcement learning. In: International Conference on Machine Learning, pp. 1928–1937. PMLR (2016)
15. Papernot, N., McDaniel, P., Goodfellow, I., Jha, S., Celik, Z.B., Swami, A.: Practical black-box attacks against machine learning. In: Proceedings of the 2017 ACM on Asia Conference on Computer and Communications Security, pp. 506–519 (2017)
16. Paszke, A., et al.: Pytorch: an imperative style, high-performance deep learning library. In: Advances in Neural Information Processing Systems, vol. 32, pp. 8026–8037 (2019)
17. Rauber, J., Brendel, W., Bethge, M.: Foolbox: A python toolbox to benchmark the robustness of machine learning models. arXiv preprint arXiv:1707.04131 (2017)
18. Schulman, J., Wolski, F., Dhariwal, P., Radford, A., Klimov, O.: Proximal policy optimization algorithms. arXiv preprint arXiv:1707.06347 (2017)
19. Shapiro, A.: Monte carlo sampling methods. Handbooks Oper. Res. Manag. Sci. **10**, 353–425 (2003)
20. Simonyan, K., Zisserman, A.: Very deep convolutional networks for large-scale image recognition. arXiv preprint arXiv:1409.1556 (2014)
21. Szegedy, C., et al.: Intriguing properties of neural networks. arXiv preprint arXiv:1312.6199 (2013)

Detecting and Mitigating Backdoor Attacks with Dynamic and Invisible Triggers

Zhibin Zheng[1], Zhongyun Hua[1,2(✉)], and Leo Yu Zhang[3]

[1] School of Computer Science and Technology, Harbin Institute of Technology, Shenzhen, Shenzhen 518055, China
20s151088@stu.hit.edu.cn
[2] Guangdong Provincial Key Laboratory of Novel Security Intelligence Technologies, Harbin Institute of Technology, Shenzhen, Shenzhen 518055, China
huazhongyun@hit.edu.cn
[3] School of Information Technology, Deakin University, Victoria 3216, Australia
leo.zhang@deakin.edu.au

Abstract. When a deep learning-based model is attacked by backdoor attacks, it behaves normally for clean inputs, whereas outputs unexpected results for inputs with specific triggers. This causes serious threats to deep learning-based applications. Many backdoor detection methods have been proposed to address these threats. However, these defenses can only work on the backdoored models attacked by static trigger(s). Recently, some backdoor attacks with dynamic and invisible triggers have been developed, and existing detection methods cannot defend against these attacks. To address this new threat, in this paper, we propose a new defense mechanism that can detect and mitigate backdoor attacks with dynamic and invisible triggers. We reverse engineer generators that transform clean images into backdoor images for each label. The generated images by the generator can help to detect the existence of a backdoor and further remove it. To the best of our knowledge, our work is the first work to defend against backdoor attacks with dynamic and invisible triggers. Experiments on multiple datasets show that the proposed method can effectively detect and mitigate the backdoor with dynamic and invisible triggers in deep learning-based models.

Keywords: AI security · Backdoor attack · Backdoor detection

1 Introduction

Deep neural networks (DNNs) have been widely used in many applications such as image classification [7], object detection [15], and language processing [1]. The performance of a DNN model relies on the high complexity of the model and the large amount of training data. It is thus acknowledged that common DNN end-users do not possess the capability to train a well-performed model by themselves. Therefore, users often outsource the training process of DNNs to a third

© The Author(s), under exclusive license to Springer Nature Switzerland AG 2023
M. Tanveer et al. (Eds.): ICONIP 2022, LNCS 13625, pp. 216–227, 2023.
https://doi.org/10.1007/978-3-031-30111-7_19

party or directly use a well-trained model from a model-sharing platform [6]. Either option introduces the attack surface of DNN backdoor.

Backdoor attacks aim to embed a backdoor into a DNN model [6]. The backdoor does not degrade the model's performance on clean inputs, but outputs the wrong results expected by the attackers when a specific trigger appears in the inputs [6]. Backdoor attacks are stealthy because the backdoor triggers are kept secret by the attacker, and model users with only clean inputs cannot activate the backdoor. In this regard, it is hard for model users to realize the existence of the backdoor [14]. Therefore, backdoor attacks cause a serious threat to the application of deep learning.

Various backdoor defense methods have been proposed to defend against backdoor attacks [2,4,20]. Most of these methods are empirical methods that are developed based on the characteristics of existing attack methods. For example, the trigger-synthesis-based defense methods [3,20,23] first reverse engineer a trigger for each label of the models and then run anomaly detection to determine the attacked target label. The saliency-map-based defense method [2] utilizes model visualization techniques to locate the potential trigger regions by finding common saliency regions in different input images. Besides, the input-filtering-based defense method [4] first adds different perturbations to the input image, and then calculates the randomness of the model's prediction to detect backdoor. However, all these defense methods are based on the assumption that a static trigger is involved in different images to launch an attack. They are ineffective against the newly proposed backdoor attacks [13,14] with dynamic and invisible triggers, since these attacks add different triggers for different clean images.

In this paper, we propose a backdoor defense method to detect and mitigate backdoor attacks with dynamic and invisible triggers. We reverse engineer the trigger pattern from the backdoored model. Different from existing trigger-synthesis-based methods that can reverse engineer only a static trigger, we reverse engineer a generator to transform clean images into backdoor images, which allows us to add different triggers for different images and capture more complex trigger patterns. The contributions of this paper are summarized as follows:

1. We design a new backdoor defense method that can defend against the backdoor attacks with dynamic and invisible triggers. We design a reverse engineering process of backdoor image generators using the characteristics of backdoor attacks.
2. The reversed generators are used for anomaly detection to detect the backdoor target. When a backdoor target is found, we use the generated backdoor images and a small subset of clean data to unlearn the backdoor.
3. We evaluate our proposed method on MNIST, CIFAR10, and GTSRB datasets with the most recent backdoor attack WaNet [14]. Experimental results show that the proposed method can effectively detect the backdoor and greatly reduce the attack success rate with only a small drop in clean accuracy.

2   Related Works

2.1   Backdoor Attacks

Backdoor attacks are aimed at embedding a backdoor associated with a trigger pattern into a DNN model. The model with a backdoor should preserve performance on clean inputs. However, when the inputs are patched with the trigger pattern, the model will output the results expected by the attackers.

Trigger pattern design is the core of backdoor attacks, where better trigger patterns can make backdoor attacks stealthy and effective. According to whether different triggers are applied to different images, backdoor attacks can be divided into static and dynamic backdoor attacks. For the case of static attacks, the injection of a static trigger into a clean image x can be formulated as

$$x' = (1 - m) \cdot x + m \cdot \Delta, \tag{1}$$

where Δ is the trigger pattern specified by the attacker, m is the mask to decide the location for the trigger to stamp on clean images. Gu et al. [6] proposed the first backdoor attack that use a small patch as the trigger pattern. Later works use optimized triggers [10], smooth triggers [21] for better stealthy.

For the case of dynamic backdoor attacks, Nguyen et al. [13] firstly suggested input-aware triggers. They optimize a generator to generate different triggers for different images with a diversity loss. When a trigger generated for one image is added to another image, the resulting image cannot activate the backdoor. WaNet [14] is another dynamic backdoor attack, which creates backdoor images using a small and smooth warping field. And the backdoor images contain triggers that vary from image to image. As the distortion caused by warping is slight, it is difficult to distinguish the backdoor images from the original images. Thus, the triggers in the WaNet attack are dynamic and invisible. Backdoor attacks with such triggers are more powerful than previous attacks as they not only break assumptions of various defense methods but also evade manual inspection.

2.2   Backdoor Defense Methods

Existing backdoor defense methods can be classified as directly reducing backdoor attack success rates (i.e., mitigation) or detecting the backdoor for a victim model (i.e., detection). For the case of mitigation, Liu et al. [12] proposed to prune neurons not useful for clean images and fine-tune the pruned model on clean images. Li et al. [11] used attention distillation to guide the fine-tuning process and showed a better result compared to pruning and fine-pruning. The major limitation of these methods is that they are blind to the existence of backdoor: if they are used for clean models, they also cause a decrease in model's accuracy.

For detection, trigger synthesis is the most popular method for its capability to not only detect the existence of the backdoor but also remove/unlearn the backdoor. Wang et al. [20] proposed the first trigger synthesis based defense method Neural Cleanse (NC), which reverses triggers for each label and uses the L1 norm of trigger masks to detect backdoor target labels. Once a backdoor is detected, they mitigate the backdoor by filtering inputs or patching the backdoored model. Zhu et al. [23] proposed GangSweep which uses GAN [5] to better reconstruct triggers. Dong et al. [3] extended it to the black-box setting where the weights of the model is not accessible.

Our work also falls into the same category of backdoor detection and mitigation. But existing methods are based on the same assumption that a universal trigger is used, so they are not able to defend against attacks with dynamic triggers.

3 Proposed Method

3.1 Threat Model and Defense Goal

We consider that a user obtains a pre-trained model from an untrusted third party. The model could be backdoored. In particular, a stronger backdoor attack with dynamic and invisible triggers [14] instead of a static trigger [6,10,21] is possible.

Following the literature studies [11,20], the defender has full access to the model and has a subset of clean data. The set of clean data can be the well-labeled data used to test the performance of the model. The defender aims to detect whether the model is backdoored and find out what the attacker's target label is. Upon detection, the defender will also try to mitigate the backdoor.

3.2 Intuition and Overview

Backdoor attacks with dynamic and invisible triggers use different triggers for clean images. The backdoored model has learned to classify images with such triggers toward the attacker specified target label. We model the process of transforming clean images into backdoor images as a generator, taking clean images as input and outputting backdoor images, and try to reverse engineer such a generator from the backdoored model. The proposed method consists of three steps:

1. Reverse engineering backdoor image generator. Given a classifier model, we reverse engineer a backdoor image generator for every label by optimizing the generator to output images that not only close to the input images but also change the prediction of the model to a specific label regardless of the input images' ground truth labels.
2. Backdoor detection. We use an outlier detection algorithm to judge if there is a generator that can generate backdoor images with small modifications and achieve a high probability of misleading the model to predict them into a specific label.
3. Backdoor mitigation. We leverage the reversed generator to generate backdoor images and combine them with a set of clean images to fine-tune the backdoored model.

3.3 Reverse Engineering Backdoor Image Generator

We define the transformation of a clean image x to a backdoor image x' via a generator model G as $x' = G(x)$. The generator G is required to perform an image-to-image transformation, thus we choose U-Net [16] as the generator model. Compared with the basic encoder-decoder model, U-Net concatenates high-level features and low-level features to propagate context information and is easier to train on a small dataset, which is suitable for our task as we assume the defender only has a small set of clean data.

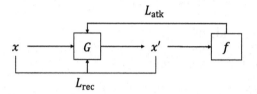

Fig. 1. The framework to reverse engineer the backdoor image generator.

As shown in Fig. 1, the reverse engineering of G from a classifier model f is the optimization for two objectives: effectiveness and stealthiness. The effectiveness goal requires the reversed generator to transform images from other classes into images that will be classified as a given class y_t with a high probability. The stealthiness goal requires that the transformed images be as similar to the original images as possible.

For the effectiveness of G to mimic trigger pattern, we use the cross-entropy loss L_{atk} to encourage the generated images to be classified as the target label y_t by model f, i.e.,

$$L_{atk} = \lambda_1 \cdot \text{CrossEntropy}(y_t, f(G(x)). \qquad (2)$$

For the stealthiness of the G's output, we use L_{rec} to encourage G to reconstruct images from inputs. It is a combination of mean square error (MSE) and learned perceptual image patch similarity (LPIPS) [22] to measure reconstruction loss, which can be formulated as

$$L_{rec} = \lambda_2 \cdot \text{MSE}(x, G(x)) + \lambda_3 \cdot \text{LPIPS}(x, G(x)). \qquad (3)$$

Here, the MSE loss measures the pixel value difference between the generated image and the input image, and the LPIPS loss measures the feature difference between the generated image and the input image.

The final loss function to train G is the combination of L_{atk} and L_{rec}:

$$L = L_{atk} + L_{rec}. \qquad (4)$$

For each label of model f, we optimize the parameters of a generator using the above loss functions. We use a binary search to find the hyper-parameters λ_1, λ_2, λ_3 to achieve a high attack success rate.

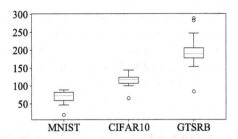

Fig. 2. Norm value distribution of clean and target labels on three datasets.

3.4 Backdoor Detection

With all the reversed generators associated with each label available, we then use them to detect the backdoor. Intuitively, for the attacker's target label, if the generator can capture the trigger pattern, it would be able to generate backdoor images with fewer modifications. To achieve a high attack success rate on clean labels, the generator needs to generate backdoor images with larger modifications. As there is not a universal trigger, the modification for label y_t is defined as the mean L1 norm of the residuals between the generated images and the clean input images:

$$m_{y_t} = \frac{1}{n} \sum_{i=1}^{n} |G_{y_t}(x_i) - x_i|, \text{ for } x_i \in X_{/y_t}, \tag{5}$$

where G_{y_t} is the generator reversed for label y_t, X is the set of clean images the defender has access to, $X_{/y_t}$ is the set of clean images that are not belonging to label y_t.

To validate our intuition, we calculate the average L1 norm value on test dataset. Figure 2 shows the distributions of residuals norm of different labels on 3 datasets. For each dataset, the box-plot shows the norm value distribution of clean labels, while the dot under the box-plot is the norm value of the target label, which is much smaller than that of clean labels. Therefore, we can leverage this characteristic to detect the target label. By taking MNIST as an example, Fig. 3 visualizes this observation. The bright and dark spots on the residual images are indications of large image modifications. The residual image for the target label 0 is grayer and more smooth than those of other labels, which means that there are fewer modifications.

We use the Median Absolute Deviation (MAD)-based outlier detection method adopted in NC [20] over each label y_t with its residual m_{y_t} obtained from Eq. (5). The anomaly index of y_t is calculated as follows:

$$AI_{y_t} = \frac{|m_{y_t} - \tilde{m}|}{c \cdot \text{median}_{i \in \{1, \cdots, K\}}(|m_{y_i} - \tilde{m}|)}, \tag{6}$$

where $\tilde{m}$ is the median of all residuals, K is the number of classes of the dataset, and c is a constant to normalize the anomaly index. For the purpose of detecting backdoor, we only need to focus on small residuals, so only labels with a residual smaller than the median residual value are considered. Similar to NC, we assume that the data satisfy normal distribution, c is thus set to 1.4826 and any label with an anomaly index larger than 2 will be considered as an attack target label with a 95% confidence level. Moreover, $\max_{t \in \{1, \cdots, K\}}(AI_{y_t})$ is called the model anomaly index.

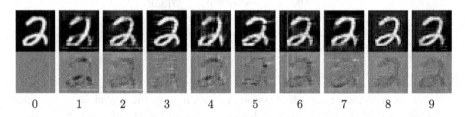

Fig. 3. Comparison of images generated by generators reversed for different labels. The first row is the generated images and the second row is the residuals between the generated image with the original images. The residuals are normalized from $[-1, 1]$ to $[0, 1]$. The label '0' is the target label.

3.5   Backdoor Mitigation

Once a backdoor is detected, we use the generator reversed for the detected target label to remove the backdoor. We use a subset of clean data and transform these clean images into backdoor images, label them with their original labels, and combine them with the subset of clean data to form a new dataset. We use the new dataset to fine-tune the backdoored model for a few epochs (empirically, 10 is enough).

4   Experiments

In this section, we describe our experiments on multiple classification tasks attacked by state-of-the-art backdoor attack method WaNet [14].

4.1   Experiment Setup

To evaluate the effectiveness of our defense method, we use three classification tasks: MNIST [9], CIFAR10 [8], GTSRB [19]. We use the default DNN models used in WaNet, namely, a simple classifier consisting of 3 ConvBlocks + 3 fcs for MNIST, and pre-activation Resnet-18 [7] for CIFAR10 and GTSRB. We also use VGG16 [18] and MobileNetV2 [17] for CIFAR10 and GTSRB. For each dataset and model architecture, we also train a clean model using the entire clean training dataset.

The hyper-parameters λ_1, λ_2, λ_3 are set to 0.1, 1, 0.1 respectively. We use an Adam optimizer with a learning rate of 0.001 to train the generator for each label. Instead of training the generators with random initial weights, we first optimize the generator using L_{rec} for 20 epochs to get a pretrained generator. The pretained weights are then loaded as initial weights to optimize generators for each label using the final loss L. For backdoor detection, we use 1% of the clean training data for each dataset. For backdoor mitigation, we use 5% of the training data.

4.2 Detection Performance

For each clean and backdoored model, we repeat the experiments 10 times with different random seeds, and the averaged model anomaly index is reported. Table 1 shows the tested anomaly indices. For backdoored MobileNetV2 models, their model anomaly indices are around 3. For other backdoored models, the model anomaly indices are larger than 4. For clean models, the anomaly indices are smaller than 2. This shows that our method can accurately detect the backdoored models.

Table 1. The backdoor detection results.

Dataset	Model	Method	Anomaly Index		Detailed Detection Results			
			Clean	Backdoor	Case 1	Case 2	Case 3	Case 4
MNIST	3ConvBlocks+3fcs	NC	1.10	0.75	0/10	0/10	0/10	10/10
		Ours	1.07	5.41	9/10	1/10	0/10	0/10
CIFAR10	PreActRes18	NC	2.52	2.32	0/10	0/10	6/10	4/10
		Ours	1.66	5.06	9/10	1/10	0/10	0/10
	VGG16	NC	1.91	4.00	0/10	4/10	3/10	3/10
		Ours	1.94	4.01	8/10	1/10	0/10	1/10
	MobileNetV2	NC	1.30	1.59	0/10	0/10	1/10	9/10
		Ours	1.30	2.98	9/10	1/10	0/10	0/10
GTSRB	PreActRes18	NC	2.65	1.52	0/10	3/10	1/10	6/10
		Ours	1.99	4.92	9/10	1/10	0/10	0/10
	VGG16	NC	1.94	1.94	0/10	0/10	3/10	7/10
		Ours	1.89	6.36	6/10	4/10	0/10	0/10
	MobileNetV2	NC	2.42	3.98	2/10	8/10	0/10	0/10
		Ours	1.95	3.24	5/10	4/10	0/10	1/10

To further assess the proposed algorithm's capability in detecting the target labels, we compare it with NC [20] by considering the following four cases:

- Case 1: The backdoor is detected and only the target label is detected.
- Case 2: The backdoor is detected and the target label is detected, but at least one clean label is identified as target label.
- Case 3: The backdoor is detected but the target label is not detected.
- Case 4: The backdoor is not detected.

The results are listed in Table 1. From this table, for the backdoored model on MNIST dataset, the target label is detected accurately 9 times with one exception: a clean label is mistakenly identified as the target label. There are similar results for all backdoored models on CIFAR10 dataset. For GTSRB dataset, the detection algorithm can detect the backdoor but has more false positive alarms for target labels on VGG16 and MobileNetV2 models than other models. This may be because there are more classes on GTSRB and the attack has side effects on other classes. While NC fails to detect the target label except for VGG16 model on CIFAR10 and MobileNetV2 model on GTSRB. The detection accuracy on these two models is still much lower than our method.

Visual Similarity. Figure 4 compares the original backdoor images and backdoor images generated by reversed generators. The generated backdoor images are close to the original backdoor images. The residuals with the clean image of the original backdoor image and the generated backdoor image both show the texture of the original images. It indicates that the optimization of the generator is capable of getting close to the trigger pattern.

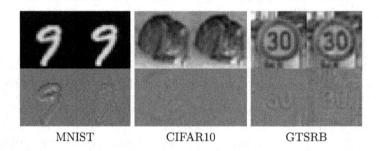

| MNIST | CIFAR10 | GTSRB |

Fig. 4. Comparison of backdoor images, generated backdoor images, and their residuals with the clean images of three datasets.

4.3   Backdoor Mitigation Performance

We use a Stochastic Gradient Descent (SGD) optimizer with a learning rate of 0.001 and a momentum of 0.9 to fine-tune the backdoored model for 10 epochs. We compare the mitigation results with basic fine-tuning and NAD [11].

The results of backdoor mitigation with different amount of clean data are shown in Table 2, where ACC stands for accuracy on clean inputs and ASR stands for attack success rate. The results show that on the MNIST and GTSRB datasets, even with only 1% clean training data, our method reduces the attack success rate to less than 1%, while the drop of the accuracy on clean data is less than 1%. On the CIFAR10 dataset, our method reduces the attack success rate to 2.58% using 5% clean data, while the accuracy on clean data drops by 1.18%. Without generated backdoor images of the reversed generator, fine-tuning has little effect on backdoor mitigation.

Figure 5 compares the mitigation performance of different methods with different learning rates when 5% of the clean training data is available. It shows that fine-tuning can be more effective when the learning rate is increased. However, fine-tuning the model with a larger learning rate has a risk of decreasing its accuracy on clean data. Similar to fine-tuning, NAD is sensitive to learning rate settings. The accuracy on clean data would drop severely when the learning rate is larger than a certain value, which is unstable. Compared to fine-tuning and NAD, our method shows a more stable mitigation effect with a small learning rate. It may be because tuning model with only clean data mitigates the backdoor through forgetting the trigger pattern, while we explicitly force the model to unlearn the trigger pattern.

Table 2. Mitigation performance with different percentages of clean training data.

Dataset	# of data	w/o reversed		w/ reversed	
		ACC	ASR	ACC	ASR
MNIST	1%	99.37	99.77	99.09	0.07
	3%	99.34	99.55	99.16	0.08
	5%	99.38	98.49	99.26	0.09
CIFAR10	1%	93.74	97.66	93.88	19.20
	3%	94.08	83.28	93.93	9.45
	5%	92.50	82.04	92.96	2.58
GTSRB	1%	98.76	93.60	98.89	0.79
	3%	99.03	94.63	99.00	0.43
	5%	99.03	94.65	99.03	0.19

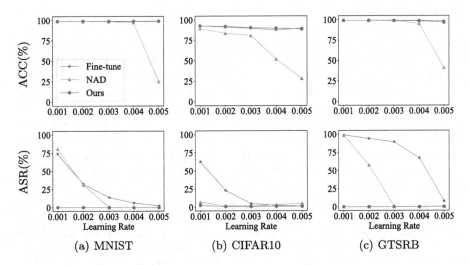

Fig. 5. Mitigation performance with different learning rates when 5% of clean training data is available.

4.4   Ablation Study

In Eq. (3), we use a combination of MSE and LPIPS. LPIPS uses a trained deep neural network to extract features to calculate features' difference and pays more attention to the image's overall changes. We conduct an ablation study without using the LPIPS loss. As shown in Fig. 6, when only the MSE loss is used, there are chances that the training falls into a local optimum. Although the generated image is basically the same as the original image in texture, there is a large change in color. This is because in an image with three channels, scattering the modifications to images across three channels may have equal MSE loss as centralizing the modifications to one channel. The LPIPS loss can help avoid this issue and better reconstruct the original image.

Fig. 6. Comparison results with and without LPIPS loss. Images from left to right are clean image, backdoor image, generated backdoor image with LPIPS loss, and generated backdoor image without LPIPS loss.

5   Conclusion

In this paper, we propose to defend against backdoor attacks with dynamic and invisible triggers by reverse-engineering the backdoor image generator. By carefully designing the optimization objectives, we can effectively reverse engineer a generator that can capture the embedded trigger pattern. The target label can be thus detected and the reversed generators can help eliminate the backdoor. We conduct experiments on multiple datasets under the WaNet attack. The experimental results show that our method can detect backdoors more effectively than the existing trigger-synthesis-based methods. Compared to fine-tuning using only clean data, fine-tuning aided by the reversed generator has a more stable backdoor mitigation performance, thus avoiding the trade-off between preserving accuracy on clean data and reducing attack success rate.

Acknowledgements. This work was supported in part by the National Natural Science Foundation of China under Grants 62071142, by the Guangdong Basic and Applied Basic Research Foundation under Grants 2021A1515011406, by the Shenzhen College Stability Support Plan under Grant GXWD20201230155427003-20200824210638001, by the Guangdong Provincial Key Laboratory of Novel Security Intelligence Technologies under Grant 2022B1212010005.

References

1. Brown, T., et al.: Language models are few-shot learners. Adv. Neural Inf. Process. Syst. **33**, 1877–1901 (2020)
2. Chou, E., Tramèr, F., Pellegrino, G.: Sentinet: detecting localized universal attacks against deep learning systems. In: IEEE S&P Workshops, pp. 48–54 (2020)
3. Dong, Y., et al.: Black-box detection of backdoor attacks with limited information and data. In: ICCV, pp. 16482–16491 (2021)
4. Gao, Y., Xu, C., Wang, D., Chen, S., Ranasinghe, D.C., Nepal, S.: Strip: a defence against trojan attacks on deep neural networks. In: ACSAC, pp. 113–125 (2019)
5. Goodfellow, I.J., et al.: Generative adversarial nets. In: NeurIPS, pp. 1–9 (2014)
6. Gu, T., Liu, K., Dolan-Gavitt, B., Garg, S.: Badnets: evaluating backdooring attacks on deep neural networks. IEEE Access **7**, 47230–47244 (2019)
7. He, K., Zhang, X., Ren, S., Sun, J.: Deep residual learning for image recognition. In: CVPR, pp. 770–778 (2016)
8. Krizhevsky, A., Hinton, G., et al.: Learning multiple layers of features from tiny images. University of Toronto, Technical Report (2009)
9. LeCun, Y., et al.: Learning algorithms for classification: a comparison on hand-written digit recognition. Neural Netw. Stat. Mech. Perspect. **261**(276), 2 (1995)
10. Li, S., Xue, M., Zhao, B.Z.H., Zhu, H., Zhang, X.: Invisible backdoor attacks on deep neural networks via steganography and regularization. IEEE Trans. Dependable Secure Comput. **18**(5), 2088–2105 (2021)
11. Li, Y., Lyu, X., Koren, N., Lyu, L., Li, B., Ma, X.: Neural attention distillation: erasing backdoor triggers from deep neural networks. In: ICLR, pp. 1–12 (2021)
12. Liu, K., Dolan-Gavitt, B., Garg, S.: Fine-pruning: defending against backdooring attacks on deep neural networks. In: Research in Attacks, Intrusions, and Defenses, pp. 273–294 (2018)
13. Nguyen, T.A., Tran, A.: Input-aware dynamic backdoor attack. In: NeurIPS, pp. 3454–3464 (2020)
14. Nguyen, T.A., Tran, A.T.: Wanet - imperceptible warping-based backdoor attack. In: ICLR, pp. 1–11 (2021)
15. Ren, S., He, K., Girshick, R., Sun, J.: Faster r-cnn: towards real-time object detection with region proposal networks. IEEE Trans. Pattern Anal. Mach. Intell. **39**(6), 1137–1149 (2016)
16. Ronneberger, O., Fischer, P., Brox, T.: U-net: convolutional networks for biomedical image segmentation. In: International Conference on Medical Image Computing and Computer-Assisted Intervention, pp. 234–241 (2015)
17. Sandler, M., Howard, A., Zhu, M., Zhmoginov, A., Chen, L.C.: Mobilenetv 2: inverted residuals and linear bottlenecks. In: CVPR, pp. 4510–4520 (2018)
18. Simonyan, K., Zisserman, A.: Very deep convolutional networks for large-scale image recognition. In: ICLR, pp. 1–14 (2015)
19. Stallkamp, J., Schlipsing, M., Salmen, J., Igel, C.: Man vs. computer: benchmarking machine learning algorithms for traffic sign recognition. Neural Netw. **32**, 323–332 (2012)
20. Wang, B., et al.: Neural cleanse: identifying and mitigating backdoor attacks in neural networks. In: S&P, pp. 707–723 (2019)
21. Zeng, Y., Park, W., Mao, Z.M., Jia, R.: Rethinking the backdoor attacks' triggers: a frequency perspective. In: ICCV, pp. 16473–16481 (2021)
22. Zhang, R., Isola, P., Efros, A.A., Shechtman, E., Wang, O.: The unreasonable effectiveness of deep features as a perceptual metric. In: CVPR, pp. 586–595 (2018)
23. Zhu, L., Ning, R., Wang, C., Xin, C., Wu, H.: Gangsweep: sweep out neural backdoors by gan. In: ACM MM, pp. 3173–3181 (2020)

NAS-StegNet: Lightweight Image Steganography Networks via Neural Architecture Search

Zhixian Wang, Guoqing Wang$^{(\boxtimes)}$, and Yang Yang

University of Electronic Science and Technology of China, Chengdu 610000, China
gqwang0420@hotmail.com

Abstract. Deep steganography describes the task of hiding a full image in another for secret communication, and such a model usually consists of hide (H) network for secret hiding followed by revealing (R) network for secret revealing. To guarantee the hiding effect for the secret communication applications (e.g., watermarking and light field messaging), most of existing deep steganography models design complex network architecture for H and R, increasing the challenge for model deployment. To achieve a better trade-off between steganography effect and model complexity, in this paper, we explore the idea of neural architecture search to learn a more practical deep steganography network, which is able to produce powerful steganography results but with much less parameters. Specifically, our automatically-learned network, termed as NAS-StegaNet, has 26× fewer parameters and requires 2× fewer GFLOPs when compared with the most powerful model. Codes are available at https://github.com/wang-MIG-CFM-UESTC/nas_stegan.git.

Keywords: Deep Steganography · Neural Architecture Search · Lightweight Network

1 Introduction

Steganography is a data hiding technology, which attempts to embed information into an inconspicuous medium and can be detected by only intended recipient [1]. Traditional steganography methods focus on the design of information encoding algorithm to seek optimal location for secret information hiding, and representative methods inlcude Least Significant Bit (LSB) for spatial domain [19], Discrete Cosine Transform for transition domain [3]. With the rapid development of deep learning techniques, a new concept has been proposed: deep steganography [2].

Instead of manually designing an algorithm to find a suitable location for secret information hiding, deep steganography uses a deep neural network to

Supported by the National Natural Science Foundation of China under grant 62102069, U20B2063 and 62220106008, the Sichuan Science and Technology Program under grant 2022YFG0032, and the Dongguan Songshan Lake Introduction Program of Leading Innovative and Entrepreneurial Talents.

M. Tanveer et al. (Eds.): ICONIP 2022, LNCS 13625, pp. 228–239, 2023.
https://doi.org/10.1007/978-3-031-30111-7_20

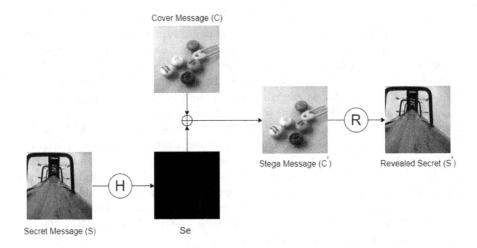

Fig. 1. General Framework of Deep Image Steganography(UDH [10]).

better embed secret messages. The key challenges for the research of steganography is to design a model with high capacity, and being secure and robust for deployment [4]. Compared with the traditional designs, deep steganography is a much better choice with very high hiding capacity. Take the image steganography as example, deep steganography can hide one or more images into another image of the same size, enjoying the hiding capacity of 24 bits per pixel (bpp) which is far beyond the traditional method of 0.5 bpp [5].

By formulating the deep image steganography task as an image-to-image translation process [20], both the convolutional neural network (CNN) and generative adversarial network (GAN) are commonly used as for designing a powerful image hiding network [2,6,7,9–12] and very promising results have been obtained. However, these models directly take the existing architecture for network design, thus leaving a space for seeking better results by exploring better task-specific network architecture designs. For example, steganalysis is a technique to analyze whether there is hidden information in a message. To produce better analytical results, some researchers focus on improving analysis networks by optimizing the pooling layer or kernel size of the convolution layer [21,22], and this may also apply to steganography tasks. Despite producing the SOTA results, it is still difficult for the deployment of the existing U-Net based steganography model, which is highly complex with a total of 16.6M parameters thus imposing large parameter storage requirement.

Imagine such a situation, when one needs to send secret information to the other, it is likely that he is acting outside or is being monitored, making it is difficult to use large memory electronic equipment. At this time, small embedded devices or mobile phones can solve his problems. However, these devices have small memory are difficult to accommodate large models. Therefore, it is important for research on lightweight steganography models.

Aimed at this, we propose to seek a much more lightweight network architecture that is specifically suitable for the deep image steganography task. However, without any prior knowledge to guide the structure design, it is impractical to try different lightweight architecture that are manually designed. Recently, Neural Architecture Search (NAS) tasks in the field of automatic machine learning have begun to emerge for solving the challenging task of large-scale image classification. Specifically, NAS firstly evaluated sample different sub-architectures, and then the candidate architectures are trained for solving the task at hand.

With such an automatic architecture searching strategy, we make an attempt on finding out a lightweight yet powerful deep steganography network architecture. Specifically, by using the metric results as optimization goal and constraining the searching space with a series of lightweight modules, we are able to obtain a more lightweight U-Net like model, which is able to produce excellent results while reducing the number of parameters to 4% of traditional U-Net [23]. In summary, our contributions are as follows:

- To our knowledge, our paper introduces the very first attempt to apply NAS for optimal network design for deep image steganography task.
- We propose different search spaces for the down-sampling and up-sampling structure, quarantining a better process for finding out the best network design.
- Extensive experiments have been carried out, and comparable results to the powerful model (i.e., UDH [10]) have been produced by our model, which is the most lightweight deep image steganography model by far.

2   Related Work

2.1   Deep Image Steganography

The network structure for the deep image steganography task can be briefly classified into three types: One is the encoder-decoder structure based on CNN, one uses GAN [4] and the other applied Invertible Neural Network (INN) [8] for secret image hiding and revealing. Shumeet first proposed the concept of deep steganography in 2017 with CNN [2]. Specifically, the proposed framework consists of three modules including the preparation network, the hiding network, and the revealing network. To reduce the network complexity and further enhance the generalization ability, StegNet [12] merges the preparation network and the hidden network into one, which solves the problem that the residual can see the prototype of the secret image, but requires the secret image and the cover image to be the same size. Zhang et al. proposed a Universal Deep Steganography model (UDH) [10], which removes the hidden Network and directly designs the preparation network to add the secret image to the cover image after preprocessing. Zhang et al. [27] use GAN to constrain the container image similar to the cover image with a discriminator. Followed CycleGAN [28], khan et al. [7] use two discriminators to constrain the container image and the extracted secret image, respectively. HiNet [29] and ISN [30] introducing Inversible Neural

Networks(INN) to Complete Steganography Tasks. These works achieved nice results all with a large number of parameters. We apply NAS for steganography and designed search pools with lightweight operations to reduce network parameters.

2.2 Neural Architecture Search

The goal of NAS is to use the high computing speed of the computer to obtain a network structure with excellent performance. Typical design includes three components: search space, search strategy, and performance estimation [25].

The early search strategy mainly included reinforcement learning [24] and evolutionary algorithms [26]. They would consume a lot of time and GPU to conduct simulation experiments. Recent work studies the one-shot idea, which is a two-stage algorithm. In the first stage, a supernet is trained, parameters are shared among all subnets, and the subnet with the highest score is finally identified as the searched network in the second stage. In 2019, Liu et al. proposed a one-shot-based differentiable optimization algorithm, DARTS [13]. The search space becomes continuous through the softmax operation, therefore the selection of the operations can be optimized through gradient descent. Xu et al. [14] optimized the calculation ratio of the number of channels, and the selection algorithm of the edge operation in the search stage, a faster search speed and a better structure can be found.

In recent work, NAS mainly uses image classification tasks to evaluate the performance of search results. Liu et al. [16] introduced NAS into the field of semantic segmentation, using the encoder-decoder structure, but the main search is the encoder, and the decoder directly uses fixed double Linear interpolation and ASPP method. Weng et al. [15] introduced NAS into the field of medical image segmentation, and defined a special search pool for the up-sampling structure. Inspired by these semantic segmentation work, we follows the differentiable NAS formulation [13,14] to prevent excessive resource consumption and extends it to steganography task.

3   Method

In this section we will explain in detail how we use NAS to search for steganography networks. First, we will introduce our search space for steganography tasks in Sect. 3.1, then we will describe the basic network structure we use in Sect. 3.2. Finally, we will introduce our search strategy in Sect. 3.3.

3.1   Search Space

We define a cell as a directed acyclic graph composed of N nodes, and its structure is shown in Fig. 2. Each node $n_i \in N$ represents an intermediate node, and the edge $e_{i,j}$ between two nodes i, j represents the operation O in the search space $\mathcal{O}$. We define that the function of each node is to convert two input tensors

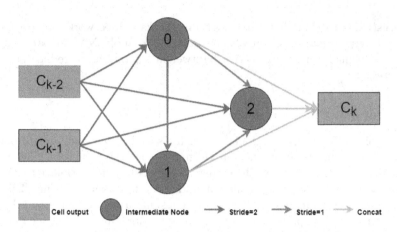

Fig. 2. Example of cell architecture. The output of the two cells is used as the input tensor to get an output tensor. The red arrow represents the operation with a stride of 2, the green arrow indicates the operation with a stride of 1, and the blue arrow represents the concatenate operation. (Color figure online)

into one output tensor, finally, concatenate the output of intermediate nodes as the output for cell.

$$\text{Cell}_k^{out} = \text{concat}(n_0^{out}, \cdots, n_{N-1}^{out}) \tag{1}$$

We divide the cell into two types: down-sampling cells (DownCell) and up-sampling cells (UpCell). The DownCell reduces the height and width of the feature map to half. The UpCell is in contrast to the DownCell. The height and width of the feature map are all multiplied by 2. To reduce parameters and save memory, we keep the number of channels equally with initial numbers in DownCell and UpCell.

Due to the encoder and decoder requiring different operations when the stride is 2, we define the DownCell and the UpCell as the Down-sampling Pool (DownPo), and the Up-sampling Pool (UpPo), as shown in Table 1.

Compared with the UpPo, the DownPo pool has two more operations: max pooling and average pooling. For the convolution operation, we use depthwise-separable convolution (SepConv) and dilated convolution (DilConv).

3.2 Architecture

In the image classification framework via the NAS used in DARTS [13], the structure of the entire network is predefined, and only cell blocks are searched to form the final network. We follow this common practice, not to search the entire network, but to make slight changes to the existing basic network.

The generally-used deep steganaography network, UDH, consists of two sub-networks, including the hiding network (H) for secret image generation followed by the revealing network (R) for secret image recovery. We follow this structure and simply set the extraction network as a superposition of several Conv-Bn-Relu blocks. We only search the structure for the H network.

Table 1. The operations included in the search pool of the two cells. Note that the number in parentheses indicates the size of the kernal used.

DownPo	UpPo
skip connection	skip connection
no connection	no connection
SepConv (3,5,7)	SepConv (3,5,7)
DilConv (3,5)	DilConv (3,5)
average pooling (3)	-
max pooling (3)	-

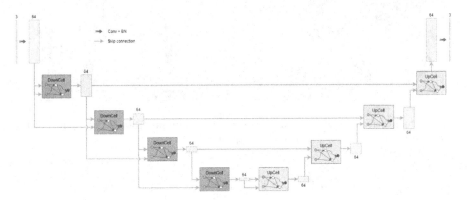

Fig. 3. Backbone network based on modified U-Net. The height and width of the yellow rectangle represent the size of the feature map and the number of channels, respectively. (Color figure online)

Unlike image classification tasks that obtain results from high-level semantics, steganography tasks need to obtain pixel-level (low-level semantic) results. Therefore, we use U-Net, an encoder-decoder network, as our backbone to reduce the memory. To make is better suitable for the task at hand, some basic modifications are made, as shown in Fig. 3. We define the input of the DownCell as the output of the last two cells, and for the UpCell, is the output of the previous cell and the output of the DownCell with the same resolution:

$$\text{DownCell}_k^{\text{input}} = (\text{Cell}_{k-1}^{\text{out}}, \text{Cell}_{k-2}^{\text{out}})$$

$$\text{UpCell}_k^{\text{input}} = (\text{Cell}_{k-1}^{\text{out}}, \text{Cell}_{l-k-1}^{\text{out}})$$

where l represents the total number of cells.

3.3 Search Strategy

As Fig. 1 shows, we respectively define cover image, secret image, container image and extracted image as C, S, C', S', the loss function of steganography can be defined as:

Fig. 4. Qualitative results for NAS-StegaNet. The columns from left to right indicate C, C', $S_e = C' - C$, S, S', and $S' - S$ respectively.

$$\mathcal{L}(S, S_e, S') = \|S_e\| + \beta \|S' - S\| \qquad (2)$$

where $S_e = C' - C$, $\|\cdot\|$ is ℓ_2 norm and β is a hyperparameter representing the reconstruction weight .

For the optimization of our model, the gradient based strategy in [14] was applied. In the search phase, we load the training set and validation set jointly, and then only optimize the network weights ω in the previous epoch:

$$\text{Update}_\omega = \text{argmin}_\omega \ \mathcal{L}_{\text{train}}(\omega, \alpha) \qquad (3)$$

After obtaining a network with better weights, we update the architecture weights α in each batch additional 1:

$$\text{Update}_\alpha = \min_\alpha \ \mathcal{L}_{\text{val}}(\text{Update}_\omega, \alpha) \qquad (4)$$

where $\mathcal{L}_{train}, \mathcal{L}_{val}$ are the $\mathcal{L}(S, S_e, S')$ obtained on the training datasets and validation datasets.

4   Experiments

4.1   Datasets and Metrics

Datasets: CIFAR10 [18] and ImageNet-1K [17] are widely used datasets for NAS research. CIFAR10 has a total of 60,000 color images, including 10,000

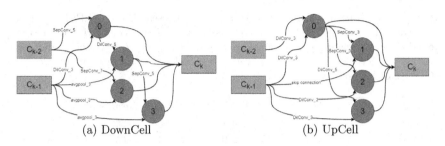

(a) DownCell   (b) UpCell

Fig. 5. Cells searched on CIFAR10. The number after the operation represents the kernel size used.

images in the test set and 50,000 images in the train set and all images have a resolution of 32×32. We can search on the low-resolution datasets to reduce the GPU days for searching an optimal structure. ImageNet-1K with 1.28 million average 469×387 color images is also used, and 50,000 are used for validation and 100,000 are used for testing. We resize image size to 128×128 and use approximately 90,000 images for training.

Metrics: Following the common setup in other deep image steganography researches [2, 10, 12], we use Peak Signal-to-Noise Ratio (PSNR), Structural Similarity (SSIM) and Perceptual Similarity (LPIPS), average pixel discrepancy (APD) to evaluate the effect of our model, and the APD are calculated as the ℓ_1 norm. The quantitative evaluations are carried out on both the container image (Ours-C in Table 2) and the revealed image (Ours-S in Table 2).

4.2   Experimental Settings

We first searched our network for a total of 50 epochs on CIFAR10. Because the network performance obtained by the search between the two is similar, but it takes 57 h to search on ImageNet-1K, and only 41 h on CIFAR10, which greatly reduces the search consumption time. The details of the experimental settings are as follows: Our network uses a total of $l = 8$ cells, of which DownCell and UpCell are 4, and each cell has $N = 4$ intermediate nodes. We set the batchsize to 200 and the learning rate to 0.005. In the first 15 epochs, we only update the network weights use the Adam optimizer. After that, the architecture weights are updated with Adam optimizer for each batch additional by reducing the learning rate from 6e-4, and with 1e-3 as the weight decay for architecture. Following [14], the hyperparameters K in the partial channel connection are set to 4 on datasets during the architecture searching.

After that, we retrained the searched structure on ImageNet-1K for 65 sepochs. We followed the setting of [10], set the learning rate to 0.001, and multiply it by 0.1 every 30 epochs. Meanwhile, the reconstruction weight hyperparameter β is set as 0.75 for all the experiments.

Table 2. Quantitative evaluation of the structure we searched on CIFAR10 and the results trained on ImageNet-1K are compared with the results of UDH and DDH [10]. The hiding and revealing performance are measured on the cover image C and secret image S, respectively. Note that for reveal network own the same parameters and FLOPs. (**Bold**: best)

	$APD \downarrow$	$PSNR \uparrow$	$SSIM \uparrow$	$LPIPS \downarrow$	$Params(M)$	$FLOPs(G)$
DDH-C	2.68	35.87	0.977	0.0046	16.66	359.67
UDH-C	2.35	39.13	**0.985**	**0.0001**	16.65	358.57
Ours-C(CIFAR10)	**2.26**	**39.23**	0.984	**0.0001**	**0.63**	**179.05**
DDH-S	3.50	34.72	**0.981**	**0.0071**	-	-
UDH-S	3.56	35.00	0.976	0.0136	-	-
Ours-S(CIFAR10)	**3.26**	**35.48**	0.978	0.0124	-	-

4.3   Results

Figure 5 visualizes the best architecture found on CIFAR10. And our evaluation results of the structures searched are shown in the Table 2. Our models have similar performance with UDH [10], while reducing the parameters from $16.65M$ to $0.69M$. These results suggest that our NAS-StegaNet can hide and reveal secret message more effectively. We observe several interesting phenomena. 1) For DownCell, edges with a stride of 2 (edges for input operations) tend to use the average pooling layer, while for UpCell, DilConv is more favored. This may mean that steganography prone retain the background features of secret information. 2) For all types of cells, have a higher output usage rate for the previous cell. We can see that all intermediate nodes in the cell structure searched on cifar10 will use C_{k-1} as an input. This may be because the deeper the network output contains more condensed global information.

4.4   Ablation Study

We first evaluated the influence of the SepConv on U-Net. The results are summarized in the Table 3. For the experimental, we simply convert the primitive convolutional layer in U-Net into SepConv. We found that although SepConv can reduce the parameters of the network, it will also cause the network to lose partial information, which leads to the deterioration of the secret information extraction. This represents the network structure we searched for is more suitable for steganography tasks.

Table 3. Ablation study for SepConv (w/o indicates without).

	PSNR-C	PSNR-S	Params(M)
U-Net(w/o SepConv)	39.13	35.00	16.65
U-Net(w/ SepConv)	39.31	33.30	1.10

Then we showed the effect on our models with network depth l (number of cells), and the network width ch (number of channels) in Table 4. Limited by time, we only evaluated the l of 4, 8 and the ch of 16, 64. The results demonstrate that superior performance can be obtained when $l = 8$ and $ch = 64$. Moreover, using a l of 4, we can further reduce the parameters while decreasing the value of PSNR slightly. Furthermore, a dramatic performance drop will occur when setting the value of ch as 16.

Table 4. Ablation study on CIFAR10 datasets for search and ImageNet-1K datasets for train. Depth represents the number of network layers l, and width represents the number of initial network channels ch.

Depth	Width	CIFAR10		
		PSNR-C	PSNR-S	Params(M)
8	16	35.95	31.56	0.06
4	64	38.87	35.19	0.30
8	64	39.10	35.39	0.63

4.5 Hiding Multiple Images

Figure 6 mean that our model can hide multiple images into one without great degradation. Therefore, the network we searched can be adapted to flexible hiding schemes which demonstrates the effectiveness of our lightweight model.

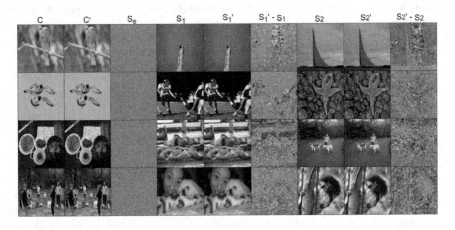

Fig. 6. Example of the result of hiding two S into one C on ImageNet-1K

5 Conclusions

Existing deep image steganography network (e.g., UDH) suffers from very large amount of parameters, making them impractical for deployment. We have solved this problem by making the first attempt on introducing NAS for a lightweight yet effective network design. By using the U-Net as the basic structure and performing cell-level searches, a novel NAS-StegaNet is searched and demonstrated to produce results that are close to or even better than the powerful UDH, but with much less parameter that are only 4% of UDH. To our knowledge, our NAS-StegaNet is the most lightweight image steganography model, and can be used for both hiding one image in another and hiding M images in N images. At present, the reveal network is not searched together with the hide network, which may cause network inconsistency, resulting in the degradation of model performance, and the security of the model is not verified. We propose undertake further research on these two weak points.

References

1. Mondal, S., Ling, Y., Ambikapathi, A.: H-Stegonet: a hybrid deep learning framework for robust steganalysis. In: 2021 IEEE International Conference on Multimedia and Expo (ICME), pp. 1–6 (2021)
2. Baluja, S.: Hiding images in plain sight: deep steganography. In: Advances in Neural Information Processing Systems (NIPS), vol. 30 (2017)
3. Hussain, M., Wahab, A., Idris, Y., Ho, A., Jung, K.: Image steganography in spatial domain: a survey. In: Signal Processing: Image Communication, vol. 65, pp. 46–66 (2018)
4. Byrnes, O., La, W., Wang, H., Ma, C., Xue, M., Wu, Q.: Data hiding with deep learning: a survey unifying digital watermarking and steganography. In: arXiv preprint, arXiv:2107.09287 (2021)
5. Pevný, T., Filler, T., Bas, P.: Using high-dimensional image models to perform highly undetectable steganography. In: Böhme, R., Fong, P.W.L., Safavi-Naini, R. (eds.) IH 2010. LNCS, vol. 6387, pp. 161–177. Springer, Heidelberg (2010). https://doi.org/10.1007/978-3-642-16435-4_13
6. Das, A., Wahi, J.S., Anand, M., Rana, Y.: Multi-image steganography using deep neural networks. In: arXiv preprint, arXiv:2101.00350 (2021)
7. Khan, N., Haan, R., Boktor, G., McComas, M., Daneshi, R.: Steganography GAN: cracking steganography with cycle generative adversarial networks. In: arXiv preprint, arXiv:2006.04008 (2020)
8. Dinh, L., David, K., Yoshua, B.: Nice: non-linear independent components estimation. In: arXiv preprint, arXiv:1410.8516 (2014)
9. Lu, S., Wang, R., Zhong, T., Rosin, P.: Large-capacity image steganography based on invertible neural networks. In: Proceedings of the IEEE/CVF Conference on Computer Vision and Pattern Recognition (CVPR), pp. 10816–10825 (2021)
10. Zhang, C., Benz, P., Karjauv, A., Sun, G., Kweon, I.: UDH: universal deep hiding for steganography, watermarking, and light field messaging. In: Advances in Neural Information Processing Systems (NIPS), vol. 33, pp. 10223–10234 (2020)
11. Zhu, J., Kaplan, R., Johnson, J., Li, F., Jung, K.: HiDDeN: hiding data with deep networks. In: Proceedings of the European Conference on Computer Vision (ECCV), pp. 657–672 (2018)

12. Wu P., Yang Y., Li, X.: StegNet: mega image steganography capacity with deep convolutional network. In: Future Internet, vol. 10 (2018)
13. Liu, H., Simonyan, K., Yang, Y.: DARTS: differentiable architecture search. In: arXiv preprint, arXiv:1806.09055 (2018)
14. Xu, Y., et al.: PC-DARTS: partial channel connections for memory-efficient architecture search. In: arXiv preprint, arXiv:1907.05737 (2019)
15. Weng, Y., Zhou, T., Li, Y., Qiu, X.: NAS-Unet: neural architecture search for medical image segmentation. In: IEEE Access, vol. 7, pp. 44247–44257 (2019)
16. Liu, C., et al.: Auto-deeplab: hierarchical neural architecture search for semantic image segmentation. In: Proceedings of the IEEE/CVF Conference on Computer Vision and Pattern Recognition (CVPR), pp. 82–92 (2019)
17. Deng, J., Dong, W., Socher, R., Li, J., Li, K., Li, F.: ImageNet: a large-scale hierarchical image database. In: Proceedings of the IEEE/CVF Conference on Computer Vision and Pattern Recognition (CVPR), pp. 248–255 (2009)
18. Alex, K.: Learning multiple layers of features from tiny images. In: Citeseer (2009)
19. Kessler, G., Hosmer, C.: An overview of steganography. In: Advances in Computers, vol. 83, pp. 51–107 (2011)
20. Isola, P., Zhu, J., Zhou, T., Efros, A.: Image-to-image translation with conditional adversarial networks. In: Proceedings of the IEEE/CVF Conference on Computer Vision and Pattern Recognition (CVPR), pp. 1125–1134 (2017)
21. Boroumand, M., Mehdi, C., Fridrich, J.: Deep residual network for steganalysis of digital images. In: IEEE Transactions on Information Forensics and Security (TIFS), vol. 14, pp. 1181–1193 (2018)
22. Zhang, R., Zhu, F., Liu, J., Liu, G.: Depth-wise separable convolutions and multi-level pooling for an efficient spatial CNN-based steganalysis. In: IEEE Transactions on Information Forensics and Security (TIFS), vol. 15, pp. 1138–1150 (2019)
23. Ronneberger, O., Fischer, P., Brox, T.: U-Net: convolutional networks for biomedical image segmentation. In: International Conference on Medical Image Computing and Computer-Assisted Intervention, pp. 234–241 (2015)
24. Pham, H., Guan, M., Zoph, B., Le, Q., Dean, J.: Efficient neural architecture search via parameters sharing. In: International Conference on Machine Learning (ICML), pp. 4095–4104 (2018)
25. Elsken, T., Metzen, J., Hutter, F.: Neural architecture search: a survey. In: The Journal of Machine Learning Research, vol. 20, pp. 1997–2017 (2019)
26. Real, E., Aggarwal, A., Huang, Y., Le, Q.: Regularized evolution for image classifier architecture search. In: Proceedings of the Association for the Advancement of Artificial Intelligence (AAAI), vol. 33, pp. 4780–4789 (2019)
27. Zhang, K., Alfredo, C., Xu, L., Veeramachaneni, K.: SteganoGAN: high capacity image steganography with GANs. In: arXiv preprint, arXiv:1901.03892 (2019)
28. Zhu, J., Park, T., Isola, P., Efros, A.: Unpaired image-to-image translation using cycle-consistent adversarial networks. In: Proceedings of the IEEE International Conference on Computer Vision (ICCV), pp. 2223–2232 (2017)
29. Jing, J., Deng, X., Xu, M., Wang, J., Guan, Z.: HiNet: deep image hiding by invertible network. In: Proceedings of the IEEE International Conference on Computer Vision (ICCV), pp. 4733–4742 (2021)
30. Lu, S., Wang, R., Zhang, T., Rosin, P.: Large-capacity image steganography based on invertible neural networks. In: Proceedings of the IEEE/CVF Conference on Computer Vision and Pattern Recognition (CVPR), pp. 10816–10825 (2021)

FIT: Frequency-Based Image Translation for Domain Adaptive Object Detection

Siqi Zhang[1,2], Lu Zhang[1], Zhiyong Liu[1,2(✉)], and Hangtao Feng[1,2]

[1] State Key Laboratory of Management and Control for Complex Systems,
Institute of Automation, Chinese Academy of Sciences, Beijing 100190, China
{zhangsiqi2020,lu.zhang,zhiyong.liu,fenghangtao2018}@ia.ac.cn
[2] School of Artificial Intelligence, University of Chinese Academy of Sciences,
Beijing 100190, China

Abstract. Domain adaptive object detection (DAOD) aims to adapt the detector from a labelled source domain to an unlabelled target domain. In recent years, DAOD has attracted massive attention since it can alleviate performance degradation due to the large shift of data distributions in the wild. To align distributions between domains, adversarial learning is widely used in existing DAOD methods. However, the decision boundary for the adversarial domain discriminator may be inaccurate, causing the model biased towards the source domain. To alleviate this bias, we propose a novel Frequency-based Image Translation (FIT) framework for DAOD. First, by keeping domain-invariant frequency components and swapping domain-specific ones, we conduct image translation to reduce domain shift at the input level. Second, hierarchical adversarial feature learning is utilized to further mitigate the domain gap at the feature level. Finally, we design a joint loss to train the entire network in an end-to-end manner without extra training to obtain translated images. Extensive experiments on three challenging DAOD benchmarks demonstrate the effectiveness of our method.

Keywords: Unsupervised Domain Adaptation · Object Detection · Frequency Domain · Image Translation · Adversarial Learning

1 Introduction

In recent years, object detectors [1–3] based on deep convolutional neural networks have demonstrated outstanding performance on a variety of datasets. However, existing object detection models still face serious challenges when deployed in practice such as autonomous driving and robotic manipulation, due to various changes in weather, illumination, object appearance, *etc.* These changes may lead to domain gaps between the training and testing data, which has been

This work was supported in part by the National Key Research and Development Plan of China under Grant 2020AAA0108902 and the Strategic Priority Research Program of Chinese Academy of Science under Grant XDB32050100.

(a)   (b)   (c)

Fig. 1. Visualization of frequency decomposition of source image: (a), (b) and (c) show original image, low-pass and high-pass filtered image.

observed to cause dramatic drops in the performance of the trained detector [6]. Although we can annotate for each new data to mitigate the problem, it is costly and even infeasible because of the countless situations in the real world. Therefore, adaptive object detectors that can bridge the domain gap from the source to the target domain are highly desirable.

Domain adaptive object detection (DAOD), which trains with labelled source datasets and unlabelled target datasets, aims to tackle domain shift to get better performance on the visually distinct target domain. Many previous works [4,6,7,20,22–24] attempt to utilize adversarial feature learning [8] to align feature distributions to extract domain-invariant features. But the adversarial training process could be unstable [22,24], which makes the decision boundary for the adversarial domain discriminator inaccurate, causing the model biased towards the source domain. To alleviate this problem, some methods [5,9,16,21] utilize the image translation model GANs, like CycleGAN [10] to translate source images to target-like images or vice versa to further mitigate the domain gap and make the detector perform better on the target datasets. However, GANs for domain adaption object detection have two following limitations. First, GANs could fail to keep semantic consistency and tend to lose important structural characteristics [25]. Second, GANs-based methods need extra training to prepare translated images before training the adaptive detector, which is time-consuming.

To address the above limitations, we propose a novel Frequency-based Image Translation method to mitigate the input-level domain gap without extra time-consuming training. Inspired by digital signal processing theories [17], we exploit the frequency information to translate the image style and maintain semantic consistency. Intuitively, the low-frequency component largely captures domain-specific information, such as colours and illuminations [26], while the high-frequency component mainly obtains domain-invariant information, such as edges and shapes, which are important details of objects [19], as shown in Fig. 1. Motivated by this, we present the Frequency-based Image Translation (FIT)

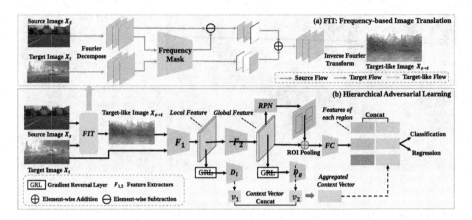

Fig. 2. Overview of the proposed framework. (a) illustrates Frequency-based Image Translation (FIT) module, where green arrows represent the flow of source data and orange arrows represent the flow of target data. The overall pipeline is illustrated in (b). The source X_s and target X_t images are fed into FIT to obtain target-like images $X_{s\rightarrow t}$, and $X_{s\rightarrow t}$ and X_t are as the inputs for the object detector. We align the local and global feature by a local domain classifier D_l and a global domain classifier D_g. F_1 and F_2 denote the different level feature extractors. The context vectors $v_{1,2}$ are extracted by the domain classifiers and concatenated with features of regions before the final fully connected layer. (Color figure online)

module, which decomposes the image into multiple frequency components, keeps domain-invariant frequency components unchanged and swaps domain-specific ones. Moreover, a novel module called Frequency Mask is designed to identify whether the frequency component is domain-specific in FIT. Then, hierarchical adversarial feature learning is utilized to further boost the performance. The entire network can be optimized in an end-to-end manner under the supervision of a joint loss function. The contributions of this work can be summarized as follows:

– A novel Frequency-based Image Translation (FIT) method is presented for DAOD, which leverages frequency information to mitigate the domain shift at the input level. To further boost the adaptation performance, we introduce hierarchical adversarial learning to align distributions at the feature level.
– Different from traditional GANs-based methods, the entire network can be trained in an end-to-end manner without extra time-consuming training, since the proposed frequency-based image translation is embedded as a module in the detection network.
– We conduct extensive experiments on three challenging DAOD benchmarks and our FIT achieves favorable performance under various domain-shift scenarios, demonstrating the effectiveness of the proposed method.

2 Proposed Method

2.1 Overview

Problem Definition. The domain adaptation [8] task typically considers two domains: the source domain S and target domain T. Specifically, we have access to a labelled source dataset $\mathcal{D}_s = \{(x_i^s, y_i^s)\}_{i=1}^{n_s}$ and a target dataset $\mathcal{D}_t = \{x_j^t\}_{j=1}^{n_t}$ with no ground-truth annotations. Here, x_i^s denotes i^{th} source image, y_i^s denotes the corresponding label and n_s denotes the number of source images. Similarly, x_j^t denotes j^{th} target image and n_t denotes the number of target images. The source and target domains with different data distributions share the same label space, and the goal of domain adaptive object detection is to train an detector with $\mathcal{D}_s$ and $\mathcal{D}_t$, which performs well on the target dataset $\mathcal{D}_t$. Following the mainstream domain adaptive object detection methods [4,6,7,20,22–24], the proposed method is based on the Faster RCNN [1] framework.

Overall Framework. The overall framework of the proposed method is shown in Fig. 2. We first transform source images X_s to target-like images $X_{s \to t}$ via frequency-based image translation (FIT), as shown in Fig. 2(a). The key idea is to decompose the image into multiple frequency components and then feed them to the Frequency Mask to identify domain-specific frequency components. Then we replace the domain-specific components of the source image with the corresponding ones of the target image and get the target-like image $X_{s \to t}$ via the Inverse Fourier Transform. Afterwards, we put target-like images $X_{s \to t}$ and target images X_t into object detector and align the local and global feature by hierarchical adversarial learning, as shown in Fig. 2(b). Through this framework, the domain gap at both input and feature level can be mitigated. The details of the proposed method are given in the following sections.

2.2 Frequency-Based Image Translation

In order to mitigate the domain gap at the input level, a novel frequency-based image translation is presented to obtain translated images without changing their semantic structures. The framework of frequency-based image translation is shown in Fig. 2(a).

First, Fourier transform $\mathcal{F}(\cdot)$ is performed on the image x of size $H \times W$:

$$\mathcal{F}(x)(a, b) = \sum_{h=0}^{H-1} \sum_{w=0}^{W-1} x(h, w) e^{-i2\pi \cdot \left(\frac{ha}{H} + \frac{wb}{W}\right)}, \tag{1}$$

for $a = 0, \ldots, H-1$, $b = 0, \ldots, W-1$.

Then, we decompose the frequency space representation $\mathcal{F}(x)$ of the image into N components $\{x^1, x^2, \ldots, x^N\}$ of equal bandwidth via band-pass filter $\mathcal{B}(\cdot; \cdot)$:

$$x^{fs} = \mathcal{B}(\mathcal{F}(x); N) = \{x^1, x^2, \ldots, x^{N-1}, x^N\}, \tag{2}$$

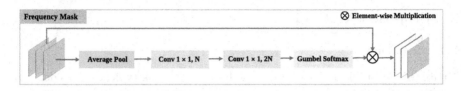

Fig. 3. Structure of Frequency Mask.

$$x^n = \begin{cases} \mathcal{F}(x)(i,j), & \text{if } \frac{n-1}{N} < d\left((i,j),(c_i,c_j)\right) < \frac{n}{N} \\ 0, & \text{otherwise} \end{cases}, \tag{3}$$

where c_i and c_j denote the image centroid, $d(\cdot,\cdot)$ denotes the Euclidean distance, and N is the number of components. In our experiments, we set $N = 64$.

To identify which frequency component is domain-specific, we design a module called Frequency Mask and its structure is shown in Fig. 3. Motivated by the Squeeze-and-Excitation Networks [15], which model the interdependencies between the channels and recalibrate the channel-wise feature responses adaptively, we design similar structure and add Gumbel-Softmax [18] to make the value close to one-hot vector. '1' means the frequency component is domain-specific, while '0' denotes it is domain-invariant. By Frequency Mask, we find the domain-specific components $DS(x_s^{fs})$:

$$DS(x_s^{fs}) = M\left(x_s^{fs}\right) \cdot x_s^{fs}, \tag{4}$$

where $M(x_s^{fs})$ represents the output of Gumbel-Softmax in Fig. 3. Then, we replace the domain-specific components of the source image with the corresponding ones of the target image:

$$\hat{x}_{s \to t}^{fs} = x_s^{fs} - DS(x_s^{fs}) + DS(x_t^{fs}), \tag{5}$$

After replacing components, we combine all frequency components and perform Inverse Fourier transform $\mathcal{F}^{-1}(\cdot)$. Finally, we obtain the target-like image:

$$x_{s \to t} = \mathcal{F}^{-1}\left(\sum \hat{x}_{s \to t}^{fs}\right). \tag{6}$$

In order to keep the consistency of semantic information, we regulate the reconstruction loss:

$$\mathcal{L}_{rec}(X) = \left\| H(X) - H(\hat{X}) \right\|_1, \tag{7}$$

where X and $\hat{X}$ represent the original and translated image. $H(\cdot)$ represents the band-pass filter that extracts the middle and high-frequency components, which largely capture the semantic information.

2.3   Hierarchical Adversarial Feature Learning

After the frequency-based image translation, we put target-like and target images into the object detector and further mitigate the feature-level domain gap by the domain classifier and gradient reversal layer (GRL) [8]. Since different domains could have distinct scene layouts, fully matching the entire distributions of source and target images at the global image-level may fail [4,20]. Therefore, we adopt different strategies on the local and global features.

The global feature alignment module consists of a global domain classifier D_g and a GRL. The GRL connects the global domain classifier and the backbone, which reverses the gradients that flow through the backbone, as shown in Fig. 2(b). It means that the global domain classifier D_g aims to distinguish which domain the global feature comes from, whereas the backbone attempts to confuse the classifier. Here, the source images are given the domain label $d = 0$ and the label is 1 for the target images. The loss of the global feature alignment module is calculated as follows,

$$\mathcal{L}_{glb_s} = -\frac{1}{n_s} \sum_{i=1}^{n_s} D_g \left(F_2(F_1\left(x_i^s\right))\right)^\gamma \cdot \log \left(1 - D_g \left(F_2(F_1\left(x_i^s\right))\right)\right), \quad (8)$$

$$\mathcal{L}_{glb_t} = -\frac{1}{n_t} \sum_{i=1}^{n_t} \left(1 - D_g \left(F_2 \left(F_1 \left(x_i^t\right)\right)\right)\right)^\gamma \cdot \log \left(D_g \left(F_2 \left(F_1 \left(x_i^t\right)\right)\right)\right), \quad (9)$$

$$\mathcal{L}_{glb} = \frac{1}{2} \left(L_{glb_s} + L_{glb_t}\right), \quad (10)$$

where n_s and n_t represent the number of source and target images, x^s and x^t are the target-like and target images, and F_1 and F_2 denotes the first seven convolutional layers of the backbone VGG16 and the rest convolutional layers. The detailed structure of global domain classifier D_g is shown in Fig. 4(a).

Similar with the adversarial training in global alignment, the local domain classifier D_l and shallow layers of the backbone are connected by the GRL. The loss function of local alignment can be written as:

$$\mathcal{L}_{loc_s} = \frac{1}{n_s HW} \sum_{i=1}^{n_s} \sum_{w=1}^{W} \sum_{h=1}^{H} D_l \left(F_1\left(x_i^s\right)\right)_{wh}^2, \quad (11)$$

$$\mathcal{L}_{loc_t} = \frac{1}{n_t HW} \sum_{i=1}^{n_t} \sum_{w=1}^{W} \sum_{h=1}^{H} \left(1 - D_l \left(F_1 \left(x_i^t\right)\right)_{wh}\right)^2, \quad (12)$$

$$\mathcal{L}_{loc} = \frac{1}{2} \left(\mathcal{L}_{loc_s} + \mathcal{L}_{loc_t}\right), \quad (13)$$

where $D_l \left(F_1 \left(x_i\right)\right)_{wh}$ represents the output of the local domain classifier D_l in each location. The detailed structure of local domain classifier D_l is shown in Fig. 4(b).

To achieve better adaptation, we regularize the domain discriminator. Previous work has shown that it is effective for stabilizing the adversarial training by regularizing the domain classifier with the segmentation loss in domain adaptive segmentation [27]. Similar with this approach, we regularize the domain discriminator with the detection loss. Formally, we extract the different levels of

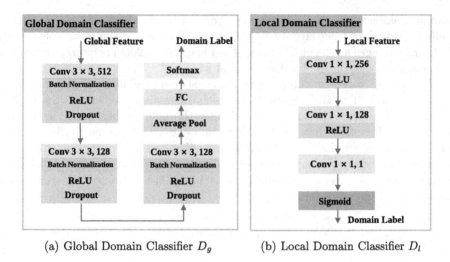

(a) Global Domain Classifier D_g (b) Local Domain Classifier D_l

Fig. 4. Structure of Domain Classifiers.

context vectors v_1 and v_2 from the middle layers of the domain classifiers D_l snd D_g respectively. Each context vector has 128 dimensions. Then, we concatenate the vectors to obtain the aggregated context vector and all region-wise features are concatenated with the aggregated context to train the domain classifiers to minimize the detection loss and domain classification loss, as illustrated in Fig. 2(b).

2.4 Overall Objective

We denote the loss of Faster RCNN [1] as $\mathcal{L}_{det}$ and the overall loss function $\mathcal{L}_{total}$ can be summarized as:

$$\mathcal{L}_{total} = \mathcal{L}_{det} + \mathcal{L}_{rec} + \lambda(\mathcal{L}_{glb} + \mathcal{L}_{loc}), \tag{14}$$

where λ is the hyper-parameter to balance the detection, reconstruction loss and hierarchical alignment losses.

3 Experiments

3.1 Datasets

We extensively evaluate our approach on three challenging domain adaptive object detection tasks with distinct domain shifts, including adaptation under different weather (Cityscapes [11] → Foggy Cityscapes [12]), adaptation from the synthetic to the real scene (Sim10K [13] → Cityscapes) and adaptation under different cameras (KITTI [14] → Cityscapes). Cityscapes [11] is a dataset of urban street scenes with 8 categories captured with on-board cameras, which

Table 1. Results (%) on the adaptation from Cityscapes to Foggy Cityscapes. 'No DA' indicates the model is only trained with the source images and directly tested on the target images without any domain adaptation. The best results are in **bold**, and the second best results are underlined.

Methods	person	rider	car	truck	bus	train	moto	bicycle	mAP
No DA	23.3	27.9	32.8	11.4	23.5	9.3	12.2	25.2	20.7
DA [6]$_{CVPR'2018}$	25.0	31.0	40.5	22.1	35.3	20.2	20.0	27.1	27.6
DivMatch [9]$_{CVPR'2019}$	30.8	40.5	44.3	27.2	38.4	34.5	28.4	32.2	34.6
SWDA [4]$_{CVPR'2019}$	29.9	42.3	43.5	24.5	36.2	32.6	30.0	35.3	34.3
HTCN [5]$_{CVPR'2020}$	33.2	**47.5**	47.9	31.6	47.4	**40.9**	32.3	37.1	39.8
CDN [7]$_{ECCV'2020}$	35.8	45.7	50.9	30.1	42.5	29.8	30.8	36.5	36.6
ATF [20]$_{ECCV'2020}$	34.6	<u>47.0</u>	50.0	23.7	43.3	38.7	33.4	<u>38.8</u>	38.7
Progressive [16]$_{WACV'2020}$	36.0	45.5	**54.4**	24.3	44.1	25.8	29.1	35.9	36.9
VDD [22]$_{ICCV'2021}$	33.4	44.0	51.7	**33.9**	**52.0**	34.7	34.2	36.8	<u>40.0</u>
CDTD [21]$_{IJCV'2021}$	31.6	44.0	44.8	30.4	41.8	40.7	<u>33.6</u>	36.2	37.9
RPA [23]$_{CVPR'2021}$	33.4	44.3	50.1	29.9	44.8	<u>39.1</u>	29.9	36.3	38.5
DDF [24]$_{TMM'2022}$	**37.2**	46.3	51.9	24.7	43.9	34.2	33.5	**40.8**	39.1
FIT-DA (Ours)	<u>36.6</u>	45.8	<u>52.2</u>	<u>32.2</u>	<u>48.1</u>	34.6	**34.7**	37.2	**40.2**

has 2975 training images and 500 validating images. Foggy Cityscapes [12] is the synthetic foggy version of Cityscapes. Sim10K [13] is a virtual dataset including 10000 images generated by the Grand Theft Auto gaming engine. KITTI [14] is an autonomous driving dataset that has 7481 images, which is captured by a standard station wagon with two high-resolution video cameras. In the test, we use mean average precision (mAP) metrics for evaluation.

3.2 Implementation Details

Our detector is original Faster R-CNN [1] without extra modules. We adopt VGG-16 [28] pre-trained on ImageNet [29] as our backbone. In our experiments, the shorter side of the image is resized to 600. Each batch is composed of one source image and one target image. The networks are trained with a learning rate of 0.001 for 50K iterations, then with a learning rate of 0.0001 for 20K more iterations. We use a momentum of 0.9 and a weight decay of 0.0005. N is 64 in Eq. (2). For Sim10K $\rightarrow$ Cityscapes, we set $\lambda = 0.1$ in Eq. (14). For the rest two tasks, we set $\lambda = 1$. Our method is implemented with PyTorch.

3.3 Comparison Experiments

Adaptation Under Different Weather. Table 1 shows the performance of our method on Cityscapes $\rightarrow$ Foggy Cityscapes. We can see that our method alleviates the domain gap across different weather conditions and outperforms all competitors in Table 1. Compared with GANs-based methods: DivMatch [9], Progressive [16] and CDTD [21], our method improves the result by +5.6%, +3.3% and +2.3% in mAP, which demonstrates the advantage of the proposed Frequency-based Image Translation for domain adaptive object detection.

Table 2. Sim10K to Cityscape.

Methods	mAP
Source Only	34.2
DA [6]$_{CVPR'2018}$	39.0
SWDA [4]$_{CVPR'2019}$	40.1
HTCN [5]$_{CVPR'2020}$	42.5
ATF [20]$_{ECCV'2020}$	42.8
CDTD [21]$_{IJCV'2021}$	42.6
RPA [23]$_{CVPR'2021}$	45.7
DDF [24]$_{TMM'2022}$	44.3
FIT-DA (Ours)	**48.6**

Table 3. KITTI to Cityscapes.

Methods	mAP
Source Only	32.2
DA [6]$_{CVPR'2018}$	38.5
SWDA [4]$_{CVPR'2019}$	43.1
CDN [7]$_{ECCV'2020}$	44.9
ATF [20]$_{ECCV'2020}$	42.1
Progressive [16]$_{WACV'2020}$	43.9
DDF [24]$_{TMM'2022}$	46.0
FIT-DA (Ours)	**46.3**

Table 4. Ablation analysis of our method. LA is local feature alignment and GA is global feature alignment. CTV represents the context vector and FIT denotes the frequency-based image translation.

Methods	LA	GA	CTV	FIT	C → F	S → C
Source only					20.7	34.2
FIT-DA				✓	31.8	40.5
FIT-DA		✓	✓	✓	38.5	46.4
FIT-DA	✓		✓	✓	35.7	42.2
FIT-DA	✓	✓		✓	37.4	44.3
FIT-DA	✓	✓	✓		34.5	40.3
FIT-DA (Ours)	✓	✓	✓	✓	**40.2**	**48.6**

Adaptation from the Synthetic to Real Scene. We evaluate the detection performance on car on Sim10K to Cityscapes benchmark. As we can see the results in Table 2, our method has a significant performance boost over other methods, further indicating the effectiveness of our method.

Adaptation Under Different Cameras. There exists a domain gap between datasets captured through different cameras due to the diversity of hardware devices. We conduct the cross-camera adaptation from KITTI to Cityscapes. The results are presented in Table 3, and our method has competitive performance among all the comparison methods.

3.4   Ablation Study

Effectiveness of Each Component. We conduct the ablation experiments on Cityscapes → Foggy Cityscapes (C → F) and Sim10K → Cityscapes (S → C) to validate the effectiveness of each module in our framework. The results in Table 4 show that all the modules contribute to the performance improvement

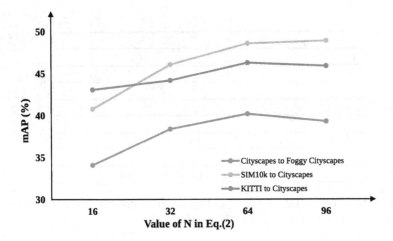

Fig. 5. Detection performances on the three benchmarks with different values of N.

Table 5. Performances with different choices of domain-specific frequency components.

Settings	Choice	C → F	S → C
Non-learnable	FC[1]	38.3	44.2
	FC[1, 2]	37.3	45.9
	FC[1, 2, 3]	36.1	41.8
	FC[1, 2, 3, 4]	33.9	40.0
Learnable	Frequency mask	**40.2**	**48.6**

(especially FIT module), which indicates the effectiveness of each component in our method.

Method of Choosing Domain-Specific Frequency Components. In FIT framework, Frequency Mask is the core module, which choose the domain-specific frequency component in a learnable way. We compare the adaptation performance of using Frequency Mask with using fixed low-frequency components to determine domain-specific components in Table 5. The results suggest the Frequency Mask captures the domain-specific information better. Although low-frequency components largely captures domain-specific information, the distributions of domain-specific components are not completely consistent for images from different domains, making it difficult to capture these components preciously just using fixed low-frequency components.

Value of N in Eq. 2. N is the number of frequency components after Fourier decomposing. Figure 5 shows the influence on adaptation performance with different N. As N is related to the division of frequency bands, which is critical for finding domain-specific frequency components, it affects the quality of translated images. In our experiments, $N = 64$ is the best choice considering the performance on the three benchmarks.

$$\text{C} \rightarrow \text{F} \qquad\qquad \text{C} \rightarrow \text{F} \qquad\qquad \text{S} \rightarrow \text{C} \qquad\qquad \text{S} \rightarrow \text{C}$$

Fig. 6. Example results on Cityscapes to Foggy Cityscapes (C → F) and Sim10K to Cityscapes (S → C). The fist row is the results of SWDA and the second row is the results of our FIT-DA. The class and score predictions are at the top left corner of the bounding box. Zoom in to visualize the details.

3.5   Visualization

Figure 6 illustrates some examples of detection results on Cityscapes to Foggy Cityscapes and Sim10K to Cityscapes. Obviously, our method produces more accurate bounding box predictions and has a stronger ability to detect obscured instances.

4   Conclusion

In this paper, a novel Frequency-based Image Translation (FIT) method for DAOD is presented to reduce domain shift at the input level. Compared to other image translation methods for DAOD, it is embedded in the detection network and does not need extra time-consuming training. Additionally, we introduce hierarchical adversarial feature learning to further mitigate the domain gap at the feature level. Meanwhile, a joint loss function is designed to optimize the entire network in an end-to-end manner. Extensive experiments on three challenging DAOD benchmarks validate the effectiveness of our method. In the future, we will utilize the frequency information in the feature space to investigate the feature augmentation for DAOD.

Acknowledgements. This work was supported in part by the National Key Research and Development Plan of China under Grant 2020AAA0108902 and the Strategic Priority Research Program of Chinese Academy of Science under Grant XDB32050100.

References

1. Ren, S., He, K., Girshick, R., Sun, J.: Faster R-CNN: towards real-time object detection with region proposal networks. Adv. Neural. Inf. Process. Syst. **28**, 91–99 (2015)

2. Redmon, J., Divvala, S., Girshick, R., Farhadi, A.: You only look once: unified, real-time object detection. In: Proceedings of the IEEE Conference on Computer Vision and Pattern Recognition, pp. 779–788 (2016)
3. Tian, Z., Shen, C., Chen, H., He, T.: FCOS: fully convolutional one-stage object detection. In: Proceedings of the IEEE International Conference on Computer Vision, pp. 9627–9636 (2019)
4. Saito, K., Ushiku, Y., Harada, T., Saenko, K.: Strong-weak distribution alignment for adaptive object detection. In: Proceedings of the IEEE Conference on Computer Vision and Pattern Recognition, pp. 6956–6965 (2019)
5. Chen, C., Zheng, Z., Ding, X., Huang, Y., Dou, Q.: Harmonizing transferability and discriminability for adapting object detectors. In: Proceedings of the IEEE Conference on Computer Vision and Pattern Recognition, pp. 8869–8878 (2020)
6. Chen, Y., Li, W., Sakaridis, C., Dai, D., Van Gool, L.: Domain adaptive faster R-CNN for object detection in the wild. In: Proceedings of the IEEE Conference on Computer Vision and Pattern Recognition, pp. 3339–3348 (2018)
7. Su, P., et al.: Adapting object detectors with conditional domain normalization. In: Vedaldi, Andrea, Bischof, Horst, Brox, Thomas, Frahm, Jan-Michael. (eds.) ECCV 2020. LNCS, vol. 12356, pp. 403–419. Springer, Cham (2020). https://doi. org/10.1007/978-3-030-58621-8_24
8. Ganin, Y., Lempitsky, V.: Unsupervised domain adaptation by backpropagation. In: International Conference on Machine Learning, pp. 1180–1189. PMLR (2015)
9. Kim, T., Jeong, M., Kim, S., Choi, S., Kim, C.: Diversify and match: a domain adaptive representation learning paradigm for object detection. In: Proceedings of the IEEE Conference on Computer Vision and Pattern Recognition, pp. 12456–12465 (2019)
10. Zhu, J.Y., Park, T., Isola, P., Efros, A.A.: Unpaired image-to-image translation using cycle-consistent adversarial networks. In: Proceedings of the IEEE International Conference on Computer Vision, pp. 2223–2232 (2017)
11. Cordts, M., et al.: The cityscapes dataset for semantic urban scene understanding. In: Proceedings of the IEEE Conference on Computer Vision and Pattern Recognition, pp. 3213–3223 (2016)
12. Sakaridis, C., Dai, D., Van Gool, L.: Semantic foggy scene understanding with synthetic data. Int. J. Comput. Vis. **126**(9), 973–992 (2018). https://doi.org/10. 1007/s11263-018-1072-8
13. Johnson-Roberson, M., Barto, C., Mehta, R., Sridhar, S.N., Rosaen, K., Vasudevan, R.: Driving in the matrix: can virtual worlds replace human-generated annotations for real world tasks? In: 2017 IEEE International Conference on Robotics and Automation, pp. 746–753. IEEE (2017)
14. Geiger, A., Lenz, P., Stiller, C., Urtasun, R.: Vision meets robotics: the KITTI dataset. Int. J. Robot. Res. **32**(11), 1231–1237 (2013)
15. Hu, J., Shen, L., Sun, G.: Squeeze-and-excitation networks. In: Proceedings of the IEEE Conference on Computer Vision and Pattern Recognition, pp. 7132–7141 (2018)
16. Hsu, H.K., et al.: Progressive domain adaptation for object detection. In: Proceedings of the IEEE Winter Conference on Applications of Computer Vision, pp. 749–757 (2020)
17. Oppenheim, A.V.: Discrete-Time Signal Processing. Pearson Education India (1999)
18. Jang, E., Gu, S., Poole, B.: Categorical reparameterization with Gumbel-Softmax. arXiv preprint arXiv:1611.01144 (2016)

19. Li, J., Duan, L.Y., Chen, X., Huang, T., Tian, Y.: Finding the secret of image saliency in the frequency domain. IEEE Trans. Pattern Anal. Mach. Intell. **37**(12), 2428–2440 (2015)
20. He, Zhenwei, Zhang, Lei: Domain adaptive object detection via asymmetric tri-way faster-RCNN. In: Vedaldi, Andrea, Bischof, Horst, Brox, Thomas, Frahm, Jan-Michael. (eds.) ECCV 2020. LNCS, vol. 12369, pp. 309–324. Springer, Cham (2020). https://doi.org/10.1007/978-3-030-58586-0_19
21. Shen, Z., et al.: CDTD: a large-scale cross-domain benchmark for instance-level image-to-image translation and domain adaptive object detection. Int. J. Comput. Vis. **129**(3), 761–780 (2021). https://doi.org/10.1007/s11263-020-01394-z
22. Wu, A., Liu, R., Han, Y., Zhu, L., Yang, Y.: Vector-decomposed disentanglement for domain-invariant object detection. In: Proceedings of the IEEE International Conference on Computer Vision, pp. 9342–9351 (2021)
23. Zhang, Y., Wang, Z., Mao, Y.: RPN prototype alignment for domain adaptive object detector. In: Proceedings of the IEEE Conference on Computer Vision and Pattern Recognition, pp. 12425–12434 (2021)
24. Liu, D., et al.: Decompose to adapt: cross-domain object detection via feature disentanglement. IEEE Trans. Multimed. (2022)
25. Chen, Y., Li, G., Jin, C., Liu, S., Li, T.: SSD-GAN: measuring the realness in the spatial and spectral domains. In: Proceedings of the AAAI Conference on Artificial Intelligence, vol. 35, pp. 1105–1112 (2021)
26. Piotrowski, L.N., Campbell, F.W.: A demonstration of the visual importance and flexibility of spatial-frequency amplitude and phase. Perception **11**(3), 337–346 (1982)
27. Sankaranarayanan, S., Balaji, Y., Jain, A., Lim, S.N., Chellappa, R.: Learning from synthetic data: addressing domain shift for semantic segmentation. In: Proceedings of the IEEE Conference on Computer Vision and Pattern Recognition, pp. 3752–3761 (2018)
28. Simonyan, K., Zisserman, A.: Very deep convolutional networks for large-scale image recognition. arXiv preprint arXiv:1409.1556 (2014)
29. Deng, J., Dong, W., Socher, R., Li, L.J., Li, K., Fei-Fei, L.: ImageNet: a large-scale hierarchical image database. In: Proceedings of the IEEE Conference on Computer Vision and Pattern Recognition, pp. 248–255. IEEE (2009)

Single Image Dehazing Using Frequency Attention

Bin Hu, Zhuangzhuang Yue, Yuehua Li, Lili Zhao, and Shi Cheng[✉]

School of Information Science and Technology, Nantong University,
Nantong, Jiangsu, China
{hubin,lyh,ylzh,chenshi}@ntu.edu.cn, 2010320052@stmail.ntu.edu.cn

Abstract. Image degradation is a negative impact on computer vision tasks, and single image dehazing methods based on data-driven have witnessed the continuously growing and achieved great success. However, most data-driven methods are based on convolution neural network (CNN) which can be considered as be an implicit modeling of frequency domain. While discrete cosine transform (DCT) can be used in the CNN to model the features in frequency domain explicitly. Therefore, we propose an end-to-end image dehazing network with frequency attention (FA) based on DCT. Then, we select the top-K low-frequency components as the output of DCT layer, so the FA module can extract information from different frequency components. The experiments on the benchmark RESIDE demonstrate that our methods achieve better results than the previous state-of-art methods.

Keywords: Image dehazing · discrete cosine transform · frequency attention

1 Introduction

The quality of the captured image from a camera can be reduced by the particles such as fog in the air. The differences between images will decrease, and the edges and textures of the objects in the image will become blurred. So the image degradation is a negative impact on computer vision tasks such as image segmentation, object detection.

In order to reduce the impact of image degradation, lots of research has been done to recover the clear images from hazy images. Before deep learning has widely used in computer vision tasks, most methods solve the problem by the priors obtained from the statistics in the images such as the optical model [1], atmospheric scattering model (ASM) [2] and dark channel prior (DCP) [3]. However, they cannot handle the complex real-world images.

With the development of deep learning, image dehazing methods based on convolution neural network (CNN) have prevailed, and lots of networks have

Supported by Nantong Science and Technology Program Project (JC2020065).

been proposed. In order to restore a higher quality output from a hazy image, different CNN blocks are proposed as plugins used in network architectures, such as dilated convolution [4,5], multi-stage fusion [6–8], attention modules [9,10] and et al. However, CNN extract features from low level to high level, which can be considered as be an implicit modeling of frequency domain. While discrete cosine transform (DCT) is widely used in image and video compression including JPEG, MPEG and H.26x in signal processing. DCT can compress data with high energy [11], and it can be used in the CNN to model the features in frequency domain explicitly.

Therefore, we propose an end-to-end image dehazing network with frequency attention (FA) based on DCT. DCT has a great energy compaction in signal processing, and most image information is concentrated in low-frequency components, so we adopt a DCT based module to restore the image with high quality in frequency domain. The qualitative and quantitative experimental results on bench mark show that the proposed method is effective compared with the state-of-the-art methods.

2 Related Work

2.1 Single Image Dehazing

Early works mainly based on ASM and different priors such as DCP. ASM provides a simple yet solid theoretical basis for image dehazing. The formula is defined as.

$$I(x) = J(x)t(x) + A(1 - t(x)) \tag{1}$$

where $I(x)$ and $J(x)$ denote the hazy image and clear image respectively, and $t(x)$ stands for the medium transmission map, and A means the atmospheric light. The methods based on ASM try to solve formula (1) to get the clear output by unsupervised [12], semi-supervised [13] and supervised [14] algorithms. DCP [3] proposed a dark channel prior to estimate A and $t(x)$ in formula (1). DCP is simple yes reliable, and has a great impact on image dehazing research. However, the model based methods cannot handle the complex real-world images, and DCP is found to be unreliable when the color of the scene objects are similar to the atmospheric light.

Recently, with the development of deep learning, CNN based methods have occupied the main current. Li et al. [15] first designed an end-to-end network to output the clear image by using standard convolution layers, it is simple yet effective. Then various convolution blocks are proposed for image dehazing including skip connection [16,17], dense connection [18], multi-scale convolution [19] and et al. In order to increase the receptive field, dilated convolution layers are used to improve the performance [4,5]. Also, various attention mechanisms are proposed such as channel attention and spatial attention [9,10,20,21]. Channel attention can earn more weight from different channels of the features, and spatial attention can exploit the different spatial attention map from the input feature map. Zhang et al. [22] designed a multi-scale feature fusion structure to enhance the information on same and different spatial scales. Besides, unsupervised methods based on GAN are used to transfer the hazy domains to haze-free domains [23,24]. However, those methods are mainly adopt convolution neural networks and lack in extracting feature in frequency domain, which can model the feature in different frequency.

2.2 Learning in the Frequency Domain

Frequency learning has been widely used in signal processing, and there are rich features in low frequency domain for image processing tasks [25] such as image classification [26]. With the great power of energy compaction, DCT also is used for model pruning [27]. However, there are few works using DCT for image dehazing.

3 Proposed Method

In this section, we mainly introduce the detail of the proposed method. Firstly, we discuss the architecture of the network, which is shown in Fig. 1. Firstly, the hazy image is processed by a 3×3 convolution layer. Second, a down-sampling layer is used to resize the size of feature map, and the cost of GPU memory can be reduced. And then several basic units are used. Finally, an up-sampling layer and a 3×3 convolution layer are used to restore the image. The encoder-decoder network with a series of residual connections can reduce the semantic gap and facilitate the optimizer to deal with the dehazing task.

The basic unit of the network consists of several residual block (RB) [28] which is used in many works and FA. The residual connection in the unit can pass the image details from shallow layers to deep layers.

3.1 Frequency Attention

In this section, we introduce the frequency attention based on discrete cosine transform (DCT). According to two-dimensional DCT, the frequency spectrum can be alculated according to the following formula:

$$f_{h,w}^{dct} = \sum_{i=0}^{H-1} \sum_{j=0}^{W-1} x_{i,j} B_{h,w}^{i,j}, \quad s.t. h = 0, 1, ..., H-1, w = 0, 1, ..., W-1 \quad (2)$$

where $x \in \mathbb{R}^{H \times W}$ means the input, H and W are height and width of the input, respectively. Correspondingly, x can be recovered from the frequency spectrum f by:

$$x_{i,j} = \sum_{i=0}^{H-1} \sum_{w=0}^{W-1} f_{h,w} B_{h,w}^{i,j} \quad (3)$$

FcaNet [29] proves that global average pooling (GAP) is a special case of DCT when $h = 0$ and $w = 0$ in Eq. (2). When h and w are 0, $f_{0,0}^{dct}$ means the lowest

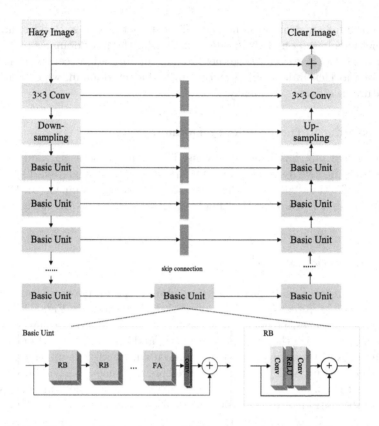

Fig. 1. The architecture of the network, which is an encoder-decoder like network. The network consists of several basic units, which is built by frequency attention.

frequency information of DCT, and the higher frequency components are lost, which also provide useful information for image dehazing.

Base on the DCT, we propose a frequency attention module to extract information from different frequency components. Let X be the input feature map, 2D DCT is applied to get the frequency components as Fig. 2 shows. Previous works show that low-frequency components are better for CNNs [20,25], so we select the top-K low-frequency components as the output.

Inspired by convolutional block attention module (CBAM) [30], we design the frequency attention module (FA) as Fig. 3 shows. Firstly, the input feature map is transformed to frequency components as introduced in Fig. 2, then a multilayer perception (MLP) is used to get the descriptor, and after a sigmoid function, the output of first stage is get. And the architecture of second stage is similar to the first stage, two convolution layers and a DCT layer are used to get the final output.

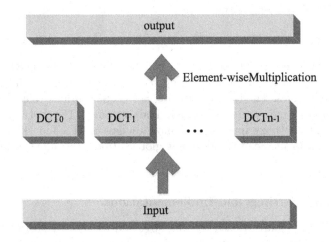

Fig. 2. DCT layer.

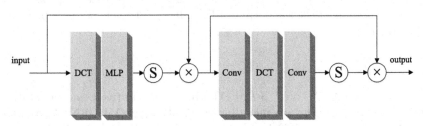

Fig. 3. Frequency attention module.

3.2 Loss Function

Various loss functions are proposed for image dehazing, such as smooth L_1 loss, L_2 loss, mean squared error (MSE), perceptual loss and et al. According to previous research, the smooth L_1 loss is better than others for image restoration tasks [28], so L_1 loss is adopted:

$$L_s = \frac{1}{N} \sum_{x=1}^{N} \sum_{i=1}^{3} \times F_s(\hat{J}_\iota(x) - J_i(x)) \tag{4}$$

where

$$F_s(e) = \begin{cases} 0.5e^2, & if\ |e| < 1, \\ |e| - 0.5, & otherwise. \end{cases} \tag{5}$$

$\hat{J}_\iota(x)$ and $J_i(x)$ stand the intensity of the ith color channel of pixel x in the de-hazed image and hazy image, respectively, and N is the pixel count of the image.

4   Experiment Results

4.1   Dataset

We evaluate the proposed method on the benchmark dataset RESIDE [31]. RESIDE is large-scale synthetic training sets, and it contains both in-door and outdoor hazy images. The Indoor Training Set (ITS) contains 1399 clean images and 13990 hazy images, and the Outdoor Training Set (OTS) contains 8477 clean images and 296695 hazy images. The hazy images of ITS and OTS are generated by the clean images with different global atmosphere light A and scatter parameters.

4.2   Training Settings and Implementations

The training images are resized to a fix size, and rotated by 90,180,270°C randomly and horizontal flipped for data augmentation. The Adam optimizer with default parameters is used. The number of Basic Unit in the network is set to 13, and each basic unit contains 7 residual block, and we select the top-16 low-frequency components as the output of DCT layer. The network is implemented on PyTorch.

4.3   Results

We compare the results on ITS and OTS both quantitatively and qualitatively with previous state of the arts including DCP, AOD-Net [15], GFN [32], GCANet [33], DehazeNet [34], GridDehazeNet [10] and FFA-Net [9]. As Table 1 shows, DCP is a prior based method and it performs worst. AOD-Net is a simple network, so its performance is not good either. And our results perform better than others.

Also, we give the visual results as Fig. 4 shows. The first column is the input hazy image, and the last column is the ground truth. The details which in the rectangular box are shown in the even rows. From the results in Fig. 4, we find that DCP has a serious problem on color distortion especially in the blue sky and sun image. AOD-net is a simple network, and it cannot remove all the hazy in the image, also the restored image has a low brightness. GCA-net also cannot remove all the hazy, and the details of the tower in the first row are lost. There are some white spot in the results of Grid-Dehazenet. Our results are almost entirely in line with the ground truth, especially, the restoration of blue sky in first image and sun in second image is much better.

Table 1. Quantitative comparison on ITS and OTS.

Method	Indoor		Outdoor	
	PSNR	SSIM	PSNR	SSIM
DCP[3]	16.62	0.8179	19.13	0.8148
AOD-Net[15]	19.06	0.8504	20.29	0.8765
GFN[32]	22.30	0.8800	21.55	0.8444
GCANet[33]	30.23	0.9800	28.68	0.9712
DehazeNet[34]	21.14	0.8472	22.46	0.8514
GridDehazeNet[10]	32.16	0.9836	30.86	0.9819
FFA-Net[9]	36.39	0.9886	33.57	0.9849
Ours	36.62	0.9901	34.71	0.9904

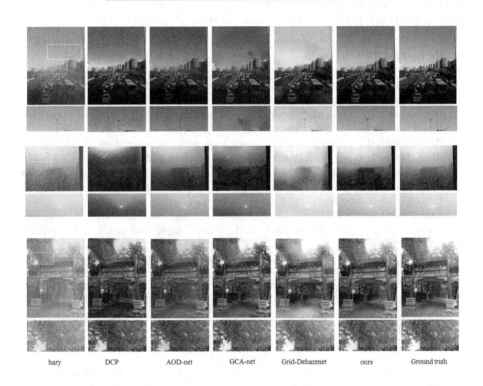

hazy   DCP   AOD-net   GCA-net   Grid-Dehazenet   ours   Ground truth

Fig. 4. Visual results on OTS. The regions in the rectangular box are detailed in the even rows.

4.4   Ablation Studies

In the DCT layer, we select top-K low-frequency components as the output. We evaluate the results on different values of K in this section. We can find that when $K = 16$, the result achieves the highest PSNR value, and it proves that every frequency component is useful for the task.

Table 2. Ablation studies on different value of K.

K	1	2	4	8	16
PSNR	36.12	36.31	36.48	36.59	36.62

5 Conclusion

We propose an end-to-end single image dehazing method with frequency attention (FA) based on DCT. The encoder-decoder network with a series of residual
connections can reduce the semantic gap and facilitate the optimizer to deal
with the dehazing task. We select the top-K low-frequency components from
DCT to design the FA module, which can restore the image with high quality
in frequency domain. The experimental results on bench mark show that the
proposed network performs better than the state-of-the-art methods.

References

1. Koschmieder, H.: Theorie der horizontalen sichtweite. Beitrage zur Physik der
   freien Atmosphare, pp. 33–53 (1924)
2. McCartney, E.J.: Optics of the atmosphere: scattering by molecules and particles.
   New York (1976)
3. He, K., Sun, J., Tang, X.: Single image haze removal using dark channel prior.
   IEEE Trans. Pattern Anal. Mach. Intell. **33**(12), 2341–2353 (2010)
4. Zhang, S., He, F., Ren, W.: Photo-realistic dehazing via contextual generative
   adversarial networks. Mach. Vis. Appl. **31**(5), 1–12 (2020). https://doi.org/10.
   1007/s00138-020-01082-5
5. Zhang, S., He, F.: DRCDN: learning deep residual convolutional dehazing networks. Vis. Comput. **36**(9), 1797–1808 (2020). https://doi.org/10.1007/s00371-
   019-01774-8
6. Li, H., Li, J., Zhao, D., Xu, L.: DehazeFlow: multi-scale conditional flow network for
   single image dehazing. In: Proceedings of the 29th ACM International Conference
   on Multimedia, pp. 2577–2585 (2021)
7. Zamir, S.W., et al.: Multi-stage progressive image restoration. In: Proceedings
   of the IEEE/CVF Conference on Computer Vision and Pattern Recognition, pp.
   14821–14831 (2021)
8. Hu, B., Gu, M., Li, Y.: A multistage with multiattention network for single image
   dehazing. Scientific Programming 2022 (2022)
9. Qin, X., Wang, Z., Bai, Y., Xie, X., Jia, H.: FFA-Net: feature fusion attention
   network for single image dehazing. In: Proceedings of the AAAI Conference on
   Artificial Intelligence, vol. 34, pp. 11908–11915 (2020)
10. Liu, X., Ma, Y., Shi, Z., Chen, J.: GridDehazeNet: attention-based multi-scale
    network for image dehazing. In: Proceedings of the IEEE/CVF International Con-
    ference on Computer Vision, pp. 7314–7323 (2019)
11. Rao, K.R., Yip, P.: Discrete Cosine Transform: Algorithms, Advantages, Applica-
    tions. Academic Press, San Diego (2014)
12. Li, B., Gou, Y., Gu, S., Liu, J.Z., Zhou, J.T., Peng, X.: You only look yourself:
    unsupervised and untrained single image dehazing neural network. Int. J. Comput.
    Vis. **129**(5), 1754–1767 (2021). https://doi.org/10.1007/s11263-021-01431-5

13. Liu, Y., et al.: From synthetic to real: image dehazing collaborating with unlabeled real data. In: Proceedings of the 29th ACM International Conference on Multimedia, pp. 50–58 (2021)

14. Zhang, H., Patel, V.M.: Densely connected pyramid dehazing network. In: Proceedings of the IEEE Conference on Computer Vision and Pattern Recognition, pp. 3194–3203 (2018)

15. Li, B., Peng, X., Wang, Z., Xu, J., Feng, D.: AOD-Net: all-in-one dehazing network. In: Proceedings of the IEEE International Conference on Computer Vision, pp. 4770–4778 (2017)

16. Zhang, Z., Zhao, L., Liu, Y., Zhang, S., Yang, J.: Unified density-aware image dehazing and object detection in real-world hazy scenes. In: Proceedings of the Asian Conference on Computer Vision (2020)

17. Zhao, D., Xu, L., Ma, L., Li, J., Yan, Y.: Pyramid global context network for image dehazing. IEEE Trans. Circuits Syst. Video Technol. **31**(8), 3037–3050 (2020)

18. Zhang, S., et al.: Semantic-aware dehazing network with adaptive feature fusion. IEEE Trans. Cybern. **53**(1), 454–467 (2021)

19. Zhang, X., Li, J., Hua, Z.: MFFE: multi-scale feature fusion enhanced net for image dehazing. Signal Process.: Image Commun. **105**, 116719 (2022)

20. Hu, J., Shen, L., Sun, G.: Squeeze-and-excitation networks. In: Proceedings of the IEEE Conference on Computer Vision and Pattern Recognition, pp. 7132–7141 (2018)

21. Zhang, X., Wang, J., Wang, T., Jiang, R.: Hierarchical feature fusion with mixed convolution attention for single image dehazing. IEEE Trans. Circuits Syst. Video Technol. **32**(2), 510–522 (2021)

22. Zhang, T., Li, J., Fan, H.: Progressive edge-sensing dynamic scene deblurring. Comput. Vis. Media **8**(3), 495–508 (2022). https://doi.org/10.1007/s41095-021-0246-4

23. Liu, W., Hou, X., Duan, J., Qiu, G.: End-to-end single image fog removal using enhanced cycle consistent adversarial networks. IEEE Trans. Image Process. **29**, 7819–7833 (2020)

24. Dudhane, A., Murala, S.: CDNet: single image de-hazing using unpaired adversarial training. In: 2019 IEEE Winter Conference on Applications of Computer Vision (WACV), pp. 1147–1155. IEEE (2019)

25. Xu, K., Qin, M., Sun, F., Wang, Y., Chen, Y.K., Ren, F.: Learning in the frequency domain. In: Proceedings of the IEEE/CVF Conference on Computer Vision and Pattern Recognition, pp. 1740–1749 (2020)

26. Ehrlich, M., Davis, L.S.: Deep residual learning in the JPEG transform domain. In: Proceedings of the IEEE/CVF International Conference on Computer Vision, pp. 3484–3493 (2019)

27. Liu, Z., Xu, J., Peng, X., Xiong, R.: Frequency-domain dynamic pruning for convolutional neural networks. Adv. Neural Inf. Process. Syst. **31** (2018)

28. Lim, B., Son, S., Kim, H., Nah, S., Mu Lee, K.: Enhanced deep residual networks for single image super-resolution. In: Proceedings of the IEEE Conference on Computer Vision and Pattern Recognition Workshops, pp. 136–144 (2017)

29. Qin, Z., Zhang, P., Wu, F., Li, X.: FcaNet: frequency channel attention networks. In: Proceedings of the IEEE/CVF International Conference on Computer Vision, pp. 783–792 (2021)

30. Woo, S., Park, J., Lee, J.Y., Kweon, I.S.: CBAM: convolutional block attention module. In: Proceedings of the European Conference on Computer Vision (ECCV), pp. 3–19 (2018)

31. Li, B., et al.: Benchmarking single-image dehazing and beyond. IEEE Trans. Image Process. **28**(1), 492–505 (2018)
32. Ren, W., et al.: Gated fusion network for single image dehazing. In: Proceedings of the IEEE Conference on Computer Vision and Pattern Recognition, pp. 3253–3261 (2018)
33. Chen, D., et al.: Gated context aggregation network for image dehazing and deraining. In: 2019 IEEE Winter Conference on Applications of Computer Vision (WACV), pp. 1375–1383. IEEE (2019)
34. Cai, B., Xu, X., Jia, K., Qing, C., Tao, D.: DehazeNet: an end-to-end system for single image haze removal. IEEE Trans. Image Process. **25**(11), 5187–5198 (2016)

A Recurrent Point Clouds Selection Method for 3D Dense Captioning

Shinko Hayashi, Zhiqiang Zhang, and Jinja Zhou[✉]

Hosei University, Tokyo, Japan
shinko.hayashi.3r@stu.hosei.ac.jp, zhou@hosei.ac.jp

Abstract. 3D dense captioning provides descriptions for corresponding objects in 3D scenes represented as RGB-D scans and point clouds. However, when generating a description, existing methods select points randomly from a point cloud regardless of importance, which degrades the quality of the description by removing important points or including low-value points. To solve the above problem, we propose a recurrent point clouds selection (RPCS) method to mitigate descriptive deficiencies in 3D dense captioning by iteratively checking the caption results of the different point clouds. Our method is divided into two steps. On step 1, this work randomly selects cloud points and uses objectness score to evaluate the generated description. The objectness score indicates whether the proposed object is close to the ground truth; the higher the score, the closer the proposed object is to the ground truth in the positive value. On step 2, if the objectness score is lower than the threshold, step 1 is processed to generate another group of cloud points and evaluate the results. This loop stops when the objectness score is no longer reduced. The loop termination conditions are configurable according to the requirement of accuracy and processing time. As a result, our work can decrease the deficient descriptions and outperforms previous state-of-the-art methods by a large margin (**6.58%~35.70% CiDEr, BLUE-4, METEOR, ROUGE improvement**).

Keywords: 3D Dense Captioning · Point Clouds · Recurrent Point Clouds Selection

1 Introduction

In computer vision research, there are many deep neural networks for visual scene understanding with natural language description, such as image captioning [1,8,15,21,23] and dense captioning [11,13,22,24] which generates sentences to describe how people or animals behave, or an event in the input image. However, 2D images have a restricted field of vision, making it impossible to grasp the physical size and locations of objects and can not achieve more accurate scene descriptions. In contrast, 3D dense captioning uses 3D scenes, which allows the

© The Author(s), under exclusive license to Springer Nature Switzerland AG 2023
M. Tanveer et al. (Eds.): ICONIP 2022, LNCS 13625, pp. 263–274, 2023.
https://doi.org/10.1007/978-3-031-30111-7_23

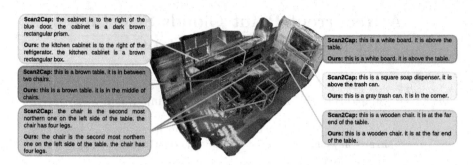

Scan2Cap: the cabinet is to the right of the blue door. the cabinet is a dark brown rectangular prism.

Ours: the kitchen cabinet is to the right of the refrigerator. the kitchen cabinet is a brown rectangular box.

Scan2Cap: this is a brown table. it is in between two chairs.

Ours: this is a brown table. it is in the middle of chairs.

Scan2Cap: the chair is the second most northern one on the left side of the table. the chair has four legs.

Ours: the chair is the second most northern one on the left side of the table. the chair has four legs.

Scan2Cap: this is a white board. it is above the table.

Ours: this is a white board. it is above the table.

Scan2Cap: this is a square soap dispenser. it is above the trash can.

Ours: this is a gray trash can. it is in the corner.

Scan2Cap: this is a wooden chair. it is at the far end of the table.

Ours: this is a wooden chair. it is at the far end of the table.

Fig. 1. The descriptions corresponding to the objects in validation scene "scene0011". In our method, the description can be better than Scan2Cap [5], however, there are still some poor descriptions, such as misidentifying "tv" as "whiteboard" and some chairs surrounding the table with the same description.

actual size and location relation to be determined and described more accurately. Scan2Cap [5], the first 3D dense captioning research, adapts PointNet++ [19] backbone and VoteNet [18] to detect objects in 3D scenes and introduces the Relational Graph module and Context-aware Captioning module to address the issue that object relations are often ignored when describing detected objects in 2D images.

Although Scan2Cap [5] can generate promising results, however, when generating or evaluating descriptions, Chen et al. [5] randomly select 40,000 points from a point cloud, which may generate poor descriptions since some of those points may not be important in the point cloud. For this issue, we propose a recurrent point clouds selection (RPCS) method, which uses the high-level overall objectness scores to achieve more accurate descriptions. Since the points used in each loop are different, some descriptions of objects in the scene may also differ, so we store good descriptions based on the objectness score in each loop. By doing so, low objectness scores regardless of random points in the point cloud become less and corresponding objects are more correctly described, thus reducing the number of deficient descriptions.

Also, the objectness score is obtained by adapting VoteNet [18] in the neural network. From the voting module in the VoteNet [18], a vote cluster $\mathcal{C} = \{w_i\}$ with $i = 1, ..., n$ and its cluster center w_j is given, where $w_i = [z_i; h_i]$ with $z_i \in \mathbb{R}^3$ as the vote location and $h_i \in \mathbb{R}^C$ as the voting feature. Vote locations are transformed to a local normalized coordinate system by $z_i' = (z_i - z_j)/r$, then the object proposal for this cluster $p(\mathcal{C})$ is generated as follows:

$$p(\mathcal{C}) = \text{MLP}_2 \left\{ \max_{i=1,...,n} \{\text{MLP}_1([z_i'; h_i])\} \right\}, \tag{1}$$

where votes from each cluster are independently processed by a MLP_1 before being max-pooled to a single feature vector and passed to MLP_2 where information from different votes are further combined. The proposal p is represented as a multidimensional vector with an objectness score, bounding box parameters, and semantic classification scores. MLP_2 outputs 2 objectness scores, including positive and negative values, which indicate the object is how close to the ground truth object center (within 0.3 m) or how far from any center of a ground truth object (greater than 0.6 m). In this work, we focus on the positive objectness score, the higher the objectness score, the object is correctly detected, to consider whether the randomly selected points contain something unimportant or not contain less unimportant ones for the corresponding object. Figure 1 shows some example descriptions in a 3D scene; the higher the objectness scores, the better descriptions are generated. In summary, our contributions are threefold:

- We propose a recurrent point clouds selection (RPCS) method for 3D dense captioning, which can make less use of low-value points for proposal objects regardless of randomly selected points from the point cloud.
- RPCS mitigates deficient descriptions for each proposal object by iteratively checking the caption results of the different point clouds.
- Our method outperforms the based state-of-the-art method by a significant margin (**6.58%~35.70%**).

2 Related Works

In the field of image captioning, various methods have been proposed and have attracted much attention in the past few years [1,8,12,15,21,23], and some methods, over grid regions [15,23] and over detected objects [1,16], adapt attention mechanisms to obtain useful information in the image and focus on particular regions in the image. Also, some methods combine attention with graph neural networks [9,10,25,26] or transformers [6] to obtain relationships between objects.

To further improve the usefulness of image captioning, dense captioning is proposed. The task of dense captioning is first proposed by Johnson et al. [11] and this task is the expansion category of image captioning and generates descriptions for all detected objects. Yang et al. [24] utilize the global image feature as context to consider the context outside of the salient image regions. Kim et al. [13] also propose the object relations between detected regions in the image. However, 2D images have a restricted field of vision, the 2D-image-based captioning methods could not grasp the information available in the 3D context.

To solve the problems of 2D dense captioning, 3D dense captioning is proposed. The task of 3D dense captioning is first proposed by Chen et al. [5] and this task is to understand 3D scenes with dense natural language descriptions of color and spatial information for all detected objects. Chen et al. [4] combines the 3D visual grounding and dense captioning tasks and succeeded in mutually improving the performance of these tasks. Zhihao et al. [27] propose a cross-modal knowledge transfer using Transformer that adapts a teacher-student framework and successfully improves the performance of 3D dense captioning. However, existing methods generate descriptions by randomly selecting points from the point cloud, which may include unimportant points, degrading the quality of descriptions.

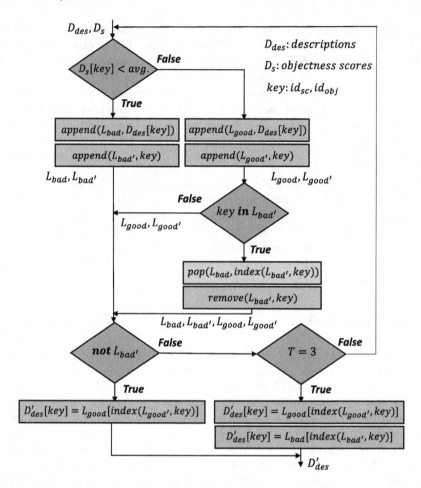

Fig. 2. The flowchart of the recurrent point clouds selection (RPCS). D_{des} and D_s are the dictionaries containing descriptions and objectenss scores, respectively, and *key* in these dictionaries contains the object information. T represents as the minimum number of loops to be aborted if the elements have not been reduced by that number of times. The objectness score threshold and T can be changed according to the requirement of accuracy and processing time.

3   Proposed Approach

3.1   Proposed Recurrent Point Clouds Selection Strategy (RPCS)

To mitigate the problem of generating deficient descriptions due to using random points that may contain unimportant points in the point cloud as mentioned in Sect. 1, we propose an RPCS that sets the overall objectness scores to a higher level so as not to use as much information as possible including unimportant points. Figure 2 shows a flowchart of the proposed RPCS. When generating descriptions, the objectness score threshold is used to consider whether the randomly selected points contain something unimportant or not contain less unimportant ones for the detected object, and store the corresponding description and the object information in the bad or good list, respectively. If the objectness score corresponding to the object is less than the threshold, which is the average value of the whole object in the scene as a baseline, the points for the object are considered to contain something unimportant, then the corresponding description and object information are stored in the bad list. If the objectness score is more than or equal to the threshold, the points for the object are considered not to contain less unimportant ones, then the description and corresponding object information are stored in the good list, and if the previous object information is in the bad list, it is removed along with its corresponding description from the bad list. The goal is to repeat this process until all the elements in the bad list have been removed and to evaluate descriptions in the good list, however, since there is a possibility that the elements may not be reduced from one loop, as a baseline, we interrupt the loop in the middle and interrupt after 3 loops, in which evaluates the descriptions in the good and bad lists.

In order to explain the RPCS in more detail, the pseudo code is also provided in Algorithm 1. The steps of explaining algorithm are as follows:

1. The input *scores* and *candidates* are dictionaries, which contain objectness scores corresponding to the objects and the object information such as scene ids, object ids, object names, and corresponding descriptions, respectively.
2. In the first step (lines 2 to 9), if objectness scores are less than *average*, scene ids, object ids, and object names are stored in the **bad_obj** and corresponding descriptions in the **bad_des**; if objectness scores are *average* or more, stored in the **good_obj** and **good_des**, respectively. **bad_obj**, **bad_des**, **good_obj**, and **good_des** are all lists, empty at first loop beginning. However, in the next loop, some objectness scores corresponding to the objects contained in **good_obj** in the previous loop may become less than *average*, to ignore it we only focus on what is not contained in **good_obj** in the previous loop.
3. Next, some objectness scores corresponding to the objects contained in **bad_obj** in the previous loop may become *average* or more, store corresponding object information and descriptions in **good_obj** and **good_des**, respectively, and then remove previous ones from **bad_obj** and **bad_des** (lines 10 to 12).
4. Before the final step, create an empty dictionary **final_candidates** instead of input *candidates* to prepare for evaluating stored descriptions (line 16).
5. As the final step, if all elements are removed from **bad_obj**, descriptions in **good_des** are stored to **final_candidates** (line 19), if elements are not reduced 3 times, descriptions in **bad_des** are also stored to **final_candidates** (line 27). If these conditions are not met, start the next loop with other random 40,000 points (line 31). Then the loop is finished, obtain descriptions from **final_candidates**.

Algorithm 1. Recurrent Point Clouds Selection (RPCS)

Input: scores, candidates
Output: final_candidates
1: **for** key **in scores do**   ▷ objectness score for each object
2:     **if** score[key] < $avg.$ **then**
3:         **if** (key not in good_obj) and (key not in bad_obj) **then**
4:             **bad_obj**.append(key)   ▷ store the object information
5:             **bad_des**.append(**candidates**[key])   ▷ store the description
6:         **end if**
7:     **else**
8:         **good_obj**.append(key)   ▷ store the object information
9:         **good_des**.append(candidates[key])   ▷ store the description
10:         **if** key **in bad_obj then**
11:             **bad_des**.pop(**bad_obj**.index(key))   ▷ remove the bad description
12:             **bad_obj**.remove(key)   ▷ remove the bad object information
13:         **end if**
14:     **end if**
15: **end for**
16: final_candidates = {}   ▷ prepare for evaluating descriptions
17: **if not bad_obj then**
18:     **for** key **in good_obj do**
19:         final_candidates[key] = **good_des**[**good_obj**.index(key)]   ▷ store descriptions from good_obj
20:     **end for**
21:     **end the loop**
22: **else if bad_obj is not reduced 3 times then**
23:     **for** key **in good_obj do**
24:         final_candidates[key] = **good_des**[**good_obj**.index(key)]
25:     **end for**
26:     **for** key **in bad_obj do**
27:         final_candidates[key] = **bad_des**[**bad_obj**.index(key)]   ▷ store descriptions from bad_obj
28:     **end for**
29:     **end the loop**
30: **else**
31:     **start the next loop**
32: **end if**

3.2 Combination with Existing 3D Dense Captioning Works

Our method can be applied to the 3D dense captioning network based on point clouds. The recent networks consist of the 3D object detection area and the own framework or modules area with a caption decoder. In the object detection area, almost all networks apply VoteNet [18] to obtain 3D proposals for point clouds, and our method is suitable for such networks. In this paper, we adapt RPCS to a network proposed by Chen et al. [5], as shown in Fig. 3. The RPCS is applied in the caption decoder, then uses random points to evaluate descriptions for each proposal object, and processes it again with other random points if the corresponding objectness score is less than the threshold. In the network proposed by Chen et al. [5], the point cloud is represented as $\mathcal{P} = \{(p_i, f_i)\} \in \mathcal{R}^{N_P \times 135}$, where $p_i \in \mathcal{R}^3, i = 1, ..., N_P$ are the coordinates and $f_i \in \mathcal{R}^{132}$ are the additional features, PointNet++ [19] backbone and VoteNet [18] are used to detect 3D objects, and the relation graph module and a captioning module to enhance object features and output object relation features, and generate descriptive tokens for the each object proposal.

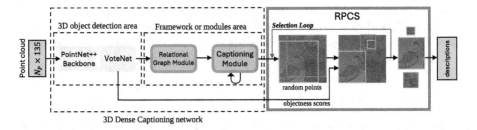

Fig. 3. 3D dense captioning network based on Scan2Cap [5] with the proposed recurrent point clouds selection (RPCS), which mitigates deficient descriptions. The color rectangles represent as the randomly selected points and the corresponding proposal objects whose objectness score is above the threshold (the orange color is in first loop, the green color is after the second loop), and the red rectangle represents as the proposal object whose objectness score is less than the threshold. (Color figure online)

4 Experimental Results and Comparison

4.1 Dataset and Metrics

Dataset. We use the ScanRefer [3] dataset consisting of 51,583 descriptions in 800 scenes inside the room from ScanNet [7]. ScanNet [7] point clouds consist of geometric coordinates as well as additional features representing appearance and height from the ground, and we split the data in the same way as in [5], composing a train set and a val set.

Metrics. We combine the standard metrics for image captioning, CiDEr [20], BLEU-4 [17], METEOR [2] and ROUGE [14], with Intersection-over-Union (IoU) scores between the predicted bounding boxes and the ground truth bounding boxes to measure the quality of generated descriptions. We also follow [5] and define the combined metrics as $m@k = \frac{1}{N}\sum_{i=1}^{N} m_i u_i$, where $u_i \in \{0,1\}$ is set to 1 if the IoU score for the i-th box is above k, otherwise 0. We use m to represent the above captioning metrics, CiDEr [20], BLEU-4 [17], METEOR [2] and ROUGE [14]. N is the number of detected object bounding boxes. For simplicity, we omit CiDEr [20], BLEU-4 [17], METEOR [2] and ROUGE [14] as C, B-4, M, R, respectively.

4.2 Comparisons with the Previous Works

Quantitative Results. We compare our proposal with Scan2Cap [5] on the validation split of ScanRefer [3]. We adapt our RPCS to Scan2Cap [5] when generating descriptions, and compare the results on 3 validation scenes, "scene0011", "scene0222" and "scene0704". As shown in Table 1, our proposal has better results, some of them with have a significant performance boost (54.33% compared to 31.01% on C@0.5). Table 2 shows the comparison of captioning results obtained by our proposal and Scan2Cap [5]. Our proposal also outperforms Scan2Cap [5] in average captioning scores, however, it is time-consuming since it loops many times per scene.

Table 1. Comparision of the example results for 3 validation scenes "scene0011", "scene0222", and "scene0704" obtained by Scan2Cap [5] and our baseline. The IoU percentage of the predicted bounding boxes with the ground truth is above 0.5. Our method outperforms Scan2Cap [5] on all metrics.

	sc0011				sc0222				sc0704			
	C@0.5	B-4@0.5	M@0.5	R@0.5	C@0.5	B-4@0.5	M@0.5	R@0.5	C@0.5	B-4@0.5	M@0.5	R@0.5
Scan2Cap [5]	31.01	24.99	24.65	45.21	37.60	23.54	20.45	42.81	50.40	22.37	23.27	46.28
Ours	**54.33**	**36.59**	**30.03**	**60.13**	**51.25**	**35.69**	**24.83**	**55.94**	**62.11**	**26.21**	**26.50**	**53.75**

Table 2. Comparision of 3D dense captioning average results obtained by Scan2Cap [5] and our baseline. The IoU percentage of the predicted bounding boxes with the ground truth is above 0.25 and 0.5. Our method also outperforms Scan2Cap [5] on average for all metrics.

	C@0.25	B-4@0.25	M@0.25	R@0.25	Time(s)	C@0.5	B-4@0.5	M@0.5	R@0.5	Time(s)
Scan2Cap [5]	56.82	34.18	26.29	55.27	**287**	39.08	23.32	21.97	44.78	**286**
Ours	**64.77**	**36.05**	**28.53**	**59.01**	15497	**44.92**	**26.41**	**23.78**	**47.16**	961

Since the loop termination conditions of the RPCS are configurable, we compare the results of our proposal under different termination conditions. "scene0011", "scene0222", and "scene0704" have average objectness scores of about 3~4, respectively, so we change the conditions to the objectness threshold 5 with interrupting the loop after elements are not reduced 10 times to compare these 3 scenes. As shown in Table 3, except ROUGE [14] metric on scene "scene0011", almost results better than previous termination conditions. Since using random points in the point cloud, captioning results are different every time when generating descriptions for each scene, however, in stricter conditions, it is possible to consistently obtain high captioning results. Also, we change the condition to interrupting the loop after elements are not reduced 10 times and 15 times, and the objectness threshold to 3 with interrupting the loop after elements are not reduced 3 times to compare captioning results. As shown in Table 4, the more permissive the termination conditions are, the shorter the processing time, and the more stringent the conditions, the better the results for the captioning results, but the longer the processing time. Therefore, good results can be obtained by setting strict termination conditions, except for being time-consuming.

Furthermore, in Table 5, we compare the results with other recent methods. *3D-2D Proj.* is a baseline method proposed in [5], which also applies VoteNet [18] to extract 3D proposals, projects them onto a 2D image, then adopts the 2D proposals to [23] to generate descriptions. $\mathcal{X}$-Trans2Cap [27] proposes a cross-modal knowledge transfer using Transformer with a teacher-student framework to produce faithful descriptions. As shown in Table 5, the 2D-based method, *3D-2D Proj.*, obtains the lowest results and it is clear that it is considerably inferior to 3D-based methods. While Scan2Cap [5] achieves better results than the 2D-based method, $\mathcal{X}$-Trans2Cap [27] further improves the performance. Also, our

Table 3. Comparision of the example results for 3 validation scenes "scene0011", "scene0222", and "scene0704" obtained by our proposal with the different termination conditions which are the baseline conditions and the objectness threshold of 5 with interrupting the loop after elements are not reduced 10 times (marked "Ours $(D_s < 5, T = 10)$"), respectively. The IoU percentage of the predicted bounding boxes with the ground truth is above 0.5. Almost results become better than previous conditions.

	sc0011				sc0222				sc0704			
	C@0.5	B-4@0.5	M@0.5	R@0.5	C@0.5	B-4@0.5	M@0.5	R@0.5	C@0.5	B-4@0.5	M@0.5	R@0.5
Ours (baseline)	54.33	36.59	30.03	**60.13**	51.25	35.69	24.83	55.94	62.11	26.21	26.50	53.75
Ours $(D_s < 5, T = 10)$	**69.68**	**37.80**	**30.10**	60.11	**72.43**	**35.95**	**26.09**	**56.82**	**72.85**	**34.31**	**29.54**	**60.83**

Table 4. Comparision of the 3D dense captioning average results obtained by our proposal with the different termination conditions which are the objectness threshold of 3 with interrupting the loop after elements are not reduced 3 times (marked "Ours $(D_s < 3, T = 3)$"), the baseline conditions, and the objectness threshold of $avg.$ with interrupting the loop after elements are not reduced, 10 times (marked "Ours $(D_s < avg., T = 10)$") and 15 times (marked "Ours $(D_s < avg., T = 15)$"). The IoU percentage of the predicted bounding boxes with the ground truth is above 0.25 and 0.5. If we do not care about time consumption, the results become better.

	C@0.25	B-4@0.25	M@0.25	R@0.25	Time(s)	C@0.5	B-4@0.5	M@0.5	R@0.5	Time(s)
Ours $(D_s < 3, T = 3)$	63.15	35.16	28.25	58.28	**4333**	44.89	26.38	23.82	47.16	**858**
Ours (baseline)	64.77	36.05	28.53	59.01	15497	44.92	26.41	23.78	47.16	961
Ours $(D_s < avg., T = 10)$	65.48	36.33	28.70	59.44	28694	50.78	29.64	25.27	50.71	3656
Ours $(D_s < avg., T = 15)$	**65.73**	**36.43**	**28.77**	**59.54**	52578	**53.03**	**30.90**	**25.92**	**52.28**	7615

Table 5. Comparison of the 3D dense captioning average results obtained by the based method introducing our proposal and other state-of-the-art methods. The IoU percentage of the predicted bounding boxes with the ground truth is above 0.25 and 0.5. All of these methods apply VoteNet [18] to obtain 3D proposals. Our method is based on Scan2Cap [5], with the termination condition, the objectness threshold of $avg.$ with interrupting the loop after elements are not reduced 15 times, which give better results than other methods.

	Detection	C@0.25	B-4@0.25	M@0.25	R@0.25	C@0.5	B-4@0.5	M@0.5	R@0.5
Scan2Cap [5]	VoteNet [18]	56.82	34.18	26.29	55.27	39.08	23.32	21.97	44.78
3D-2D Proj. [23]	VoteNet [18]	19.73	17.86	19.83	40.68	11.47	8.56	15.73	31.65
X-Trans2Cap [27]	VoteNet [18]	61.83	35.65	26.61	54.70	43.87	25.05	22.46	45.28
Ours (Scan2Cap based)	VoteNet [18]	**65.73**	**36.43**	**28.77**	**59.54**	**53.03**	**30.90**	**25.92**	**52.28**

method based on Scan2Cap [5] gives better results than these state-of-the-art methods, and since these methods also extract 3D proposals using VoteNet [18], the proposed RPCS is expected to be applied and mitigate deficient descriptions.

GT: this is a black refrigerator. it sets in the corner of the room, by the counter.
Scan2Cap: this is a white refrigerator. it is to the right of the stove.
Scan2Cap+Ours: this is a stainless steel refrigerator. it is to the right of the kitchen cabinet.

GT: there is a window with blinds covering it. it has a baseboard heater below it.
Scan2Cap: the object is a drawer. it is located under the bed on the right side of the bed.
Scan2Cap+Ours: this is a window with blinds. it is to the left of the desk.

GT: the sink is below the mirror and right of the toilet. it is below and to the left of the black soap dispenser.
Scan2Cap: this is a white sink. it is to the left of the toilet.
Scan2Cap+Ours: this is a white sink. it is to the left of a soap dispenser.

GT: this is a ottoman. the ottoman is near the wall.
Scan2Cap: this is a brown hamper. it is to the left of a chair.
Scan2Cap+Ours: this is a ottoman. the ottoman is near the wall.

GT: the coffee table is in the center of the room surrounded by a sofa. the coffee table is a brown rectangle.
Scan2Cap: the couch is surrounding the coffee table. the couch is a long, gray, and rectangular.
Scan2Cap+Ours: the coffee table is in the center of the room. the coffee table is a brown rectangle.

GT: the object is the bed with a green spread, with a laptop computer resting on it. the bed is to the left of the nightstand.
Scan2Cap: there is a bed with a brown headboard. it is to the right of the desk.
Scan2Cap+Ours: the bed is in the middle of the room. the bed is to the left of the nightstand.

GT: the trash can is right of the blue covered trash can. the trash can is a gray rectangular prism.
Scan2Cap: the trash can is the second rightmost one out of a row of trash cans. the trash can is a cylinder.
Scan2Cap+Ours: the trash can is the farthest right one out of a row of trash cans. the trash can is a gray cylinder.

GT: to the right of this armchair is a black table. this gray armchair is located near the left wall.
Scan2Cap: this is a brown ottoman. it is to the left of the chair.
Scan2Cap+Ours: this is a brown armchair. it is to the left of the table.

Fig. 4. Comparisons of sample descriptions. Correct descriptions are in green color and incorrect in red color. Our proposal can mitigate deficient descriptions, however, there are still some poor descriptions. (Color figure online)

Qualitative Results. Figure 4 shows the qualitative results of Scan2Cap [5] and our proposal. Higher overall objectness scores can mitigate deficient descriptions of the corresponding objects, such as object misrecognition and misidentified relationships to surrounding objects. For instance, the object "window" is incorrectly described as a drawer under the bed, and the object "coffee table" is incorrectly described as the couch surrounding the coffee table, as shown in Fig. 4. However, in our proposal, they can be described the same as the ground truth. And modifier for the object is also can be mitigated; for instance, a "black refrigerator" in the ground truth is represented as a "white refrigerator", but in our proposal, it can be described as a "stainless steel refrigerator".

5   Conclusion

In this work, we propose the recurrent point clouds selection (RPCS) method to solve the problem in 3D dense captioning which produces poor descriptions by randomly selecting points from the point cloud regardless of their importance. Our method can keep the overall objectness scores at a higher level so as not to use as much information as possible including unimportant points, thus mitigating descriptive deficiencies for corresponding proposal objects. Our method also outperforms state-of-the-art methods and can be applied to the network of those methods, it is expected to obtain more accurate descriptions. Furthermore, our method achieves excellent captioning results, however, there is a problem that it takes time to generate descriptions, so we will continue to work on solving this problem.

References

1. Anderson, P., et al.: Bottom-up and top-down attention for image captioning and visual question answering. In: CVPR, pp. 6077–6086 (2018)
2. Banerjee, S., Lavie, A.: METEOR: an automatic metric for MT evaluation with improved correlation with human judgments. In: ACL Workshop, pp. 65–72 (2005)
3. Chen, D.Z., Chang, A.X., Nießner, M.: ScanRefer: 3D object localization in RGB-D scans using natural language. In: Vedaldi, A., Bischof, H., Brox, T., Frahm, J.-M. (eds.) ECCV 2020. LNCS, vol. 12365, pp. 202–221. Springer, Cham (2020). https://doi.org/10.1007/978-3-030-58565-5_13
4. Chen, D.Z., Wu, Q., Nießner, M., Chang, A.X.: D3Net: a speaker-listener architecture for semi-supervised dense captioning and visual grounding in RGB-D scans. arXiv preprint arXiv:2112.01551 (2021)
5. Chen, Z., Gholami, A., Nießner, M., Chang, A.X.: Scan2Cap: context-aware dense captioning in RGB-D scans. In: CVPR, pp. 3193–3203 (2021)
6. Cornia, M., Stefanini, M., Baraldi, L., Cucchiara, R.: Meshed-memory transformer for image captioning. In: CVPR, pp. 10578–10587 (2020)
7. Dai, A., Chang, A.X., Savva, M., Halber, M., Funkhouser, T., Nießner, M.: ScanNet: richly-annotated 3D reconstructions of indoor scenes. In: CVPR, pp. 5828–5839 (2017)
8. Donahue, J., et al.: Long-term recurrent convolutional networks for visual recognition and description. In: TPAMI, vol. 39, pp. 677–691 (2017)
9. Gao, L., Wang, B., Wang, W.: Image captioning with scene-graph based semantic concepts. In: ICMLC, pp. 225–229 (2018)
10. Jingpeng, H., Zhuo, L., Zhihong, C., Zhen, L., Xiang, W., Tsung-Hui, C.: Graph enhanced contrastive learning for radiology findings summarization. In: ACL (2022)
11. Johnson, J., Karpathy, A., Fei-Fei, L.: DenseCap: fully convolutional localization networks for dense captioning. In: CVPR, pp. 4565–4574 (2016)
12. Karpathy, A., Fei-Fei, L.: Deep visual-semantic alignments for generating image descriptions. In: CVPR, pp. 3128–3137 (2015)
13. Kim, D.J., Choi, J., Oh, T.H., Kweon, I.S.: Dense relational captioning: triple-stream networks for relationship-based captioning. In: CVPR, pp. 6271–6280 (2019)

14. Lin, C.Y.: Rouge: a package for automatic evaluation of summaries. In: Text Summarization Branches Out, pp. 74–81 (2004)
15. Lu, J., Xiong, C., Parikh, D., Socher, R.: Knowing when to look: Adaptive attention via a visual sentinel for image captioning. In: CVPR, pp. 3242–3250 (2017)
16. Lu, J., Yang, J., Batra, D., Parikh, D.: Neural baby talk. In: CVPR, pp. 7219–7228 (2018)
17. Papineni, K., Roukos, S., Ward, T., Zhu, W.J.: Bleu: a method for automatic evaluation of machine translation. In: The 40th Annual Meeting of ACL, pp. 311–318 (2002)
18. Qi, C.R., Litany, O., He, K., Guibas, L.J.: Deep hough voting for 3D object detection in point clouds. In: ICCV, pp. 9277–9286 (2019)
19. Qi, C.R., Yi, L., Su, H., Guibas, L.J.: Pointnet++: deep hierarchical feature learning on point sets in a metric space. arXiv preprint arXiv:1706.02413 (2017)
20. Vedantam, R., Zitnick, C.L., Parikh, D.: Cider: consensus-based image description evaluation. In: CVPR, pp. 4566–4575 (2014)
21. Vinyals, O., Toshev, A., Bengio, S., Erhan, D.: Show and tell: a neural image caption generator. In: CVPR, pp. 3156–3164 (2015)
22. Xiangyang, L., Jiang, S., Han, J.: Learning object context for dense captioning. In: AAAI, pp. 8650–8657 (2019)
23. Xu, K., et al.: Show, attend and tell: neural image caption generation with visual attention. arXiv preprint arXiv:1502.03044 (2015)
24. Yang, L., Tang, K., Yang, J., Li, L.J.: Dense captioning with joint inference and visual context. In: CVPR, pp. 2193–2202 (2017)
25. Yang, X., Tang, K., Zhang, H., Cai, J.: Auto-encoding scene graphs for image captioning. In: IEEE Conf. Comput. Vis. Pattern Recog., pp. 10677–10686 (2019)
26. Yao, T., Pan, Y., Li, Y., Mei, T.: Exploring visual relationship for image captioning. In: ECCV, pp. 684–699 (2018)
27. Yuan, Z., et al.: X-Trans2Cap: cross-modal knowledge transfer using transformer for 3D dense captioning. In: CVPR, pp. 8563–8573 (2022)

Multi-domain Feature Fusion Neural Network for Electrocardiogram Classification

Jiao Nan, Zhanquan Sun[✉], Lin Sun, Min Huang, and Zifan Li

School of Optical-Electrical and Computer Engineering, University of Shanghai for Science and Technology, Shanghai, China
sunzhq@usst.edu.cn

Abstract. Computerized electrocardiogram (ECG) interpretation technology is of great significant in detecting cardiovascular disease. Recently, many research studies information mining of ECG signals from multiple domain. However, these models of automatic arrhythmia detection extract the features in time-domain and time-frequency domain separately, and merge the features in the last layer. The information of middle layers of multi-domain is not used fully. In this study, we develop an ECG classifier based on Multiple Domain Features Fusion Network with Lead Attention (MDFF-LA), which can realize feature fusion in time domain and frequency domain features respectively. Filter useful lead information through the attention module prior to data entry. In the process of feature extraction, the feature weights of one domain are optimized in multiple stages by the attention module generated by another domain, where the multi-domain information constantly complements each other at the middle layers. Finally, we conduct comprehensive experiments on three multi lead ECG databases to test the performance. The results demonstrate that the fusion features in multi domain extracted by MDFF-LA can obtain more valuable information, which can provide supported diagnosis for clinicians in practical.

Keywords: Electrocardiogram · Convolutional neural network · Multi-domain · Feature fusion

1 Introduction

Cardiovascular diseases is a fairly common problem among critically ill patients [1]. Electrocardiogram (ECG) is one of the most widely used medical measure to detect cardiovascular diseases, and it is a non-invasive, computerized technique [2]. However, the interpretation of ECG waveform is a challenging

---

Supported by Medical Engineering Cross Project of USST (10-21-302-413), National Defence Basic Research Program (JCKY2019413 D001), National Natural Science Foundation of China (61374040).

© The Author(s), under exclusive license to Springer Nature Switzerland AG 2023
M. Tanveer et al. (Eds.): ICONIP 2022, LNCS 13625, pp. 275–286, 2023.
https://doi.org/10.1007/978-3-031-30111-7_24

problem, which requires rich special knowledge and experience. Hence, an accurate solution for automatic detection of arrhythmia needs to be constructed.

Machine learning and deep learning become the mainstream methods of ECG automatic diagnosis. Machine learning usually consist of two modules, i.e. feature extraction and classifier training, which have some common shortcomings: (1) The feature selection and extraction are complex comparatively [3,4]; (2) The classifier needs to supply strong modeling capability for these features [5,6]. However, deep learning technology breaks through the limitations of machine learning. As a data-driven technology, it can automatically extract and select features. The features are obtained through convolution, and the classification function is realized by the full connection layer.

Recent years, some research focuses on obtaining the characteristics of ECG signal in time-frequency domain [7]. Comparing the slightly varying waveform, the sensitivity of ECG in time-frequency domain is higher than that of routine ECG [8]. There are many common time-frequency analysis methods, such as Short-Time Fourier Transform (STFT), Continuous Wavelet Transform (CWT) and Hilbert Huang Transform (HHT) [9,10] etc. However, only extracting the information in the time-frequency domain is too fragmented to get all the ECG information. To solve this problem, some models extract the information in time domain and time-frequency domain at the same time. Jin et al. [11] developed an AF detector based on a twin-attentional convolutional long short-term memory neural network (TAC-LSTM), aiming at the influence of different input. However, in these researches, only the last layer features extracted from different domains are spliced, which does not fuse the information more deeply. Minh et al. [12] refined the architecture of feature fusion, which semantically fused the information by a transformer encoder. However, the information of middle layers of multi-domain is not used fully. These models isolated the relationship between ECG signal in time domain and ECG signal in time-frequency domain, so that the information between two domains could not complement each other.

By examining neural network models for single domain feature extraction, we find that models only pay attention to the features of one lead, ignoring the relationship between leads and arrhythmia classes. For example, atrial fibrillation changes noticeably in lead II and lead VI. Due to the timing of ECG signal, many research works have stacked Convolutional Neural Network (CNN) and Recurrent Neural Network (RNN) to extract local and global features in single lead [13,14]. With the emergence of attention module, using spatial and temporal attention mechanisms to emphasize necessary information and suppresse unimportant features in single lead [15].

To address the above limitations, we designed an ECG classifier based on Multiple Domain Features Fusion Network with Lead Attention (MDFF-LA). The highlights of our proposed method are as follows: (1) Our method combines multi-domain information, i.e. the time domain and time-frequency domain, and fuse the information in multi-domain at different stages. Multi-domain information is constantly complementing each other at the middle layer, allowing full integration of multi-domain information. (2) The attention module is applied to the raw ECG leads to emphasize the leads selectively, so that convolution can be targeted to extract information from the effective leads. (3) We conduct the extensive experiments based on the proposed method. Experimental results show that the proposed method has an excellent performance in three multi leads databases through comparing with current deep learning methods.

2   Methods

2.1   Dataset Description

Three databases are used to demonstrate the efficiency of the proposed sorting methods, i.e. the China Physiological Signal Challenge 2018 (CPSC 2018), Alibaba Cloud Tianchi ECG Competition 2019 (ACTEC 2019) and PTB-XL. The CPSC 2018 contains 6877 12-lead ECG records with varied lengths ranging from 6 to 60 s and it contains 9 ECG types, i.e. 8 arrhythmia classes and a normal rhythm. ACTEC 2019 includes 40000 medical ECG samples with 34 classification results, provided by the engineering research center of the Ministry of education of mobile health management system of Hangzhou Normal University. The PTB-XL ECG dataset is a large dataset of 21837 clinical 12-lead ECGs from 18885 patients of 10 s lengths. The in total 71 different ECG statements conform to the SCP-ECG standard and cover diagnostic, form, and rhythm statements. All the three ECG databases are multi label.

2.2   Overview of the Proposed Method

The proposed method consists of four aspects. Firstly, the ECG signals is preprocessed to obtain time sequences and time-frequency sequences with Short-Time Fourier transform (STFT). Secondly, before convolution operation, the attention module is added to screen for leads that are more relevant to the disease. Then, a convolutional neural network is utilized to classify arrhythmia, which extract the features in time domain and time-frequency domain. When extracting features at middle layers of the convolution model, the information in multi domain is fused for several times. Finally, the fused features are used to classify the type of arrhythmia.

2.3   Pre-processing and Short-Time Fourier Transform

In this study, the ECG information of the input model is divided into two parts: (I) time series; (II) time-frequency series. ECG signal is a kind of weak electronic signal which can be easily interfered by many kinds of noises, such as powerline interference, electromyography interference, environment noise and so on. In this experiment, median filter algorithm is used to remove baseline drift. A low pass filter with 50HZ cut-off frequency is used to filter powerline interferences. The ECG data is normalized to [0,1].

For time-frequency sequence, we adopt STFT to convert the original ECG wave signal into ECG spectrum. In contrast to Fourier transform, STFT reduces the loss of temporal information and maintains the capture of time-varying sensitive information to a certain extent [15]. Mathematically, it can be described as follows:

$$S(t, f) = \int_{-\infty}^{\infty} z(\tau)h(\tau - t)e^{-j2\pi f\tau} d\tau \qquad (1)$$

where $z(\tau)$ is the signal and $h(\tau - t)$ the window. We use Hamming window and set the window size to 128. The time sequence and time-frequency sequence of one lead ECG signals are shown in Fig. 1 respectively.

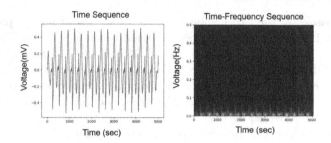

Fig. 1. Time sequence of one lead ECG signals and time-frequency sequence of the ECG signals.

2.4   The Architecture of MDFF-LA

As shown in Fig. 2, the model consists of two similar parts of attention convolutional neural network, which are used to deal with ECG in time series and time-frequency series respectively. The model used for processing each domain ECG signals consists of attention modules, convolution layers and pooling layers. The first attention module is added to enhance the weight of important leads and suppress the weight of unimportant leads. Then, two-dimensional convolution is used to extract the features of multiple leads in time domain or time-frequency domain. Overall, the model is called Multiple Domain Features Fusion Network with Lead Attention (MDFF-LA).

The two domain features can complement each other. We perform multiple feature fusion at the middle layers. Through Conv block1 and Conv block2, the extracted features of one domain model are crossed to generate attention module for the other domain model. Information from multiple domains can be fully integrated in different stages, which include shallow features and deep features. Besides, we combine the output of time series with the output of time-frequency series furtherly, and concatenated features are transformed to dense layer for classification. Finally, the feature vectors generated by the pooling operation are concatenated and passed to the subsequent layers for classification.

2.5   Convolutional Neural Network Block

A convolutional neural network inspired by VGGNet is applied to extract features. The convolution neural network consists of 3 convolutional blocks, i.e. 12

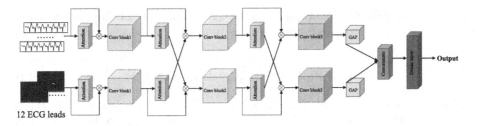

Fig. 2. The architecture of MDFF-LA.

2-dimension (2D) convolution layers, as shown in Fig. 3. Conv3 × 64 means that the kernel size of the convolution layer is 3 and the kernel number of the convolution layer is 64. Conv3 × 128 and Conv3 × 256 are the same meaning. Each convolution layer is followed by a batch normalization layer and a rectified linear unit layer. After performing two convolution operations, dropout with a rate of 0.2 is set to prevent over fitting of the neural network. Because the length and width of ECG signal in time domain and time-frequency domain are different, the size of pooling is different. For the time domain, considering that the width is equal to the number of leads 12, the maximum pooling kernel size 5 is only carried out in the width direction. For the time-frequency domain, a pooling kernel size 3 is used to reduce the length by 3 times in both length and width.

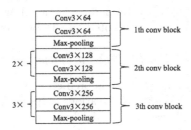

Fig. 3. Layer configuration for the convolutional neural network.

2.6 Attention Module

In order to extract more features of important leads and fuse more features of ECG information in different domain in the process of convolution operation, the attention module is introduced in view of reference [17]. Attention module can highlight some leads with strong relationship with arrhythmia. In the process of feature extraction, attention is embedded many times and used alternately, so that the features in multi-domain are fully fused. Feature is used as the input of attention layer. Firstly, through the spatial maximum pooling operation and average pooling operation from the shared feature F, and two spatial features F_{max} and F_{avg} are obtained. Average pooling is easier to capture global changing information while maximum pooling is easier to capture local changing information. Then, the aggregated information is input into the shared dense layers to generate attention A.

$$A = \sigma[W_2 ReLU(W_1 F_{\max}) + W_2 ReLU(W_1 F_{avg})] \tag{2}$$

where σ is a sigmoid function to compresses the attention weight to [0,1], $W_1 \in R^{C/r \times C}$ and $W_1 \in R^{C/C \times r}$ are the weights of the shared dense layers. r is the ratio to reduce the number of characteristic channels to save network parameters, and we set its experience to 0.3.

After obtaining the learned attention weight A, we multiply it with the original feature map F to produce a new feature map F_{new} after the feature weight is corrected.

$$F_{new} = F \otimes A \tag{3}$$

where, $\otimes$ represents element-wise multiplication. And the attention weights are broadcasted along the feature channels.

2.7 Training

The preprocessed ECG signals are input into convolutional neural network models in batches and batch size is set 32, which can balance the available graphics processing unit (GPU) memory and training speed. Stochastic Gradient Descent (SGD) is applied to update the weights iteratively due to its ability to accelerate the training process of deep network model. The learning rate is set to 0.01. The binary cross entropy function is adopted as the loss function of the network, defined as:

$$H(p,q) = -\sum p(x) log q(x) \tag{4}$$

where $p(x)$ is the true result, $q(x)$ is the prediction result.

Due to the different proportion of various arrhythmias, using the accuracy as the stop standard will lead the model being conducive to the category of arrhythmias with many samples. F1 score is fairer and objective than accuracy to judge the classification ability of the model. Therefore, in the experiment, we used the F1 score as the stop criterion. When the F1 score on the validation set does not improve in 200 epochs, the model is set to terminate the training. The experiment is operated on GPU server with two 1080i GPU cards, 8 kernel CPU, 128G memory.

2.8 Evaluation Metrics

In this study, we used five experiments to verify the accuracy and necessity of the network. To evaluate the performance of classifiers, Hamming Loss, Jaccard Index, Accuracy, Precision, Recall and F1 score are adopted as the evaluation indices.

Hamming Loss is used to investigate the misclassification of samples on a single marke. Jaccard Index represents the similarity and difference between limited sample sets. The accuracy means that predicted samples correctly predicted account for all samples. Precision means the probability of the actual positive sample in all the predicted positive samples, representing the prediction accuracy of the positive sample. Recall measures the ability of the classifier to recognize positive examples. Micro-F1 score is a typical metric for measuring the classification performance of a model.

3 Experiment and Result

3.1 Experiments of Generality for Model

We experimented on three multi lead ECG databases (CPSC 2018, ACTEC 2019 and PTB-XL) to prove the rationality of model MDFF-LA, which include three aspects, i.e. weight of important leads with arrhythmia, feature extraction and fusion in multi domain. Experimental data show that the model MDFF-LA is universal and suitable for ECG classification, showed in Table 1.

Table 1. Classification performance of MDFF-LA in three ECG databases.

Databases	Hamming loss	Jaccard index	Accuracy	Precision	Recall	F1
CPSC 2018	0.035	0.737	0.800	0.853	0.845	0.849
ACTEC 2019	0.010	0.781	0.751	0.920	0.837	0.877
PTB-XL	0.021	0.549	0.382	0.776	0.653	0.709

3.2 Experiments of Different Domain Information for Model

We compared the performance of different network models (Convolutional Neural Network in Time-Frequency Domain with Lead Attention (CNN-TFD-LA), Convolutional Neural Network in Time Domain with Lead Attention (CNN-TD-LA) and Convolutional Neural Network in Multi-Domain with Lead Attention (CNN-MD-LA), respectively). Table 2 shows the prediction results of CNN-TFD-LA, CNN-TD-LA and CNN-MD-LA. CNN-MD-LA combines feature in multi domain to obtain better classification performance. Therefore, multi-domain information facilitates the classification performance of the model.

Table 2. Comparison of different domain information tests of Arrhythmia detection in CPSC 2018.

Algorithm	Hamming loss	Jaccard index	Accuracy	Precision	Recall	F1
CNN-TFD-LA	0.043	0.686	0.755	0.833	0.796	0.814
CNN-TD-LA	0.040	0.709	0.763	0.835	0.824	0.829
CNN-MD-LA	**0.038**	**0.723**	**0.775**	**0.843**	**0.835**	**0.839**

3.3   Experiments of Lead Attention for Model

To verify the role of lead attention, we compared performance of different network models (Convolutional Neural Network in Time-Frequency Domain (CNN-TFD), CNN-TFD-LA, Convolutional Neural Network in Time Domain (CNN-TD), CNN-TD-LA, Convolutional Neural Network in Multi-Domain (CNN-MD), CNN-MD-LA, Multiple Domain Features Fusion Network (MDFF), and MDFF-LA, respectively), showed in Table 3. Lead attention has achieved better classification results in time domain, time-frequency domain, multi domain and multi domain of multiple feature fusion. The experiment demonstrates the effectiveness of the initiative to use attention to screen for leads with strong arrhythmia relationship before the convolution operation. Lead attention makes the convolution to be more targeted to obtain information about valid leads in depth.

Table 3. Classification performance with different structures about lead attention in CPSC 2018.

Algorithm	Hamming loss	Jaccard index	Accuracy	Precision	Recall	F1
CNN-TFD	0.045	0.674	**0.759**	0.820	0.791	0.805
CNN-TFD-LA	**0.043**	**0.686**	0.755	**0.833**	**0.796**	**0.814**
CNN-TD	0.043	0.686	0.741	0.830	0.798	0.814
CNN-TD-LA	**0.040**	**0.709**	**0.763**	**0.835**	**0.824**	**0.829**
CNN-MD	0.040	0.712	0.771	0.826	**0.837**	0.832
CNN-MD-LA	**0.038**	**0.723**	**0.775**	**0.840**	0.835	**0.837**
MDFF	0.037	0.727	0.788	0.843	0.841	0.842
MDFF-LA	**0.035**	**0.737**	**0.800**	**0.853**	**0.845**	**0.849**

3.4   Experiments of Features Fusion for Model

To illustrate the effectiveness of multi-domain feature fusion in the middle layer, we have also carried out multiple feature fusion in the process of feature extraction. In order to fuse the features more fully, we undertake the research with some models which feature fusion in different stages (after Conv block 1 (MDFF-LA -CB1), after Conv block 2 (MDFF-LA-CB2), after Conv block 3 (MDFF-LA -CB3), after Conv block 1 and Conv block 2 (MDFF-LA-CB12, MDFF-LA), after Conv block 1 and Conv block 3 (MDFF-LA -CB13), after Conv block 2 and Conv block 3 (MDFF-LA-CB23), after Conv block 1, Conv block 2 and Conv block 3 (MDFF-LA-CB123)), showed in Table 4. Feature fusion performs one or more times at different stages, which reduces or improves the classification performance of the model. Feature fusion occurs after Conv block 1 or Conv block 2, and the classification performance of the model degrades. When feature fusion added to Conv block 3, the classification ability of the model is improved. For feature fusion in any two stages, the classification ability of the model is improved. The neural network of feature fusion after Conv block 1 and

Conv block 2, achieving the best classification results. However, the classification performance of model for three times feature fusion will be degraded. Different positions of multi-domain information in the middle layer have different effects on model information extraction.

Table 4. Classification performance with different structures about features fusion in CPSC 2018.

Algorithm	Hamming loss	Jaccard index	Accuracy	Precision	Recall	F1
CNN-MD-LA	0.038	0.723	0.775	0.084	0.835	0.839
MDFF-LA-CB1	0.041	0.705	0.762	0.816	0.839	0.827
MDFF-LA-CB2	0.038	0.721	0.785	0.834	0.839	0.836
MDFF-LA -CB3	0.036	0.730	0.789	**0.859**	0.830	0.844
MDFF-LA-CB12	**0.035**	**0.737**	**0.800**	0.853	**0.845**	**0.849**
MDFF-LA-CB13	0.036	0.730	0.783	0.858	0.830	0.844
MDFF-LA-CB23	0.036	0.731	0.791	0.846	0.843	0.844
MDFF-LA-CB123	0.039	0.712	0.774	0.837	0.827	0.832

3.5   Classification Performance Comparison with Other Works

In this paper, we employ experimental conditions similar to those of existing methods in order to fairly evaluate the performance of the proposed method. The designed model is compared with other classification models in three multi label ECG databases. The Table 5, Table 6 and Table 7 show the results of some existing research methods. The network structures of [18] and [14] were built on the basis of CNN and RNN to extract local and global features. He et al. modified the structure of CNN and added the residual element through the short-circuit mechanism [14]. Wang et al. proposed a record level ECG classification method by combining 1-D deep convolutional neural network and long short-term memory network (LSTM) [19]. Yao et al. improved the original model design [18], adding the attention module behind LSTM to focuses on the characteristics of the local waveform of a single lead [15]. Zhang et al. also used the temporal

Table 5. Classification performance comparison with other works and the proposed model in CPSC 2018.

Author	Hamming loss	Jaccard index	Accuracy	Precision	Recall	F1
Yao et al. [18]	0.043	0.686	0.741	0.830	0.798	0.814
Wang et al. [19]	0.040	0.705	0.770	0.847	0.808	0.827
He et al. [14]	0.038	0.720	0.797	0.847	0.827	0.837
Yao et al. [15]	0.038	0.721	0.798	0.845	0.831	0.838
Zhang et al. [20]	0.036	0.733	0.786	0.848	0.844	0.846
Ours	**0.035**	**0.737**	**0.800**	**0.853**	**0.845**	**0.849**

Table 6. Classification performance comparison with other works and the proposed model in ACTEC 2019.

Author	Hamming loss	Jaccard index	Accuracy	Precision	Recall	F1
Yao et al. [18]	0.012	0.748	0.719	0.883	0.830	0.855
Wang et al. [19]	0.011	0.753	0.727	0.887	0.832	0.859
He et al. [14]	0.011	0.755	0.730	0.886	0.836	0.860
Yao et al. [15]	0.011	0.758	0.733	0.910	0.819	0.862
Zhang et al. [20]	**0.010**	0.772	**0.751**	**0.936**	0.815	0.871
Ours	**0.010**	**0.781**	**0.751**	0.920	**0.837**	**0.877**

Table 7. Classification performance comparison with other works and the proposed model in PTB-XL.

Author	Hamming loss	Jaccard index	Accuracy	Precision	Recall	F1
Yao et al. [18]	0.023	0.509	0.331	0.725	0.631	0.675
Wang et al. [19]	0.023	0.516	0.345	0.715	0.650	0.681
He et al. [14]	0.022	0.522	0.357	0.745	0.636	0.686
Yao et al. [15]	0.023	0.524	0.346	0.723	0.656	0.688
Zhang et al. [20]	**0.021**	0.530	**0.393**	**0.781**	0.623	0.693
Ours	**0.021**	**0.549**	0.382	0.776	**0.653**	**0.709**

attention and spatial attention after convolution operation for many times [20]. The attention module improved the classification performance of the network. The network model we designed has achieved good classification results in multi classification.

4   Conclusion

Arrhythmia has serious harmful effects on human beings and can cause sudden death. Therefore, it is of great significance to build a high-precision and stable arrhythmia detection system. In this study, we develop a MDFF-LA method for arrhythmia detecting. MDFF-LA can more fully extract and fuse the features of ECG signals in multi domain after emphasizing important leads. Multi-stage feature fusion of different domains is performed at the middle layer causing information features of different domains to complement each other. Classification performance of the method proposed by us exceed other methods on three multi lead databases. In addition, further experiments show the effectiveness and rationality of our model design. It can help cardiologists diagnose arrhythmias in clinical practice and to obtain scientific basis for prevention and diagnosis of arrhythmias with patients. We will continue to study the related field knowledge of ECG automatic diagnosis and improve the accuracy of ECG classification.

References

1. Gordon, P., Kerton, M.: Cardiac arrhythmias in the critically ill. Anaesth. Intensive Care Med. **22**(5), 301–321 (2021)
2. Barcos, J.C., Humphreys, J.D., Santacruz, I.A.T., et al.: Enhancing electrocardiographic analysis by combining a high-resolution 12-lead ECG with novel software tools. J. Electrocardiol. **70**, 70–74 (2022)
3. De Lannoy, G., Francois, D., Delbeke, J., et al.: Weighted conditional random fields for supervised interpatient heartbeat classification. IEEE Trans. Biomed. Eng. **59**(1), 241–247 (2012)
4. Daamouche, A., Hamami, L., Alajlan, N., et al.: A wavelet optimization approach for ECG signal classification. Biomed. Signal Process. Control **7**(4), 342–349 (2012)
5. Sharma, P., Dinkar, S.K., Gupta, D.V.: A novel hybrid deep learning method with cuckoo search algorithm for classification of arrhythmia disease using ECG signals. Neural Comput. Appl. **33**(19), 13123–13143 (2021)
6. Barni, M., Failla, P., Lazzeretti, R., et al.: Privacy-preserving ECG classification with branching programs and neural networks. IEEE Trans. Inf. Forensics Secur. **6**(2), 452–468 (2011)
7. Alqudah, A.M., Qazan, S., Al-Ebbini, L., Alquran, H., Qasmieh, I.A.: ECG heartbeat arrhythmias classification: a comparison study between different types of spectrum representation and convolutional neural networks architectures. J. Ambient. Intell. Humaniz. Comput. **13**, 4877–4907 (2021). https://doi.org/10.1007/s12652-021-03247-0
8. Haberl, R., Jilge, G., Pulter, R., et al.: Comparison of frequency and time domain analysis of the signal-averaged electrocardiogram in patients with ventricular tachycardia and coronary artery disease: methodologic validation and clinical relevance. J. Am. Coll. Cardiol. **12**(1), 150–8 (1988)
9. Zhang, J., Tian, J., Cao, Y., et al.: Deep time-frequency representation and progressive decision fusion for ECG classification. Knowl.-Based Syst. **7**, 92871–92880 (2019)
10. Huang, J.S., Chen, B.Q., Yao, B., et al.: ECG arrhythmia classification using STFT-based spectrogram and convolutional neural network. IEEE Access **190**, 92871–92880 (2020)
11. Jin, Y.R., Qin, C.J., Huang, Y.X., et al.: Multi-domain modeling of atrial fibrillation detection with twin attentional convolutional long short-term memory neural networks. Knowl.-Based Syst. **193**, 105460 (2020)
12. Minh, D.L., Singh, R.V., Quang, S.T., et al.: Multi-module recurrent convolutional neural network with transformer encoder for ECG arrhythmia classification. In: 2021 IEEE EMBS International Conference on Biomedical and Health Informatics (BHI). EMB, IEEE, Athens, Greece (2021). https://doi.org/10.1109/BHI50953.2021.9508527
13. Lih, O.S., Jahmunah, V., San, T.R., et al.: Comprehensive electrocardiographic diagnosis based on deep learning. Artif. Intell. Med. **103**, 101789 (2020)
14. He, R.N., Liu, Y., Wang, K.Q., Zhao, N., et al.: Automatic cardiac arrhythmia classification using combination of deep residual network and bidirectional LSTM. IEEE Access **7**, 102119–102135 (2019)
15. Yao, Q.H., Wang, R.X., Fan, X.M., et al.: Multi-class arrhythmia detection from 12-lead varied-length ECG using attention-based time-incremental convolutional neural network. Inf. Fusion **53**, 174–182 (2020)
16. Malikiosis, R.D.: Spark deficient Gabor. Pac. J. Math. **294**(1), 159–180 (2018)

17. Woo, S., Park, J., Lee, J.Y., Kweon, I.S.: Cbam: convolutional block attention module. In: Proceedings of the European Conference on Computer Vision (ECCV), pp. 3–19 (2018)
18. Yao, Q.H., Fan, X.M., Cai, Y.P., et al.: Time-incremental convolutional neural network for arrhythmia detection in varied-length electrocardiogram. In: 6th IEEE Int. Conf. on Dependable, Autonom and Secure Comp/16th IEEE Int. Conf. on Pervas Intelligence and Comp/4th IEEE Int. Conf. on Big Data Intelligence and Comp/3rd IEEE Cyber Sci. and Technol Congress, pp. 754–761. IEEE, Athens, GREECE (2018). https://doi.org/10.1109/DASC/PiCom/DataCom/CyberSciTec
19. Wang, Y.C., Xiao, B., Bi, B.X., et al.: Pay attention and watch temporal correlation: a novel 1-D convolutional neural network for ECG record classification, p. 4. Singapore (2019). https://doi.org/10.23919/CinC49843.2019.9005505
20. Zhang, J., Liu, A.P., Gao, M., et al.: ECG-based multi-class arrhythmia detection using spatio-temporal attention-based convolutional recurrent neural network. Artif. Intell. Med. **106**, 101856 (2020)

Graph-Based Contextual Attention Network for Single Image Deraining

Bin Hu, Mingcen Gu, Yuehua Li, Lili Zhao, and Shi Cheng[✉]

School of Information Science and Technology, Nantong University, Nantong,
Jiangsu, China
{hubin,lyh,ylzh,chenshi}@ntu.edu.cn, 2010320047@stmail.ntu.edu.cn

Abstract. Rain streaks degrade the images and badly affect the outdoor
vison tasks, and deep learning based single image deraining approach has
witnessed the continuously growing and achieved great success. However, traditional convolution operation which uses a slide window can
only extract the local feature patch, most of them ignore the correlation
among the features for image deraining. We propose a graph-based contextual attention network for single image deraining. Firstly, we project
the input feature map to a latent global context representation, and then
the global information flow to spatial graph attention and channel graph
attention, respectively. After the two graph attention, a grouped linear layer is adopted to mine the correlation among the information and
enhance the global contextual information. The experiments on several
datasets demonstrate that our methods achieve better results than the
previous state-of-art methods.

Keywords: Image deraining · contextual attention · graph
convolutional

1 Introduction

Rain streaks degrade the images captured from outdoor, and have badly affected
the outdoor vison tasks such as object detection and image segmentation [1,2].
In order to solve this problem, lots of algorithms have been proposed for image
deraining. However, single image deraining is an ill-posed inverse problem, for a
rainy image corresponds to lots of possible solutions.

Early works on single image deraining are model-based approach, which consider that the rain image O can be declared as a linear sum of the background
image B and a rain layer R. These methods including image decomposition [3],
sparse coding [4] and priors based Gaussian mixture models [5] are popular
before 2017. The model of background and rain layer can be expressed as:

$$O = B + R \tag{1}$$

Supported by Nantong Science and Technology Program Project (JC2020065).

M. Tanveer et al. (Eds.): ICONIP 2022, LNCS 13625, pp. 287–297, 2023.
https://doi.org/10.1007/978-3-031-30111-7_25

The model-based methods often try to obtain the background image B rely on the statistical analysis by Eq. (1).

Since 2017, lots of data-driven approach (deep learning approach) is proposed because of the fast development of deep learning in computer vision tasks. Most approach are indicated by deep convolution neural network (CNN) [6], generative adversarial network (GAN) [7] and semi/unsupervised methods [8]. Compared to the model-driven approaches, the data-driven approaches based on deep learning achieve more superior performance with robustness.

Most deep learning methods use CNN as the backbones, although great progress has been made in single image deraining. In CNN, a convolution layer usually uses a slide window kernel W to extract local features, and the mechanism limits the network to capture global contextual information, which is import for image deraining. To address the limits of CNN, lots of methods are proposed to utilize global information such as attention mechanism [9], non-local network [10], graph convolutional network [11,12] and et al. However, most methods ignore the correlation among the features for image deraining.

Motivated by the research above, we propose a graph-based contextual attention network for single image deraining. We project the input feature map to a latent global context representation, and then the global information flow to spatial graph attention and channel graph attention, respectively. After the two graph attention, a grouped linear layer is adopted to mine the correlation among the information and enhance the global contextual information.

Above all, the contributions of the work are follows:

1) We propose a graph-based contextual attention network for single image deraining. With the spatial graph attention with spatial coherence computing and channel graph attention with channel correlation computing, the convolution kernel is modulated to extract global contextual information.
2) We propose to mine the correlation among the global information by project the information into a bag of latent features, which enhance the descriptors.
3) Our network is an end-to-end network and is easy to implement. The experiment results show that our work achieves superior performance in comparison with the state-of-the-art methods on serval datasets.

2   Related Work

2.1   Single Image Deraining

Single image deraining methods can be simply categorized into model-driven (prior-based) approach and data-driven (deep learning based) approach.

Before 2017, most methods are model-driven approach, and those methods are rely on the statistical analysis on the rain layer and background layer of the rainy images. The methods adopt various prior knowledges to separate the rain layer and background layer. Sparse coding [13] is used reconstruct the rain streaks and background signals. Kang et al. [3] represent the vector of rainy image as a combination of ow-frequency component and high-frequency component via sparse coding and dictionary learning, and then remove the rain layers by image decomposition. Luo et al. [4] propose the discriminative sparse coding to learn the dictionary of rain streaks and background layers. Zhu et al. [14] remove the rain streaks by iterative layer separation via layer-specific priors. Gaussian mixture models (GMMs) are used to model the rain layer and background layer [5].

Since deep learning has achieved significant success in computer vision, the methods based on data-driven approach have been popular after 2017. Lots of new network structure and loss functions are proposed to improve the performance of image deraining. Yang et al. [15] design a network to detect and remove rain streaks. The deep detail network [16] takes only high frequency details as input and then output the clear images. Following those methods, many network architectures are proposed, such as multi-stage network [17] and encoder-decoder network [18]. Different attention mechanisms [19,20] are introduced to design the network. Besides CNN, researchers also introduce GANs [21], recurrent neural networks [22] and graph convolutional networks (GCNs) [12] to design the network. But methods based on GAN are lack in fine-grained details.

2.2 Graph Convolutional Networks

Most deep learning based approaches use convolutional operations to design a CNN based deraining method. However, CNN can only extracts the local spatial image feature but lacks in broad contextual information because of its limited receptive field. While graph based methods have shown great success in computer vision tasks such as semi-supervised classification [11] and detection [23] by modeling contextual information. The non-local networks [10] which can capture the spatial information between all pixels in the image is the most widely used GCN form. Also the GCNs have been applied in single image deraining to capture the global contextual information [12,18].

3 Graph-Based Contextual Attention Network

In CNN, the input rainy image can be extracted as features by convolution layers. Let the input feature map be $F_{in} \in \mathbb{R}^{c_{in} n \times w \times h}$, where c_{in}, w, h denote the number of channel, width and height of the input, respectively, and the convolution layer uses a slide window to extract the local feature patch by a convolution kernel $W \in \mathbb{R}^{c_{out} \times c_{in} \times k_1 \times k_2}$, where c_{out}, k_1, k_2 are the output number of channel, width and height of the convolution kernel. So only local features are extracted from the convolution layers.

3.1 Context Feature Extraction

To address the limits of the traditional convolution operation, a graph-based contextual attention convolution is proposed to extract the global information from the input feature map. As Fig. 1 shows, firstly, a pooling layer is adopted to

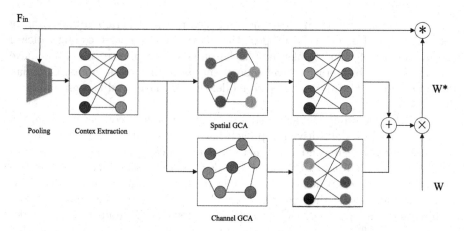

Fig. 1. Graph-based contextual attention convolution (GCAC), spatial GCA denotes spatial graph attention and channel GA denotes channel graph attention.

get the global spatial information and the size of the input feature map is reduced to $w' \times h'$. Then, following the previous work [10,24], a linear layer with weight $W_e \in \mathbb{R}^{w' \times h' \times e}$ is used to project the input feature to a latent global context representation with size e, where $e = (k_1 + k_2)/2$. And the global information $F_g \in \mathbb{R}^{c_i n \times e}$ is obtained, we reform $F_g = [\mathbf{f}_1, \dots, \mathbf{f}_e]$ as bag of global features.

3.2   Graph-Based Contextual Attention

Recently, graph-based contextual attention have been introduced to study the correlation among the features [11,25,26]. Following those works, we denote the graph $G(F_g, R)$ to learn the correlation among the global context information $F_g = [\mathbf{f}_1, \dots, \mathbf{f}_e]$ extracted from the linear layer, where F_g and R are the nodes and edges of the graph. The bag of global context features $F_g = [\mathbf{f}_1, \dots, \mathbf{f}_e]$ flow to two paths, one is the spatial graph attention (SGA), and the other is the channel graph attention (CGA). The two graph-based attention modules learn the global contextual information from spatial and channel relationship. The spatial graph attention [10] is designed to explore global spatial information between pixels, while the channel graph attention [12] is derived from the channels of feature map to explore the global information between channels.

According to Non-local network [10], the GCN operation can be defined as:

$$F_{GCN} = AF_{in}W \tag{2}$$

where $F_{in} \in \mathbb{R}^{w'h' \times e}$ is the input feature map, A and W are the adjacency matrix and the weight matrix, respectively. The SGA module is designed to model the global spatial correlations from F_{in}:

$$F_{sGA} = F_{in} + A_{sGA}F_s W_{sGA} \tag{3}$$

where F_s is generated from F_{in}, $F_s = conv(F_{in})$, and $conv(\cdot)$ means a convolution layer with 1×1 kernel size. W_{sGA} is the weight matrix, which is implemented by a 1×1 convolution layer. A_{sGA} is the adjacency matrix, which is calculated by matrix multiplication operation:

$$A_{sGA} = conv(F_{in}) \otimes conv(F_{in})^{\mathrm{T}} \tag{4}$$

where $conv(\cdot)$ is a convolution layer with 1×1 kernel size.

Similar to SGA module, the CGA module is designed as:

$$F_{cGA} = F_{in} + \phi(\widehat{F_c}) = F_{in} + \phi(A_{cGA}F_c W_{cGA}) \tag{5}$$

where $\phi(\cdot)$ is a hidden-to-output operation which is implemented by a 1×1 convolution layer, and A_{cGA} and W_{cGA} are adjacency matrix and weight matrix. F_c is generated from F_{in}, and F_c is calculated by:

$$F_c = softmax(\kappa(F_{in})^{\mathrm{T}} \otimes \varsigma(F_{in})) \tag{6}$$

where $\varsigma(F_{in}) \in \mathbb{R}^{w'h' \times e/2}$ and $\kappa(F_{in}) \in \mathbb{R}^{w'h' \times e/4}$ are two 1×1 convolution layers, and $softmax$ operation is used to avoid numerical instabilities. And $F_c \in \mathbb{R}^{e/4 \times e/2}$ contains $e/4$ nodes whose dimension is $e/2$, an identity matrix I is used to propagate the nodes, and the $\widehat{F_c}$ can be calculated by:

$$\widehat{F_c} = (I + A_{cGA})(\kappa(F_{in})^{\mathrm{T}} \otimes \varsigma(F_{in}))W_{cGA} \tag{7}$$

A_{cGA} and W_{cGA} are implemented by two convolution layers. And the shape of $\widehat{F_c}$ is reshaped to $w'h' \times e$ by $\phi(\cdot)$.

Inspired by [24], the grouped linear layer is adopted after the two graph attention, respectively. Consequently, two tensors $F_s = F_{sGA}W_s$ and $F_c = F_{cGA}W_c$ are generated, where W_s and W_c are two weights of the two linear layers, respectively.

After the generation of F_s and F_c, we element-wise add the two tensors to get the final global attention mask $F_A = F_s + F_c$. Then we form the graph-based contextual attention convolution (GCAC) weight via:

$$W^* = W \odot F_A \tag{8}$$

With the GCAC, the global context information is captured from the input feature maps.

3.3 Graph-Based Context Attention Network for Image Deraining

The proposed GCAC can be simply used to design a network. We design a symmetrical network for single image deraining as Fig. 2 shows. The network takes the rainy image X as the input and the clear image Y as the predicted result. The network consists of two convolution layers for pre-processing, several basic units and two convolution layers for reconstructing output. The pre-processing and reconstruction layers are designed standard 3×3 convolutional operations.

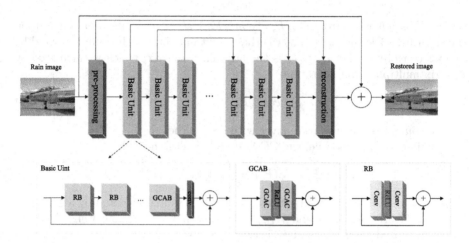

Fig. 2. The architecture of the proposed network. RB, GCAB, GCAC are residual block, graph-based contextual attention bock and graph-based contextual attention convolution, respectively.

The skip connection used in the symmetrical network can pass the image details from shallow layers to deep layers. Following the previous residual block, we use a simplified residual block (RB) [27] which used in many works, and design a graph-based contextual attention bock (GCAB) to form the basic unit. The GCAB is designed as:

$$F_o = W_{A2}\psi(W_{A1}F_i) + F_i \tag{9}$$

where F_o and F_i are output and input feature map, respectively, W_{A1} and W_{A2} are the proposed GCAC layer, and $\psi(\cdot)$ is the ReLU activation function.

4 Experiment Results

4.1 Datasets and Implementation

We evaluate our method on five benchmark datasets including DID-Data [6], Rain100L [28], Rain100H [28], Rain800 [21] and SPA-Data [19]. DID-Data contains 12,000 pair training images and 1,200 pair testing images. Rain100L and Rain100H are two testing subsets from the same dataset, and they contains testing images with light rain streaks and heavy rain streaks, respectively. SPA-Data is a real world rainy dataset that consists of about 640 thousand training images and 1,000 testing images.

We implement and train the network in PyTorch. The number of basic module of the network is set to 11, and each basic module contains 19 RBs and 1 GCAB. Adam optimizer with the default parameters is used for training, and cosine annealing strategy [29] is adopted to adjust the learning rate.

4.2   Comparisons

We compare our method on peak signal to noise ratio (PSNR) and structure similarity (SSIM) with the recent state-of-the-art algorithms including DDN [16], DID-MDN [6], RESCAN [30], JORDER-E [28], SIRR [31], PReNet [32], SPANet [18], RCDNet [33], DualGCN [12], MPRNet [17] and RLNet [34], and the results are list in Table 1. Compared with the recent state-of-the-art algorithms, we achieve a better results on those data sets. The best two methods DualGCN and ours are both graph based, it improve that the global information is import for image deraing. Compared with the DualGCN, the PSNR of our method increased by about 2.6%, it improves the effectiveness of the information projection, which models the correlation among the global information.

Table 1. Quantitative comparison on different data sets. Bold and underline indicate top 1st and 2nd rank, respectively.

Method	Datasets				
	DID-Data	Rain100L	Rain100H	Rain800	SPA-Data
DDN (TIP'17)	27.33/0.853	33.50/0.944	20.12/0.635	21.16/0.732	34.67/0.945
DID-MDN (CVPR'18)	27.93/0.861	23.79/0.773	15.54/0/520	21.22/0.750	36.65/0.963
RESCAN (ECCV'18)	29.12/0.880	36.58/0.970	25.92/0.823	23.90/0.828	36.62/0/961
JORDER-E (TPAMI'19)	32.89/0.931	38.97/0.983	28.95/0.908	28.17/0.8897	38.72/0.973
SIRR (CVPR'19)	29.75/0.847	38.71/0.987	28.79/0.881	24.68/0.829	36.12/0.938
PReNet (CVPR'19)	30.40/0.891	36.28/0.979	27.64/0.884	22.83/0.790	38.61/0.966
SPANet (CVPR'19)	22.96/0.720	27.85/0.881	13.27/0.412	21.22/0.687	38.66/0.968
RCDNet (CVPR'20)	29.81/0.859	38.60/0.983	28.83/0.886	24.59/0.821	41.12/0.977
DualGCN (AAAI'21)	32.98/0.934	39.05/0.991	29.06/0.912	28.28/0.896	41.56/0.981
MPRNet (CVPR'21)	31.73/0.910	36.69/0.972	27.65/0.878	27.86/0.856	40.17/0.974
RLNet (CVPR'21)	32.62/0.917	37.38/0.980	28.87/0.902	27.95/0.870	40.21/0.978
Ours	**33.84/0.937**	**40.01/0.994**	**30.33/0.917**	**29.42/0.904**	**41.73/0.990**

We give some visualization results of deraining examples in Fig. 3. We can observe that RESCAN and JORDER_E cannot remove the rain completely. We

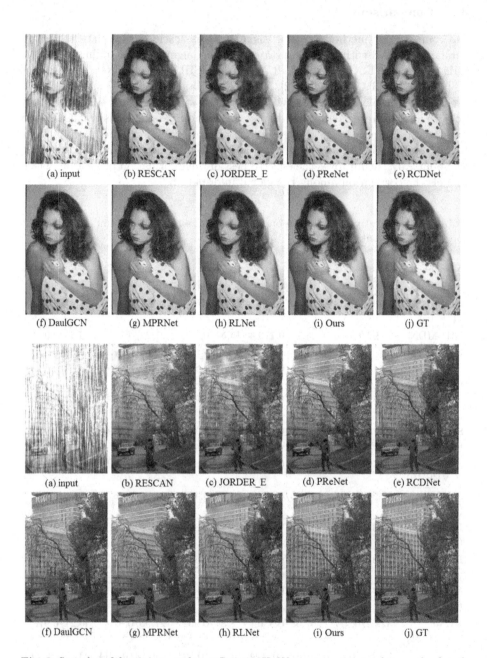

Fig. 3. Samples of deraining results on Rain100H. We can zoom in to observe the detail of the girl's hair and arm in the first image, and the text and building in the second image, the results of ours is better than others obviously.

can zoom in to observe the detail of the girl's hair and the arm in the first image, all the methods lose the details expect ours especially the joint part of arm and body. For the second image, we can zoom in to observe the text on the building and the details of the building, the result of ours is better than others obviously.

4.3 Ablation Studies

In this section, we mainly discuss the effectiveness of the graph-based contextual attention convolution. We use PSNR to evaluate the performance, and the PSNR reduced by 0.5 to 1 when GCAC is not used. The results are list in Table 2.

Table 2. Ablation studies on different data sets

Method	Datasets				
	DID-Data	Rain100L	Rain100H	Rain800	SPA-Data
w/o GCAC	33.23	39.37	29.86	28.28	40.77
Ours	**33.84**	**40.01**	**30.33**	**29.42**	**41.73**

5 Conclusion

In this paper, a graph-based contextual attention network for single image deraining. With the spatial graph attention with spatial coherence computing and channel graph attention with channel correlation computing, the convolution kernel is modulated to extract global contextual information. And we project the global information into a bag of latent features to mine the correlation among the information, which enhance the descriptors. The experiment results on several datasets demonstrate that the proposed method perform well than the state-of-arts.

References

1. Garg, K., Nayar, S.K.: Vision and rain. Int. J. Comput. Vis. **75**(1), 3–27 (2007). https://doi.org/10.1007/s11263-006-0028-6
2. Yang, W., Tan, R.T., Wang, S., Fang, Y., Liu, J.: Single image deraining: from model-based to data-driven and beyond. IEEE Trans. Pattern Anal. Mach. Intell. **43**(11), 4059–4077 (2020)
3. Kang, L.W., Lin, C.W., Fu, Y.H.: Automatic single-image-based rain streaks removal via image decomposition. IEEE Trans. Image Process. **21**(4), 1742–1755 (2011)
4. Luo, Y., Xu, Y., Ji, H.: Removing rain from a single image via discriminative sparse coding. In: Proceedings of the IEEE International Conference on Computer Vision, pp. 3397–3405 (2015)

5. Li, Y., Tan, R.T., Guo, X., Lu, J., Brown, M.S.: Rain streak removal using layer priors. In: Proceedings of the IEEE Conference on Computer Vision and Pattern Recognition, pp. 2736–2744 (2016)
6. Zhang, H., Patel, V.M.: Density-aware single image de-raining using a multi-stream dense network. In: Proceedings of the IEEE Conference on Computer Vision and Pattern Recognition, pp. 695–704 (2018)
7. Goodfellow, I., et al.: Generative adversarial networks. Commun. ACM **63**(11), 139–144 (2020)
8. Yue, Z., Xie, J., Zhao, Q., Meng, D.: Semi-supervised video deraining with dynamical rain generator. In: Proceedings of the IEEE/CVF Conference on Computer Vision and Pattern Recognition, pp. 642–652 (2021)
9. Qin, X., Wang, Z., Bai, Y., Xie, X., Jia, H.: FFA-Net: feature fusion attention network for single image dehazing. In: Proceedings of the AAAI Conference on Artificial Intelligence, vol. 34, pp. 11908–11915 (2020)
10. Wang, X., Girshick, R., Gupta, A., He, K.: Non-local neural networks. In: Proceedings of the IEEE Conference on Computer Vision and Pattern Recognition, pp. 7794–7803 (2018)
11. Kipf, T.N., Welling, M.: Semi-supervised classification with graph convolutional networks. arXiv preprint arXiv:1609.02907 (2016)
12. Fu, X., Qi, Q., Zha, Z.J., Zhu, Y., Ding, X.: Rain streak removal via dual graph convolutional network. In: Proceedings of the AAAI Conference on Artificial Intelligence, vol. 35, pp. 1352–1360 (2021)
13. Elad, M., Aharon, M.: Image denoising via learned dictionaries and sparse representation. In: 2006 IEEE Computer Society Conference on Computer Vision and Pattern Recognition (CVPR 2006), vol. 1, pp. 895–900. IEEE (2006)
14. Zhu, L., Fu, C.W., Lischinski, D., Heng, P.A.: Joint bi-layer optimization for single-image rain streak removal. In: Proceedings of the IEEE International Conference on Computer Vision, pp. 2526–2534 (2017)
15. Yang, W., Tan, R.T., Feng, J., Liu, J., Guo, Z., Yan, S.: Deep joint rain detection and removal from a single image. In: Proceedings of the IEEE Conference on Computer Vision and Pattern Recognition, pp. 1357–1366 (2017)
16. Fu, X., Huang, J., Zeng, D., Huang, Y., Ding, X., Paisley, J.: Removing rain from single images via a deep detail network. In: Proceedings of the IEEE Conference on Computer Vision and Pattern Recognition, pp. 3855–3863 (2017)
17. Zamir, S.W., et al.: Multi-stage progressive image restoration. In: Proceedings of the IEEE/CVF Conference on Computer Vision and Pattern Recognition, pp. 14821–14831 (2021)
18. Li, G., He, X., Zhang, W., Chang, H., Dong, L., Lin, L.: Non-locally enhanced encoder-decoder network for single image de-raining. In: Proceedings of the 26th ACM International Conference on Multimedia, pp. 1056–1064 (2018)
19. Wang, T., Yang, X., Xu, K., Chen, S., Zhang, Q., Lau, R.W.: Spatial attentive single-image deraining with a high quality real rain dataset. In: Proceedings of the IEEE/CVF Conference on Computer Vision and Pattern Recognition, pp. 12270–12279 (2019)
20. Hu, X., Fu, C.W., Zhu, L., Heng, P.A.: Depth-attentional features for single-image rain removal. In: Proceedings of the IEEE/CVF Conference on Computer Vision and Pattern Recognition, pp. 8022–8031 (2019)
21. Zhang, H., Sindagi, V., Patel, V.M.: Image de-raining using a conditional generative adversarial network. IEEE Trans. Circuits Syst. Video Technol. **30**(11), 3943–3956 (2019)

22. Yu, F., Koltun, V.: Multi-scale context aggregation by dilated convolutions. arXiv preprint arXiv:1511.07122 (2015)
23. Krishna, R., et al.: Visual genome: connecting language and vision using crowd-sourced dense image annotations. Int. J. Comput. Vis. **123**(1), 32–73 (2017). https://doi.org/10.1007/s11263-016-0981-7
24. Lin, X., Ma, L., Liu, W., Chang, S.-F.: Context-gated convolution. In: Vedaldi, A., Bischof, H., Brox, T., Frahm, J.-M. (eds.) ECCV 2020. LNCS, vol. 12363, pp. 701–718. Springer, Cham (2020). https://doi.org/10.1007/978-3-030-58523-5_41
25. Zhang, Y., Li, K., Li, K., Fu, Y.: MR image super-resolution with squeeze and excitation reasoning attention network. In: Proceedings of the IEEE/CVF Conference on Computer Vision and Pattern Recognition, pp. 13425–13434 (2021)
26. Zhou, B., Andonian, A., Oliva, A., Torralba, A.: Temporal relational reasoning in videos. In: Proceedings of the European Conference on Computer Vision (ECCV), pp. 803–818 (2018)
27. Lim, B., Son, S., Kim, H., Nah, S., Mu Lee, K.: Enhanced deep residual networks for single image super-resolution. In: Proceedings of the IEEE Conference on Computer Vision and Pattern Recognition Workshops, pp. 136–144 (2017)
28. Yang, W., Tan, R.T., Feng, J., Guo, Z., Yan, S., Liu, J.: Joint rain detection and removal from a single image with contextualized deep networks. IEEE Trans. Pattern Anal. Mach. Intell. **42**(6), 1377–1393 (2019)
29. He, T., Zhang, Z., Zhang, H., Zhang, Z., Xie, J., Li, M.: Bag of tricks for image classification with convolutional neural networks. In: Proceedings of the IEEE/CVF Conference on Computer Vision and Pattern Recognition, pp. 558–567 (2019)
30. Li, X., Wu, J., Lin, Z., Liu, H., Zha, H.: Recurrent squeeze-and-excitation context aggregation net for single image deraining. In: Proceedings of the European Conference on Computer Vision (ECCV), pp. 254–269 (2018)
31. Wei, W., Meng, D., Zhao, Q., Xu, Z., Wu, Y.: Semi-supervised transfer learning for image rain removal. In: Proceedings of the IEEE/CVF Conference on Computer Vision and Pattern Recognition, pp. 3877–3886 (2019)
32. Ren, D., Zuo, W., Hu, Q., Zhu, P., Meng, D.: Progressive image deraining networks: a better and simpler baseline. In: Proceedings of the IEEE/CVF Conference on Computer Vision and Pattern Recognition, pp. 3937–3946 (2019)
33. Wang, H., Xie, Q., Zhao, Q., Meng, D.: A model-driven deep neural network for single image rain removal. In: Proceedings of the IEEE/CVF Conference on Computer Vision and Pattern Recognition, pp. 3103–3112 (2020)
34. Chen, C., Li, H.: Robust representation learning with feedback for single image deraining. In: Proceedings of the IEEE/CVF Conference on Computer Vision and Pattern Recognition, pp. 7742–7751 (2021)

ADTR: Anomaly Detection Transformer with Feature Reconstruction

Zhiyuan You[1], Kai Yang[2], Wenhan Luo[3], Lei Cui[4], Yu Zheng[1],
and Xinyi Le[1(✉)]

[1] Shanghai Jiao Tong University, Shanghai, China
zhiyuanyou@foxmail.com, lexinyi@sjtu.edu.cn
[2] SenseTime Research, Hong Kong, China
[3] Sun Yat-sen University, Guangzhou, China
[4] Tsinghua University, Beijing, China

Abstract. Anomaly detection with only prior knowledge from normal samples attracts more attention because of the lack of anomaly samples. Existing CNN-based pixel reconstruction approaches suffer from two concerns. First, the reconstruction source and target are raw pixel values that contain indistinguishable semantic information. Second, CNN tends to reconstruct both normal samples and anomalies well, making them still hard to distinguish. In this paper, we propose Anomaly Detection TRansformer (ADTR) to apply a transformer to reconstruct pre-trained features. The pre-trained features contain distinguishable semantic information. Also, the adoption of transformer limits to reconstruct anomalies well such that anomalies could be detected easily once the reconstruction fails. Moreover, we propose novel loss functions to make our approach compatible with the normal-sample-only case and the anomaly-available case with both image-level and pixel-level labeled anomalies. The performance could be further improved by adding simple synthetic or external irrelevant anomalies. Extensive experiments are conducted on anomaly detection datasets including MVTec-AD and CIFAR-10. Our method achieves superior performance compared with all baselines.

Keywords: Anomaly Detection · Transformer · Attention Mechanism

1 Introduction

Unsupervised anomaly detection [4,8,14] aims to identify anomalies using prior knowledge from only normal samples. Due to the extreme lack of anomalies in production lines, anomaly detection is attracting more and more interests.

From the view of statistics, anomalies may be seen as distribution outliers of normal samples. In this setting, CNN-based reconstruction models like Auto-Encoder (AE), Variational Auto-Encoder (VAE), and Generative Adversarial

---

This work is sponsored by the Shanghai Foundation for Development of Science and Technology (21SQBS01502) and National Natural Science Foundation of China (62176152).

M. Tanveer et al. (Eds.): ICONIP 2022, LNCS 13625, pp. 298–310, 2023.
https://doi.org/10.1007/978-3-031-30111-7_26

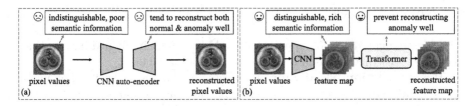

Fig. 1. (a) **CNN-based pixel reconstruction methods** tend to reconstruct both normal samples and anomalies well, making them still hard to distinguish. Also, the pixel values contain indistinguishable semantic information. (b) **Our method** reconstructs features with distinguishable semantic information. Besides, the adoption of transformer limits the reconstruction of anomalies.

Network (GAN) are usually adopted to model the distribution of normal samples [8,12,15,18]. These methods train a model with only normal samples based on the assumption of generalization gap, which means that the reconstruction succeeds with only normal samples but fails with anomalies. The anomaly detection is performed with a distance metric between a sample and its reconstruction.

As show in Fig. 1a, one concern about these approaches is the poor representation ability. The reconstruction targets are raw pixel values with poor semantic information. Therefore, these pixel reconstruction approaches usually fails when normal and anomalous regions share similar pixel values but different semantic information like different textures. In another aspect, it has been verified that the feature extractor pre-trained on large public datasets could extract distinguishable features for normal samples and anomalies [5,29]. Thus we propose to reconstruct pre-trained features instead of raw pixel values.

Taking CNN as the reconstruction model brings another issue (Fig. 1a). CNN tends to take shortcuts to learn a somewhat "identical mapping", which means the anomalous regions are also reconstructed quite well [15]. The great success of transformer in computer vision inspires us to propose a transformer-based reconstruction model. The query embedding in attention layer of transformer could limit the tendency of "identical mapping", which helps distinguish normal samples and anomalies (See Sect. 3.2).

Besides, more anomaly samples are available with the runs of production lines [5], bringing anomaly detection the demands of compatibility with both the normal-sample-only case (only normal samples are available) and the anomaly-available case (normal samples and a few anomalies are available). Therefore, a unified approach that is compatible with both cases would be a better solution.

In this paper, we propose a concise but powerful transformer-based anomaly detection approach. As shown in Fig. 1b, a frozen pre-trained CNN backbone is adopted to extract features, then a transformer is used for feature reconstruction. The proposed approach has strong representation abilities, and could limit the tendency of "identical mapping". Moreover, novel loss functions are proposed for the compatibility with the anomaly-available case. The performance could be further improved by adding simple synthetic or external irrelevant anomalies. Our approach achieves state-of-the-art anomaly detection performance in anomaly detection datasets including MVTec-AD [4] and CIFAR-10 [17].

2   Related Work

Existing anomaly detection approaches could be generally divided into two categories: reconstruction-based ones and projection-based ones.

Reconstruction-based approaches assume that the reconstruction model trained with normal samples has a generalization gap with anomalies, thus fails to reconstruct anomalies. AE [6,12,15,24] and GAN [25,28,38] are intuitive choices of reconstruction models. Zhou et al. [39] and Xia et al. [34] respectively adopt the structural information and semantic segmentation information for better reconstruction. Zaheer et al. [38] utilize a discriminator to distinguish good or bad quality of reconstruction, and the predicted possibility of bad quality serves as an anomaly score. Gong et al. [15] and Park et al. [24] introduce a memory module to select the most similar embedding in embedding storage of normal samples to restrict the generalization on anomalies. Dehaene et al. [12] refine the selection method with an iterative gradient-based approach.

Projection-based approaches project samples into an embedding space, where normal samples and anomalies are more distinguishable. SVDD [27] extracts feature representation with the one-class classification objective. Yi and Yoon [36] propose a patch-based SVDD with multiple kernels. Liu et al. [20] and Kwon et al. [18] find that the back-propagated gradients of normal samples and anomalies are more distinguishable. FCDD [21] is trained to enlarge the embedding differences between normal samples and anomalies, where the mapped samples are themselves an explanation heat map. Bergmann et al. [5] utilize a teacher-student network, assuming that the embedding differences between normal samples and anomalies would be enlarged through knowledge distillation. Salehi et al. [29] extend the knowledge distillation to multi-layer, multi-scale scheme, enlarging the distillation gap between normal samples and anomalies. PaDiM [11] models normal distribution using pre-trained features, then utilize a distance metric to measure the anomalies. Wang et al. [33] compare the embeddings of local pattern and global pattern to detect anomalies.

Transformer in Anomaly Detection. Transformer [32] has been successfully used in computer vision [9]. Some attempts also try to utilize transformer for anomaly detection. InTra [26] adopts transformer to recover the image by recovering all masked patches one by one. VT-ADL [23] and AnoVit [37] both apply transformer encoder to reconstruct images. However, these methods mainly focus on indistinguishable raw pixels, and do not figure out why transformer brings improvement. In contrast, we reconstruct pre-trained features instead of raw pixels. We also confirm the efficacy of the query embedding in attention layer to prevent the "identical shortcut".

3   Method

In this part, we first introduce the architecture of ADTR, followed by the analysis of why transformer could limit to reconstruct anomalies well. Finally, we propose two loss functions to extend our approach compatible with available anomalies.

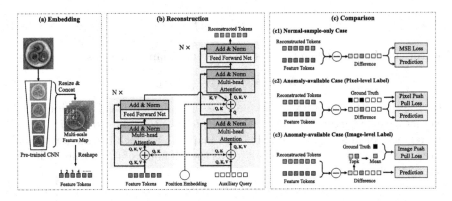

Fig. 2. Overview of our method. (a) Embedding: a pre-trained CNN backbone is applied to extract the multi-scale features. (b) Reconstruction: a transformer is utilized to reconstruct the feature tokens with an auxiliary learnable query embedding. (c) Comparison: our approach is compatible with both normal-sample-only case and anomaly-available case. The anomaly score maps are obtained through the differences between extracted and reconstructed features.

3.1 Architecture

Embedding. A frozen pre-trained CNN backbone is first utilized for feature extraction (Fig. 2a). Here we use EfficientNet-B4 [31] pre-trained on ImageNet. The features from *layer1* to *layer5* are resized to the same size, then concatenated together to form a multi-scale feature map, $f \in \mathbb{R}^{C \times H \times W}$. Note that here we define *layer* as the combination of stages with the same size of features. We adopt multi-scale feature map because feature maps from different layers have different levels of receptive fields thus are sensitive to different anomalies.

Reconstruction. The reconstruction stage is shown in Fig. 2b. The feature map, $f \in \mathbb{R}^{C \times H \times W}$, is first split to $H \times W$ feature tokens. To reduce the computation consumption, a 1×1 convolution is applied to reduce the dimension of these tokens before they are fed into the transformer. Also, their dimensions are recovered by another 1×1 convolution when output by transformer. The transformer encoder embeds the input feature tokens into a latent feature space. Each encoder layer follows the standard architecture [32] with multi-head attention, feed forward network (FFN), residual connection, and normalization. The transformer decoder follows the standard architecture [32] with an auxiliary query embedding. The auxiliary query is a learned embedding with the same size of the input feature tokens. The transformer decoder transforms this learned query embedding to reconstruct the feature tokens using multi-head self-attention and encoder-decoder attention mechanisms. The learned position embedding [9] is included because transformer is permutation-invariant.

Comparison. In normal-sample-only case, the model is trained with the MSE loss, $\mathcal{L}_{norm}$, between the backbone extracted features, f, and the reconstructed features, $\hat{f} \in \mathbb{R}^{C \times H \times W}$, as follows,

$$\mathcal{L}_{norm} = \frac{1}{H \times W} ||\boldsymbol{f} - \hat{\boldsymbol{f}}||_2^2. \tag{1}$$

Inference. We first define the feature difference map, $\boldsymbol{d}(i, u)$, as,

$$\boldsymbol{d}(i, u) = \boldsymbol{f}(i, u) - \hat{\boldsymbol{f}}(i, u), \tag{2}$$

where i represents the index of channel, u is the index of spatial position (height together with width for simplicity). *Anomaly localization* aims to localize anomalous regions, producing an anomaly score map, $\boldsymbol{s}(u)$, which assigns an anomaly score for each pixel, u. $\boldsymbol{s}(u)$ is calculated as the $L2$ norm of the feature difference vector, $\boldsymbol{d}(:, u)$.

$$\boldsymbol{s}(u) = ||\boldsymbol{d}(:, u)||_2. \tag{3}$$

Anomaly detection aims to detect whether an image contains anomalous regions. We intuitively take the maximum value of the averagely pooled $\boldsymbol{s}(u)$ as the anomaly score of the whole image.

3.2   Preventing "Identical Mapping" with Transformer

We suspect that, compared with CNN, the query embedding in attention layer makes transformer difficult to learn an "identical mapping". We denote the features in a normal image as $\boldsymbol{x}^+ \in \mathbb{R}^{K \times C}$, where K is the feature number, C is the channel dimension. The features in an anomalous image are denoted as $\boldsymbol{x}^- \in \mathbb{R}^{K \times C}$. We take a 1-layer network as the reconstruction net, which is trained on $\boldsymbol{x}^+$ with the MSE loss and tested to detect anomalous regions in $\boldsymbol{x}^-$.
Convolutional Layer in CNN. We first visit a fully-connected layer, whose weights and bias are denoted as $\boldsymbol{w} \in \mathbb{R}^{C \times C}, \boldsymbol{b} \in \mathbb{R}^C$, respectively. When using this layer as the reconstruction model of normal samples, it can be written as,

$$\hat{\boldsymbol{x}} = \boldsymbol{x}^+ \boldsymbol{w} + \boldsymbol{b} \in \mathbb{R}^{K \times C}. \tag{4}$$

with the MSE loss pushing $\hat{\boldsymbol{x}}$ to $\boldsymbol{x}^+$, the model may take shortcut to regress $\boldsymbol{w} \rightarrow \boldsymbol{I}$ (identity matrix), $\boldsymbol{b} \rightarrow \boldsymbol{0}$. Ultimately, this model could also reconstruct $\boldsymbol{x}^-$ well, failing in anomaly detection. A convolutional layer with 1×1 kernel is equivalent to a fully-connected layer. Besides, An $n \times n$ $(n > 1)$ kernel has more parameters and larger capacity, and can complete whatever 1×1 kernel can. Thus, the convolutional layer also has the chance to learn a shortcut.

Transformer with query embedding contains an attention layer with a learnable query embedding, $\boldsymbol{q} \in \mathbb{R}^{K \times C}$. This attention layer can be denoted as,

$$\hat{\boldsymbol{x}} = \mathtt{softmax}(\boldsymbol{q}(\boldsymbol{x}^+)^T/\sqrt{C})\boldsymbol{x}^+ \in \mathbb{R}^{K \times C}. \tag{5}$$

To push $\hat{\boldsymbol{x}}$ to $\boldsymbol{x}^+$, the attention map, $\mathtt{softmax}(\boldsymbol{q}(\boldsymbol{x}^+)^T/\sqrt{C})$, should approximate $\boldsymbol{I}$ (identity matrix), so $\boldsymbol{q}$ must be highly related to $\boldsymbol{x}^+$. Considering that $\boldsymbol{q}$ in the trained model is relevant to normal samples, the model could not reconstruct $\boldsymbol{x}^-$ well. The ablation study in Sect. 4.4 shows that without the attention layer or the query embedding, the performance of transformer respectively drops by 2.4% or 3%, which is almost the same as CNN. This reflects that the query embedding in attention layer helps prevent transformer from learning an "identical shortcut".

3.3 Adaptation with Anomaly-Available Case

In practice, anomalies gradually increase with the runs of production lines, which brings the demands of compatibility with these increasing anomalies. Thus we adapt ADTR to ADTR+ for compatibility with the anomaly-available case.

Adaptation with Pixel-Level Labels. Inspired by [21], we firstly calculate a pseudo-Huber loss, $\phi(u)$, using the feature difference map, $d(i, u)$.

$$\phi(u) = ((\frac{1}{C} \sum_i^C |d(i, u)|)^2 + 1)^{\frac{1}{2}} - 1. \tag{6}$$

The pseudo-Huber loss, $\phi(u)$, is designed as a difference map, which is easy to train and extend. Then the reconstruction loss function with pixel-level labels is denoted as $\mathcal{L}_{px}$ and could be described as a "push-pull loss" as,

$$\mathcal{L}_{px} = \frac{1}{HW} \sum_u^{HW} (1 - y(u))\phi(u) - \alpha \log(1 - \exp(-\frac{1}{HW} \sum_u^{HW} y(u)\phi(u))), \tag{7}$$

where the first term pulls the reconstructed normal features to the extracted features, and the second term pushes the reconstructed anomalous features away from the original features, $y(u)$ is the pixel-level label (0 for normal sample and 1 for anomaly) and α is a weight term.

Adaptation with Image-Level Labels. Since anomaly samples could contain both anomalous and normal regions, simply treating all regions of anomaly samples as anomalous regions confuses the model. Considering that larger values in $\phi(u)$ are more likely to be anomalous regions, we firstly collect k maximum values of $\phi(u)$, then calculate their mean as the anomaly score of the image.

$$q = \frac{1}{k} \sum \text{top\_k}(\phi). \tag{8}$$

Then the image-level loss, $\mathcal{L}_{img}$, could be calculated as,

$$\mathcal{L}_{img} = (1 - y)q - \alpha y \log(1 - \exp(-q)), \tag{9}$$

where y is the image-level label (0 for normal sample and 1 for anomaly) and α is a weight term. In $\mathcal{L}_{img}$, the first term pulls the reconstructed features of normal samples towards the extracted features, while the second term pushes the reconstructed features of anomalies away from the extracted features.

4 Experiment

4.1 Dataset

MVTec-AD [4] is a multi-category, multi-defect, industrial anomaly detection dataset with 15 categories. The ground-truth includes both image labels and anomaly segmentation. In *normal-sample-only case*, we follow the original setting to use normal samples for training, and test on both normal and anomaly sam-

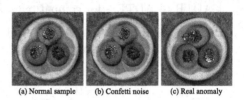

(a) Normal sample   (b) Confetti noise   (c) Real anomaly

Fig. 3. Synthetic anomalies by adding confetti noise on normal samples.

ples. In *anomaly-available case*, following [21], we synthesize anomalies by adding confetti noise on normal samples (Fig. 3).

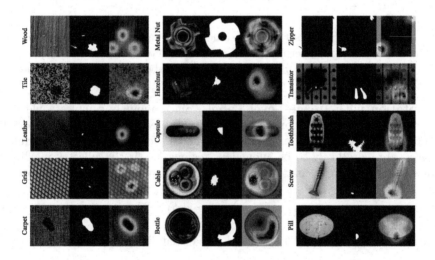

Fig. 4. Anomaly detection results on MVTec-AD [4]. From left to right: the anomaly sample, the ground-truth, and the anomaly score map of ADTR.

CIFAR-10 [17] is a classical classification dataset with 10 classes. Each class has 5000 images for training and 1000 images for testing. In *normal-sample-only case*, following [18], the training set of one class is used for training, and the test set contains normal images of the same class and the same number of anomaly images randomly sampled from other classes. In *anomaly-available case*, an irrelevant dataset, CIFAR-100 [17], is used as an auxiliary dataset. We randomly select the same number of images from CIFAR-100 as anomalies.

4.2 Anomaly Detection on MVTec-AD

The performance of our method is evaluated on anomaly detection and localization tasks of MVTec-AD [4].

Setup. The sizes of the image and feature map are selected as 256×256 and 16×16, respectively. The numbers of the encoder layer and decoder layer (N in Fig. 2) in transformer are both set as 4. The features from *layer1* to *layer5* of EfficientNet-B4 [31] are resized and concatenated to form a 720-channel feature map. The reduced channel dimension is set as 256. AdamW optimizer [22] with weight decay 1×10^{-4} is used for training with batch size 16. In *normal-sample-only case*, models are trained with $\mathcal{L}_{norm}$ in Eq. (1) for 500 epochs. The learning rate is 1×10^{-4} initially, and dropped by 0.1 after 400 epochs. In *anomaly-available case*, the pixel-level loss, $\mathcal{L}_{px}$, in Eq. (7) is adopted for training, where α is chosen as 0.003. The trained model in normal-sample-only case is firstly loaded. Then the model is trained for 300 epochs with the learning rate of 1×10^{-4} for first 200 epochs and 1×10^{-5} for last 100 epochs.

Qualitative results on MVTec-AD are shown in Fig. 4. Our approach successfully detects different kinds of anomalies with high localization accuracy. Especially, for the shown "Metal Nut" example, where the anomaly is a flipped normal sample, our approach detects the "flip" anomaly successfully though there are no obvious vision anomalies like texture disorder nor color change.

Quantitative results of anomaly localization are given in Table 1. Our approach is compared with SSIM-AE [6], AnoGAN [30], VEVAE [20], SMAI [19], GDR [12], P-Net [39], FCDD [21], SCADN [35], PSVDD [36], SPADE [10], KDAD [29], Loc-Glo [33]. With pure normal samples, ADTR stably outperforms the best baseline, SPADE [10], by 1.2%. With merely simple synthetic anomalies, the performance of ADTR+ is further improved by 0.3%.

Table 1. Anomaly localization results under pixel-level AUROC metric on MVTec-AD [4].

	Texture					Object										
	Carp.	Grid	Leat.	Tile	Wood	Bott.	Cable	Caps.	Haze.	Meta.	Pill	Screw	Toot.	Tran.	Zipp.	Mean
SSIM-AE [6]	87	94	78	59	73	93	82	94	97	89	91	96	92	90	88	86
AnoGAN [30]	54	58	64	50	62	86	78	84	87	76	87	80	90	80	78	74
VEVAE [20]	78	73	95	80	77	87	9	74	98	94	83	97	94	93	78	86
SMAI [19]	88	97	86	62	80	86	92	93	97	92	92	96	96	85	90	89
GDR [12]	74	96	93	65	84	92	91	92	98	91	93	95	99	92	87	89
P-Net [39]	57	**98**	89	**97**	**98**	**99**	70	84	97	79	91	**1.00**	99	82	90	89
FCDD [21]	96	91	98	91	88	97	90	93	95	94	81	86	94	88	92	92
SCADN [35]	64.9	79.6	76.3	67.7	67.2	69.6	81.4	68.7	88.4	75.4	74.7	87.6	90.1	68.9	67.0	75.2
PSVDD [36]	92.6	96.2	97.4	91.4	90.8	98.1	96.8	95.8	97.5	98.0	95.1	95.7	98.1	97.0	95.1	95.7
SPADE [10]	97.5	93.7	97.6	87.4	88.5	98.4	**97.2**	99.0	**99.1**	**98.1**	96.5	98.9	97.9	94.1	96.5	96.0
KDAD [29]	95.6	91.8	98.1	82.8	84.8	96.3	82.4	95.9	94.6	86.4	89.6	96.0	96.1	76.5	93.9	90.7
Loc-Glo [33]	96	78	90	80	81	93	94	90	84	91	93	96	96	**1.00**	99	91
ADTR (ours)	98.7	**95.0**	98.1	93.8	91.2	98.0	96.8	**99.1**	98.6	97.0	98.3	99.3	98.5	97.9	97.2	97.2
ADTR+ (ours)	**98.8**	94.2	**98.6**	95.9	93.0	98.0	97.0	**99.1**	98.8	96.8	**98.7**	99.3	**99.2**	97.8	97.6	**97.5**

Table 2. Anomaly detection results under image-level AUROC metric on MVTec-AD [4].

	Texture					Object										Mean
	Carp.	Grid	Leat.	Tile	Wood	Bott.	Cable	Caps.	Haze.	Meta.	Pill	Screw	Toot.	Tran.	Zipp.	
GANomaly [2]	69.9	70.8	84.2	79.4	83.4	89.2	75.7	73.2	78.5	70.0	74.3	74.6	65.3	79.2	74.5	76.2
ArNet [13]	70.6	88.3	86.2	73.5	92.3	94.1	83.2	68.1	85.5	66.7	78.6	**100**	**100**	84.3	87.6	83.9
SPADE [10]	–	–	–	–	–	–	–	–	–	–	–	–	–	–	–	85.5
SCADN [35]	50.4	98.3	65.9	79.2	96.8	95.7	85.6	76.5	83.3	62.4	81.4	83.1	98.1	86.3	84.6	81.8
PSVDD [36]	98.6	90.3	76.7	92.9	94.6	92.0	90.9	**94.0**	86.1	81.3	97.8	**100**	91.5	96.5	**97.9**	92.1
TS [5]	95.3	**98.7**	93.4	95.8	95.5	96.7	82.3	92.8	91.4	94.0	86.7	87.4	98.6	83.6	95.8	92.5
KDAD [29]	79.3	78.0	95.1	91.6	94.3	99.4	89.2	80.5	98.4	73.6	82.7	83.3	92.2	85.6	93.2	87.7
ADTR (ours)	**100**	97.5	**100**	**100**	99.9	**100**	**92.5**	93.1	**100**	**94.9**	92.1	94.0	93.1	97.6	95.8	96.4
ADTR+ (ours)	**100**	97.8	**100**	**100**	99.9	**100**	**92.5**	92.5	99.9	94.5	**93.3**	94.2	93.9	**98.0**	97.0	**96.9**

Table 3. Anomaly detection results under image-level AUROC metric on CIFAR-10 [17].

	Airplane	Automobile	Bird	Cat	Deer	Dog	Frog	Horse	Ship	Truck	Mean
KDE [7]	65.8	52.0	65.7	49.7	72.7	49.6	75.8	56.4	68.0	54.0	61.0
VAE [3]	63.4	44.2	64.0	49.7	74.3	51.5	74.5	52.7	67.4	41.6	58.3
LSA [1]	73.5	58.0	69.0	54.2	76.1	54.6	75.1	53.5	71.7	54.8	64.1
AnoGAN [30]	67.1	54.7	52.9	54.5	65.1	60.3	58.5	62.5	75.8	66.5	61.8
DSVDD [27]	61.7	65.9	50.8	59.1	60.9	65.7	67.7	67.3	75.9	73.1	64.8
OCGAN [25]	75.7	53.1	64.0	62.0	72.3	62.0	72.3	57.5	82.0	55.4	65.7
GradCon [18]	76.0	59.8	64.8	58.6	73.3	60.3	68.4	56.7	78.4	67.8	66.4
GT [14]	76.2	84.8	77.1	73.2	82.8	84.8	82.0	88.7	89.5	83.4	82.3
TS [5]	78.9	84.9	73.4	74.8	85.1	79.3	89.2	83.0	86.2	84.8	82.0
Loc-Glo [33]	79.1	70.3	67.5	56.1	73.9	63.8	73.2	67.4	81.4	72.2	70.5
KDAD [29]	90.5	90.4	80.0	77.0	86.7	91.4	89.0	86.8	91.5	88.9	87.2
ADTR(ours)	94.1	97.4	92.3	89.0	93.2	94.4	97.4	95.8	96.3	96.7	94.7
ADTR+(ours)	**96.2**	**98.0**	**94.5**	**91.7**	**95.1**	**95.6**	**98.0**	**97.1**	**98.0**	96.9	**96.1**

Quantitative results of anomaly detection are shown in Table 2. Our approach is compared with GANomaly [2], ArNet [13], SPADE [10], SCADN [35], PSVDD [36], TS [5], KDAD [29]. ADTR considerably exceeds all baseline methods ($\geq$ 3.9%) with only normal samples. The performance of ADTR+ is improved by 0.5% with simple synthetic anomalies.

4.3 Anomaly Detection on CIFAR-10

To further validate the anomaly detection ability, we evaluate our model in the unsupervised one-class classification task of CIFAR-10 [17].

Setup. The setup is the same as that in Sect. 4.2 except the followings. First, the sizes of the image and feature map are 32×32 and 8×8, respectively. Second, in anomaly-available case, the model is trained with the image-level loss, $\mathcal{L}_{img}$, in Eq. (9), where α and k are selected as 0.003 and 20, respectively.

Quantitative results on CIFAR-10 are shown in Table 3. The competitors include: KDE [7], VAE [3], LSA [1], AnoGAN [30], DSVDD [27], OCGAN [25], GradCon [18], GT [14], TS [5], Loc-Glo [33], KDAD [29]. ADTR surpasses KDAD [29] by a great margin (7.5%) when training in normal-sample-only case. In anomaly-available case, the performance of ADTR+ is further improved by 1.4% with the help of external irrelevant dataset, reflecting the effectiveness of the designed image-level loss function, $\mathcal{L}_{img}$.

4.4   Ablation Study

Extensive ablation studies with pixel-level AUROC metric are conducted on anomaly localization task of MVTec-AD [4].

Attention and Auxiliary Query Embedding. As shown in Table 4a, a CNN revised from ResNet [16] is firstly included as the baseline of the reconstruction model. (1) The replacement of the attention layer is a concatenation followed by projection. If we remove the attention layer (w/o Attn) from the transformer, the performance shows no obvious superiority to CNN. (2) Without the auxiliary query embedding (w/o Query), meaning that only the encoder embedding is input to the decoder, the performance is even worse than CNN. (3) Equipped with both attention and auxiliary query embedding (Attn+Query), transformer stably outperforms CNN by 2.8%. This proves our assertion in Sect. 3.2 that the auxiliary query embedding in attention layer helps prevent transformer from reconstructing anomalies well.

Reconstructed Target. In Table 4b, reconstructing features surpasses pixel values substantially, indicating that the features extracted by pre-trained backbone are more distinguishable for normal samples and anomalies than raw pixels.

Backbone and Multi-scale Features. (1) As shown in Table 4c, four different backbones all achieve quite good performance, reflecting that our method could cooperate with different types of backbones. (2) In Table 4d, multi-scale features obviously outperform last-layer feature, because multi-scale features contain different levels of receptive fields thus are sensitive to different anomalies.

Table 4. Ablation study on (a) attention & auxiliary query embedding, (b) reconstructing pixels *vs.* features, (c) backbone, and (d) multi-scale features under pixel-level AUROC metric on anomaly localization of MVTec-AD [4].

(a) Attention & auxiliary query embedding					(b) Reconstructing pixels *vs.* features		
	CNN	w/o Attn	w/o Query	Attn+Query		Pixels	Features
Pixel AUROC	94.4	94.8	94.2	**97.2**	Pixel AUROC	91.3	**97.2**
(c) Backbone					(d) Multi-scale features		
	Res-18	Res-34	Efficient-B0	Efficient-B4		Last-layer	Multi-scale
Pixel AUROC	95.3	95.7	96.4	**97.2**	Pixel AUROC	96.0	**97.2**

4.5   Visualization of Feature Difference Vectors

We visualize the feature difference vectors $\boldsymbol{d}(:,u)$ in Eq. (2) to better interpret our approach. Specifically, we randomly sample 600 feature difference vectors (normal : anomaly = 1:1) from MVTec-AD [4]. Then t-SNE is utilized to visualize the high dimensional vectors in a 2D space, as shown in Fig. 5. Firstly, normal samples and anomalies are mostly colored with blue and red, respectively, indicating good

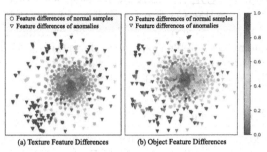

(a) Texture Feature Differences   (b) Object Feature Differences

Fig. 5. visualization of feature difference vectors by t-SNE. Circles and triangles respectively represent normal samples and anomalies. The color map indicates the predicted anomaly possibility. Our method brings large generalization gap between normal samples and anomalies.

anomaly detection ability. Secondly, normal samples are well clustered, and there is a wide gap between the normal samples and anomalies. These observations indicate that our approach brings a large generalization gap between normal samples and anomalies.

5   Conclusion

In this paper, we propose anomaly detection transformer to utilize a transformer to reconstruct pre-trained features. First, the pre-trained features contain distinguishable semantic information. Second, the adoption of transformer prevents reconstructing anomalies well such that anomalies could be detected easily once the reconstruction fails. Our method brings a large generalization gap between normal samples and anomalies. Moreover, we propose novel loss functions to extend our approach from normal-sample-only case to anomaly-available case with both image-level labeled and pixel-level labeled anomalies, further improving the performance. Our approach achieves the state-of-the-art performance on anomaly detection benchmarks including MVTec-AD and CIFAR-10.

References

1. Abati, D., Porrello, A., Calderara, S., Cucchiara, R.: Latent space autoregression for novelty detection. In: CVPR (2019)
2. Akcay, S., Atapour-Abarghouei, A., Breckon, T.P.: GANomaly: semi-supervised anomaly detection via adversarial training. In: Jawahar, C.V., Li, H., Mori, G., Schindler, K. (eds.) ACCV 2018. LNCS, vol. 11363, pp. 622–637. Springer, Cham (2019). https://doi.org/10.1007/978-3-030-20893-6_39
3. An, J., Cho, S.: Variational autoencoder based anomaly detection using reconstruction probability. Spec. Lect. IE **2**(1), 1–18 (2015)

4. Bergmann, P., Fauser, M., Sattlegger, D., Steger, C.: MVTec AD: a comprehensive real-world dataset for unsupervised anomaly detection. In: CVPR (2019)
5. Bergmann, P., Fauser, M., Sattlegger, D., Steger, C.: Uninformed students: student-teacher anomaly detection with discriminative latent embeddings. In: CVPR (2020)
6. Bergmann, P., Lwe, S., Fauser, M., Sattlegger, D., Steger, C.: Improving unsupervised defect segmentation by applying structural similarity to autoencoders. In: International Conference on Computer Vision Theory and Applications (2019)
7. Bishop, C.M.: Pattern recognition and machine learning. Springer, New York (2006). https://link.springer.com/book/9780387310732
8. Borghesi, A., Bartolini, A., Lombardi, M., Milano, M., Benini, L.: Anomaly detection using autoencoders in high performance computing systems. In: AAAI (2019)
9. Carion, N., Massa, F., Synnaeve, G., Usunier, N., Kirillov, A., Zagoruyko, S.: End-to-end object detection with transformers. In: Vedaldi, A., Bischof, H., Brox, T., Frahm, J.-M. (eds.) ECCV 2020. LNCS, vol. 12346, pp. 213–229. Springer, Cham (2020). https://doi.org/10.1007/978-3-030-58452-8_13
10. Cohen, N., Hoshen, Y.: Sub-image anomaly detection with deep pyramid correspondences. arXiv preprint arXiv:2005.02357 (2020)
11. Defard, T., Setkov, A., Loesch, A., Audigier, R.: PaDiM: a patch distribution modeling framework for anomaly detection and localization. In: Del Bimbo, A., et al. (eds.) ICPR 2021. LNCS, vol. 12664, pp. 475–489. Springer, Cham (2021). https://doi.org/10.1007/978-3-030-68799-1_35
12. Dehaene, D., Frigo, O., Combrexelle, S., Eline, P.: Iterative energy-based projection on a normal data manifold for anomaly localization. In: ICLR (2020)
13. Fei, Y., Huang, C., Jinkun, C., Li, M., Zhang, Y., Lu, C.: Attribute restoration framework for anomaly detection. IEEE Trans. Multimed. 24, 116–127 (2020)
14. Golan, I., El-Yaniv, R.: Deep anomaly detection using geometric transformations. In: Bengio, S., Wallach, H.M., Larochelle, H., Grauman, K., Cesa-Bianchi, N., Garnett, R. (eds.) NIPS (2018)
15. Gong, D., et al.: Memorizing normality to detect anomaly: memory-augmented deep autoencoder for unsupervised anomaly detection. In: ICCV (2019)
16. He, K., Zhang, X., Ren, S., Sun, J.: Deep residual learning for image recognition. In: CVPR (2016)
17. Krizhevsky, A.: Learning multiple layers of features from tiny images. Master's thesis, University of Tront (2009)
18. Kwon, G., Prabhushankar, M., Temel, D., AlRegib, G.: Backpropagated gradient representations for anomaly detection. In: Vedaldi, A., Bischof, H., Brox, T., Frahm, J.-M. (eds.) ECCV 2020. LNCS, vol. 12366, pp. 206–226. Springer, Cham (2020). https://doi.org/10.1007/978-3-030-58589-1_13
19. Li, Z., et al.: Superpixel masking and inpainting for self-supervised anomaly detection. In: BMVC (2020)
20. Liu, W., et al.: Towards visually explaining variational autoencoders. In: CVPR (2020)
21. Liznerski, P., Ruff, L., Vandermeulen, R.A., Franks, B.J., Kloft, M., Müller, K.: Explainable deep one-class classification. In: ICLR (2021)
22. Loshchilov, I., Hutter, F.: Decoupled weight decay regularization. In: ICLR (2019)
23. Mishra, P., Verk, R., Fornasier, D., Piciarelli, C., Foresti, G.L.: VT-ADL: a vision transformer network for image anomaly detection and localization. In: International Symposium on Industrial Electronics (2021)
24. Park, H., Noh, J., Ham, B.: Learning memory-guided normality for anomaly detection. In: CVPR (2020)

25. Perera, P., Nallapati, R., Xiang, B.: OCGAN: one-class novelty detection using GANs with constrained latent representations. In: CVPR (2019)
26. Pirnay, J., Chai, K.: Inpainting transformer for anomaly detection. arXiv preprint arXiv:2104.13897 (2021)
27. Ruff, L., et al.: Deep one-class classification. In: ICML (2018)
28. Sabokrou, M., Khalooei, M., Fathy, M., Adeli, E.: Adversarially learned one-class classifier for novelty detection. In: CVPR (2018)
29. Salehi, M., Sadjadi, N., Baselizadeh, S., Rohban, M.H., Rabiee, H.R.: Multiresolution knowledge distillation for anomaly detection. In: CVPR (2021)
30. Schlegl, T., Seeböck, P., Waldstein, S.M., Schmidt-Erfurth, U., Langs, G.: Unsupervised anomaly detection with generative adversarial networks to guide marker discovery. In: Niethammer, M., et al. (eds.) IPMI 2017. LNCS, vol. 10265, pp. 146–157. Springer, Cham (2017). https://doi.org/10.1007/978-3-319-59050-9_12
31. Tan, M., Le, Q.: EfficientNet: rethinking model scaling for convolutional neural networks. In: ICML (2019)
32. Vaswani, A., et al.: Attention is all you need. In: NIPS (2017)
33. Wang, S., Wu, L., Cui, L., Shen, Y.: Glancing at the patch: anomaly localization with global and local feature comparison. In: CVPR (2021)
34. Xia, Y., Zhang, Y., Liu, F., Shen, W., Yuille, A.L.: Synthesize then compare: detecting failures and anomalies for semantic segmentation. In: Vedaldi, A., Bischof, H., Brox, T., Frahm, J.-M. (eds.) ECCV 2020. LNCS, vol. 12346, pp. 145–161. Springer, Cham (2020). https://doi.org/10.1007/978-3-030-58452-8_9
35. Yan, X., Zhang, H., Xu, X., Hu, X., Heng, P.A.: Learning semantic context from normal samples for unsupervised anomaly detection. In: AAAI (2021)
36. Yi, J., Yoon, S.: Patch SVDD: patch-level SVDD for anomaly detection and segmentation. In: ACCV (2020)
37. Yunseung, L., Pilsung, K.: AnoViT: unsupervised anomaly detection and localization with vision transformer-based encoder-decoder. arXiv preprint arXiv:2203.10808 (2022)
38. Zaheer, M.Z., Lee, J.H., Astrid, M., Lee, S.I.: Old is gold: redefining the adversarially learned one-class classifier training paradigm. In: CVPR (2020)
39. Zhou, K., et al.: Encoding structure-texture relation with P-Net for anomaly detection in retinal images. In: Vedaldi, A., Bischof, H., Brox, T., Frahm, J.-M. (eds.) ECCV 2020. LNCS, vol. 12365, pp. 360–377. Springer, Cham (2020). https://doi.org/10.1007/978-3-030-58565-5_22

SCIEnt: A Semantic-Feature-Based Framework for Core Information Extraction from Web Pages

Zhiwei Wang[1,2], Yan Guo[1(✉)], Yangyuanxiang Xu[2], Yuanhai Xue[1], Yue Liu[1], Huawei Shen[1], and Xueqi Cheng[1]

[1] Institute of Computing Technology, Chinese Academy of Sciences, Beijing, China
{wangzhiwei20s,guoy,xueyuanhai,liuyue,shenhuawei,cxq}@ict.ac.cn
[2] University of Chinese Academy of Sciences, Beijing, China
xuyangyuanxiang20@mails.ucas.ac.cn

Abstract. Core Information Extraction (CIE) from web pages aims to extract valuable text to provide data for downstream Text Data Mining (TDM) tasks. Web page representations in existing CIE methods are either based on HTML structural features or visual features. Neither of these representations really understands the semantic associations inherent in the web page, leading to poor extraction quality. This paper proposes a new web page representation method based on semantic features from the perspective of readers' reading and understanding of web pages. In this method, we introduce a new concept of web page skeleton to parse and represent the web page from a semantic point of view. To observe the relationship between the various parts of the skeleton, we project the skeleton onto the DOM tree and get the skeleton tree. Based on this new web page representation, we propose SCIEnt, a semantic-feature-based web page CIE framework. SCIEnt consists of four modules, i.e. Skeleton Tree Construction, Node Splitting, Node Classification, Semantic Aggregation and Correction. Algorithms in each module can be flexibly replaced according to the requirement of downstream TDM task. We evaluate SCIEnt in terms of three well-studied datasets. Results show that SCIEnt far outperforms baseline methods, and the semantic-feature-based web page representation have superiority in web page CIE.

Keywords: Web page extraction framework · Web page representation · Semantic features · Web page skeleton tree

1 Introduction

Web pages contain a wealth of mixed information and it is difficult to extract specific information in bulk. Some information mixed in web pages can be an important data source for many data mining tasks. Consider the Text Data Mining (TDM) task, many intelligent TDM models today are data-driven. Getting a

This work is supported by the Strategic Priority Research Program of Chinese Academy of Sciences, Grant No. XDC02060200.

good TDM model often requires training with large-scale high-quality text data. Therefore, how to extract relevant information from web pages for downstream data mining tasks has become an urgent issue.

The value of information in the web page is determined by the specific downstream data mining task. We name the information in the web page, applicable as input data for a specific downstream task, as the core information of the web page. And the process of obtaining core information from the web page is called Core Information Extraction (CIE). For example, for downstream tasks such as News Summary, the input data is the content of news, not the ads or copyright statements. So for this task, the goal of web page CIE is to extract the news content, and make the extraction results free of text with other semantics such as ads. This paper intends to study how to effectively extract the core information related to downstream tasks from the web page, so as to provide large-scale high-quality data for different downstream TDM tasks.

A web page has three forms: HTML source code, DOM (Document Object Model) tree, and snapshot, which provide different perspectives for web page representation. According to different representations, the existing web page CIE methods can be divided into HTML-structure-feature-based methods and visual-feature-based methods. The method based on HTML structural features generally takes the HTML source code of the web page as input, parses the HTML into a DOM tree, and further performs calculations on the DOM tree. Nowadays, product placement in the web article is very common. And these ads are often written directly into a paragraph of the article by the author. In the DOM tree, often a paragraph is in a single node. As shown in Fig. 1a and Fig. 1d, Node A and B are mixed with two semantics of *news* and *ad*. Directly classifying the entire node cannot separate these two semantics in A and B. The visual-feature-based method partitions the page into different visual blocks based on the visual associations of web page elements, rather than the semantic relations. In Fig. 1b, different semantics are distinguished with different colors. Figure 1c shows the result of visual block partitioning. As you can see, this partitioning method cannot separate different semantics.

In this paper, we stand in the perspective of readers' reading and understanding web pages, re-parse the web page structure, propose a new web page representation method based on semantic features. On this basis, we propose SCIEnt, a semantic-feature-based web page CIE framework. We divide the process of reading a web page into three parts: before, during, and after reading. Before reading, people partition elements according to their arrangement on the web page, and the method based on visual features is to simulate this process. During reading, people obtain the text semantics of the web page by processing and analyzing the text. After reading, people divide and reorganize the web page text according to their own understanding of the text semantics, so that semantically similar texts will be organized together. Our method simulates these processes. We propose the concept of web page skeleton to summarize the main content of the web page from a semantic point of view. Compared to the original web page, the skeleton of the web page removes styles and visual elements, but retains the semantics expressed by the original web page to the greatest extent possible. In order to separate different semantics, we project the skeleton

(a) The original web page   (b) Representation based on semantic features   (c) Representation based on visual features

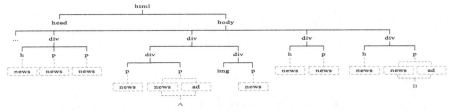

(d) Representation based on HTML structural features

Fig. 1. Different representations of the web page. (Color figure online)

onto the DOM tree and get the skeleton tree. The skeleton tree depicts the relationships between each part of the skeleton. We split the nodes of the skeleton tree to separate the semantics. Then we make fine-grained classifications, and finally organize the nodes of the same semantics together to form a semantic block. The web page CIE is carried out on these semantic blocks. As shown in Fig. 1b, different colors identify different semantics: rose red for *news*, orange for *ad*, and green for *other* semantics. The text of the same semantics can be visually dispersed and has nothing to do with their position on the DOM tree. Our semantic-feature-based web page representation can achieve effects like Fig. 1b very well. The main contributions of this work are as follows:

- We analyze the web page from a semantic perspective, and propose the web page representation based on semantic features.
- We propose SCIEnt, a web page CIE framework based on semantic features. SCIEnt adopts a modular architecture that algorithms in each module can be flexibly replaced according to the requirement of downstream TDM task.
- We design Core Accuracy, a new metric to measure the availability of extraction results in downstream TDM tasks.
- We conduct extensive experiments on both public and real datasets. Results show that SCIEnt outperforms baseline methods.

2   Related Work

Methods based on HTML structural features are widely used. Some methods [1,2] reconstruct DOM trees to generate extraction rules or templates to extract

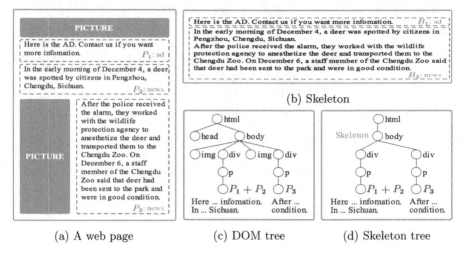

Fig. 2. Construction of the skeleton tree.

some DOM tree nodes. Other methods [3,4] carry out statistical algorithms on DOM tree nodes combined with HTML sequences. Some works [5–7] model the task as a classification problem on HTML tags. Traditional machine learning methods such as SVM, LR, and DT are used for node classification. Some studies [8,9] model the DOM tree nodes as sequences, and use HMM or neural sequence labeling model CNN to mark sequences. Lately, attentions between HTML tokens and text are constructed in [10] for single attribute extraction in the web page.

The method based on visual features was first proposed by Cai et al. [11]. The proposed VIPS extracts all suitable visual blocks, calculates separators, and reconstructs the structure of the web page according to these visual blocks and separators. Since then, a large number of web page information extraction methods based on visual features have been proposed [12,13]. rExtractor [14] extracts data records based on the visual and content similarities between displayed data records. Some deep learning structures such as CNN are also used in visual-feature-based methods [15,16].

3 Semantic-Feature-Based Web Page Representation

3.1 Web Page Skeleton

Let $W_i, i = 1, \cdots, N$, be a web page, where N presents the number of web pages. We define plain text content in a web page, not related to formats or styles, as skeleton, noted as $S_i, i = 1, \cdots, N$. The skeleton of a web page reflects its unique semantic expression as well as its main idea. Figure 2b is the skeleton of the web page shown in Fig. 2a. As you can see, although the style and pictures are removed, it has little effect on the meaning of the web page. We can clearly tell which sentence in the skeleton is *news* and which sentence is *advertising*.

3.2   Skeleton Tree

The text of multiple semantics is intertwined in the skeleton, but it is difficult to see the relationship between the semantics. Therefore, we need to observe and analyze the skeleton from different perspective. It is easy to find that the text in the skeleton is all distributed in leaf nodes of the DOM tree. We correspond the text in the skeleton and the text node of the DOM tree one by one, and project the skeleton onto the DOM tree. In this perspective we can see the structural information of the skeleton, and we call the skeleton of this perspective the skeleton tree. Figure 2c shows the DOM tree in Fig. 2a, and Fig. 2d is the skeleton tree. In order not to be distracted by the visual elements and layout of the web page, we kept all the text nodes in the DOM tree and removed the other leaf nodes. To maintain the order of text paragraphs, we ensure the relative position of the text nodes in the skeleton tree to be the same as that in the DOM Tree. Compared with the original perspective of the skeleton, the structural information of the skeleton is clear at a glance from the perspective of the skeleton tree. The tree structure depicts the relationship between different parts of the skeleton.

3.3   Skeleton Node

We call the leaf node of the skeleton as the skeleton node, denoted as $s_{ij}, j = 1, \cdots, N_i$. Where N_i represents the number of skeleton nodes in the skeleton tree. Since there is a consistent one-to-one match between the skeleton node and the text of the web page W_i, we use the sequence of skeleton nodes to represent the skeleton, denoted as $S_i = [s_{i1}, s_{i2}, \cdots, s_{iN_i}], i = 1, \cdots, N$, where $s_{ij}, j \in [0, N_i]$ represents the jth skeleton node ordered by the post-order traversal.

The text semantic in each skeleton node is still mixed up and the granularity of which is uneven at present. If we directly classify the skeleton nodes, like the HTML-structure-based method dose, it is impossible to distinguish between different semantics in a single node. As shown in Fig. 2d, P_1 expresses the semantics of *ad* and P_2 expresses the semantics of *news*, which are located in the same skeleton node. To solve this problem, we need fine-grained skeleton nodes. We split a skeleton node into multiple sub-nodes until each sub-node only contains one semantic. The skeleton node can be split by paragraph or by sentence, etc. according to the task requirements. We define the sub-node split from the skeleton node s_{ij} as skeleton sub-node, noted as $s_{ij}^k, k \in [0, M_{ij}]$, where M_{ij} represents the number of the skeleton sub-nodes split from s_{ij}. Hence, the skeleton can be represented not only by the sequence of skeleton nodes, but also by the sequence of skeleton sub-nodes, i.e. $S_i = [s_{i1}, s_{i2}, \cdots, s_{iN_i}] = [s_{i1}^1, s_{i1}^2, \cdots, s_{i1}^{M_{i1}}, s_{i2}^1, s_{i2}^2, \cdots, s_{i2}^{M_{i2}}, \cdots, s_{iN_i}^1, s_{iN_i}^2, \cdots, s_{iN_i}^{M_{iN_i}}]$.

3.4   Semantic Block

When the appropriate split granularity is chosen, each skeleton sub-node contains text with single semantics. The text of skeleton sub-nodes that belongs to the same semantics forms the same semantic block. Meanwhile, the skeleton is also divided into different semantic blocks. As shown in Fig. 2b, B_1 is the *ad* semantic block and B_2 is the *news* semantic block. Note the semantic collection of web page W_i as $Q = \{q_{i1}, q_{i2}, \cdots, q_{iP}\}$, P is the number of semantics that the web page contains. Then web page W_i can be expressed as a collection of semantic blocks, i.e. $W_i = \{B_{i1}, B_{i2}, \cdots, B_{iP}\}$, $B_{ip} = \{s_{ij}^k | Sem(s_{ij}^k) = q_{ip}\}$, where $Sem(x) = y$ indicates that the semantics of x is closest to y.

The number of semantics (or semantic blocks) of web pages depends on the downstream task of web page CIE. When the web page is decomposed into different semantic blocks, since each semantic block corresponds to a unique semantic label, we can extract the corresponding semantic block or aggregate multiple semantic blocks as the result of the web page CIE, according to the needs of the downstream task.

4   SCIEnt Framework

On the basis of the semantic-feature-based web page representation, we further design Semantic-feature-based web page Core Information Extraction framework, or SCIEnt for short. As shown in Fig. 3, SCIEnt adopts a modular architecture and consists of four modules: Skeleton Tree Construction, Node Splitting, Node Classification, and Semantic Aggregation and Correction. Each module can adapt specific algorithms to the different needs of downstream TDM tasks.

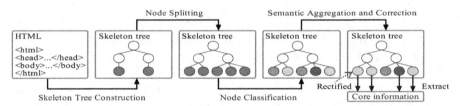

Fig. 3. The framework of SCIEnt.

4.1   Skeleton Tree Construction

Since the web skeleton is completely distributed in the leaf nodes of the DOM tree, we directly parse the web page HTML into a DOM tree, and prune the DOM tree to build the skeleton tree. Before pruning, considering some text like the bold text is wrapped by *span* tags, which will have a great impact on the DOM tree structure and text node semantics, we remove all *span* tags, such that a sentence is completely in one text node. During the pruning, we remove the subtree without text nodes until all leaf nodes in the tree are text nodes. To make the text in the skeleton tree consistent with the text that readers see in the rendered web page, we also removed the node with *hidden* attribute.

4.2   Node Splitting

A skeleton node may contain text with different semantics. If the extraction is directly made according to the granularity of the skeleton node, the extraction result will be mixed with different semantics. Therefore, we split the skeleton node so that each skeleton sub-node contains only one semantic. According to different need of semantic granularity, we divide the text in each skeleton node by paragraph, sentence or phrase, and split a skeleton node into one or more skeleton sub-nodes. Skeleton sub-nodes split from the same skeleton node inherit the characteristics of the skeleton node, have the same ancestral nodes, and are siblings of each other. The order of the siblings is the same as the order of the text in the skeleton node.

4.3   Node Classification

After choosing the appropriate split granularity, each skeleton sub-node contains only one semantic. The next thing is to identify these semantics to form different semantic blocks, that is, to classify the skeleton sub-nodes. Note the classifier as $f(\theta, s_{ij}^k)$, where θ is the model parameter. Then the probability that the skeleton sub-node s_{ij}^k expresses the semantic q is expressed as $P(y = q|\theta, s_{ij}^k) = \frac{e^{f(\theta^q, s_{ij}^k)}}{\sum_{l=1}^{Q} e^{f(\theta^l, s_{ij}^k)}}$, where Q represents the number of semantic categories. Get the probability that the skeleton sub-node s_{ij}^k is predicted as either semantic. Take the semantic with the highest probability as the final prediction of the skeleton sub-node. The skeleton sub-node is assigned to corresponding semantic blocks, i.e. $s_{ij}^k \in B_{\arg\max_q P(y=q|\theta, s_{ij}^k)}$.

In the Node Classification module, the classifier can be traditional machine learning methods, such as LR, GBDT, etc., or deep learning methods such as CNN and RNN. Traditional machine learning methods require manually selected characteristics, and only consider the statistical characteristics of the text. For two sentences with the same semantics but different structures, the machine learning classifier may give the opposite answers. So we need a classifier that can model the semantics of the text. In SCIEnt, we use the pre-trained language model BERT [17] to represent the semantics of the input text. As shown in Fig. 4, the classifier consists of BERT and a fully connection (FC) layer. Input the text of the skeleton sub-node into the BERT, then the output of CLS of BERT is fed into a follow-up FC layer, and finally the text score for each semantics is output.

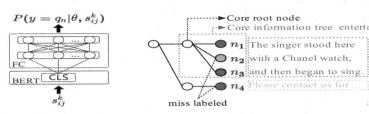

Fig. 4. The classifier.   **Fig. 5.** Semantic Correction.

4.4 Semantic Aggregation and Correction

After classifying the skeleton sub-nodes and obtaining different semantic blocks, the next thing is to select and output one or more of these semantic blocks as the core information according to the specific task requirements of web page CIE. We call this process Semantic Aggregation.

The semantic blocks can be aggregated by rules or unsupervised learning. In practice, output aggregation results directly may cause problems. At Node Classification stage, it is possible that some phrases or sentences in the paragraph are misclassified because the granularity of the sub-node is too small. Taking Fig. 5 as an example, the node n_2 originally belongs to the semantics of *entertainment*, but the classifier mistakenly marks it as *ad*, because of words like *Chanel* with strong *ad* semantic properties. The partially missing returned text is not semantically coherent and even makes no sense, which is not in line with human writing logic. Such text is unqualified for downstream TDM tasks. Therefore, before directly extracting the aggregated result, it is also necessary to correct the misclassified part.

Taking news extraction as an example, we found that in the result of Node Classification, if a few nodes marked as *ad* are distributed among a large of continuous *news* nodes, these nodes are most likely misclassified. If nodes marked as *ad* appear continuously in the *news* block, these nodes are most likely product placement. Therefore, we proposed a semantic correction method named Path Clustering, to correct the error caused by Node Classification. We introduce Path Clustering Coefficient $\alpha, \alpha \in [0, 1]$ in the method, which is a hyperparameter to control the threshold of semantic correction. The smaller α is, the greater the probability that the continuous *ad*-labeled nodes will be corrected as *news*, and vice versa.

Algorithm 1. Path Clustering Algorithm

Input: Path sequence P
Parameter: Path clustering coefficient α
Output: Core root node path C

```
 1: w ← min(lengthOf(path) for each path in P);
 2: h ← lengthOf(P);
 3: C ← [ ];
 4: pIdx ← [1, · · · , h];
 5: for all i ∈ [1, w] do
 6:     nDict ← dict();
 7:     for all j ∈ pIdx do
 8:         nDict[Ps[j][i]] ← nDict.get(Ps[j][i], 0) + 1;
 9:     end for
10:     maxKey ← max(nDict, key = nDict.get);
11:     if nDict[maxKey] < α∗lengthOf(pIdx) then
12:         break;
13:     end if
14:     C.append(maxKey);
15:     newpIdx ← [ ];
16:     for all j ∈ pIdx do
17:         if maxKey == Ps[j][i] then
18:             newpIdx.append(j);
19:         end if
20:     end for
21:     pIdx ← newpIdx;
22: end for
23: return C
```

The pseudocode of Path Clustering algorithm is shown in Algorithm 1. After Semantic Aggregation, we get all the skeleton sub-nodes that labeled as core information. The paths of these skeleton sub-nodes are input to the algorithm.

The output is the path of the core root node. SCIEnt marks all skeleton sub-nodes in the subtree of core root node as core information, while other skeleton sub-nodes are marked as non-core information. As shown in Fig. 5, after applying this algorithm to p_1, p_3, p_4, p_4 is filtered out, p_2 is corrected. Hence, the final output core information contains n_1, n_2, and n_3.

5   Experiments

Web page CIE is mainly to provide high-quality data for downstream TDM tasks. So in practical application, we require the web page CIE methods of high extraction efficiency. The method based on visual features often needs to parse web pages to images, which is time consuming. It is also highly dependent on the layout of web pages. When slightly changing the layout of a web page, the extraction result may vary greatly. Moreover, this method only roughly chunk the web page according to the visual features of web elements, cannot distinguish between different semantics. Because of its relatively low stability and low extraction efficiency, this method has not been widely used in the industry. Therefore, we do not conduct experiments of visual-feature-based methods. We selected CETD [3] and BoilerNet [9], two web page CIE methods based on HTML structural features, and tested them together with SCIEnt on three datasets.

5.1   Datasets

CleanEval [18] is a competitive evaluation that contains Chinese and English web pages, of which we use only English pages. We divide the dataset into development set and test set. The test set of CleanEval contains 136 web pages. The GoogleTrends-2017 (GT2017) [9] dataset consists of 180 pages randomly selected from a larger pool of web sites retrieved from the top search of Google's queries. We re-divide the GT2017 into development set and test set, and obtain a test set of a total of 103 pages. In order to evaluate the ability of generalization of SCIEnt on the Chinese web pages, we create the NewsCN dataset. NewsCN contains a total of 3,000 news pages published from Chinese news web sites in the last two years. We randomly take 2,000 pages as the development set and the remaining 1,000 pages as the test set.

5.2   Performance Metrics

We take P, R and F1 as the evaluation indicators as ever [3]. In addition, we define Core Accuracy of web page CIE, noted as CA. Given an extraction quality threshold t, when $F1_i > t$, we believe the extraction result of the W_i meets the quality requirements of the downstream TDM task, and the extraction result of W_i can be used as input data for this downstream TDM task. So the ith extraction result is considered as a success, denoted as $a_i = 1$, otherwise it is considered as a failure, denoted as $a_i = 0$. Then we have $CA = \frac{1}{N} \sum_{i=1}^{N} a_i$, where N is the total number of test pages. CA measures the extraction quality of models at the page level. It provides the intuition of the proportion of results that meet the requirements of extraction quality.

5.3   Comparison Results

As can be seen in Table 1, SCI-Ent has a high recall. Because it would rather make the extraction results contain some non-core information, than distort semantics by the incompleteness of core information. The CA of CETD and BoilerNet indicates, although F1 is high, the extraction quality of a majority of pages is inferior. Results show that SCIEnt outperforms CETD and BoilerNet

Table 1. The performance of models on three datasets ($t = 0.9$)

Dataset	Method	P	R	F1	CA
CleanEval	CETD	**0.95**	0.88	0.91	0.77
	BoilerNet	0.89	0.90	0.90	0.54
	SCIEnt	0.94	**0.96**	**0.95**	**0.96**
GT2017	CETD	0.90	0.90	0.89	0.68
	BoilerNet	0.78	0.83	0.80	0.30
	SCIEnt	**0.91**	**0.95**	**0.93**	**0.74**
NewsCN	CETD	0.70	**0.98**	0.80	0.31
	SCIEnt	**0.98**	**0.98**	**0.97**	**0.94**

on both CleanEval and GT2017 in both F1 and CA. It confirms that the web representation based on semantic features is superior to the method based on HTML structure features in web page CIE.

The rule-based model CETD is language-independent and is theoretically applicable to the extraction of any language web page. Hence, we retrain and evaluate SCIEnt on the Chinese dataset NewsCN, and evaluate the model CETD on that dataset to compare the generalization ability on different languages. As you can see, SCIEnt performs just as well on the Chinese dataset NewsCN as on two English datasets, and better than CETD on Chinese dataset, especially in CA. Results show that SCIEnt has strong generalization ability. On the one hand, this is because SCIEnt adopts modular architecture, the classifier of Node Classification module can be replaced according to the requirements of different downstream TDM tasks, so that SCIEnt can extract web pages in different languages. On the other hand, SCIEnt models the semantics of the web page. The web page semantics do not change with the structure, so SCIEnt achieves better extraction results on datasets with different structures.

5.4   Sensitivity Analysis

Path Clustering algorithm is an unsupervised method, which has a hyperparameter named Path Clustering Coefficient, noted as α. In order to study the influence of parameter α, we carried out the parameter sensitivity analysis experiment on three datasets: CleanEval, GT2017 and NewsCN. As shown in Fig. 6, with the increase of α, F1 shows a trend of rising first and falling later. The sensitivity of α varies from one dataset to another. In terms of overall trend, when α is set between 0.5 and 0.7, the model performs better, and on unknown datasets, α can be set in this interval and should be adjusted appropriately to achieve the best extraction results.

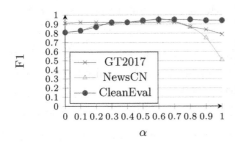

Fig. 6. Effect on F1 when α ranges from 0 to 1, where the step size is 0.1.

Table 2. Ablation study on dataset CleanEval. ($t = 0.9$)

Classifier	P	R	F1	CA
BERT	**0.94**	0.96	**0.95**	**0.96**
RoBERTa	0.93	**0.97**	0.94	0.93
ALBERT	0.93	**0.97**	0.94	0.94
XLM	0.93	**0.97**	0.94	0.93
NB	0.88	0.92	0.90	0.90
SVM	0.88	0.92	0.90	0.89
LR	0.88	0.92	0.90	0.89

5.5 Ablation Study

To verify the superiority of using the pre-trained language model as the classifier, we replaced the classifier of SCIEnt to make comparison. We use four different pre-trained models, i.e., BERT, RoBERTa, ALBERT and XLM, and three machine learning models, i.e., Naive Bayes (NB), Support Vector Machine (SVM) and Logistic Regression (LR) as the classifier. They are all evaluated on the dataset CleanEval. Table 2 reveals that the results of using machine learning models as the classifier is not as good as using pre-trained language models, indicating pre-trained language models capture more semantics than machine learning models. In addition, it also shows that no matter what kind of pre-trained language model is used, there is not much difference in extraction results, illustrating that the web page representation based on semantic features can stably represent the semantics of web pages. Whether a traditional machine learning model or a pre-trained language model is used as the classifier, the extraction results are better than CETD and BoilerNet, which confirms the superiority of SCIEnt framework.

6 Conclusion and Future Work

Extracting core information from web pages can provide effective structured data for the downstream TDM task. Traditional web page CIE methods represent the web page either based on HTML structural features or visual features. Neither of these web page representations can separate different semantics in a web page. From the perspective of readers' reading and understanding web pages, this paper proposes a new web page representation method based on semantic features. This method re-parse and construct the web page based on semantics. It is not affected by visual interference such as web page layout and style, nor is it disturbed by the underlying coding structure of the web page. Based on it, we propose SCIEnt, a semantic-feature-based web page CIE framework. SCIEnt adopts a modular architecture, and the algorithms in the module can be flexibly replaced according to the requirements of the downstream TDM task. Thus it has good generalization ability.

In the future, we will continue to optimize and explore the algorithms in each module of SCIEnt. In addition, we will explore the ability of semantic-feature-based web page representation on other web page comprehension tasks, such as web page classification and web page retrieval.

References

1. Yu, X., Jin, Z.: Web content information extraction based on DOM tree and statistical information. In: 2017 IEEE 17th International Conference on Communication Technology (ICCT), pp. 1308–1311. IEEE (2017)
2. Yuliana, O.Y., Chang, C.H.: A novel alignment algorithm for effective web data extraction from singleton-item pages. Appl. Intell. **48**(11), 4355–4370 (2018). https://doi.org/10.1007/s10489-018-1208-0
3. Sun, F., Song, D., Liao, L.: DOM based content extraction via text density. In: Proceedings of the 34th International ACM SIGIR Conference on Research and Development in Information Retrieval, pp. 245–254 (2011)
4. Tan, Z., He, C., Fang, Y., et al.: Title-based extraction of news contents for text mining. IEEE Access **6**, 64085–64095 (2018)
5. Yao, J., Zuo, X.: A machine learning approach to webpage content exraction (2013)
6. Utiu, N., Ionescu, V.S.: Learning web content extraction with DOM features. In: 2018 IEEE 14th International Conference on Intelligent Computer Communication and Processing (ICCP), pp. 5–11. IEEE (2018)
7. Yang, H., Lu, H., Li, S., Li, M., Sun, Y.: Research on content extraction of rich text web pages. In: Sun, X., Pan, Z., Bertino, E. (eds.) ICAIS 2019. LNCS, vol. 11635, pp. 279–287. Springer, Cham (2019). https://doi.org/10.1007/978-3-030-24268-8_26
8. Vogels, T., Ganea, O.-E., Eickhoff, C.: Web2Text: deep structured boilerplate removal. In: Pasi, G., Piwowarski, B., Azzopardi, L., Hanbury, A. (eds.) ECIR 2018. LNCS, vol. 10772, pp. 167–179. Springer, Cham (2018). https://doi.org/10.1007/978-3-319-76941-7_13
9. Leonhardt, J., Anand, A., Khosla, M.: Boilerplate removal using a neural sequence labeling model. In: Companion Proceedings of the Web Conference, pp. 226–229 (2020)
10. Wang, Q., Fang, Y., Ravula, A., et al.: WebFormer: the web-page transformer for structure information extraction. In: Proceedings of the ACM Web Conference, pp. 3124–3133 (2022)
11. Cai, D., Yu, S., Wen, J.-R., Ma, W.-Y.: Extracting content structure for web pages based on visual representation. In: Zhou, X., Orlowska, M.E., Zhang, Y. (eds.) APWeb 2003. LNCS, vol. 2642, pp. 406–417. Springer, Heidelberg (2003). https://doi.org/10.1007/3-540-36901-5_42
12. Zhao, H., Meng, W., Wu, Z., et al.: Fully automatic wrapper generation for search engines. In: Proceedings of the 14th International Conference on World Wide Web, pp. 66–75 (2005)
13. Lu, Y., He, H., Zhao, H., et al.: Annotating search results from web databases. IEEE Trans. Knowl. Data Eng. **25**(3), 514–527 (2011)
14. Anderson, N., Hong, J.: Visually extracting data records from the deep web. In: Proceedings of the 22nd International Conference on World Wide Web, pp. 1233–1238 (2013)

15. Liu, J., Lin, L., Cai, Z., et al.: Deep web data extraction based on visual information processing. J. Ambient Intell. Human. Comput. 1–11 (2017). https://doi.org/10. 1007/s12652-017-0587-0
16. Kumar, A., Morabia, K., Wang, J., et al.: CoVA: context-aware visual attention for webpage information extraction. arXiv preprint arXiv:2110.12320, 2021
17. Devlin, J., Chang, M.W., Lee, K., et al.: BERT: pre-training of deep bidirectional transformers for language understanding. arXiv preprint arXiv:1810.04805 (2018)
18. Baroni, M., Chantree, F., Kilgarriff, A., et al.: Cleaneval: a competition for cleaning web pages. In: Lrec (2008)

Hierarchical Down-Sampling Based Ultra High-Resolution Image Inpainting

Haixin Wang, Jian Yang, and Jinja Zhou[✉]

Hosei University, Tokyo, Japan
haixin.wang.8v@stu.hosei.ac.jp, zhou@hosei.ac.jp

Abstract. In recent years, image inpainting techniques have received a great deal of attention in the field of image processing. Many useful inpainting algorithms have been proposed, such as automatically removing some objects or repairing damaged images. Since the image inpainting prediction hole part from an already known area, the original input information was needed, then down-sampling it easy to tract. But most methods suffer from image quality degradation at the down-sampling part because the general average or maximum value-based down-sampling loses pixel information. In addition, Performing a large calculation at once time costs a lot of memory that makes them only can handle the input smaller than 1 K. In this paper, we propose a new ultra-high resolution image inpainting method that fully utilizes and densifies pixels by down-sampling the images without extra processing such as calculation, adapting it to existing models resulting in higher quality inpainting with ultra high-resolution images (more than 2 K images), and also can adjust the down-sampling to an arbitrary level for your need. The experiments demonstrate that the proposed HdCRAE quantitatively and qualitatively outperforms state-of-the-art.

Keywords: Image inpainting · Hierarchical down-sampling · ultra high-resolution · Image enhancement

1 Introduction

Since image inpainting was proposed by Bertalmio [3], it has become a popular topic in the field of computer vision and is attracting more and more researchers. Image inpainting is to restore missing or damaged portions of the work, making it more clear and restoring a sense of unity to the work. The algorithm can also be used to remove unwanted objects or writing on the image as well as to repair damaged photographs as shown in Fig. 1. Traditional methods can be broadly divided into two types, the one is a diffusion based [1,14] approach that uses variational methods and partial differential equations to diffuse the information from the surrounding regions to the interior of the missing regions, propagating local image appearance surrounding the target holes. For example, Pyramid model based Down-sampling Inpainting (PDI) model proposed by Wang et al.

© The Author(s), under exclusive license to Springer Nature Switzerland AG 2023
M. Tanveer et al. (Eds.): ICONIP 2022, LNCS 13625, pp. 324–334, 2023.
https://doi.org/10.1007/978-3-031-30111-7_28

[15] used a novel method, which excellent effect on regular image patterns, but did not give the desired results for repairing irregular or unknown objects. The other is texture synthesis techniques which replenish holes by borrowing from the known area [2,4,5,9,13] or finding the pixel closest to the hole boundary from the background area and gradually filling in the unknown area [6–8,16]. These methods are sufficient for the visual effect of repairing a small missing area. However, the quality of performance is greatly reduced when the missing area increases with image size. Furthermore, the data used for training is also costly and very memory intensive. These problems can be improved by Contextual Residual Aggregation (CRA) [17]. The method includes aggregated weighted high-frequency residuals from know areas and then is used to generate high-frequency residuals for unknown regions, finally adding these high-frequency residuals to the generated inpainted result. All processes work on low-resolution images, so memory usage and costs have decreased considerably but it still has some problems like the down-sampling methods. Down-sampling takes the average or maximum of several pixels which do not give the best effect because some pixels are vanished and can not be fully deployed.

Fig. 1. The inpainting results from proposed HdCRAE on high-resolution images. (a) Image restoration, (b) Remove object.

To solve this issue, we propose Hierarchical down-sampling based ultra high-resolution image inpainting. In order to improve the efficiency and accuracy of image restoration algorithms when inpainting images, we focused on down-sampling based on the pyramid model. By extracting pixels along the rows and columns the image is transformed into smaller images, it is obvious that the image is smaller and more pixel-dense twice than once down-sampling. Furthermore, compared to other down-sampling methods, our methods down-sampling the image in the situation where pixel value does not change then the information will not be lost, all pixels are available. And the smaller size results in denser distribution of pixels and more peripheral pixels remain. Therefore, adapting hierarchical down-sampling to the CRA model improves the performances of aggregating weighted high-frequency residuals from known areas and more effective high-frequency residuals for unknown areas, enhancing the effectiveness of inpainting.

When using CRA to predict and generate unknown portions, visible noises will be inevitably mixed in with the generated image, to make it better we implemented image enhancement denoising convolutional neural networks (DnCNN) [18], which is able to easily be extended to general image denoising not only tasks make results more natural and integrated, but also low computational complexity and low demands on equipment, by this enhancement, we can obtain a promising effect on goal achievement. To remove the noise for a superior image x from noisy observation y, we follow an image degradation model, which is formulated as:

$$y = x + v \tag{1}$$

where v is additive white Gaussian noise (AWGN) with standard deviation σ.

In a summary, the contributions of this paper are shown as follows: 1) Hierarchical down-sampling based Contextual Residual Aggregation is proposed to make image inpainting models perform to the fullest potential. The hierarchical down-sampling technique enables fully utilizing the pixels, which can gain more information from known parts around the hole making the results better. 2) Extensive experimental results show that the additional enhancement module makes the results significantly surpass CRA while using masks of different scales, thus demonstrating the robustness of the entire framework. 3) Our proposed framework successfully combined the image inpainting and image denoising. In addition, as a texture-based inpainting method, HdCRAE requires less computation and is adaptable to the ultra high-resolution image, making it easy to spread.

2   Related Work

Contextual Residual Aggregation (CRA). The contextual residual aggregation [17] work first down-sampling the input image by an average of 6 patches after that the neural network predicts a low-resolution inpainted result and upsample it to yield a large blurry image by Bicubic. Then produce the high-frequency residuals for in-hole patches by aggregating weighted high-frequency residuals from contextual patches, the weight means the attention scores which are computed by obtaining region affinity between patches inside and those outside the holes. At last, add the aggregated residuals to the large blurry image to obtain a sharp result finished inpainting. Our image inpainting mechanism is based on that, and improved down-sampling method protects from more information loss.

Denoising Convolutional Neural Networks (DnCNN). Denoising convolutional neural networks [18] can be successfully combined with the proposed methods since contextual residual aggregation input the contextual residual into generator attain the deficiency part will occur some noise. The DnCNN model not only has the capacity to handle the unknown noise make our framework perform quantitatively and qualitatively but also promising run time implementation, keeping our proposal's advantage about calculations are infinitely less.

3   Proposed HdCRAE

3.1   Overall Framework

For image inpainting, the texture-based inpainting method often has some problems like high computational, time-consuming, and difficulty in handling high-resolution datasets. The CRA model improved considerably on these issues. However, about down-sampling, we observe that CRA model like many existing works often follows a similar approach. For example, taking the average or maximum of some pixels, those methods lose a certain number of pixels thereby preventing the model from performing as well as it should. Down-sampling is an indispensable technique to obtain more accurate and effective algorithms, we propose hierarchical down-sampling which reduced pixel loss, leading the inpainting model to perform better.

On the other hand, our experiments results show that the addition of hierarchical downsampling increased CRA, but the image quality due to generator and synthesis still have space to improve. Therefore, we connect image enhancement module to achieve superior performance. The overall framework is shown in Fig. 2, the experimental results proof that image enhance module can make the results more visually pleasing and further improved the CRA model.

3.2   Hierarchical Down-Sampling Algorithm

Before feeding the input image to the generator, we subtract the input image and the down-sampled image to obtain the contextual residual. Figure 3 illustrates

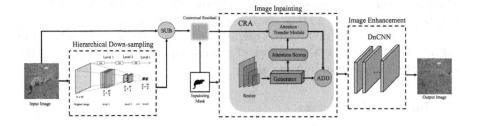

Fig. 2. The overview of proposed HdCRAE framework.

the proposed hierarchical down-sampling. The hierarchical down-sampling down-sampled image without the pixel loss by giving a tensor $I \in \mathbb{R}^{H \times W}$ such that:

$$I = \begin{bmatrix} I(0,0) & \cdots & I(0,W) \\ \vdots & \ddots & \vdots \\ I(H,0) & \cdots & I(H,W) \end{bmatrix} \tag{2}$$

And our down-sampling focuses on the matrices from the upper left of $I \in \mathbb{R}^{H \times W}$, the extraction of pixels located in the elements of each matrix is repeated along the rows and columns of the entire image to obtain different dimension down-sampled images from different levels i, where $i = n - 1$, $\{n = 2, 3, 4, ...\}$. The down-sampled images $D^{nk} \in \mathbb{R}^{\frac{H}{K} \times \frac{W}{K}}$, where $K = 2^i$ and $k = 1, 2, 3, 4...$, The procedures are shown on Algorithm 1 followed by:

$$D_1^{nk}(x,y) = D^{ik}(2x, 2y) \tag{3}$$

$$D_2^{nk}(x,y) = D^{ik}(2x, 2y + 1) \tag{4}$$

$$D_3^{nk}(x,y) = D^{ik}(2x + 1, 2y) \tag{5}$$

$$D_4^{nk}(x,y) = D^{ik}(2x + 1, 2y + 1) \tag{6}$$

The x and y means each element of I. k is the number of down-sampling images $k \in 2^K$. D is the matrix down-sampled from the previous level.

After inpainting we are able to derive input image with same dimension from inpainted $D'^{nk} \in \mathbb{R}^{\frac{H}{K} \times \frac{W}{K}}$ as:

$$\sum_{k=1}^{K} \sum_{n=2}^{N} \sum_{\alpha=1}^{4} D_\alpha' nk \tag{7}$$

By repeating this process we can attain the inpainted image without pixel loss and dimension change, thus reach the promising performance.

$I(x,y)$ means pixel value of the original image, h, w denotes the height and width of the image. n is the number of down-sampling, and m represents the number of down-sampled images.

3.3   Image Enhancement

When using the generator to predict and create the lacked area, that is difficult to avoid the noise. That caused the results to become poor image quality. We defined the generated down-sampled image as D and the noise as φ, the output image Y can represent as:

$$Y = \sum_{k=1}^{K} D_k + \Phi \tag{8}$$

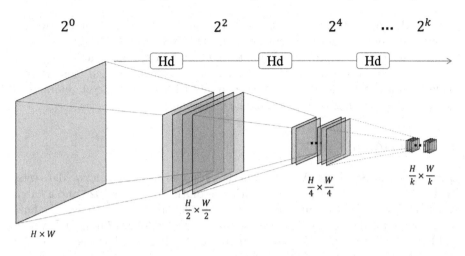

Fig. 3. Hierarchical down-sampling algorithm. The bottom side represent the level of down-sampling. The top side represent the number of images obtained by down-sampling.

Algorithm 1. Hierarchical down-sampling Algorithm

BEGIN

$H \leftarrow \frac{h}{2}$

$W \leftarrow \frac{w}{2}$

n denotes the nth down-sampling, m $\leftarrow 2^{2(n-1)}$

$$D^{(n=0)} = \{(x,y)|x, \ni I\}$$

initialization;

for $x = 1, ..., H$ **do**

  for $y = 1, ..., W$ **do**

    $D_1^m = \{(x,y)|x = 2x, y = 2y\}$

    $D_2^m = \{(x,y)|x = 2x, y = 2y + 1\}$

    $D_3^m = \{(x,y)|x = 2x + 1, y = 2y\}$

    $D_4^m = \{(x,y)|x = 2x + 1, y = 2y + 1\}$

  end

end

To reach the goal of recovering Y to a clean image X with no effect on lightweight calculations, our method implemented the DnCNN after merging the down-sampled image thus improving results to the high image quality and comparing to enhancement before merging that decreasing the calculation and memory cost made the output image Y more nearly to clean image X and yet guaranteed treatable the ultra high-resolution image inpainting.

4   Experiments

Experimental Settings. We evaluate the proposed HdCRAE on two data sets, FFHQ [11] and DIV2K [13]. Images with a resolution of 1024×1024 and a batch size of 512 were used to train the model on the GeForce RTX 2080 Ti. The final model was implemented on TensorFlow v1.13 with CUDNN v7.6, CUDA v10.0, and Pytorch v1.0.1.post2. We also compared with CRA, which is the base of the proposed methodology. In order to figure out the effect of image enhancement, we conduct the ablation study model without enhancement module, dubbed as HdCRA.

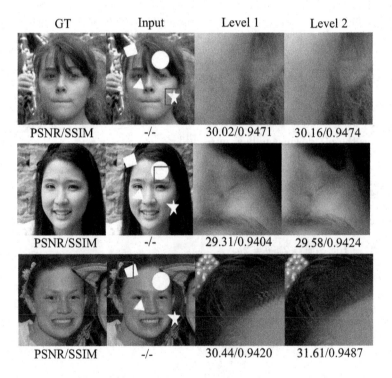

Fig. 4. Comparisons between down-sampling level 1 and level 2. From results the level 2 obtain the better evaluation metrics.

Table 1. The average index of performance comparison between down-sampling level 1 and level 2.

Down-sampling	FID↓	PSNR↑	SSIM↑
Level 1	57.95	28.41	0.9385
Level 2	38.92	28.71	0.9406

Fig. 5. Comparisons with irregular mask holes under 25% and over 25% area of the high-resolution images from FFHQ (1K images). Both under 25% and over 25% perform equally well.

Experimental Results. We compare our method with the CRA and Edge Connect (EC) [12] method, to ensure a fair comparison, the same settings were used in the experiment as much as possible. The generator was used in the same way as the CRA and no special treatment was added. The image enhancement model was trained for $2K$ iterations with a batch size of 32. Trained on 1024 × 1024 images with irregular holes area of the whole image. Finally, because the higher and higher resolution images have more pixels, we can modify details many times in high-resolution images.

Since there are very few existing methods for processing images with ultra-high-resolution (more than 2K), and even fewer with self-defined masks, only CRA were compared with the irregular mask. The Globally and locally consistent image completion (GLCIC) [10] method even if tractable the 2 K image but the mask should be random quadrangle, we the results in Fig. 6. And Fig. 5 represent our visual effect of the model results with the addition of enhancement in 1 K images was also remarkably clean. The improvements in FID, PSNR and SSIM values can also be observed.

Table 2 Table 3, and Table 4 report the average valuation index slightly better than without hierarchical down-sampling regardless of the size of the mask. Moreover, we compared down-sampling level 2 with down-sampling level 1 in Fig. 4 and Table 1. It was observed through the PSNR and SSIM that level 2 was slightly better than level 1.

Table 2. Average qualitative evaluation using mask holes over 25% area of the image. Note that EC models cause out of memory error when tested on 2 K images, thus the corresponding cells are left empty.

Mask scale	>25%					
Image size	$1K$(FFHQ [11])			$2K$(DIV2K [13])		
Metrics	FID↓	PSNR↑	SSIM↑	FID↓	PSNR↑	SSIM↑
EC [12]	112.703	21.08	0.771	–	–	–
CRA [17]	68.158	20.15	0.8082	61.141	19.62	0.7862
HdCRA (Ours)	56.166	20.45	0.8080	53.351	19.88	0.7799
HdCRAE (Ours)	**45.728**	**21.10**	**0.8166**	**46.984**	**20.66**	**0.7933**

Table 3. Average qualitative evaluation using mask holes under 25% area of the image. Note that EC models cause out of memory error when tested on 2 K images, thus the corresponding cells are left empty.

Mask scale	<25%					
Image size	$1K$(FFHQ [11])			$2K$(DIV2K [13])		
Metrics	FID↓	PSNR↑	SSIM↑	FID↓	PSNR↑	SSIM↑
EC [12]	84.093	27.75	0.934	–	–	–
CRA [17]	49.227	28.88	0.9387	28.598	28.31	0.9460
HdCRA (Ours)	34.652	29.19	0.9384	26.544	28.50	0.9445
HdCRAE (Ours)	**20.926**	**30.03**	**0.9440**	**21.472**	**28.83**	**0.9476**

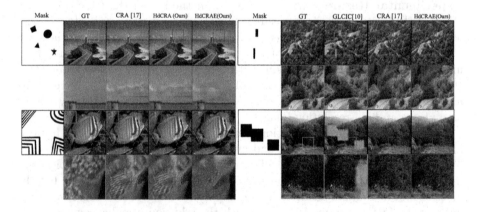

Fig. 6. Comparisons with irregular mask holes under 25% and over 25% area of the image from DIV2K (2K images). Note that with few processing images with self-defined masks ultra-high-resolution (more than 2K) we compared CRA and our no addition with the enhancement model. And for a fair comparison with the GLCIC model, the random quadrangle mask was used.

Table 4. Average qualitative evaluation using mask holes under 25% and over 25% area of the image.

Image size	>2K(DIV2K [13])							
Mask scale	<25%				>25%			
Metrics	FID↓	PSNR↑	SSIM↑	L1(%)↓	FID↓	PSNR↑	SSIM↑	L1(%)↓
GLCIC [10]	88.254	14.13	0.2964	17	162.251	12.65	0.2471	20
CRA [17]	16.42	31.71	0.9625	0.84	96.3217	22.45	0.8589	3
HdCRAE (Ours)	**13.94**	**31.89**	**0.9629**	**0.08**	**88.5134**	**22.56**	**0.8597**	**0.1**

5 Conclusion

In this paper, we proposed hierarchical down-sampling based ultra high-resolution image inpainting that works well at high-resolution images and presented the down-sampling without loss of pixel information and low equipment requirements image inpainting methods, which successfully improved contextual residual aggregated by hierarchical down-sampling. The maximum FID value is increased by 60% while comparing with state-of-the-art methods. Furthermore, the higher level of down-sampling better the results can be confirmed from our experiments. On the other hand, the necessity for enhancement was proved by our experiment results, both the evaluation metrics and the visual effect of presented HdCRAE were greater than without enhancement. Regarding to the future work, investigating and developing a more effective image enhancement approach to achieve prominent amelioration is taken into account.

References

1. Ballester, C., Bertalmio, M., Caselles, V., Sapiro, G., Verdera, J.: Filling-in by joint interpolation of vector fields and gray levels. IEEE Trans. Image Process. **10**(8), 1200–1211 (2001)
2. Barnes, C., Shechtman, E., Finkelstein, A., Goldman, D.B.: PatchMatch: a randomized correspondence algorithm for structural image editing. ACM Trans. Graph. **28**(3), 24 (2009)
3. Bertalmio, M., Sapiro, G., Caselles, V., Ballester, C.: Image inpainting. In: Proceedings of the 27th Annual Conference on Computer Graphics and Interactive Techniques, pp. 417–424 (2000)
4. Buyssens, P., Daisy, M., Tschumperlé, D., Lézoray, O.: Exemplar-based inpainting: technical review and new heuristics for better geometric reconstructions. IEEE Trans. Image Process. **24**(6), 1809–1824 (2015)
5. Darabi, S., Shechtman, E., Barnes, C., Goldman, D.B., Sen, P.: Image melding: combining inconsistent images using patch-based synthesis. ACM Trans. Graph. **31**(4), 1–10 (2012)
6. Drori, I., Cohen-Or, D., Yeshurun, H.: Fragment-based image completion. In: ACM SIGGRAPH 2003 Papers, pp. 303–312 (2003)
7. Efros, A.A., Freeman, W.T.: Image quilting for texture synthesis and transfer. In: Proceedings of the 28th Annual Conference on Computer Graphics and Interactive Techniques, pp. 341–346 (2001)

8. Efros, A.A., Leung, T.K.: Texture synthesis by non-parametric sampling. In: Proceedings of the Seventh IEEE International Conference on Computer Vision, vol. 2, pp. 1033–1038. IEEE (1999)

9. Huang, J.B., Kang, S.B., Ahuja, N., Kopf, J.: Image completion using planar structure guidance. ACM Trans. Graph. **33**(4), 1–10 (2014)

10. Iizuka, S., Simo-Serra, E., Ishikawa, H.: Globally and locally consistent image completion. ACM Trans. Graph. **36**(4), 1–14 (2017)

11. Karras, T., Laine, S., Aila, T.: A style-based generator architecture for generative adversarial networks. In: Proceedings of the IEEE/CVF Conference on Computer Vision and Pattern Recognition, pp. 4401–4410 (2019)

12. Nazeri, K., Ng, E., Joseph, T., Qureshi, F.Z., Ebrahimi, M.: EdgeConnect: generative image inpainting with adversarial edge learning. arXiv preprint arXiv:1901.00212 (2019)

13. Timofte, R., Gu, S., Wu, J., Van Gool, L.: NTIRE 2018 challenge on single image super-resolution: methods and results. In: Proceedings of the IEEE Conference on Computer Vision and Pattern Recognition Workshops, pp. 852–863 (2018)

14. Tschumperlé, D., Deriche, R.: Vector-valued image regularization with PDEs: a common framework for different applications. IEEE Trans. Pattern Anal. Mach. Intell. **27**(4), 506–517 (2005)

15. Wang, M., Yan, B., Gharavi, H.: Pyramid model based down-sampling for image inpainting. In: 2010 IEEE International Conference on Image Processing, pp. 429–432. IEEE (2010)

16. Wilczkowiak, M., Brostow, G.J., Tordoff, B., Cipolla, R.: Hole filling through photomontage. In: Proceedings of the British Machine Vision Conference (BMVC 2005) (2005)

17. Yi, Z., Tang, Q., Azizi, S., Jang, D., Xu, Z.: Contextual residual aggregation for ultra high-resolution image inpainting. In: Proceedings of the IEEE/CVF Conference on Computer Vision and Pattern Recognition, pp. 7508–7517 (2020)

18. Zhang, K., Zuo, W., Chen, Y., Meng, D., Zhang, L.: Beyond a Gaussian denoiser: residual learning of deep CNN for image denoising. IEEE Trans. Image Process. **26**(7), 3142–3155 (2017)

Vision Transformer with Depth Auxiliary Information for Face Anti-spoofing

Shenyuan Li, Jiwen Dong, Junting Chen, Xizhan Gao, and Sijie Niu$^{(\boxtimes)}$

University of Jinan, Jinan, China
sjniu@hotmail.com

Abstract. Face anti-spoofing (FAS) is an important part of the face recognition system. Although methods based on convolutional neural networks (CNN) have achieved great success, CNN may not be able to make good use of global information, resulting in the degradation of classification performance. Because Vision transformer (ViT) can use attention mechanisms to aggregate global information, some ViT based methods have been proposed. But most of these works treat the FAS problem as binary classification task, making it difficult to capture spoofing cues. In this work, we use ViT as our backbone. Then, we design an auxiliary supervised branch to exploit the depth information of the face image so that the algorithm can take the depth information into account when classifying. Cross domain experiments between CASIA-MFSD and Replay-Attack and intra experiments on OULU-NPU demonstrate the effectiveness of our method.

Keywords: Face anti-spoofing · Vision Transformer · Convolution · Depth map supervision

1 Introduction

Face recognition has brought great convenience to people's life. However, deceivers will cheat the face recognition system through presentation attacks (PA). PA includes print attack (printing face on paper), video replay attack(replay face on digital devices) and 3D mask attack. These attacks may cause serious harm to public security. Therefore, it is very important to design presentation attack detection (PAD) algorithms which can deal with various attacks in different scenarios.

In order to deal with various attacks, a large number of PAD algorithms have been proposed. Both the traditional methods [1,2,9,15,18,19] and the deep learning methods [13,16,22–26] show the effectiveness of detecting face attacks. At first, researchers use hand-crafted features for classification, such as LBP [1,2,9], HOG [15], DoG [19], or SIFT [18]. Then, due to the strong ability of

Thanks to No. 62101213, No. ZR2020QF107, No. ZR2020MF137, No. ZR2019MF040 for funding.

M. Tanveer et al. (Eds.): ICONIP 2022, LNCS 13625, pp. 335–346, 2023.
https://doi.org/10.1007/978-3-031-30111-7_29

feature extraction of CNN, many CNN based methods are proposed. Most of these works regard FAS problem as a binary or multi classification problem. However, taking presentation attack detection as simple classification problem may make the network unable to learn robust features, resulting in the weak generalization ability of the network.

Therefore, many FAS algorithms with auxiliary supervision have been proposed. There are three main kinds of auxiliary information: rPPG signal, face depth information and reflection information. Due to the blood flow, the real face can show a periodic change, while the masked face will not change significantly, so **Remote Photoplethysmography (rPPG) signal** can be regarded as a kind of temporal information; The real face has **depth information**, while the printed and replayed face attacks are planar, so the **depth information** can be used as a clue to distinguish the real and fake face; Under the illumination, face skin and human material will have different **reflection information**, so it also can be used as a cue to distinguish the real and fake face. These auxiliary information are widely used as a prior information to guide the network to learn robust features in most of the state-of-the-art methods, such as BCN [24], CDCN [25], DCN [26], and auxiliary [16]. All of these CNN based methods have achieved promising results.

Recently, vision transformer has been widely used in many fields. On the one hand, when trained with large amounts of data, the Vision Transformer model outperformed state-of-the-art methods in many vision benchmarks. On the other hand, the weak inductive bias of ViT means that the algorithm is more malleable, which may be beneficial to the generalization performance of the algorithm. So, some transformer's work in the field of FAS has been proposed [10,12]. ViTranZFAS [10] fix the pre-training parameters of ViT on the imagenet and modifies only the last layer to explore the performance of ViT on the FAS task. ViTF [12] explores the FAS task through a few-shot approach.

Unlike previous transformer work, we are the first single frame level work to use depth auxiliary information to supervise the transformer. On the one hand, the depth information can provide fine-grained supervision to the FAS task making the features learned by the model more robust. On the other hand, ViT may be an important help in reconstructing the depth of a human face because it can use attention mechanism to capture global contextual information to establish a long-range dependency. For example, when reconstructing the depth of a face, the eyes and nose or chin of a person should have feature dependencies. So we designed an auxiliary supervised branch to exploit the face depth information.

In this paper, we use ViT to explore FAS task. Furthermore, in order to make the transformer model more generalizable and stable in FAS task, we use depth information as auxiliary supervision.

The contributions of our work can be summarized as follows:

1) We use depth auxiliary information to supervise the transformer to explore FAS task.
2) A novel encoder-decoder structure is designed to generate the depth map so that the depth information is considered into classification.

3) Our algorithm shows good performance in cross-domain experiments between the CASIA-MFSD dataset and the Replay-Attack dataset, as well as intra experiments on the OULU-NPU dataset.

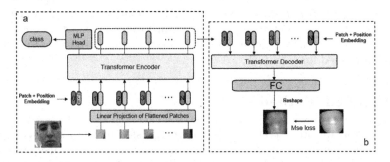

Fig. 1. The overview of DE-ViT. (a) vision transformer. (b) depth estimation module.

2 Related Works

2.1 Face Anti-spoofing Methods

Researchers have recently made great progress in the field of face anti-spoofing. Early researchers used hand-crafted descriptors such as LBP [1,2,9], HOG [15], DOG [19], SIFT [18], etc. to train a classifier to distinguish between real and spoofing faces. With the development of deep learning, more and more researchers choose to use convolutional neural networks to explore the field of FAS. The original convolutional neural network-based approaches viewed the FAS task as a binary or multi classification task. This may lead the algorithm to learn features that are not discriminative (borders of photos, etc.), which affects the generalization performance of the algorithm. Some recent work [16,24–26] has used auxiliary information (depth, reflection, etc.) to guide the network to ignore such features to learn more robust ones.

Depth and reflection information can provide fine-grained supervision for the FAS algorithm, allowing the FAS algorithm to learn more reasonable spoofing cues. Depth labels usually supervise the FAS algorithm to regress live faces to facial depth and spoofing faces to zero map. The reflection label, on the other hand, supervises the FAS algorithm to regress the spoofing face to the reflection map and the real face to the zero map. We found that the generation of pseudo-labels for reflection maps is strongly influenced by illumination, so we used pseudo facial depth as auxiliary information in this work.

2.2 Transformer

The original transformer [21] was proposed for machine translation. Transformer's attention layer uses the attention mechanism to aggregate information from the entire input sequence. The advantages of transformer are that it can use the attention mechanism to capture global contextual information to establish a long-range dependency on the target and extract more powerful features. Recently Dosovitskiy et al. [7] proposed the vision transformer (ViT), which applies the transformer to image classification task. The Vision Transformer has achieved great success on many vision tasks when a large amount of data is available for training. This motivated us to use transformer to explore the FAS field. To the best of our knowledge, no depth supervised transformer-based methods has ever proposed for FAS task.

3   Proposed Method

In this section we present DE-ViT, a vision transformer with supervised with depth auxiliary information. As shown in Fig. 1, our model is an encoder-decoder structure, which consists of two main modules: 1) ViT (Fig. 1 (a)). 2) Depth estimation module (Fig. 1 (b)). We use the ViT for classification and the depth estimation module as an auxiliary supervised branch.

3.1   ViT

Our encoder part is the same as the Vision Transformer [7]. The flow of the encoder part is shown in Fig. 1 (a). Given an image as input, we first reshape the image into a sequence of flattened $2D$ patches, where P is the patch size and $N = \frac{HW}{P}$ is the number of patches. These flatten patches will map to D dimensions with a trainable linear projection. Then a learnable embedding $F_0^0 = f_{class}$ will be spliced with these patches. In order to maintain the position information of each patches the learnable position embedding E_{pos} will be added, and the obtained results will be used as the input of the transformer encoder. Finally, the y_{class} obtained from the transformer encoder is input into the MLP Head to get the classification result. The calculation of the encoder part can be formulated as:

$$
\begin{aligned}
F_0 &= [f_{class}; f_p^1; f_p^2; \cdots ; f_p^N] + E_{pos}, \\
q_l &= k_l = v_l = LN(F_{l-1}), \\
F_l' &= MSA(q_l, k_l, v_l) + F_{l-1}, \\
F_l &= MLP(LN(F_l')) + F_l', \qquad l = 1, ..., L \\
y_{class} &= LN(F_L^0), \\
F_{feature} &= [F_L^1, F_L^2, \cdots , F_L^N],
\end{aligned}
\tag{1}
$$

where LN denotes the layernorm layer, MLP denotes two full connection (FC) layers with a GELU activation function, MSA denotes the multi-head self-attention. More details can be seen at [7].

3.2   Depth Estimation Module

The traditional deep learning based FAS algorithm regards the presentation attack detection (PAD) as a binary or multi classification problem. This may make the model learn some non-robust features, which will affect the generalization ability of the model. Depth map supervision provides pixel-level supervision and priori information for the FAS task, which can improve the generalization ability of the FAS algorithm.

Inspired by IPT's [4] approach to using transformer, we designed a module to learn face depth information. As shown in Fig. 1 (b), our depth estimation module is composed of transformer decoder and a FC layer, and it is used only in the training phase. The calculation of our depth estimation module can be formulated as:

$$
\begin{aligned}
D_0 &= F_{feature} = [F_L^1, F_L^2, \cdots, F_L^N], \\
q_i &= k_i = LN(D_{i-1}) + E_{task}, v_i = LN(D_{i-1}), \\
D_i' &= MSA(q_i, k_i, v_i) + D_{i-1}, \qquad\qquad i = 1, ..., K \\
q_i' &= LN(D_i') + E_{task}, k_i' = LN(D_{i-1}) + E_{pos}, \\
v_i' &= LN(D_{i-1}), \\
D_i'' &= MSA(q_i', k_i', v_i') + D_i', \\
D_i &= MLP(LN(D_i'')) + D_i'', \\
D_{feature} &= [D_K^1, D_K^2, \cdots, D_K^N],
\end{aligned}
\tag{2}
$$

where $D_{feature}$ denotes the feature obtained by transformer decoder and E_{task} is task embedding which learn to decoder the features. The size of patch D_K^i is $P^2 \times C$ and changes to $\frac{h}{16} \times \frac{w}{16}$ after passing through the full connection layer. Then reshaped into a depth map with the size of $h \times w$, where h and w are the height and width of the depth map in pixels. The detail of transformer encoder and decoder of our method are introduced in Fig. 2.

3.3   Loss Function

In consideration of the depth information may not contain all the important information of classification, our method uses two loss functions. One is the cross entropy loss for classification, and the other is the MSE loss for auxiliary supervision. The loss used in our method can be formulated as:

$$
L_{BCE} = \frac{1}{N} \sum_{n=1}^{N} -B_{gt} log(f_{MLPhead}(F_{class})),
\tag{3}
$$

$$
L_{MSE} = \frac{1}{N} \sum_{n=1}^{N} ||D_{gt} - D_{pre}||^2,
\tag{4}
$$

$$
L_{overall} = (1 - \alpha) L_{BCE} + \alpha L_{MSE},
\tag{5}
$$

where B_{gt} is the binary label, $f_{MLPhead}$ denotes one hidden layer and one soft-max layer, D_{gt} and D_{Pre} are the ground-truth depth map and the generated depth map. α is the parameter used to balance the two loss.

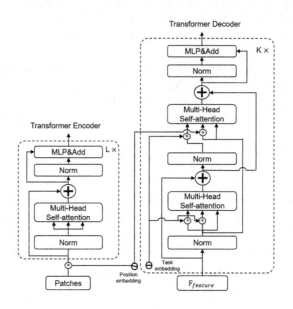

Fig. 2. The detail of transformer encoder and decoder.

4   Experiments

4.1   Dataset and Metric

Databases. Three databases CASIA-MFSD [27], Replay-Attack [5] and OULU-NPU [3] are used in our experiments. CASIA-MFSD and Replay-Attack are used for cross testing. CASIA-MFSD was taken by 50 people, with a total of 600 videos, including three types of attacks: wrap photo attack, cropp photo attack and replay video attack. Replay-Attack consists of 1,300 video clips of photo and video attack attempts of 50 clients, under different lighting conditions. Oulu-NPU contains four protocols to verify the generalization of the model. We use CASIA-MFSD and Replay-Attack for cross testing, and OULU-NPU for intra testing.

Evaluation Metrics. According to the ISO standardization, we use Attack Presentation Classification Error Rate APCER, Bona Fide Presentation Classification Error Rate BPCER and ACER in intra testing, and HTER for cross testing. The calculation of ACER and HTER can be formulated as:

$$ACER = \frac{APCER + BPCER}{2}, \tag{6}$$

$$HTER = \frac{FRR + FAR}{2}, \tag{7}$$

Where FRR is False Rejection Rate and FAR is False Acceptance Rate. All of these metrics details can be found at [6].

4.2 Implementation Details

Preprocessing. In order to facilitate the generation of depth map and avoid the influence of environment, we cut out the face area first. All images are cropped to a size of 256 × 256. We use PRNet [8] to generate depth map for living face, then resize them to the size of 64 × 64 and normalize them to [0,1]. To distinguish living faces from spoofing faces, the depth map of spoofing faces is set to 0. Note that the size of input image of all models in our experiments is 256 × 256, but the transformer models we use is loaded with pre-trained weights 'vit-b_16-244'.

Table 1. The cross testing results between CASIA-MFSD and Replay-Attack. (HTER%)

Method	Train	Test	Train	Test
	CASIA-MFSD	Replay-Attack	Replay-Attack	CASIA-MFSD
Motion-Mag [9]	50.1		47.0	
CNN [22]	48.5		45.5	
STASN [23]	31.5		30.9	
Face-des [13]	28.5		41.1	
Auxiliary [16]	27.6		**28.4**	
BCN [24]	16.6		36.4	
CDCN [25]	15.5		32.6	
DCN [26]	**15.3**		29.4	
Ours	18.9		36.4	

Settings. Our method(DE-ViT) is implemented by Pytorch [17] and trained with AdamW [14] optimizer for 55 epochs. We use a cosine decay learning rate scheduler and 3 epochs of linear warm-up. A batch size of 32, an initial learning rate of 2e-5, and a weight decay of 1e-2 are adopted. The number of encoder layer L is 12 and decoder layer K is 3. The balance loss parameter α is set to 0.5.

4.3   Cross-database Testing

Table 2. Intra testing results on Oulu-NPU four protocols.

Protocol	Method	APCER(%)	BPCER(%)	ACER(%)
1	Face-des [13]	1.2	1.7	1.5
	STASN [23]	1.2	2.5	1.9
	Auxiliary [16]	1.6	1.6	1.6
	CDCN [25]	0.4	1.7	1.0
	BCN [24]	0.0	1.6	0.8
	DCN [26]	1.3	0.0	0.6
	Ours	0.9	0.1	**0.5**
2	Face-des [13]	4.2	4.4	4.3
	STASN [23]	4.2	0.3	2.2
	Auxiliary [16]	2.7	2.7	2.7
	CDCN [25]	1.5	1.4	**1.5**
	BCN [24]	2.6	0.8	1.7
	DCN [26]	2.2	2.2	2.2
	Ours	3.0	0.3	1.7
3	Face-des [13]	4.0±1.8	3.8±1.2	3.6±1.6
	STASN [23]	4.7±3.9	0.9±1.2	2.8±1.6
	Auxiliary [16]	2.7±1.3	3.1±1.7	2.9±1.5
	CDCN [25]	2.4±1.3	2.2±2.0	2.3±1.4
	BCN [24]	2.8±2.4	2.3±2.8	2.5±1.1
	DCN [26]	2.3±2.7	1.4±2.6	1.9±1.6
	Ours	1.4±1.0	1.9±3.5	**1.7±1.5**
4	Face-des [13]	5.1±6.3	6.1±5.1	5.6±5.7
	STASN [23]	6.7±10.6	8.3±8.4	7.5±4.7
	Auxiliary [16]	5.0±4.5	10.4±6.0	9.5±6.0
	CDCN [25]	4.6±4.6	9.2±8.0	6.9±2.9
	BCN [24]	2.9±4.0	7.5±6.9	5.2±3.7
	DCN [26]	6.7±6.8	0.0±0.0	**3.3±3.4**
	Ours	5.7±4.8	1.5±3.2	3.5±3.4

In this experiment, we use HTER to evaluate the generalization ability of the model. It can be seen from Table 1 that we have achieved very competitive results in both the experiments from CASIA-MFSD to Replay-Attack and the experiments from Replay-Attack to CASIA-MFSD. However, we noticed that our algorithm performance decreased slightly from low resolution dataset (replay attack) to high resolution dataset (CASIA-MFSD). This may be because it is difficult for the vision transformer to capture sufficient information to generate depth when the low resolution image is used as input. As demonstrated in Sect. 4.4, our method achieves good results on high-resolution data sets (Oulu-NPU).

4.4 Intra-database Testing

The Oulu NPU dataset is divided into four protocols to measure the generalization ability of the model. Protocols 1, 2 and 3 were designed to verify the generalization ability of the PAD method under different lighting conditions and backgrounds, different presentation attack in structures (PAI) (e.g., different displays and prints), and different devices, respectively. Protocol 4 takes into account all the above factors and is most similar to the real scenario. As shown in Table 2, our proposed method achieves the best results on Oulu NPU protocols 1 and 3 (0.5% and 1.7%ACER), and is very close to the best results on protocols 2 and 4, which proves the good generalization ability of our method.

4.5 Contrast Experiments

ResFASNet. We used resnet101 [11] loaded with pre-trained model and replaced the final fully connected layer to make it suitable for binary classification.

EfficientFASNet. Efficient net [20] is the most popular and powerful convolutional neural network. It has achieved great success in many tasks. Here we use efficientnet-b4 loaded with the pre-trained model. Similar to ResFASNet, we replaced the final fully connected layer.

ViT-FAS. Here we use ViT-Base [7] and replace the final layer with a new full connection layer. In particular, we load the pre-trained weights "vit-b_16-244" and adjust model to make it available for an image size of 256×256.

Table 3. Ablation study of baseline models.(HTER%)

Method	Train	Test
	CASIA-MFSD	Replay-Attack
ResFASNet	30.3	
EfficientFASNet	28.5	
ViT-FAS	26.5	
DE-ViT	18.9	

Experiment Results. In order to explore the generalization ability of the model, all contrast experiments are conducted on cross testing (train on CASIA-MFSD, test on Replay-Attack). The experimental results are shown in Table 3. ViT-FAS performed better than ResFASNet and EfficientFASNet in cross

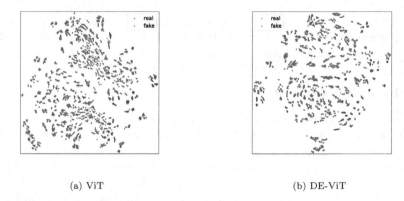

(a) ViT   (b) DE-ViT

Fig. 3. The t-SNE visualizations of the extracted features by (a) vit method and (b) DE-ViT from Replay database, in the cross testing from CASIA to Replay.

domain experiment. This proves that transformer is better than convolutional neural network in FAS task. Our method obtains 19% HTER results, which is 7.5 points lower than ViT-FAS.

Moreover, Fig. 3 shows the t-SNE visualization of 5000 random samples features from Replay database extracted by ViT and DE-ViT. It can be seen that some of the fake face features and real face features extracted by ViT are mixed together. In contrast, the features extracted by our method are more separable. This demonstrate the good generalization ability of our method.

5   Conclusion

In this paper, we use transformer to explore FAS task and show the effectiveness of transformer in FAS task. The proposed method estimates face depth information through encoder-decoder structure, thus the depth information can be considered in the classification. Extensive experiments show superior performance of our method. In the future work, we will explore a more lightweight transformer model for FAS task.

References

1. Boulkenafet, Z., Komulainen, J., Hadid, A.: Face anti-spoofing based on color texture analysis. In: 2015 IEEE International Conference on Image Processing (ICIP), pp. 2636–2640. IEEE (2015)
2. Boulkenafet, Z., Komulainen, J., Hadid, A.: Face spoofing detection using colour texture analysis. IEEE Trans. Inf. Forensics Secur. **11**(8), 1818–1830 (2016)
3. Boulkenafet, Z., Komulainen, J., Li, L., Feng, X., Hadid, A.: OULU-NPU: a mobile face presentation attack database with real-world variations. In: 2017 12th IEEE International Conference on Automatic Face & Gesture Recognition (FG 2017), pp. 612–618. IEEE (2017)

4. Chen, H., et al.: Pre-trained image processing transformer. In: Proceedings of the IEEE/CVF Conference on Computer Vision and Pattern Recognition, pp. 12299–12310 (2021)
5. Chingovska, I., Anjos, A., Marcel, S.: On the effectiveness of local binary patterns in face anti-spoofing. In: 2012 BIOSIG-Proceedings of the International Conference of Biometrics Special Interest Group (BIOSIG), pp. 1–7. IEEE (2012)
6. Chingovska, I., Mohammadi, A., Anjos, A., Marcel, S.: Evaluation methodologies for biometric presentation attack detection. In: Marcel, S., Nixon, M.S., Fierrez, J., Evans, N. (eds.) Handbook of Biometric Anti-Spoofing. ACVPR, pp. 457–480. Springer, Cham (2019). https://doi.org/10.1007/978-3-319-92627-8_20
7. Dosovitskiy, A., et al.: An image is worth 16×16 words: transformers for image recognition at scale. arXiv preprint arXiv:2010.11929 (2020)
8. Feng, Y., Wu, F., Shao, X., Wang, Y., Zhou, X.: Joint 3D face reconstruction and dense alignment with position map regression network. In: Proceedings of the European Conference on Computer Vision (ECCV), pp. 534–551 (2018)
9. de Freitas Pereira, T., Anjos, A., De Martino, J.M., Marcel, S.: Can face anti-spoofing countermeasures work in a real world scenario? In: 2013 International Conference on Biometrics (ICB), pp. 1–8. IEEE (2013)
10. George, A., Marcel, S.: On the effectiveness of vision transformers for zero-shot face anti-spoofing. In: 2021 IEEE International Joint Conference on Biometrics (IJCB), pp. 1–8. IEEE (2021)
11. He, K., Zhang, X., Ren, S., Sun, J.: Deep residual learning for image recognition. In: Proceedings of the IEEE Conference on Computer Vision and Pattern Recognition, pp. 770–778 (2016)
12. Huang, H.P., et al.: Adaptive transformers for robust few-shot cross-domain face anti-spoofing. arXiv preprint arXiv:2203.12175 (2022)
13. Jourabloo, A., Liu, Y., Liu, X.: Face De-spoofing: anti-spoofing via noise modeling. In: Proceedings of the European Conference on Computer Vision (ECCV), pp. 290–306 (2018)
14. Kingma, D.P., Ba, J.: Adam: a method for stochastic optimization. arXiv preprint arXiv:1412.6980 (2014)
15. Komulainen, J., Hadid, A., Pietikäinen, M.: Context based face anti-spoofing. In: 2013 IEEE Sixth International Conference on Biometrics: Theory, Applications and Systems (BTAS), pp. 1–8. IEEE (2013)
16. Liu, Y., Jourabloo, A., Liu, X.: Learning deep models for face anti-spoofing: binary or auxiliary supervision. In: Proceedings of the IEEE Conference on Computer Vision and Pattern Recognition, pp. 389–398 (2018)
17. Paszke, A., et al.: Automatic differentiation in pytorch (2017)
18. Patel, K., Han, H., Jain, A.K.: Secure face unlock: spoof detection on smartphones. IEEE Trans. Inf. Forensics Secur. 11(10), 2268–2283 (2016)
19. Peixoto, B., Michelassi, C., Rocha, A.: Face liveness detection under bad illumination conditions. In: 2011 18th IEEE International Conference on Image Processing, pp. 3557–3560. IEEE (2011)
20. Tan, M., Le, Q.: EfficientNet: rethinking model scaling for convolutional neural networks. In: International Conference on Machine Learning, pp. 6105–6114. PMLR (2019)
21. Vaswani, A., et al.: Attention is all you need. In: Advances in Neural Information Processing Systems, pp. 5998–6008 (2017)
22. Yang, J., Lei, Z., Li, S.Z.: Learn convolutional neural network for face anti-spoofing. arXiv preprint arXiv:1408.5601 (2014)

23. Yang, X., et al.: Face anti-spoofing: model matters, so does data. In: Proceedings of the IEEE/CVF Conference on Computer Vision and Pattern Recognition, pp. 3507–3516 (2019)
24. Yu, Z., Li, X., Niu, X., Shi, J., Zhao, G.: Face anti-spoofing with human material perception. In: Vedaldi, A., Bischof, H., Brox, T., Frahm, J.-M. (eds.) ECCV 2020. LNCS, vol. 12352, pp. 557–575. Springer, Cham (2020). https://doi.org/10.1007/978-3-030-58571-6_33
25. Yu, Z., et al.: Searching central difference convolutional networks for face anti-spoofing. In: Proceedings of the IEEE/CVF Conference on Computer Vision and Pattern Recognition, pp. 5295–5305 (2020)
26. Zhang, K.Y., et al.: Structure destruction and content combination for face anti-spoofing. In: 2021 IEEE International Joint Conference on Biometrics (IJCB), pp. 1–6. IEEE (2021)
27. Zhang, Z., Yan, J., Liu, S., Lei, Z., Yi, D., Li, S.Z.: A face antispoofing database with diverse attacks. In: 2012 5th IAPR International Conference on Biometrics (ICB), pp. 26–31. IEEE (2012)

Dynamically Connected Graph Representation for Object Detection

Shuyu Miao[1]([✉]), Lin Zheng[1], Hong Jin[1], and Rui Feng[2]

[1] Ant Group, Tokyo, China
`miaoshuyu.msy@antgroup.com`
[2] School of Computer Science, Shanghai Key Laboratory of Intelligent Information Processing, Fudan University, Shanghai, China

Abstract. Graph models have been applied to object detection task to implement relational reasoning in many works, achieving a tremendous advancement in performance. However, all these models focus on graph representation with shallow layers and the performance decreases with more graph layers. The most crucial problem lies in that deep graph models encounter the vanishing gradient problem, which causes over-smoothing and no diversity among the features of graph nodes. This paper proposes a novel *Dynamically Connected Graph (DCG)* representation mechanism to break the limitation of the vanishing gradient problem in deep graph-based detection models and make the graph models deeper and better. DCG adaptively learns the feature re-usage and new feature exploration for better optimization of the deep graph. Deep graph models still meet another general problem, i.e., the class imbalance of the original data distribution causes an imbalance in graph node learning. Thus, a *Node Balanced Loss* is further introduced to address this issue by normalizing and reassigning the weights of losses of various classes. Our method can be easily integrated into existing detection models without extra modification. Experiments prove that DCG significantly improves the performance of various baseline models on the challenging MS-COCO and PascalVOC.

Keywords: Object Detection · Vanishing Gradient Problem · Dynamically Connected Graph Representation · Node Balanced Loss

1 Introduction

Object detection aims to analyze and describe the locations and classes of the objects in an image, as an important but challenging task. it benefits many visual tasks or multimedia tasks, like object tracking [12,29,40], temporal action proposal generation [17,21,31], video caption [32,35] and video detection [9,30,39]. Most of the classical object detection models focus on making the feature representation more valuable to boost the performance, while ignoring the powerful reasoning ability when humans see and discriminate the objects in an image.

For example, the co-occurrence probability of *a mouse* and *a computer display* appearing in an image is significantly greater than that of *a tile* and *a computer display*. Fortunately, graph models have aroused wide public concern in object detection to realize the relational reasoning for modeling the relationship between objects due to their strong non-euclidean representation capabilities [5, 23, 33]. However, it can not be ignored that such graph models in these works are mostly shallow with no more than 3 or 4 layers, which is far from the powerful reasoning ability of humans. As the number of graph model layers deepens, the model performance decreases instead. Exploring the deep graph models with better reasoning ability like humans sheds new light on the development of graph-based object detection.

Besides the object detection task, considerable literature has grown up around the theme of modeling graph representation based on shallow layers. Graph Convolutional Network (GCN) [10] trained a graph convolutional model for semi-supervised classification and showed the performance degraded when using more than 3 layers. A multi-label recognition model [4] was proposed based on the graph convolution network for image recognition to learn the potential relationships among image categories. [13] introduced a spatial-temporal graph convolution for skeleton-based action recognition to model the natural connection between two key points. However, all these methods are based on shallow graph models. Several works [16, 37] have explored that deep graph models meet the vanishing gradient problem like primitive CNNs without residual connection [7] or dense connection [8]. It causes over-smoothing during the backpropagation and no diversity among the graph node features [16]. Limited by these issues, most graph-based models are not deeper than 3 or 4 layers. Borrowing concepts from CNNs, residual/dense connections, and dilated convolutions are specifically adopted to design deep graph architectures to alleviate the vanishing gradient problem [14]. Residual connection contributes to feature re-usage, while dense connection facilitates new feature exploration. Enjoying the benefits from both connections would make an important impact on the optimization of deep graphs in object detection.

In this paper, we present a novel *Dynamically Connected Graph*, called **DCG**, to extend the shallow graph models to deep graph models for unstructured graph representation with better reasoning ability and break the limitations of the vanishing gradient problem. DCG models dynamic connection representation for better feature re-usage and new feature exploration via fusing the residual connection and dense connection. It solves the gradient propagation problem of deep graph models successfully and makes the graph models deeper and better. Besides, the class imbalance of the original data distribution leads to the imbalanced learning of the graph node features, which brings that the propagation of many graph nodes is easily ignored. Thus, we further introduce a *Node Balanced Loss* (**NBloss**) to balance the graph optimization. NBloss normalizes and reassigns the weights of losses of various classes, rewards the losses of classes with fewer samples, and restrains the losses of classes with more samples. Based on this, the corresponding graph nodes can be optimized accordingly. Our method is plug and play to existing detection models for better graph relational reasoning. Extensive experiments on the challenging MS-COCO and PascalVOC validate that DCG improves the performance of existing detection models by about 3%. The main contributions of our paper are summarized as:

- A novel *dynamically connected graph* representation mechanism is proposed to address the vanishing gradient problem of deep graph models and facilitate better feature re-usage and new feature exploration.
- *Node balanced loss* is presented to balance the graph optimization brought by the class imbalance of the original data distribution via normalizing and reassigning the weights of losses of various classes.
- *Deep graph representation* is introduced to object detection for the higher ceiling of the graph performance and better relational reasoning like humans *for the first time*, which promotes the detection performance significantly.

2   Related Works

2.1   Graph Representation for Object Detection

Relational reasoning ability of object detection has recently witnessed rapid progress due to the increasing emergence of graph representation models [5,15,16,22,23,33,37]. A class-based graph representation [23] was proposed to learn the relation between various classes to make the object detection have relation reasoning ability. Xu et al. [33] introduced a spatial-aware graph relation network to properly encode high-order object relations and leverage the information between co-occurrence and locations of objects for better relational reasoning. The idea of modeling the relationships between proposals for object detection from the graph learning perspective was explored [5]. However, these works are based on the shallow graph models, and the performance degrades with more graph layers. As pointed out by the research works [16,37], deep graph models meet the vanishing gradient problem. Thus, our work attempts to provide new insights into designing deep graph models far away from the vanishing gradient problem in object detection for the first time.

2.2   Vanishing Gradient Problem

Vanishing gradient problem exists not only in deep graph models but also in the early days of CNNs [28,34]. In the early development of CNNs, with the increase of network depth, the performance of the model is greatly limited due to vanishing gradient. ResNet [7] innovatively introduced an identity shortcut connection, which skipped one or more layers directly and moved a big step forward in the design of deep CNNs. It broke the constraints of vanishing gradient in deep models and made better feature re-usage through identity mapping. DenseNet [8] utilized a dense connection to directly connect all layers. In this architecture, the input feature of each layer was composed of all the previous layer feature maps, and its output feature was transmitted to each subsequent layer. It alleviated the problem of vanishing gradient, strengthened feature propagation, and enabled new feature exploration. A highly efficient and modularized dual-path network was further presented [3] to explore more effective connection styles. Inspired by these CNN concepts, we consider merging the residual and dense connection in one block to learn dynamically connected graph representation to make graph models much deeper. We focus on designing deeper graph models for existing detection models and significantly improving their performance.

2.3 Imbalance in Object Detection

One of the greatest challenge of object detection is that the samples are imbalanced. An Online Hard Example Mining (OHEM) algorithm [27] was proposed to automatically select hard examples to join in the training process, so that the training was more effective and fast. Although OHEM increased the weights of misclassified samples, it ignored the samples that were easy to be classified. To address this issue, Focal loss [18] reduced the weight of easy-to-classify samples, making the model focus more on hard-to-classify samples during training. Easy-to-classify samples should not be paid too much attention, but particularly hard-to-classify samples (outliers) also should not be paid too much attention. Gradient Harmonizing Mechanism (GHM) [1] balanced the sample weights based on sample density to balance various kinds of samples. However, all the above methods rely on the manual adjustment of hyper-parameters. Chen et al. [2] introduced the sample-free-based loss mechanism by combining optimal bias initialization, guided loss, and class-adaptive threshold. In this paper, we point out that class imbalance leads to an imbalance of graph node representation, causing partial graph nodes to be ignored. To solve this problem, we propose a dedicated weight balanced loss for balancing the optimization of the graph model. Extensive experiments validate the effectiveness of our proposed loss function.

3 Methodology

3.1 Overview of Our Proposed DCG

The overall architecture flow of DCG-based detection models is shown in Fig. 1. The original object detection model and our proposed DCG are decoupled. It means DCG can be easily embedded in these models to improve detection performance without extra modification. The existing detection models can be roughly divided into feature extraction backbone and detection head process, and DCG can be seamlessly inserted between such two processes.

3.2 Initial Definition of Graph Representation

Graph Definition. A graph $\mathcal{G}$ is defined by a tuple $\mathcal{G} := (\mathcal{V}, \mathcal{E})$. $\mathcal{V} := \{v_i\}_{i=1:N^v}$ represents the set of graph nodes, where N^v is the number of graph nodes.

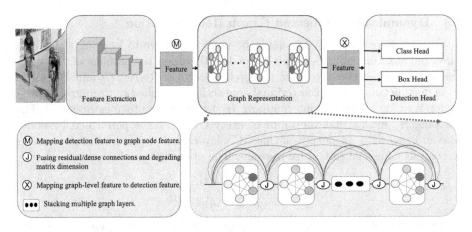

Fig. 1. An overview of our novel *dynamically connected graph* representation mechanism for object detection. Our proposed DCG is inserted between the feature extraction backbone and detection head of the detection model. *(top)* The overall detection framework includes feature extraction backbone, graph representation, and detection head. *(bottom)* Dynamically connected graph representation is a specific form of graph representation.

$\mathcal{E} := \{e_{a,b}^j\}_{j=1:N^e}$ represents the set of graph edges, where N^e is the number of graph edges and $e_{a,b} \in \mathcal{E}$ denotes the nodes v_a and v_b are connected by an edge $e_{a,b}$. Node features and edge information will be updated in each iteration.

Graph Convolution Network. Our **DCG** can be applied to different graph-based detection models like [23] by extending shallow graph models to deep graph models. We take GCN [10] as an example. GCN aims to extract richer features from graph data at a node by aggregating the features of its neighborhood nodes and updating self-features via learning a non-linear function for optimizing the graph $\mathcal{G}$. The feature of every node v_i is described as $x_i \in \mathbb{R}^D$, where D is the dimension of the node features. All the node features are stacked by $X = [x_{v_1}, x_{v_2}, \cdots, x_{v_{N^v}}]^T \in \mathbb{R}^{N^v \times D}$. The graph structure is described in the form of an adjacency matrix $\mathcal{A} \in \mathbb{R}^{N^v \times N^v}$. A node-level output $\mathcal{Z} \in \mathbb{R}^{N^v \times E}$ is produced by the graph, where E is the feature dimension of the graph output. The propagation process of graph network layer can be formulated as the following non-linear function:

$$\mathcal{G}^{(l)} = \mathcal{F}(\mathcal{G}^{(l-1)}, \mathcal{W}) \quad s.t. \quad l \in \{0, 1, 2, \cdots, \mathcal{L}\}, \tag{1}$$

where $\mathcal{L}$ is the number of the layers of the graph; $\mathcal{G}^{(l)} \in \{\mathcal{V}_l, \mathcal{E}_l\}$ and $\mathcal{G}^{(l+1)} \in \{\mathcal{V}_{l+1}, \mathcal{E}_{l+1}\}$ are the input and output of the l-th graph layer; $\mathcal{W}$ is the learnable parameter of the graph; and $\mathcal{V}_0 = X$, $\mathcal{E}_0 = \mathcal{A}$, and $\mathcal{Z} = \mathcal{V}_{\mathcal{L}}$. Furthermore, limited by the vanishing gradient problem, it is no more than 3 or 4 layers in most graph models. Thus, we explore to expand the graph models much deeper as 28 or 56 like [7,8].

3.3 Dynamically Connected Graph Representation

Intuitively, more graph layers should be more conducive to learning richer features, but stacking multiple graph layers directly brings the problem of vanishing gradient [16,37]. Benefiting from the great success of ResNet [7] and DenseNet [8] in deep networks [14], the residual connection is applied to deep graph models for learning an underlying mapping $\mathcal{H}_1$ by fitting another mapping $\mathcal{F}$. The mapping $\mathcal{H}_1$ means that the input of every GCN layer is performed by node-wise addition with the output of this GCN layer. Such a graph residual learning strategy is beneficial to feature re-usage. The operation can be written as:

$$
\begin{aligned}
\mathcal{G}^{(l+1)} &= \mathcal{H}_1(\mathcal{G}^{(l)}, \mathcal{A}) \\
&= \mathcal{F}(\mathcal{G}^{(l)}, \mathcal{A}) + \mathcal{G}^{(l)}.
\end{aligned}
\tag{2}
$$

Dense connection is adopted to deep graph networks for exploiting information flow from different GCN layers, which is operated as the underlying mapping $\mathcal{H}_2$. The mapping denotes that every graph layer output is concatenated with the input graph $\mathcal{G}^{(0)}$ and all the intermediate GCN layer outputs. The operation can be written as:

$$
\begin{aligned}
\mathcal{G}^{(l+1)} &= \mathcal{H}_2(\mathcal{G}^{(l)}, \mathcal{A}) \\
&= T(\mathcal{F}(\mathcal{G}^{(l)}, \mathcal{A}), \mathcal{G}^{(l)}) \\
&= T(\mathcal{F}(\mathcal{G}^{(l)}, \mathcal{A}), \cdots, \mathcal{F}(\mathcal{G}^{(1)}, \mathcal{A}), \mathcal{G}^{(0)}),
\end{aligned}
\tag{3}
$$

where T is node-wise concatenation.

Based on the above two connections, we further propose *dynamically connected graph* representation as the underlying mapping $\mathcal{H}$. Residual connection and dense connection are merged to attack the vanishing gradient problem. To this end, the graph-level outputs of $\mathcal{G}^{(l+1)}$ obtained by $\mathcal{H}_1$ and $\mathcal{H}_2$ are operated by the node-wise concatenation function. The mapping dynamically learns the free combination of long-connection graph representation and short-connection graph representation, maximizes the retention of valuable features, and explores potential feature relations. Therefore, $\mathcal{G}^{(l+1)}$ can be gained by:

$$
\begin{aligned}
\mathcal{G}^{(l+1)} &= \mathcal{H}(\mathcal{G}^{(l)}, \mathcal{A}) \\
&= T(\mathcal{F}(\mathcal{G}^{(l)}, \mathcal{A}) + \mathcal{G}^{(l)}, T(\mathcal{F}(\mathcal{G}^{(l)}, \mathcal{A}), \mathcal{G}^{(l)})) \\
&= T(\mathcal{F}(\mathcal{G}^{(l)}, \mathcal{A}) + \mathcal{G}^{(l)}, T(\mathcal{F}(\mathcal{G}^{(l)}, \mathcal{A}), \cdots, \mathcal{A}), \mathcal{G}^{(0)}).
\end{aligned}
\tag{4}
$$

Simply merging two outputs would cause a sharp increase in the amount of computation. The main reason is that dense connection leads to a higher matrix dimension. To alleviate this issue, a special linear function is designed to degrade the matrix dimension as the original matrix dimension, which is a fully connected function denoted as $\mathcal{J}(\cdot)$. We define a learnable parameter $W \in \mathbb{R}^{a \times b}$, where a is the original dimension of the matrix, b is the degraded dimension of the matrix, and $a > b$. Therefore, we expand Eq. (4) to Eq. (5).

$$
\begin{aligned}
\mathcal{G}^{(l+1)} &= \mathcal{J}(\mathcal{G}^{(l+1)}) \\
&= \mathcal{G}^{(l+1)} \cdot W
\end{aligned}
\tag{5}
$$

$\mathcal{J}(\cdot)$ reduces the complexity of the model and increases the training speed. The performance of the model is unaffected because it retains the most effective features based on full feature mapping. Because all the operations are based on matrices, the weights are shared by the adjacency matrix, and the unique design of $\mathcal{J}(\cdot)$ reduces the matrix dimension, this strategy has a lower computation cost compared with the simple residual and dense connection.

3.4   Node Balanced Loss

Because each graph node corresponds to one class, the sample distribution of the dataset plays a great impact on the learning of node features. The dataset $\mathcal{D}$ is defined by $\mathcal{D} := \{d_i\}_{i=1:N^d}$, where d_i is the i-th class and N^d is the number of classes in $\mathcal{D}$. Most of the numbers of samples of various classes $\{d_1, d_2, \cdots, d_{N^d}\}$ are imbalanced, which weakens the representation of some graph nodes with fewer samples. Node Balanced loss (NBloss) is proposed to tackle this issue. $\mathcal{L}_{d_i}$ denotes the classification loss of the i-th class, mapping the graph node representation. The total classification loss can be formulated as:

$$\mathcal{L} = \frac{1}{[\mathcal{D}]} \sum_{i=1}^{N^d} \mathcal{L}_{d_i}. \tag{6}$$

To address the problem of graph node representation imbalance due to class imbalance, we normalize and reassign the weights of various class losses. $[d_i]$ denotes the number of the i-th class samples. α_i is the weight of the i-th loss of class classification, which is firstly obtained by $\frac{[\mathcal{D}]}{[d_i]}$. To make the model robust, the mean value and standard deviation of $\{\alpha_i\}_{i=1:N^d}$ are calculated by $mean(\cdot)$ and $std(\cdot)$. The final weight α_i can be obtained as:

$$\alpha_i = \frac{\frac{[\mathcal{D}]}{[d_i]} - mean(\frac{[\mathcal{D}]}{[d_i]})}{std(\frac{[\mathcal{D}]}{[d_i]})}. \tag{7}$$

Furthermore, α_i is performed by $softmax(\cdot)$ to get the final weights, which is formulated as:

$$\alpha_i = \frac{exp(\alpha_i)}{\sum_{i=1}^{N^d} exp(\alpha_i)}, \tag{8}$$

where $\{\alpha_i\}_{i=1:N^d}$ corresponds to $\{\mathcal{L}_{d_i}\}_{i=1:N^d}$. By our NBloss function, the influence of the classes with a large number of samples will be suppressed, and the influence of the classes with a small number of samples will be encouraged during training. The final loss $\mathcal{L}$ in Eq. (6) can be reformulated as:

$$\mathcal{L} = \frac{1}{[\mathcal{D}]} \sum_{i=1}^{N^d} \alpha_i \mathcal{L}_{d_i}. \tag{9}$$

Table 1. Experiments of our DCG embedded into YOLOv3 and Libra R-CNN on MS-COCO test-dev set.

Method	Backbone	AP	AP_{50}	AP_{75}	AP_S	AP_M	AP_L
Faster R-CNN [26]	VGG16	21.9	42.7	-	-	-	-
RFB Net512 [19]	VGG16	33.8	54.2	35.9	16.2	37.1	47.4
Refinedet512 [36]	VGG16	33.0	54.5	35.5	16.3	36.3	44.3
RetinaNet800 [18]	ResNet101-FPN	39.1	59.1	42.3	21.8	42.7	50.2
FSAF [38]	ResNet101	–	54.0	33.6	17.8	35.4	46.5
YOLOv3 [25]	Darknet53	33.0	57.9	34.4	18.3	35.4	41.9
Libra R-CNN [24]	ResNet50-FPN	38.7	59.9	42.0	22.5	41.1	48.7
YOLOv3 [25]+ClassGCN28 [23]	Darknet53	33.6	58.5	35.1	19.0	36.0	42.3
YOLOv3+DCG28	Darknet53	35.3	60.3	36.6	20.8	37.5	43.9
YOLOv3+DCG56	Darknet53	36.0	60.8	37.6	21.6	38.5	44.4
Libra R-CNN [24]+ClassGCN28 [23]	ResNet50-FPN	39.2	60.6	42.3	23.3	41.5	49.2
Libra R-CNN+DCG28	ResNet50-FPN	41.0	62.2	44.0	25.0	43.4	51.1
Libra R-CNN+DCG56	ResNet50-FPN	**41.8**	**62.6**	**44.9**	**26.0**	**44.2**	**51.8**

4 Experiments

4.1 Datasets

MS-COCO and *PascalVOC* are adopted in experiments. To verify the effectiveness of our method, we choose classic detection models as our baselines, like RefineDet [36], YOLOv3 [25] and Libra R-CNN [24]. Objectively, the same strategies are used as baselines.

4.2 Overall Performance

Experimental Results on MS-COCO. It can be seen from Table 1, that rigidly stacking graph models into deeper networks (+ClassGCN) only improves the performance by a little bit because of vanishing gradient problem. Contrarily, for the YOLOv3 model with DCG, we can obtain 2.3%, 2.4%, and 2.2% improvement of AP, AP_{50}, and AP_{75} separately with 28 graph layers. When the number of graph layers is 56, AP, AP_{50}, and AP_{75} are significantly elevated by 3.0%, 2.9%, and 3.2%. For the Libra R-CNN model, DCG boosts the AP, AP_{50}, and AP_{75} values by 2.3%, 2.3%, and 2.0% with 28 graph layers, and 3.1%, 2.7%, and 2.9% with 56 graph layers.

Experimental Results on PascalVOC. Table 2 shows the experimental results of our method on PascalVOC. We validate the performance of DCG with the baseline RefineDet [36]. Closer inspection of the table shows that for the 320×320 input size, DCG significantly improves the mAP of RefineDet by 1.3% from 80.0% to 81.3% with 28 graph layers, and 1.8% from 80.0% to 81.8% with 56 graph layers. For the 512×512 input size, the mAP is increased by 1.2% from 81.8% to 83.0% with 28 graph layers, and 1.8% from 81.8% to 83.6%. The results fully demonstrate the effectiveness of our method. and indicate that the

Table 2. Experiments of our DCG embedded into RefineDet on the *PascalVOC*2007 test.

Method	Backbone	mAP
Fast R-CNN [6]	VGG16	70.0
Faster R-CNN [26]	VGG16	73.2
RON384++ [11]	VGG16	77.6
SSD300 [20]	VGG16	77.5
RFBNet300 [19]	VGG16	80.5
SSD512 [20]	VGG16	79.5
RFBNet512 [19]	VGG16	82.2
RefineDet320 [36]	VGG16	80.0
RefineDet512 [36]	VGG16	81.8
RefineDet320+DCG28	VGG16	81.3
RefineDet320+DCG56	VGG16	81.8
RefineDet512+DCG28	VGG16	83.0
RefineDet512+DCG56	VGG16	**83.6**

Table 3. Experiments on Node Balanced loss.

Component	AP	AP_{50}	AP_{75}	AP_S	AP_M	AP_L
w/o NBloss	34.7	59.6	36.0	20.2	37.1	43.8
w/o $softmax(\cdot)$	35.1	59.8	36.9	20.7	37.7	44.0
w NBloss	36.0	60.8	37.6	21.6	38.5	44.4

proposed DCG is effective for multi-scale inputs. The vanishing gradient problem in deep graph models has been well solved by our dynamically connected graph representation.

4.3   Ablation Study

To verify that every component of our method is effective, we conduct detailed ablation studies. The training is on the MS-COCO trainval35k set, and the testing on the MS-COCO test-dev set with the YOLOv3 model with 56 graph layers.

Experiment on Node Balanced Loss. Node Balanced loss (NBloss) aims to balance the graph propagation, due to the imbalance of class samples in the data distribution. The related experimental results are shown in Table 3. It can be viewed that when adopting DCG without NBloss, the AP value is decreased by 1.3% from 36.0% to 34.7%; and when utilizing DCG with WRloss but no $softmax(\cdot)$, the value AP is reduced by 0.9% from 36.0% to 35.1%.

Table 4. Experiments on different connections.

Component	AP	AP_{50}	AP_{75}	AP_S	AP_M	AP_L
Baseline [25]	33.0	57.9	34.4	18.3	35.4	41.9
w/o residual/dense [23]	33.2	58.0	34.6	18.6	35.5	42.1
w residual	34.9	58.8	35.8	19.8	37.5	43.6
w dense	34.8	58.5	36.0	19.6	37.5	43.7
w/o $\mathcal{J}(\cdot)$	36.2	60.9	37.8	21.7	38.6	44.7
w $\mathcal{J}(\cdot)$	36.0	60.8	37.6	21.6	38.5	44.4

Experiment on Different Connections. We verify the performance of the graph model by different connections, and validate the rationality of the operation of degrading the matrix dimension in Eq. (5) that alleviates the problem of a sharp increase in computation amount caused by the increase of matrix dimensions. The related experimental results are shown in Table 4. It can be seen from the table that rigidly stacking graph models into deeper networks by [23] only improves the performance by a little bit because of the vanishing gradient problem. The table presents that the performance of DCG is better than that with the residual connection or dense connection. When DCG with or without degrading the matrix dimension operation, the *AP* values are almost equal.

5 Conclusion

In this paper, considering that the vanishing gradient problem restricts the development of deep graph models in object detection, a novel Dynamically Connected Graph (DCG) representation mechanism is proposed to make the graph models as deeper as CNNs with better reasoning ability like humans. Our DCG also contributes to feature re-usage and new feature exploration benefiting from dynamic long and short connections. We further present a Node Balanced loss to address the issue of the imbalance of graph node feature representation caused by the class imbalance of the original data distribution. Our method can be easily embedded with existing detection models to make them have better relational reasoning ability with deeper graph layers for the first time.

References

1. Buyu, L.I., Liu, Y., Wang, X.: Gradient harmonized single-stage detector. In: AAAI, pp. 8577–8584 (2019)
2. Chen, J., et al.: Is sampling heuristics necessary in training deep object detectors? (2019)
3. Chen, Y., Li, J., Xiao, H., Jin, X., Yan, S., Feng, J.: Dual path networks. In: NeurIPSW, pp. 4467–4475 (2017)

4. Chen, Z.M., Wei, X.S., Wang, P., Guo, Y.: Multi-label image recognition with graph convolutional networks. In: IEEE Conference on Computer Vision and Pattern Recognition (2019)
5. Du, X., Shi, X., Huang, R.: Repgn: object detection with relational proposal graph network. arxiv (2019)
6. Girshick, R.B.: Fast R-CNN. In: IEEE International Conference on Computer Vision (ICCV), pp. 1440–1448 (2015)
7. He, K., Zhang, X., Ren, S., Sun, J.: Deep residual learning for image recognition. In: IEEE Conference on Computer Vision and Pattern Recognition, pp. 770–778 (2016)
8. Huang, G., Liu, Z., Der Maaten, L.V., Weinberger, K.Q.: Densely connected convolutional networks. In: IEEE Conference on Computer Vision and Pattern Recognition, pp. 2261–2269 (2017)
9. Kang, K., Ouyang, W., Li, H., Wang, X.: Object detection from video tubelets with convolutional neural networks. In: IEEE Conference on Computer Vision and Pattern Recognition, pp. 817–825 (2016)
10. Kipf, T.N., Welling, M.: Semi-supervised classification with graph convolutional networks. In: International Conference on Learning Representations (2017)
11. Kong, T., Sun, F., Yao, A., Liu, H., Lu, M., Chen, Y.: Ron: reverse connection with objectness prior networks for object detection. In: IEEE Conference on Computer Vision and Pattern Recognition, pp. 5244–5252 (2017)
12. Li, B., Yan, J., Wu, W., Zhu, Z., Hu, X.: High performance visual tracking with siamese region proposal network. In: CVPR (2018)
13. Li, C., Cui, Z., Zheng, W., Xu, C., Yang, J.: Spatio-temporal graph convolution for skeleton based action recognition. In: Proceedings of the Thirty-Second AAAI Conference on Artificial Intelligence, pp. 3482–3489 (2018)
14. Li, G., Muller, M., Thabet, A.K., Ghanem, B.: DeepGCNs: can GCNs go as deep as CNNs? In: IEEE International Conference on Computer Vision, pp. 9267–9276 (2019)
15. Li, H., Miao, S., Feng, R.: DG-FPN: learning dynamic feature fusion based on graph convolution network for object detection. In: 2020 IEEE International Conference on Multimedia and Expo (ICME), pp. 1–6. IEEE (2020)
16. Li, Q., Han, Z., Wu, X.: Deeper insights into graph convolutional networks for semi-supervised learning. In: Proceedings of the Thirty-Second AAAI Conference on Artificial Intelligence, pp. 3538–3545 (2018)
17. Lin, T., Liu, X., Li, X., Ding, E., Wen, S.: BMN: boundary-matching network for temporal action proposal generation. In: IEEE International Conference on Computer Vision, pp. 3889–3898 (2019)
18. Lin, T., Goyal, P., Girshick, R.B., He, K., Dollár, P.: Focal loss for dense object detection. In: IEEE International Conference on Computer Vision, pp. 2999–3007 (2017)
19. Liu, S., Huang, D., Wang, Y.: Receptive field block net for accurate and fast object detection. In: ECCV, pp. 404–419 (2018)
20. Liu, W., et al.: SSD: single shot multibox detector. In: IEEE Conference on Computer Vision and Pattern Recognition, pp. 21–37 (2015)
21. Liu, Y., Ma, L., Zhang, Y., Liu, W., Chang, S.F.: Multi-granularity generator for temporal action proposal. In: IEEE Conference on Computer Vision and Pattern Recognition, pp. 3604–3613 (2019)
22. Miao, S., Feng, R.: Object-oriented relational distillation for object detection. In: ICASSP 2021–2021 IEEE International Conference on Acoustics, Speech and Signal Processing (ICASSP), pp. 1510–1514. IEEE (2021)

23. Miao, S., Feng, R., Zhang, Y., Fan, W.: Learning class-based graph representation for object detection. In: European Conference on Artificial Intelligence, vol. 325, pp. 2752–2759 (2020)

24. Pang, J., Chen, K., Shi, J., Feng, H., Ouyang, W., Lin, D.: Libra R-CNN: towards balanced learning for object detection. In: IEEE Conference on Computer Vision and Pattern Recognition, pp. 821–830 (2019)

25. Redmon, J., Farhadi, A.: Yolov3: an incremental improvement (2018). https://pjreddie.com/media/files/papers/YOLOv3.pdf

26. Ren, S., He, K., Girshick, R.B., Sun, J.: Faster R-CNN: towards real-time object detection with region proposal networks. In: Conference and Workshop on Neural Information Processing Systems (NeurIPS), pp. 1–10 (2015)

27. Shrivastava, A., Gupta, A., Girshick, R.: Training region-based object detectors with online hard example mining. In: IEEE Conference on Computer Vision and Pattern Recognition, pp. 761–769 (2016)

28. Simonyan, K., Zisserman, A.: Very deep convolutional networks for large-scale image recognition. In: IEEE Conference on Computer Vision and Pattern Recognition (2014)

29. Wang, Q., Zhang, L., Bertinetto, L., Hu, W., Torr, P.H.: Fast online object tracking and segmentation: a unifying approach. In: IEEE Conference on Computer Vision and Pattern Recognition (2019)

30. Wang, S., Zhou, Y., Yan, J., Deng, Z.: Fully motion-aware network for video object detection. In: European Conference on Computer Vision, pp. 542–557 (2018)

31. Wang, X., Zhang, S., Qing, Z., Shao, Y., Gao, C., Sang, N.: Self-supervised learning for semi-supervised temporal action proposal (2021)

32. Tang, X., Gao, X., Liu, J., Zhang, H.: A spatial-temporal approach for video caption detection and recognition. IEEE Trans. Neural Netw. **13**(4), 961–971 (2002). https://doi.org/10.1109/TNN.2002.1021896

33. Xu, H., Jiang, C., Liang, X., Li, Z.: Spatial-aware graph relation network for large-scale object detection. In: IEEE Conference on Computer Vision and Pattern Recognition, pp. 9298–9307 (2019)

34. Zeiler, M.D., Fergus, R.: Visualizing and understanding convolutional networks. In: European Conference on Computer Vision (2013)

35. Zhang, D., Chang, S.F.: Event detection in baseball video using superimposed caption recognition. In: Proceedings of the Tenth ACM International Conference on Multimedia, pp. 315–318. MULTIMEDIA 2002 (2002)

36. Zhang, S., Wen, L., Bian, X., Lei, Z., Li, S.Z.: Single-shot refinement neural network for object detection. In: IEEE Conference on Computer Vision and Pattern Recognition, pp. 4203–4212 (2018)

37. Zhou, J., et al.: Graph neural networks: a review of methods and applications. arXiv (2018)

38. Zhu, C., He, Y., Savvides, M.: Feature selective anchor-free module for single-shot object detection. In: IEEE Conference on Computer Vision and Pattern Recognition, pp. 840–849 (2019)

39. Zhu, X., Dai, J., Yuan, L., Wei, Y.: Towards high performance video object detection. In: IEEE Conference on Computer Vision and Pattern Recognition, pp. 7210–7218 (2018)

40. Zhu, Z., Wang, Q., Bo, L., Wu, W., Yan, J., Hu, W.: Distractor-aware siamese networks for visual object tracking. In: European Conference on Computer Vision (2018)

Multi-Class Anomaly Detection

Suresh Singh$^{(\boxtimes)}$ (iD), Minwei Luo, and Yu Li

Portland State University, Portland 97207, OR, USA
singh@cs.pdx.edu
http://www.cs.pdx.edu/ singh

Abstract. We study anomaly detection for the case when the *normal* class consists of more than one object category. This is an obvious generalization of the standard one-class anomaly detection problem. However, we show that jointly using multiple one-class anomaly detectors to solve this problem yields poorer results as compared to training a single one-class anomaly detector on all normal object categories together. We further develop a new anomaly detector called DeepMAD that learns compact distinguishing features by exploiting the multiple normal objects categories. This algorithm achieves higher AUC values for different datasets compared to two top performing one-class algorithms that either are trained on each normal object category or jointly trained on all normal object categories combined. In addition to theoretical results we present empirical results using the CIFAR-10, fMNIST, CIFAR-100, and a new dataset we developed called RECYCLE.

Keywords: Out of Distribution · Anomaly · Multi-Class Anomaly

1 Introduction

To motivate this formulation of anomaly detection, let us consider several applications: imagine a roadside garbage container that automatically separates any discarded trash into recycles (glass, cans, plastic, etc.) or garbage; machines used to separate normal blood cells (monocytes, neutrophils, basophils, etc.) from bacteria in the blood (sepsis); generic classification problems such as separating all species of cats from any other animal; separating sounds of different mosquito species from any other sound; and many, many more. In all these examples, the normal class in fact consists of several different object categories.

It is trivial to see that one-class anomaly detection algorithms [2,10,11,13, 17,20,26,28,30,33] can be easily applied to this problem. In a typical one-class formulation, the classifier is only trained on in-distribution samples and it learns a probability density function P that captures normal behavior. Points that then map to a low probability region are classified as anomalous. Generalizing to the case when m different object categories are all considered normal, we can train m separate one-class anomaly detectors $P_1, \cdots, P_m$ and use them together to classify any new object. However, we show that this approach is not as good as

M. Tanveer et al. (Eds.): ICONIP 2022, LNCS 13625, pp. 359–371, 2023.
https://doi.org/10.1007/978-3-031-30111-7_31

training a single one-class anomaly detector on all the m categories combined. This result is demonstrated empirically as well as theoretically. The intuition is that errors in classification of each of the m one-class detectors are cumulative thus resulting in poor joint performance.

Novelty detection [19,24,34] and out-of-distribution (OOD) detection [12, 15,16,18,25,35] are related to anomaly detection but our model has significant differences. A common model used in OOD is to consider, for example, the fMNIST data set as normal and MNIST as OOD or CIFAR-10 as normal and natural images as OOD, etc. This model for defining what constitutes OOD has recently come under criticism [3,25]. As [3] notes, *different datasets are created and curated differently* and thus it is likely that OOD algorithms are learning to identify these idiosyncrasies rather than meaningful features. Similarly, they argue that the context of the OOD task also matters – what is OOD in one context may not be OOD in another. [25] also argues for more realistic benchmarks for OOD (they use a bacteria genome database) observing that distributional shifts learned in traditional OOD tasks may have learned the background and thus have high error when used for realistic tasks. Finally, [3] proposes testing OOD algorithms by sticking to the same dataset. They study the case when 9 out of 10 CIFAR-10 classes are normal and the tenth is treated as anomalous. We note that our model follows this approach. We study cases when 2, 5, 9 classes of CIFAR-10, MNIST, and fMNIST are normal while the others are anomalous.

In novelty detection, a data point is considered an outlier if it differs significantly from a collection of normal data points (e.g., a mammogram with a lump is an outlier in a collection of normal mammograms). The outlier and the normal data points can be seen as being similar and what distinguishes the outlier is its distance from a majority of normal points. A multi-class novelty model is an extension where the number of outliers is large enough to form a separate cluster but all the points are still similar. *In our multi-class model, on the other hand, the normal classes can be arbitrary* (e.g., the normal classes may be dog and truck images).

There are two variations of the multi-class anomaly detection problem – the class labels of normal samples may either be unknown or known during training. We call these two formulations *inseparable* and *separable* respectively and we study both in this paper. The major results of our paper can be summarized as follows:

- We prove that training a one-class classifier in the inseparable case yields a higher AUC (Area Under the Curve) value than training multiple single-class classifiers for the separable case and combining their outputs. Our proof uses the formulation for combining probabilities from the theory of *belief functions*, Lemma 1.
- We illustrate these results using two recent one-class anomaly detection algorithms DROCC [11] and DeepSVDD [27] for the task of multi-class anomaly detection. We did not use OOD algorithms such as [12] because, as noted previously [3], these algorithms train on one dataset and test on another and likely learn dataset idiosyncrasies rather than meaningful image features.

– We present a new algorithm called DeepMAD for the *separable case* and compare the accuracy of DeepMAD and two versions of each of DROCC and DeepSVDD (in one version we train m separate one-class classifiers and in the other we train a single one-class classifier by combining training data from all m normal classes). These five algorithms are compared using CIFAR-10, fMNIST, CIFAR-100, and RECYCLE. The RECYCLE dataset is one we created to study the problem of classifying recycles described in the first paragraph of this section. We show that DeepMAD performs well above all other algorithms by a wide margin.

2   Related Work

A variety of one-class novelty detection algorithms exist [1,5,19,23,29,36] but, as noted previously [21], these algorithms are not directly applicable to the single-class anomaly detection problem. In the latter case we assign labels whereas in the former we learn a latent space that separates the normal class effectively from outliers. There have been a few additional works that consider a multi-class novelty detection problem [22] where the dataset is split into two subsets of classes with one treated as normal. A large reference labeled dataset is used in conjunction with normal samples during training. Our problem formulation differs from this in that we do not use any reference dataset as a comparator to extract features. We also consider cases when the split of normal/anomaly classes is variable.

In OOD detection, out of distribution samples are detected via the basic predictive confidence of the classifier (MSP score) [12] or measures such as temperature-scaled softmax scores (ODIN) [18] or confidence calibration [8]. Approaches as in [13] use other datasets as proxies for OOD examples while [16] uses a GAN to generate negative examples for training. Other methods to improve the representation of in-distribution data includes using self-supervised approaches such as contrastive learning [5], alternative training strategies such as margin loss [35], or metric learning [19]. [4] splits normal data into M and learns features to separate them.

Recently, deep learning based methods have been used for anomaly detection. For instance [26,27] describes DeepSVDD in which a deep neural network is trained to map the in-class data into a sphere of minimal volume, while [9] maps to multiple spheres. DROCC [11] trains the network to distinguish manufactured out of distribution points that are perturbations of in-distribution points as a way to learn more discriminating features. Another approach has been to train auto encoders to learn a lower dimensional representation [17,20,30]. An entirely different way of looking at the problem is to use self-supervision by learning *transformations* where a new sample is classified as normal if transforms applied to it can be correctly identified [10,14,32].

3 Generalizing Single-Class Anomaly Detection to the Multi-Class Case

Table 1. Probability estimates for sample $x \in \mathcal{X}$ provided by m classifiers.

	$\{1\}$	$\{2\}$	$\cdots$	$\{m\}$	$\neg\{1\}$	$\neg\{2\}$	$\cdots$	$\neg\{m\}$
$(\mathcal{F}_1, D_1)$	$P_1(1\|x)$	0	$\cdots$	0	$1 - P_1(1\|x)$	0	$\cdots$	0
.	.	.	$\cdots$	.	.	.	$\cdots$	.
$(\mathcal{F}_m, D_m)$	0	0	$\cdots$	$P_m(m\|x)$	0	0	$\cdots$	$1 - P_m(m\|x)$

Typically, anomaly detection methods learn a mapping $\mathcal{F}$ from the input space $\mathcal{X}$ to some lower dimensional feature space $\mathcal{Y}$. Inputs are then assigned a probability of being normal or anomalous via some function D. $\mathcal{F}$ is learned using the normal class examples while D may be static (e.g., a distance measure of points in feature space) or jointly learned. If we consider using one-class anomaly detectors for the multi-class case where there are $1 \leq i \leq m$ normal classes, for each class we learn one classifier giving us a set $\{(\mathcal{F}_i, D_i)\}_{i=1}^{m}$ of one-class anomaly detectors. The question then is how to use these m detectors for the multi-class case.

We can use the theory of *belief functions* in risk analysis [6] to provide a framework for combining these m classifiers. In this theory, the assumption is that m experts provide their own assessment of risk among a set of choices. In order to develop a final estimate of risk, these m estimates are combined. In the standard formulation of the problem, as applied to our case, let $U = \{1, 2, \cdots, m, \Lambda\}$ denote the set of m normal classes (1 to m) and a catchall class labeled Λ which represents all *anomalies*. A general classifier assigns a probability $P(u|x)$ for the probability that an image x belongs to the *set of classes* u, where $u \in 2^U$ is an element of the powerset of U. For example if $u = \{1, 5\}$ then this is the probability that x belongs to either of the classes 1 or 5.

This general formulation simplifies considerably for our case because the one-class classifiers will only assign a non-zero probability to two instances, as illustrated in Table 2. In other words, $P_i(u|x) = 0$ for all cases except when $u = \{i\}$ or $u = \neg\{i\} = \{1, 2, \cdots, i - 1, i + 1, \cdots, m, \Lambda\}$. In order to combine the predictions of the m one-class classifiers to compute the probability that x is anomalous, we need to determine $P(\Lambda|x)$. According to the Dempster-Schafer Theory [7,31],

$$P(\Lambda|x) = \frac{1}{K} \sum_{\neg\{1\} \cap \neg\{2\} \cap \cdots \neg\{m\} = \Lambda} \prod_{i=1}^{m} P_i(\neg\{i\}|x) \tag{1}$$

which simplifies to,

$$P(\Lambda|x) = \frac{1}{K} \prod_{i=1}^{m} P_i(\neg\{i\}|x) \tag{2}$$

since there is just one case when the intersection of the predictions of the m classifiers yields Λ. K is given by,

$$K = 1 - \sum_{u_1 \cap u_2 \cap \cdots u_m = \phi} \prod_{i=1}^{m} P_i(u_i|x) \tag{3}$$

where $u_i \in \{\{i\}, \neg\{i\}\}$ (because all other probabilities are zero from Table 1, these are the only non-trivial cases). The above equation simplifies to,

$$K = \prod_{i=1}^{m} P_i(\neg\{i\}|x) + \sum_{i=1}^{m} P_i(\{i\}|x) \prod_{j=1, j \neq i}^{m} P_j(\neg\{j\}|x) \tag{4}$$

From the above formulation it follows that an input x is classified as normal if *any* of the m classifiers say it is normal and it is classified as anomalous if *all* classifiers classify it as anomalous. Thus, it is easy to construct a multi-class anomaly detection algorithm using m one-class anomaly detectors as shown in Algorithm 1.

Algorithm 1: (Using m Single-Class Anomaly Detectors)
Training: Let $(\mathcal{F}, D)$ be any one-class anomaly detection algorithm. Given training set $X = \{X_1, X_2, \cdots, X_m\}$ consisting of training examples from m classes $X_i \subset \mathcal{X}_i$, train $(\mathcal{F}, D)$ separately on each class producing m classifiers, $(\mathcal{F}_i, D_i) 1 \leq i \leq m$.
Testing: Given $x \in \mathcal{X}$, classify it with each of the m classifiers. *Declare x anomalous if all classifiers classify it as anomalous.*

Another approach for training a one-class classifier for the m class case is to simply combine the training data from all m classes and treat that as a single class. Doing so gives us *Algorithm 2*.

Lemma 1. *Algorithm 2 has a higher AUC value than Algorithm 1 when the same one-class anomaly detection algorithm is used in both cases.*

Proof. Recall that the AUC value is the integral of the True Positive Rate (TPR) vs False Positive Rate (FPR) curve (each point on the curve corresponds to a different value for the detection threshold T). For Algorithm 1 we can write the TPR and FPR as,

$$\begin{aligned}
\text{TPR}_1 &= 1 - \text{False Negative Rate} \\
&= 1 - \frac{1}{K} \sum_{x_j \in \cup_{i=1}^{m} \mathcal{X}_i} p(x_j) \prod_{i=1}^{m}(1 - P_i(\{i\}|x_j)) \\
\text{FPR}_1 &= 1 - \text{True Negative Rate} \\
&= 1 - \frac{1}{K} \sum_{x_j \in \mathcal{X} \setminus \cup_{i=1}^{m} \mathcal{X}_i} p(x_j) \prod_{i=1}^{m}(1 - P_i(\{i\}|x_j))
\end{aligned}$$

Note that the difference in the two expressions is in the sets that x_j is selected from. The similar expressions for Algorithm 2 are,

$$\text{TPR}_2 = 1 - \text{False Negative Rate}$$
$$= 1 - \frac{1}{K} \sum_{x_j \in \cup_{i=1}^m \mathcal{X}_i} p(x_j)(1 - P(\text{Normal}|x_j))$$
$$= 1 - \frac{1}{K} \sum_{x_j \in \cup_{i=1}^m \mathcal{X}_i} p(x_j) \sum_{i=1}^m \frac{1}{m}(1 - P_i(\{i\}|x_j))$$
$$\text{FPR}_2 = 1 - \text{True Negative Rate}$$
$$= 1 - \frac{1}{K} \sum_{x_j \in \mathcal{X} \setminus \cup_{i=1}^m \mathcal{X}_i} p(x_j)(1 - P(\text{Normal}|x_j)$$
$$= 1 - \frac{1}{K} \sum_{x_j \in \mathcal{X} \setminus \cup_{i=1}^m \mathcal{X}_i} p(x_j) \sum_{i=1}^m \frac{1}{m}(1 - P_i(\{i\}|x_j))$$

The difference in the TPR (and FPR) values of this set of expressions is that for Algorithm 1 we take a product of the form $\prod_i (1 - P_i(\{i\}|x_j))$ whereas for Algorithm 2 this is a mean of the same probabilities. As m increases, we would expect the product term to decrease rapidly. Indeed, the values of TPR and FPR for Algorithm 1 approach 1 as m increases resulting in a purely random classifier. On the other hand, the TPR and FPR for Algorithm 2 remain relatively unchanged since they use an arithmetic mean of the probability rather than a geometric mean as in Algorithm 1.

Corollary 3.1.1: As m increases, the AUC value for Algorithm 1 decreases.

Proof. This is easy to see since we note that the FPR and TPR values approach 1 as m increases.

Note that this corollary is not true for Algorithm 2.

Algorithm 2: (Training a *single* classifier by combining all m classes)
Training: Given a training set $X = X_1 \cup X_2 \cup \cdots \cup X_m$, train a single classifier $(\mathcal{F}, D)$ on this set (see *Notes* in Algorithm 1).
Testing: An example x is classified as normal or anomalous by this classifier (just like any single-class classifier).

Algorithms Studied

In this paper, we use two recent and high performing single-class anomaly detection algorithms – DROCC [11] and DeepSVDD [27]. For each algorithm we consider two variations corresponding to Algorithms 1 & 2 above. Thus, DROCC(m) and DeepSVDD(m) correspond to Algorithm 1 while DROCC and DeepSVDD correspond to Algorithm 2. We use the author provided code for our experiments.

4   DeepMAD: Deep Multi-class Anomaly Detection

Our algorithm DeepMAD is based on Algorithm 1 above, but with important changes. The key insights we applied are the following:

- For any single-class algorithm, AUC values increase if the *variance* of the pdf of the normal class is reduced because it increases the TPR (True Positive Rate) without affecting the FPR (False Positive Rate), assuming the means of the normal and anomaly classes do not shift.
- In the absence of anomalies, the features learned represent what is *common* among the examples. DROCC attempts to learn *discriminating* features by using the manufactured examples. Similarly, algorithms that add noise to training examples or that learn transformations etc. are all attempting to reduce variance by learning discriminating features.
- When moving to the multi-class case, observe that Algorithm 2 will likely learn features common among the normal classes in addition to features common to examples within each class, potentially resulting in larger variance.

In DeepMAD, we train m one-class anomaly detectors. However, rather than train P_i only on the ith normal class, we train it on all m classes where we use the training examples from classes $\neg\{i\}$ as anomalous.

Algorithm 3: DeepMAD
Training: Randomly initialize m autoencoders A_i; For every A_i, train A_i on provided examples X_i; Using the encoder part E_i of the autoencoder, identify a point c_i, the "center" for this class;
For every encoder E_i, create *labeled* training data $\{(x, l)|$ if $x \in X_i$ then $l = +1$ else if $x \in \bigcup_{j=1, j\neq i}^{m} X_j$ then $l = -1\}$. Then train E_i on this data using loss function $\mathcal{L}$
Result: m trained encoders E_i
Testing: Given x to classify, compute $d(x) = \min_{i=1}^{m} ||c_i - E_i(x; \theta_i)||_2$; If $d(x) < \gamma$ then x is *normal* else x is anomalous

The algorithm for training proceeds in two steps. For every class i, we first train an autoencoder A_i using MSE between the original and reconstructed image as the loss. Next, we train only the encoder E_i using the other $m - 1$ classes as anomalies and class i as normal. For this, we identify a "center" c_i for this class and learn a representation that maps points in X_i to a sphere around it and maps other points from the remaining classes far away. We use the following loss function,

$$\mathcal{L}(\theta_i) = \frac{1}{N} \sum_{j=1}^{N_i} ||E_i(x_j; \theta_i) - c_i||^2 +$$
$$\frac{\eta}{N} \sum_{x_k \in (\bigcup_j X_j)\setminus X_i} (\max(0, \delta - ||E_i(x_k; \theta_i) - c_i||_2)^2)$$
$$+ \frac{\lambda}{2} \sum_{l=1}^{L} ||W^l||_F^2$$

where $N = |X_1| + \cdots + |X_m|$ is the total number of normal class samples and $N_i = |X_i|$ is the number of examples from normal class i. δ, η and λ are hyperparameters where η controls the weight given to the *constrastive loss* term, δ is a distance from the identified center beyond which we would like to map points not in X_i, and λ is a regularization parameter. The first term above attempts to learn a small sphere about the center for normal points in class i while the second term attempts to maximize the distance of the remaining normal points (all except points in X_i) far away. The ensemble method in [35] also trains multiple classifiers and uses a mean of their softmax outputs for OOD detection. Unlike out work, they train on one dataset and test on another (recall the criticism from [3] which notes that detecting anomalies within the same dataset is much harder than between dastasets). It is also unclear how well their method would scale down to few normal classes (say 2).

5   Empirical Evaluation

Table 2. AUC range for CIFAR-10 (10 repetitions for each).

	2-in, 8-out	5-in, 5-out	9-in, 1-out
DROCC	0.4728 ⟷ 0.7252	0.4316 ⟷ 0.7219	0.4107 ⟷ 0.7146
	±0.0119 ± 0.0081	±0.0257 ± 0.0039	±0.0454 ± 0.0079
Outlier	(0, 8)0.8359 ± 0.0117		
DROCC(m)	0.4216 ⟷ 0.6912	0.3806 ⟷ 0.7023	0.3439 ⟷ 0.6896
	±0.0424 ± 0.0188	±0.0047 ± 0.0648	±0.1034 ± 0.0453
Outlier	(0, 8)0.8255 ± 0.0137		
DeepSVDD	0.4088 ⟷ 0.7623	0.3382 ⟷ 0.7105	0.3058 ⟷ 0.6844
	±0.0068 ± 0.0193	±0.0076 ± 0.0077	±0.0169 ± 0.0144
DeepSVDD(m)	0.4147 ⟷ 0.7516	0.3482 ⟷ 0.6909	0.3580 ⟷ 0.5864
	±0.0129 ± 0.0093	±0.0123 ± 0.0133	±0.0166 ± 0.0167
DeepMAD	**0.5396⟷ 0.7647**	**0.4929 ⟷ 0.7738**	**0.5437 ⟷ 0.7230**
	±0.0031 ± 0.0014	±0.0046 ± 0.0022	±0.0028 ± 0.0084

We compare the performance of five algorithms – DROCC, DROCC(m), DeepSVDD, DeepSVDD(m), and DeepMAD using four data sets: CIFAR-10, fMNIST, RECYCLE, and CIFAR-100. For DROCC and DeepSVDD we used the code provided by the authors while for the m normal classes case we simply replicated the code and modified the main() function appropriately. For all the experiments with DROCC, we used parameters radius $r = 8$, $\mu = 1$, learning rate = 0.001, ascent step = 0.001 and the Adam optimizer. The reason for this choice is that the ablation study reported in [11] shows a fairly stable AUC value for this parameter setting across classes (though the specific value for r that achieves optimal AUC for the single-class case can vary by class). The parameter settings for DeepSVDD used $\eta = 1$, learning rate of 0.0001 and the Adam optimizer as well. We consider three multi-class anomaly detection cases for CIFAR-10 and fMNIST – (2/8), (5/5), and (9/1) which correspond to 2, 5, 9 normal classes respectively. For CIFAR-100 we used the 20 super-classes and studied the (2/18) case. For the RECYCLE dataset (described below) we have 5 classes of recyclables and one that is assorted trash. We considered two cases:

in one, we study the (4/1) case using only the recycles; we then study the case when the 5 recycles are all normal and the trash is the anomaly. In all cases we repeated the experiments with random seeds and report the 95% confidence intervals of the achieved AUC values.

Table 3. AUC range for fMNIST (10 repetitions each).

	2-in, 8-out	5-in, 5-out	9-in, 1-out
DROCC	0.6873 ↔ 0.9774	0.5738 ↔ 0.9260	0.5408 ↔ 0.8247
	±0.0937 ± 0.0049	±0.0397 ± 0.0307	±0.0961 ± 0.0507
Mean	0.8161	0.7448	0.6992
Deep	0.6622 ↔ 0.9871	0.5438 ↔ 0.9279	0.4551 ↔ 0.8825
SVDD	±0.0502 ± 0.0033	±0.0274 ± 0.0325	±0.0285 ± 0.0137
Mean	**0.8538**	0.7269	0.6523
Deep	0.6434 ↔ 0.9714	0.5732 ↔ 0.8832	0.4860 ↔ 0.9395
MAD	±0.0640 ± 0.0011	±0.0485 ± 0.0137	±0.0267 ± 0.3466
Mean	0.8329	**0.7739**	**0.7613**

Table 4. AUC range for RECYCLE, and CIFAR-100 (10 repetitions each).

	RECYCLE		CIFAR-100
	4-in, 1-out	5-in, 1-out	2-in, 18-out
DROCC	0.4447 ↔ 0.7997	0.9056	0.3548 ↔ 0.7329
	±0.0176 ± 0.0719		±0.0006 ± 0.0971
Mean	0.6128		0.5638
Deep	0.3703 ↔ 0.8728	0.9012	0.4196 ↔ 0.7185
SVDD	±0.0207 ± 0.0079		±0.0077 ± 0.0180
Mean	0.5791		0.5559
Deep	0.5906 ↔0.8283	**0.9838**	0.5384 ↔ 0.8213
MAD	±0.0035 ± 0.0073		±0.0018 ± 0.0012
Mean	**0.6966**		**0.6580**

Fig. 1. Examples of objects from the five (5) recycle classes.

The RECYCLE dataset contains 5 classes of recycles – glass bottles, plastic bottles, cans, crushed cans, and cardboard boxes. The original images were 3008

Fig. 2. Examples of trash (images are all random trash and cannot be divided into classes).

× 2000 × 3 but we downsampled them to 32 × 32 × 3. There are a total of 11,000 images evenly divided among the five classes. 10,000 are used for training and 1,000 for testing. Samples of these images are shown in Fig. 1. We also created 300 *trash* images to use for testing. Some examples are shown in Fig. 2.

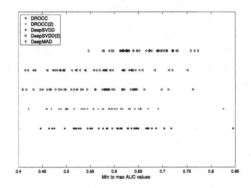

Fig. 3. 2 in and 8 out case (CIFAR-10).

Table 2 summarizes the AUC values achieved by each of the five algorithms on CIFAR-10. For the (2/8) case, there are 45 combinations, for (5/5) there are 252 and for (9/1) there are ten. In the table we report on the range of values with their confidence intervals. For all three cases we note that DeepMAD has the best AUC values (with the exception of an outlier for DROCC). But more importantly, observe that DROCC performs better than DROCC(m) and DeepSVDD performs better than DeepSVDD(m), which is what the theoretical results from the previous section predicted. Figure 3 provides a scatter plot of all 45 combinations for the five algorithms for the case when two classes are normal. It is clear that which classes are considered normal has a huge impact on AUC values. For all but DeepMAD, some cases have an AUC value below 0.5 which is very poor. Its noteworthy that the case when classes 0 and 8 are normal DROCC has a very high AUC value (labeled an outlier in table). Figure 4 plots the AUC values for the ten best and ten worst cases (out of 252) for when 5 classes are normal. The specific class combinations that have the best and worst values are different for the algorithms. Aside from the fact that DeepMAD performs better, it also shows considerably low variance. This points to it learning very good discriminating features.

We ran DROCC, DeepSVDD, and DepMAD on the fMNIST, RECY-CLE, and CIFAR-100 datasets as well. We did not run DROCC(m) and DeepSVDD(m) because, as we have seen, training over all classes jointly performs better. Tables 3, 4 provides AUC values for CIFAR-100 where we see that DeepMAD has about 10% improvement over the other algorithms. In this case we considered the 20 super-classes and considered just a single case when 2 of the 20 classes are considered normal. The tables also provides AUC values for fMNIST. Here the picture is more mixed. For the case when 2 classes are normal, all algorithms are very similar with DeepSVDD holding a 2% edge when averaged over 45 combinations. DeepMAD performs better on average for the other two cases. For the RECYCLE dataset, DeepMAD performs well above the other two algorithms. The (4/1) case is when we only consider the five recycle classes and use four of them as defining normal with the fifth being the anomaly. We also considered the case when the five recycle classes were all normal and *trash* was considered as the anomaly. This is the (5/1) case in the table. All algorithms perform considerably better because trash images are very different from recycles. However, DeepMAD is about 8% better than the other two.

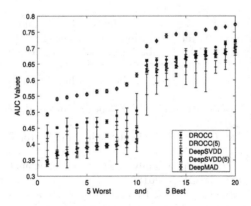

Fig. 4. 5 in 5 out case (CIFAR-10).

6   Conclusions

The problem of multi-class anomaly detection occurs naturally in many settings, making it an important model to study. In this paper we adapt two one-class anomaly detection algorithms to the multi-class setting and develop a new algorithm DeepMAD. Our paper reports two significant results. First, we show that jointly learning a classifier (using single-class anomaly detection algorithms) for all the normal classes provides higher classification accuracy as compared to using several single-class classifiers. And second, we show that using supervised learning by treating $m - 1$ out of m normal classes as anomalous results in much compact representations and hence uniformly higher accuracy. We compare our algorithm against two recent single-class classifiers adapted to the multi-class scenario and show improvements in AUC values for almost all cases for CIFAR-10, fMNIST, and CIFAR-100. However, we still see AUC values below 0.7 for many cases and addressing this constitutes a major thrust of future work. We also take a first step in analyzing the feature vectors as a way to understand where the algorithms fail. We hope that the methodology we use for this analysis of feature vectors can be extended by others. Our code is available at https:// anonymous.4open.science/r/DeepMAD-16D4/.

References

1. Abati, D., Porrello, A., Calderara, S., Cucchiara, R.: And: autoregressive novelty detectors. In: Proceedings of the CVPR (2019)
2. Ackay, A., Atapour-Abarghouei, A., Breckopn, T.P.: Ganomaly: semi-supervised anomaly detection via adversarial training. In: Proceedings of the ACCV, pp. 622–637 (2018)
3. Ahmed, F., Courville, A.: Detecting semantic anomalies. In: Proceedings of the AAAI, vol. 34, pp. 3154–3162 (2020)
4. Bergman, L., Hoshen, Y.: Classification-based anomaly detection for general data. In: Proceedings of the ICLR (2020)
5. Chen, T., Kornblith, S., Norouzi, M., Hinton, G.: A simple framework for contrastive learning of visual representations. In: Proceedings of the ICML (2020)
6. Clemen, R.T., Winkler, R.L.: Combining probability distributions from experts in risk analysis. Risk Anal. **19**(2), 187–203 (1999)
7. Dempster, A.P.: Upper and lower probabilities induced by a multivalued mapping. Ann. Math. Stat. **38**(2), 325–339 (1967)
8. DeVries, T., Taylor, G.W.: Learning confidence for out-of-distribution detection in neural networks (2018)
9. Ghafoori, Z., Leckie, C.: Deep multi-sphere support vector data description. In: SIAM SDM (2020)
10. Golan, I., El-Yaniv, R.: Deep anomaly detection using geometric transformations. In: Proceedings of the NeurIPS (2018)
11. Goyal, S., Raghunathan, A., Simhadri, M., Jain, P.: DROCC: deep robust one-class classification. In: Proceedings of the ICML (2020)
12. Hendrycks, D., Gimpel, K.: A baseline for detecting misclassified and out-of-distribution examples in neural networks. In: Proceedings of the ICLR (2017)
13. Hendrycks, D., Mazeika, M., Dietterich, T.: Deep anomaly detection with outlier exposure. In: Proceedings of the ICLR (2019)
14. Hendrycks, D., Mazeika, M., Kadavath, S., Song, D.: Using self-supervised learning can improve model robustness and uncertainity. In: Proceedings of the NeurIPS (2019)
15. Lakshminarayanan, B., Pritzel, A., Blundell, C.: Simple and scalable predictive uncertainity estimation using deep ensembles. In: 31st Conference on Neural Information Processing Systems (2017)
16. Lee, K., Lee, H., Lee, K., Shin, J.: Training confidence-calibrated classifiers for detecting out-of-distribution samples (2018)
17. Li, D., Chen, D., Goh, J., Ng, S.K.: Anomaly detection with generative adversarial networks for multivariate time series. In: 7th International Workshop on Big Data, Streams and Heterogeneous Source Mining: Algorithms, Systems, Programming Models and Applications (ACM KDD), August 2018

18. Liang, S., Li, Y., Srikant, R.: Enhancing the reliability of out-of-distribution image detection in neural networks. In: Proceedings of the International Conference on Learned Representations (ICLR 2018), February 2018
19. Masana, M., Ruiz, I., Serrat, J., van de Weijer, J., Lopez, A.M.: Metric learning for novelty and anomaly detection. In: British Machine Vision Conference (2018)
20. Nguyen, D.T., Lou, Z., Klar, M., Brox, T.: Anomaly detection with multiple-hypothesis predictions. In: Proceedings of the ICML (2019)
21. Perera, P., Nallapati, R., Xiang, B.: Ocgan: one-class novelty detection using gans with constrained latent representations. In: Proceedings of the CVPR (2019)
22. Perera, P., Patel, V.M.: Deep transfer learning for multiple class novelty detection. In: Proceedings of the CVPR (2019)
23. Pidhorskyi, S., Almohsen, R., Adjeroh, D.A., Doretto, G.: Generative probabilistic novelty detection with adversarial autoencodres. In: Proceedings of the NeurIPS (2018)
24. Pimentel, M.A.F., Clifton, D.A., Clifton, L., Tarassenko, L.: A review of novelty detection. Signal Processing, pp. 215–249 (2014)
25. Ren, J., et al.: Likelihood ratios for out-of-distribution detection. In: 33rd Conference on Neural Information Processing Systems (2019)
26. Ruff, L., et al.: Deep one-class classification. In: Proceedings 35th International Conference on Machine learning (ICML). Stockholm, Sweeden (2018)
27. Ruff, L., et al.: Deep semi-supervised anomaly detection. In: Proceedings of the ICLR (2020)
28. Ruff, L., et al.: A unifying review of deep and shallow anomaly detection, September 2020. arXiv:2009.11732v2
29. Sabokrou, M., Khalooei, M., Fathy, M., Adeli, E.: Adversarially learned one-class classifier for novelty detection. In: Proceedings of the CVPR, pp. 3379–3388 (2018)
30. Schlegl, T., Seebock, P., Waldstein, S.M., Schmidt-Erfurth, U., Langs, G.: Unsupervised anomaly detection with generative adversarial networks to guide marker discovery. In: International Conference on Information Processing in Medical Imaging, pp. 146–157 (2017)
31. Shafer, G.: Mathematical Theory of Evidence. Princeton University Press, Princeton (1976)
32. Sohn, K., Li, C.L., Yoon, J., Jin, M., Pfister, T.: Learning and evaluating representations for deep one-class classification. In: Proceedings of the ICLR (2021)
33. Song, H., Jiang, Z., Men, A., Yang, B.: A hybrid semi-supervised anomaly detection model for high-dimensional data. Computational Intelligence and Neuroscience (2017)
34. Tack, J., Mo, S., Jeong, J., Shin, J.: CSI: novelty detection via contrastive learning on distributionally shifted instances. In: Proceedings of the NeurIPS (2020)
35. Vyas, A., Jammalamadaka, N., Zhu, X., Das, D., Kaul, B., Wilke, T.L.: Out-of-distribution detection using an ensemble of self-supervised leave-out classifiers. In: Proceedings of the ECCV, pp. 550–564 (2018)
36. Zhang, H., Li, A., Guo, J., Guo, Y.: Hybrid models for open set recognition (2020)

Understanding Graph and Understanding Map and Their Potential Applications

Gangli Liu[✉]

Tsinghua University, Bejing 100084, China
`gl-liu13@mails.tsinghua.edu.cn`

Abstract. Based on a previously proposed data structure, Understanding Tree, this paper introduces two data structures: Understanding Graph and Understanding Map, and explores their potential applications. Understanding Graph and Understanding Map are special cases of semantic network. They are constructed according to concept definitions. Potential applications of them include quantitatively measuring a concept's complexity degree, measuring a concept's importance degree in a domain, and computing an optimized learning sequence for comprehending a concept etc.

Keywords: Mind Map · Semantic Network · Concept Map · Meaningful Learning

1 Introduction

In a previous paper [11], Understanding Tree is introduced for evaluating a person's understanding degree to a piece of knowledge. This paper introduces two more data structures, Understanding Graph and Understanding Map, and explores their potential applications. We first recap several concepts formulated previously. Section 2 discusses Understanding Graph and its applications. Section 3 gives an account of Understanding Map and its applications. Section 4 compares Understanding Graph and Understanding Map with other knowledge organizing and representing tools. Section 5 concludes the paper.

1.1 Recap of Several Definitions

Here we recap several concepts mentioned in previous papers.

Knowledge Point. A Knowledge Point is a piece of knowledge which is explicitly defined and has been widely accepted. A Knowledge Point's definition may be differently phrased, however, they should be consistent.

Basic Knowledge Point (BKP). A BKP is a Knowledge Point that is simple enough so that it is not interpreted by other Knowledge Points. In a sense, they are like axioms in mathematics, serving as premises or starting points for further reasoning and arguments. In practice, it is subjective which set of Knowledge Points should be categorized as BKPs. It can be decided by a group of experts empirically.

M. Tanveer et al. (Eds.): ICONIP 2022, LNCS 13625, pp. 372–383, 2023.
https://doi.org/10.1007/978-3-031-30111-7_32

Familiarity Measure. A Familiarity Measure is a score that depicts a person's familiarity degree to a Knowledge Point at a particular time. It is calculated with the formulas mentioned in [11].

Understanding Tree. An Understanding Tree is a treelike data structure which compiles the background Knowledge Points that are essential to understand the root Knowledge Point.

Fully Extended Understanding Tree (FEUT). A fully extended Understanding Tree is a tree like Fig. 3 of [9]. It can be constructed manually by some experts of a domain, or automatically with Algorithm 1 of [9]. Its construction is based on a collection of definitions that are self-contained. 'Self-contained' means for every Knowledge Point mentioned in a definition, either it is a BKP, or it has at least one definition in the collection. The root of an FEUT is the Knowledge Point at which the construction starts.

Standard Understanding Tree (SUT). A standard Understanding Tree is a tree that is generated from an FEUT, by removing identical nodes from it. Figure 1 is an example of SUT. It is similar to Fig. 6 of [11], except that the BKPs are labeled with shading and the Knowledge Points are not tagged with Familiarity Measures. Figure 2 is another example of SUT. It is constructed according to the definitions of Table 1.

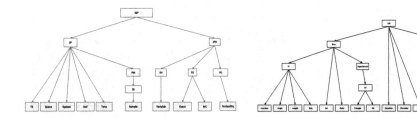

Fig. 1. SUT of SSP   **Fig. 2.** SUT of The Law of Sines

2   Understanding Graph

An Understanding Graph is a rooted graph that is also generated from an FEUT. Instead of removing identical nodes from an FEUT, it merges them. The merging of identical nodes keeps the connection relations (links and their directions) between Knowledge Points unchanged; only the nodes that representing the same Knowledge Point are merged into one node. E.g., Fig. 3 is an Understanding Graph constructed based on Fig. 3 of [9]. Figure 4 is another Understanding Graph constructed based on the definitions of Table 2. The definitions listed in Table 1 and 2 are cited from Wikipedia and other authoritative websites like

Table 1. A set of self-contained definitions

Knowledge Point	Definition	Involved Knowledge Points
law of cosines (LoC)	The law of cosines relates the lengths of the sides of a triangle to the cosine of one of its angles. the law of cosines states: $c^2 = a^2 + b^2 - 2ab\cos\gamma$, where γ denotes the angle contained between sides of lengths a and b and opposite the side of length c.	length, side, triangle, cosine, angle
cosine	The trigonometric function that is equal to the ratio of the side adjacent to an acute angle (in a right triangle) to the hypotenuse.	trigonometric function (TF) , ratio, side, acute angle (AA), right triangle (RT), hypotenuse
Pythagorean theorem (PT)	The Pythagorean theorem states that the square of the hypotenuse is equal to the sum of the squares of the other two sides.	square, hypotenuse, sum, side
hypotenuse	A hypotenuse is the longest side of a right triangle, the side opposite the right angle.	side, right triangle (RT), right angle (RA)
right triangle	A triangle with a right angle.	triangle, right angle (RA)
law of sines (LoS)	The law of sines is an equation relating the lengths of the sides of a triangle to the sines of its angles. According to the law, $$\frac{a}{\sin A} = \frac{b}{\sin B} = \frac{c}{\sin C} = d$$ where a, b, and c are the lengths of the sides of a triangle, and A, B, and C are the opposite angles, while d is the diameter of the triangle's circumcircle.	equation, length, side, triangle, sine, angle, opposite angle (OA), diameter, circumcircle
sine	The sine is a trigonometric function of an angle. The sine of an acute angle is defined in the context of a right triangle: for the specified angle, it is the ratio of the length of the side that is opposite that angle to the length of the longest side of the triangle (the hypotenuse).	trigonometric function (TF), angle, acute angle (AA), right triangle (RT), ratio, length, side, triangle, hypotenuse
circumcircle	The circumcircle of a polygon is a circle which passes through all the vertices of the polygon.	polygon, circle, vertex
trigonometric function (TF)	The trigonometric functions are functions of an angle. They relate the angles of a triangle to the lengths of its sides.	function, angle, triangle, length, side

Wolfram MathWorld. The shaded nodes in Fig. 3 and 4 are BKPs. In Understanding Graph, each Knowledge Point is represented by one and only one node. If two Knowledge Points have the same name but with different meanings (such as homonyms), or a concept has a particular meaning in a specific field (such as the term 'compact' in Topology), they are differentiated with different nodes. If a Knowledge Point has multiple names, such as the law of sines, they are incorporated into one node.

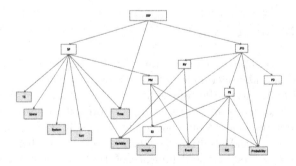

Fig. 3. Understanding Graph of Strictly Stationary Process (SSP)

Table 2. Another set of self-contained definitions

Knowledge Point	Definition	Involved Knowledge Points
Poincare conjecture	Every simply connected, closed 3-manifold is homeomorphic to the 3-sphere.	simply connected, closed, 3-manifold, homeomorphic, 3-sphere
simply connected	A topological space is called simply-connected if it is path-connected and every path between two points can be continuously transformed, staying within the space, into any other such path while preserving the two endpoints in question.	topological space, path-connected, path, point, endpoint
closed manifold	A closed manifold is a type of topological space, namely a compact manifold without boundary.	topological space, compact manifold, boundary
3-manifold	A topological space X is a 3-manifold if it is a second-countable Hausdorff space and if every point in X has a neighborhood that is homeomorphic to Euclidean 3-space.	topological space, second-countable, Hausdorff space, point, neighborhood, homeomorphic, Euclidean 3-space
homeomorphic	A homeomorphism is a continuous function between topological spaces that has a continuous inverse function.	continuous function, topological space, inverse function
the 3-sphere	A 3-sphere with center $(C0, C1, C2, C3)$ and radius r is the set of all points $(x0, x1, x2, x3)$ in real, 4-dimensional space $(R4)$ such that $\sum_{i=0}^{3}(x_i - C_i)^2 = (x_0 - C_0)^2 + (x_1 - C_1)^2 + (x_2 - C_2)^2 + (x_3 - C_3)^2 = r^2$	center, radius, point, real, 4-dimensional space
topological space	A topological space is an ordered pair (X, τ), where X is a set and τ is a collection of subsets of X, satisfying the following axioms: (1) The empty set and X itself belong to τ . (2) Any (finite or infinite) union of members of τ still belongs to τ. (3) The intersection of any finite number of members of τ still belongs to τ.	ordered pair, set, subset, axiom, empty set, finite, infinite, union, intersection
path-connected	A topological space is said to be path-connected if given any two points on the topological space, there is a path starting at one point and ending at the other.	topological space, point, path
compact manifold	A compact manifold is a manifold that is compact as a topological space.	manifold, compact, topological space
boundary	The boundary of a subset S of a topological space X is the set of points which can be approached both from S and from the outside of S.	subset, topological space, point
Hausdorff space	The Hausdorff space is a topological space in which distinct points have disjoint neighborhoods.	topological space, point, disjoint, neighborhood
second-countable	A topological space is second countable if it has a countable topological basis.	topological space, topological basis
neighborhood	A neighborhood of a point is a set of points containing that point where one can move some amount away from that point without leaving the set.	Point, set
continuous function	A continuous function is a function for which sufficiently small changes in the input result in arbitrarily small changes in the output.	function, change, input, output
inverse function	An inverse function is a function that undoes the action of another function.	function, action
manifold	A manifold is a second countable Hausdorff space that is locally homeomorphic to Euclidean space.	second countable, Hausdorff space, homeomorphic, Euclidean space
compact	A topological space is compact if every open cover of X has a finite subcover.	topological space, open cover, subcover
topological basis	A topological basis is a subset B of a set T in which all other open sets can be written as unions or finite intersections of B.	subset, set, open set, union, finite, intersection
open cover	A collection of open sets of a topological space whose union contains a given subset.	open set, topological space, union, subset
subcover	Let S be a set. Let U be a cover for S. A subcover of U for S is a set $V \subseteq U$ such that V is also a cover for S.	set, cover
open set	Let S be a subset of a metric space. Then the set S is open if every point in S has a neighborhood lying in the set.	subset, metric space, set, point, neighborhood
intersection	The intersection of two sets A and B is the set of elements common to A and B.	set, element
subset	A subset is a portion of a set.	portion, set
cover	A cover of a set X is a collection of sets whose union contains X as a subset.	set, union, subset
metric space	A metric space is a set S with a global distance function that, for every two points x, y in S, gives the distance between them as a nonnegative real number $g(x, y)$. A metric space must also satisfy (1) $g(x, y) = 0$ iff $x = y$, (2) $g(x, y) = g(y, x)$, (3) The triangle inequality $g(x, y) + g(y, z) >= g(x, z)$.	set, function, point, distance, nonnegative real number, triangle inequality

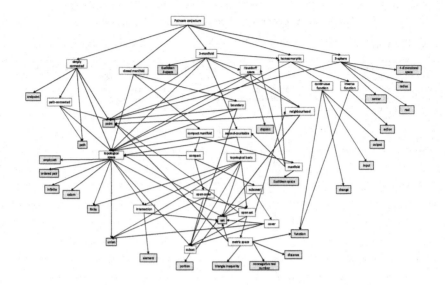

Fig. 4. Understanding Graph of the Poincare Conjecture

2.1 Application of Understanding Graph

Several potential applications have been devised for Understanding Graph.

Extending Keywords in Information Retrieval. Information Retrieval
(IR) is the activity of obtaining information resources relevant to an information
need from a collection of information resources. When a person has an information need, he inputs some keywords to a search engine. Usually, the more
keywords are submitted, the more a search engine can know the user's information need. However, more keywords require more cognitive burden for the user
to formulate and input. With Understanding Graph, we can extend the user's
keywords automatically. E.g., if the user inputs *'Poincare Conjecture'*, it seems
the user wants to know what the Poincare Conjecture is, then the keyword is
extended into *"Poincare Conjecture, simply connected, closed manifold, homeomorphic, topological space, Hausdorff space, neighborhood ..."* The extension can
be accomplished according to a Knowledge Point's SUT, from low level nodes
to high level nodes. A document or a series of documents containing all the
extended keywords are good candidates for the information need.

**Indicating the Minimum Information for Understanding a Knowledge
Point.** Because an Understanding Graph is just a transformation of an Understanding Tree. The Understanding Tree of a Knowledge Point is constructed
based on the definitions of itself and its descendants. It compiles the most essential information for understanding a Knowledge Point. Knowing other Knowledge Points may help understand it, but they are not the most necessary ones.
E.g., knowing the Pythagorean Theorem is a special case of the Law of Cosines

(LoC) helps understand the LoC, but it is not essential for understanding it. Thus it indicates the minimum information for understanding a Knowledge Point.

Indicating an Optimized Learning Sequence for Comprehending a Knowledge Point. In meaningful learning, the learners are 'integrating' new information into old information [14]. Meaningful learning is opposed to rote learning and refers to a learning method where the new knowledge to acquire is related with previous knowledge [1]. If a person is facing a piece of information full of concepts he does not understand, he cannot practice a meaningful learning. E.g., if we tell a person who knows little about Mathematics, the content of Poincare Conjecture (now it is a theorem): *"Every simply connected, closed 3-manifold is homeomorphic to the 3-sphere."* He will be perplexed by what a manifold is and what do you mean by saying *'simply connected', 'closed',* and *'homeomorphic'*. To practice a meaningful learning, a person should have understood these concepts before facing this information. Understanding Graph provides a way to comprehend a complicated Knowledge Point in a from-easy-to-difficult order.

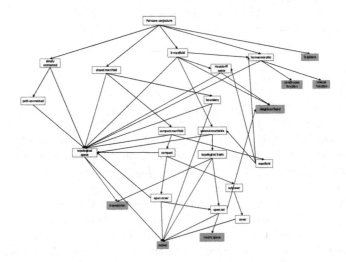

Fig. 5. Understanding Graph of the Poincare Conjecture without BKPs

Taking the Poincare Conjecture as an example, Fig. 4 is its Understanding Graph. To learning it meaningfully, a person can lean the BKPs first. After comprehended them, they are removed from the Understanding Graph. The graph becomes Fig. 5. The next learning target is determined by checking the out-degrees of the nodes of Fig. 5. The nodes with the smallest out-degrees are candidates (the light blue ones). When a node is learned and understood, it is removed from the Understanding Graph, then compares the out-degrees of the rest nodes to determine the next target. The learning process ends until the last node is removed. Because of decaying of human memory, what have been learned may be forgotten, a person can review previously learned information at any time during the learning process.

Algorithm 1 calculates an optimized learning sequence for comprehending a Knowledge Point, based on its Understanding Graph. The subject is supposed having understood the BKPs before learning, so the learning sequence does not consider BKPs.

Algorithm 1. Calculating an optimized learning sequence for comprehending a Knowledge Point based on its Understanding Graph

---

Input:
    Knowledge Point k_i's Understanding Graph, G_i;
Output:
    An optimized learning sequence for comprehending k_i, S;
1: Set $S = NULL$;
2: Remove the BKPs and their edges from G_i;
3: Check whether G_i is $NULL$. If so, go to step 6;
4: Check the out-degrees of the nodes of G_i. Select the node with the smallest out-degree, record it as n_i (if there are more than one such nodes, select one arbitrarily);
5: Insert n_i into S. Remove n_i and its edges (to or from n_i) from G_i. Go to step 3;
6: **return** S.

According to Algorithm 1 and Understanding Graph Fig. 4, one of the suggested learning sequences for comprehending Poincare Conjecture is: *intersection, subset, metric space, neighborhood, continuous function, inverse function, 3-sphere, open set, cover, subcover, topological space, topological basis, open cover, compact, path-connected, simply connected, homeomorphic, boundary, Hausdorff space, second-countable, manifold, compact manifold, closed manifold, 3-manifold, Poincare Conjecture.*

The sequence is 'optimized' in that it recommends the easier ones first, then the harder ones, and comprehending the harder ones relies on understanding the easier ones. In a sense, the learning process is similar to let a group of sappers clear a minefield first, before troops pass through the field. An Understanding Graph is like a map tagging where the mines are located and how they are connected; Algorithm 1 calculates an optimized sequence for clearing the mines. Therefore, it facilitates a meaningful learning path for comprehending the root Knowledge Point.

Measuring a Knowledge Point's Complexity Degree. In Understanding Graph, a Knowledge Point is represented as a bag-of-concepts and their relations. By examining the size of a Knowledge Point's Understanding Graph, its complexity degree can be obtained. E.g., by estimating empirically, we know that the Poincare Conjecture is more complicated than the Pythagorean Theorem. Understanding Graph facilitates a quantitative description of this fact. A Knowledge Point's complexity degree is defined as the number of vertices plus the number of edges of its Understanding Graph (Eq. 1). Hence according to Poincare Conjecture and Pythagorean Theorem's Understanding Graphs (Fig. 4 and Fig. 6), the complexity degree of Poincare Conjecture equals 148; the complexity degree of Pythagorean Theorem equals 17. Based on this definition, Poincare Conjecture is indeed more complicated than Pythagorean Theorem.

$$C_i = |V_i| + |E_i| \tag{1}$$

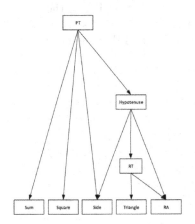

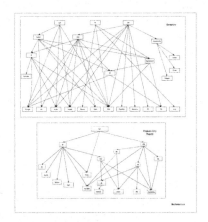

Fig. 6. Understanding Graph of the Pythagorean Theorem

Fig. 7. An exemplary Understanding Map

3   Understanding Map

Suppose we have constructed Understanding Graphs for a set of Knowledge Points in a domain, a map can be generated by merging the graphs. An Understanding Map is defined as a graph which is constructed by merging a set of Understanding Graphs (at least one). The merging combines identical nodes into one node and keeps the link relations unchanged. E.g., Fig. 7 is an Understanding Map constructed based on four Understanding Graphs: LoC, PT, LoS, and SSP. The Understanding Graphs of LoC, PT, and LoS are constructed based on the definitions of Table 1. Figure 3 is the Understanding Graph of SSP. Figure 4 is the Understanding Graph of Poincare Conjecture. It can also be deemed as an Understanding Map. Understanding Graph and Understanding Map are devised for different applications. The main difference is that an Understanding Graph is constructed for depicting a concept; an Understanding Map is for depicting a domain. Understanding Map usually contains large scale of Knowledge Points. If we combine all Knowledge Points' Understanding Graphs, a very large Understanding Map is generated. It is called the Global Understanding Map (GUM), like a world map.

3.1   A Node's N-Level Neighborhood

A node's n-level open neighborhood in an Understanding Map is defined as a set of nodes whose undirected distance to the node is less than or equal to n, excluding the node itself. The n-level closed neighborhood is defined in the same way but also includes the node itself. The n-level neighborhood is an extension of the term 'neighborhood' in graph theory. Before counting distances, the edges in Understanding Map are converted to be undirected. The distance between any two vertices in a graph is the length of the shortest path having the two vertices as its endpoints. E.g., in the Understanding Map of Fig. 8, the 1-level open neighborhood of the node 'topological basis' is a set that is comprised of the green nodes; the 2-level open neighborhood is comprised of the green nodes plus the yellow nodes.

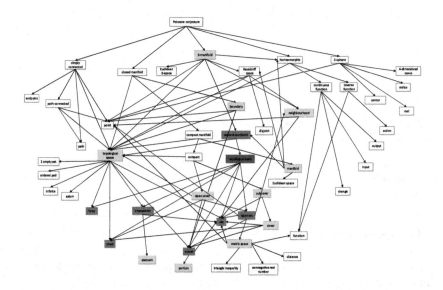

Fig. 8. A node's n-level neighborhood in an Understanding Map

3.2   Application of Understanding Map

Following are some potential applications of Understanding Map.

Another Way for Calculating Understanding Degree. In [11], we propose a method for calculating a person's understanding degree to a Knowledge Point, by examining his Familiarity Measures to the Knowledge Point's SUT. Here we propose an alternative way for calculating the understanding degree, based on an Understanding Map (preferably the GUM). It calculates the subject's average Familiarity Measure to a Knowledge Point's n-level closed neighborhood (n is a parameter to be determined), then uses the average as the understanding degree.

Illustrating the Knowledge Characteristics of a Person or a Corpus.
An Understanding Map is analogous to a real map. E.g., in a sense, Fig. 7 is like
Fig. 9. A Knowledge Point is like a building; closely related Knowledge Points
constitute a district, a city, or a country, just like a domain or discipline. A
person lives in a city, he must be familiar with the buildings of the city than
other cities. It is analogous to a person having expertise in a domain, and being
familiar with the topics of the domain. In [11], we propose a formula for calculating a person's familiarity degree to a Knowledge Point at a particular time. If
we display the familiarity degrees in an Understanding Map, by setting a node's
brightness being proportional to a person's familiarity degree to the node, a night
light map like Fig. 10 can be obtained. It illustrates a person's knowledge characteristics. A corpus can be illustrated similarly in an Understanding Map. A
node's brightness can be set being proportional to its term frequency in the corpus. The Understanding Map Supervised Topic Model (UM-S-TM) is a further
exploration of this application [10].

Measuring a Knowledge Point's Importance Degree in a Domain. In
Understanding Graph and Understanding Map, an edge from node A to node B
means understanding A relies on understanding of B. If a node is relied on by a
lot of nodes, it implies the node is very important for understanding other nodes.
Therefore, if we have an Understanding Map that contains all the Knowledge
Points in a domain, by counting a node's in-degree, its importance rank in the
domain is obtained (Eq. 2). E.g., suppose we have a domain called "Poincare
Conjecture's definition", also suppose the Understanding Map of Fig. 4 contains
all the Knowledge Points of this domain, by counting each node's in-degree, the
top 3 important Knowledge Points in this domain are: topological space (11),
point (9), and set (9). If we ignore the BKPs when ranking importance, the top
4 are: topological space (11), subset (6), homeomorphic (3), and neighborhood
(3).

$$I_i = |E_{i\_in}|$$ (2)

4 Related Work

A variety of tools have been devised for organizing concepts and their relationships, such as mind map, spider diagram, semantic network, and concept
map etc. A mind map is a diagram used to visually organize information. It is
hierarchical and shows relationships among pieces of the whole [2,3,19]. Mind
maps are considered to be a type of spider diagram. A spider diagram is a
boolean expression involving unitary spider diagrams and the logical symbols
conjunction, disjunction, and negation. A unitary spider diagram adds exis-
tential points to an Euler or a Venn diagram [8,17]. A semantic network is a
network that represents semantic relations between concepts. It is a directed
or undirected graph consisting of vertices, which represent concepts, and edges,
which represent semantic relations between concepts [4–6,16]. Typical standard-
ized semantic networks are expressed as semantic triples. A semantic triple is a

Fig. 9. A part of a real map

Fig. 10. A part of a night light map

set of three entities that codifies a statement about semantic data in the form of subject-predicate-object expressions (e.g. "a bear is a mammal", or "fish lives in water"). A concept map is a diagram that depicts suggested relationships between concepts [7,12,13,18]. It typically represents ideas and information as boxes or circles, which it connects with labeled arrows in a downward-branching hierarchical structure. The relationship between concepts can be articulated in linking phrases such as causes, requires, or contributes to [15].

Understanding Graph and Understanding Map can be considered as special cases of the tools listed above. The two main differences are: Firstly, the data sources for constructing Understanding Maps and Understanding Graphs are distinctive and simple. The sufficient and necessary condition for constructing an Understanding Graph is a set of self-contained definitions of Knowledge Points. Merging a set of Understanding Graphs (at least one) produces an Understanding Map. Secondly, the relations between concepts in Understanding Graph and Understanding Map are monotonous. An edge from node A to node B always means understanding A relies on understanding of B. Since the relation is simplex, it is omitted in the graphs.

5   Conclusion

Understanding Graph and Understanding Map provide a way for organizing concepts, exploring their relationships according to their definitions. Besides proposing notions, we propose potential applications of the two data structures, such as quantitatively measuring a concept's complexity degree according to its Understanding Graph, quantitatively measuring a concept's importance degree in a domain according to the domain's Understanding Map, and computing an optimized learning sequence for comprehending a concept based on its Understanding Graph etc. Further study is necessary for evaluating their performances in these applications. In addition, an algorithm is devised for calculating an optimized learning sequence for comprehending a concept.

References

1. Ausubel, D.P.: The Acquisition and Retention of Knowledge: A Cognitive View. Springer Science & Business Media, Berlin, Heidelberg (2012). https://doi.org/10.1007/978-94-015-9454-7
2. Buzan, T., Buzan, B.: How to mind map. Thorsons, London (2002)
3. Dhindsa, H.S., Roger Anderson, O., et al.: Constructivist-visual mind map teaching approach and the quality of students' cognitive structures. J. Sci. Educ. Technol. **20**(2), 186–200 (2011)
4. Doerfel, M.L., Barnett, G.A.: A semantic network analysis of the international communication association. Hum. Commun. Res. **25**(4), 589–603 (1999)
5. Drieger, P.: Semantic network analysis as a method for visual text analytics. Procedia Soc. Behav. Sci. **79**, 4–17 (2013)
6. Fellbaum, C.: A semantic network of English verbs. WordNet Electron. Lex. Database **3**, 153–178 (1998)
7. Hager, P.J., Scheiber, H.J., Corbin, N.C.: Designing & Delivering: Scientific, Technical, and Managerial Presentations. John Wiley & Sons, Hoboken (1997)
8. Howse, J., Stapleton, G., Taylor, J.: Spider diagrams. LMS J. Comput. Math. **8**, 145–194 (2005)
9. Liu, G.: Understanding tree: a tool to estimate one's understanding of knowledge. arXiv preprint arXiv:1612.07714 (2016)
10. Liu, G.: Topic model supervised by understanding map. arXiv preprint arXiv:2110.06043 (2021)
11. Liu, G.: ICKEM: a tool for estimating one's understanding of conceptual knowledge. In: Memmi, G., Yang, B., Kong, L., Zhang, T., Qiu, M. (eds.) Knowledge Science, Engineering and Management. KSEM 2022. LNCS, vol. 13368, pp. 44–57. Springer, Cham (2022). https://doi.org/10.1007/978-3-031-10983-6_4
12. Markham, K.M., Mintzes, J.J., Jones, M.G.: The concept map as a research and evaluation tool: further evidence of validity. J. Res. Sci. Teach. **31**(1), 91–101 (1994)
13. McClure, J.R., Sonak, B., Suen, H.K.: Concept map assessment of classroom learning: reliability, validity, and logistical practicality. J. Res. Sci. Teach. Off. J. Natl. Assoc. Res. Sci. Teach. **36**(4), 475–492 (1999)
14. Novak, J.D.: Meaningful learning: the essential factor for conceptual change in limited or inappropriate propositional hierarchies leading to empowerment of learners. Sci. Educ. **86**(4), 548–571 (2002)
15. Novak, J.D., Cañas, A.J.: The theory underlying concept maps and how to construct and use them (2008)
16. Sowa, J.F.: Semantic networks (1987)
17. Stapleton, G., Howse, J., Taylor, J., Thompson, S.: The expressiveness of spider diagrams. J. Log. Comput. **14**(6), 857–880 (2004)
18. Tseng, S.S., Sue, P.C., Su, J.M., Weng, J.F., Tsai, W.N.: A new approach for constructing the concept map. Comput. Educ. **49**(3), 691–707 (2007)
19. Zubaidah, S., Fuad, N.M., Mahanal, S., Suarsini, E.: Improving creative thinking skills of students through differentiated science inquiry integrated with mind map. J. Turkish Sci. Educ. **14**(4), 77–91 (2017)

BBSN: Bilateral-Branch Siamese Network for Imbalanced Multi-label Text Classification

Jiangjiang Zhao[1,2], Jiyi Li[2(✉)], and Fumiyo Fukumoto[2]

[1] Hangzhou Dianzi University, Hangzhou, China
zhaojiangjiang@hdu.edu.com
[2] University of Yamanashi, Kofu, Japan
{jyli,fukumoto}@yamanashi.ac.jp

Abstract. In multi-label text classification, the numbers of instances in different categories are usually extremely imbalanced. How to learn good models from imbalanced data is a challenging task. Some existing works tackle it through class re-balancing strategies or imbalanced loss objectives, but their performance remains limited in the cases of imbalanced distributed data. In this work, we propose a model, which combined Siamese Network and Bilateral-Branch Network to deal with both representation learning and classifier learning simultaneously. In the siamese network component, we propose a category-specific similarity strategy to improve the representation learning and adapt a novelty dynamic learning mechanism to make the model end-to-end trainable, and in the bilateral-branch network, we adopt the cumulative learning strategy to shift the learning focus from universal pattern to tail learning. In general, we adopt a multi-task architecture to ensure that both the head categories and the tail categories are adequately trained. The experiments on two benchmark datasets show that our method can improve the performance on the entire and tail categories, and achieves competitive performance compared with existing approaches.

Keywords: Multi-label Text Classification · Imbalanced Data

1 Introduction

In the real world, large-scale data tend to exhibit a long-tailed distribution, where a few categories occupy most of the data (a.k.a head category), while most categories have rarely few instances (a.k.a tail category). Figure 1 illustrates the categories distribution of the RCV1 [8] dataset. Multi-label classification becomes complicated when there are extremely imbalanced distributions and label co-occurrences, which causes a tremendous challenge to the multi-label classification performance, especially for the tail categories.

Most existing works tackle the data imbalance issue by data re-sampling [15,20] and data augmentation [2,13] methods, which are effective in promoting

© The Author(s), under exclusive license to Springer Nature Switzerland AG 2023
M. Tanveer et al. (Eds.): ICONIP 2022, LNCS 13625, pp. 384–396, 2023.
https://doi.org/10.1007/978-3-031-30111-7_33

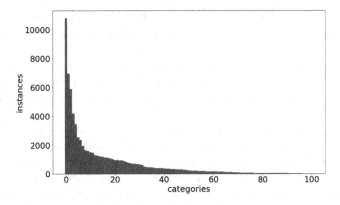

Fig. 1. The distribution of instance numbers of categories for the RCV1 training data, the category number of instances follows the long-tailed distribution, where many tail categories have only a small number of instances.

the classifier learning directly and mitigating the data shortage in tail categories. However, there is the problem of label co-occurrence in multi-label text classification, and simply increasing the number of tail categories does not solve the data imbalanced problem well. At the same time, these methods will also damage the model's representation ability to learn deep features. Specifically, re-sampling and data augmentation methods are at risk of overfitting on the tail categories.

Recently, more studies have used loss re-weighting methods [3,12,21] that directly modify the classification loss and consider the long-tailed distribution and label co-occurrence. However, this scheme is inconsistent with the principle of cross-entropy loss and naturally brings along with contradiction between learning generalizable representations and facilitating learning for tail categories. Meanwhile, these methods also have difficulties in practical applications. For example, focal loss [12] can handle hard samples well in theory, but in practice, we have difficulty in setting the appropriate weights for hard samples, which often needs to be fine-tuned with the real task.

In this paper, to address the problem of multi-label text classification for imbalanced data, we propose a unified **B**ilateral-**B**ranch **S**iamese **N**etwork (BBSN) model, which is based on a multi-task architecture that can deal with both representation learning and classifier learning. On the one hand, we adapt the Siamese network [1] and propose a novel contrastive learning method to improve representation ability especially for the tail categories. At the same time, we use one branch of the Siamese network for the classification task, which allows the model to learn the classification ability while also preventing overfitting on contrastive learning. In addition, we develop a novel dynamic learning strategy in the Siamese network that allows the network to adjust from classification learning to contrastive learning gradually, making the model end-to-end trainable. On the other hand, we refer the idea of the Bilateral-Branch network [27] from the computer vision domain to improve the model's learning of tail categories. There are two branches in the Bilateral-Branch network responsible for conventional representation learning and classifier learning respectively, we further use the cumulative learning strategy in the Bilateral-Branch network that can shift the learning focus from universal pattern to tail learning. To alleviate the problems in the traditional re-sampling methods, we considered both uniform sampling and re-balanced sampling. The uniform sampling leads to the original long-tailed distribution, while the re-balanced sampling expects to achieve a balanced distribution, but yields another biased distribution due to label-occurrence. Compared to previous approaches, our model avoids overfitting and take care of both representation learning and classifier learning.

We verify the proposed method based on two benchmark datasets with imbalanced data, the experiment results show that our model can significantly improve the performance on the entire and tail categories. The main contributions of this paper can be summarized as follows.

- We propose an end-to-end model BBSN to tackle the data imbalanced problem in the multi-label text classification task with multi-task architecture.
- We propose a category-specific similarity strategy in the Siamese network, and a novel dynamic learning mechanism is developed that make the learning focus of the Siamese network adjust from classification learning to contrastive learning.
- We evaluate our model on two benchmark datasets for multi-label text classification with extremely imbalanced data, and our proposed model achieves competitive performance compared with previous approaches.

2   Related Work

Multi-label Text Classification. Text classification is an important technique for diverse practical classification-related NLP tasks, e.g., scientific claim verification [26], metaphor detection [19], academic document scoring [9,10,16]. Recently, deep learning approaches were proposed to learn better text representations for multi-label text classification task. For example, Liu et al. [14] proposed XML-CNN to use a convolution neural network for text encoding. Shimura et al. [17] proposed a method to learn a local classifier at each hierarchical category level that can effectively utilize the data in the upper levels to contribute to categorization in the lower levels. Yang et al. [25] proposed a hierarchical attention network approach with the attention mechanism to capture the most relevant part of the input text for each label. Shimura et al. [18] presented an approach to text categorization by leveraging a predominant sense of a word depending on the domain. Xiao et al. [22] proposed label-specific attention network to learn new document representation for multi-label text classification. However, such methods ignore the data imbalanced problem, which we believe is crucial for accurate multi-label text classification.

Data Imbalance Distribution in Classification. The imbalanced data is a common problem in the classification task. Most of the existing works are presented in the computer vision domain. For exmaple, Zhou et al. [27] proposed

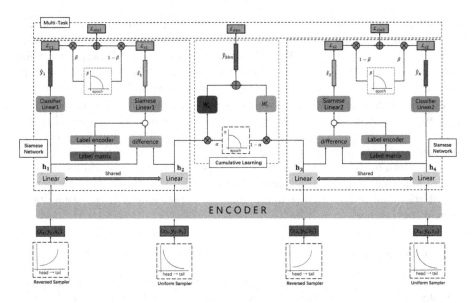

Fig. 2. The framework of our BBSN model.

BBN for the long-tailed visual recognition task. Lin et al. [12] addressed this imbalance problem by reshaping the standard cross-entropy loss. Guo et al. [5] proposed an approach to train on both uniform and re-balanced samplings collaboratively. In recent years, many methods have also been proposed to deal with the long tail problem in the natural language processing domain. For example, Yang et al. [24] proposed HSCNN with a multi-task architecture based on Single and Siamese networks for improving the performance on tail or entire categories. Huang et al. [7] compared the application of a series of balancing loss functions to address the imbalance problem in multi-label classification. In contrast, our approach integrates the Bilateral-Branch network with the Siamese network in a multi-task structure and can handle both representation learning and classifier learning.

3   Methodology

3.1   Definitions and Notations

We denote the dataset as $\mathcal{D} = (x, y)$, here $x = \{x_1, x_2, ..., x_n\}$ is the text content and $y_i \in \{0, 1\}^{|\mathcal{C}|}$ is the corresponding label, we definite $y_i^j = 1$ if category j belongs to instance x_i, where $|\mathcal{C}|$ is the total number of labels. The purpose of a multi-label text classification task is assigning relevant labels to the instances.

3.2   Overview of the Model

Figure 2 illustrates the proposed BBSN model with multi-task architecture that consists of four branches. We use these four branches to form a Bilateral-Branch network and two Siamese networks. Concretely, from left to right, the first and second branches make up one Siamese network to improve the representation learning of the model, meanwhile, the third and fourth branches compose another Siamese network. In addition, we combine the second branch with the third branch as the Bilateral-Branch network, we term the second branch as the "conventional learning branch" for universal pattern learning and the third branch as the "re-balancing branch" for the tail data learning. Each branch has the same network structure and shares the encoder parameters. Further more, the sampler is an important part of our method which controls a focus on the universal pattern or tail data in the learning process. We consider uniform sampling and re-balanced sampling that uniform sampling leads to the original long-tailed distribution and re-balanced sampling follows another biased distribution due to label co-occurrence.

3.3   Data Samplers

We consider two sampling methods, terms the uniform sampler and reversed sampler respectively. The uniform sampler focuses on universal pattern learning and retains the original data distribution that each instance in the training dataset is sampled only once. To give more attention to the tail data, we apply the reversed sampler. For the reversed sampler, the sampling possibility of each category is proportional to the reciprocal of its sample size. In formulations, we denote the number of samples for category j is N_j and the maximum number of all the categories is N_{max}. We calculate the sampling possibility P_j for category j as

$$P_j = \frac{w_j}{\sum_{k=1}^{|C|} w_k} \tag{1}$$

where $w_j = \frac{N_{max}}{N_j}$. We randomly sample a category according to P_j, and we select an instance that contains category j, by repeating this reversed sampler process, training data is obtained.

We sample (x_1, y_1, s_1) and (x_2, y_2, s_1) as the input data for the Siamese network from the uniform sampler and reversed sampler respectively, here the $s_1 \in \mathbb{R}^{|C|}$ is the similarity vector of two branch instances of the Siamese network on every category, and we definite $s_1^j = 1$ if the two branch instances all have the category j, else $s_1^j = 0$.

3.4   Siamese Network Component

In our model, we have two Siamese networks. In each Siamese network, the backbone parameters will be shared. The structure and operation of the two Siamese networks are identical, and here we just introduce one of the Siamese networks as the example.

We denote the (x_1, y_1, s_1) and (x_2, y_2, s_1) as the input data for the Siamese network, where s_1 means the similarity of the two input on each category, x_1 is sampled by the uniform sampler, and x_2 come from the reversed sampler. We fed x_1 and x_2 into the encoder, and next used an extra linear layer to acquire the document feature vectors $\mathbf{h_1} \in \mathbb{R}^d$ and $\mathbf{h_2} \in \mathbb{R}^d$, where d is the hidden size of the feature vector.

On the one hand, we add a classifier in one branch of the Siamese network to learn the classification ability and prevent overfitting. We can get the prediction vectors $\hat{y}_1 \in \mathbb{R}^{|\mathcal{C}|}$ by a classifier

$$\hat{y}_1 = \mathbf{W}_1^\top \mathbf{h}_1 + \mathbf{b}_1 \tag{2}$$

where $\mathbf{W}_1 \in \mathbb{R}^{d \times |\mathcal{C}|}$ and $\mathbf{b}_1 \in \mathbb{R}^{|\mathcal{C}|}$ are randomly initialized and then continuously update during the training process. We apply binary cross entropy as our loss function

$$BCE(\hat{y}, y) = ylog(\hat{y}) + (1 - y)log(1 - \hat{y}) \tag{3}$$

therefore, the loss between $\hat{y}_1$ and ground truth y_1 can be calculated as

$$\mathcal{L}_{c_1} = BCE(\hat{y}_1, y_1) \tag{4}$$

Category-Specific Similarity. On the other hand, we aim to improve the representation ability of the model. The category-specific similarity in HSCNN [24] just calculates the similarity of two instances on a certain category. In the inference stage, the HSCNN model compares each test instance with the sample instances from tail categories to classify the test instances. However, we need an end-to-end structure from the input instance to the output labels in our model. We thus propose a strategy that calculates the similarity of two instances on each category, and the similarity vector $\hat{s}_1$ calculated by

$$\mathbf{H_c} = ReLU(\frac{\mathbf{W}_l^\top \mathbf{Q_c}}{\sqrt{|\mathcal{C}|}})^\top \tag{5}$$

$$\mathbf{H} = |\sigma(\mathbf{H_1}) - \sigma(\mathbf{H_2})| \circ \mathbf{H_c} \tag{6}$$

$$\hat{s}_1 = \mathbf{W_{s_1}} \mathbf{H}^\top + \mathbf{b_{s_1}} \tag{7}$$

here $\mathbf{Q_c} \in \mathbb{R}^{|\mathcal{C}| \times |\mathcal{C}|}$ is the eye matrix about the one-hot encoding of every category, $\mathbf{W}_l \in \mathbb{R}^{|\mathcal{C}| \times d}$ is the parameter weights of the label encoder. The $\mathbf{H_1} \in \mathbb{R}^{|\mathcal{C}| \times d}$ and $\mathbf{H_2} \in \mathbb{R}^{|\mathcal{C}| \times d}$ is the matrix generated by repeating $\mathbf{h_1}$ and $\mathbf{h_2}$ respectively, and $\mathbf{W_{s_1}} \in \mathbb{R}^{1 \times d}$, $\mathbf{b_{s_1}} \in \mathbb{R}^{|\mathcal{C}|}$ are the weights and bias of siamese linear, $\sigma(\cdot)$ means the sigmoid function, $\circ$ means the element-wise multiplication. We use binary cross entropy loss function to calculate the similarity loss, therefore, the loss between $\hat{s}_1$ and s_1 is illustrated as

$$\mathcal{L}_{s_1} = BCE(\hat{s}_1, s_1) \tag{8}$$

Dynamic Learning Mechanism. Additional, different from the staged training method in HSCNN [24], we propose a dynamic learning mechanism that can adjust the training focus from classification learning to contrastive learning gradually by a trade-off parameter

$$\beta = 1 - \left(\frac{T}{T_{max}}\right)^2 \tag{9}$$

where T is the current epoch and T_{max} is the total training epochs, which makes our method an end-to-end trainable model, and gives the model more freedom to automatically adjust to the data.

We add up the classification loss and the similarity loss as the total loss of a Siamese network:

$$\mathcal{L}_{sia_1} = \beta \mathcal{L}_{c_1} + (1 - \beta)\mathcal{L}_{s_1} \tag{10}$$

and the other Siamese network has a symmetric structure, we can obtain the feature vectors $\mathbf{h_3} \in \mathbb{R}^d$ and $\mathbf{h_4} \in \mathbb{R}^d$ just like $\mathbf{h_1}$ and $\mathbf{h_2}$, and calculate the loss $\mathcal{L}_{c_2}$ and $\mathcal{L}_{s_2}$ which is equivalent to $\mathcal{L}_{c_1}$ and $\mathcal{L}_{s_1}$ respectively, and we can get the loss of the second Siamese network:

$$\mathcal{L}_{sia_2} = \beta \mathcal{L}_{c_2} + (1 - \beta)\mathcal{L}_{s_2} \tag{11}$$

3.5 Bilateral-Branch Network Component

The BBN [27] is originally proposed in the computer vision domain with the conventional learning branch and re-balancing branch for the long-tailed visual recognition task. In this work, we see the vector $\mathbf{h_2} \in \mathbb{R}^d$ as the feature input for the conventional learning branch produced by the uniform sampler, which is responsible for learning universal patterns. At the same time, we see the vector $\mathbf{h_3} \in \mathbb{R}^d$ as the feature input for the re-balancing learning branch that comes from the reversed sampler and is responsible for modeling the tail data.

We apply the cumulative learning strategy to shift the learning focus between the conventional learning branch and re-balancing learning branch follow by original BBN method, which controls the weights for $\mathbf{h_2}$ and $\mathbf{h_3}$ with an adaptive parameter α

$$\alpha = 1 - \left(\frac{T}{T_{max}}\right)^2 \tag{12}$$

the weighted feature vectors $\alpha \mathbf{h_2}$ and $(1 - \alpha)\mathbf{h_3}$ will be sent into the classifier $\mathbf{W_c} \in \mathbb{R}^{d \times |\mathcal{C}|}$ and $\mathbf{W_r} \in \mathbb{R}^{d \times |\mathcal{C}|}$ respectively, and the output will be integrated by element-wise addition. The output logits of the Bilateral-Branch network is illustrated as

$$\hat{y}_{bbn} = \alpha \mathbf{W_c}^\top \mathbf{h_2} + (1 - \alpha)\mathbf{W_r}^\top \mathbf{h_3} \tag{13}$$

Table 1. The data statistics about RCV1 and AAPD dataset.

Dataset	Train	Test	$|\mathcal{C}|$	$\overline{N_c}$	N_c^{max}	N_c^{min}
RCV1	23149	781265	103	225	10787	1
AAPD	54840	1000	54	2340	17152	350

In addition, the cumulative learning strategy also acts on the loss function, which is designed to first learn the universal patterns and then pay attention to the tail data gradually that can exhaustively improve long-tailed classification performance. The weighted cross-entropy classification loss of the Bilateral-Branch network is illustrated as

$$\mathcal{L}_{bbn} = \alpha BCE(\hat{y}_{bbn}, y_2) + (1 - \alpha)BCE(\hat{y}_{bbn}, y_3) \tag{14}$$

3.6 Multi-task Learning Architecture

We adopt multi-task learning architecture in our model that combined the Siamese network and the Bilateral-Branch network, which can both take care of representation learning and classifier learning simultaneously. Briefly, we obtain the total loss of our model:

$$\mathcal{L} = \mathcal{L}_{sia1} + \mathcal{L}_{sia2} + \mathcal{L}_{bbn} \tag{15}$$

and the whole network is unified and end-to-end trainable. During inference, we fix the $\alpha = 0.5$ and use the $\hat{y}_{bbn}$ as the model prediction for test samples.

4 Experiment

4.1 Dataset

We use two benchmark datasets for long-tailed multi-label text classification, including RCV1 [8] and AAPD [23]. Table 1 lists the dataset statistics. There are many categories, and the categories are extremely imbalanced on the instance numbers. We randomly split the raw training data into 80% for training and 20% for validation. For RCV1 dataset, we set the tail threshold $N_\phi = 100$, and there are 35 categories to be considered as tail categories, and we get 23 tail categories in AAPD dataset while we set the tail threshold $N_\phi = 1000$.

4.2 Implementation Details

We use BERT [4] as our model's encoder and use the bert-base-cased version for all experiments. We take Adam as our optimizer and set the initial learning rate = 5e-5 with the warmup linear scheduler, and set the feature vector hidden size $d = 1024$. According to existing work [6], we train our models for 50 epochs with a batch size of 32. We use F1-score as our metric, and we use both micro-F1 and

Table 2. Main results on the RCV1 dataset.

Model	Micro-F1	Macro-F1	Micro-P@1	Micro-P@3	Micro-P@5	Macro-P@1	Macro-P@3	Macro-P@5
Results1: Entire Categories Metrics on the RCV1 Dataset								
BERT	87.11	65.03	**97.64**	82.24	54.20	88.22	64.97	39.13
WCELoss	86.49	64.22	97.26	81.90	53.94	90.46	69.41	44.13
FocalLoss	**87.15**	66.72	97.57	82.16	54.10	91.15	66.51	38.06
DBLoss	85.08	65.68	97.51	82.57	55.00	**93.00**	69.29	43.53
CTURS	85.69	66.12	97.26	81.20	53.11	85.47	63.86	38.61
BBSN (ours)	86.58	**70.48**	95.72	**82.66**	**55.75**	82.66	**69.43**	**44.69**
Results2: Tail Categories Metrics on the RCV1 Dataset								
BERT	58.84	40.69	45.36	17.02	10.21	34.16	17.86	10.46
WCELoss	57.59	39.14	44.29	16.42	9.85	35.35	17.70	10.10
FocalLoss	61.76	45.55	48.51	18.13	10.88	36.40	17.68	10.27
DBLoss	62.77	50.75	55.76	21.50	13.03	43.90	19.40	12.44
CTURS	66.03	52.68	54.84	20.39	12.23	42.23	20.33	12.22
BBSN (ours)	**75.10**	**63.25**	**68.76**	**25.81**	**15.49**	**60.86**	**26.80**	**17.37**

macro-F1 to provide a more analytical perspective. We also use the precision at top-k (P@k) as the evaluation metric. Because the numbers of instances of tail categories are much smaller than those of the head categories, the micro average is not exactly fair to the tail categories. Therefore, similar to the micro and macro average to the F1-score, we also use both micro and macro average to the P@k.

4.3   Baselines

We compare our proposed approach with the following existing works.

BERT [4]: The powerful pre-train language model in natural language processing domain, for multi-label text classification task, which only uses the text as input. In this paper, we use two linear networks as a classifier after BERT.

WCE Loss [12]: The weighted cross-entropy loss function, we apply this loss function based on the BERT baseline.

Focal Loss [12]: A re-weight loss approach that focuses training on a sparse set of hard examples. We apply this re-weight loss based on the BERT baseline.

DBLoss [7]: The distribution-balanced loss approach, which is critical in the multi-label scenario and explicitly assigns lower weight on easy-to-classify negative instances. We apply this loss function based on the BERT baseline.

CTURS [5]: It was originally proposed for solving long-tailed visual recognition task, which trains both uniform and re-balanced samplings in a collaborative way. For a fair comparison, we use the same backbone with our method.

4.4   Results and Discussion

Firstly, The RCV1 dataset results are listed in Table 2. We can see our model achieves competitive performance compared to other approaches in the entire

Table 3. Main results on the AAPD dataset.

Model	Micro-F1	Macro-F1	Micro-P@1	Micro-P@3	Micro-P@5	Macro-P@1	Macro-P@3	Macro-P@5
Result1: Entire Categories Metrics on the AAPD Dataset								
BERT	72.90	58.18	84.00	55.43	33.92	72.98	42.87	27.16
WCELoss	71.95	56.65	82.70	54.70	33.32	70.92	44.20	26.91
FocalLoss	72.78	58.19	83.70	54.93	33.44	72.04	42.31	26.00
DBLoss	72.17	57.68	83.80	57.40	35.40	74.91	46.76	**30.45**
CTURS	72.08	57.10	83.80	52.43	31.76	74.53	39.75	24.94
BBSN (ours)	**74.05**	**62.20**	**84.50**	**59.80**	**37.20**	**79.86**	**48.45**	30.17
Result2: Tail Categories Metrics on the AAPD Dataset								
BERT	55.58	50.32	42.28	16.53	9.92	41.34	21.97	14.71
WCELoss	50.70	47.50	37.80	14.77	8.86	37.97	19.11	12.30
FocalLoss	55.78	50.84	44.72	16.67	10.00	42.63	20.06	13.21
DBLoss	53.45	48.68	45.12	16.80	10.08	43.18	19.01	12.45
CTURS	54.02	51.22	37.80	14.53	8.78	37.19	17.46	11.01
BBSN (ours)	**65.04**	**60.13**	**55.28**	**21.68**	**13.01**	**52.57**	**25.68**	**16.14**

categories. The results of the tail categories demonstrate that BERT does not perform as well as other approaches which focus on the imbalanced distributed data. Additionally, our model consistently achieved the best performance on the tail categories. Specifically, the tail micro-F1 score gains 12.33% from DBLoss and 9.07% from CTURS method.

Second, Table 3 demonstrates the results of the AAPD dataset. For this dataset, the other approaches did not work efficiently compared with BERT. On the contrary, our method has an advantage of 1.15% and 4.02% in the entire micro-F1 and macro-F1 score compared with BERT. Notably, the improvement of the tail categories is more prominent than the entire categories. Specifically, our approach outperformed 9.46% in the tail micro-F1 score and 9.81% in the tail macro-F1 score compare to BERT.

Among these methods, BERT is the simplest baseline which directly used in the long-tailed datasets. As expected, BERT is worse than our model in terms of average performance, especially in the tail categories. The WCE loss, Focal loss and DB loss are re-weight loss approaches for tackling imbalanced data, although they have improved in average performance compared with BERT, the effect is still not as good as our method, because they do not comply with the principle of cross-entropy loss. The collaborative training approach of CTURS increases the uniformity of the two branches, but also affects the representation ability of the model. In general, our method takes care of both representation learning and classifier learning, which has a significant advantage over other baseline methods for tacking the imbalanced data.

4.5 Ablation Study

To evaluate the effectiveness of each component, we conduct a detailed ablation study on the proposed architecture. Firstly, we remove the Bilateral Branch network which is represented as w/o BBN. Secondly, the Siamese network is

Table 4. The ablation study results.

Model	Micro-F1	Macro-F1	Micro-P@1	Micro-P@3	Micro-P@5	Macro-P@1	Macro-P@3	Macro-P@5
Results1: Entire Categories Metrics on the RCV1 Dataset								
w/o BBN	85.67	67.11	94.77	81.20	53.79	81.67	62.28	38.05
w/o Siamese	85.10	64.45	93.66	81.96	55.19	**88.71**	68.39	44.07
Our method	**86.58**	**70.48**	**95.72**	**82.66**	**55.75**	82.66	**69.43**	**44.69**
Results2: Tail Categories Metrics on the RCV1 Dataset								
w/o BBN	65.74	53.04	54.90	20.46	12.28	45.65	22.11	14.10
w/o Siamese	69.99	52.01	65.32	25.73	15.45	48.88	22.38	14.17
Our method	**75.10**	**63.25**	**68.76**	**25.81**	**15.49**	**60.86**	**26.80**	**17.37**
Results3: Entire Categories Metrics on the AAPD Dataset								
w/o BBN	70.61	57.56	83.10	56.83	35.16	73.76	44.29	27.63
w/o Siamese	72.03	58.24	83.00	57.67	35.60	73.45	45.68	29.95
Our method	**74.05**	**62.20**	**84.50**	**59.80**	**37.20**	**79.86**	**48.45**	**30.17**
Results4: Tail Categories Metrics on the AAPD Dataset								
w/o BBN	61.38	57.31	51.63	20.46	12.28	**53.56**	23.61	15.05
w/o Siamese	60.71	56.53	50.41	19.24	11.54	48.24	23.64	15.30
Our method	**65.04**	**60.13**	**55.28**	**21.68**	**13.01**	52.57	**25.68**	**16.14**

clipped, and the model is called w/o Siamese. Table 4 lists the ablation performance results.

We can see that the removal of the Bilateral-Branch network (w/o BBN) significantly hampers the model's performance. The Bilateral-Branch network is divided into a conventional learning branch and a re-balancing branch, which focuses on the universal pattern of the original dataset and the tail learning on the new dataset generated by reversed sampling. The experimental results show that our model drops the entire micro-F1 score by at least 0.91% without the Bilateral-Branch network, and this shortcoming is magnified to 3.67% on the tail categories.

We also eliminate the Siamese network component (w/o Siamese) to verify its effectiveness. The Siamese network is responsible for the representation learning by contrastive learning with the category-specific similarity strategy and dynamic learning mechanism. This decreases the model average performance which implies that contrastive learning is important for representation learning. This also proves that representation learning is crucial for imbalanced multi-label text classification.

5 Conclusion and Future Work

In this paper, we propose a unified model BBSN with a multi-task architecture to tackle the imbalanced distribution problem in the multi-label text classification task. In addition, we propose a category-specific similarity strategy to improve the representation learning and apply a novelty dynamic learning mechanism to make the model end-to-end trainable. Compared to previous approaches, our method can improve the performance on the entire categories and tail categories, and achieves a competitive result.

Acknowledgements. This work was partially supported by JKA and by Research Grant for Young Scholars funded by Yamanashi Prefecture.

References

1. Bromley, J., Guyon, I., LeCun, Y., Säckinger, E., Shah, R.: Signature verification using a Siamese time delay neural network. Adv. Neural Inf. Process. Syst. **6** (1993)
2. Chu, P., Bian, X., Liu, S., Ling, H.: Feature space augmentation for long-tailed data. In: European Conference on Computer Vision, pp. 694–710 (2020)
3. Cui, Y., Jia, M., Lin, T.Y., Song, Y., Belongie, S.: Class-balanced loss based on effective number of samples. In: Proceedings of the IEEE/CVF Conference on Computer Vision and Pattern Recognition, pp. 9268–9277 (2019)
4. Devlin, J., Chang, M.W., Lee, K., Toutanova, K.: BERT: pre-training of deep bidirectional transformers for language understanding. In: Proceedings of the 2019 Conference of the North American Chapter of the Association for Computational Linguistics, pp. 4171–4186 (2019)
5. Guo, H., Wang, S.: Long-tailed multi-label visual recognition by collaborative training on uniform and re-balanced samplings. In: Proceedings of the IEEE/CVF Conference on Computer Vision and Pattern Recognition, pp. 15089–15098, June 2021
6. Hu, T., Li, J., Fukumoto, F., Zhou, R.: A multi-task based bilateral-branch network for imbalanced citation intent classification. In: 16th International Conference on Ubiquitous Information Management and Communication, pp. 1–8 (2022)
7. Huang, Y., Giledereli, B., Köksal, A., Özgür, A., Ozkirimli, E.: Balancing methods for multi-label text classification with long-tailed class distribution. In: Proceedings of the 2021 Conference on Empirical Methods in Natural Language Processing, pp. 8153–8161 (2021)
8. Lewis, D.D., Yang, Y., Russell-Rose, T., Li, F.: RCV1: a new benchmark collection for text categorization research. J. Mach. Learn. Res. **5**, 361–397 (2004)
9. Li, J., Fukumoto, F.: Multi-task neural shared structure search: a study based on text mining. In: International Conference on Database Systems for Advanced Applications, pp. 202–218 (2021)
10. Li, J., Sato, A., Shimura, K., Fukumoto, F.: Multi-task peer-review score prediction. In: Proceedings of the First Workshop on Scholarly Document Processing, pp. 121–126 (2020)
11. Li, Y., Wang, T., Kang, B., Tang, S., Wang, C., Li, J., Feng, J.: Overcoming classifier imbalance for long-tail object detection with balanced group softmax. In: Proceedings of the IEEE/CVF Conference on Computer Vision and Pattern Recognition, pp. 10991–11000 (2020)
12. Lin, T.Y., Goyal, P., Girshick, R., He, K., Dollár, P.: Focal loss for dense object detection. In: Proceedings of the IEEE International Conference on Computer Vision, pp. 2980–2988 (2017)
13. Liu, J., Sun, Y., Han, C., Dou, Z., Li, W.: Deep representation learning on long-tailed data: a learnable embedding augmentation perspective. In: Proceedings of the IEEE/CVF Conference on Computer Vision and Pattern Recognition, pp. 2970–2979 (2020)
14. Liu, J., Chang, W.C., Wu, Y., Yang, Y.: Deep learning for extreme multi-label text classification. In: Proceedings of the 40th International ACM SIGIR Conference on Research and Development in Information Retrieval, pp. 115–124 (2017)
15. More, A.: Survey of resampling techniques for improving classification performance in unbalanced datasets. arXiv preprint arXiv:1608.06048 (2016)

16. Muangkammuen, P., Xu, S., Fukumoto, F., Saikaew, K.R., Li, J.: A neural local coherence analysis model for clarity text scoring. In: Proceedings of the 28th International Conference on Computational Linguistics, pp. 2138–2143 (2020)
17. Shimura, K., Li, J., Fukumoto, F.: HFT-CNN: learning hierarchical category structure for multi-label short text categorization. In: Proceedings of the 2018 Conference on Empirical Methods in Natural Language Processing, pp. 811–816 (2018)
18. Shimura, K., Li, J., Fukumoto, F.: Text categorization by learning predominant sense of words as auxiliary task. In: Proceedings of the 57th Annual Meeting of the Association for Computational Linguistics, pp. 1109–1119 (2019)
19. Su, C., Fukumoto, F., Huang, X., Li, J., Wang, R., Chen, Z.: DeepMet: a reading comprehension paradigm for token-level metaphor detection. In: Proceedings of the Second Workshop on Figurative Language Processing, pp. 30–39 (2020)
20. Wallace, B.C., Small, K., Brodley, C.E., Trikalinos, T.A.: Class imbalance, redux. In: 2011 IEEE 11th International Conference on Data Mining, pp. 754–763 (2011)
21. Wu, T., Huang, Q., Liu, Z., Wang, Y., Lin, D.: Distribution-balanced loss for multi-label classification in long-tailed datasets. In: European Conference on Computer Vision, pp. 162–178 (2020)
22. Xiao, L., Huang, X., Chen, B., Jing, L.: Label-specific document representation for multi-label text classification. In: Proceedings of the 2019 Conference on Empirical Methods in Natural Language Processing and the 9th International Joint Conference on Natural Language Processing, pp. 466–475 (2019)
23. Yang, P., Sun, X., Li, W., Ma, S., Wu, W., Wang, H.: SGM: sequence generation model for multi-label classification. In: Proceedings of the 27th International Conference on Computational Linguistics, pp. 3915–3926 (2018)
24. Yang, W., Li, J., Fukumoto, F., Ye, Y.: HSCNN: a Hybrid-Siamese convolutional neural network for extremely imbalanced multi-label text classification. In: Proceedings of the 2020 Conference on Empirical Methods in Natural Language Processing, pp. 6716–6722 (2020)
25. Yang, Z., Yang, D., Dyer, C., He, X., Smola, A., Hovy, E.: Hierarchical attention networks for document classification. In: Proceedings of the 2016 Conference of the North American Chapter of the Association for Computational Linguistics: Human Language Technologies, pp. 1480–1489 (2016)
26. Zhang, Z., Li, J., Fukumoto, F., Ye, Y.: Abstract, rationale, stance: a joint model for scientific claim verification. In: Proceedings of the 2021 Conference on Empirical Methods in Natural Language Processing, pp. 3580–3586 (2021)
27. Zhou, B., Cui, Q., Wei, X.S., Chen, Z.M.: BBN: bilateral-branch network with cumulative learning for long-tailed visual recognition. In: Proceedings of the IEEE/CVF Conference on Computer Vision and Pattern Recognition, pp. 9719–9728 (2020)

Deep Hierarchical Semantic Model for Text Matching

Chongxu Hu[1,2], Xiaoyan Gongye[1(✉)], and Xiaohu Zhang[1,3]

[1] School of Cyber Science and Engineering, Qufu Normal University, Jining, China
`yigongsd@qfnu.edu.cn`
[2] School of Computer Science and Engineering, Tianjin University of Technology, Tianjin, China
[3] School of Control and Computer Engineering, North China Electric Power University, Baoding, China

Abstract. In recent years, some effective textual matching models are proposed for solving NLP-related tasks. However, these models have the following issues: they cannot extract semantic information at different levels from the words of text pairs; they cannot integrate the low-level information to fine-tune the high-level information. To address these, this paper proposes a novel deep learning neural network, namely deep hierarchical semantic model (DHSM), for text matching, which consists of multiple semantic processing layers, a pooling layer, and a prediction layer. Specifically, each semantic processing layer consists of three parts: encoding part, interaction part and fusion part; and it can well represent the semantic information and enable the information interaction at different levels through attention mechanism. Moreover, the pooling layer uses pooling method to extract key information of the text pairs, based on which the prediction layer determines the relationship between text pairs.

Keywords: Text Matching · Deep Hierarchical Semantics Model · Deep Learning · Neural network · Attention Mechanism

1 Introduction

Text matching, as a core task in natural language processing(NLP), has been widely used in related tasks including natural language inference [12], textual entailment [13,26] and paraphrase identification [4,5]. Text matching aims to predict the semantic relationship that exists between two texts.

A great text matching model needs to effectively capture the rich semantic information in the text as well as the interaction information during the matching process. Traditional text matching models rely on features of manual tagging which cannot capture the deep semantics of text. In contrast, deep learning models can better capture and represent the rich semantic information of text through deep neural networks [2]; and effectively capture the semantic

M. Tanveer et al. (Eds.): ICONIP 2022, LNCS 13625, pp. 397–408, 2023.
https://doi.org/10.1007/978-3-031-30111-7_34

information of the interaction between two texts through an attention-based interaction module [25]. However, many deep learning models ignore the impact of different levels of representative and interactive information about the words in a text pair on text matching.

Representative information is different from words with different lengths of contextual information in the text, and by the same token, interaction information generated when words with different representative information interact is also different. Take the text matching of the following two texts as an example:

Text1: Last week, teachers and students saw notable places.

Text2: Last week, students and teachers visited famous sites.

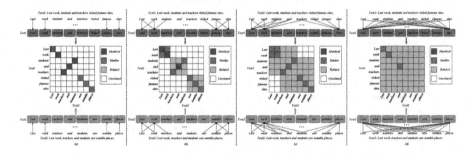

Fig. 1. Interaction between different semantic information of a word.

Figure 1 reflects that the interaction of words containing different semantic information in a text pair yields different relationship matrices, and the interactive information reflected in the relationship matrix is also different. The information contained in the words in Fig. 1(a)(b)(c)(d) are self-semantic information, short-distance local semantic information, long-distance local semantic information, and global semantic information, respectively. The interactive information focuses on the interactions between words in Fig. 1(a). The interactive information can reflect the interaction relations between phrases in Fig. 1(b). The interactive information can further enhance the interactive relationship between phrases in Fig. 1(c). The interactive information can reflect the interactive relationship between sentences in Fig. 1(d). It can be seen that with the increasing level of representative information of words, the interactive information reflected by the interaction between words in two texts also shows the characteristic of hierarchical structure. In Fig. 1, the high-level interactive information cannot clearly reflect the low-level interactive information, it is necessary to include the low-level information when constructing the high-level information.

Based on the above analysis, an effective text matching model faces three additional challenges. Firstly, existing matching models ignore the hierarchical relationship between the different representative and interactive information of words. Secondly, existing matching models cannot effectively capture the information at different levels of words. Finally, existing models lack an effective way to transfer information between different levels. To address the above challenges, this paper attempts to propose a novel model, and the specific contributions of this paper are as follows.

1) In this paper, we propose a novel model, namely DHSM, which makes full use of representative information and interactive information of different levels of words in the text to provide an effective method for realizing the text matching, and verify the effectiveness of the model through several text matching tasks.

2) In this paper, we have trained multiple semantic processing layers with encoding part, interaction part and fusion part from the perspective of hierarchy. Encoding part of each layer uses attention-based methods to capture and express representative information of words at different levels. Interaction part of each layer obtains interactive information at different levels through attention-based methods. And fusion part of each layer transfers the information from the lower-level to the higher-level information.

3) In this paper, extensive experiments are conducted to evaluate the performance of DHSM on three tasks, i.e., natural language inference, textual entailment and paraphrase identification. The experimental results show the effectiveness of the DHSM model on these datasets.

2   Related Work

Existing text matching models can be divided into traditional text matching model approaches and deep learning matching models.

Traditional text matching models mainly include the following models: constructed probability models based on word frequency distribution determine the relationship between text pairs by high frequency words in the text [6,17], e.g., TF-IDF and BM25. LDA-based [16,28] text matching models determine relationships by extracting topics from the text.

Traditional text matching relies on hand-crafted features that fail to capture the deep semantics of text. Existing deep learning text matching models are as follows. Hu et al. [7] proposed an ARCI model using CNN to fuse the semantic information of neighboring words in a text. Marco et al. [14] and Shen et al. [23] proposed improved methods such as Tree-LSTM to obtain the long-range dependency information of the text. With the proposed self-attentive mechanism, Liu et al. [15] use the attention mechanism to capture and express the semantic information of text instead of the traditional LSTM. ESIM [1] encodes the semantic information using BiLSTM and the alignment information obtained from the attention mechanism. With the development of preprocessing models such as BERT, Sun et al. [24] used BERT preprocessing techniques to solve the text matching problem.

3   Proposed Method

This section describes DHSM architecture, as shown in Fig. 2(a). It consists of the following layers: embedding layer, semantic processing layers, pooling layer, and prediction layer. The text pairs inputs are defined as $T_a = \{a_1, a_2, \cdots a_{la}\}$ and $T_b = \{b_1, b_2, \cdots b_{lb}\}$, where a_i is the i-th word in text T_a, la is the length of text T_a, b_j is the j-th word in text T_b, and lb is the length of text T_b. The structure and functions of each layer are described in detail below.

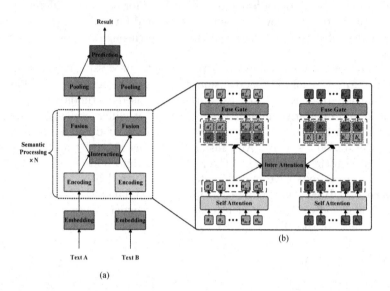

Fig. 2. The structure diagram of DHSM, (a) is the structure of DHSM, (b) is the detail of semantic processing layer.

3.1 Embedding Layer

The role of the embedding layer is to map words to a higher dimensional space. In this paper, word embedding is used to represent words. In word embedding, each word is represented as a high-dimensional vector by using a pre-trained word embedding method (e.g. GloVe). Through the embedding layer, the two texts are transformed into their respective embedding matrices represented as $A \in \mathbb{R}^{la \times d}$ and $B \in \mathbb{R}^{lb \times d}$, where d denotes the dimensionality of the embedding vector.

3.2 Semantic Processing Layer

There are several semantic processing layers in the model, and the role of different semantic processing layers is to extract the representative information and interactive information of words at different levels and fuse them to pass them to the next layer. As show in Fig. 2(b), the semantic processing layer is composed of three parts: the encoding part, the interaction part and the fusion part.

Encoding Part. The role of the encoding part is to extract the representative information of words. Given that the encoding parts of different semantic processing layers need to obtain representative information of different length of words, the encoding part uses a window-based multi-head self-attention mechanism to calculate representative information of each word, where only the attention within the window is calculated for each word. By adjusting the size of the words' windows in the encoding parts of the different semantic processing layers, we obtain representative information about words of different length; in particular, the words' windows in the bottom layer contain only their own words, the words' windows in the top layer contain all the words in the sentence. The size of the words' window is increased layer by layer, and the increment of the words' window(δ_I) is determined by the text length(L) and the number of layers(N), as shown by:

$$\delta_I = Ceiling\left((L-1)/2 \times (N-1)\right) \tag{1}$$

Ceiling denotes upward rounding function. The word is centered left and right to increase δ_I. The encoding parts T_a and T_b of each semantic processing layer process the representative information of words by the same method, and they share the same encoding part structure. As an example, the operation of T_a in the encoding part is as follows:

$$A^e = SelfAttention\left(A\right) \tag{2}$$

where $A \in \mathbb{R}^{la \times d}$ represents the input of the semantic processing layer, $SelfAttention()$ represents the window-based multi-head self-attention operation, and $A^e \in \mathbb{R}^{la \times d}$ represents the window-based multi-head self-attention output. Similarly, for the text T_b, we get the output representative information $B^e \in \mathbb{R}^{lb \times d}$.

Interaction Part. The role of the interaction part is to obtain the interactive information between a pair of texts. Firstly, The interaction matrix obtained by the attention-based interaction method, and then each row of the interaction matrix is normalized using the *softmax* function as the attention weight of T_a. Similarly, each column in the interaction matrix is processed by a similar method as the attention weight of T_b. The interaction equations are as follows:

$$E = A^e B^{eT} \tag{3}$$

$$A^i = softmax\left(E\right)B^e \tag{4}$$

$$B^i = softmax\left(E^T\right)A^e \tag{5}$$

where $E \in \mathbb{R}^{la \times lb}$ denotes the interaction matrix; $A^i \in \mathbb{R}^{la \times d}$ and $B^i \in \mathbb{R}^{lb \times d}$ represent the interactive information of words between two texts from the perspective of T_a and T_b, respectively.

Fusion Part. The role of the fusion part is to fuse the representative information (e.g.: A^e, B^e) and the interactive information (e.g.: A^i, B^i), and the fused information is used as the output of the semantic processing layer. In this paper, the fusion of A^e and A^i is implemented using a fusion gate with the following equation:

$$\alpha = \left[A^e; A^i; A^e A^i\right] \tag{6}$$

$$z_k = sigmoid\left(\alpha w_k + b_k\right)(k = 1, 2, 3) \tag{7}$$

$$\theta = [\tanh\left(\alpha w_4\right) + b_4] \tag{8}$$

$$A^f = \theta * z_1 + A^e * z_2 + A^i * z_3 \tag{9}$$

where $w_i \in \mathbb{R}^{3d \times d}$ and $b_i \in \mathbb{R}^{3d \times d}$ $(i = 1, 2, 3, 4)$ are the parameters of the fusion gate; $A^f \in \mathbb{R}^{la \times d}$ denotes the output matrix of T_a; $\left[A^e; A^i; A^e A^i\right]$ denotes the join operation; * denotes the element-by-element multiplication method. In the same way, T_b has fused $B^e \in \mathbb{R}^{lb \times d}$ and $B^i \in \mathbb{R}^{lb \times d}$ to obtain the output matrix $B^f \in \mathbb{R}^{lb \times d}$.

3.3 Pooling Layer

The role of the pooling layer is to extract the key information of the semantic processing layer at the top layer. In this paper, maximum pooling and average pooling are used to process the representative information of the semantic processing layer at the top layer. The final pooling vector is the concatenation of the maximum pooling and the average pooling, operations as follows:

$$P_a = \left[MaxPooling\left(A^f\right); MeanPooling\left(A^f\right)\right] \tag{10}$$

$$P_b = \left[MaxPooling\left(B^f\right); MeanPooling\left(B^f\right)\right] \tag{11}$$

$$P = [P_a; P_b] \tag{12}$$

where P_a and P_b denote the connection vectors of the maximum pooling and average pooling vectors of T_a and T_b, respectively, and P is used as the connection vector of the output of the layer.

3.4 Prediction Layer

The prediction layer gets the matching result of two texts by a two-layer feedforward neural network, as shown in equation:

$$\hat{r} = softmax\left(relu\left(relu\left(P * w_5 + b_5\right) * w_6 + b_6\right)\right) \tag{13}$$

where w_5, w_6 and b_5, b_6 denote the parameters of each layer of the feedforward neural network, respectively. Since this paper studies classification tasks, $softmax$ is used as the final activation function.

3.5 Loss Function

The loss function used in this paper for the classification task is cross entropy. Its formula is expressed as follows:

$$Loss = -\sum_{i=1}^{k} r_i \log \hat{r}_i \tag{14}$$

where r_i denotes the true value of the sample for the i-th classification, $\hat{r}_i$ denotes our model output value of the sample for the i-th classification, and k is the number of classes.

4 Experiments

To verify the effectiveness of the DHSM, we have tested the proposed architecture on three tasks: natural language inference, and textual entailment and paraphrase identification.

4.1 Experimental Setup

During training, Adam method [10] and L2 conditioning method were used for optimization with batchsize of 128. Dropout was applied to all feedforward connections with a rate of 0.9. Word embeddings are initialized using a 300-dimensional GloVe word vector (GloVe 840B 300D) [21], and out-of-vocabulary (OOV) word embeddings were initialized by sampling from $[-.01, .01]$, and all word embeddings are fixed during training. Sequence lengths were padded with zeros to the batch maximum.

4.2 Natural Language Inference

The task of natural language inference is to infer logical relations between two texts. The SciTail dataset [9] is the classic dataset for natural language inference tasks, it consists of 27K text pairs, with 23596 text pairs in the training set, 1304 text pairs in the test set, and 2126 text pairs in the validation set.

To verify the validity of the model, we compared the following five baseline data: Chen et al. [1], Parikh et al. [20] and Li et al. [12] introduce an attention-based alignment to model texts. DGEM and DGEM w/o edges proposed by Khot et al. [9] are powerful models on the SciTail dataset.

Table 1 shows that DHSM has 82.5% accuracy on the SciTail test set, outperforming Chen et al. [1], Parikh et al. [20] and Li et al. [12] by 10.2%, 8.8% and 3%, respectively. DHSM improves the best performance compared to DGEM [9] and DGEM w/o edges [9] before by 11.7% and 5.2%, respectively.

Table 1. Comparison with other models on the SciTail test set.

Model	Acc.(%)
Chen et al.(ESIM) [1]	72.3
Parikh et al.(DecAtt) [20]	73.7
Li et al.(SA- ESIM) [12]	79.5
Khot et al.(DGEM) [9]	70.8
Khot et al.(DGEM w/o edges) [9]	77.3
DHSM	82.5

4.3 Textual Entailment

The textual entailment task aims to determine whether there is an entailment relationship between two texts. The SemEval 2014 [18] task evaluates the system's prediction of the textual entailment relationship for text pairs in the SICK dataset [19]. The three class are entailment, contradiction and neutral. The sizes of text pairs from the SICK training, validation, and test sets were 4439, 495, and 4906, respectively.

To verify the validity of the model, we compared the following five baseline data: Alice et al. [11], Jiang et al. [27] and Sergio et al. [8] are the top three systems in the SemEval task. Shen et al. [23] and Marco et al. [14] are based on the improved Tree-LSTM structure for text matching.

Table 2 shows that DHSM has an accuracy of 85.0% on the SICK test set, and compared to the top three models in the SemEval task, DHSM improves the performance by 0.4%, 1.4%, and 1.9%, respectively. DHSM outperforms the Shen et al. [23] and Marco et al. [14] by 4.8% and 1.4%, respectively.

Table 2. Comparison with other models on the SICK test set.

Model	Acc.(%)
Alice et al.(Illinois-LH) [11]	84.6
Jiang et al.(ECNU) [27]	83.6
Sergio et al.(UNAL-NLP) [8]	83.1
Shen et al.(TG-HTreeLSTM) [23]	83.6
Marco et al.(DC-TreeLSTM) [14]	80.2
DHSM	85.0

4.4 Paraphrase Identification

The main task of the paraphrase identification is to determine whether two sentences have the same meaning, and this task is considered an important task for text matching. We use the Microsoft Research Paraphrase (MSRP) corpus [3]. The train set contains 2753 true/1323 false pairs and the test set contains 1147 true/578 false paraphrase pairs.

To verify the validity of the model, we compared the following five baseline data: Rus et al. [22], El-Alfy et al. [4] and Ferreira et al. [5] are traditional text matching model and have achieved good results on paraphrase identification tasks. Hu et al. [7] and Hu et al. [7] are both deep models for extracting interaction features by CNN.

Table 3 shows that DHSM has an accuracy of 74.3% on the MSRP test set, Compared with Rus et al. [22], El-Alfy et al. [4] and Ferreira et al. [5], DHSM improves by 4%, 0.4% and 0.2%, respectively. DHSM outperforms Hu et al. [7] and Hu et al. [7] with 2.68% and 2.38%, respectively.

Table 3. Comparison with other models on the MSRP test set.

Model	Acc.(%)
Rus et al. [22]	70.3
El-Alfy et al. [4]	73.9
Ferreira et al. [5]	74.1
Hu et al.(ARC-I) [7]	69.6
Hu et al.(ARC-II) [7]	69.9
DHSM	74.3

4.5 Analysis of the Number of Semantic Processing Layers and the Increment of the Words' Window

In DHSM, the smaller the number of semantic processing layers, the larger the words' window increment will be, resulting in the model not being able to access the semantic and interaction information of rich words; conversely, too many layers will result in the model's lower-level information not being effectively transferred to the higher-level information. To investigate the effects of window increments and the number of semantic processing layers on model performance,

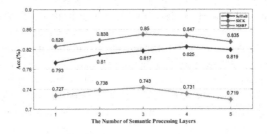

Fig. 3. Influence of the numbere of semantic processing layers.

this paper conducted experiments on the SciTail, SICK and MSPC datasets with the values of semantic processing layers while keeping other parameters constant, and the experimental results are shown in Fig. 3.

The experimental results demonstrate that the accuracy of the test set of SciTail, SICK, and MSRP shows a trend of increasing and then decreasing with the increase of the number of semantic processing layers. This indicates that multiple semantic processing layers can extract and transfer the semantic information of words at different levels in the text. However, the performance of DHSM decreases when the model is too complex.

4.6   Visualization Analysis

To verify that the interaction module in each semantic processing layer can effectively obtain interaction information and the effectiveness of the multi-layer interaction module, this paper extracts a text pair from the SciTail dataset and analyzes the word-to-word attention distribution of different semantic processing layers. Figure 4 in (a) (b) (c) (d) represents the heat maps of interaction modules in semantic processing layers of layers 1,2,3,4, respectively.

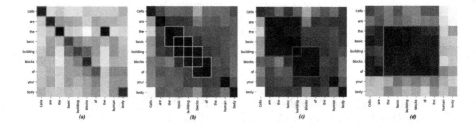

Fig. 4. Visualization of attention weights.

In Fig. 4(a), the matrices in the first layer reflect mainly the interactions between words in a text pair, framed in yellow respectively. In Fig. 4(b), the matrix at the second level reflects the interactions between phrases in the text pair, respectively framed in pink. In Fig. 4(c), the third level of the matrix reflects the interactions between phrases in the text pair, respectively framed in red. In Fig. 4(d), the matrix in the fourth layer further enhances the ability of the text to reflect the interaction relationships, as can be seen in the green boxes. In addition, each layer contains the interactions from the previous layer. The heat map demonstrates that the interaction relationships between text pairs increase as the semantic processing layer increases, and that the interaction information at higher levels can reflect the interaction relationships at lower levels, demonstrating the effectiveness of DHSM interaction method and information transfer.

5   Conclusion

In this paper, we discuss the hierarchical relationships between different semantic information in the text and the impact on text matching. Based on this, DHSM proposed in this paper implements the extraction of representative and interactive information about different levels of words, as well as the fusion and transfer of information between different levels, and determines the relationships between text pairs based on this information. In this paper, we have conducted extensive experiments on natural language inference tasks, text nesting tasks and paraphrase recognition tasks, and DHSM outperforms most existing models on many tasks and has good generalisation capabilities.

In the further, we will apply the preprocessing model to DHSM to enhance the performance of the model further; meanwhile, based on the existing research, we will build a model to optimize the encoding and analysis of text by using semantic trees or semantic graphs. Moreover, we plan to apply DHSM to more natural language related tasks to validate the model's capability.

References

1. Chen, Q., Zhu, X., Ling, Z., Wei, S., Jiang, H., Inkpen, D.: Enhanced lstm for natural language inference. arXiv preprint arXiv:1609.06038 (2016)
2. Chen, Z., et al.: Information retrieval: a view from the Chinese IR community. Front. Comp. Sci. **15**(1), 1–15 (2021)
3. Dolan, B., Brockett, C.: Automatically constructing a corpus of sentential paraphrases. In: Third International Workshop on Paraphrasing (IWP2005) (2005)
4. El-Alfy, E.S.M., Abdel-Aal, R.E., Al-Khatib, W.G., Alvi, F.: Boosting paraphrase detection through textual similarity metrics with abductive networks. Appl. Soft Comput. **26**, 444–453 (2015)
5. Ferreira, R., Cavalcanti, G.D., Freitas, F., Lins, R.D., Simske, S.J., Riss, M.: Combining sentence similarities measures to identify paraphrases. Comput. Speech Lang. **47**, 59–73 (2018)
6. He, B., Huang, J.X., Zhou, X.: Modeling term proximity for probabilistic information retrieval models. Inf. Sci. **181**(14), 3017–3031 (2011)
7. Hu, B., Lu, Z., Li, H., Chen, Q.: Convolutional neural network architectures for matching natural language sentences. Adv. Neural Inf. Process. Syst. **27** (2014)
8. Jimenez, S., Duenas, G., Baquero, J., Gelbukh, A.F., Bátiz, A.J.D., Mendizábal, A.: UNAL-NLP: combining soft cardinality features for semantic textual similarity, relatedness and entailment. In: SemEval@ COLING, pp. 732–742 (2014)
9. Khot, T., Sabharwal, A., Clark, P.: SciTail: a textual entailment dataset from science question answering. In: Thirty-Second AAAI Conference on Artificial Intelligence (2018)
10. Kingma, D.P., Ba, J.: Adam: a method for stochastic optimization. arXiv preprint arXiv:1412.6980 (2014)
11. Lai, A., Hockenmaier, J.: Illinois-LH: a denotational and distributional approach to semantics. In: SemEval@ COLING, pp. 329–334 (2014)
12. Li, P., Yu, H., Zhang, W., Xu, G., Sun, X.: SA-NLI: a supervised attention based framework for natural language inference. Neurocomputing **407**, 72–82 (2020)

13. Liu, P., Qiu, X., Chen, X., Wu, S., Huang, X.J.: Multi-timescale long short-term memory neural network for modelling sentences and documents. In: Proceedings of the 2015 Conference on Empirical Methods in Natural Language Processing, pp. 2326–2335 (2015)

14. Liu, P., Qiu, X., Huang, X.: Dynamic compositional neural networks over tree structure. arXiv preprint arXiv:1705.04153 (2017)

15. Liu, Y., Sun, C., Lin, L., Wang, X.: Learning natural language inference using bidirectional LSTM model and inner-attention. arXiv preprint arXiv:1605.09090 (2016)

16. Liu, Y., Tang, A., Sun, Z., Tang, W., Cai, F., Wang, C.: An integrated retrieval framework for similar questions: word-semantic embedded label clustering-LDA with question life cycle. Inf. Sci. **537**, 227–245 (2020)

17. Marchesin, S., Purpura, A., Silvello, G.: Focal elements of neural information retrieval models. An outlook through a reproducibility study. Inf. Process. Manag. **57**(6), 102109 (2020)

18. Marco, M., Luisa, B., Raffaella, B., Stefano, M., Roberto, Z., et al.: Semeval-2014 task 1: evaluation of compositional distributional semantic models on full sentences through semantic relatedness and textual entailment. In: Proceedings of the SemEval, pp. 1–8 (2014)

19. Marelli, M., Menini, S., Baroni, M., Bentivogli, L., Bernardi, R., Zamparelli, R.: A sick cure for the evaluation of compositional distributional semantic models. In: Proceedings of the Ninth International Conference on Language Resources and Evaluation (LREC 2014), pp. 216–223 (2014)

20. Parikh, A.P., Täckström, O., Das, D., Uszkoreit, J.: A decomposable attention model for natural language inference. arXiv preprint arXiv:1606.01933 (2016)

21. Pennington, J., Socher, R., Manning, C.D.: Glove: global vectors for word representation. In: Proceedings of the 2014 Conference on Empirical Methods in Natural Language Processing (EMNLP), pp. 1532–1543 (2014)

22. Rus, V., McCarthy, P.M., Lintean, M.C., McNamara, D.S., Graesser, A.C.: Paraphrase identification with lexico-syntactic graph subsumption. In: FLAIRS Conference, pp. 201–206 (2008)

23. Shen, G., Deng, Z.H., Huang, T., Chen, X.: Learning to compose over tree structures via POS tags for sentence representation. Expert Syst. Appl. **141**, 112917 (2020)

24. Sun, C., Huang, L., Qiu, X.: Utilizing bert for aspect-based sentiment analysis via constructing auxiliary sentence. arXiv preprint arXiv:1903.09588 (2019)

25. Xu, S., Shijia, E., Xiang, Y.: Enhanced attentive convolutional neural networks for sentence pair modeling. Expert Syst. Appl. **151**, 113384 (2020)

26. Zhao, H., Lu, Z., Poupart, P.: Self-adaptive hierarchical sentence model. In: Twenty-fourth International Joint Conference on Artificial Intelligence (2015)

27. Zhao, J., Zhu, T., Lan, M.: ECNU: one stone two birds: ensemble of heterogenous measures for semantic relatedness and textual entailment. In: SemEval@ COLING, pp. 271–277 (2014)

28. Zhou, G., Zhao, J., He, T., Wu, W.: An empirical study of topic-sensitive probabilistic model for expert finding in question answer communities. Knowl. Based Syst. **66**, 136–145 (2014)

Multimodal Neural Network for Demand Forecasting

Kumar Dheenadayalan$^{(\boxtimes)}$, Nitesh Kumar, Suprabath Reddy,
and Sumant Kulkarni

Zenlabs, Zensar Technologies Ltd., Pune, India
d.kumar@iiitb.org

Abstract. Demand forecasting applications have immensely benefited from the state-of-the-art Deep Learning methods used for time series forecasting. Traditional uni-modal models are predominantly seasonality driven which attempt to model the demand as a function of historic sales along with information on holidays and promotional events. However, accurate and robust sales forecasting calls for accommodating multiple other factors, such as natural calamities, pandemics, elections, etc., impacting the demand for products and product categories in general. We propose a multi-modal sales forecasting network that combines real-life events from *news articles* with traditional data such as historical sales and holiday information. Further, we fuse information from general product trends published by Google trends. Empirical results show statistically significant improvements in the SMAPE error metric with an average improvement of **7.37%** against the existing state-of-the-art sales forecasting techniques on a real-world supermarket dataset.

Keywords: multimodal · CNN · news encoding · time series · forecasting

1 Introduction

Demand forecasting deals with predicting the number of goods and services that might be consumed under prevailing market conditions [1]. It can affect other aspects of the supply chain, like raw material procurement, human resource estimation, manufacturing, inventory management by retailers, etc. Hence, accurate forecasting enables businesses to be effective, reduce financial losses, and increased consumer satisfaction. Many visible events like social unrest, pandemic, economic recession, extreme weather activity which have an impact on the forecasting, cannot be easily quantified [2]. Such events are mostly observed through multimodal data (text - news/policy articles, image - weather report, video - news recordings; time-series data like weather recordings) [3] which necessitates its inclusion for more accurate demand forecasting. For example, air purifiers demand in California during forest fires [4].

K. Dheenadayalan and N. Kumar—Both the authors have made equal contribution towards the paper.

One of the major challenges in multimodal networks is the difficulty in combining different modalities to achieve a particular goal without facing generalization issues. Ideally, the addition of newer input sources should lead to more useful information for the forecasting task but training an end-to-end multimodal model is not as competitive as their unimodal counterparts in many cases [5]. The main causes for this as reported in [6] are:

- Multimodal models may overfit due to their increased model capacity.
- Different modalities overfit or generalize at varying rates, so training them jointly with a single optimization strategy is sub-optimal.

To deal with the aforementioned issues of multimodal networks, we use the Overfitting to Generalization Ratio (OGR) [6] as a criterion for weight updates that helps in alleviating the issues in a multimodal network. In our proposed model, we use historical sales data, google trends related to some of the product categories, and news articles as inputs sources. To embed event-related information for sales forecasting, we use a Multimodal AI (MAI) model. Here, multiple independent models operate on different sources/modes of input that are eventually combined using fusion strategies like mid-fusion or late-fusion [7] to make the final forecast. Our contributions in this paper include:- **1:** Using Multi-source and multimodal data to show statistically significant forecast accuracy compared to the existing state-of-the-art models like DeepGLO [8] and Filternet [9]. **2:** We show the importance of different modes of information based on the events occurring in a geographic region. Events and their impact encoded through news articles and their influence on estimating the sales numbers are demonstrated. **3:** We also show the advantage of monitoring Overfitting to Generalization Ratio (OGR) [6] for successive weight updates that are necessary to ensure better generalization of multimodal networks.

2   Related Work

Deep Learning for Time-Series: Time series is a sequence of physical measurements over uniform/non-uniform time intervals. Traditional [10] and modern statistical models [11] have shown varying degrees of performance for sales forecasting. Statistical forecast models are mostly linear [12], and forecasting at scale is better achieved by Prophet [11], which introduces non-linearity and is robust to outliers. In the past decade, Deep learning has advanced the research in forecasting by a big margin [13], which can be attributed to the ability of deep learning models to automate the feature engineering process and learn the underlying complex representations [14].

Several architectures have shown promise in sequence modeling tasks [13]. The successful application of Convolutional Neural Networks (CNN) [15], initially designed for computer vision applications, by scaling down the CNNs from 2D to 1D extends its applicability for time series datasets, both in univariate and multivariate form [16]. Similar to spatial invariance of 2D CNNs, temporal invariance is also true in the 1D CNNs. Another significant improvement was proposed in Filternet [9], where multiple layers of stacked 1D CNN or LSTM units were used. The stacking at different layers enables input length invariance which makes it suitable for variable-length time series. DeepGLO presented in [8], uses global information to aid forecasting in an ensemble fashion. The process of combining multiple models dealing with different input sources in a deep learning setting offers a major advantage over existing statistical and deep learning techniques which are explored in this paper.

Representation Learning for News Articles: The idea of incorporating information from news articles has been a well-explored area in the context of stock market predictions. Deep sequential models such as LSTMs and GRUs have been extensively used to model time series data. Encoding events extracted from news articles using Deep Learning Framework [17, 18] and then using them for stock prediction is a generic theme across the literature. Predicting future stock prices by modelling transcripts of the earnings call was proposed in [19]. A hybrid attention network (HAN) was introduced in [18] to classify stock increase/decrease based on news articles related to the problem at hand. We adopt the HAN model to estimate the '+'ve/'-'ve influence of news on sales of products.

Multimodal AI: The existing literature in multimodal AI deals with identical tasks like the Kinetics video dataset where video clips, audio, and optical flow are used as data modalities for action recognition [20]. Similarly, the mini-sports dataset consists of audio and video for sports classification [21]. All these problems majorly deal with sourcing data and modalities with correlated information directly dealing with the problem at hand. However, the uniqueness of sales forecasting is the need to extract relevant events that can affect the sales and associate these events with a subset of categories which adds to the complexity. Identifying and associating relevant news with product sales is a unique and challenging task. Applications like stock market trend predictions have explored diverse sources like utilizing online news [22]. We adapt the ideas presented in [18] to embed news as a mode of data for sales forecasting.

3 Problem Formulation

To formalize the sale forecasting problem, let the sales data $S = (s_t^p : t \in T)$ and google trends data $G = (g_t^q : t \in T)$ be two time series for each product category $p \in P$ and trend category $q \in Q$ respectively. Let $n_t \in N$ represent the set of news articles on day 't' and $\phi(.)$ be a transformation function. We aim to use a multimodal network to approximate the function

$$y_t^p = f(s_{t-1:t-w}^p, g_{t-1:t-w}^q, \phi(n_{t-1:t-w}))$$

Here, the forecast for time step 't' is estimated by a fusion of inputs from multi-source (sales, google trends) and multimode (time-series and unstructured text like news) at various frequencies. 'w' represents the window size indicating the number of past time steps to be used for all the data sources.

4   Data Collection and Pre-processing

We used the sales forecasting dataset published by a supermarket company, Corporación Favorita[1]. The daily sales number for each item is provided from 2013 to 2017 with each item belonging to 21 different product categories, but the data for several products are sparse. Hence, we aggregate the product category-wise sales and consider 20 categories in our experiment, while omitting 1 category (LAWN AND GARDEN) due to extreme sparsity of sales numbers. A subset of these will be affected by climatic and geopolitical events in Ecuador. A total of 1,610 days' data was aggregated for each of the 20 sales product categories of which 1,300 instances were used for training and the rest were used for testing. This roughly amounts to 3 years of data for training and 1 one year for testing.

News Articles: We collect news articles from popular news websites that operate in Ecuador and other South American regions for the time for which we have the sales data. Overall, we collected ≈198K articles from 3 websites - USNews[2], ElCommercio[3], and BBC News[4]. Since a lot of these news articles were in native South American languages, we use Google Translate[5] to translate them into the English language. We release this dataset[6] for future research purposes. The pipeline consists of 3 phases as described below.

Keyword Generation: A list of candidate keywords are generated for each of the product categories such that the news corpus can be queried with these keywords and later ranked according to their relevance towards each category. To accomplish this, firstly, we manually map each of the product categories C_1 in the sales dataset to the appropriate entities in Amazon[7] and Google taxonomy[8] Secondly, we consider all sub-categories (descendants) of the mapped entities from both the taxonomies to create uni-gram, bi-gram, and tri-gram tokens for each product category. This served as the first set of candidate keywords - keywords coming from taxonomy entities.

Further, we enriched the keyword set S_i for category C_i by extending the keyword set to all the brands associated with the mapped entity in Amazon taxonomy. This gave us a larger keyword set so that we don't miss out on news articles that might be relevant to our categories. A final round of manual pass was done to remove any generic keyword. By the end of this exercise, for each of the 20 categories C_1–C_{20} in our sales dataset, we had keyword sets S_1–S_{20}.

Relevancy Scoring: Not all articles will carry information pertinent to the product categories under consideration. We term these articles as noisy articles.

---

[1] https://www.kaggle.com/c/favorita-grocery-sales-forecasting.
[2] https://www.usnews.com/topics/locations/ecuador.
[3] https://elcomercio.pe/.
[4] https://www.bbc.com/news [2014-2017].
[5] https://pypi.org/project/googletrans/.
[6] https://www.kaggle.com/reviewerh/news-dataset.
[7] http://jmcauley.ucsd.edu/data/amazon/.
[8] https://www.google.com/basepages/producttype/taxonomy-with-ids.en-US.txt.

The proportion of noisy articles in the acquired news dataset is quite large compared to meaningful articles. Using the keywords generated, on a majority of days, we were able to extract a minimum of 20 articles and a maximum of 178 news articles for less than 10 days. Although we see that there are 10's to 100's of news articles collected each day, upon manual exploration we found that only a few articles per day are relevant to at least one of the 20 categories. This motivated us to design a relevancy scoring module that would filter out noisy articles. We aim to assign a probability distribution over the 20 categories for each news article if it is relevant to at least one category, else mark it as irrelevant. As a first step, we use TextRank [23] to find the importance score for every word in a given news article N_j. For any category i, we find the keywords which are present in the article N_j as well as in the keyword set S_i. Let this intersection have 'l' keywords, denoted by $F_{ij} = w_1, w_2, \ldots, w_l$, and let the TextRank scores of these keywords be denoted by $r_1, r_2 \ldots, r_l$. We further reduce the size of F_{ij} by considering top 50% keywords according to their descending order of TextRank scores. We denote this new set by $\hat{F}_{ij} = w_1, w_2, \ldots, w_h$, where $h = \frac{l}{2}$. Now the relevancy score of an article N_j towards a category i is defined by -

- How important is the kth keyword $w_k \in \hat{F}_{ij}$ to the category i. This is denoted by $\frac{1}{m_k}$, where m is the number of categories that have the keyword w_k
- How important is a keyword $w_k \in \hat{F}_{ij}$ to the article N_j. This is given by $r_k * \frac{1}{rank(w_k)}$, where $rank(w_k)$ is the index of keywords sorted descending by TextRank scores.

The combined score is given by $score_{ij} = \sum_{k=1}^{h} \frac{1}{m_k} * r_k * \frac{1}{rank(w_k)}$.

We then calculate this score for every category i and news article j. We review the top scoring articles for each category to define a threshold t_i above which the articles showed significant relevance towards category i. Let $\triangle_{ij} = score_{ij} - t_i$ as the actual relevance score between category i and article j. Articles with $\triangle_{ij} \leq 0$ for every category i were deemed irrelevant/noisy. For articles which had $\triangle_{ij} > 0$ for at least one category were deemed relevant. The maximum number of relevant articles in a day turned out to be 5 in our dataset.

News Encoding: To generate encoding for the news articles, we use the HAN proposed in [18] for classification, with a few modifications. The 64 dimension vector from the second attention layer was used as the news embedding for the last day of the input sequence.

- The input fed to the network was a tensor of shape $n_{articles} * n_{days} * n_{dim}$, where $n_{articles} = 5$/day, $n_{days} = 7$ days, and $n_{dim} = 100$ is the dimension of doc2vec [24] trained on a larger corpus of news articles.
- Unidirectional GRU is used with Batch normalization.
- The output for the classification task had 40 classes -2 binary variables for each of the 20 classes constructed as follows :

$$C_i^+ = \begin{cases} 1 & \text{sales increased by } 5\% \\ 0 & \text{otherwise} \end{cases} \qquad C_i^- = \begin{cases} 1 & \text{sales decreased by } 5\% \\ 0 & \text{otherwise} \end{cases}$$

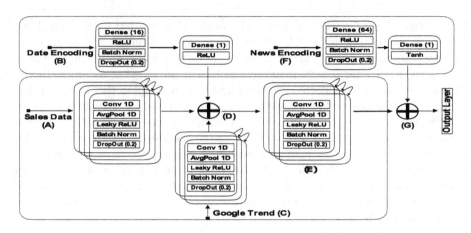

Fig. 1. Multimodal network architecture for demand forecasting.

4.1 Date Encoding

We use cyclic feature encoding for 4 pieces of date information, namely, the day of the month, day of the year, week of the month, and month of the year using $x_{sin} = \sin(\frac{2*\pi*x}{\max(x)})$ and $x_{cos} = \cos(\frac{2*\pi*x}{\max(x)})$.

Maximum value of x in the equations is = $\{(28\text{--}31)$ days, $(365\text{--}366)$ days, $(4\text{--}5)$ weeks, 12 months$\}$ for the 4 different date information that we want to encode. This gives us a 8-dimensional date encoding for each day. To forecast the sales for the day 't', we use the encoding only for day t instead of encoding for all days in window w. This ensures a minimal increase in the feature dimensionality.

4.2 Google Trends

Use of google trends[9] and search keywords to better predict the interest of consumers in products has been of interest in the past. The search volume of a particular product can be associated with the interest in the product and the same can be translated to the sales volume as well. [25] presented a similar line of work for box-office revenue. For sales forecasting, not all product categories in the sales dataset have the corresponding google trends available. We manually skimmed through the available categories and identified 12 google trend categories (*Food Drink, Internet, Computer Electricals, Business, Shopping, Books, Hobbies, Auto, Home Garden, Art, Beauty, Health*) that were either exactly matching or closely related to the categories present in the sales data.

Correlation Analysis: Since we have 12 different trend categories (q) and 20 product categories (p), we perform a pairwise Pearson's correlation for each pair of trends time series and product sales time series. The pairs which have a correlation above 0.4 are retained. From these shortlisted pairs, we derive a set of

---

[9] http://trends.google.com/trends, 2020.

unique trend categories for each product category. These sets are not disjoint for each product category, i.e. a trend category can be relevant for multiple product categories. The purpose of setting a high threshold is to use only meaningful google trends for our task for sales forecasting. Consequently, only a few product categories will eventually get trend categories data in the modeling phase. These are marked with (*) and boldfaced in Sect. 6.

5 Multimodal Forecast Model

Filternet [9] is a novel architecture inspired by the wavenet model which includes a stack of identical blocks of Conv1D layers. It was initially proposed for a time-series classification problem but we have extended the idea of stacking blocks of Conv1D in multiple layers to the forecasting problem. Our architecture as shown in Fig. 1 consists of sales network (A), date encoding network (B), trend data network (C) and news encoding network (F). (A), (C), and (E) are Filternet blocks having 4, 3, and 4 Conv1D stacks respectively with kernel sizes varying from 1 to ∥stack size∥. The 8 dimensional encodings for the date are passed to (B), consisting of a fully connected network with 16 hidden units followed by a single output node. The output of the Date Encoding network (B) and the flattened output of trend network (C) are concatenated with the output of each stack in Sales network (A) as represented by the $\oplus$ operation (D) in Fig. 1. This is fed as an input to the second Filternet block (E).

The 64 dimensional news encodings (described in Sect. 4) for the past 'w' days are flattened and fed into the News Network (F), which is a fully connected network with one hidden layer of 64 units and an output node with tanh activation. The output acts as a gating unit and enables the overall network to learn if the news articles need to be used for prediction. The output of the news network is concatenated to the flattened output of the stacks in (E) as shown by $\oplus$ operation (G). This is fed into the output layer whose dimension depends on the forecasting window {1,7}. Linear activation is used in the output layer to generate the final forecast.

All the Conv1D blocks in Filternet use a leaky ReLU activation function along with average pooling and the fully connected layers use the ReLU activation function. Additionally, all the layers have batch normalization, dropout, and L2 regularization. This is to ensure that the model capacity is limited and issues of multimodal networks as listed in Sect. 1 are alleviated.

5.1 Overfitting-Generalization-Ratio (OGR)

OGR, which is the ratio of overfitting ($\triangle O$) to generalization ($\triangle G$) was first introduced in [6], and used as a means of blending gradients with successful application in video classification. A rough estimation of generalization is given by $\triangle G = (\text{val\_loss}_e - \text{val\_loss}_{e-1})$ and overfitting is estimated using $\triangle O = (\text{val\_loss}_e - \text{train\_loss}_e) - (\text{val\_loss}_{(e-1)} - \text{train\_loss}_{(e-1)})$

Intuition is that a small value of OGR indicates that the model is learning well with good generalization and less overfitting. It can also be interpreted as the quality of learning between epochs in the training phase. A similar idea is used in our work for successive weight updates. Instead of using OGR for gradient blending, we compare the OGR values in successive epochs in the training procedure and update the weights only if the OGR is less than a small number (10^{-2}). It must be noted that a positive OGR value indicates overfitting. We implement this simple yet effective check of measuring the improvements in generalization and overfitting for the current epoch with the previous epoch as the reference. This is contrary to the popular way of weight updates where successive epochs' training loss or validation loss is used. Using OGR, we ensure that the multimodal network does not get into the overfitting zone and ensure optimal updates only when the generalization of the model occurs with every weight update.

6   Results

In the network shown in Fig. 1, we used Adam optimizer with Mean Absolute Error(MAE) as the loss metric, a learning rate and decay factor of 0.01. The experiments were run for 200 epochs, and one model was trained per product category. A window size of 30 was set to consider the last 30 days' inputs as a sequence. The models provided in [8] and [9] were used to benchmark DeepGLO and Filternet respectively on the sales data. We generate forecasts for each of the product categories. The results of daily and weekly forecasts are discussed next.

The experiments were run over 10 iterations on our dataset discussed in Sect. 4. Benchmarks are created against DeepGLO, Filternet as well as the Non-OGR version of our model. We refrain from presenting results on statistical models like SARI-MAX, TABTS, FB-Prophet as the results were not satisfactory. Symmetric Absolute Mean Percentage Error(**SMAPE**) [26] is a relative error metric with specified upper and lower limits widely used for forecasting tasks, and the same is used in the evaluation of our algorithm. To discuss our work in line with the existing literature, we only present the results related to SMAPE values. The mean and variance of SMAPE for the daily forecast for each product category are presented in Table 1.

Wilcoxon Signed-Ranks Test: For statistical comparison of machine learning algorithms, research surveys have suggested the use of Wilcoxon Signed-Ranks Test as the most suitable statistical test when compared to other tests like paired t-test, averaging over multiple runs, etc., [27]. The Wilcoxon signed-ranks test [27] is a non-parametric equivalent to the paired t-test, used to rank the difference in the performance of two models in a multi-dataset scenario. Since we train one model per product category, we have 20 datasets across which the proposed model is tested. Keeping the model architecture constant, and treating each of the product categories as an independent dataset, we can use the signed-rank test to conclude if the results are statistically significant or not. We define the following hypothesis: H_0 : there isn't any significant difference in SMAPE of the proposed and benchmark models, and the difference between each pair of

Table 1. Benchmarking of Daily forecast SMAPE results

Name	DeepGLO	Filternet	Multimodal Network (with OGR)	Multimodal Network (Non-OGR)
AUTOMOTIVE*	18.92 ± 4.26	12.93 ± 0.43	12.17 ± 0.08	12.41 ± 1.71
BEAUTY*	20.94 ± 3.64	15.55 ± 0.36	14.25 ± 0.16	18.63 ± 4.98
BEVERAGES*	17.05 ± 4.79	10.79 ± 0.51	10.47 ± 0.82	11.31 ± 2.78
BREAD/BAKERY*	16.80 ± 5.40	8.21 ± 0.19	7.08 ± 0.17	7.69 ± 1.17
CLEANING	17.83 ± 4.96	11.18 ± 0.38	9.97 ± 0.20	10.45 ± 0.74
DAIRY*	17.50 ± 4.76	9.82 ± 0.30	9.04 ± 0.14	9.47 ± 0.91
DELI	17.50 ± 4.84	10.27 ± 0.30	8.71 ± 0.13	10.30 ± 1.85
EGGS	19.02 ± 4.32	11.16 ± 0.46	10.29 ± 0.40	10.60 ± 0.86
FROZEN FOOD*	22.85 ± 5.50	12.39 ± 0.43	12.66 ± 1.01	13.33 ± 2.08
GROCERY	17.23 ± 5.18	10.35 ± 0.42	9.17 ± 0.30	9.59 ± 0.85
HARDWARE	21.20 ± 3.39	15.96 ± 0.16	15.90 ± 0.14	18.90 ± 5.66
HOME APPLIANCES	30.11 ± 2.87	25.93 ± 0.17	26.02 ± 0.13	27.43 ± 4.02
LINGERIE	18.96 ± 4.57	12.57 ± 0.17	11.34 ± 0.05	12.01 ± 1.75
LIQUOR,WINE,BEER	28.86 ± 4.04	21.09 ± 0.41	21.99 ± 0.50	22.81 ± 2.87
MEATS	18.26 ± 4.56	9.70 ± 0.21	9.06 ± 0.17	9.59 ± 1.10
PERSONAL CARE	18.63 ± 4.27	12.26 ± 0.33	10.80 ± 0.09	11.70 ± 1.82
POULTRY	18.87 ± 3.91	11.29 ± 0.88	9.78 ± 0.17	10.49 ± 1.53
PREPARED FOODS	17.09 ± 5.90	8.45 ± 0.29	7.37 ± 0.10	8.15 ± 1.10
PRODUCE*	19.86 ± 4.31	10.94 ± 1.02	10.23 ± 0.58	10.95 ± 1.70
SEAFOOD	20.17 ± 4.24	12.65 ± 0.41	11.55 ± 0.05	12.30 ± 1.15

SMAPE values follows a symmetric distribution around zero. H_1 : there is a significant difference in SMAPE, and the difference between the pairs is not normally distributed around zero.

For a significance level(α) of 0.05 and $n = 20$, the signed-rank test for the best model in literature, i.e., Filternet and the proposed multimodal network, we observed a $z - score$ of -3.7333 and the p-value was 0.0002. Owing to the low p-value, and a significantly smaller $z - score$, we reject the null hypothesis (H_0). Hence, proving the generalization of the multimodal network. Two key takeaways from this experiment were:

- Results of our Multimodal Forecasting model *without OGR* are competitive and comparable with the Filternet model. However, the statistical significance cannot be established as there are nearly equal proportions of product categories where either of them perform better than the other.
- The variance in the results obtained without OGR is relatively high. *However, enabling OGR ensures that the model performs well across the dataset.*

The weekly forecast is also statistically significant (z-score is -3.9199 & p-value is 0.00008), and the SMAPE errors for the same is captured in Table 2 in columns 1 & 2. The weekly forecasts were generated for 40 weeks in 2017. To understand the use of modalities in identifying demand shifts, we drill down on the forecast results obtained for product categories during a major flood that hit South America between 3rd week of March 2017 to 2nd week of April 2017. The SMAPE errors on the weekly forecast during this interval are shown in Table 2 in columns 3 & 4. In line with the comparison, w.r.t. overall forecast accuracy, the SMAPE values during the event are reasonably good with additional information through news and google trends having a positive impact on the product

Table 2. Weekly results for entire dataset along with results during a major event

Name	Filternet (weekly)	Multimodal (weekly)	Filternet(event)	Multimodal Network(event)
AUTOMOTIVE*	7.73 ± 0.36	6.82 ± 0.39	5.71 ± 0.61	5.20 ± 1.71
BEAUTY*	10.65 ± 0.41	9.92 ± 1.14	7.28 ± 0.47	6.50 ± 1.05
BEVERAGES*	7.93 ± 1.08	7.17 ± 0.72	9.04 ± 1.08	6.75 ± 0.98
BREAD/BAKERY*	4.58 ± 0.42	3.93 ± 0.36	3.75 ± 0.63	2.71 ± 0.68
CLEANING	8.96 ± 0.69	7.74 ± 0.33	7.59 ± 1.60	5.91 ± 1.02
DAIRY*	5.27 ± 0.90	5.09 ± 0.74	5.83 ± 1.12	6.35 ± 0.73
DELI	4.61 ± 0.26	4.63 ± 0.18	3.79 ± 0.75	3.38 ± 0.65
EGGS	5.15 ± 0.21	3.80 ± 0.20	5.51 ± 0.60	5.05 ± 0.71
FROZEN FOOD*	12.69± 2.63	12.88± 2.04	27.64 ± 11.25	22.67 ± 5.21
GROCERY	9.97 ± 1.35	8.97 ± 0.68	17.42 ± 1.81	14.25 ± 1.77
HARDWARE	9.05 ± 0.48	8.14 ± 0.87	4.20 ± 0.76	2.05 ± 1.47
HOME APPLIANCES	10.48± 1.38	11.09± 1.09	29.26 ± 6.28	26.73 ± 4.41
LINGERIE	6.23 ± 0.45	5.54 ± 0.58	4.25 ± 1.41	3.09 ± 1.78
LIQUOR WINE BEER	14.59± 1.20	13.16± 0.26	16.20 ± 4.45	15.26 ± 2.53
MEATS	6.82 ± 0.38	5.48 ± 0.56	4.07 ± 0.33	3.06 ± 0.44
PERSONAL CARE	11.45± 0.74	7.88 ± 0.19	9.04 ± 0.98	6.79 ± 2.09
POULTRY	6.20 ± 0.63	4.47 ± 0.37	4.13 ± 1.02	2.64 ± 0.62
PREPARED FOODS	4.59 ± 0.44	4.09 ± 0.18	6.45 ± 0.61	5.59 ± 0.60
PRODUCE*	3.31 ± 0.56	3.64 ± 0.17	2.31 ± 0.37	2.42 ± 0.32
SEAFOOD	6.39 ± 0.39	5.89 ± 0.34	0.95 ± 0.66	2.84 ± 0.38

categories. For a similar statistical test setting of $\alpha = 0.05$ and $n = 20$, the value of z is -3.1733 the p-value is 0.00152, thus validating the significance of the improved forecasts during the event.

6.1   DeepShap Feature Importance

To understand the importance of using additional sources and modes of data, we use DeepShap [28] to estimate Shapley values which evaluate the contribution of each feature in model prediction. For interpretation purposes, it must be noted that the larger the absolute Shapley values of a feature, the greater is its influence on the forecast made by the model. The x-axis represents the Shapley values and the y-axis represents the features. The jittered representation for each feature gives an indication of the distribution of Shapley values for each feature. The pink and blue colors represent a positive and negative influence on the forecasts respectively. The summary plot in Fig. 2a & 2b shows the most important features identified by DeepShap. In Fig. 2a the model was trained with data collected during major events (i.e., social unrest and earthquake during 2015 to 2016) and we see 5 features related to news are influencing the predictions. The extent of the impact is relatively small as *Shapley values are averaged over the entire data* and hence the influence of events on sales lasts for a relatively shorter duration. However, the influence of events is still positive, as seen by the pink-colored representation of Shapley values. In Fig. 2b, the summary plot is shown for a model trained using data without any significant event (i.e., 2013–2014). We do not observe news encoding related features but 5 trend-based features appear to have an impact on the forecast. The top 6 features identified in both the scenarios discussed consists of the sales numbers recorded on 7, 14, 21 and 28 days in the past which shows that the weekly trends have the most impact along with *date encoding #4* and previous days sales.

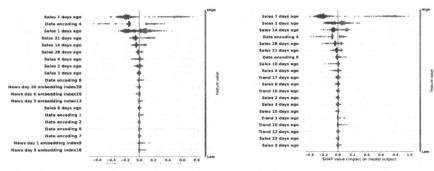

(a) Feature importance when trained on a subset of data with major events.

(b) Feature importance when trained on a subset of data with no events.

Fig. 2. Feature importance using DeepShap

7 Conclusion

In this work, we have proposed a novel approach to automate the process of encoding events that directly influence the prediction of a model. We demonstrate improved forecast accuracy with the multimodal architecture under general conditions and significant improvements under event-specific conditions. There are no limitations in the applicability of the proposed network in other applications or domains. Exploring use of additional modality is also an option with the caveat of carefully preventing overfitting. Future work will explore better models with newer data sources along with a wider applications of multimodal networks to different domains.

References

1. Armstrong, J.S., Green, K.C.: Demand forecasting: evidence-based methods. SSRN Electron. J. (2005)
2. Thomassey, S.: Sales forecasts in clothing industry: the key success factor of the supply chain management. Int. J. Prod. Econ. **128**(2), 470–483 (2010). Supply Chain Forecasting Systems
3. Sawhney, R., Agarwal, S., Wadhwa, A., Shah, R.: Deep attentive learning for stock movement prediction from social media text and company correlations. In: Proceedings of the 2020 Conference on Empirical Methods in Natural Language Processing (EMNLP), pp. 8415–8426, November 2020
4. De Livera, A.M., Hyndman, R.J.: California Air Cleaning Units Mraket. Technical report, TechScience Research (2019)
5. Goyal, Y., Khot, T., Summers-Stay, D., Batra, D., Parikh, D.: Making the v in vqa matter: elevating the role of image understanding in visual question answering. In: 2017 IEEE Conference on Computer Vision and Pattern Recognition (CVPR), pp. 6325–6334 (2017)

6. Wang, W., Tran, D., Feiszli, M.: What makes training multi-modal classification networks hard? In :2020 IEEE/CVF Conference on Computer Vision and Pattern Recognition (CVPR), pp. 12692–12702, Los Alamitos, CA, USA (2020). IEEE
7. Owens, A., Efros, A.A.: Audio-Visual Scene Analysis with Self-Supervised Multisensory Features. arXiv e-prints, page arXiv:1804.03641 (2018)
8. Sen, R., Yu, H.-F., Dhillon, I.: Think globally, act locally: a deep neural network approach to high-dimensional time series forecasting (2019)
9. Chambers, R.D., Yoder, N.C.: Filternet: a many-to-many deep learning architecture for time series classification. Sensors 20(9) (2020)
10. Box, G.E.P., Jenkins, G.: Time Series Analysis, Forecasting and Control. Holden-Day Inc., USA (1990)
11. Taylor, S.J., Letham, B.: Forecasting at scale. Am. Stat. 72(1), 37–45 (2018)
12. Makridakis, S., Wheelwright, S.C., Hyndman, R.J.: Forecasting: Methods and Applications, 3rd ed. John Wiley & Sons, USA (1997)
13. Lim, B., Zohren, S.: Time series forecasting with deep learning: a survey. ArXiv, abs/2004.13408 (2020)
14. Bengio, Y., Courville, A., Vincent, P.: Representation learning: a review and new perspectives. IEEE Trans. Pattern Anal. Mach. Intell. 35(8), 1798–1828 (2013)
15. Lecun, Y., Bottou, L., Bengio, Y., Haffner, P.: Gradient-based learning applied to document recognition. Proc. IEEE 86(11), 2278–2324 (1998)
16. Kiranyaz, S., Avci, O., Abdeljaber, O., Ince, T., Gabbouj, M., Inman, D.J.: 1D convolutional neural networks and applications: a survey. Mech. Syst. Signal Process. 151, 107398 (2021)
17. Nassirtoussi, A.K., Aghabozorgi, S., Wah, T.Y., Ngo, D.C.L.: Text mining of news-headlines for forex market prediction. Expert Syst. Appl. 42(1), 306–324 (2015)
18. Hu, Z., Liu, W., Bian, J., Liu, X., Liu, T.Y.: Listening to chaotic whispers: a deep learning framework for news-oriented stock trend prediction. In: Proceedings of the Eleventh ACM International Conference on Web Search and Data Mining, WSDM 2018, pp. 261–269, New York, NY, USA (2018)
19. Wang, W.Y., Hua, Z.: A semiparametric Gaussian copula regression model for predicting financial risks from earnings calls. In: Proceedings of the 52nd Annual Meeting of the Association for Computational Linguistics (Volume 1: Long Papers), pp. 1155–1165, Baltimore, Maryland. Association for Computational Linguistics (2014)
20. Zisserman, A., et al.: The kinetics human action video dataset (2017)
21. Karpathy, A., Toderici, G., Shetty, S., Leung, T., Sukthankar, R., Fei-Fei, L.: Large-scale video classification with convolutional neural networks. In: 2014 IEEE Conference on Computer Vision and Pattern Recognition, pp. 1725–1732 (2014)
22. Li, Q., Tan, J., Wang, J., Chen, H.: A multimodal event-driven LSTM model for stock prediction using online news. IEEE Trans. Knowl. Data Eng. 1 (2020)
23. Mihalcea, R., Tarau, P.: TextRank: bringing order into text. In: Proceedings of the 2004 Conference on Empirical Methods in Natural Language Processing, pp. 404–411, Barcelona, Spain. Association for Computational Linguistics (2004)
24. Le, Q., Mikolov, T.: Distributed representations of sentences and documents. In: Proceedings of the 31st International Conference on International Conference on Machine Learning, vol. 32, ICML 2014. JMLR.org (2014)
25. Kulkarni, G., Kannan, P.K., Moe, W.: Using online search data to forecast new product sales. Decis. Support Syst. 52(3), 604–611 (2012)
26. Schnaars, S.P.: Long-range forecasting: from crystal ball to computer: J. scott armstrong, 2nd ed. Wiley, New York (1985). [uk pound] 22.95 (paper), pp. 689. International Journal of Forecasting, 2(3), 387–390, 1986

27. Demšar, J.: Statistical comparisons of classifiers over multiple data sets. J. Mach. Learn. Res. **7**, 1–30 (2006)
28. Lundberg, S.M., Lee, S.I.: A unified approach to interpreting model predictions. In: Guyon, I., et al. (eds.) Advances in Neural Information Processing Systems, vol. 30, pp. 4765–4774. Curran Associates Inc. (2017)

Image Super-Resolution Based on Adaptive Feature Fusion Channel Attention

Qizhang Song, Baodi Liu[(✉)], and Weifeng Liu

College of Control Science and Engineering,
China University of Petroleum (East China), Qingdao, China
thu.liubaodi@gmail.com, liuwf@upc.edu.cn

Abstract. Since the advent of SENet, existing image super-resolution models based on deep learning have been keen to improve networks' cross-channel understanding and characterization capabilities by introducing channel attention into the network. However, we note that previous channel attention uses one or two global pooling methods as preprocessing methods to extract channel description scalars and treat each channel description scalar equally. Doing so will undoubtedly lose some of the feature information and bring some inflexibility, resulting in the network's inability to distinguish each feature channel very well. To address it, we propose adaptive feature fusion channel attention(AFFCA). Specifically, we use diverse global pooling methods to extract feature description scalars and adaptively adjust each scalar's weights to merge them into a representative feature description scalar. Our method can extract enough information from the feature map to further enhance the ability to distinguish between channels so as to optimize the channel attention performance. We have done some experiments and compared them with the existing methods. The results show that our AFFCA achieves better accuracy and visual improvements against current methods.

Keywords: Super-resolution · Channel attention · Adaptive feature fusion

1 Introduction

Resolution is the primary criterion used to evaluate the quality of an image. Higher resolution images tend to have more details, clearer outlines, and more information. However, in natural scenes, image acquisition is easily interfered by

This research was supported by the Natural Science Foundation of Shandong Province under Grants ZR2019MF073, the Fundamental Research Funds for the Central Universities, China University of Petroleum (East China) under Grant 20CX05001A, the Major Scientific and Technological Projects of CNPC under Grant ZD2019-183-008, and the Creative Research Team of Young Scholars at Universities in Shandong Province under Grant 2019KJN019.

M. Tanveer et al. (Eds.): ICONIP 2022, LNCS 13625, pp. 422–434, 2023.
https://doi.org/10.1007/978-3-031-30111-7_36

factors such as network transmission loss and the quality of camera equipment, making it challenging to obtain ideal high-resolution images directly. Because the cost of improving image resolution from the hardware level is too high, how to achieve image super-resolution from the software level has become a hot research direction in the field of computer vision. The key to this task is restoring the original image's high-frequency features, colors, and spatial relationships as undistorted as possible. Image super-resolution has been applied in many fields, such as target detection, image quality enhancement, and medical imaging.

The deep CNN network has been widely used in image super-resolution reconstruction tasks recently. People are keen to add a channel attention mechanism to the CNN network to give different weights to the channels in the feature map to increase the differences between channels so that limited resources are allocated to places with higher information density and ultimately the performance of the super-resolution reconstructed network is improved. Generally speaking, the first step to achieving channel attention is to compress the characteristic graph into a descriptive scalar. At the beginning, SENet [5] gets this scalar through global average pooling(GAP). In fact, the GAP is simple to calculate, but this compression method can not fully represent the characteristic information of the whole channel, some high-frequency and complex details will be lost, and this lost high-frequency information is very important for the task of super-resolution reconstruction. Therefore, we need a more effective method to compress the channel. CBAM [19] and SRM [9] used global maximum pooling and global standard deviation pooling to enhance the performance of GAP. FCANet [14] later proved that gap is a particular case of DCT. FCANet expanded the gap from the perspective of frequency, and assigned specific frequency components to the channels in each group by grouping, thus retaining the information in the characteristic map to a certain extent. However, among the methods mentioned above, CBAM and SRM only use one or two pooling methods to extract features, and treat the importance of different pooling methods equally, which results in a lack of flexibility. In addition, because the grouping of feature channels and the allocation of frequency components are fixed in FCANet, the features extracted from each group are also fixed, so the network cannot flexibly select the appropriate frequency components. In the task of super-resolution reconstruction, RCAN [21] has achieved good performance. However, it still uses gap to calculate channel attention, which undoubtedly loses some feature information and cannot effectively distinguish each channel.

To solve the above problems, we propose an image super-resolution reconstruction model based on adaptive feature fusion attention. Its main contributions are as follows: (1) We propose the adaptive feature fusion attention mechanism(AFFCA). This method uses a variety of global pooling methods to extract feature description scalars, and adaptively adjusts the weight of each scalar to fuse them into a representative feature description scalar, so as to fully extract the information in the feature map, and further strengthen the discrimination ability between channels to optimize the performance of channel attention. (2) We introduce spatial attention(SA) on the basis of CA, which further improves the learning and characterization ability of CNN network, enabling the network to focus on more valuable areas at the spatial level.

2   Related Work

In recent years, super-resolution reconstruction model based on deep learning is very popular. The earliest SRCNN model introduced a three-layer CNN network, which can learn and recover more detailed features than the traditional "interpolation" algorithm. However, the training of SRCNN is time-consuming and difficult to converge. To speed up the training and testing of SRCNN, a fast training network structure, FSRCNN [1], was proposed.

Ledig et al. [8] then acquired SRResNet by introducing ResNet to the super-resolution task, which improves network depth. They also proposed SRGAN [8] on the basis of combining generative adversarial network(GAN) and perceptual loss. The model was then introduced in the EnhanceNet [15] network, which combines perceptual loss with texture synthesis techniques. In fact, although SRGAN and Enhancenet networks can better improve the blurring and artifacts of reconstructed images and perform better in visual perception, the performance of PSNR and SSM indices is not as good as that of super-resolution reconstruction models based on CNN networks. The subsequent VDSR [6] and DRCN [7] models introduced a 20-layer deep network, which improved considerably compared to SRCNNs with only 3-layer networks. Later, with the advent of the deep residual network [3] (ResNet), the number of network layers could increase to 1,000 layers. EDSR and MDSR proposed by Lim et al. [11] are the early models to apply ResNet in this field. They have considerable network width and network depth (over 165 layers) and the author simplifies the residual blocks in the model to make them more suitable for super-resolution tasks. RCAN [21] subsequently proposed the network structure of Residual in Residual (RIR), which further improved the network depth and learning efficiency of deep networks, and pioneered the introduction of channel attention in the super-resolution reconstruction network, which significantly improved the model performance.

On the other hand, people have never stopped exploring the field of visual attention. Following the channel attention proposed by SENet, GE puts forward spatial attention, which enables the network to focus on the spatial characteristics of information. BAM [13], CBAM [19], and CoordAttention [4] then explored how space and channel attention combine. In addition, CBAM mentions that GAP can result in the loss of some feature information, so it uses GAP and global maximum pooling to compress feature maps and achieves performance improvements. Incidentally, a combination of GAP and global standard deviation pooling is also proposed in SRM [9]. In addition, GsoP proposed building dense spatial feature maps. SkNet [10] introduces signature attention through two network branches with different convolution sizes. ResNeSt [20] proposed a separate attention module based on the ideas of SkNet and GoogleNet [17]. ECANet [18] simplifies the full connection layer in SENet, using only two full connection layers, and uses one-dimensional convolution to turn global interaction into local interaction in channel attention, which improves training efficiency and model performance. It is worth noting that almost all of the above work focuses on classification and detection tasks, whereas attention mechanisms a re not widely used in super-resolution reconstruction tasks.

3   Method

This section focuses on the principles of AFFCA and spatial attention, as well as the network architecture for super-resolution reconstruction.

3.1   Adaptive Feature Fusion Channel Attention

The overall structure of AFFCA is shown in Fig. 1. First, we convert each feature channel into a scalar describing the channel using four pooling methods: global average pooling, global maximum pooling, global variance pooling, and global standard deviation pooling. For example, for an image $x = [x_1, x_2, ..., x_c]$ with a size of $H \times W$, it consists of C feature channels. We use N global pooling methods to obtain C-dimensional characteristic scalars $S_N = [s_{N,1}, s_{N,2}, ..., s_{N,c}]$, each representing one of the characteristics of the channel. Taking $N = 4$ as an example, S_N can be formulated as

$$S_0 = \mu(x_c) = \frac{1}{H \times W} \sum_{i=1}^{H} \sum_{j=1}^{W} x_c(i, j), \tag{1}$$

$$S_1 = MAX(x_c) = MAX(x(i, j))\ i \in [0, H], j \in [0, W], \tag{2}$$

$$S_2 = \sigma^2(x_c) = \frac{1}{H \times W} \sum_{i=1}^{H} \sum_{j=1}^{W} (x(i, j) - \mu(x_c))^2, \tag{3}$$

$$S_3 = \sigma(x_c) = \sqrt{\frac{1}{H \times W} \sum_{i=1}^{H} \sum_{j=1}^{W} (x(i, j) - \mu(x_c))^2}. \tag{4}$$

After extracting N scalars describing the feature channels, the output size becomes $X \in R^{C \times N \times 1 \times 1}$. Then we fuse the N feature description scalars by one-dimensional convolution:

$$S_{fuse} = Conv_{1d}[S_0 \cdots S_N], \tag{5}$$

Finally, we get the fused characteristic $X_{fuse} \in R^{C \times 1 \times 1 \times 1}$, and then we choose to use sigmoid functions and gating mechanisms to learn the correlation between channels, which can be formulated as

$$Weight_{fuse} = sigmoid(C_U \delta(C_D X_{fuse})), \tag{6}$$

where $\delta(\cdot)$ represents the ReLU [12] function, C_D and C_U are convolution layers responsible for channel reduction and amplification respectively, and $Weight_{fuse}$ is the final channel weight. This part of the processing is consistent with RCAN, that is, learning the correlation between channels through two convolution layers with a convolution core size of 1. Each of the two feature maps will interact during the above process. Finally, we multiply the weight information of channel c by the corresponding channel x, which can be formulated as

$$CA = x_c \cdot Weight_{fuse}. \tag{7}$$

Through the above operations, we realized the feature fusion channel attention.

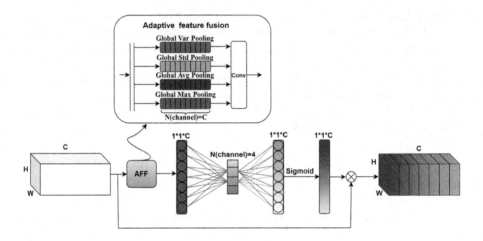

Fig. 1. Overall structure of AFFCA.

3.2   Spatial Attention

In order for the model to focus on learning areas with rich detail, we introduced spatial attention into each residual block. Since the sequential effects of channel and spatial attention have been demonstrated in CBAM [19], the output of channel attention is directly used as the input of spatial attention in this work. The structure of spatial attention is shown in Fig. 2, and spatial attention can be formulated as

$$SA = Sigmoid\,(M_{spatial}) \cdot X_{in}. \tag{8}$$

where $M_{spatial}$ represents spatial attention map, which is obtained by fusing average pooling feature F_{avg} and maximum pooling feature F_{max}.

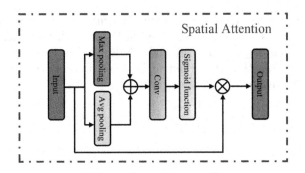

Fig. 2. Structure of spatial attention.

3.3 Network Architecture

As shown in Fig. 3, the network architecture in our model is mainly divided into four parts: the shallow feature extraction module, the deep feature extraction module based on multiple residual groups, the up-sampling module, and the reconstruction module. These four modules are described in this section. First, the shallow feature extraction module, which extracts the feature F_{SF} through a single convolution layer N_{SF}, which can be formulated as

$$F_{SF} = N_{SF}(I_{LR}). \tag{9}$$

After the F_{SF} is obtained, we input it into the deep feature extraction network N_{RG}, then there are

$$F_{DF} = N_{RG}(F_{SF}), \tag{10}$$

where N_{RG} in formula 13 represents a network composed of several residual groups and long skip connections, and the residual groups include several residual blocks and short skip connections. In order to fully extract deep features, this part has a deep network. Network depths exceed 400 layers in the baseline model alone. The resulting feature F_{DF} needs to be amplified by the up-sampling module, that is

$$F_{UP} = N_{UP}(F_{DF}), \tag{11}$$

where N_{UP} in formula 14 is an up-sampling module. Generally, the three common methods of up-sampling are deconvolution layer [1], up-nearest neighbor sampling + convolution [2], and ESPCN [16]. In this task, the ESPCN method is used, which can not only avoid the artificial traces of deconvolution, but also solve the checkerboard effect, so it is widely used in the field of super-resolution reconstruction.

After upsampling, the final output is obtained from a single convolution layer reconstruction module as

$$O_{SR} = N_{RBD}(F_{UP}). \tag{12}$$

Finally, we need to select a reasonable loss function for training. Since the advantages of L1 loss have been analyzed in EDSR [11], this task continues to follow L1 loss training. We use LR and HR images as inputs and outputs to optimize the loss function with random gradient descent with the objective of minimizing L1, which can be formulated as

$$L(\Theta) = \frac{1}{N} \sum_{i=1} \|H_{net}(I_{LR}^i) - I_{HR}^i\|. \tag{13}$$

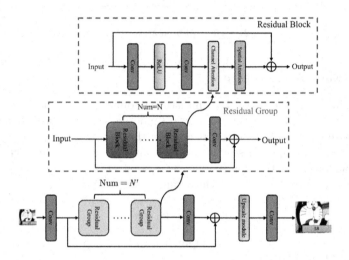

Fig. 3. Network architecture of AFFSR.

4 Experiment

4.1 Experimental Settings

Training Settings. We selected 800 images from the DIV2K dataset for training, and due to limitations in graphics card performance, we did not use any data augmentation method, and we used bicubic downsampling to obtain the input low-resolution images. In each batch of training, we selected two low-resolution images of size as input for a total of 15 rounds of training. Our model is trained by L1 loss function and ADAM optimizor with $\beta_1 = 0.9$, $\beta_2 = 0.999$ and $\varepsilon = 10^{-8}$. The initial learning rate is set to 1×10^{-4}. We use Pytorch to implement our model with GTX1080 GPU.

Test Settings. We selected standard datasets Set5, Set14, B100, and Urban100 for testing, and used peak singnal-to-noise ratios(PSNR) and structural similarity(SSIM) to quantify the results on the Y channel in the YCbCr space of the output image.

Network Parameter Settings. The number of residual blocks and residual groups is set to 20 and 10, respectively. The convolution kernel size in the residual blocks is both, the number of characteristic channels is 64, and the step size is 1. In channel attention, the convolution core size is set to 1×1, the number of channels first decreases to 4 and then increases to 64 dimensions, the convolution core size is set to 7×7 in spatial attention, and the channel is reduced from 2 to 1 dimension.

4.2   Hyperparametric Experiment

Fusing the features from multiple global pooling to improve performance is the core content of our method, therefore the selection of global pooling method is a very important superparameter in this paper. This section will analyze the performance impact of different global pooling methods to obtain the best set of hyperparameters. First, keep the other parameters unchanged, we don't introduce spatial attention for the time being. Then, we select one of the four pooling methods: global average pooling, global maximum pooling, global variance pooling, and global standard deviation pooling to train. Each group of experiments is trained for 15 rounds under the magnification of, and select the best performance model to test on Set5 and Set14, as shown in Table 1.

Table 1. Comparison of different global pooling methods

Global pooling method	Data Set			
	Set5		Set14	
	PSNR	SSIM	PSNR	SSIM
Avg	30.3622	0.85730	26.3447	0.72347
Avg+Max	30.3635	0.85924	**26.3777**	0.72631
Avg+Var+Std	30.3756	0.86064	26.3575	**0.72689**
Avg+Max+Var+Std	**30.4196**	**0.86075**	26.3271	0.72420

The results show that when multiple global pooling methods are used to fuse features, the performance of the model is significantly improved. As shown in the third row of Table 1, when the features obtained from the three global pooling are combined, the PSNR on set5 and set14 test sets is increased by 0.0133 db and 0.0128 db compared with the single global average pooling. Similarly, as shown in the forth row of Table 1, when the two global pooling features are combined, the PSNR on the two test sets is increased by 0.0013 db and 0.0019 db. When the four global pooled features are fused, it can be seen from the fourth row of Table 1 that the PSNR of set5 data set is increased by 0.0574 db, but decreased by 0.0176 db on set14. We consider that this is due to the increase in the number of features and the difficulty of weight training in a short time.

4.3   Ablation Experiment

In this section, to verify the effectiveness of each method, RCAN is considered a Baseline, and AFFCA and spatial attention are introduced on the basis of baseline. In this group of experiments, the magnification is still set to 4. The final test results on set5 and set14 are as follows.

Table 2. Ablation experiments on Set5 and Set14

Global pooling method	Spatial attention	AFFCA	Data Set			
			Set5		Set14	
			PSNR	SSIM	PSNR	SSIM
Avg	✗	✗	30.3622	0.85730	26.3447	0.72347
Avg	✓	✗	30.2968	0.86036	26.3419	0.72682
Avg+Var+Std	✗	✓	30.3756	0.86064	26.3575	0.72689
Avg+Var+Std	✓	✓	**30.4418**	**0.86229**	**26.3927**	**0.72747**

As shown in rows 1 and 3 of Table 2, after the introduction of AFFCA, the PSNR increased by 0.0133 dB and 0.0128 dB compared with the original model. In addition, as shown in rows 2 and 4, the PSNR increased by 0.145 dB and 0.0509 dB after the introduction of AFFCA, which fully illustrates the importance of AFFCA. On the other hand, SA can bind AFFCA well.

As shown in rows 3 and 4 of Table 2, after the introduction of spatial attention, PSNR increased by 0.0663 dB and 0.0353 dB compared with the original model. PSNR decreased by 0.0654 dB and 0.0028 dB compared with baseline in rows 1 and 2. In view of this phenomenon, we do not deny the validity of SA. We think this phenomenon is due to the increase of training difficulty after the introduction of SA. This phenomenon also shows that AFFCA can be better combined with SA and has better expandability. It is also worth noting that after the introduction of SA, SSIM has been significantly improved in all models and all test sets, especially in AFFSA models. As shown in row 4 in Table 2, we found that the SSIM values improved by 0.00165 and 0.00059 compared with the original model and 0.00499 and 0.00401 compared with baseline, which shows that SA is critical for performance improvement.

Finally, as shown in row 4 of Table 2, the models with both AFFCA and SA achieved maximum performance on both PSNR and SSIM, with PSNR values increased by 0.0796 dB and 0.0481 dB and SSIM by 0.00499 and 0.00401 compared with baseline. The above results fully demonstrate the validity of our method.

4.4   Model Comparison

Comparison of Quantitative Results. In the previous section, we mentioned that both FCANet and CBAM have improved the global pooling of channel attention. We name the model that uses the pooling method in CBAM as

CBAMP, and the model that introduces AFFCA and SA as AFFSR. This section compares our model with previous advanced methods at different magnifications. The experimental settings are consistent with previous work, and the number of training rounds for each model is consistent at the same magnification. The results are shown in Tables 3, 4 and 5.

Table 3. Quantitative comparison for $4 \times SR$

Model	Data Set							
	Set5		Set14		B100		Urban100	
	PSNR	SSIM	PSNR	SSIM	PSNR	SSIM	PSNR	SSIM
RCAN [21]	30.362	0.8573	26.345	0.7235	26.087	0.6901	23.401	0.6857
FCANet_Top32 [14]	30.310	0.8561	26.329	0.7237	26.026	0.6907	23.347	0.6856
FCANet_Top16 [14]	30.158	0.8499	26.249	0.7189	26.023	0.6877	23.292	0.6799
CBAMP [19]	29.894	0.8444	26.131	0.7166	26.027	0.6865	23.211	0.6753
AFFSR(Ours)	**30.442**	**0.8623**	**26.393**	**0.7275**	**26.095**	**0.6937**	**23.414**	**0.6902**

Table 4. Quantitative comparison for $3 \times SR$

Model	Data Set							
	Set5		Set14		B100		Urban100	
	PSNR	SSIM	PSNR	SSIM	PSNR	SSIM	PSNR	SSIM
RCAN [21]	29.424	0.8654	25.661	0.7495	27.447	0.7659	23.879	0.7385
FCANet_Top32 [14]	29.259	0.8639	25.629	0.7499	27.438	0.7666	23.870	0.7386
CBAMP [19]	29.132	0.8563	25.608	0.7443	27.263	0.7595	23.750	0.7284
AFFSR(Ours)	**29.548**	**0.8685**	**25.726**	**0.7531**	**27.485**	**0.7686**	**23.910**	**0.7427**

Table 5. Quantitative comparison for $2 \times SR$

Model	Data Set							
	Set5		Set14		B100		Urban100	
	PSNR	SSIM	PSNR	SSIM	PSNR	SSIM	PSNR	SSIM
RCAN [21]	36.299	0.9485	30.846	0.8891	27.215	0.8034	27.755	0.8675
FCANet_Top32 [14]	34.886	0.9513	30.222	0.8909	27.073	**0.8064**	27.533	0.8683
CBAMP [19]	35.603	0.9464	30.531	0.8864	27.229	0.8004	27.283	0.8570
AFFSR(Ours)	**36.528**	**0.9523**	**30.945**	**0.8924**	**27.256**	0.8052	**27.829**	**0.8703**

The key to FCANet [14] in Tables 3, 4 and 5 is to consider GAP as a special case of Discrete Cosine Transform (DCT), thereby extending a single GAP in the frequency domain. This method can result in significant performance improvements in classification tasks. In this paper, the best 16 and 32 frequency components in [4] are selected for experimentation under the magnification of 4×. The results show that the Top32 frequency component of FCANet is lower than RCAN in PSNR value, and the performance degradation is more obvious in Top16 frequency component. However, this method has a slight improvement in the SSM value.

The above results show that FCANet is not effective in super-resolution reconstruction tasks and can not distinguish channel characteristics very well. In addition, CBAM points out that channel weights can be computed using global average pooling and global maximum pooling respectively and added together to improve channel attention performance. When we reference the pooling method in CBAM, the results show that the performance is not as good as RCAN at all scales. We believe that this phenomenon is due to the equal treatment of the two pooled features by CBAM. At the same time, we find that the speed of loss convergence is slow during CBAMP training, but this paper uses short-term small-scale training, which results in poor performance. AFFSR replaces the original channel attention with adaptive multifeature fusion based on RCAN and introduces spatial attention to capture important feature information.

As shown in the last row of Tables 3, 4 and 5, the performance of each index of the proposed AFFSR is optimal for each dataset and each magnification. Especially at 2× scale, PSNR value is increased by 0.2294 dB, 0.0983 dB, 0.0410 dB and 0.0735 dB compared with RCAN, while it is increased by 0.0038, 0.0033, 0.0018 and 0.0028 on SSMA. This result fully demonstrates the effectiveness of AFFSR in super-resolution reconstruction tasks.

Contrast of Visual Effects. This section compares the visual effects of each method. As shown in Fig. 4, we find it difficult for Bicubic, a traditional interpolation-based method, to recover the outline of the hand and to perform very poorly in the visual effect. RCAN, a method based on deep CNN network, is much better than Bicubic. This method can restore the outline of the hand to some extent, but the result of this method brings very obvious stripe noise, which affects the visual perception. CBAMP and FCANet suppress the stripe noise to some extent, but it is still difficult to remove the fine noise. Our method not only eliminates most of the noise, but also reconstructs the most delicate contour.

Figure 5 is the results of each method on the building dataset. Compared to other methods, our method avoids the checkerboard effect and reconstructs the finest textures and lines.

5   Conclusion

In this paper, aiming at the problem that channel attention in the existing super-resolution model cannot sufficiently extract channel characteristics in the pre-processing section, we propose adaptive feature fusion channel attention. Our method uses a variety of global pooling methods to extract feature description scalars, and adaptively adjusts the weights of each scalar to merge them into a representative feature description scalar. Our method can extract enough information from the feature map to further enhance the ability to distinguish between channels. Ultimately optimizes channel attention performance. We have done some experiments and compared them with the existing methods. The experimental results show that the model is superior to the existing model in both quantitative and visual results.

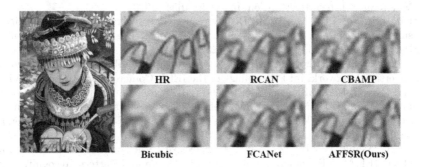

Fig. 4. Visual comparison of comic(2×) in Set14.

Fig. 5. Visual comparison of img019(2×) in Urban100.

References

1. Dong, C., Loy, C.C., Tang, X.: Accelerating the super-resolution convolutional neural network. In: Leibe, B., Matas, J., Sebe, N., Welling, M. (eds.) ECCV 2016. LNCS, vol. 9906, pp. 391–407. Springer, Cham (2016). https://doi.org/10.1007/978-3-319-46475-6_25
2. Dumoulin, V., Shlens, J., Kudlur, M.: A learned representation for artistic style. arXiv preprint arXiv:1610.07629 (2016)
3. He, K., Zhang, X., Ren, S., Sun, J.: Deep residual learning for image recognition. In: Proceedings of the IEEE Conference on Computer Vision and Pattern Recognition, pp. 770–778 (2016)

4. Hou, Q., Zhou, D., Feng, J.: Coordinate attention for efficient mobile network design. In: Proceedings of the IEEE/CVF Conference on Computer Vision and Pattern Recognition, pp. 13713–13722 (2021)
5. Hu, J., Shen, L., Sun, G.: Squeeze-and-excitation networks. In: Proceedings of the IEEE Conference on Computer Vision and Pattern Recognition, pp. 7132–7141 (2018)
6. Kim, J., Lee, J.K., Lee, K.M.: Accurate image super-resolution using very deep convolutional networks. In: Proceedings of the IEEE Conference on Computer Vision and Pattern Recognition, pp. 1646–1654 (2016)
7. Kim, J., Lee, J.K., Lee, K.M.: Deeply-recursive convolutional network for image super-resolution. In: Proceedings of the IEEE Conference on Computer Vision and Pattern Recognition, pp. 1637–1645 (2016)
8. Ledig, C., et al.: Photo-realistic single image super-resolution using a generative adversarial network. In: Proceedings of the IEEE Conference on Computer Vision and Pattern Recognition, pp. 4681–4690 (2017)
9. Lee, H., Kim, H.E., Nam, H.: SRM: a style-based recalibration module for convolutional neural networks. In: Proceedings of the IEEE/CVF International Conference on Computer Vision, pp. 1854–1862 (2019)
10. Li, X., Wang, W., Hu, X., Yang, J.: Selective kernel networks. In: Proceedings of the IEEE/CVF Conference on Computer Vision and Pattern Recognition, pp. 510–519 (2019)
11. Lim, B., Son, S., Kim, H., Nah, S., Mu Lee, K.: Enhanced deep residual networks for single image super-resolution. In: Proceedings of the IEEE Conference on Computer Vision and Pattern Recognition Workshops, pp. 136–144 (2017)
12. Nair, V., Hinton, G.E.: Rectified linear units improve restricted Boltzmann machines. In: ICML (2010)
13. Park, J., Woo, S., Lee, J.Y., Kweon, I.S.: Bam: Bottleneck attention module. arXiv preprint arXiv:1807.06514 (2018)
14. Qin, Z., Zhang, P., Wu, F., Li, X.: FcaNet: frequency channel attention networks. In: Proceedings of the IEEE/CVF International Conference on Computer Vision, pp. 783–792 (2021)
15. Sajjadi, M.S., Scholkopf, B., Hirsch, M.: Enhancenet: single image super-resolution through automated texture synthesis. In: Proceedings of the IEEE International Conference on Computer Vision, pp. 4491–4500 (2017)
16. Shi, W., et al.: Real-time single image and video super-resolution using an efficient sub-pixel convolutional neural network. In: Proceedings of the IEEE Conference on Computer Vision and Pattern Recognition, pp. 1874–1883 (2016)
17. Szegedy, C., et al.: Going deeper with convolutions. In: Proceedings of the IEEE Conference on Computer Vision and Pattern Recognition, pp. 1–9 (2015)
18. Wang, Q., Wu, B., Zhu, P., Li, P., Hu, Q.: ECA-NET: efficient channel attention for deep convolutional neural networks. In: 2020 IEEE/CVF Conference on Computer Vision and Pattern Recognition (CVPR) (2020)
19. Woo, S., Park, J., Lee, J.Y., Kweon, I.S.: CBAM: convolutional block attention module. In: Proceedings of the European Conference on Computer Vision (ECCV), pp. 3–19 (2018)
20. Zhang, H., et al.: ResNeST: split-attention networks. In: Proceedings of the IEEE/CVF Conference on Computer Vision and Pattern Recognition, pp. 2736–2746 (2022)
21. Zhang, Y., Li, K., Li, K., Wang, L., Zhong, B., Fu, Y.: Image super-resolution using very deep residual channel attention networks. In: Proceedings of the European Conference on Computer Vision (ECCV), pp. 286–301 (2018)

SGFusion: Camera-LiDAR Semantic and Geometric Fusion for 3D Object Detection

Xuhua Chen[1] , Xinhua Zeng[1]([✉]), Chengxin Pang[2], and Xin Hu[3]

[1] Academy for Engineering and Technology, Fudan University,
Shanghai 200433, China
zengxh@fudan.edu.cn

[2] Electronics and Information Engineering, Shanghai University of Electric Power,
Shanghai 200000, China
chengxin.pang@shiep.edu.cn

[3] School of Optical-Electrical and Computer Engineering, University of Shanghai
for Science and Technology, Shanghai 200093, China
huxin@usst.edu.cn

Abstract. Camera and lidar are considered as important sensors to achieve higher-level autonomous driving. And the complementary information provided by these sensors offer more opportunities for improving performance. However, it is difficulty to fuse them because of different representation of both. In this work, we propose a novel fusion framework SGFusion, which fuse the image and point clouds at semantic and geometric level. The SGFusion framework is divided in sequence into two stages: semantic fusion stage and geometric fusion stage. First, the point clouds are painted with the object-level semantic information obtained from an 2D object detector on the semantic fusion stage. At the same time, the output candidates of 2D object detector will be saved. Then, the painted point clouds fed to one LiDAR-based detector for obtaining high-quality 3D detection candidates. Finally, on the geometric fusion stage, these 3D detection candidates and the 2D detection candidates just saved are combined, and the more accurate detection results can be obtained by using their geometric consistence. The experimental results on KITTI detection benchmark show that our SGFusion achieve up to 3.64% AP improvement on car class compared to three different baselines. Furthermore, our method outperforms prior state-of-the-art works on 3D detection of car class on KITTI testing benchmark.

Keywords: Object detection · Multimodal fuison · Autonomous driving

1 Introduction

With development of autonomous driving and domestic robots and intelligence transaction, 3D object detection, as the crucial and indispensable task is extensively studied. Currently, the LiDAR sensors are the most commonly used 3D

M. Tanveer et al. (Eds.): ICONIP 2022, LNCS 13625, pp. 435–446, 2023.
https://doi.org/10.1007/978-3-031-30111-7_37

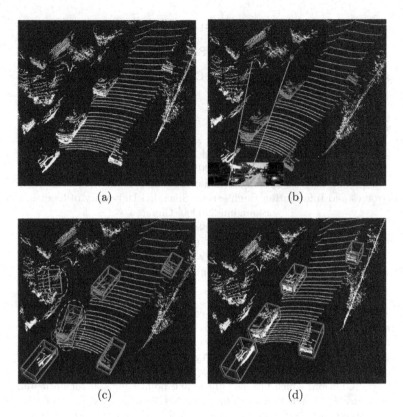

Fig. 1. (a) The raw pointcloud; (b) Painted pointcloud with the 2D object detection results; (c) give the object detection results with False Positive (FP) based on semantic fusion stage; (d) show the results come from the proposed two stages (sematic fusion stage and geometric fusion stage) fusion framework.

sensors in autonomous driving because it can provide 3D point clouds to capture the 3D structures of the scenes. Benefited from development of the deep learning techniques on point-cloud based representation, various LiDAR-based 3D object detection approaches have been proposed. However, performances of these LiDAR-based detectors are considerably limited by inherent defects of point clouds, such as its sparsity, texture and color information missing. On the other hand, camera can provide density texture and color information. The combination of both sensors provide a feasible way of improving 3d detection performance. But how to find a excellent fusion method is a challenging task. Inspired by this, Many multimodal fuse meths is proposed. MV3D [3] proposes a ROI feature fusion strategy which use a simple network to fuse the features of 2d image RoI and 3d RoI. CLOCs [12] obtain more precise confidences of 3D bounding boxes by utilizing correspondence association of 2D and 3D detection candidates. PointPainting [19] is a simple yet effective sequential fusion framework, and achieve outstanding performance on different benchmarks of 3D detection.

By utilizing a existing semantic segmentation network, this method can obtain pixel-wise semantic segmentation results and adds them into a point-cloud-based 3D detector with the 2D-3D projection algorithm. The work of PointPainting indicates that the combination of the semantic mask and raw point cloud can be used for improving detection accuracy of point-cloud-based 3D object detection approaches, especially point sparse objects.

However, the submitted results of painted PointRCNN on the KITTI testing split for car class show a drop in 3D mAP about 3%+. Which indicates over-reliance on accurate pixel-wise semantic mask will weaken the 3D detector's ability to extract information from point clouds. We point out the coarse object-wise semantic information offered by a object detector instead of pixel-wise sematic segmentation can make 3D detector maintain a more reasonable balance between semantic information and point clouds. Besides, The method of painting points with image semantic information have the problem of blurred boundaries of obstacles and many-to-one correspondence of 3D points and 2D image pixels. Figure 1(b) shows an example of the point cloud painted by 2D semantic mask, where the point cloud in a large frustum area behind and around the car have been mis-classified as foreground. Which impact 3D object detector to generate some FP detection results (shown in Fig. 1(c)). This problem of object detection is even more serious, but the bounding box from object detection contain explicit boarder information. Inspired by CLOCs, we can leverage the geometric consistence of same class 2D detections and 3D detections to filter out False Positive 3D detections. Therefore, Using the object detection network to do semantic fusion and geometric fusion not only improves speed and save comptution, but also greatly relieved the problem of boundary blurring, finally refine the detection results. Based this idea, we present a novel fusion framework, named SGFusion, which can simple yet effective fuse different types of sensors in semantic level and geometric level respectively. First, any off-the-shelf image detector can be used to get course semantic estimation (compare with pixel-level semantic segmentation) from RGB image. Then, painted points obtained by semantic fusion fuse semantic information and raw point cloud fed to one lidar only detector and get high-quality 3D box candidates. Finally, The geometric fusion combine output candidates of 3D detector and 2D detector to get more accurate detection results. The main contributions of our work can be summarized:

1) We first propose a general two stages fusion framework "SGFusion", which fuse the different modal information in the semantic level and geometric level and significantly improve the 3D object detection performance.

2) We point out the coarse object-level semantic mask can offer sufficient semantic information for 3D object detection and avoid overfitting problem. Besides, the 2D detection can be directly used by geometric fusion and further improve detection accuracy.

3) We demonstrate the superiority of the proposed fusion framework with experimental results on the KITTI. Specially, the results show our method clearly outperforming the prior published works.

2 Related Work

3D Object Detection in Point Cloud. Currently 3D object detection methods mostly focus on the basis of pure point cloud. PointPillar [8] convert point cloud into pillars and utilize a 2d convolution network to get detection results. 3DSSD [22] suggest that the detection on less representative points is feasible, and make a trade-off between accuracy and speed. VoxelRCNN [4] proposes the heigh performance of 3d object detection just need coarse voxels but not need precise positioning of raw points. SECOND [21] provides a sparse 3D convolution layer and significantly improve inference speed. PV-RCNN [17] benefit from both point-based and voxel-based methods. CenterPoint [23] present a anchor-free framework based in CenterNet [5] and achieve excellent performance.

Multi-modal Fusion. MV3D [3] proposes a ROI feature fusion strategy which utilize a 2d image RoI refine 3d RoI. MMF [9] realizes a multi-sensor fusion framework with an multi-task learning strategy. [25] use cascade architecture consist of image detector and lidar detector to exploit multi-modal information. ContFuse [10] benefits from the continuous fusion of image feature maps and BEV feature maps. 3D-CVF [24] fuses features from multi-view images and LiDAR. F-PointNet [13] uses the image to produce proposals and refines the final 3D bounding box in frustums with PointNet. CLOCs [12] utilizes 2D candidates to refine the confidences of 3D candidates with a learnable network.

PointPainting [19] is a simple yet effective sequential fusion method, which assigns a class score vector to each point by projecting points into image semantic segmentation mask, and feds the semantic to lidar-only 3d detector.

3 Method

The overall architecture of our "SGFusion" framework is illustrated in Fig. 2 the proposed multi-modaly fusion framework mainly divide to two stages: stage semantic fusion stage and geometric fusion stage. First, any off-the-shelf image detector can be used to get course semantic estimation (compare with pixel-level semantic segmentation) from RGB image. Then, painted points obtained by semantic fusion fuse semantic information and raw point cloud fed to one lidar only detector and get high-quality 3D box candidates. Finally, The geometric fusion combine output candidates of 3D detector and 2D detector to get more accurate detection results.

3.1 Semantic Fusion

2D images contain dense texture and color information, which can complementary to point cloud and thus improve detection accuracy. PointPating and PatingFusion [20] utilize a semantic segmentation network to get precise sematic information from 2D image by generating pixel-wise sematic labels. But, we propose a different viewpoint, the coarse semantic estimation obtained by a object detector can offer approximate complementary information to point cloud and

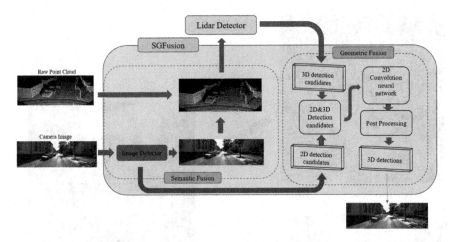

Fig. 2. Overview of our "SGFusion". First, any off-the-shelf image detector can be used to get coarse semantic estimation (compare with pixel-level semantic segmentation) from RGB image. Then, painted points obtained by semantic fusion stage fuse semantic information and raw point cloud fed to one lidar only detector and get high-quality 3D box candidates. Finally, The geometric fusion combine output candidates of 3D detector and 2D detector to get more accurate detection results.

has potential for further fusion. Besides, due to our framework do not rely on the specific object detectors, we can use any state-of-the-art image object detectors, such as Cascade RCNN [2], Faster RCNN [16], YOLOs [1,14,15], Swin transformer [11] etc. The 2D image is fed to an object detector and output a bounding boxes set, we denote it as $B \in R^{(n \times 6)}$, where n is the number of objects and the 6 is the number of single bounding box parameters, including the center coordinates (x, y), the width w, height h, the class c and the confidence f. Then we convert boxes set B to semantic mask ,denoted as $S \in R^{(w \times h \times m)}$, where w, h are the width and height of the image, m is the number of classes. Inspired by the fusion strategy in PointPainting, we leverage the camera project matrices to reproject the semantic mask into corresponding 3D point. The transform of lidar point is projected from 3D coordinate to camera penal coordinate and add semantic information is:

$$P_{painted} = Paint(M, T, S, P) \tag{1}$$

where $P_{painted}$ is the painted point which contain additional semantic information, the "Paint" represents the transform process, the $M \in R^{(3 \times 4)}$ is the camera matrix, the $T \in R^{(4 \times 4)}$ is the homogenous matrix and $P \in R^{(n \times 3)}$ is raw lidar points. After "Paint" painted, the painted points $P_{painted} \in R^{(n \times (3+m))}$ can be obtained and fed to one lidar only detector.

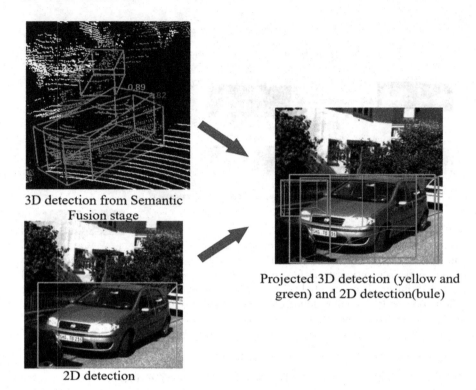

3D detection from Semantic Fusion stage

Projected 3D detection (yellow and green) and 2D detection(bule)

2D detection

Fig. 3. 3D and 2D detection. 3D detection of an object accurately detected by the 3D detector has greater overlap with correct 2D detection in image plane. While The overlap between incorrect 3D detection and accurate 2D detection usually is smaller.

3.2   Geometric Fusion

Although the semantic fusion based on semantic masks can significantly improve the detection accuracy (especially obstacles with sparse point clouds), some 3D points are misclassified due to the unavoidable ambiguity of the boundary, which produce some negative impact in the detection performance. For example, some point clouds is misclassified on around and behind the car in Fig. 3. It is precisely due to the influence of incorrect painted points that the lidar detector can not accurately predict the position and orientation of some obstacles and even generates a 3D bounding box in the background point cloud. To address this problem, we trained a convolution neural network to learn relation of 2D detections and 3D detections for geometric fusion. The geometric fusion utilize an image-based IOU between 2D detections and projected 3D detections to filter out some false positive detections. In other word, The geometric fusion suppressing confidence score of these 3D detections which have smaller overlap with the 2D detection in the image plane. The architecture is shown in Fig. 4 the process can be described as: First, n high-quality 3D detection candidates and k 2D detection candidates

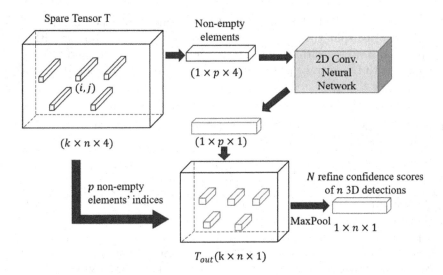

Fig. 4. The architecture of Geometric Fusion module. The non-empty elements of input sparse tensor T are fed to the convolution neural network to learn a refine confidence scores map Tout. Then, the final n available confidence scores can be obtained by utilizing maxpooling operation.

are converted to a $k \times n \times 5$ tensor T. For each element T_{ij}, there are 4 channels denoted as follows:

$$T_{ij} = \{IoU_{ij}, s_{2d}, s_{3d}, d_j\} \qquad (2)$$

where IoU_{ij} is the IoU between i_{th} 2D detection and j_{th} projected 3D detection, s_{2d} and s_{3d} are the confident scores for i_{th} 2D detection and j_{th} 3D detection respectively. d_j represents the normalized distance between the j_{th} 3D bounding box and the LiDAR in image plane. Elements T_{ij} with zero IoU are eliminated due to geometric inconsistence, thus the tensor T is sparse. Then the p non-empty elements are fed to the 2D convolution neural network to generate a fused confidence score for each association. By aggrating p non-empty elements' fused confidence score and their indices, we can obtain the output tensor Tout of shape $k \times n \times 1$. Finally the output tensor is mapped to the refine confidence map for n 3D detection through a maxpooling operation in the first dimension.

4   Experiment

4.1   Dataset and Evaluation Metrics

Our fusion method is evaluated on the 3D object detection benchmark KITTI dataset which consist of 7481 training samples and 7518 testing samples. Each instance in dataset is classified to three difficulty levels: easy, moderate and hard, according to its object size, truncation condition and occlusion condition. The testing set doesn't provide ground truth labels, so we need to submit our detection results to KITTI online server. For convenience of experimental studies, we follow [6] to split the original training samples into a train set with 3712 samples and a val set with 3769 samples. The results on val set and test set are evaluated with the average precision calculated by 0.7 3D IoU threshold and 40 recall positions.

Table 1. Evaluation results on KITTI validation dataset. Our SGFusion fusion methods outperform the baseline methods.

Method	mAP(3D)	3D AP			BEV AP		
		Easy	Moderate	Hard	Easy	Moderate	Hard
SECOND [21]	81.49	88.62	79.26	76.59	92.35	88.44	86.02
SECOND+SGFusion	84.22	92.42	82.90	77.33	95.41	89.60	86.37
Improvement	+2.73	+3.8	+3.64	+0.74	+3.06	+1.16	+0.35
PointRCNN [18]	84.68	91.53	82.35	80.15	93.01	88.91	88.50
PointRCNN+SGFusion	86.33	92.06	85.56	81.38	93.95	89.43	87.39
Improvement	+1.65	+0.53	+3.21	+1.23	+0.94	+0.52	-1.11
VoxelRCNN [4]	86.55	92.15	85.01	82.48	93.55	91.18	88.92
VoxelRCNN+SGFusion	87.89	92.25	86.75	84.68	95.28	92.41	92.56
Improvement	+1.34	+0.1	+1.74	+2.2	+1.73	+1.23	+3.64

4.2   Implementation Details

We use Cascade RCNN [2] to obtain 2D object detection results, due to its outstanding object detection ability. The Cascade-RCNN is trained on KITTI image dataset. We implement SGFuison in Pytorch using the open-sourced Open-PCdet. Our training consists of two stages: 1) we first train the Lidar detector for 80 epoch, which need the painted point cloud as input and generates high-quality 3D bounding box candidates. 2)We then train the geometric fusion network for 6 epochs. We observe that the geometric fusion network convergence fast due to its simple yet effective structure.

4.3   Evaluation Results

Evaluation on Baseline Methods: We using three different lidar detectors: SEC-OND [21], PointRCNN [18] and VoxelRCNN [4] as the baseline methods to eval Our SGFusion framework. Especially, We retrained the three baseline methods according to the OpenPCDet for ensuring the comparison conditions are consistent. The experiment result suggests that our fusion framework achieved significant improvements on mAP and APs of different difficulty levels compared to all baselines. Table 1 shows detailed results. From the table, we can observed that the proposed fusion strategy obtain improvement gain on mAP (1.5+) and BEV AP (only PointRCNN+SGFusion has slight drop on hard level). Interestingly, The detectors with slightly weaker detection ability (i.e. SECOND, PointRCNN) achieve greater improvement (+3.64 and +3.21 on moderate level, respectively) with the support of SGFusion. And the VoxelRCNN with good detection performance still obtain improvement (+1.74 on moderate level) from our fusion strategy.

Table 2. Comparison with state-of-the-art methods on the KITTI test set for car 3D detection, with average precisions of 40 sampling recall points evaluated on the KITTI server.

Method	Modality	3D AP				BEV AP		
		mAP	Easy	Moderate	Hard	Easy	Moderate	Hard
SECOND [21]	LiDAR	73.9	83.34	72.55	65.82	89.39	83.77	78.59
PointPillars [8]	LiDAR	75.29	82.58	74.31	68.99	90.07	86.56	82.81
PointRCNN [18]	LiDAR	84.68	86.96	75.64	70.70	92.13	87.39	82.72
3DSSD [22]	LiDAR	84.68	88.36	79.57	74.55	92.66	89.02	85.86
PV-RCNN [17]	LiDAR	82.83	90.25	81.43	76.82	94.98	90.65	86.14
Voxel-RCNN [4]	LiDAR	83.19	90.90	81.62	77.06	94.85	88.83	86.13
MV3D [3]	LiDAR+RGB	64.20	74.97	63.63	54.00	86.62	78.93	69.80
F-PointNet [13]	LiDAR+RGB	70.86	82.19	69.79	60.59	91.17	84.67	74.77
PointPainting [19]	LiDAR+RGB	73.63	82.11	71.70	67.08	92.45	88.11	83.36
AVOD [7]	LiDAR+RGB	73.52	83.07	71.76	65.73	89.75	84.95	78.32
3D-CVF [24]	LiDAR+RGB	80.79	89.20	80.05	73.11	93.52	89.56	82.45
CLOCs PVCas [12]	LiDAR+RGB	82.25	88.94	80.67	77.15	93.05	89.80	**86.57**
SGFusion (ours)	LiDAR+RGB	**83.76**	**91.13**	**82.64**	**77.53**	**94.76**	**91.11**	86.25

4.4 Comparison with State-of-the-Arts

To compare our SGFusion with other sate-of-the-art methods, we submit our best results to KITTI online test server. The result is shown in Table 2, our fusion method achieve outperform results. From this table, we can find that our SGFusion offers 1% improvement on the moderate level compared with the baseline VoxelRCNN [4]. And our method achieve comparable results with other state-of-the-art lidar-based or multimodal fusion methods.

4.5 Ablation Study

We design a series of ablation studies to verify effectiveness of different fusion stages. All the experiments are executed on the validation split of KITTI. Table 3 shows all the results, from experiment (b) we can see that semantic fusion stage marks 2.01%, 1.86% improvement on easy and moderate levels, respectively. And the experiment (c) demonstrate geometric fusion stage contributes 2.6%, 1.46% on easy AP and moderate AP, respectively. The higher improvement on moderate level from semantic fusion stage benefit from better detection ability for object of the point cloud is sparse. The geometric fusion offers advantages in easy level detection. But the combination of semantic fusion and geometric fusion provide the most improvement in each difficulty level, 3.8%, 3.6%, 0.73%, respectively (Fig. 5).

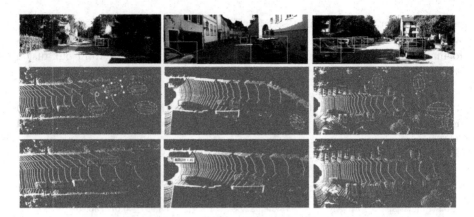

Fig. 5. Visualization of detection results on the KITTI dataset. The first row is detection results of image, the second row is detection results obtained by SECOND and the last row is the detection results obtained by SECOND+SGFusion. The bounding boxes of the ground-truth and prediction are in the color red and green respectively. Especially, these false detection generated by SECOND but removed by SECOND+SGFusion are marked by yellow dash ellips. (Color figure online)

Table 3. Ablation study on SGFusion. "SF": Semantic Fusion. "GF": Geometric Fusion. The results are calculated by 40 recall positions for car class.

Experiment	SF	GF	3D AP		
			Easy	Moderate	Hard
(a)			88.62	79.26	76.59
(b)	✓		90.63	81.12	76.62
(c)		✓	91.23	80.76	77.24
(d)	✓	✓	92.42	82.90	77.33

4.6   Conclusion

In this paper, we present a simple yet efficient multi-modal fusion framework, "SGFusion" which can fuse detection results from 2D/3D object detection networks in semantic level and geometric level. Especially, we are the first to use an object detection network to do semantic fusion and geometric fusion, respectively. The effectiveness of the proposed fusion strategy is proved by experimental results on KITTI dataset. Furthermore, the proposed "SGFusion" does not rely on specific 2D/3D object detector. Which means our strategy will continue to benefit from the development of 2D/3D object detection in the future.

References

1. Bochkovskiy, A., Wang, C., Liao, H.M.: YOLOv4: optimal speed and accuracy of object detection. CoRR abs/2004.10934 (2020)
2. Cai, Z., Vasconcelos, N.: Cascade R-CNN: delving into high quality object detection. In: 2018 IEEE Conference on Computer Vision and Pattern Recognition, pp. 6154–6162. Computer Vision Foundation/IEEE Computer Society (2018)
3. Chen, X., Ma, H., Wan, J., Li, B., Xia, T.: Multi-view 3D object detection network for autonomous driving. In: 2017 IEEE Conference on Computer Vision and Pattern Recognition, CVPR 2017, pp. 6526–6534. IEEE Computer Society (2017)
4. Deng, J., Shi, S., Li, P., Zhou, W., Zhang, Y., Li, H.: Voxel R-CNN: towards high performance voxel-based 3D object detection. In: Proceedings of the AAAI Conference on Artificial Intelligence, pp. 1201–1209. AAAI Press (2021)
5. Duan, K., Bai, S., Xie, L., Qi, H., Huang, Q., Tian, Q.: CenterNet: keypoint triplets for object detection. In: International Conference on Computer Vision, pp. 6568–6577. IEEE (2019)
6. Geiger, A., Lenz, P., Stiller, C., Urtasun, R.: Vision meets robotics: the KITTI dataset. Int. J. Robot. Res. **32**(11), 1231–1237 (2013)
7. Ku, J., Mozifian, M., Lee, J., Harakeh, A., Waslander, S.L.: Joint 3D proposal generation and object detection from view aggregation. In: 2018 IEEE/RSJ International Conference on Intelligent Robots and Systems, IROS 2018, pp. 1–8. IEEE (2018)
8. Lang, A.H., Vora, S., Caesar, H., Zhou, L., Yang, J., Beijbom, O.: PointPillars: fast encoders for object detection from point clouds. In: IEEE Conference on Computer Vision and Pattern Recognition, CVPR 2019, pp. 12697–12705. Computer Vision Foundation/IEEE (2019)
9. Liang, M., Yang, B., Chen, Y., Hu, R., Urtasun, R.: Multi-task multi-sensor fusion for 3D object detection. In: IEEE Conference on Computer Vision and Pattern Recognition, CVPR 2019, pp. 7345–7353. Computer Vision Foundation/IEEE (2019)
10. Liang, M., Yang, B., Wang, S., Urtasun, R.: Deep continuous fusion for multi-sensor 3D object detection. In: Ferrari, V., Hebert, M., Sminchisescu, C., Weiss, Y. (eds.) ECCV 2018. LNCS, vol. 11220, pp. 663–678. Springer, Cham (2018). https://doi.org/10.1007/978-3-030-01270-0_39
11. Liu, Z., et al.: Swin transformer: hierarchical vision transformer using shifted windows. In: 2021 IEEE/CVF International Conference on Computer Vision, pp. 9992–10002. IEEE (2021)
12. Pang, S., Morris, D.D., Radha, H.: CLOCs: camera-lidar object candidates fusion for 3D object detection. In: IEEE/RSJ International Conference on Intelligent Robots and Systems, pp. 10386–10393. IEEE (2020)
13. Qi, C.R., Liu, W., Wu, C., Su, H., Guibas, L.J.: Frustum pointnets for 3D object detection from RGB-D data. In: 2018 IEEE Conference on Computer Vision and Pattern Recognition, pp. 918–927. Computer Vision Foundation/IEEE Computer Society (2018)
14. Redmon, J., Divvala, S.K., Girshick, R.B., Farhadi, A.: You only look once: unified, real-time object detection. In: 2016 IEEE Conference on Computer Vision and Pattern Recognition, CVPR 2016, pp. 779–788. IEEE Computer Society (2016)
15. Redmon, J., Farhadi, A.: YOLOv3: an incremental improvement. CoRR abs/1804.02767 (2018)

16. Ren, S., He, K., Girshick, R.B., Sun, J.: Faster R-CNN: towards real-time object detection with region proposal networks. IEEE Trans. Pattern Anal. Mach. Intell. **39**(6), 1137–1149 (2017)
17. Shi, S., et al.: PV-RCNN: point-voxel feature set abstraction for 3D object detection. In: 2020 IEEE/CVF Conference on Computer Vision and Pattern Recognition, CVPR 2020, pp. 10526–10535. Computer Vision Foundation/IEEE (2020)
18. Shi, S., Wang, X., Li, H.: PointRCNN: 3D object proposal generation and detection from point cloud. In: IEEE Conference on Computer Vision and Pattern Recognition, CVPR 2019, pp. 770–779. Computer Vision Foundation/IEEE (2019)
19. Vora, S., Lang, A.H., Helou, B., Beijbom, O.: PointPainting: sequential fusion for 3D object detection. In: 2020 IEEE/CVF Conference on Computer Vision and Pattern Recognition, CVPR 2020, pp. 4603–4611. Computer Vision Foundation/IEEE (2020). https://doi.org/10.1109/CVPR42600.2020.00466
20. Xu, S., Zhou, D., Fang, J., Yin, J., Zhou, B., Zhang, L.: FusionPainting: multimodal fusion with adaptive attention for 3D object detection. In: 24th IEEE International Intelligent Transportation Systems Conference, ITSC 2021, pp. 3047–3054. IEEE (2021)
21. Yan, Y., Mao, Y., Li, B.: SECOND: sparsely embedded convolutional detection. Sensors **18**(10), 3337 (2018)
22. Yang, Z., Sun, Y., Liu, S., Jia, J.: 3DSSD: point-based 3D single stage object detector. In: 2020 IEEE/CVF Conference on Computer Vision and Pattern Recognition, pp. 11037–11045. Computer Vision Foundation/IEEE (2020)
23. Yin, T., Zhou, X., Krähenbühl, P.: Center-based 3D object detection and tracking. In: IEEE Conference on Computer Vision and Pattern Recognition, pp. 11784–11793. Computer Vision Foundation/IEEE (2021)
24. Yoo, J.H., Kim, Y., Kim, J., Choi, J.W.: 3D-CVF: generating joint camera and LiDAR features using cross-view spatial feature fusion for 3D object detection. In: Vedaldi, A., Bischof, H., Brox, T., Frahm, J.-M. (eds.) ECCV 2020. LNCS, vol. 12372, pp. 720–736. Springer, Cham (2020). https://doi.org/10.1007/978-3-030-58583-9_43
25. Zhao, X., Liu, Z., Hu, R., Huang, K.: 3D object detection using scale invariant and feature reweighting networks. In: Proceedings of the AAAI Conference on Artificial Intelligence, pp. 9267–9274. AAAI Press (2019)

SATNet: Captioning with Semantic Alignment and Feature Enhancement

Wenhui Bai[1], Canlong Zhang[1,2(✉)], Zhixin Li[1,2], Peiyi Wei[1],
and Zhiwen Wang[3]

[1] School of Computer Science and Engineering, Guangxi Normal University,
Guilin 541004, China
zcltyp@163.com
[2] Guangxi Key Lab of Multi-source Information Mining and Security,
Guilin 541004, China
[3] School of Computer Science and Technology, Guangxi University of Science
and Technology, Liuzhou 545006, China

Abstract. The fusion of region and grid features based on location alignment can make the utilization of image features better to a certain extent, thus improving the accuracy of image captioning. However, it still inevitably introduces semantic noise because of spatial misalignment. To address the problem, this paper proposes a novel image captioning model based on semantic alignment and feature enhancement, which contains a Visual Features Adaptive Alignment Module (VFAA) and a Features Enhancement Module (FEM). The VFAA module, at the encoder layer, utilizes Visual Semantic Graph (VSG) to generate pure semantic information for more accurately guiding the alignment and fusion of the region and grid features, thus further reducing the semantic noise caused by spatial dislocation. In addition, to ensure that the features that eventually enter the decoder layer do not lose their specific attributes, we design the FEM module to fuse the original region and grid features. To validate the effectiveness of the proposed model, we conduct extensive experiments on the MS-COCO dataset and test it on the online test server. The experimental results show that our model is superior to many state-of-the-art methods.

Keywords: Image captioning · Semantic alignment · Feature enhancement · Visual semantic graph

1 Introduction

Image captioning [24], as one of the core research topics in the field of cross-modal, has attracted the attention of related researchers in recent years. Different from classification and detection, image captioning aims to use image features and language models to generate a fluent and diversified natural language caption consistent with the content of the image.

© The Author(s), under exclusive license to Springer Nature Switzerland AG 2023
M. Tanveer et al. (Eds.): ICONIP 2022, LNCS 13625, pp. 447–458, 2023.
https://doi.org/10.1007/978-3-031-30111-7_38

Inspired by machine translation [4], researchers began to use the encoder-decoder structure [20] as the main framework of the image captioning model. At first, the encoder used CNN to process the image to obtain its grid features [17]. Later, [2] proposed using the feature of the complete objects extracted from the image as the final image feature vectors, named the region features. After that, the self-attention network [18] gradually replaced the traditional network used in the framework. Recently, the Dual-Layer Collaborative Transformer [13] introduced the geometric alignment graph as a necessary condition for constructing a 0/1 (False/True) attention mask to fuse the region and grid features.

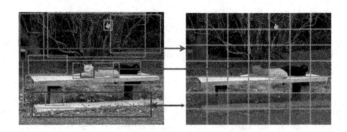

Fig. 1. Alignment graph of region and grid features. The left picture is region features and the right picture is grid features.

However, there are still problems in the fusion of the two features. The region features are extracted by the detector, and they are inconsistent with the ones obtained by semantic segmentation. The former uses the region specified by the anchor box as the detected region, so the shape of the region is regular. The shape of the grid is also regular, so the region and grid features will be aligned well by location only when the size of the object is close to the one of the grid, just as the cat head detected by the object detector in Fig. 1. However, in most cases, the size of objects is larger or smaller than its anchor box, and the space occupied by the individual object is not the whole region of the anchor boxes, so when using the absolute 0/1 (False/True) attention mask based on the position of the anchor boxes to align the grid features, it will inevitably introduce noisy information. Such as the signboard and slate detected by the object detector in Fig. 1, the grid region corresponding to the location information occupy much more space than the object's real region, which will introduce semantic noise.

To address the problem, this paper proposes a new fusion method, which uses semantic alignment weight (soft-attention) instead of the absolute 0/1 attention mask (hard-attention) to more carefully and accurately align two visual features. We calculate the semantic weights between the region and grid features and use the semantic weights as a "bridge" to guide the fusion and alignment of the two features, which can obtain more detailed image captioning. In our case, we only utilize pure semantic information to calculate the weight of semantic content, which is integrated into the joint processing of features. Considering that the traditional fusion of two features may ignore their original attributes, we fuse the processed initial features and the aligned features so that we can not only get the complementary two kinds of feature attributes but also ensure that the unique attributes are contained in the original features will not be lost.

Therefore, we design the VFAA and FEM in our Semantic Alignment Transformer (SATNet) to achieve the above two goals respectively, and conduct adequate experiments on the MS-COCO [12] dataset. The quantitative and qualitative analysis show that our proposed method outperforms many existing models.

Our contributions can be summarized as follows:

- We propose the visual features adaptive alignment module and generate visual semantic information by constructing a visual semantic graph between image features, which can guide the alignment and fusion of the region and grid features.
- We propose the features enhancement module to fuse the original features and aligned features to ensure that the unique attributes contained in the original features will not be lost.
- We construct a novel image captioning model based on the above two modules, which can generate more accurate and comprehensive captions for images, and show excellent performance on the MS-COCO dataset.

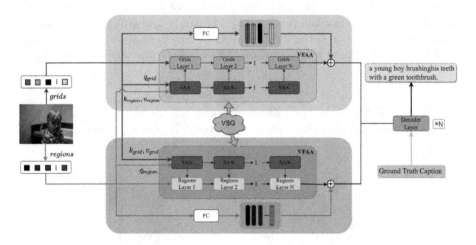

Fig. 2. The design of our SATNet framework.

2   Proposed Method

In this section, we briefly introduce the transformer model [18] and position encoding in Sect. 2.1. Then, we show visual semantic graph representation in Sect. 2.2. Next, we will present the VFAA and FEM in Sect. 2.3 and 2.4. Finally, we will give the training details of our model in Sect. 2.5. Figure 2 shows the overall design of our model.

2.1   Transformer Model and Position Encoding

The transformer core module is the self-attention module, and the Fig. 3(b) shows its structure. The input consists of Q, K, and V matrices, usually regarded as a mapper between a set of queries and features key/value pairs. Given the input features X, the self-attention module can obtain the permutation-invariant encoding of X. This process can be expressed as follows:

$$Z = Attention(Q, K, V) = softmax(\frac{QK^T}{\sqrt{d_k}}). \qquad (1)$$

$$Q = XW_q, \quad K = XW_k, \quad V = XW_v. \qquad (2)$$

where Q, K, V have the same dimension, W_q, W_k, and W_v are learnable weight matrices. The transformer network uses sine and cosine functions of different frequencies to represent position information. The formula is shown as follows:

$$PE_{(pos,2i)} = sin(pos/10000^{(2i/d_{model})}), \qquad (3)$$

$$PE_{(pos,2i+1)} = cos(pos/10000^{(2i/d_{model})}), \qquad (4)$$

where pos and i are the position and dimension, the d_{model} is the dimension of the input. In our model, the sine and cosine functions of different frequencies represent the absolute position information of grid features and ground truth text features. For region features, the 4-d bounding box $B_i = (x_{min}, y_{min}, x_{max}, y_{max})$ is used as the embedded representation of its absolute position information:

$$PE_i^r = B_i W_{emb}. \qquad (5)$$

where i is the index of the bounding box, (x_{min}, y_{min}) and (x_{max}, y_{max}) denote the top-left and bottom-right corners of the box, W_{emb} is a parameter matrix.

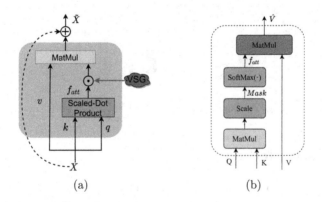

(a)   (b)

Fig. 3. (a) Details of the SAA module. (b) Scaled Dot-Product Attention.

2.2 Visual Semantic Graph Representation

Given an image, we first obtain the region and grid features of the image according to the operation provided by [9], denoted as $V_R = (r_1, r_2, \cdots, r_i)$ and $V_G = (g_1, g_2, \cdots, g_i)$, where $(r_i, g_i) \in \mathbb{R}^N$ is the dimension of the image features.

To guide the alignment and fusion of two visual features more accurately, we design the VSG (see Fig. 4). VSG is essentially a directed graph, the circles and rhombus represent grid (g_i) and region (r_j) features vertexes respectively, similar region and grid vertexes are connected by a directed edge, α on the directed edge represents the weight of similar semantic alignment. Although connected edges indicate that they are similar, the weight of the edge at the same position is different due to different arrow directions.

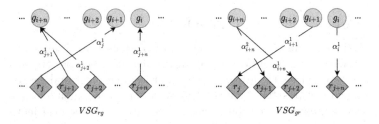

Fig. 4. Example of the visual semantic graph.

When obtaining the semantic alignment information, the two visual features V_R and V_G are first normalized to V_R^{ln} and V_G^{ln}. Then the normalized vectors are subjected to dot-product operation and divided by the parameter γ. Finally, the weight representation is obtained through softmax function operation:

$$VSG_{rg} = softmax(V_R^{ln}(V_G^{ln})^T/\gamma), \quad VSG_{gr} = softmax(V_G^{ln}(V_R^{ln})^T/\gamma), \quad (6)$$

where $(V_R^{ln}, V_G^{ln}) \in \mathbb{R}^{i \times N}$, the value of N is 2048. We set the parameter γ to smooth the size of the results because the dimensions of the features are large.

2.3 Visual Features Adaptive Alignment

VFAA consists of Semantic Adaptive Alignment (SAA) and Cross Fusion Attention (CFA). Given two groups of input image features V_R and V_G, their dimensions are first projected to $d_{model} = 512$, and then are input into the SAA module (see Fig. 3(a)) to guide their alignment by using semantic alignment weights, which can obtain aligned features, finally sends the aligned features to the CFA for further alignment and fusion. In our SAA module, different self-attention layers in the transformer structure obtain an attention weight representation f_{att} when processing image features, which represents the alignment relationship between region and grid features. However, it still inevitably introduces semantic noise due to spatial misalignment. So we weight f_{att} using the purely semantic alignment weights obtained from the VSG, which can obtain a more accurate alignment. Above processing can be formulated as:

$$f_{att}^{rg} = f_{att} \odot VSG_{rg}, \quad f_{att}^{gr} = f_{att} \odot VSG_{gr}, \tag{7}$$

where f_{att} is the weight of the scaled dot-product attention, VSG_{rg}, and VSG_{gr} represent pure semantic alignment weights, and $\odot$ is the Hadamard Product. Note that in the CFA module we do not apply VSG, nor do we use other additional masks:

$$X^{l+1} = Attention(Q', K', V'), \tag{8}$$

$$Q' = [X^l W_Q], \quad K' = [X^l W_K; FFN'_{SAA}], \quad V' = [X^l W_V; FFN'_{SAA}]. \tag{9}$$

where X^l represents two visual features, and $X^l \in \{V_R, V_G\}$, FFN'_{SAA} is the output of the SAA, W_Q, W_K, and W_V are learnable parameter matrices.

2.4   Features Enhancement Module

After analysis, we found that the two aligned visual features through the VFAA module will lose a few attributes of the original features because the main function of above module is to provide more accurate alignment and fusion for the region and grid features. Through the VFAA module, the region features can be enhanced by context information from grid features, and the grid features can also be enhanced by object information from region features. However, this may lead to neglecting some details of the original features that belong to themselves.

Therefore, inspired by the idea of the classic residual network [6] we first use a fully-connected network followed by the dropout and ReLU to project the feature's dimension to 512-D, and then respectively sum them with corresponding the output of the VFAA module and normalize the result, finally splice the normalized features together as the image feature input of the decoder:

$$F_R = LN_1(\tilde{Z}_R + FC_1(V_R)), \quad F_G = LN_2(\tilde{Z}_G + FC_2(V_G)). \tag{10}$$

where F_R and F_G are the features of the input decoder, $\tilde{Z}$ is the output of the VFAA, LN and FC are the normalization layer and the fully connected layer.

2.5   Training Details

Following the standard practice of image captioning [2,16], we first train our model with a word-level cross-entropy loss (XE):

$$L_{XE}(\theta) = -\sum_{t=1}^{T} log(p_\theta(\omega_t^* | \omega_{1:t-1}^*)), \tag{11}$$

After that, we continuously optimize the non-differentiable evaluation metrics through self-critical sequence training [16] following [5]:

$$\nabla_\theta L_{RL}(\theta) = -\frac{1}{k} \sum_{i=1}^{k} ((r(y_{1:T}^i) - b)\nabla_\theta log p_\theta(y_{1:T}^i)), \quad b = \frac{1}{k} \sum_{i=1}^{k} r(y_{1:T}^i). \tag{12}$$

where k is the number of beam searches, $r(\cdot)$ is the reward function, $y^i_{1:T}$ is the i-th sentence in the sampling sequence, and b is the baseline.

3   Experiments

3.1   Dataset and Evaluation Metrics

We evaluate our model on the standard MS-COCO dataset [12], which contains 123,287 images. Each picture contains 5 different captions. In offline evaluation, we follow the splits method provided by Karpathy et al. [10], where 5000 images are used for validation, 5000 images are used for testing, and the remaining images are used for training. Then, we downloaded the unlabeled test dataset officially provided by COCO for online evaluation, which contains 40,775 images. We convert all sentences to lowercase, remove punctuation in all sentences, and remove words that occur less than five times, resulting in a vocabulary of 10201 words. Following the standard evaluation protocol, we use all the classic metrics to evaluate the performance of our proposed model on image captioning, including BLEU [15], METEOR [3], ROUGR [11], CIDEr [19], and SPICE [1].

3.2   Implementation Details

We use the backbone provided by [9] to extract the image features. For the sentences, we linearly project each word to the input dimension of the model using one-hot encoding. In our implementation, we set the d_{model} to 512, the number of heads to 8, the number of layers for both encoder and decoder 3, the probability of dropout to 0.1, and the value of γ to 64. During the whole training process, we employ the Adam optimizer to help train model, the beam size is set to 5, and during the cross-entropy training, we set the batch size to 50, and the batch size is changed from 50 to 100 during the self-critical sequence training.

Table 1. Ablation experiments on our model, where B-1, B-4, M, R, C and S are short for BLEU-1, BLEU-4, METEOR, ROUGE, CIDEr, and SPICE scores.

VFAA	FEM	B-1	B-4	M	R	C	S
✗	✗	81.5	39.5	29.4	59.2	131.6	23.0
✗	✓	81.6	39.6	29.5	59.2	132.2	23.1
✓	✗	82.1	**40.3**	29.6	**59.7**	133.6	23.2
✓	✓	**82.2**	**40.3**	**29.7**	59.6	**134.2**	**23.3**

3.3   Ablation Analysis

To fully demonstrate the effectiveness of our methods, We conduct ablation experiments on our model and the DLCT model [13], the image features used in our experiments are all extracted with ResNext101 as the backbone network.

Table 2. Ablation experiments on the DLCT model.

CBG	GAG	FEM	VSG	B-1	B-4	M	R	C	S
✗	✗	✗	✗	81.2	39.2	29.2	58.6	132.6	22.8
✓	✗	✗	✗	80.8	38.7	29.0	58.7	130.8	22.7
✗	✓	✗	✗	81.4	39.8	29.5	59.1	133.8	23.0
✗	✓	✓	✗	82.0	40.2	29.6	**59.5**	134.3	23.0
✗	✗	✗	✓	**82.2**	**40.4**	**29.7**	59.5	**134.5**	**23.2**

VFAA. From Table 1 we can see that using the semantic alignment information contained in the VSG to help guide the alignment and fusion of two features can improve the performance of the model, increasing the CIDEr score from 131.6% to 133.6%, and the scores of other evaluation metrics also have different improvements. In addition, from the ablation experiments on DCLT in Table 2, it can also be seen that our proposed VSG improves the CIDEr score by 1.9% and 0.7% compared respectively with the baseline model and the original model. It is higher than the model using CBG.

Table 3. Performance comparisons on the COCO Karpathy test split.

Model	B-1	B-4	M	R	C	S
SCST [16]2017	-	34.2	26.7	57.7	114.0	-
Up-Down [2]2018	79.8	36.3	27.7	56.9	120.1	21.4
GCN-LSTM [24]2018	80.5	38.2	28.5	58.3	127.6	22.0
SGAE [23]2019	80.8	38.4	28.4	58.6	127.8	22.1
ORT [7]2019	80.5	38.6	28.7	58.4	127.3	22.6
AOA [8]2019	80.2	38.9	29.2	58.8	129.8	22.4
M^2 [5]2020	80.8	39.1	29.2	58.6	131.2	22.6
X-Transformer [14]2020	80.9	39.7	29.5	59.1	132.8	**23.4**
RSTNet [25]2021	81.1	39.3	29.4	58.8	133.3	23.0
DGET [22]2022	81.3	**40.3**	29.2	59.4	132.4	23.3
Ours(ResNeXt-101)	**82.2**	40.3	**29.7**	**59.6**	**134.2**	23.3

FEM. It can be seen from Table 1 that FEM improves the CIDEr score from 131.6% to 132.2%, and in addition to the ROUGR evaluation metric, other metrics also have different improvements. In addition, we can observe that the FEM module has improved the scores of all metrics and has good performance in the two metrics of B-4 and ROUGR from Table 2. The results of the ablation experiments on two different models can prove that the FEM is beneficial.

3.4   Quantitative Analysis

Offline Evaluation. We show the performance comparison of our model with some recent models on the offline "Karpathy" test split in Table 3, the models we compared include SCST [16], Up-Down [2], GCN-LSTM [24], SGAE [23], ORT [7], AOA [8], M^2 [5], X-Transformer [14], RSTNet [25]and DGET [22]. It can be seen from Table 3 that our model outperforms all models on BLEU-1, METEOR and CIDEr scores, the performance of BLEU-4 and ROUGE is comparable with DGET. On the whole, the model we proposed is significantly better than other methods in most evaluation indicators.

Table 4. Leaderboards of various methods on the online MS-COCO test server, where S-T, U-D, H-N, G-L, S-E, X-T and R-N are short for SCST, Up-Down, HAN, GCN-LSTM, SGAE, X-Transformer, and RSTNet.

Model	B-1		B-2		B-3		B-4		M		R		C	
	c5	c40	c5	c40	c5	c40	c5	c40	c5	c40	c5	c40	c5	c40
S-T [16]	78.1	93.7	61.9	86.0	47.9	75.9	35.2	64.5	27.0	35.5	56.3	70.7	114.7	116.7
U-D [2]	80.2	95.2	64.1	88.8	49.1	79.4	36.9	68.5	27.6	36.7	57.1	72.4	117.9	120.5
H-N [21]	80.4	94.5	63.8	87.7	48.8	78.0	36.5	66.8	27.4	36.1	57.3	71.9	115.2	118.2
G-L [24]	80.8	95.2	65.5	89.3	50.8	80.3	38.7	69.7	28.5	37.6	58.5	73.4	125.3	126.5
S-E [23]	81.0	95.3	65.6	89.5	50.7	80.4	38.5	69.7	28.2	37.2	58.6	73.6	123.8	126.5
AOA [8]	81.0	95.0	65.8	89.6	51.4	81.3	39.4	71.2	29.1	38.5	58.9	74.5	126.9	129.6
M^2 [5]	81.6	96.0	66.4	90.8	51.8	82.7	39.7	72.8	29.4	39.0	59.2	74.8	129.3	132.1
X-T [14]	81.3	95.4	66.3	90.0	51.9	81.7	39.9	71.8	29.5	39.0	59.3	74.9	129.3	131.4
R-N [25]	81.7	96.2	66.5	90.9	51.8	82.7	39.7	72.5	29.3	38.7	59.2	74.2	130.1	132.4
Ours	**82.6**	**96.6**	**67.2**	**91.4**	**52.5**	**83.5**	**40.5**	**73.8**	**29.6**	**39.4**	**59.8**	**75.3**	**131.5**	**133.7**

Online Evaluation. We report the performance of our proposed model on the official online COCO test server in Table 4, and the official test images are 5 reference captions (c5) and 40 reference captions (c40), respectively. The model used for our online evaluation is an ensemble of four models trained on the "Karpathy" training split.

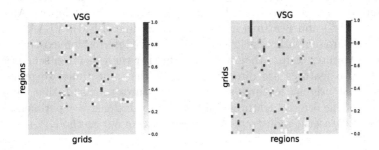

Fig. 5. Visualization example of the visual vemantic graph.

3.5 Qualitative Analysis and Visualization

For a more intuitive understanding of our proposed Visual Semantic Graph, we show a visualization example of semantic weights used in the alignment of two visual features Fig. 5, which can be understood corresponding to Fig. 4 shown above. From the figure, we can observe the details of the weight distribution; compared with the absolute 0/1 (False/True), the semantic information weight is more "soft," which can make the alignment of image features more accurate.

Figure 6 shows several examples of captions generated by our model and the original Transformer model under the same conditions. From the results, we can see that our model can generate more accurate and more vivid captions with more details.

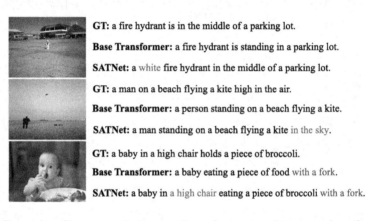

Fig. 6. Examples of image captioning results and corresponding ground truth captions generated by our proposed method and the baseline Transformer model.

4   Conclusion

This paper proposes a novel image captioning model based on semantic alignment and features enhancement. Including Visual Features Adaptive Alignment module (VFAA) and Features Enhancement Module (FEM), VFAA uses VSG to obtain a pure semantic relationship and uses the information existing between two visual features to fuse them to reduce further the generated semantic noise when the two features are directly fused. At the same time, we propose the FEM fuse the original features and aligned features to ensure that the unique attributes of the original features will not be lost. Extensive experiments on the MS-COCO dataset demonstrate the effectiveness of our method.

Acknowledgements. This work is supported by National Natural Science Foundation of China (Nos. 62266009, 61866004, 62276073, 61966004, 61962007), Guangxi Natural Science Foundation (Nos. 2018GXNSFDA281009, 2019GXNSF DA245018, 2018GXNSFDA294001), Guangxi "Bagui Scholar" Teams for Innovation and Research

Project, Innovation Project of Guangxi Graduate Education (No. JXXYYJSCXXM-2021-013), and Guangxi Collaborative Innovation Center of Multi-source Information Integration and Intelligent Processing.

References

1. Anderson, P., Fernando, B., Johnson, M., Gould, S.: SPICE: semantic propositional image caption evaluation. In: Leibe, B., Matas, J., Sebe, N., Welling, M. (eds.) ECCV 2016. LNCS, vol. 9909, pp. 382–398. Springer, Cham (2016). https://doi.org/10.1007/978-3-319-46454-1_24

2. Anderson, P., et al.: Bottom-up and top-down attention for image captioning and visual question answering. In: Proceedings of the IEEE Conference on Computer Vision and Pattern Recognition, pp. 6077–6086 (2018)

3. Banerjee, S., Lavie, A.: Meteor: an automatic metric for MT evaluation with improved correlation with human judgments. In: Proceedings of the ACL Workshop on Intrinsic and Extrinsic Evaluation Measures for Machine Translation and/or Summarization, pp. 65–72 (2005)

4. Cho, K., et al.: Learning phrase representations using RNN encoder-decoder for statistical machine translation. arXiv preprint arXiv:1406.1078 (2014)

5. Cornia, M., Stefanini, M., Baraldi, L., Cucchiara, R.: Meshed-memory transformer for image captioning. In: Proceedings of the IEEE/CVF Conference on Computer Vision and Pattern Recognition, pp. 10578–10587 (2020)

6. He, K., Zhang, X., Ren, S., Sun, J.: Deep residual learning for image recognition. In: Proceedings of the IEEE Conference on Computer Vision and Pattern Recognition, pp. 770–778 (2016)

7. Herdade, S., Kappeler, A., Boakye, K., Soares, J.: Image captioning: transforming objects into words. In: Advances in Neural Information Processing Systems, vol. 32 (2019)

8. Huang, L., Wang, W., Chen, J., Wei, X.Y.: Attention on attention for image captioning. In: Proceedings of the IEEE/CVF International Conference on Computer Vision, pp. 4634–4643 (2019)

9. Jiang, H., Misra, I., Rohrbach, M., Learned-Miller, E., Chen, X.: In defense of grid features for visual question answering. In: Proceedings of the IEEE/CVF Conference on Computer Vision and Pattern Recognition, pp. 10267–10276 (2020)

10. Karpathy, A., Fei-Fei, L.: Deep visual-semantic alignments for generating image descriptions. In: Proceedings of the IEEE Conference on Computer Vision and Pattern Recognition, pp. 3128–3137 (2015)

11. Lin, C.Y.: Rouge: a package for automatic evaluation of summaries. In: Text Summarization Branches Out, pp. 74–81 (2004)

12. Lin, T.-Y., et al.: Microsoft COCO: common objects in context. In: Fleet, D., Pajdla, T., Schiele, B., Tuytelaars, T. (eds.) ECCV 2014. LNCS, vol. 8693, pp. 740–755. Springer, Cham (2014). https://doi.org/10.1007/978-3-319-10602-1_48

13. Luo, Y., et al.: Dual-level collaborative transformer for image captioning. In: Proceedings of the AAAI Conference on Artificial Intelligence, vol. 35, pp. 2286–2293 (2021)

14. Pan, Y., Yao, T., Li, Y., Mei, T.: X-linear attention networks for image captioning. In: Proceedings of the IEEE/CVF Conference on Computer Vision and Pattern Recognition, pp. 10971–10980 (2020)

15. Papineni, K., Roukos, S., Ward, T., Zhu, W.J.: Bleu: a method for automatic evaluation of machine translation. In: Proceedings of the 40th Annual Meeting of the Association for Computational Linguistics, pp. 311–318 (2002)

16. Rennie, S.J., Marcheret, E., Mroueh, Y., Ross, J., Goel, V.: Self-critical sequence training for image captioning. In: Proceedings of the IEEE Conference on Computer Vision and Pattern Recognition, pp. 7008–7024 (2017)

17. Simonyan, K., Zisserman, A.: Very deep convolutional networks for large-scale image recognition. arXiv preprint arXiv:1409.1556 (2014)

18. Vaswani, A., et al.: Attention is all you need. In: Advances in Neural Information Processing Systems, vol. 30 (2017)

19. Vedantam, R., Lawrence Zitnick, C., Parikh, D.: Cider: consensus-based image description evaluation. In: Proceedings of the IEEE Conference on Computer Vision and Pattern Recognition, pp. 4566–4575 (2015)

20. Vinyals, O., Toshev, A., Bengio, S., Erhan, D.: Show and tell: a neural image caption generator. In: Proceedings of the IEEE Conference on Computer Vision and Pattern Recognition, pp. 3156–3164 (2015)

21. Wang, W., Chen, Z., Hu, H.: Hierarchical attention network for image captioning. In: Proceedings of the AAAI Conference on Artificial Intelligence, vol. 33, pp. 8957–8964 (2019)

22. Xian, T., Li, Z., Zhang, C., Ma, H.: Dual global enhanced transformer for image captioning. Neural Netw. **148**, 129–141 (2022)

23. Yang, X., Tang, K., Zhang, H., Cai, J.: Auto-encoding scene graphs for image captioning. In: Proceedings of the IEEE/CVF Conference on Computer Vision and Pattern Recognition, pp. 10685–10694 (2019)

24. Yao, T., Pan, Y., Li, Y., Mei, T.: Exploring visual relationship for image captioning. In: Proceedings of the European Conference on Computer Vision (ECCV), pp. 684–699 (2018)

25. Zhang, X., et al.: RSTNet: captioning with adaptive attention on visual and non-visual words. In: Proceedings of the IEEE/CVF Conference on Computer Vision and Pattern Recognition, pp. 15465–15474 (2021)

Halyomorpha Halys Detection Using Efficient Neural Networks

Alexandru Dinca[1], Nicoleta Angelescu[2], Loretta Ichim[1], and Dan Popescu[1]($\boxtimes$)

[1] University POLITEHNICA Bucharest, Bucharest, Romania
marius.dinca1411@stud.acs.upb.ro, {loretta.ichim,dan.popescu}@upb.ro
[2] University VALAHIA Targoviste, Târgoviște, Romania
nicoleta.angelescu@valahia.ro

Abstract. In previous years, there has been a strong increase in working models in the areas of artificial intelligence development based on high-performance hardware and software resources. The agricultural industry was also placed in this area of development, which benefited greatly from the research and tools that emerged. The pest detection area was the main aspect considered and the attachment of image processing modules motivated the adoption of such techniques. After analyzing the existing solutions, the paper implemented two convolutional neural network architectures adapted for the automatic detection of Halyomorpha Halys pests by classifying images taken from orchards. The new version of the Python programming language, its attached libraries, and architectures adapted from EfficientNet and Google Inception V3 was used to define, train, and test the proposed architectures. The good results obtained (accuracy between 0.92 and 0.95) make it possible to implement an efficient system for detecting and monitoring harmful insects in a complex environment such as orchard trees.

Keywords: Image Processing · Neural Networks · Orchard Monitoring · Insect Detection

1 Introduction

A recent report by Stanford University (Artificial Intelligence Index Report) [1] shows the impact that artificial intelligence has had in a multitude of domains and the fact that it targets highly diversified areas. In the interval 2019–2020, the publications on the artificial intelligence area increased considerably by a percentage of 34.5%, considering the increase of 19.6% from the previous interval 2018–2019. In this sense, the area of artificial intelligence has also had a major impact in the field of agriculture or biology - they are rapidly developing tools for data analysis, complex representations of them, or pest detection.

For agricultural areas, the algorithms of pest detection and monitoring have enhanced automation processes and provided stakeholders or farmland owners with an overview of the areas they are considering. Automatic pest detection

M. Tanveer et al. (Eds.): ICONIP 2022, LNCS 13625, pp. 459–470, 2023.
https://doi.org/10.1007/978-3-031-30111-7_39

and monitoring using remote-controlled imaging systems are key steps in these areas. Following the evolution of ImageNet systems that provide a graph of performance over time, there has been an upward trend in research areas that focus on increasing the efficiency of the training of image recognition and classification systems. Although the ImageNet – based module is very popular and developed accordingly, existing computer vision systems do not always provide the desired results. In this case, there are a lot of research works that look at the robustness and performance of image classifiers based on hand-built and custom datasets.

The integration of revolutionary technologies has helped agricultural landowners to reduce monitoring costs and carefully introduce non-invasive tools to detect the issues they have in mind. As the authors in [2] pointed out, the monitoring areas were limited to the automatic acquisition and analysis of digital images to identify pests for reference agricultural ecosystems. It has been observed that the integration of computer vision modules and artificial intelligence can meet the need for fast, timely, and accurate detection of areas of interest, compared to manual identification which requires execution time, is prone to errors, and is often done by experts in the field.

Given the automatic acquisition of digital images, research areas were later developed for the implementation of algorithms capable of extracting useful information from these images using deep learning methods based on convolutional neural networks (CNNs). They have proven useful in satisfying these aspects of image classification or segmentation, used in practical work as input data sets for specific algorithms. The methodologies and architectures are diverse and present various values of performance. The CPAFNet neural network model developed in [3] was used to identify common pests of agricultural ecosystems. The CNN was modeled accordingly, and the key parameters were gradually transformed based on the data set to obtain a high-performance model. To validate the proposed model, the test experiments were based on VGG16, Inception V3, and ResNet50 architectures.

In the same trend, the authors in [4] proposed a solution for automatic pest detection based on images taken from mustard and bean crops. Regions of interest are properly extracted using Wavelet transformation and image fusion techniques. At the same time, the research area allowed the integration of the proposed solution with real-time monitoring networks based on IoT or wireless sensor networks.

Because of its characteristics of very high mobility and increased reproductive potential, Halyomorpha Halys (HH) is a harmful pest, spreading today throughout the European continent. They are causing significant damage to agricultural crops and especially to orchards. In this context, the population feeds on fruits and seeds, bringing significant mutations to plant products, being almost impossible to market them. Although solutions have been adopted, strongly based on the fight against HH using insecticides, they do not bring satisfactory results, such solutions being unfriendly to the environment and affecting the credibility or trust of the production staff. This paper aims to introduce a less invasive automation solution that will allow the acquisition of data from the orchards and then the identification and recognition of the target population for this type of pest. Thus, we proposed a solution for detecting harmful insects like HH in images using performant CNNs and, especially, a modified Single Shot Detector (SSD) [5]. The paper's goal was to implement and compare two efficient CNN architectures modified for the automatic detection of HH by classifying images taken from orchards. The new version of the Python programming language, its attached libraries, and architectures adapted from EfficientNet and Google Inception V3, was used to define, train, and test the proposed architectures. The CNNs used were trained and tested on a new dataset created by the authors.

2   Materials and Methods

2.1   Convolutional Neural Networks Used

As is well known, a CNN is structured using several convolution layers, pooling (mean-pooling or max-pooling) layers, fully connected layers, and normalization layers [6]. The convolutional network type is a key element for many computer vision algorithms. We chose, modified, and tested two performant neural networks, EfficientNet and Inception V3, to detect HH in an orchard context.

EfficientNet [7] represented the network model that rethought the scaling mode of the architecture to optimize the performances. In general, the three scaling dimensions are represented by depth, width, and resolution. The authors of this paper provided a systematic research model for concrete balancing in their scaling to achieve notable performances. At the same time, the authors in [8] introduced new methods of network optimization through the efficient use of computing resources, increasing the number of input data, and the use of graphics processing units for training.

On the other hand, we discuss an alternative presented by Google, called Inception, that was proposed in the 2014 ImageNet Visual Recognition Challenge [9]. The Inception architecture brought to the forefront a revolutionary, high-performance technique for image recognition and detection algorithms. The complex architecture of the CNN Inception network features various techniques to increase performance in terms of both speed and model accuracy. Over time, it has seen steady growth, which has been reduced to several versions, V1 [10], V2-V3 (with quite similar structures) [11], and V4 [12], each new version showing considerable improvements compared to the previous one.

To increase the performance and relevance of a convolutional network, the techniques usually involve adding extra layers and computing areas that will ultimately increase the so-called depth of the proposed network [13]. In the case of Inception architectures, the number of layers practically does not increase but goes wider, implementing several convolution areas of different sizes in the same layer. Therefore, choosing a standard kernel size is obviously difficult. For such variable dimensions, the need is introduced to create kernels of different sizes for the analysis of the distribution of areas of interest, because in a set of images the area occupied by the object of interest can be considerably different.

The larger core is preferred for globally distributed areas of interest, and the smaller core is chosen for locally distributed information in the image.

As the first objective of the paper was to propose and implement a CNN modified from the EfficientNet architecture (CNN-1) and a modified architecture that includes the Inception V3 architecture (CNN-2). For the CNN-1 network, we introduced new layers that are shown in Fig. 1. The new layers for the network CNN-2, modified and based on Inception V3, are shown in Fig. 2. Using bold text, the changes and techniques brought to create the architectures for this work are shown.

Layer	Remarks
Input	Input image 224x224x3
Rescaling	Normalization
EfficientNet	EfficientNet B0, no pre-trained weights, classes=2
GlobalAveragePooling2D	
Dropout	
Dense + Softmax	Probability of belonging to the class

Fig. 1. CNN-1 EfficientNet modified

Layer	Remarks
Input	Input image 128x128x3
Rescaling	Normalization
Inception V3	Tensor-type input due to normalization operations, no pre-trained weights, classes=2
GlobalAveragePooling2D	
Dense (256) + Relu	256 neurons
Dropout (0.5)	50% of neurons eliminated
Dense (64) + Relu	64 neurons
Dense (2) + Softmax	Probability of belonging to the class

Fig. 2. CNN-2 Inception V3 modified

The proposed Inception architecture has been tested on color images with a resolution of 299×299 pixels. Each Inception module performs 4 types of operations: convolution with kernels of size 1×1, 3×3, and 5×5, followed by a max-pooling layer to reduce the resolution of the feature map. The usefulness of 1×1 convolution reduces the depth, and the results of the operations are sent to form the block called Filter Concatenation. The mentioned global characteristics are captured by the 5×5 convolution layer, and the distributed ones are captured by the 3×3 layer. Compared to the EfficientNet network architecture, the drive time is similar for both architectures.

2.2 Dataset Used

Images with HH and without HH were hand-cut into patches (sub-images) of
128 × 128 size. These were made on different orchards, and the images were generated manually by tracking and photographing the individual insects, especially
HH. The data set with such images were created by the authors. Examples of
such patches for the learning phase are given in Fig. 3.

Fig. 3. Examples of learning images (patches): a) images containing HH, b) images
without HH.

At the testing phase, each CNN receives 128 × 128-pixel color RGB patches
at the input. These patches are generated by a sliding box algorithm. For both
networks analyzed, the original image set was divided into two sets: training and
testing. The patches were resized in the network training/testing process. The
original image set (without augmentation) includes the following sets: training
set: 160 images with HH and 262 images without HH (nonHH); testing set: 72
images with HH and 60 images without HH.

To increase the robustness of the network, the number of images in the
original set for the training and validation stage was increased by data augmentation [14]. The operations selected for image pre-processing are adaptive
histogram equalization (CLAHE - Contrast Limited Adaptive Histogram Equalization), Gaussian noise generation, median filter, optical distortion, blur, etc.
The second objective of this paper was to describe and implement a robust,

well-structured data set that would become a solid entry point for classification algorithms. In this regard, images of the HH were taken manually and integrated with images of insects belonging to other classes. In this way, the classification classes denoted HH for the appearance of the harmful stink bug in the image and nonHH for the images without HH or insects of any kind, were constructed theoretically. Furthermore, these classes have been transposed and implemented practically at the software level. The dataset is an original one and no other insect databases were used for this work. Furthermore, the data set was annotated manually, with increased care, using the popular bounding-box structures and then subsequently prepared for the training area.

2.3   Software Used

From the software point of view, the CNN network used to classify HH and nonHH was implemented in Python version 3.9, using the TensorFlow version 2.7 library developed by Google [15]. CNN's implementation used the Keras module [16] in Tensorflow.

The number of network parameters is small, which reduces the complexity of the calculation in the network drive phase, respectively in the prediction phase. As mentioned earlier, the network input layer is a 128×128 pixels RGB color image, each color level is represented by 8 bits. Figure 4 shows the details that is representative of the implementation of augmentation operations. Through the augmentation operation, a larger number of images is obtained: 2110 images as a training set and 500 images as a test set.

In the proposed network the convolution layers have a kernel size of 3×3 for each plane. The max-pooling operation operates on a 2×2 size window. The image is partitioned into 2×2 blocks, and each block is replaced by a pixel whose value is the maximum in the block (max-pooling operation). The first Conv2D layer generates 8 feature maps. Equation (1) shows the number of parameters (No) to be learned.

$$No = 8 * (3 * 3 * 3 + 1) = 224 \tag{1}$$

For the first feature map, we learn a matrix of weights measuring $3 \times 3 \times 3 +$ bias. The third parameter in the multiplication operation is the number of planes in the image. Before entering, the pixels in each color plane are normalized in the range [0, 1] by dividing by 255.

From the mathematical point of view, a feature map is the result of the activation function f applied to the convolution operation between the RGB image and the convolution kernel K to which we add the bias b to each pixel resulting from the convolution operation. It is important to note that if the RGB image has P planes, the result of the convolution operation will be an image with the same number of planes. If we add the same bias b to each pixel, the result is a plane P image. The activation function reduces the number of planes to 1 and we get a feature map. The feature map will become the entry for the next layer of the network.

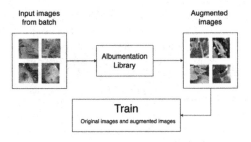

Fig. 4. The organigram for the augmentation process.

To reduce the number of calculation operations for the convolution operation the MaxPooling 2D operation was used. This reduced the resolution of the feature map. It can be observed the increase of the number of feature maps as the resolution decreases, which leads to a controlled increase in the number of network parameters. The Flatten operation turns a feature map into a vector. Then we apply through the Dense type of operation, and we build a fully connected layer of multilayer perceptron type (Multi-layer Perceptron - abbreviation MLP). This MLP network will be used for classification. The activation function used is SoftMax. Because we have two classes (HH and nonHH), the function selected for optimization is binary cross-entropy. The lower its value, the better the classification. The number of epochs for the convolutional network is 25 epochs. The optimization algorithm is Adam, with a learning rate of 0.0001.

To assess the algorithm performance, the aim was to calculate the representative indices using the confusion matrix. This is a visual indicator of performance and provides an overview of the errors and performance of the classification algorithm [17]. A test dataset with the expected results is required to calculate the confusion matrix (Fig. 5). Then a prediction is made for each row in the dataset. Subsequently, the correct predictions for each class and the number of incorrect predictions, organized by the predicted class, are noted. Each row in the matrix corresponds to an existing class and each column to a predicted class. Finally, the number of correct and incorrect classifications is completed, with better visualization and interpretation of the data.

3 Experimental Results and Discussions

The evolution of the accuracy function for CNN-1 EfficientNet is shown in Fig. 6a and for CNN-2 Inception V3 in Fig. 6b. The evolution of the loss function for CNN-1 EfficientNet is shown in Fig. 6c and for CNN-2 Inception V3 in Fig. 6d.

		Predicted class: Positive	Predicted class: Negative
True class	Positive	True Positive (TP)	False Negative (FN)
	Negative	False Positive (FP)	True Negative (TN)

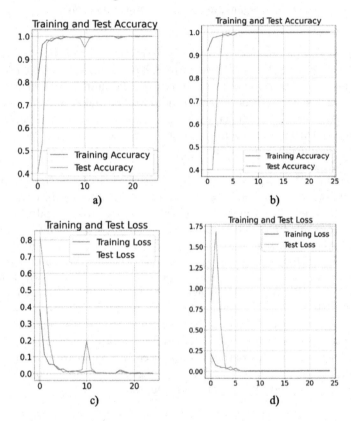

Fig. 5. The confusion matrix

Fig. 6. Performance indicators for CNNs. a) Accuracy graph for CNN-1, b) Accuracy graph for CNN-2, c) Loss function graph for CNN-1, d) Loss function graph for CNN-2.

For each figure, the blue line shows the evolution of the functions in the training area, and the orange one reflects the evolution of the functions in the testing phase. Examples from the testing dataset are shown in Fig. 7.

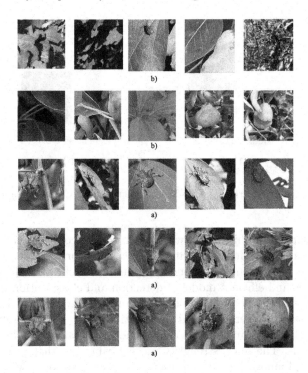

Fig. 7. Examples of testing images: a) images with HH, b) images without HH.

For the efficient visualization of the data and the performances, the implementation of the confusion matrix for each network was used. For CNN-1 EfficientNet, the non-normalized confusion matrix is attached in Fig. 8a. For CNN-2 Inception V3, the non-normalized confusion matrix is attached in Fig. 8b. For their implementation, 32 images with bugs (HH) were considered, respectively 61 images without HH from the data set.

Based on the confusion matrix, the performance evaluation was limited to the calculation of its representative indices: precision, sensitivity, specificity, accuracy, and F1 score. These indicators are presented for each network in Table 1. The first disadvantage of the proposed implementation for this paper is the small to medium size of the data set with images from orchards. For the development of an impact solution, a considerable data set of representative images for the population of HH may be entered, described, and annotated accordingly. Secondly, the images in the major set to be introduced may exemplify enough hypostases for the reference stink bug. The discussion and this implementation want to propose a set of images of various sizes, scaled or rotated, and with a considerable arrangement of the referenced bug. Very important that the acquisition of images is not perfect many times, the object of interest in the image may be small, cropped, or with reduced clarity.

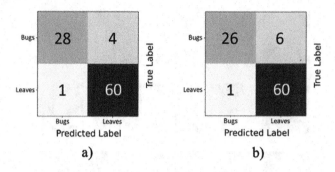

Fig. 8. Confusion matrices: a) CNN-1 EfficientNet, b) CNN-2 Inception V3.

The experiment in this paper was based on the comparison between two CNN architectures in terms of classification accuracy for HH. The contributions that can define the present paper and study are those of laying the foundations of an automatic and efficient model of detection and classification in a complex environment, as illustrated, and represented in the case of orchards and associated crops. The experiment adopted the creation of a basic data set and passed through the well-known augmentation operations for training and validating the proposed models. The main strengths and contributions that can validate the performance and innovation of the present work are based on new working models in this context of pest detection or classification, in natural settings, with fast, non-invasive, automatically monitored solutions. The modification of the presented networks was done in accordance with the context we are presenting, for effective pest detection. The obtained results show the advantages of using deep-learning models and algorithms. The evaluation for the present work was done with a series of state-of-the-art architectures so that it could pave the way for much more laborious research. It was clearly seen that complex architecture and parameter optimization can bring a considerable impact on the results. Considering the database created, it can be concluded that errors and experimental results are strongly influenced by the poses in which the insects are in the image (small object issue, image background being complex, species similarity). In this sense, training on a robust and well-anchored database in relation to these image details is a future point of research and improvement. On the other hand, starting from the components of the present architectures, convincing results were obtained. However, from what we observed in this paper, fine-tuning techniques or testing using decision systems can improve the architectures we use.

Table 1. Performance evaluation.

Network	Precision	Sensitivity	Specificity	Accuracy	F1
CNN-1 EfficientNet	1.00	0.87	1.00	0.95	0.93
CNN-2 InceptionV3	0.96	0.81	0.98	0.92	0.87

4 Conclusions

Following the experimental results described in this paper and considering the two proposed networks, a comparative study shows that the CNN-1 EfficientNet network performs better than the CNN-2 Inception V3 network, due to the complexity of the associated structure. The increased performance of the CNN-1 EfficientNet network is also visible in the associated confusion matrix. Sensitivity and accuracy are higher (so better) with CNN-1 EfficientNet.

At the same time, considerable information and results were presented, which allows the development of this work in the future by enlarging the image dataset and by implementing more robust classification solutions. Given these strategies, for the present paper, it is possible to considerably improve the entry point and the learning model for the presented algorithms and architectures, the subject of the paper being one of interest. However, we can see from the data presented in this paper, that the advantages of the existing data set are the fact that the implemented architectures presented a good classification score. Also in this sense, the increase of the data set would imply the further development of the classification solutions.

Finally, it was studied how the introduction of a convolutional network with considerable dimensions improved the metrics and values of performance indicators. As further work, we intend to use the individual selected neural networks as subjective classifiers inside a combined multi-network system (as a collective intelligence) to better detection of HH in different hypostases inside the trees.

Acknowledgement. This work was supported by HALY.ID project. HALY.ID is part of ERA-NET Co-fund ICT-AGRI-FOOD, with funding provided by national sources [Funding agency UEFISCDI, project number 202/2020, within PNCDI III] and co-funding by the European Union's Horizon 2020 research and innovation program, Grant Agreement number 862665 ERA-NET ICT-AGRI-FOOD (HALY-ID 862671).

References

1. Zhang, D., et al.: The AI Index 2021 Annual Report. AI Index Steering Committee, Human-Centered AI Institute, Stanford University, Stanford, CA (2021)
2. De Cesaro, T., Jr., Rieder, R.: Automatic identification of insects from digital images: a survey. Comput. Electron. Agric. (178), 105784, 1–7 (2020)

3. Wang, J., Li, Y., Feng, H., Ren, L., Du, X., Wu, J.: Common pests image recognition based on deep convolutional neural network. Comput. Electron. Agric. (179), 105834, 1–9 (2020)
4. Nagar, H., Sharma, R.S.: Pest detection on leaf using image processing. In: 2021 International Conference on Computer Communication and Informatics (ICCCI), pp. 1–5 (2021)
5. Trufelea, R., Dimoiu, M., Ichim, L., Popescu, D.: Detection of harmful insects for orchard using convolutional neural networks. U.P.B. Sci. Bull. Ser. C **83**(4), 85–96 (2021)
6. Escontrela, A.: Convolutional Neural Networks from the ground up (2018). https://towardsdatascience.com/convolutional-neural-networks-from-the-ground-up-c67bb41454e1
7. Tan, M., Le, Q.: EfficientNet: rethinking model scaling for convolutional neural networks. In: International Conference on Machine Learning. PMLR (2019)
8. Krizhevsky, A., Sutskever, I., Hinton, G.E.: ImageNet classification with deep convolutional neural networks. Commun. ACM **60**(6), 84–90 (2017)
9. Raj, B.: A Simple Guide to the Versions of the Inception Network (2018). https://towardsdatascience.com/a-simple-guide-to-the-versions-of-the-inception-network-7fc52b863202
10. Szegedy, C., et al.: Going deeper with convolutions. In: 2015 IEEE Conference on Computer Vision and Pattern Recognition (CVPR), pp. 1–9 (2015)
11. Szegedy, C., Vanhoucke, V., Ioffe, S., Shlens, J., Wojna, Z.: Rethinking the inception architecture for computer vision. In: 2016 IEEE Conference on Computer Vision and Pattern Recognition (CVPR), pp. 2818–2826 (2016)
12. Szegedy, C., Ioffe, S., Vanhoucke, V., Alemi, A.: Inception-v4, inception-ResNet and the impact of residual connections on learning. In: Proceedings of the Thirty-First AAAI Conference on Artificial Intelligence (AAAI 2017), pp. 4278–4284. AAAI Press (2017)
13. Anwar, A.: Difference between AlexNet, VGGNet, ResNet, and Inception (2019). https://towardsdatascience.com/the-w3h-of-alexnet-vggnet-resnet-and-inception-7baaaecccc96
14. Albumentations Library. https://albumentations.ai. Accessed Jan 2020
15. Abadi, M., et al.: TensorFlow: large-scale machine learning on heterogeneous systems (2015). Software is available from https://tensorflow.org/
16. Chollet, F., et al.: Keras. GitHub (2015/2020). https://github.com/fchollet/keras
17. Orellana, E.: Breakdown Confusion Matrix (2020). https://emiliaorellana44.medium.com/breakdown-confusion-matrix-2cf25842f1ae

HPointLoc: Point-Based Indoor Place Recognition Using Synthetic RGB-D Images

Dmitry Yudin[1,4(✉)] ⓘ, Yaroslav Solomentsev[1,2] ⓘ, Ruslan Musaev[1] ⓘ, Aleksei Staroverov[1,4] ⓘ, and Aleksandr I. Panov[1,3,4] ⓘ

[1] Moscow Institute of Physics and Technology, Moscow Region, Dolgoprudny 141700, Russia
{yudin.da,panov.ai}@mipt.ru, {solomentsev.yak,musaev.rv}@phystech.edu
[2] LLC Integrant, Moscow 127495, Russia
[3] Federal Research Center "Computer Science and Control" of the Russian Academy of Sciences, Moscow 117312, Russia
[4] AIRI (Artificial Intelligence Research Institute), Moscow 105064, Russia

Abstract. We present a novel dataset named as HPointLoc, specially designed for exploring capabilities of visual place recognition in indoor environment and loop detection in simultaneous localization and mapping. The loop detection sub-task is especially relevant when a robot with an on-board RGB-D camera can drive past the same place ("Point") at different angles. The dataset is based on the popular Habitat simulator, in which it is possible to generate photorealistic indoor scenes using both own sensor data and open datasets, such as Matterport3D. To study the main stages of solving the place recognition problem on the HPointLoc dataset, we proposed a new modular approach named as PNTR. It first performs an image retrieval with the Patch-NetVLAD method, then extracts keypoints and matches them using R2D2, LoFTR or SuperPoint with SuperGlue, and finally performs a camera pose optimization step with TEASER++. Such a solution to the place recognition problem has not been previously studied in existing publications. The PNTR approach has shown the best quality metrics on the HPointLoc dataset and has a high potential for real use in localization systems for unmanned vehicles. The proposed dataset and framework are publicly available: https://github.com/metra4ok/HPointLoc.

Keywords: Visual Place Recognition · Indoor Localization · Synthetic Image · RGB-D Image · Deep Learning · Dataset

1 Introduction

Place recognition based on camera images is an important task for navigation of a robot or an unmanned vehicle [28]. This can significantly reduce the cost and simplify the determining of the agent spatial pose in a 3D scene.

This work was supported by the Russian Science Foundation (Project No. 20-71-10116).

Generally, visual localization has three key stages. The first stage is retrieving for a given query image the most similar image from a previously known database images based on global embeddings [3,10,22]. At the second stage, key points are extracted on the query image and on the retrieved image and matching them. Some modern methods combine these procedures into a single model [30]. Camera pose optimization is performed in the third step so that key points on the retrieved image coincide in location with the key points on the query image. Sometimes this is done in 2D [15,18], sometimes directly in 3D [38,41].

The modern trend is learning neural network-based methods that allow these stages to be performed. For them, it is important to have a diverse and large-scale dataset. It should contains, in addition to images, information about the exact camera position with 6 degrees of freedom (6DoF), and also data about distances corresponding to each pixel (depth map).

This paper focuses on the development of the dataset for indoor localization that can later be used for robot's intelligent navigation in the photorealistic simulator Habitat [27]. This will help to investigate quantitatively and qualitatively loop detection methods of a robot's movement when it enters the vicinity already visited place ("Points", see Fig. 1).

Another contribution is the development of a new modular approach named as PNTR. It first performs an image retrieval with the Patch-NetVLAD method, then extracts keypoints and matches them using R2D2, LoFTR or SuperPoint with SuperGlue, and finally performs a camera pose optimization step with TEASER++.

2   Related Work

Image Retrieval. The search problem for the closest image in the database can be reformulated as a ranking problem. The solution requires finding informative and compact local and global descriptors of the images.

The common "classical" approaches obtain global features (embeddings) by aggregating the local descriptors using the bag of words (BoW) scheme (DBoW2, DBoW3, FBoW) or vectors of locally aggregated descriptors (VLAD).

In the last few years, new approaches based on deep neural networks have been released: NetVLAD [3], distilled model HF-Net [25], Ap-Gem [22] approach with differentiable rank loss function, graph-based approach GraphVLAD [40]. They had surpassed the classical ones by feature learning for the specific problems. The similar image candidates can be also re-ranked by analyzing statistics of geometrically correct matchings of local descriptors for image patches as in Patch-NetVLAD [11].

To improve the matching of embeddings, some approaches utilize semantic information. They demonstrate good performance on popular benchmarks [20, 21,37]. However, in spite of notable achievements of these neural networks, there still is a problem with the extraction of invariant semantic descriptors of images which have only low-quality semantic/instance segmentation. Such segmentation is often inherent in real-time neural networks that generate a lot of noise in the resulting masks.

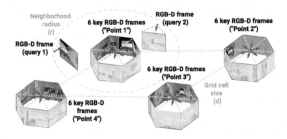

Fig. 1. Illustration of the localization problem by RGB-D query image. It is necessary to determine the pose of the corresponding camera relative to the camera poses from known database, consisting of a regular grid of groups of 6 cameras ("Points")

Local Feature Extraction and Matching. In the next stage of visual localization, we should find (detect and describe) common local features (keypoints) on pairs of images. The difficulties of this problem include getting informative and compact local descriptors and providing geometrical verification of detected keypoints.

The classical local features extraction approaches utilize scale-invariant feature transform based on local gradients [17], BoW [10], etc. These methods have achieved remarkable performance, but analogously to image retrieval methods, modern neural network approaches have outperformed them by finding more robust local features. The recent approaches demonstrate good results due to appropriate loss functions, neural network architectures (CNN, Transformers), and training schemes (Siamese networks). For example, the advantage of end-to-end training allows us to simultaneously train keypoint detectors and descriptors: SuperPoint [8], R2D2 [23], D2-Net [9], etc.

The goal of the feature matching is to find the same keypoint descriptors in both query and retrieved images. Metric learning is the most common training scheme for this purpose. Models using attention mechanisms (LoFTR [30]) and graphs (SuperGlue [26]) provide highly accurate keypoint matching. In the inference mode, models can use the k-nearest neighbors algorithm for fast and robust matching (Faiss [12]).

Camera Pose Optimization. is the final stage of the visual localization. The main problem is to find such transformation (SE3 rotation matrix and translation vector), which minimizes the pairwise distance between point clouds. The most popular methods for this task are the classical optimization approaches such as easy PnP + Ransac [18], graph-based g2o [15], 3D point cloud-based ICP [41], reliable to outliers TEASER++ [38], neural network-based methods PoseNet [13], DCP [34], etc.

Table 1. Open datasets for visual place recognition and localization

Dataset	Year	Scene	Synth.	Pose info.	RGB	Depth info.	Sem. segm.	Inst. segm.	Frames
Pittsburgh [19]	2015	outdoor		GPS	×				278k
Landmarks [19]	2016	outdoor		Label	×				10k
Google-Landmarks [19]	2017	outdoor		GPS	×				1.2M
Nordland [19]	2018	outdoor		GPS	×				143k
CMU-Seasons [1]	2018	outdoor		6DoF Pose	×	laser scan			82.5k
Aachen Day-Night [1]	2018	outdoor		6DoF Pose	×	stereo			7.5k
RobotCar Seasons [1]	2018	outdoor		6DoF Pose	×	laser scan			38k
Tokyo 24/7 [19]	2018	outdoor		GPS	×	stereo			2.8M
Argoverse Stereo [7]	2019	outdoor		6DoF Pose	×	stereo/laser scan			6,6k
NuScenes-lidarseg [5]	2020	outdoor		6DoF Pose	×	laser scan	×		40k point clouds
Waymo Perception [31]	2020	outdoor		6DoF Pose	×	stereo/laser scan			390k
KITTI360 [36]	2020	outdoor		6DoF Pose	×	laser scan	×	×	4×83k
Mapillary SLS [19]	2020	outdoor		6DoF Pose	×				1.68M
TUM Indoor [16]	2012	indoor		6DoF Pose	×	laser scan			7k
7-scenes [19]	2013	indoor		6DoF Pose	×	RGB-D cam			17k
Baidu [39]	2017	indoor		6DoF Pose	×	laser scan	×		2k
Matterport3D [6]	2017	indoor	×	Need to gen	×	3D Models	×	×	194k
TUM-LSI [16]	2017	indoor		6DoF Pose	×	laser scan			220
ScanNet [16]	2017	indoor		6DoF Pose	×	RGB-D cam	×	×	2.4M
2D-3D-Semantics [4]	2017	indoor		6DoF Pose	×	RGB-D cam	×		70k
Gibson [35]	2017	indoor	×	Need to gen	×	3D Models			572 3D Scenes
Inloc [1]	2018	indoor		6DoF Pose	×	3D point clouds			10k
Replica [29]	2019	indoor	×	Need to gen	×	3D Models	×	×	18 3D Scenes
ROI10 [33]	2020	indoor		6DoF Pose	×	RGB-D cam			250k
HM3D [2]	2021	indoor	×	Need to gen	×	3D Models			1000 3D scenes
Naver Labs [16]	2021	indoor		6DoF Pose	×	laser scan	×		100k
HPointloc (our)	2021	indoor	×	6DoF Pose	×	RGB-D cam	×	×	76k

Complex Localization Approaches. One of the most popular three-stage approach using neural networks is Hierarchical-Localization [25]. HF-Net [25] is an image retrieval method in this pipeline, SuperPoint [8] is a keypoint extracting method, SuperGlue [26] is used for keypoint matching and PnP [14] with RANSAC [18] are for pose optimization. Another state-of-the-art method on long-term visual localization benchmark [1] is Kapture framework using Ap-GeM [22] and R2D2 [23].

One of the most popular classic non-neural network visual localization method is the combination of ORB [24] and DBoW2 [10], which is widely used in popular SLAM methods, in particular, ORB-SLAM2, OpenVSLAM, etc.

Datasets. There are many known datasets used to solve the visual localization problem (see Table 1). The most popular of them are presented in the long-term visual localization challenge [1], as well as in the survey publications [16, 19]. One of the drawbacks of most of them is the lack of data on instance or semantic segmentation, or depth maps, while this information is very important for improving existing localization approaches.

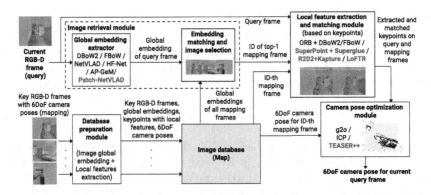

Fig. 2. Scheme of the proposed framework for indoor localization using RGB-D images. Green text refers to the methods used in the PNTR approach. (Color figure online)

If we take the InLoc dataset [32] as an example, then it includes not dense depth maps, but sparse point clouds. Using them we can obtain an estimate of the depths for the found keypoints, but in our formulation of the problem, we do not consider such a case.

Of particular note are datasets that contain 3D models of a scene, rather than separate frames (Matterport3D [6], Gibson [35], Replica [29], HM3D [2]). They are usually integrated into simulators of the movement of intelligent agents, for example, into the popular and computationally efficient Habitat [27]. At the same time, to obtain a reproducible and useful result based on them, it is necessary to have sufficiently large and diverse samples from these datasets containing frames and the corresponding 6DoF camera poses.

In our work, we propose filling the gap among such datasets, specialized in the study of a specific problem—localization of an intelligent agent in the vicinity of some key point belonging to a regular grid in a 3D indoor environment.

3   The Point-Based Approach to Indoor Place Recognition

In this paper, we solve the problem of developing an approach to estimating the 6DoF pose P_q of an intelligent agent from the image I_q of its RGB-D camera (query) in the vicinity of the poses of the cameras P_{db}^i from the database (see Fig. 1), forming a regular grid of groups of 6 cameras (named as "Points"), the images I_{db}^i of which cover 360° environment. This formulation required the development of a new special dataset, named as HPointLoc. The dataset was generated automatically based on the Habitat simulator [27]. This will provide the perspective to use the results for navigation and learning the behavior of an intelligent agent (robot) in real time in this photorealistic environment. On such a "Point"-based dataset, it is advisable to evaluate the quality of the state-of-the-art methods of visual place recognition. For their study, we proposed the modular PNTR approach, which is based on trainable methods. Its modules are described in more detail later in the paper.

Image Retrieval Module. When developing the approach, various classical image retrieval models based on a bag of words and neural network approaches are investigated. They form global feature vectors (embeddings) E_q and E_{db}^i for the query image I_q and images from the database I_{db}^i. Based on embeddings, the most similar image in the sense of the similarity metric S is determined:

$$top1 = \underset{i}{\operatorname{argmin}} \left(S \left(E_q, E_{db}^i \right) \right). \tag{1}$$

We consider the classic approaches DBoW2 [10] and FBoW [42] based on the formation of global descriptors by image keypoints extracted with ORB method. The main attention is paid to neural network-based approaches of image retrieval: NetVLAD [3], AP-GeM [22], HF-Net [25] and the Patch-NetVLAD [11] (its speed up configuration (s) is included in our PNTR approach).

Local Feature Extraction and Matching Module. At the next stage of extraction and matching of local features, k matches of keypoints on the query image F_q^j and the most similar from the base F_{top1}^j are found, $j = 1..k$. For each of these points, the distance to the camera in meters (depth) is known: D_q^j and D_{top1}^j, $j = 1..k$ respectively.

We examined the extraction and matching of image keypoints based on the classic popular approaches ORB with DBoW2 [10] or FBoW [42] and the popular neural networks (used in our PNTR approach) SuperPoint [8] with SuperGlue [26], R2D2 [23] with feature matching based on the Kapture toolbox, as well as monolithic LoFTR [30].

Database Preparation Module. The database used to solve the localization problem is an array of $i \in [1, N]$ camera poses P_{db}^i, the corresponding images I_{db}^i, depth maps D_{db}^i, and segmentation masks M_{db}^i (the latter may not be used in the approach). In addition, extracted and pre-calculated global image embeddings E_{db}^i and information about local features F_{db}^j on them are additionally entered into the database.

Camera Pose Optimization Module. Further, the final estimation of the 4×4 matrix of the camera pose P_q is carried out. This pose corresponds to the query image I_q. In our case, the 4×4 matrix of the camera pose P_{db}^{top1} is known for retrieved image I_{db}^{top1} from database. An optimization problem is solved for calculating the relative pose of the camera P_{top1}^q (also a 4×4 matrix) based on minimizing some functional L that takes into account the 2D coordinates F and the depth D of matched keypoints in two images:

$$P_{top1}^q = \underset{P_{top1}^q, j \in [1,k]}{\operatorname{argmin}} \left(L \left(F_q^j, D_q^j, F_{top1}^j, D_{top1}^j \right) \right). \tag{2}$$

The final pose is defined as $P_q = P_{db}^{top1} P_{top1}^q$.

The widely used classical approaches g2o [15], ICP [41] and the newer and highly accurate Teaser++ [38] (used in the proposed PNTR approach) are investigated as methods for optimizing pose in our framework.

Table 2. Summary for HPointLoc Dataset.

	Points	Poses	Categories	Instances	Maps
HPointLoc-Val	23	1088	33	3266	1
HPointLoc-All	1757	86678	41	488717	49

Table 3. Statistics for semantic instances in the proposed HPointLoc-Val and HPointLoc-All datasets

Category	Instances per categ.		Category	Instances per categ.		Category	Instances per categ.	
	Val	All		Val	All		Val	All
appliances	0	1077	cushion	15	5600	shelving	18	4174
bathtub	11	621	door	1174	44500	shower	4	2190
beam	0	981	fireplace	0	1022	sink	6	1924
bed	28	5025	floor	0	52274	sofa	6	3871
blinds	0	162	furniture	8	376	stairs	15	3557
board_panel	3	145	gym equipment	10	105	stool	0	7660
cabinet	26	5383	lighting	184	8797	table	15	11886
ceiling	249	28143	mirror	16	1752	toilet	3	271
chair	48	27946	misc	180	47346	towel	17	482
chest_of_drawers	8	2123	objects	67	28902	tv_monitor	10	1190
clothes	0	191	picture	115	18945	void	10	14445
column	0	4255	plant	2	4330	wall	601	107291
counter	0	1521	railing	0	8157	window	110	19041
curtain	11	8129	seating	6	2927			

To quantitatively evaluate the quality of the overall framework, we use Recall metric with different thresholds. It is calculated as the fraction of query images which localization errors do not exceed the specified threshold, respectively for distance (translation) $\epsilon_t \in \{0.25\,\text{m}, 0.5\,\text{m}, 1\,\text{m}, 5\,\text{m}\}$ and rotation $\epsilon_r \in \{2°, 5°, 10°, 20°\}$. Such thresholds were chosen to assess the prospects for solving the problem of loop detection and place recognition in selected indoor environment using the developed approach.

4 HPointLoc Dataset

The dataset consists of 49 scenes from the Matterport3D dataset and is intended for training computer vision algorithms or testing them. It was formed according to the following algorithm.

For each scene, a regular grid with N key poses was generated $(x_{id}, y_{id}, z_{id}, yaw_{id}, pitch_{id}, roll_{id})$ with a distance of 2 m in x and y axes for one scene. A key pose has a unique id and field $status = 1$. An example of a key pose is marked in red in Fig. 3a. For each key pose, the two steps were taken:

1. Generation of five more key pose orientations with 60° step, simulating a 360° key pose camera. The view angle of each frame is 90°. An example of six key images is shown in Fig. 3d.

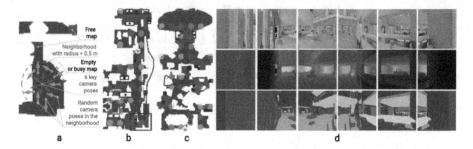

a   b   c   d

Fig. 3. Explanations to the HPointLoc dataset content: a - the generation of query and database camera poses based on maps from the Matterport3D dataset in the Habitat environment; b - scene map in the HPointLoc-Val dataset; c - another scene map in the HPointLoc-All dataset, the positions of the camera centers are shown by colored dots in the circles corresponding to the "Points"; d - examples of 6 RGB-images (upper row), depth map (middle row) and map of object instance segmentation (lower row), corresponding to one "point" of scene map of the HPointLoc dataset. Six RGB-images from the "point" cover overlapping areas with a 360-degree view (Color figure online)

2. Generation of $M = 50$ random poses in a radius $r = 0.5$ m relative to the key pose (x_{id}, y_{id}, z_{id}) with a random orientation (shown in green in Fig. 3a). If the generated pose does not fall into the free area, then it is discarded; otherwise we add a new record to the dataset.

Each item in the dataset contains the following data: RGB and depth images, instance segmentation of the frame with 41 classes (see Table 3) and GPS with Compass data (orientation of the camera is described with a quaternion).

RGB images have 256×256 size and camera viewing angle 90°. Gaussian noise model was also added with factor intensity of 0.02, a mean value of 0 and a sigma parameter of 1.

Depth images also have 256×256 size. The depth values lie in the range from 0 to 1, where 0 is a minimum possible depth (0 m) and 1—maximum possible (10 m). Such maximum depth is chosen because most of the commercial RGB-D cameras, such as Intel RealSense, and ZED, have similar restrictions. The RGB-D camera stands at 1.25 m high above the floor.

Examples of top-view scene maps with marked frame poses are shown in Fig. 3b and 3c. Thus, in each spherical "Point" with a radius of 0.5 m, there are 6 key images with the prior known data for localization methods (images in the database) and several dozen images (query images), which need to be localized. It is assumed in the experiments that the pose of query images by which it is necessary to localize is not known to the localization method, but all other data, including the depth map, are known. A summary of the dataset contents is shown in Table 2. The dataset is split into two parts: the validation HPointLoc-Val, which contains only one scene, and the complete HPointLoc-All dataset, containing all 49 scenes, including HPointLoc-Val. We use these datasets only to validate the proposed approaches, not for training.

Table 4. Localization quality based on the pose of top-1 retrieved image (without pose optimization) on the HPointLoc-Val dataset

Image retrieval method	Embedding size	(5m,20°)	(1m,10°)	(0.5m,5°)	(0.25m,2°)	(5m)	(1m)	(0.5m)	(0.25m)
FBoW+ORB	~1000000	0.527	0.269	0.133	0.029	0.911	0.864	0.864	0.476
DBoW2+ORB	~1000000	0.542	0.269	0.133	0.028	0.916	0.877	0.877	0.477
AP-GeM	2048	0.559	0.282	0.138	**0.032**	0.964	0.893	0.893	0.493
NetVLAD	32768	0.584	0.291	**0.141**	**0.032**	0.971	0.919	0.919	0.502
HF-Net	4096	**0.600**	0.291	0.138	**0.032**	0.975	0.934	0.934	0.505
Patch-NetVLAD(s)	512×100 patches	0.590	**0.296**	0.140	0.030	**0.983**	**0.963**	**0.963**	0.515
Patch-NetVLAD(p)	4096×300 patches	0.576	0.292	0.138	0.028	0.979	0.961	0.961	**0.517**

5   Experimental Results

Localization Quality Estimation. The experiments were held on the HPoint-Loc dataset. All the methods we used were already pre-trained. According to Fig. 2 and Eq. 1, the image retrieval method searches for the closest similar image. After that, we optimize the relative pose of the retrieved image to query image (Eq. 2). Knowing the absolute pose of the image from the database, we get the final pose of the query. The quality metrics we used are generally recognized and observable on the benchmark [1]. Accuracy metrics of true localizations with the optimization step on the HPointLoc-Val dataset are given in Table 4. Pose optimization in this case is not accomplished: in other words, the result pose is equivalent to the database image that was accepted as the most similar. From Table 4, we can see that at all translation error thresholds the best method is Patch-NetVLAD with speed up configuration (s). The angle error should be taken into account secondary since the frame pose is not optimized.

The localization results of different methods on the HPointLoc-Val dataset are given in the Table 5. SuperPoint is abbreviated as SP. As we can see, the highest quality for image retrieval, keypoints matching, camera pose optimization is for the proposed PNTR approach with the R2D2 feature extraction and matching based on procedure form the Kapture tool (symbol 'K').

The localization results of different methods on the HPointLoc-All are given in Table 6. We can see that all metrics are worse in comparison with the evaluation on the HPointLoc-Val dataset. This directly follows from the degradation of the image retrieval methods.

In the course of the experiments, we detect problems that sometimes arise in all of the considered image retrieval methods. In Fig. 4, they are shown using the NetVLAD method as an example and demonstrate the need to explicitly use semantics and the complexity of choosing a top-1 image containing the same scene as the query image, but the corresponding camera is located at a considerable distance from the camera for images from the database.

Time Performance Estimation. Average execution time of the main stages of localization implemented in the proposed framework is shown in Table 7. The performance was evaluated on a workstation with NVidia RTX2080Ti 11 Gb GPU, AMD Threadripper 1900X (8 cores, 3.8 GHz) CPU, 64 Gb RAM.

Table 5. Quality of various visual localization methods on the HPointLoc-Val dataset

Approach	(5m,20°)	(1m,10°)	(0.5m,5°)	(0.25m,2°)	(5m)	(1m)	(0.5m)	(0.25m)
DBoW2+ORB+g2o	**0.844**	**0.812**	**0.78**	**0.726**	0.905	0.857	0.841	0.796
FBoW+ORB+g2o	0.801	0.766	0.719	0.64	0.903	0.826	0.78	0.719
AP-GeM+R2D2(K)+ICP	**0.961**	0.862	0.805	0.733	**0.965**	0.871	0.828	0.777
AP-GeM+R2D2(K)+ TEASER++	0.939	**0.904**	**0.881**	**0.877**	0.958	**0.906**	0.882	**0.880**
Ap-GeM+LofTR+g2o	0.907	0.868	0.837	0.746	0.963	0.902	**0.891**	0.861
Ap-Gem+LofTR+ICP	0.903	0.846	0.825	0.762	0.961	0.877	0.862	0.807
AP-GeM+LoFTR+ TEASER++	0.892	0.854	0.831	0.766	0.964	0.899	0.884	0.823
NetVLAD+SP+SuperGlue+g2o	0.917	0.891	0.874	0.845	0.967	0.919	0.906	0.887
NetVLAD+SP+SuperGlue+ICP	0.944	0.893	0.868	0.794	0.969	0.909	0.893	0.842
NetVLAD+SP+SuperGlue+TEASER++	0.924	0.892	0.872	0.854	0.966	0.918	0.906	0.888
NetVLAD+R2D2(K)+ICP	**0.968**	0.901	0.855	0.788	0.969	0.907	0.877	0.827
NetVlad+R2D2+TEASER++	0.941	**0.916**	**0.907**	**0.905**	0.967	0.918	0.911	**0.908**
NetVLAD+LoFTR+g2o	0.921	0.891	0.876	0.78	0.968	**0.92**	**0.916**	0.892
NetVLAD+LoFTR+ICP	0.917	0.875	0.858	0.799	**0.971**	0.9	0.892	0.843
NetVLAD+LoFTR+ TEASER++	0.908	0.881	0.865	0.806	**0.971**	0.919	0.914	0.857
HF-Net+SP+SuperGlue+g2o	0.925	0.903	0.881	0.852	0.972	0.936	0.92	0.896
HF-Net+R2D2(K)+ICP	**0.974**	0.917	0.869	0.791	**0.975**	0.925	0.894	0.834
HF-Net+R2D2(K)+ TEASER++	0.953	**0.936**	**0.923**	**0.92**	0.973	**0.938**	**0.926**	**0.923**
Patch-NetVLAD(s)+SP+SuperGlue+ g2o	0.946	0.924	0.908	0.883	**0.982**	0.96	0.947	0.924
Patch-NetVLAD(s)+ R2D2(K)+ICP	**0.98**	0.94	0.896	0.808	0.984	0.952	0.925	0.852
PNTR with SP+SuperGlue	0.947	0.925	0.899	0.882	0.975	0.945	0.934	0.914
PNTR with R2D2(K)	0.964	**0.957**	**0.952**	**0.945**	0.98	0.958	**0.953**	**0.951**
PNTR with LoFTR	0.942	0.919	0.900	0.834	0.977	0.953	0.948	0.892

Table 6. Quality of various visual localization methods on the HPointLoc-All dataset

Approach	(5m,20°)	(1m,10°)	(0.5m,5°)	(0.25m,2°)	(5m)	(1m)	(0.5m)	(0.25m)
DBoW2+ORB (top-1 db)	0.592	0.303	0.150	0.033	0.903	0.882	0.881	0.498
FBoW+ORB2 (top-1 db)	0.547	0.285	0.142	0.032	0.825	0.799	0.797	0.464
NetVLAD (top-1 db)	0.643	0.317	0.158	**0.034**	0.957	0.877	0.876	0.487
AP-GeM (top-1 db)	0.635	0.311	0.156	**0.034**	0.944	0.813	0.812	0.452
HF-Net (top-1 db)	**0.646**	0.318	0.158	**0.034**	0.955	0.879	0.878	0.487
Patch-NetVLAD(s) (top-1 db)	0.644	**0.320**	**0.159**	0.034	**0.969**	**0.944**	**0.942**	**0.516**
DBoW2+ORB2+g2o	**0.876**	**0.857**	0.648	0.404	0.901	0.870	0.661	0.419
FBoW+ORB2+g2o	0.776	0.748	0.571	0.357	0.820	0.769	0.590	0.377
NetVLAD+SP+SuperGlue+ g2o	0.925	**0.867**	0.647	0.406	0.950	**0.877**	0.656	0.414
NetVLAD+SP+SuperGlue+ICP	0.935	0.853	0.803	0.426	**0.956**	0.876	0.844	0.476
NetVLAD+R2D2(K)+ICP	**0.944**	0.842	0.748	0.341	**0.956**	**0.877**	0.846	0.474
NetVLAD+LoFTR+ICP	0.934	0.863	**0.829**	**0.458**	**0.956**	**0.877**	0.849	**0.480**
NetVLAD+LoFTR+TEASER++	0.933	0.865	0.829	0.445	**0.956**	**0.877**	0.849	0.479
Ap-Gem+LofTR+ICP	**0.907**	0.797	**0.765**	**0.423**	0.943	0.813	**0.787**	**0.445**
AP-GeM+LoFTR+ TEASER++	**0.907**	0.799	**0.765**	0.412	0.943	0.813	**0.787**	**0.445**
HF-Net+SP+SuperGlue+g2o	0.924	**0.868**	0.647	0.405	0.949	0.878	0.656	0.413
HF-Net+SP+SuperGlue+ICP	0.94	0.861	0.817	0.439	**0.955**	**0.879**	0.85	0.479
HF-Net+SP+SuperGlue+TEASER++	0.935	0.854	0.803	0.425	**0.955**	0.877	0.845	0.475
HF-Net+R2D2(K)+ICP	**0.943**	0.845	0.752	0.342	**0.955**	**0.879**	0.848	0.475
HF-Net+LoFTR+TEASER++	0.933	0.867	**0.831**	**0.445**	**0.955**	**0.879**	**0.851**	**0.480**
Patch-NetVLAD(s)+SP+SuperGlue+g2o	0.956	0.934	0.693	0.433	0.967	0.942	0.701	0.440
Patch-NetVLAD(s)+R2D2(K)+ ICP	**0.964**	0.931	0.876	0.464	0.968	**0.944**	0.911	0.507
PNTR with SP+SuperGlue	0.952	0.919	0.863	0.451	0.968	0.942	0.905	0.502
PNTR with R2D2(K)	0.961	**0.938**	**0.906**	**0.503**	**0.969**	0.943	**0.913**	**0.509**
PNTR with LoFTR	0.959	0.936	0.895	0.472	**0.969**	**0.944**	**0.913**	0.508

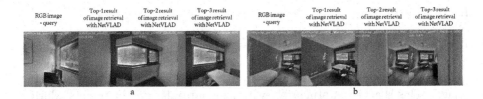

Fig. 4. Typical problems that sometimes arise in all image retrieval methods (for example, the NetVLAD) on the HPointLoc Dataset: a — the selection of an image that looks like a query occurs with an error caused by the lack of explicit usage of semantics (information about the presence of objects, about their color, etc.); b — the selected top-1 image contains the same scene as the query image, but is more than 1 m away from it and generates a localization error.

Table 7. Average execution time of different localization stages, sec

Image retrieval			Camera Pose Optimization	
Method	Embedding extr	Embedding match	Method	Avg. time
NetVLAD	0.00698	-	g2o	0.00074
AP-GeM	0.01542	0.00013	ICP	0.00158
HF-Net	0.09049	-	Teaser++	0.00826
Patch-NetVLAD(s)	0.01843	0.1039		
Patch-NetVLAD(p)	0.26415	2.28863		
Local feature extraction and matching			Overall PNTR approach	
Method	Feature extr	Feature match	Method	Avg. time
ORB+DBoW2	0.00407	0.00013	PNTR with SuperPoint+SuperGlue	0.13565
ORB+FBoW	0.00092	0.00007	PNTR with R2D2	0.19207
SuperPoint+SuperGlue	0.0025	0.00256	PNTR with LoFTR	0.15189
R2D2(K)	0.0585	0.00298		
LoFTR	0.0213	-		

It should be noted that the Patch-NetVLAD (s) configuration is more than 20 times faster than the Patch-NetVLAD (p) configuration, and their quality metrics are almost the same. NetVLAD leads in speed, which is 10 times faster than the almost identical in quality HF-Net method and than the significantly better Patch-NetVLAD (s).

Among the considered methods of feature extraction, the ORB FBoW method is the leader, which is almost five times faster than the classical ORB + DBoW2 method, but is significantly inferior in quality. LoFTR, the one-stage method for extracting and matching points, has the highest performance among neural network approaches, which slightly exceeds the total performance of the SuperPoint+SuperGlue method combination. Among the considered optimization methods, the fastest is the g2o method, which is twice as fast as ICP and almost 10 times faster than TEASER++. The total performance of the most accurate groups of methods, except the slow Patch-NetVLAD (p), is 5–10 FPS, which indicates the potential for their practical use in a parallel global localization stream to solve the problem of loop closure in SLAM methods.

6   Conclusions

In this work, we have shown that the proposed HPointLoc dataset allows us to visually assess the quality and performance of various approaches for solving the place recognition problem in a photorealistic indoor environment using RGB-D images. This is achieved primarily by creating a regular grid of "Points" during dataset preparation. This can be done for any real indoor environment whose model can be imported into the Habitat simulator.

On the small subset HPointLoc-Val, localization methods exhibit similar behavior, so it is sufficient to use it for quick evaluation and comparison of algorithms. The experiment results have revealed the limitations of the considered localization methods. It leads to the need of explicitly taking into account the semantics of the scene and the importance of a correct interpretation of the localization error when using one or the other image retrieval method.

State-of-the-art results were obtained for the proposed PNTR approach, which showed the highest quality on the developed dataset, but is rather slow and provides an image processing speed of 5 FPS. Nevertheless, it can be still used for loop detection in SLAM algorithms in a parallel stream to the main visual tracking procedure.

References

1. Long-Term Visual Localization. https://www.visuallocalization.net/
2. Habitat matterport dataset (2021). https://aihabitat.org/datasets/hm3d/
3. Arandjelović, R., Gronat, P., Torii, A., Pajdla, T., Sivic, J.: NetVLAD: CNN architecture for weakly supervised place recognition (2016)
4. Armeni, I., Sax, S., Zamir, A.R., Savarese, S.: Joint 2D-3D-semantic data for indoor scene understanding. arXiv preprint arXiv:1702.01105 (2017)
5. Caesar, H., et al.: nuScenes: a multimodal dataset for autonomous driving. arXiv:1903.11027 (2019)
6. Chang, A., et al.: Matterport3D: learning from RGB-D data in indoor environments. arXiv preprint arXiv:1709.06158 (2017)
7. Chang, M.F., et al.: Argoverse: 3D tracking and forecasting with rich maps. In: Proceedings of the IEEE/CVF Conference on Computer Vision and Pattern Recognition, pp. 8748–8757 (2019)
8. DeTone, D., Malisiewicz, T., Rabinovich, A.: Superpoint: self-supervised interest point detection and description (2018)
9. Dusmanu, M., et al.: D2-Net: a trainable CNN for joint detection and description of local features. arXiv preprint arXiv:1905.03561 (2019)
10. Gálvez-López, D., Tardos, J.D.: Bags of binary words for fast place recognition in image sequences. IEEE Trans. Rob. **28**(5), 1188–1197 (2012)
11. Hausler, S., Garg, S., Xu, M., Milford, M., Fischer, T.: Patch-NetVLAD: multiscale fusion of locally-global descriptors for place recognition. In: Proceedings of the IEEE/CVF Conference on Computer Vision and Pattern Recognition (2021)
12. Johnson, J., Douze, M., Jégou, H.: Billion-scale similarity search with GPUs. arXiv preprint arXiv:1702.08734 (2017)

13. Kendall, A., Grimes, M., Cipolla, R.: PoseNet: a convolutional network for real-time 6-DoF camera relocalization. In: Proceedings of the IEEE International Conference on Computer Vision, pp. 2938–2946 (2015)

14. Kneip, L., Scaramuzza, D., Siegwart, R.: A novel parametrization of the perspective-three-point problem for a direct computation of absolute camera position and orientation. In: CVPR 2011, pp. 2969–2976 (2011). https://doi.org/10.1109/CVPR.2011.5995464

15. Kümmerle, R., Grisetti, G., Strasdat, H., Konolige, K., Burgard, W.: G2O: a general framework for graph optimization. In: 2011 IEEE International Conference on Robotics and Automation, pp. 3607–3613 (2011). https://doi.org/10.1109/ICRA.2011.5979949

16. Lee, D., et al.: Large-scale localization datasets in crowded indoor spaces. In: Proceedings of the IEEE/CVF Conference on Computer Vision and Pattern Recognition, pp. 3227–3236 (2021)

17. Lowe, D.G.: Distinctive image features from scale-invariant keypoints. Int. J. Comput. Vis. (2004). https://doi.org/10.1023/B:VISI.0000029664.99615.94

18. Fischler, M.A., Bolles, R.C.: Random sample consensus: a paradigm for model fitting with applications to image analysis and automated cartography. Commun. ACM **24**(6), 381–395 (1981). https://doi.org/10.1145/358669.358692

19. Masone, C., Caputo, B.: A survey on deep visual place recognition. IEEE Access **9**, 19516–19547 (2021)

20. Neubert, P., Schubert, S., Schlegel, K., Protzel, P.: Vector semantic representations as descriptors for visual place recognition. In: Proceedings of Robotics: Science and Systems (RSS) (2021)

21. Peng, G., Yue, Y., Zhang, J., Wu, Z., Tang, X., Wang, D.: Semantic reinforced attention learning for visual place recognition. In: 2021 IEEE International Conference on Robotics and Automation (ICRA), pp. 13415–13422. IEEE (2021)

22. Revaud, J., Almazan, J., de Rezende, R.S., de Souza, C.R.: Learning with average precision: training image retrieval with a listwise loss (2019)

23. Revaud, J., et al.: R2D2: repeatable and reliable detector and descriptor (2019)

24. Rublee, E., Rabaud, V., Konolige, K., Bradski, G.: ORB: an efficient alternative to sift or surf. In: 2011 International Conference on Computer Vision, pp. 2564–2571 (2011). https://doi.org/10.1109/ICCV.2011.6126544

25. Sarlin, P.E., Cadena, C., Siegwart, R., Dymczyk, M.: From coarse to fine: robust hierarchical localization at large scale (2019)

26. Sarlin, P.E., DeTone, D., Malisiewicz, T., Rabinovich, A.: SuperGlue: learning feature matching with graph neural networks. In: CVPR (2020)

27. Savva, M., et al.: Habitat: a platform for embodied AI research. In: Proceedings of the IEEE/CVF International Conference on Computer Vision (ICCV) (2019)

28. Staroverov, A., Yudin, D.A., Belkin, I., Adeshkin, V., Solomentsev, Y.K., Panov, A.I.: Real-time object navigation with deep neural networks and hierarchical reinforcement learning. IEEE Access **8**, 195608–195621 (2020)

29. Straub, J., et al.: The replica dataset: a digital replica of indoor spaces. arXiv:1906.05797 (2019)

30. Sun, J., Shen, Z., Wang, Y., Bao, H., Zhou, X.: LoFTR: detector-free local feature matching with transformers (2021)

31. Sun, P., et al.: Scalability in perception for autonomous driving: Waymo open dataset. In: Proceedings of the IEEE/CVF Conference on Computer Vision and Pattern Recognition, pp. 2446–2454 (2020)

32. Taira, H., et al.: InLoc: indoor visual localization with dense matching and view synthesis. In: Proceedings of the IEEE Conference on Computer Vision and Pattern Recognition, pp. 7199–7209 (2018)

33. Wald, J., Sattler, T., Golodetz, S., Cavallari, T., Tombari, F.: Beyond controlled environments: 3D camera re-localization in changing indoor scenes. In: Vedaldi, A., Bischof, H., Brox, T., Frahm, J.-M. (eds.) ECCV 2020. LNCS, vol. 12352, pp. 467–487. Springer, Cham (2020). https://doi.org/10.1007/978-3-030-58571-6_28

34. Wang, Y., Solomon, J.M.: Deep closest point: learning representations for point cloud registration. In: Proceedings of the IEEE/CVF International Conference on Computer Vision, pp. 3523–3532 (2019)

35. Xia, F., Zamir, A.R., He, Z., Sax, A., Malik, J., Savarese, S.: Gibson ENV: real-world perception for embodied agents. In: 2018 IEEE Conference on Computer Vision and Pattern Recognition (CVPR). IEEE (2018)

36. Xie, J., Kiefel, M., Sun, M.T., Geiger, A.: Semantic instance annotation of street scenes by 3D to 2D label transfer. In: Conference on Computer Vision and Pattern Recognition (CVPR) (2016)

37. Xue, F., Budvytis, I., Reino, D.O., Cipolla, R.: Efficient large-scale localization by global instance recognition. In: Proceedings of the IEEE/CVF Conference on Computer Vision and Pattern Recognition, pp. 17348–17357 (2022)

38. Yang, H., Shi, J., Carlone, L.: Teaser: fast and certifiable point cloud registration. IEEE Trans. Rob. **37**(2), 314–333 (2020)

39. Yu, H., Yang, S., Gu, W., Zhang, S.: Baidu driving dataset and end-to-end reactive control model. In: 2017 IEEE Intelligent Vehicles Symposium (IV). IEEE (2017)

40. Zhang, C., Budvytis, I., Liwicki, S., Cipolla, R.: Lifted semantic graph embedding for omnidirectional place recognition. In: 2021 International Conference on 3D Vision (3DV), pp. 1401–1410. IEEE (2021)

41. Zhang, Z.: Iterative point matching for registration of free-form curves and surfaces. Int. J. Comput. Vis. **13**(2), 119–152 (1994)

42. Zhao, R., Mao, K.: Fuzzy bag-of-words model for document representation. IEEE Trans. Fuzzy Syst. **26**(2), 794–804 (2017)

In Situ Augmentation for Defending Against Adversarial Attacks on Text Classifiers

Lei Xu[1], Laure Berti-Equille[2], Alfredo Cuesta-Infante[3],
and Kalyan Veeramachaneni[1(✉)]

[1] Massachusetts Institute of Technology, Cambridge, USA
{leix,kalyanv}@mit.edu
[2] Institute of Research for Development (IRD), Marseille, France
laure.berti@ird.fr
[3] Universidad Rey Juan Carlos, Móstoles, Spain
alfredo.cuesta@urjc.es

Abstract. In text classification, recent research shows that adversarial attack methods can generate sentences that dramatically decrease the classification accuracy of state-of-the-art neural text classifiers. However, very few defense methods have been proposed against these generated high-quality adversarial sentences. In this paper, we propose LMAg (Language-Model-based Augmentation using Gradient Guidance), an in situ data augmentation method as a defense mechanism effective in two representative defense setups. Specifically, LMAg transforms input text during the test time. It uses the norm of the gradient to estimate the importance of a word to the classifier's prediction, then replaces those words with alternatives proposed by a masked language model. LMAg is an additional protection layer on the classifier that counteracts the perturbations made by adversarial attack methods, thus can protect the classifier from adversarial attack without additional training. Experimental results show that LMAg can improve after-attack accuracy of BERT text classifier by 51.5% and 17.3% for two setups respectively.

Keywords: Adversarial Robustness · Text Classification · Data Augmentation

1 Introduction

In the past few years, adversarial attack methods on text classifiers have been studied extensively [8,16,25,26]. The goal of this type of attack is to rewrite a sentence such that a text classifier returns an incorrect prediction. Recently proposed attack methods can drastically decrease the accuracy of state-of-the-art classifiers: the adversarial sentences they generate are semantically similar to the original sentences and are of high grammatical quality making them hard to detect and discriminate from original sentences [5,10].

As adversarial attacks can effectively degrade the accuracy of a text classifier, defending against such attacks has become a natural need. The effort to

© The Author(s), under exclusive license to Springer Nature Switzerland AG 2023
M. Tanveer et al. (Eds.): ICONIP 2022, LNCS 13625, pp. 485–496, 2023.
https://doi.org/10.1007/978-3-031-30111-7_41

defend against adversarial attacks on text classification mainly uses adversarial training [8,25]. However, adversarial training of a text classifier has several issues. First, adversarial training are limited by the existing attack methods. Once a better attack method emerges, the classifier needs to be retrained to defend against the new attack, creating extra burden for developers. Second, finding an adversarial sentence for a text classifier is inefficient and can take a few seconds because it often involves heuristic search [8,25] or inference of a neural language model [5,10]. Therefore, adversarial training of a classifier for 10k steps can take a few days, which prevents developers from efficiently deploying classifiers. To address this issue, instead of generating adversarial sentences during training, some work [8] generates a fixed set of adversarial sentences in advance and uses them to tune the classifier. This solution reduces the efficacy of adversarial training, because when the classifier is improved, new adversarial examples are needed to further robustify the classifier. Third, no consensus on the efficacy of adversarial training has been demonstrated yet [15]: some works (e.g., [8,18]) showed that adversarial training is effective whereas others (e.g., [1]) showed it is not. Beyond the differences in the benchmark datasets, we will show that the efficacy of a defense method can be measured under two different setups, making the results hard to compare (See Sect. 3).

In this paper, we propose a simple and elegant method to defend against adversarial attacks by in situ augmentation – transforming the input sentence during inference – rather than tuning the classifier. Since most attack methods modify the sentence by replacing a small portion of words in the sentence, counteracting these substitutions is one intuitive idea to defend against attacks. We can assume that words modified by the attack methods tend to have a high impact on the classifier's prediction, thus tending to increase the gradient norm. By substituting these words, we can attempt to counteract the modifications made by the attacker. As such, this paper proposes language-model-based augmentation with gradient guidance (LMAg). In LMAg, we compute the gradient of the classifier's prediction with respect to the input word embeddings. We then use the gradient norm as a weight to randomly mask words in the sentence, and employ a BERT [3] language model to fill in masked words. Since LMAg is a data-augmentation method at test time, it does not need additional training of the classifier and is easier to deploy. Our experimental results show that the proposed method is effective in defending against various attacks at a cost of slightly increasing inference time.

2   Related Work

Significant research has been done concerning adversarial attacks on text classifiers. Early works attempted to attack the classifier by injecting anomalies such as typos [4,11]. One line of research [8,25] uses synonym substitution to find adversarial sentences. Recent works including [5,9,10,24] introduce a pre-trained language model in finding substitutions so that the adversarial sentences can be more fluent. [27] provides a comprehensive survey on existing attack methods. Adversarial attack libraries have also been developed [16,26]. Adversarial training is an effective solution to protect classifiers from adversarial attacks in computer vision [14,20]. So it's not surprising that similar defending approaches have been applied to text classification. Among the attack methods mentioned above, many [8,10,24,25] use adversarial training to make the classifier resist the attacks. Adversarial training is also used in tasks such as reading comprehension [6,23] and machine translation [2]. [7] proposes certified defense, but it can not be applied on transformer-based models. [22] proposes synonym encoding (SEM). SEM constructs a synonym dictionary, and maps a cluster of synonyms to the most frequent word in that cluster to offset the adversarial perturbation.

3   Problem Formulation

In this section, we formulate the adversarial attack task and two defense setups.

Adversarial Attack on Text Classification. Given a sentence $\mathbf{x} = x_1, \ldots, x_l$ and its label y where l is the length of the sentence, a text classifier $f(\cdot)$ is supposed to make a prediction $\hat{y} = f(\mathbf{x})$ where $\hat{y} = y$ with high probability. When $f(\mathbf{x}) = y$, an adversarial attack method $\mathcal{A}(\mathbf{x}, y, f)$ generates an adversarial sentence $\mathbf{u}$ where $\mathbf{u}$ is grammatically correct and has the same semantic meaning as $\mathbf{x}$, but $f(\mathbf{u}) \neq y$. The efficacy of adversarial attack is measured by **after-attack accuracy (AAcc)** on the test set $\mathcal{D}$ such as: $\mathbb{P}_{(\mathbf{x},y) \sim \mathcal{D}}[f(\mathcal{A}(\mathbf{x}, y, f)) = y]$.

As attack methods can successfully decrease the accuracy of a classifier, defending against these attacks is necessary. The goal of the defense is to construct a classifier $f'(\cdot)$ such that it retains high classification accuracy even when it is attacked with adversarial sentences. Note that there is no constraint on how $f'(\cdot)$ is constructed; it may be constructed either by tuning the classifier's parameters or by adding additional protections, such as adversarial sentence detection and/or text transformation.

Setup I: Efficacy of Original Defense Against Adversarial Examples. We generate adversarial examples by attacking the original classifier $f(\cdot)$, then we evaluate the robustness of the robusified classifier $f'(\cdot)$ based on the absence of misclassification on these examples. In this setup, the AAcc is defined as: $\mathbb{P}_{(\mathbf{x},y) \sim \mathcal{D}}[f'(\mathcal{A}(\mathbf{x}, y, f)) = y]$. Several works [18,22] follow this setup and show significant improvement in after-attack accuracy.

Setup II: Efficacy of Boosted Defense Against Adversarial Examples. We generate adversarial examples by attacking the robustified classifier $f'(\cdot)$. In this setup, the AAcc is defined as: $\mathbb{P}_{(\mathbf{x},y) \sim \mathcal{D}}[f'(\mathcal{A}(\mathbf{x}, y, f')) = y]$. A few works [8,25] following this setup show relatively lower efficacy in defense.

The difference between the two setups is whether the adversarial examples are generated on the original classifier or the robusified classifier. We believe Setup II is prevailing in practice because Setup I underestimates the efficacy of attack methods. Most attack methods [5,8,25] stop early when an adversarial sentence is found, but this early stop only indicates that the algorithm has found an adversarial example against the original classifier. This adversarial sentence may fail on the robustified classifier. But if the attack method directly attacks the robustified classifier and runs sufficient iterations, it may still find efficient adversarial examples. Also, Setup II is more realistic. When a robustified classifier is deployed, users interact with the robustified classifier rather than the original one. Thus, it is more likely that an attacker directly attacks the robustified classifier.

4   In Situ Data Augmentation

In this section, we introduce LMAg, an in situ data augmentation to defend adversarial attacks. LMAg consists of three steps: (1) Estimate the importance of words using the gradient of the classifier; (2) Generate multiple rephrases by stochastically masking important words in the input sentence and filling in with alternative words using a masked language model; and (3) Make a prediction based on the majority of predictions on the rephrases. Figure 1 illustrates the procedure to generate one rephrase. Algorithm 1 shows the pseudo code to generate rephrases.

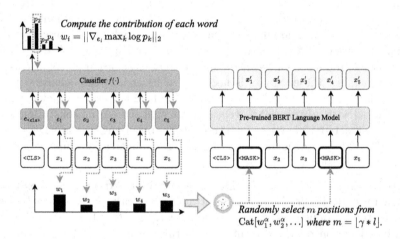

Fig. 1. Rephrase an input sentence using LMAg. The input sentence is forward and backward propagated through the classifier to compute the importance of each word. Then the input sentence is masked according to the importance weights, and a BERT language model is used to generate a rephrase.

4.1   Estimate Importance of Words Using Gradients

Gradient information has been widely used in attack methods. In white-box settings where attackers have full access to the classifier, gradient is directly used to pick candidate substitutions [11], whereas in black box settings, gradient is approximated by comparing the classifier's output with or without a word [10]. When building a defense, we assume that we have full access to the classifier; thus we directly compute gradients to identify important words that contribute

Algorithm 1: LMAg method.

Input: Sentence $\mathbf{x} = \{x_1, \ldots, x_l\}$; A classifier $f(\cdot)$ which includes the embedding layer $E(\cdot)$, and upper layers $g(\cdot)$ which takes embeddings and returns a probability distribution over classes; Number of rewrites λ; Mask ratio γ; Hyperparameter α.

Output: λ rewritten sentences.

1 results $\leftarrow$ empty list;
2 $\mathbf{e}_1, \ldots, \mathbf{e}_l \leftarrow E(\mathbf{x})$;
3 max_log_p $= \max_k g(\mathbf{e}_1, \ldots, \mathbf{e}_l)_k$;
4 $w_1, \ldots, w_l \leftarrow [\nabla_{\mathbf{e}_i} \text{max\_log\_p}]_{i=1\ldots l}$
5 $m \leftarrow \max(1, \lfloor l \times \gamma \rfloor)$;
6 **for** i in $1 \ldots \lambda$ **do**
7 $\quad$ $\mathbf{x}^{(i)} \leftarrow \mathbf{x}$;
8 $\quad$ $t_1, \ldots, t_m \sim \text{Cat}[w_1^\alpha, \ldots, w_l^\alpha]$;
9 $\quad$ **for** j in $1 \ldots m$ **do**
10 $\quad\quad$ $x_{t_j}^{(i)} \leftarrow \text{MASK}$;
11 $\quad$ **end**
12 $\quad$ $\hat{\mathbf{x}}^{(i)} \leftarrow [\arg\max \text{BERT}(\mathbf{x}^{(i)})_j]_{j=1\ldots l}$;
13 $\quad$ results.append($\hat{\mathbf{x}}^{(i)}$);
14 **end**
15 **return** results

the most to the classification. Let $\mathbf{x} = x_1, \ldots, x_l$ be an input sentence of length l. We split a text classifier into two components:

$$f(\mathbf{x}) = \arg\max_k g(E(\mathbf{x}))_k,$$

where $E(\mathbf{x}) = \mathbf{e}_1, \ldots, \mathbf{e}_l$ is the input embedding layer that converts the words x_i into embeddings $\mathbf{e}_i$, and $g(\cdot)$ is the upper layers that made prediction from word embeddings. For transformer-based models, $\mathbf{e}_i$ denotes the sum of word embedding, position embedding and token type embedding. The output of $g(\cdot)$ is a probability distribution over all classes. We use $g()_k$ to denote the probability of k-th class. We find the log probability of the most likely class predicted by the classifier (i.e., $\log \max_k g(E(\mathbf{x})_k)$), then compute the L2 norm of the gradient with respect to the input embeddings to capture the importance weight of words. Specifically, the importance weight of i-th word is

$$w_i = ||\nabla_{\mathbf{e}_i} \log \max_k g(E(\mathbf{x}))_k||_2.$$

4.2 Stochastic Multiple Rephrasing

After calculating the importance weight of each word, we have to replace the important words, hoping to counteract the adversarial attack. However, if we threshold the importance weight then mask and substitute words, it is possible to mask all important words and make the sentence generated by the language model semantically different from the original sentence. For example, in sentiment analysis, if we mask all the adjectives that express sentiment, then the language model may generate a sentence with the opposite sentiment. To overcome this problem, we used a stochastic substitute method.

We randomly sample $m = \lfloor l \times \gamma \rfloor$ positions in the sentence using w_i as weights, where γ is the masking ratio. Specifically we sample positions:

$$t_1, \ldots, t_m \sim \mathrm{Cat}(w_1^\alpha, \ldots, w_l^\alpha),$$

where Cat means a multinomial distribution. $t_1, \ldots, t_m$ represents the positions being masked in the sentence. They are sampled from the multinomial distribution without replacement. The hyperparameter α is a smoothing factor. When $\alpha = 0$, the probability density for each position being masked is uniform. When $\alpha \to \infty$, only top-m most important words will be masked. Then we replace these positions with a special MASK token and use BERT language model to impute the most likely sentence as

$$\hat{\mathbf{x}} = [\arg\max \mathrm{BERT}(\mathbf{x})_i]_{i=1\ldots l},$$

where $\mathrm{BERT}(\mathbf{x})$ is a BERT language model. Note that all l words in the rephrase $\hat{\mathbf{x}}$ are proposed by the BERT language model, although only mask m words are masked. $\hat{\mathbf{x}}$ may have more than m word substitutions.

Different mask positions result in different rephrases. To make the classifier more stable, we generate λ sentences for each adversarial sentence by selecting different mask positions. We then take the majority predictions of λ sentences as the prediction for the input sentence.

Applying LMAg leads to an increase in inference time, because it introduces extra computation of 1 backward propagation of the classifier, and λ forward propagation of BERT language model. However, the λ forward propagation can be parallelized on a GPU. So the inference time does not increase much.

5   Experiments

In this section, we compare the efficacy of LMAg with baselines under two setups discussed in Sect. 3.

Datasets. We use 5 text classification datasets: (1) **AG's News** [28]; (2) Movie Reviews (**MR**) [17]; (3) **Yelp** Reviews [28]; (4) **IMDB** Movie Reviews [13]; and the binary variation [21] of Stanford Sentiment Treebank v2 (**SST2**) [19].

Original Classifier. For all datasets, we use the BERT-base classifier [3] (#layers = 12, hidden_size = 768). We fine-tune the classifier on 20k batches (5k batches on MR and IMDB), with batch size 32. We use the AdamW optimizer [12] and learning rate 0.00002.

Metrics: We measure clean accuracy (**CAcc**) – the accuracy of the classifier on the original test, and after-attack accuracy (**AAcc**) – the accuracy of the classifier after being adversarially attacked by an attack method.

Table 1. Dataset details. #C means number of classes. Len is the average number of BERT word-pieces in a sentence. PWWS, TF, PSO, BA, BAE shows the AAcc(%) on the original classifier using the corresponding attack method.

Name	Type	#C	Train/Test	Len	CAcc	PWWS	TF	PSO	BA	BAE
AG	Topic	4	120k/7.6k	54	92.2	29.9	9.9	20.8	17.9	73.6
MR	Sentiment	2	9k/1k	24	88.1	18.4	9.4	7.5	13.8	37.0
Yelp	Sentiment	2	160k/38k	182	96.5	3.7	4.3	NA[a]	9.0	50.5
IMDB	Sentiment	2	25k/25k	305	89.8	10.0	6.3	NA[a]	18.2	46.1
SST2	Sentiment	2	67k/0.9k	54	92.4	14.7	7.5	8.1	20.8	38.6

[a]We fail to attack Yelp and IMDB datasets with PSO because it is inefficient on long sentences.

Attack Methods: We pick 5 recently proposed adversarial attack methods implemented in TextAttack [16]: (1) [18] proposes the probability weighted word saliency (**PWWS**), which determines the synonym substitution using both the word saliency and the classification probability; (2) TextFooler [8] (**TF**) is a synonym substitution algorithm with semantic similarity checker and part-of-speech checker; (3) BERT-ATTACK [10] (**BA**) and (4) **BAE** [5] both use BERT language models to propose word substitutions; and (5) SememePSO [25] (**PSO**[1]) substitutes words based on sememes – the minimum semantic units, and uses particle swarm optimization.

Details of the datasets, the CAcc, and the AAcc of the original classifier against attack methods are shown in Table 1.

Baselines: We compare our method with 2 baseline defense methods: (1) **SEM** [22]: we follow the hyper-parameters recommended by authors. We convert the training data using SEM and train the classifier using the same convention as the original classifier mentioned above; and (2) Adversarial training (**AT**): we sample 10k sentences from each of the training set, then use TF to attack the original classifier with these sentences. We use TF in adversarial training because of its efficiency and attack efficacy. We then merge the generated adversarial sentences with the original training set, then fine-tune the original classifier for another 5k batches. For each training batch, we sample half of the sentences from the original training set, and the other half from the set of adversarial sentences.

Our Method: For LMAg, we set the number of rephrases $\lambda = 10$, the mask ratio $\gamma = 0.2$, and $\alpha = 0.6$. We fine-tune the BERT language model on the training set for 5000 steps with batch size 32 and learning rate 0.00002.

5.1 Experimental Results

Does Applying Defense Methods Reduce the CAcc? Figure 2 shows the CAcc of classifiers after applying a defense. AT and LMAg both cause slight decrease in CAcc in most cases. SEM results in a greater decrease in CAcc.

---

[1] We fail to attack Yelp and IMDB datasets with PSO because it is inefficient on long sentences.

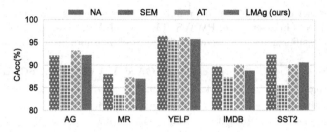

Fig. 2. CAcc of the classifier after applying defense methods. NA means the original classifier.

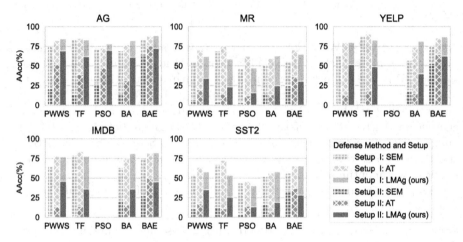

Fig. 3. AAcc of the classifier for each adversarial method (X-axis) on both setups. The translucent (taller) bars represent the AAcc of setup I – original defense. The solid (short) bars represent the AAcc of setup II – boosted defense.

Are Defense Methods Effective on Setup I – Original Defense? The translucent (taller) bars in Fig. 3 show the AAcc for this setup. All the methods including ours successfully defend against a large portion of adversarial examples, and improve AAcc by more than 45% compared to the original classifier. Our LMAg improves AAcc by 51.5% on average while AT performs slightly better with an improvement of 53.7%.

Are Defense Methods Effective on Setup II – Boosted Defense? The solid (shorter) bars in Fig. 3 show the AAcc for this setup. The AAcc is significantly lower than in Setup I, showing that this setup is more challenging. When the attack methods get access to the robustified classifiers and run more iterations, they can still find adversarial sentences. SEM does not improve AAcc whereas AT slightly improves the AAcc by 6.6% compared to the original classifier. LMAg can improve the AAcc by 17.3% on average which is significantly better than the other two baselines. Furthermore, LMAg achieves the best improvement on all 5 datasets and 4 out of 5 attack methods.

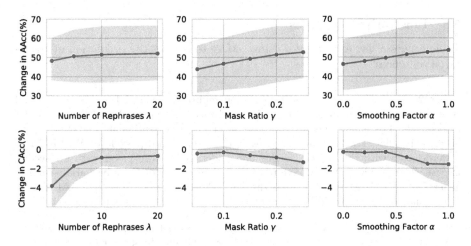

Fig. 4. The effect of hyperparameters on Setup I. The upper row shows the change of AAcc, the lower row shows the change of CAcc. The line in each figure shows the change of AAcc or CAcc averaged over all 5 datasets. The colored band shows the maximum and minimum change of 5 datasets.

5.2   Effect of Hyperparameters

We further evaluate the effect of three hyperparameters of LMAg, namely the number of rephrases λ, the mask ratio γ, and smoothing factor α. We tune one hyperparameter with the other two fixed. The results are demonstrated in Fig. 4.

Effect of Number of Rephrases λ. We measure $\lambda = 1, 5, 10, 20$ when $\gamma = 0.2$ and $\alpha = 0.6$. We observe that when increasing k from 1 to 10, the CAcc on the original test set increases significantly. When $k = 1$, the CAcc decreases as much as 6%. We interpret it as when 20% of the words are covered, the language model may generate a sentence whose label is different from the original sentence, which causes a significant drop in CAcc. Using multiple rewrites can alleviate this problem. We also observe that the average AAcc improves when λ increases.

Effect of Mask Ratio γ. We measure $\gamma = 0.05, 0.1, 0.15, 0.2, 0.25$ while $\lambda = 10$ and $\alpha = 0.6$. We observe that masking more words leads to more improvement on AAcc but leading to lower CAcc. When masking more words, it's harder for the language model to rephrase the sentence and retain the same label; meanwhile it is more likely to counteract adversarial modifications.

Effect of Smoothing Factor α. We measure $\alpha = 0.0, 0.2, 0.4, 0.6, 0.8, 1.0$ with $\lambda = 10$ and $\gamma = 0.2$. Note that when $\alpha = 0$, the mask positions are sampled uniformly. We observe that larger α leads to higher AAcc but lower CAcc. The reason for this is that when α becomes larger, the probability distribution of the selected position becomes sparser. Some positions have a high probability of being masked while others are hardly masked. In this case, the same masking positions may be selected for multiple rewrites, which is similar to setting a smaller λ.

5.3   Illustrative Examples

Table 2 gives a few examples of using LMAg to correct the prediction of adversarial sentences.

Table 2. Two adversarial sentences and their rephrases generated by LMAg. Ori and Adv indicate the original sentence and the adversarial sentence found by TextFooler respectively. Pos or Neg means the positive or negative sentiment predicted by the original classifier. The 3rd row visualizes w_i at BERT's word-piece level. We **boldface** 5 word-pieces with the largest weights and underlines 5 word-pieces with the second largest weights. The following five rows show 5 rephrases of the adversarial sentence generated by LMAg. We **boldface** the masked word-pieces. Note that LMAg may change unmasked words. In both examples, the classifier's prediction is corrected.

Ori (Neg)	without shakespeare's eloquent language, the update is dreary and sluggish.
Adv (Pos)	without shakespeare's eloquent **dialect**, the **refreshing** is **sorrowful** and **unmotivated**.
Visualize w_i	without shakespeare's el ##o ##quent dialect, the **refreshing** is sorrow ##ful and un ##mot ##ivated.
R1 (Neg)	without shakespeare's eloquent **wit**, the film is sorrowful and **unmotional**.
R2 (Pos)	**like** shakespeare's eloquent **plays**, the film is sorrowful and unmotivated.
R3 (Neg)	without shakespeare's eloquent wit, the film**ly** sorrow**ful** and **un**motivated.
R4 (Neg)	without shakespeare's **eloquent** wit, the film is sorro**wful** and unmot**ivated**.
R5 (Neg)	without shakespeare's elo**quent**ism, **miss** film is sorrowful and unmotivated.
Ori (Pos)	compelling revenge thriller, though somewhat weakened by a miscast leading lady.
Adv (Neg)	**cogent** revenge thriller, though somewhat weakened by a miscast leading lady.
Visualize w_i	co ##**gent** **revenge thriller**, though somewhat **weakened** by a mis ##**cast** leading lady.
R1 (Pos)	co**hesive** revenge thriller, though somewhat overshadowed by **its** miscast leading lady.
R2 (Neg)	cogent revenge thriller, **playedly** performance by **a** miscast leading lady.
R3 (Pos)	**a entertaining entertaining** thriller, though somewhat hampered by a miscast leading lady.
R4 (Neg)	cogent revenge **thriller, only** somewhat hampered by a miscast leading **man**.
R5 (Pos)	**a entertaining** revenge thriller, though somewhat **hampered** by a miscast leading lady.

6   Conclusion

In this paper, we laid out two different setups in defending adversarial attack, namely (1) original defense and (2) boosted defense against adversarial examples. We show that the latter is both more realistic and more challenging. We introduce LMAg, a novel in situ augmentation to defend adversarial attacks on text classifiers. LMAg achieves comparable performance on Setup I and significantly better performance on Setup II. Since LMAg is an in situ data transformation, it does not change the architecture of the classifier, so it can be easily integrated with other defense methods. Although we improved the after-attack accuracy by 17.3%, the problem of defending adversarial attack is far from being solved. In the future, we will attempt to further improve the defense method by integrating LMAg with other methods, and meanwhile try to improve the efficiency.

Acknowledgment. Alfredo Cuesta-Infante has been funded by the Spanish Government research project MICINN PID2021-128362OB-I00.

References

1. Alzantot, M., Sharma, Y., Elgohary, A., Ho, B.J., Srivastava, M., Chang, K.W.: Generating natural language adversarial examples. In: Proceedings of the 2018 Conference on Empirical Methods in Natural Language Processing (2018)
2. Belinkov, Y., Bisk, Y.: Synthetic and natural noise both break neural machine translation. In: International Conference on Learning Representations (2018)
3. Devlin, J., Chang, M.W., Lee, K., Toutanova, K.: BERT: pre-training of deep bidirectional transformers for language understanding. In: Proceedings of the Annual Conference of the North American Chapter of the Association for Computational Linguistics (2019)
4. Ebrahimi, J., Rao, A., Lowd, D., Dou, D.: HotFlip: white-box adversarial examples for text classification. In: Proceedings of the 56th Annual Meeting of the Association for Computational Linguistics (Volume 2: Short Papers) (2018)
5. Garg, S., Ramakrishnan, G.: BAE: BERT-based adversarial examples for text classification. In: Proceedings of the Conference on Empirical Methods in Natural Language Processing and the International Joint Conference on Natural Language Processing (2020)
6. Jia, R., Liang, P.: Adversarial examples for evaluating reading comprehension systems. In: Proceedings of the 2017 Conference on Empirical Methods in Natural Language Processing (2017)
7. Jia, R., Raghunathan, A., Göksel, K., Liang, P.: Certified robustness to adversarial word substitutions. In: Proceedings of the Conference on Empirical Methods in Natural Language Processing and the International Joint Conference on Natural Language Processing (2019)
8. Jin, D., Jin, Z., Zhou, J.T., Szolovits, P.: Is BERT really robust? Natural language attack on text classification and entailment. In: Proceedings of the AAAI Conference on Artificial Intelligence (2020)
9. Li, D., et al.: Contextualized perturbation for textual adversarial attack. In: Proceedings of the 2021 Conference of the North American Chapter of the Association for Computational Linguistics: Human Language Technologies, pp. 5053–5069 (2021)
10. Li, L., Ma, R., Guo, Q., Xue, X., Qiu, X.: BERT-attack: adversarial attack against BERT using BERT. In: Proceedings of the Conference on Empirical Methods in Natural Language Processing and the International Joint Conference on Natural Language Processing (2020)
11. Liang, B., Li, H., Su, M., Bian, P., Li, X., Shi, W.: Deep text classification can be fooled. In: Proceedings of the International Joint Conferences on Artificial Intelligence (2017)
12. Loshchilov, I., Hutter, F.: Decoupled weight decay regularization. In: Proceedings of the International Conference on Learning Representations (2019)
13. Maas, A.L., Daly, R.E., Pham, P.T., Huang, D., Ng, A.Y., Potts, C.: Learning word vectors for sentiment analysis. In: Proceedings of the Annual Meeting of the Association for Computational Linguistics: Human Language Technologies (2011)
14. Madry, A., Makelov, A., Schmidt, L., Tsipras, D., Vladu, A.: Towards deep learning models resistant to adversarial attacks. In: International Conference on Learning Representations (2018)

15. Morris, J., Lifland, E., Lanchantin, J., Ji, Y., Qi, Y.: Reevaluating adversarial examples in natural language. In: Proceedings of the 2020 Conference on Empirical Methods in Natural Language Processing: Findings (2020)
16. Morris, J., Lifland, E., Yoo, J.Y., Grigsby, J., Jin, D., Qi, Y.: TextAttack: a framework for adversarial attacks, data augmentation, and adversarial training in NLP. In: Proceedings of the Conference on Empirical Methods in Natural Language Processing: System Demonstrations (2020)
17. Pang, B., Lee, L.: Seeing stars: exploiting class relationships for sentiment categorization with respect to rating scales. In: Proceedings of the Annual Meeting of the Association for Computational Linguistics (2005)
18. Ren, S., Deng, Y., He, K., Che, W.: Generating natural language adversarial examples through probability weighted word saliency. In: Proceedings of the 57th Annual Meeting of the Association for Computational Linguistics, pp. 1085–1097 (2019)
19. Socher, R., et al.: Recursive deep models for semantic compositionality over a sentiment treebank. In: Proceedings of the 2013 Conference on Empirical Methods in Natural Language Processing (2013)
20. Tramèr, F., Kurakin, A., Papernot, N., Goodfellow, I., Boneh, D., McDaniel, P.: Ensemble adversarial training: attacks and defenses. In: International Conference on Learning Representations (2018)
21. Wang, A., Singh, A., Michael, J., Hill, F., Levy, O., Bowman, S.: Glue: a multi-task benchmark and analysis platform for natural language understanding. In: Proceedings of the 2018 EMNLP Workshop BlackboxNLP: Analyzing and Interpreting Neural Networks for NLP (2018)
22. Wang, X., Jin, H., Yang, Y., He, K.: Natural language adversarial defense through synonym encoding. In: The Conference on Uncertainty in Artificial Intelligence (2021)
23. Wang, Y., Bansal, M.: Robust machine comprehension models via adversarial training. In: Proceedings of the 2018 Conference of the North American Chapter of the Association for Computational Linguistics: Human Language Technologies (Volume 2: Short Papers) (2018)
24. Xu, L., Veeramachaneni, K.: Attacking text classifiers via sentence rewriting sampler. arXiv preprint arXiv:2104.08453 (2021)
25. Zang, Y., et al.: Word-level textual adversarial attacking as combinatorial optimization. In: Proceedings of the 58th Annual Meeting of the Association for Computational Linguistics (2020)
26. Zeng, G., et al.: OpenAttack: an open-source textual adversarial attack toolkit. In: Proceedings of the 59th Annual Meeting of the Association for Computational Linguistics (Demo) (2021)
27. Zhang, W.E., Sheng, Q.Z., Alhazmi, A., Li, C.: Adversarial attacks on deep-learning models in natural language processing: a survey. ACM Trans. Intell. Syst. Technol. (TIST) 11, 1–41 (2020)
28. Zhang, X., Zhao, J., LeCun, Y.: Character-level convolutional networks for text classification. In: Proceedings of the Conference on Advances in Neural Information Processing Systems (2015)

Relation-Guided Dual Hash Network for Unsupervised Cross-Modal Retrieval

Yuanchao Zheng⬛, Yan Dong⬛, and Xiaowei Zhang[✉]⬛

Qingdao University, Qingdao, China
xiaowei19870119@sina.com

Abstract. Cross-modal hashing has attracted a great deal of attention due to its unique low storage cost and high retrieval efficiency. However, these existing cross-modal retrieval methods fail to deal with global semantic relational redundancy, leading to an unsatisfactory performance on such data. In this paper, to address this issue, we propose a novel cross-modal hashing, namely Relation-guided Dual Hash Network (RDHN) for unsupervised cross-modal retrieval. It captures both long-range dependencies within modalities and enhances the relevance of semantic relations between different modalities through a heterogeneous feature fusion module. Besides, we designs a dual hash network for image modality and text modality, two for each, which can effectively highlight the useful global semantic relations while suppressing the redundant information. In the cross-modal retrieval tasks, our proposed RDHN improves by 4.9% $mAP@50$ for $I \rightarrow T$ and 9.1% $mAP@50$ for $T \rightarrow I$ in 128-bit compared to the state-of-the-art AGCH on the MIRFlickr-25k dataset, respectively. Code is available at https://github.com/Z000204/RDHN.

Keywords: Cross-modal hashing · Dual hash network · Relation-guided

1 Introduction

With the dramatic increase in multimodal data, cross-modal hash (CMH) retrieval [8,16] has received increasing attention from researchers because of its characteristic high retrieval efficiency and low storage cost. Cross-modal hashing is the mapping of data from different modalities into the common hamming space for matching. CMH is classified into unsupervised hashing methods and supervised hashing methods, depending on whether or not labels are used. Currently, supervised methods [10,12,18] achieve good performance because of the use of labels for supervised learning. However, labels are labor- and time-intensive to obtain, so supervisory methods are often not adapted to real-world requirements. As a result, unsupervised methods [11,14,15,17] which can reduce the dependence on data labels in the training process and have achieved good results, are receiving increasing attention.

Nevertheless, there are still some problems with existing methods. In many real-world scenarios, invalid or redundant global semantic relations are inevitable.

© The Author(s), under exclusive license to Springer Nature Switzerland AG 2023
M. Tanveer et al. (Eds.): ICONIP 2022, LNCS 13625, pp. 497–508, 2023.
https://doi.org/10.1007/978-3-031-30111-7_42

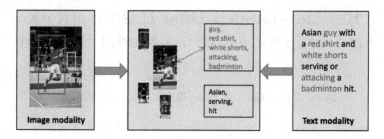

Fig. 1. A real-world image-text pair from dataset MIRFlickr. The red parts of the image and text are the primary semantic relations. The red arrow represents the red part of the text corresponding to the red part of the image. (Color figure online)

Taking Fig. 1 as an example, due to its semantic relations, i.e., "red par", "blue par", "yellow par" and "green par", the most useful visual semantic relations might be the red part. However, such relations cover only a small part of the whole image, and most of the area in that image is the background and contains many useless relations. As another example, the user can provide several imperfect text labels to describe the image, such as the text shown in Fig. 1. This text only describes key information about the corresponding image, while long-distance relational modeling generates much redundant and invalid information. At the same time, many recent approaches [6,11,14,17] perform alignment of the semantics themselves. Though these achieve good performance, there is still redundancy in the semantic relations.

Based on the above analysis, in this paper, we propose a novel unsupervised cross-modal hash retrieval method called Relation-guided Dual Hash Network (RDHN) for unsupervised cross-modal retrieval. First, we use the image transformer encoder and text transformer encoder to capture long-range dependencies within each image and text modalities to establish global dependencies. Second, we propose a heterogeneous feature fusion module to enhance the relevance of semantic relations between different modalities. Third, we introduce a dual network to improve the global semantic relations of the intra-modality by the intra-modal consistent loss. Specifically, we use the CNN-structured training network to filter the transformer-structured training network, thus enhancing the similarity of similar relations within intra-modality and suppressing dissimilar relations. Our contributions can be summarized as follows:

- We propose a novel end-to-end deep cross-modal hashing method, named Relation-guided Dual Hash Network, which deals with global semantic relational redundancy to enhance intra-and inter-modal semantic relations.
- We propose a heterogeneous feature fusion module to enhance the semantic similarity relations of different modalities by aligning the semantic relations between them. And with a designed dual-network for different modalities, two for each, RDHN can enhance the semantic relations between different modalities.

– In the cross-modal retrieval tasks, our proposed RDHN improves by 4.9% $mAP@50$ for $I \rightarrow T$ and 9.1% $mAP@50$ for $T \rightarrow I$ in 128-bit compared to the state-of-the-art AGCH [20] on the MIRFlickr-25k dataset, respectively.

2 Related Work

2.1 Cross-Modal Hashing

Cross-modal hash retrieval methods can be broadly divided into two categories: supervised methods and unsupervised methods. Supervised methods are to explore semantic information in semantic labels to supervise the generation of hash codes, such as TEACH [18], SSAH [10], DMFH [12].

Compared with the supervised methods, unsupervised methods mainly use co-occurrence information between images and texts to maximize the relationship and bridge the heterogeneous semantic gap. Based on this, several unsupervised approaches have been proposed. DSAH [17] aligns the semantic information between modalities by designing a Semantic Alignment loss and reconstructing the characteristics of different modalities. To better maintain the semantic similarity between different modalities, DGCPN [19] introduces a graph neighborhood structure to guarantee the similarity of the original data. AGCH [20] then takes a different approach to modeling the graph nodes for the original data and uses the GCN to generate the final similarity matrix.

Although these methods have achieved good results, their semantic alignment always suffers from semantic relations redundancy, which affects the performance of the retrieval.

2.2 Attention Mechanism

The attention mechanism has now been a great success and is used in a wide range of areas, such as object detection [2], person re-identification [21], natural language processing [4] and so on. The method [2] significantly improves the representation of the target detection head without increasing the computational overhead by combining multiple self-attentive mechanisms between feature levels, spatial locations, and within task-aware output channels. Relation-Aware Global Attention for Person Re-identification [21] proposes an effective relation-aware global attention (RGA) module that captures global structural information for better attentional learning. In this work, we use a transformer encoder to extract modal features and establish global dependencies in our work. We also use a self-attention mechanism to align the semantic similarity relations between different modalities to enhance their information representation.

3 Proposed Approach

3.1 Problem Definition

Assume that we have N image-text pairs which can be denoted as $O = \{I_i, T_i\}_{i=1}^{N}$. I_i and T_i represent the i-th image and the i-th text in the instance, respectively.

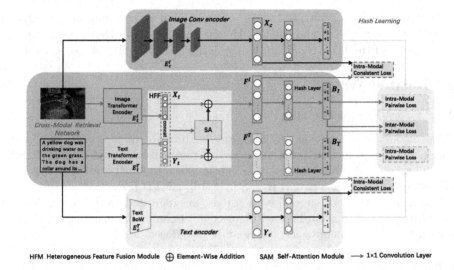

Fig. 2. The structure of our approach. With a designed dual hash network for different modalities, two for each, RDHN can effectively highlight the good global semantic relations while suppressing the redundant information.

The structure of our method is shown in Fig. 2. The aim of our method is to learn k-bit high-quality binary representations $B_I \in \{-1, +1\}^{N \times k}$ and $B_T \in \{-1, +1\}^{N \times k}$ for I and T by the dual-network. Learn two efficient hash functions that map the data of each modality into the common Hamming space.

3.2 Feature Extraction

Feature Extraction Module E_t^I and E_t^T of Cross-Modal Retrieval Network: We use vision transformer [5] as a feature extractor only, so we can capture long-distance dependencies within image modality. We follow the recent convention to use the forward extraction features of the pretrained transformer as the image feature input; the text is similar to that used the bert [3]. Denote $X_t = \{x_{t1}, ..., x_{ti}, ..., x_{tN}\} \in \mathbb{R}^{N \times d}$ and $Y_t = \{y_{t1}, ..., y_{ti}, ..., y_{tN}\} \in \mathbb{R}^{N \times d}$ as the outputs of the image network and text network respectively, where d is the numbers of channel, $x_{ti} = E_t^I \{I_i\}$ and $y_{ti} = E_t^T \{T_i\}$. We use the pre-trained feature extraction networks and fix them during the training.

Feature Extraction Module E_c^I and E_c^T of Image Conv Encoder and Text Encoder: We follow the current cross-modal retrieval methods to use the fc7 features of pretrained VGG-16 as the image feature input and the two full connects for text input. Denote $X_c = \{x_{c1}, ..., x_{ci}, ..., x_{cN}\} \in \mathbb{R}^{N \times d}$ and $Y_c = \{y_{c1}, ..., y_{ci}, ..., y_{cN}\} \in \mathbb{R}^{N \times d}$ as the outputs of the image network and text network respectively, where d is the numbers of channel, $x_{ci} = E_c^I \{I_i\}$ and $y_{ci} = E_c^T \{T_i\}$. We use the pre-trained feature extraction networks and fix them during the training.

3.3 Relation-Guided Dual Hash Network

Heterogeneous Feature Fusion. To enhance the relevance of semantic relations between different modalities, we propose a heterogeneous feature fusion (HFF) module. Specifically, the image features X_t and text features Y_t obtained from the cross-modal retrieval network are first concatenated together and then passed through a fully connected layer to obtain $F = fc(X_t : Y_t) \in \mathbb{R}^{N \times d}$. And we repeat it to $F = \{f_1, ..., f_i, ..., f_N\} \in \mathbb{R}^{N \times d}$. Then the query, key, and the value computed from the input f_i with the corresponding weights are used as input for the self-attention as:

$$query = f_i W_q, key = f_i W_k, value = f_i W_v, \tag{1}$$

where $W_q, W_k, W_v \in \mathbb{R}^{d \times D}$ are the learnable weight parameter, and D is the dimension of the learnable parameter. Self-attention(SA) is then formulated as:

$$SA(f_i) = f_i + softmax\left(\frac{f_i W_q \left(f_i W_k{}^\top\right)}{\sqrt{d}}\right). \tag{2}$$

Next, we change the dimension to get the fused features $F = \{f_1, ..., f_i, ..., f_N\} \in \mathbb{R}^{N \times d}$. Finally, the original image features X_t and the fused features F are simply combined with an element-wise sum, generating the fused features F^I for image modality. The fused features F^T of text modality are similar.

$$F^I = X_t + F, F^T = Y_t + F. \tag{3}$$

Enhancing Semantic Relations through Dual Hash. Although we have aligned the global semantic relations between the different modalities, it is difficult to discern important information because all the information is aggregated together, resulting in redundancy in the global semantic relations. In the paper, we use the instances obtained from the training of the CNN network to filter the global semantic relations obtained from the training of the transformer network. With a designed dual hash network for different modalities, two for each, RDHN can effectively highlight the useful global semantic relations of the intra-modality while suppressing the redundant semantic relations. Specifically, we exploit an intra-modal consistent loss to achieve it.

Once obtained the features from the cross-modal retrieval network and the features from the image conv encoder and text encoder, we can measure the intra-modal consistent loss, including semantic consistency loss and semantic relational consistency loss:

$$\mathcal{L}_R = \sum \|F^I - X_c\| + \sum \|F^T - Y_c\| \tag{4}$$

Thus, we can make full use of the consistency features of different networks within the same modality and are able to fully preserve important global semantic relations between different modalities.

To describe the neighbor relations, we calculate the pairwise cosine similarity matrices and define them as $\boldsymbol{S}_X^F = \boldsymbol{X}_c \boldsymbol{X}_t^\top \in [-1, +1]^{N \times N}$ and $\boldsymbol{S}_Y^F = \boldsymbol{Y}_c \boldsymbol{Y}_t^\top \in [-1, +1]^{N \times N}$. Similarly, we define matrices as $\boldsymbol{S}_{m,g}^B$, where $m, g \in \{I, T\}$ denotes the matrices are calculated by the features or hash codes of the same modality.

$$\boldsymbol{S}_{m,g}^B = \boldsymbol{B}_m \boldsymbol{B}_g^\top, \in [-1, +1]^{N \times N}. \tag{5}$$

To make the best of the co-occurrence information in data, we pay attention to the similarity matrix of the different network features. From our observations, the diagonal elements of the matrix $\boldsymbol{S}_X^F$ and $\boldsymbol{S}_Y^F$, are calculated between the pairwise features of the same images or texts from different networks. Any same images or texts must have the same label or category. Hence any diagonal element of the matrix should be equal to 1. And we can minimize the quantization error of the diagonal elements to highlight the useful information of global semantic relations while suppressing the redundant information with the same label:

$$\mathcal{L}_S = \min_{\boldsymbol{X}_c, \boldsymbol{X}_t} \sum_{i=1}^{n} \left\| 1 - \boldsymbol{S}_X^F(i, i) \right\|^2 + \min_{\boldsymbol{X}_c, \boldsymbol{X}_t} \sum_{i=1}^{n} \left\| 1 - \boldsymbol{S}_Y^F(i, i) \right\|^2 \tag{6}$$

Thus, the intra-modal consistent loss is as follows:

$$\mathcal{L}_C = \mathcal{L}_R + \mathcal{L}_s. \tag{7}$$

3.4   Hash Learning

In this paper, we use two networks trained together to generate high-quality hash codes, and the learning of both networks is consistent in the cross-modal retrieval learning part. This is illustrated below with the cross-modal retrieval network. Finally, $\boldsymbol{X}_t$ and $\boldsymbol{Y}_t$ are fed into a fully connected layer called hash layer and then binarized to generate the final binary representations $\boldsymbol{B}_I$ of image modality and $\boldsymbol{B}_T$ of text modality, respectively. As above, we obtained the similarity matrix by calculating the cosine distance, such as: $\boldsymbol{S}_{I,I}^F = \boldsymbol{X}_t \boldsymbol{X}_t^\top \in [-1, +1]^{N \times N}$, $\boldsymbol{S}_{T,T}^F = \boldsymbol{Y}_t \boldsymbol{Y}_t^\top \in [-1, +1]^{N \times N}$ and $\boldsymbol{S}_{I,T}^F = \boldsymbol{X}_t \boldsymbol{Y}_t^\top \in [-1, +1]^{N \times N}$.

Intra-modal Pairwise Loss. Pairwise loss enhances the correlation of similar instances and can reduce the similarity of semantically related unrelated instances. Therefore, it is crucial to increase an intra-modal pairwise loss for the image modality and text modality in each training network, respectively. Specifically, to ensure consistency of hash codes and features within modalities, we align the similarity information of hash codes and features from an intra-modal perspective:

$$\mathcal{L}_{\text{intra}} = \sum \left\| \gamma \boldsymbol{S}_{g,g}^F - \boldsymbol{S}_{g,g}^B \right\|^2, \quad g \in \{I, T\}, \tag{8}$$

where γ is a trade-off parameter.

Inter-modal Pairwise Loss. To further bridge the modal gap, we make alignments from the inter-modality view by inter-modal pairwise loss. We combine $\boldsymbol{S}_{I,I}^F$, $\boldsymbol{S}_{T,T}^F$ and $\boldsymbol{S}_{I,T}^F$ with a weighted summation manner:

$$\boldsymbol{S}^F = \alpha \boldsymbol{S}_{I,I}^F + \beta \boldsymbol{S}_{T,T}^F + (1 - \alpha - \beta) \boldsymbol{S}_{I,T}^F, \tag{9}$$

Table 1. The mAP@50 values of Wiki, MIRFlickr-25k, and NUS-WIDE at various code lengths. Bold data represents optimal performance.

Task	Method	Wiki				MIRFlickr-25k				NUS-WIDE			
		16bits	32bits	64bits	128bits	16bits	32bits	64bits	128bits	16bits	32bits	64bits	128bits
I→T	DBRC [7]	0.253	0.265	0.269	0.288	0.617	0.619	0.620	0.621	0.424	0.459	0.447	0.447
	UDCMH [15]	0.309	0.318	0.329	0.346	0.689	0.698	0.714	0.717	0.511	0.519	0.524	0.558
	DJSRH [14]	0.388	0.403	0.412	0.421	0.810	0.843	0.862	0.876	0.724	0.773	0.798	0.817
	JDSH [11]	0.346	0.431	0.433	0.442	0.832	0.853	0.882	0.892	0.736	0.793	0.832	0.835
	DSAH [17]	0.416	0.430	0.438	0.445	0.863	0.877	0.895	0.903	0.775	0.805	0.818	0.827
	DGCPN [19]	0.420	0.438	0.440	0.448	0.875	0.891	0.908	0.918	0.788	0.820	0.826	0.833
	AGCH [20]	0.408	0.425	0.433	0.450	0.865	0.887	0.892	0.912	**0.809**	0.830	0.831	0.852
	OURS	**0.449**	**0.459**	**0.471**	**0.475**	**0.883**	**0.913**	**0.945**	**0.957**	0.800	**0.836**	**0.868**	**0.884**
T→I	DBRC [7]	0.573	0.588	0.598	0.599	0.618	0.626	0.626	0.628	0.455	0.459	0.468	0.473
	UDCMH [15]	0.622	0.633	0.645	0.658	0.692	0.704	0.718	0.733	0.637	0.653	0.695	0.716
	DJSRH [14]	0.611	0.635	0.646	0.658	0.786	0.822	0.835	0.847	0.712	0.744	0.771	0.789
	JDSH [11]	0.630	0.631	0.647	0.651	0.825	0.864	0.878	0.880	0.721	0.785	0.794	0.804
	DSAH [17]	0.644	0.650	0.660	0.662	0.846	0.860	0.881	0.882	0.770	0.790	0.804	0.815
	DGCPN [19]	0.644	0.651	0.660	0.662	0.859	0.876	0.890	0.905	0.783	0.802	0.812	0.817
	AGCH [20]	0.627	0.640	0.648	0.658	0.829	0.849	0.852	0.880	0.769	0.780	0.798	0.802
	OURS	**0.644**	**0.651**	**0.660**	**0.663**	**0.884**	**0.914**	**0.947**	**0.960**	**0.801**	**0.837**	**0.869**	**0.886**

where $\boldsymbol{S}^F(i,j)$ represents the similarity between $i-th$ instance and $j-th$ instance, α and β are trade-off parameters. We then align the similarity of the instances with the similarity of the different modal hash codes.

$$\mathcal{L}_{\text{inter}} = \sum \left\| \gamma \boldsymbol{S}^F - \boldsymbol{S}^B_{m,g} \right\|^2, \text{ s.t. } (m,g) \in \{(I,I),(T,T),(I,T)\} \tag{10}$$

where (m,g) denotes a modality-pair. Finally, we merge the semantic relations alignment loss of the inter-modality and that of the inter-modality:

$$\mathcal{L}_{Ret} = \mathcal{L}_{inter} + \gamma \mathcal{L}_{intra}, \tag{11}$$

where γ is the trade-off parameter to balance the importance of the inter-modality and intra-modality.

We combine the losses of different modules into our final objective function, as shown below:

$$\mathcal{L}_{total} = \mathcal{L}_C + \mathcal{L}_{Ret}. \tag{12}$$

4   Experiments and Evaluations

4.1   Datasets and Evaluation

We conducted experiments on three public cross-modal retrieval datasets, including Wiki [13], MIRFlickr-25K [9] and NUS-WIDE [1], to verify the validity of our method. In our experiments, we adopt two standard retrieval evaluation metrics: $mAP@50$ and mAP to evaluate the retrieval performance.

4.2 Implementation Details

We implement the method HILN by pytorch and workstation configured with NVIDIA RTX 3090 GPU. The weight decay rate is 0.0005, and the momentum is set to 0.9. When training on the Wiki dataset, the learning rate of the network is set to 0.01, $\lambda = 0.4$, $\beta = 0.4$, $\omega = 0.2$ when training on the MIRFlickr-25k and NUS-WIDE datasets, the learning rate of the network is set to 0.001, $\lambda = 0.45$, $\beta = 0.45$, $\omega = 0.1$. The training epochs on the Wiki, MIRFLickr-25K, and NUS-WIDE datasets are set to 150, 100, and 80, respectively.

4.3 Performance

We compare our method with several representative deep unsupervised cross-modal hashing retrieval methods including DBRC [7], UDCMH [15], DJSRH [14], JDSH [11], DSAH [17], DGCPN [19], AGCH [20]. Table 1 shows the $mAP@50$ values of RDHN and other comparison methods on MIRFlickr-25k, NUS-WIDE, and Wiki in two cross-modal retrieval tasks for four lengths of hash codes. And Fig. 3 shows top-N precision curves among five comparison methods. $I \rightarrow T$ means that the query is image modality and the database is text modality. $T \rightarrow I$ is the opposite. It can be seen that our method is significantly better than the latest unsupervised cross-modal hashing methods. Specifically, compared to DSAH, for the Wiki, as shown in the results, we achieve boosts of 5.3% in average $mAP@50$ for different hash code lengths in the $I \rightarrow T$ task. Moreover, RDHN achieves boosts of 4.4% and 6.8% in average $mAP@50$ with different hash code lengths in $I \rightarrow T$ task and $T \rightarrow I$ task on MIRFlickr-25k respectively, and achieves boosts of 5.0% and 6.7% in two retrieval tasks on NUS-WIDE. The main reason for the performance improvement is due to the heterogeneous feature fusion module and dual network proposed by RDHN.

Although one performance is worse than AGCH, the rest are higher, while we ensure that the performance of $I \rightarrow T$ and $T \rightarrow I$ tasks on both MIRFlickr and NUS-WIDE data is largely consistent.

4.4 Ablation Study

We design several variants to validate the impact of our proposed modules and to demonstrate the superiority of the original RDHN.

- **RDHN-1** extracts global semantic information from images and texts using only transformer encoder (TE).
- **RDHN-2** builds on RDHN-1 using the heterogeneous feature fusion (HFF) module.
- **RDHN-3** is based on RDHN-2 and uses a training network in a dual network using only CNN as a framework to filter the semantic relations of image modalities in the main network, marked as DN-I.
- **RDHN-4** is based on RDHN-2 and uses a training network in a dual network using only CNN as a framework to filter the semantic relations of text modalities in the main network, marked as DN-T

Table 2. The mAP@50 results at 128 bits of ablation study on MIRFlickr-25k and NUS-WIDE. Bold data represents the best performance.

Model	Configuration	MIRFlickr		NUS-WIDE	
		I→T	T→I	I→T	T→I
Baseline	-	0.903	0.882	0.827	0.815
RDHN-1	Baseline+TE	0.919	0.896	0.848	0.832
RDHN-2	RDHN-1+HFF	0.941	0.941	0.875	0.874
RDHN-3	RDHN-2+(DN-I)	0.951	0.952	0.882	0.882
RDHN-4	RDHN-3+(DN-T)	0.950	0.949	0.879	0.880
RDHN-5	RDHN-4+(DN-IT)	**0.960**	**0.960**	**0.886**	**0.886**

Table 3. The mAP@50 results at 128 bits of ablation study about CNN filtering semantic relations on MIRFlickr-25k and NUS-WIDE. Bold data represents the best performance.

Model	Configuration	MIRFlickr		NUS-WIDE	
		I→T	T→I	I→T	T→I
Filter-1	CNN→TE	**0.960**	**0.960**	**0.886**	**0.886**
Filter-2	TE→TE	0.952	0.953	0.878	0.880
Filter-3	TE→CNN	0.916	0.926	0.857	0.866

- **RDHN-5** is based on RDHN-2 and uses a training network in a dual network using CNN to filter the semantic relations between image modalities and text modalities in the main network, marked as DN-IT.

We conduct $I \rightarrow T$ and $T \rightarrow I$ tasks at 128 bits on the MIRFlickr-25k and NUS-WIDE datasets. The $mAP@50$ results of six types of RDHN are shown in Table 2. From the results of Baseline, RDHN-1, and RDHN-2, We find the effectiveness of transformer encoder and heterogeneous feature fusion module. RDHN-1 improves $mAP@50$ by 1.6% and 2.1% over baseline for the $T \rightarrow I$ task on both datasets, MIRFlickr and NUS-WIDE, respectively. RDHN-1 improves $mAP@50$ by 1.8% and 2.5% over baseline for the $I \rightarrow T$ task on both datasets, MIRFlickr and NUS-WIDE, respectively. The performance improvement of the RDHN-1 attributes transformer encoder effectively captures the long-range dependencies within modalities. RDHN-2 improves $mAP@50$ by 5.0% and 5.0% over RDHN-1 for the $T \rightarrow I$ task on both datasets, MIRFlickr and NUS-WIDE, respectively. RDHN-2 improves $mAP@50$ by 2.4% and 3.1% over RDHN-1 for the $I \rightarrow T$ task on both datasets, MIRFlickr and NUS-WIDE, respectively. The performance improvement of RDHN-2 attributes heterogeneous feature fusion module, which effectively aligns semantic similarity relations between different modalities to enhance the relevance of semantic relations between different modalities.

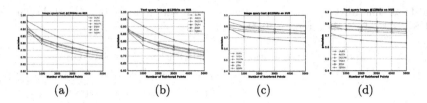

Fig. 3. Precision@top-K curves of various cross-modal hashing methods on MIRFlickr-25k and NUS-WIDE with 128 bits.

From the results of RDHN-2, RDHN-3, RDHN-4 and RDHN-5 in Table 2, We find the effectiveness of using CNN as a framework to filter the semantic relations. From the $mAP@50$ results of RDHN-2 and RDHN-3, we find that the reason for the improved performance of RDHN-3 is due to the use of a training network in a dual network using only CNN as a framework to filter the semantic relations of image modalities in the cross-modal retrieval network. From the $mAP@50$ results of RDHN-2 and RDHN-4, we find that the reason for the improved performance of RDHN-4 is due to the use of a training network in a dual network using only CNN as a framework to filter the semantic relations of text modalities in the cross-modal retrieval network. For the MIRFlickr dataset, in the $I \rightarrow T$ task, RDHN-3 improves 1.1% over RDHN-2, RDHN-4 improves 1.0% over RDHN-2, and RDHN-5 improves 1.7% over RDHN-2. In the $T \rightarrow I$ task, RDHN-3 improves 1.2% over RDHN-2, RDHN-4 improves 0.9% over DHN-2, and RDHN-5 improves 2.0% over RDHN-2. These results show that filtering both image and text modalities is better and that filtering only image modalities with a dual network is better than filtering only text modalities. And the dual network can effectively highlight the useful global semantic relations of the intra-modality while suppressing the redundant semantic relations.

Meanwhile, we take the ablation study about CNN filtering semantic relations on MIRFlickr-25k, as shown in Table 3. Specifically, Filter-1 uses CNN to filter the semantic relations between the respective modalities, where the main retrieval network uses the transformer as a feature extractor. Filter-3 is the opposite, while Filter-2 uses the transformer to filter the semantic relations. Filter-1 improves $mAP@50$ by 3.7% and 2.3% over Filter-3 for the $T \rightarrow I$ task on both datasets, MIRFlickr and NUS-WIDE, respectively. Filter-1 improves $mAP@50$ by 4.5% and 3.2% over Filter-3 for the $I \rightarrow T$ task on both datasets, MIR-Flickr and NUS-WIDE, respectively. Filter-1 improves $mAP@50$ by 0.6% and 0.7% over Filter-2 for the average retrieval task on both datasets, MIRFlickr and NUS-WIDE, respectively. From the $mAP@50$ results, we find that because the transformer can establish global dependencies and capture global semantic relationships but also adds a lot of useless information that affects the performance of the experiment, it does not achieve the same results as our method, but surpasses the original method.

5 Conclusion

This paper proposes a novel unsupervised cross-modal hashing method called Relation-guided Dual Hash Network (RDHN) for Unsupervised Cross-Modal Retrieval. To enhance the relevance of semantic relations between different modalities, we introduce a heterogeneous feature fusion module to align the semantic similarity relationship of different modalities. Moreover, we exploit a dual network that can effectively highlight the useful global semantic relations of the intra-modality while suppressing the redundant semantic relations of the intra-modality. Numerous experiments have shown the effectiveness of our method.

Acknowledgments. This work was supported in part by the National Natural Science Foundation of China (Grant No.61902204), and in part by the Natural Science Foundation of Shandong Province of China (Grant No. ZR2019BF028).

References

1. Chua, T.S., Tang, J., Hong, R., Li, H., Luo, Z., Zheng, Y.: NUS-WIDE: a real-world web image database from National University of Singapore. In: Proceedings of the ACM International Conference on Image and Video Retrieval, pp. 1–9 (2009)
2. Dai, X., et al.: Dynamic head: unifying object detection heads with attentions. In: Proceedings of the IEEE/CVF Conference on Computer Vision and Pattern Recognition, pp. 7373–7382 (2021)
3. Devlin, J., Chang, M.W., Lee, K., Toutanova, K.: BERT: pre-training of deep bidirectional transformers for language understanding. arXiv preprint arXiv:1810.04805 (2018)
4. Dong, Z., Wu, T., Song, S., Zhang, M.: Interactive attention model explorer for natural language processing tasks with unbalanced data sizes. In: 2020 IEEE Pacific Visualization Symposium (PacificVis), pp. 46–50. IEEE (2020)
5. Dosovitskiy, A., et al.: An image is worth 16 × 16 words: transformers for image recognition at scale. arXiv preprint arXiv:2010.11929 (2020)
6. Gong, H., Chen, G., Liu, S., Yu, Y., Li, G.: Cross-modal self-attention with multi-task pre-training for medical visual question answering. In: Proceedings of the 2021 International Conference on Multimedia Retrieval, pp. 456–460 (2021)
7. Hu, D., Nie, F., Li, X.: Deep binary reconstruction for cross-modal hashing. IEEE Trans. Multimed. **21**(4), 973–985 (2018)
8. Hu, H., Xie, L., Hong, R., Tian, Q.: Creating something from nothing: unsupervised knowledge distillation for cross-modal hashing. In: Proceedings of the IEEE/CVF Conference on Computer Vision and Pattern Recognition, pp. 3123–3132 (2020)
9. Huiskes, M.J., Lew, M.S.: The MIR flickr retrieval evaluation. In: Proceedings of the 1st ACM International Conference on Multimedia Information Retrieval, pp. 39–43 (2008)
10. Li, C., Deng, C., Li, N., Liu, W., Gao, X., Tao, D.: Self-supervised adversarial hashing networks for cross-modal retrieval. In: Proceedings of the IEEE Conference on Computer Vision and Pattern Recognition, pp. 4242–4251 (2018)

11. Liu, S., Qian, S., Guan, Y., Zhan, J., Ying, L.: Joint-modal distribution-based similarity hashing for large-scale unsupervised deep cross-modal retrieval. In: Proceedings of the 43rd International ACM SIGIR Conference on Research and Development in Information Retrieval, pp. 1379–1388 (2020)
12. Nie, X., Wang, B., Li, J., Hao, F., Jian, M., Yin, Y.: Deep multiscale fusion hashing for cross-modal retrieval. IEEE Trans. Circuits Syst. Video Technol. **31**(1), 401–410 (2020)
13. Pereira, J.C., Coviello, E., Doyle, G., Rasiwasia, N., Lanckriet, G.R., Levy, R., Vasconcelos, N.: On the role of correlation and abstraction in cross-modal multimedia retrieval. IEEE Trans. Pattern Anal. Mach. Intell. **36**(3), 521–535 (2013)
14. Su, S., Zhong, Z., Zhang, C.: Deep joint-semantics reconstructing hashing for large-scale unsupervised cross-modal retrieval. In: Proceedings of the IEEE/CVF International Conference on Computer Vision, pp. 3027–3035 (2019)
15. Wu, G., et al.: Unsupervised deep hashing via binary latent factor models for large-scale cross-modal retrieval. In: IJCAI, pp. 2854–2860 (2018)
16. Yan, C., Bai, X., Wang, S., Zhou, J., Hancock, E.R.: Cross-modal hashing with semantic deep embedding. Neurocomputing **337**, 58–66 (2019)
17. Yang, D., Wu, D., Zhang, W., Zhang, H., Li, B., Wang, W.: Deep semantic-alignment hashing for unsupervised cross-modal retrieval. In: Proceedings of the 2020 International Conference on Multimedia Retrieval, pp. 44–52 (2020)
18. Yao, H.L., Zhan, Y.W., Chen, Z.D., Luo, X., Xu, X.S.: Teach: attention-aware deep cross-modal hashing. In: Proceedings of the 2021 International Conference on Multimedia Retrieval, pp. 376–384 (2021)
19. Yu, J., Zhou, H., Zhan, Y., Tao, D.: Deep graph-neighbor coherence preserving network for unsupervised cross-modal hashing. In: Proceedings of the AAAI Conference on Artificial Intelligence, AAAI, pp. 4626–4634 (2021)
20. Zhang, P.F., Li, Y., Huang, Z., Xu, X.S.: Aggregation-based graph convolutional hashing for unsupervised cross-modal retrieval. IEEE Trans. Multimed. **24**, 466–479 (2021)
21. Zhang, Z., Lan, C., Zeng, W., Jin, X., Chen, Z.: Relation-aware global attention for person re-identification. In: Proceedings of the IEEE/CVF Conference on Computer Vision and Pattern Recognition, pp. 3186–3195 (2020)

Prompt-Based Learning for Aspect-Level Sentiment Classification

Guowei Li, Fuqiang Lin, Wangqun Chen, Diwen Dong, and Bo Liu[✉]

National University of Defense Technology, Changsha 410073, China
{liguowei,linfuqiang13,chenwangqun19,ddw_bak,kyle.liu}@nudt.edu.cn

Abstract. The goal of Aspect-level Sentiment Classification (ASC) is to identify the sentiment polarity towards a specific aspect of a given sentence. Mainstream methods design complicated models and require a large scale of annotated training samples, and even perform finetuning based on pre-trained language models. Therefore, those supervised methods may be impractical in real-world scenarios due to the limited availability of labeled training corpora, a.k.a. low-resource settings. To this end, we propose an aspect-specific prompt learning approach (AS-Prompt) that fully utilizes the pre-trained knowledge and aspect-related information to deal with ASC tasks, enabling pre-trained models with huge parameters to achieve considerable results under the few-shot settings. Specifically, we transfer the sentiment classification task into Masked Language Modeling (MLM) by designing appropriate prompts and searching for the ideal expression of prompts in continuous space. Meanwhile, we integrate the prompts into the input sentence, thus adapting the model to the classification task under the guidance of sentiment labels. Experimental results on SemEval-2014 Task 4 show our proposed method achieves noticeably improvement compared with the original BERT models and discrete prompt methods. In addition, we test the performance of the model's transfer on different datasets and demonstrate the superiority of prompt learning when adapting to a new domain, especially under a low-resource setting.

Keywords: Aspect-level Sentiment Classification · Prompt Learning · Few-shot Learning

1 Introduction

As an essential part of the Aspect-Based Sentiment Analysis (ABSA) task, Aspect-Level Sentiment Classification (ASC) aims to predict the sentiment polarity of each given aspect term in a sentence [28]. This aspect-level sentiment analysis can provide more fine-grained guidance in making decisions. For example, suppose a restaurant knows specific reviews for various aspects of it. In this case, they can improve management by targeting niches to better meet the needs of consumers. From the comments in Fig. 1, we can infer that consumers rate the food positively but believe that restaurants should pay more attention to their service.

M. Tanveer et al. (Eds.): ICONIP 2022, LNCS 13625, pp. 509–520, 2023.
https://doi.org/10.1007/978-3-031-30111-7_43

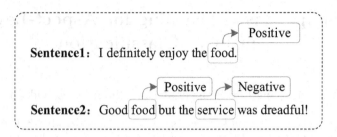

Fig. 1. Illustration examples of aspect-level sentiment classification. The aspect words and corresponding sentiment polarities are in orange and blue boxes, respectively. (Color figure online)

In recent years, researchers have proposed various neural network models for ASC tasks, which achieve promising performances [12,25]. Notably, pre-trained language models (PLMs) have been widely applied to ASC tasks considering the versatile knowledge contained in PLMs. A typical paradigm, called Fine-tuning, is to adapt PLMs to ASC tasks via additional parameters or well-designed objective function, and then optimize the whole model's parameters based on extensive task-specific training data. For example, Devlin and Sun et al. [5,19] construct a classifier to realize the ASC task based on the output of the pre-trained BERT. Nevertheless, the finetuning of such models relies on large-scale annotated corpora to ensure satisfied results, which is not easy to acquire due to costly labor. Besides, such methods suffer a marked decline in performance when transferring into a new domain [9].

Recently, researchers have realized the inefficiency of Fine-tuning under a low-resource setting, and propose a new paradigm named prompt-based learning. Instead of adapting the pre-trained LMs to the downstream tasks through target engineering, prompt-based methods reconstruct the downstream task to fit the LMs with the help of text prompts. Discrete prompts (e.g., The {aspect} made me feel [MASK]) are manually designed and bring significant improvements over the non-prompt methods [22]. Further, researchers automatically search discrete prompts and demonstrate the effectiveness of their works [11,23]. Their efforts in probing appropriate promots show that prompt-based learning methods bring considerable improvement in performance, especially under cross-domain and low-resource settings. However, those discrete prompt-based learning methods require a manual design of prompts, which is time-consuming and can only be optimized in discrete word embedding space.

To address these issues, we propose an aspect-specific prompt learning model (AS-Prompt) that automatically searches prompts with the guidance of aspect-specific information. Specifically, we design a soft prompt to adapt to the downstream ASC task by formulating the prompt as multiple learnable vectors and searching for a better representation of the prompt in continuous space. Such a way is also in line with the continuity feature of neural networks. Additionally, we leverage the sentiment label and aspect information to guide the training process by means of inserting the aspect words into the soft prompt, thus building connections between aspect and corresponding sentiment. Finally, two types of MLM tasks are utilized to optimize the pre-trained model and the soft prompt with the reformulated input by concatenating the original sentence with the prompt.

Our main contributions can be summarized as follows:

- We propose an aspect-specific prompt method (AS-Prompt) to model the sentiment classification task into an MLM task, fully utilizing the aspect and sentiment label information to adapt the model to the downstream task.
- We demonstrate the ability of the model when transferring to a new domain, and explore the impacts of various factors on prompt-based methods to provide guidance for designing and training prompts.
- Experimental results on two datasets show that our method outperforms the baselines, especially under 16-shot and 64-shot scenarios.

2   Related Work

2.1   Aspect-Level Sentiment Analysis

Unlike the traditional coarse-grained sentiment classification task, fine-grained sentiment analysis has more practical values. Hu and Liu [10] is one of the early works to analyze aspect-related sentiment within the text. Pontik [18] contributes to the benchmark datasets of customer reviews for the research of aspect-based sentiment analysis (ABSA), which contain the restaurants and laptops domains. Research like Xu et al. [27] uses pre-trained language models to achieve significant results in sentiment classification, but those models are highly dependent on large-scale training datasets.

Since manually annotating the labels of text can be time-consuming, recent developments for ABSA attempt to finetune the models with unlabeled data. Sun et al. [24] finetune the BERT using a range of domain and task-related knowledge to compromise the effectiveness of a small dataset. Beigi et al. [1] propose an approach for sentiment analysis in unknown domains by adapting sentiment information with a generated domain-specific sentiment lexicon.

Apart from the performance on the small-scale training datasets, the ability to deal with multiple domains is another concern. Beigi and Moattaar [4] propose a transfer learning framework, which uses an adaptive domain model to eliminate the difference between domains. Cao et al. [30] adopt parameter transferring and attention sharing mechanisms to establish the connection between the source domain network and the target domain network. Zhao et al. [14] utilize a sentiment analyzer that learns sentiments via domain adaptive knowledge transfer to improve the classification performance. This paper also focuses on exploring effective approaches to improve the cross-domain ASC under a few-shot setting.

2.2 Prompt-Based Learning

Prompt-based learning reformulates the downstream tasks to adapt to the original LM training task. It utilizes the pre-trained LM to predict the desired output with appropriate prompts, even without additional task-specific training [6,8]. The prompt-based learning was applied to various domains as soon as it was proposed. For example, Yin et al. [29] and Schick et al. [20] explore prompt templates in classification-based tasks, where prompts can be easily constructed. The prompt-based learning reduces or eliminates the need for large supervised data sets for training models and can be applied to few-shot or zero-shot scenarios.

Prompt templates were first created manually according to human introspection [3,21]. But manual template engineering usually fails to find optimal prompts even with rich experience [11]. To find a way that allows LM to perform a task effectively, researchers examine continuous prompts in the embedding space of the model. Li and Liang [13] prepend a sequence of continuous task-specific vectors to the input while keeping the LM parameters frozen. Hambardzumyan et al. [7] propose to initialize the search for a continuous template with discrete prompts. Jiang et al. [11] propose "P-tuning" where continuous prompts are learned by inserting trainable variables into the embedded input.

Continuous prompts formulate the prompt as additional trainable parameters and search for an appropriate prompt by gradient optimization in embedding space. Some prompt-based learning models only update the parameters in prompts during the training stage, representative examples are Prefix-Tuning [13] and WARP [7]. Different from them, Liu and David et al. [2,15] optimize both parameters in pre-trained models and prompts, which is effective especially under few-shot settings. Our work introduces soft prompts to ASC tasks and utilizes aspect-specific information and sentiment label information to improve the performance of classification.

3 Methods

This section presents the task definition and the implementation of our proposed AS-prompt model. Unlike the discrete prompts, we follow the intuition that an appropriate prompt should be explored in continuous space. The architecture of the model is shown in Fig. 2. Our method adopts the pre-trained BERT as the pre-trained model. We set the tokens in prompts as trainable parameters and automatically optimize the prompts. The input of the method is reformulated by an original sentence and designed prompt.

3.1 Overview

We transfer the ASC task to the MLM task to avoid complicated training on pre-trained models based on prompt learning. Different from optimizing the model in discrete word embedding space, we adopt trainable vectors as prompt and finetune it in continuous space. The trainable prompt is more expressive than

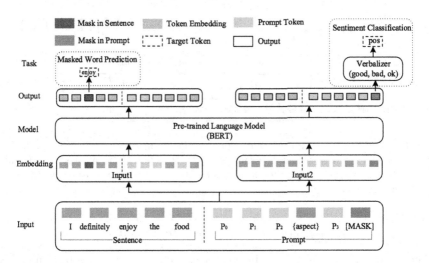

Fig. 2. The overall architecture of our model. The model design a continuous prompt to match the sentiment classification task and takes the original MLM as an auxiliary task. The {·} is the placeholder of the corresponding word. The prompt tokens are trainable vectors, while the embeddings of general tokens are fixed.

a discrete prompt formed of fixed words. To overcome the complex process of prompt selection, we directly initialize the continuous prompt with the same shape as the discrete prompt. Meanwhile, only finetuning the parameters in the prompt is far from enough for the model to learn domain-specific information. To this end, we implement the traditional MLM task of BERT as an auxiliary task to provide more semantic guidance for the model. Since the computation of training can be omitted under few-shot settings, we finetune both the model and the prompt in this paper.

3.2 Task Formulation

For a given sentence $X = [x_1, x_2, \cdots, x_n]$, the ASC task aims to identify the corresponding sentiment polarity s of each aspect a contained in the sentence, where the $s \in \{pos, neg, neu\}$. Let $\mathcal{V}$ refers to the vocabulary space of a language model $\mathcal{M}$. A prompt template T is denoted by $T = \{[p_{0:i}], a, [p_{i+1:m}], y\}$, where $p_i \in \mathcal{V}$ refers to the ith prompt token, a is the aspect token and y is the [MASK] token. The main process of the architecture can be divided into two parts, the masked word prediction task to finetune the pre-trained model and the sentiment classification task to search appropriate prompt for ASC. The Fig. 3 shows the formulation of input for the mentioned two tasks.

Given the sentence and the prompt, we concatenate the sentence with the prompt as input, and formulate the original input into the *Input1* and *Input2* to get the embeddings as follows:

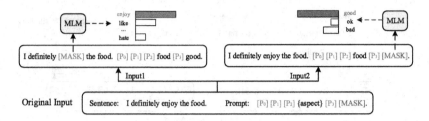

Fig. 3. An example of input formulation for the main processes of the model. $[P_i]$ in orange denotes the trainable prompt. [MASK] denotes the masked word and is differentiated by tasks. $\{\cdot\}$ is the placeholder of the corresponding aspect word. (Color figure online)

$$\{e(x_{0:j}), e(y_s), e(x_{j+1:n}), h_{0:i}, e(a), h_{i+1:m}, e(y)\}, \tag{1}$$

where $h_i (0 \le i < m)$ is a trainable vector and $e(x_i)$ is the initialization embedding of corresponding word.

For each input of MLM task, we compute the cross-entropy loss of the predictions on masked tokens. For the main task, the loss of *Input2* is:

$$\mathcal{L}_{prompt} = -\sum\sum ylog(p(y|X,T)). \tag{2}$$

Similarly, we can get the loss of *Input1* as follows:

$$\mathcal{L}_{sen} = -\sum\sum y_slog(p(y_s|X,T)). \tag{3}$$

Then we can find a suitable prompt with the downstream loss function $\mathcal{L}$ by differentially optimize the prompt h_i:

$$\hat{h}_{0:m} = argmin_h(\mathcal{L}_{prompt} + \mathcal{L}_{sen}), \tag{4}$$

where the $\mathcal{L}_{prompt}$ refers to the loss of sentiment classification and $\mathcal{L}_{sen}$ refers to the loss of masked aspect prediction.

To better adapt the prompt, we convert the groudtruth labels $\{pos, neg, neu\}$ to $\{good, bad, ok\}$ and use the new labels for predicting.

3.3   In-Domain Data Pre-training

We can get available pre-trained weights for BERT, which has been trained on large corpora before. We prepare the pre-trained model with the extra in-domain datasets to include more domain-specific information. We adopt the same way as Seoh et al. [22], we only mask adjectives, proper nouns, and nouns, which are tightly related to sentiment. The baselines execute the same operation for a fair comparison. For the laptop domain, we use the reviews written for products from the electronics category in Amazon Review Data [17]. For the restaurants domain, we extract reviews related to restaurants from Yelp Open Dataset[1].

---

[1] https://www.yelp.com/dataset.

Table 1. SemEval 2014 dataset statistics after preprocessing.

Class	Laptops		Restaurants	
	Train	Test	Train	Test
Positive	987	341	2164	728
Negative	866	128	645	196
Neutral	460	169	496	196
Total	1850	638	3602	1120

4 Experiments

4.1 Datasets

We adopt SemEval 2014 Task 4 datasets released by Pontiki et al. [18] to measure the performance of our proposed model and baselines. It contains English review sentences from laptops and restaurants. The sentiment of each aspect is labeled as positive, negative, neutral, or conflict, where neutral refers to the opinion towards the aspect that is neither positive nor negative, and conflict denotes the existence of both positive and negative sentiment for an aspect. To conduct the experiments in the same condition as early studies [27], we remove the reviews labeled as conflict and split multiple aspect-sentiment labels within one text into different sentences. We select the training data from the training dataset by random numbers for each type of few-shot. The dataset statistics after preprocessing are shown in Table 1.

4.2 Baselines

We compare our proposed model with three Bert-based methods and two prompt-based methods:

- **BERT-ADA** [19] uses both domain-specific language model finetuning and supervised task-specific finetuning to realize the ASC task.
- **BERT [CLS]** [5] inserts a [CLS] token in front of the text and takes the corresponding output vector of [CLS] as the semantic representation of the text for classification.
- **BERT NSP** [5] aims to predict whether sentence B semantically follows sentence A when entering sentence A and sentence B simultaneously.
- **BERT LM** [22] transfers the ASC task as Language Modeling and designs discrete prompts as part of the input.
- **Null Prompts** [16] sets up a prompt template in the form of input text followed by [MASK] token for all tasks and automatically searches prompt in continuous space.

Table 2. Experimental results of our method and baselines. The best results are in bold. The underlined results indicate that our model's improvements over all baselines are statistically significant in the t-test of $p < 0.05$. The results of the first four baselines are taken directly from [22].

Laptops

Model	4		16		64		256		1024		Full(1850)	
	Acc	F1	Acc	F1	Acc	F1	Acc	F1	Acc	F1	Acc	F1
BERT-ADA	-	-	-	-	-	-	-	-	-	-	79.19	74.18
BERT [CLS]	-	-	48.75	34.92	60.63	49.43	77.35	64.31	76.87	71.22	80.06	75.08
BERT NSP	-	-	48.24	31.35	60.91	49.27	72.38	64.64	76.77	71.12	80.25	75.46
BERT LM	61.60	51.29	69.05	58.60	72.80	65.54	76.59	70.65	**79.30**	**74.80**	**81.10**	**76.83**
Null Prompts	57.52	**54.15**	61.76	58.32	67.87	63.59	76.96	71.83	76.02	69.97	76.80	70.81
AS-Prompt	**62.23**	52.38	**71.79**	**64.47**	**75.39**	**69.77**	**77.74**	**72.98**	78.84	74.50	78.37	74.19

Restaurants

Model	4		16		64		256		1024		Full(3602)	
	Acc	F1	Acc	F1	Acc	F1	Acc	F1	Acc	F1	Acc	F1
BERT-ADA	-	-	-	-	-	-	-	-	-	-	87.17	80.05
BERT [CLS]	-	-	58.89	34.50	73.00	50.79	79.45	64.70	83.48	73.62	86.77	79.33
BERT NSP	-	-	61.05	32.46	74.73	53.00	79.34	65.51	83.61	74.15	87.09	79.98
BERT LM	71.78	48.16	71.99	56.65	77.79	63.30	81.10	69.27	85.12	76.60	**87.50**	**80.78**
Null Prompts	63.93	48.88	74.00	**62.85**	76.43	61.92	79.20	68.83	83.72	73.52	83.58	77.18
AS-Prompt	**71.82**	**52.09**	**74.05**	60.73	**78.16**	**63.42**	**82.09**	**69.69**	**85.59**	**76.74**	85.64	76.71

4.3 Settings

We implement our model in PyTorch and load scripts for our datasets to be compatible with the Huggingface datasets[2]. We use spaCy for POS tagging and pytokenizations[3] for tokenizer alignment. All experiments are running on NVIDIA GeForce RTX 3090 GPU. For our MLM task, we utilize the pre-trained weights obtained from the transformers library [26]. The main layers of BERT are left frozen while training and do not get any updates. We only finetune the parameters in continuous prompts to cut down computation costs. We evaluate our model with randomly re-sampled training sets of size {4, 16, 64, 256, 1024, Full}, and the dataset is split following [19] under the setting of full-shot training. The training epoch is set to 20. As for results, we perform macro F1 score and accuracy (Acc.) as metrics to measure the performance of models. The initial learning rate is set to 0.00002 and we varied this value during training to reduce the training loss below 0.00001.

4.4 Overall Results

Table 2 records the overall results of the proposed model and baselines. As we can see from the table, all the prompt-based methods generally outperform the non-prompt ones in all few-shot cases for both target domains. It indicates that the

---

[2] https://huggingface.co/docs/datasets/master/.
[3] https://github.com/tamuhey/tokenizations.

Table 3. Accuracies of proposed model trained with in-domain and cross-domain data.

In/Cross	16		Full	
	Laptops	Restaurants	Laptops	Restaurants
In	71.79	74.05	78.37	85.64
Cross	73.48	71.70	79.29	76.33

prompt-based methods can easily formulate the downstream task to MLM task and fully take advantage of the knowledge contained in the pre-trained model. All prompt-based methods but BERT-LM achieve relatively lower performance than non-prompt methods under the full-shot setting, which suggests that finetining methods are superior to prompt-based methods when there are sufficient training data.

Considering the prompt-based methods, our proposed continuous prompt method achieves better results than the discrete prompt method in majority of scenarios. It implies that a fixed prompt template is not as powerful as a continuous prompt in building the connection between the pre-trained model and the downstream task. Our method achieves significant improvement, especially in a few-shot setting. With the decrease in training data, the performance of the benchmark methods decays significantly, while our approach remains at a high level. Considerable performance in few-shot indicates the effectiveness of the proposed model. Note that results on Restaurants overall higher than those on Laptops, we can speculate that the pre-trained model contains more knowledge in Restaurants domain than Laptops.

4.5   Further Analysis

To analyze the factors that may affect the prompt-based methods and provide guidance for prompt-learning methods, we conduct four more experiments for further discussion.

Transfer Ability. To examine the transferability of the model, we train the model on the in-domain dataset and test it on the cross-domain dataset. As we can see from Table 3, our model achieves better results on cross-domain datasets for laptops dataset under both the 16-shot and the full-shot settings. Results suggest that the prompt-based method has a strong ability to adapt to a new domain with considerable performance.

In-Domain Data Pre-training. We explore the impact of in-domain information on our method. We retrain the model with the in-domain dataset (i.e., Amazon for laptops and Yelp for restaurants) and compare the performance with the original BERT. Results in Table 4 show that retrained BERT model achieves much better outcomings than the original BERT model under low-resource settings. Still, the gap between them on the full-shot dataset is neglectable. It suggests that the prompt-based method is more practical for the sizeable pre-trained model, which already contains sufficient knowledge of various domains.

Table 4. Model's performance using the original pre-trained weights (Original) and the weights further trained with domain-specific review texts (Amazon, Yelp).

Datasets	Corpora	4		16		64		256		1024		Full(1850)	
		Acc	F1	Acc	F1	Acc	F1	Acc	F1	Acc	F1	Acc	F1
Laptops	Original	55.80	46.33	64.11	59.65	68.59	61.19	75.08	68.49	77.59	72.40	77.16	72.24
	+Amazon	62.23	52.38	71.79	64.47	75.39	69.77	77.74	72.98	78.84	74.50	78.37	74.19
Restaurants	Original	68.26	44.54	73.21	59.36	77.07	61.09	79.55	68.26	84.88	74.47	84.64	76.17
	+Yelp	71.82	52.09	74.05	60.73	78.16	63.42	82.09	69.69	85.59	76.74	85.64	76.71

Table 5. The impact of finetuning of the pre-trained model on 16-shot and full shot settings.

Finetuning	16		Full	
	Laptops	Restaurants	Laptops	Restaurants
Prompt+Model	71.79	74.05	78.37	85.64
Prompt	62.23	59.73	71.47	77.73

Type of Finetuning. Our model finetunes the parameters in pre-trained BERT and continuous prompt simultaneously under the assumption that the training costs of few-shot data can be neglected. As is shown in Table 5, we further freeze the parameters in BERT and compare the results with ours. We can conclude that there is no need to adjust the parameters in the pre-trained model when the training data is limited. However, finetuning the whole parameter helps when enough training data is provided.

Impact of Aspect. Intuitively, a better-designed prompt can improve the performance to a great extent. The results in Table 6 verify our conjecture. Here, we replace the aspect word with 'things' in the prompt and compare the results with ours. Results show that well-designed prompt largely improves the performance of the model.

Table 6. The impact of aspect information in prompts on full-shot dataset.

Prompt	Laptops		Restaurants	
	Acc	F1	Acc	F1
Prompt(aspect)	78.37	74.19	85.64	76.71
Prompt(things)	74.16	70.49	78.80	67.95

5   Conclusion

In this paper, we model the ASC task as LM and test the performance of our aspect-specific prompt learning model under the few-shot and the full supervised settings. Results demonstrate that the prompt learning method can achieve considerable performance on few-shot data while reducing the training cost of large pre-trained models. Additionally, we reveal that the prompt-based approach is more practical to transfer to a new domain, and sufficient domain-specific knowledge contained in pre-trained model greatly improves the model's performance under the few-shot setting. In future work, since the prompt learning method can easily adapt to classification tasks and extraction tasks, it is possible to find a unified model to solve all subtasks of aspect-based sentiment classification based on prompt learning.

References

1. Beigi, O.M., Moattar, M.H.: Automatic construction of domain-specific sentiment lexicon for unsupervised domain adaptation and sentiment classification. Knowl. Based Syst. **213**, 106423 (2021)
2. Ben-David, E., Oved, N., Reichart, R.: PADA: a prompt-based autoregressive approach for adaptation to unseen domains. arXiv preprint arXiv:2102.12206 (2021)
3. Brown, T.B., et al.: Language models are few-shot learners. In: Proceedings of NeurIPS, pp. 1877–1901 (2020)
4. Cao, Z., Zhou, Y., Yang, A., Peng, S.: Deep transfer learning mechanism for fine-grained cross-domain sentiment classification. Connect. Sci. **33**(4), 911–928 (2021)
5. Devlin, J., Chang, M., Lee, K., Toutanova, K.: BERT: pre-training of deep bidirectional transformers for language understanding. In: Proceedings of NAACL, pp. 4171–4186 (2019)
6. Gao, T., Fisch, A., Chen, D.: Making pre-trained language models better few-shot learners. In: Proceedings of ACL, pp. 3816–3830 (2021)
7. Hambardzumyan, K., Khachatrian, H., May, J.: WARP: word-level adversarial reprogramming. In: Proceedings of ACL, pp. 4921–4933 (2021)
8. Heinzerling, B., Inui, K.: Language models as knowledge bases: on entity representations, storage capacity, and paraphrased queries. In: Proceedings of EACL, pp. 1772–1791 (2021)
9. Howard, J., Ruder, S.: Universal language model fine-tuning for text classification. In: Gurevych, I., Miyao, Y. (eds.) Proceedings of the 56th Annual Meeting of the Association for Computational Linguistics, ACL 2018, Melbourne, Australia, 15–20 July 2018, Volume 1: Long Papers, pp. 328–339. Association for Computational Linguistics (2018)
10. Hu, M., Liu, B.: Mining and summarizing customer reviews. In: Proceedings of ACM SIGKDD, pp. 168–177 (2004)
11. Jiang, Z., Xu, F.F., Araki, J., Neubig, G.: How can we know what language models know. Trans. Assoc. Comput. Linguist. **8**, 423–438 (2020)
12. Li, L., Liu, Y., Zhou, A.: Hierarchical attention based position-aware network for aspect-level sentiment analysis. In: Proceedings of CoNLL, pp. 181–189 (2018)
13. Li, X.L., Liang, P.: Prefix-tuning: optimizing continuous prompts for generation. In: Proceedings of ACL, pp. 4582–4597 (2021)
14. Li, Z., Qin, Y., Liu, Z., Wang, W.: Powering comparative classification with sentiment analysis via domain adaptive knowledge transfer. arXiv preprint arXiv:2109.03819 (2021)
15. Liu, X., et al.: GPT understands, too. arXiv preprint arXiv:2103.10385 (2021)
16. Logan, R.L., IV., Balažević, I., Wallace, E., Petroni, F., Singh, S., Riedel, S.: Cutting down on prompts and parameters: simple few-shot learning with language models. arXiv preprint arXiv:2106.13353 (2021)

17. Ni, J., Li, J., McAuley, J.J.: Justifying recommendations using distantly-labeled reviews and fine-grained aspects. In: Proceedings of EMNLP, pp. 188–197 (2019)
18. Pontiki, M., Galanis, D., Pavlopoulos, J., Papageorgiou, H., Androutsopoulos, I., Manandhar, S.: Semeval-2014 task 4: aspect based sentiment analysis. In: Proceedings of COLING, pp. 27–35 (2014)
19. Rietzler, A., Stabinger, S., Opitz, P., Engl, S.: Adapt or get left behind: domain adaptation through BERT language model finetuning for aspect-target sentiment classification. In: Proceedings of LREC, pp. 4933–4941 (2020)
20. Schick, T., Schütze, H.: Exploiting cloze-questions for few-shot text classification and natural language inference. In: Proceedings of EACL, pp. 255–269 (2021)
21. Schick, T., Schütze, H.: It's not just size that matters: small language models are also few-shot learners. In: Proceedings of NAACL, pp. 2339–2352 (2021)
22. Seoh, R., Birle, I., Tak, M., Chang, H., Pinette, B., Hough, A.: Open aspect target sentiment classification with natural language prompts. In: Proceedings of EMNLP, pp. 6311–6322 (2021)
23. Shin, T., Razeghi, Y., IV., R.L.L., Wallace, E., Singh, S.: AutoPrompt: eliciting knowledge from language models with automatically generated prompts. In: Proceedings of EMNLP, pp. 4222–4235 (2020)
24. Sun, C., Huang, L., Qiu, X.: Utilizing BERT for aspect-based sentiment analysis via constructing auxiliary sentence. In: Proceedings of NAACL, pp. 380–385 (2019)
25. Tang, H., Ji, D., Li, C., Zhou, Q.: Dependency graph enhanced dual-transformer structure for aspect-based sentiment classification. In: Proceedings of ACL, pp. 6578–6588 (2020)
26. Wolf, T., et al.: HuggingFace's transformers: state-of-the-art natural language processing. arXiv preprint arXiv:1910.03771 (2019)
27. Xu, H., Liu, B., Shu, L., Yu, P.S.: BERT post-training for review reading comprehension and aspect-based sentiment analysis. In: Proceedings of NAACL, pp. 2324–2335 (2019)
28. Yan, H., Dai, J., Ji, T., Qiu, X., Zhang, Z.: A unified generative framework for aspect-based sentiment analysis. In: Proceedings of ACL, pp. 2416–2429 (2021)
29. Yin, W., Hay, J., Roth, D.: Benchmarking zero-shot text classification: datasets, evaluation and entailment approach. In: Proceedings of EMNLP, pp. 3912–3921 (2019)
30. Zhao, C., Wang, S., Li, D., Liu, X., Yang, X., Liu, J.: Cross-domain sentiment classification via parameter transferring and attention sharing mechanism. Inf. Sci. **578**, 281–296 (2021)

Multi-knowledge Embeddings Enhanced Topic Modeling for Short Texts

Jueying He, Jiayao Chen, and Mark Junjie Li[✉]

College of Computer Science and Software Engineering, Shenzhen University, Shenzhen, China
`hejueying2020@email.szu.edu.cn`, `jj.li@szu.edu.cn`

Abstract. Topic models are widely used to extra the latent knowledge of short texts. However, due to data sparsity, traditional topic models based on word co-occurrence patterns have trouble achieving accurate results on short texts. Researchers have recently proposed knowledge-based topic models for short texts to discover more coherent and meaningful topics. Every form of knowledge aims at representing specific and oriented information but is not wide-ranging. Single knowledge-enhanced topic models only take a form of knowledge into count, which is restricted and undesirable. The more forms of knowledge we incorporate, the more comprehensive our understanding of the short text. In this paper, we propose a novel short texts topic model, named MultiKE-DMM, which combines multiple forms of knowledge and the generalized Pólya urn (GPU) model with Dirichlet Multinomial Mixture (DMM) model. The proposed approach boosts the multi-knowledge background-related words under the same topic. Access to multi-form knowledge permits the creation of an intelligent topic modelling algorithm that considers semantic and fact-oriented relationships between words, offering improved performance over four comparison models on four real-world short text datasets.

Keywords: Topic Model · Short Texts · Word Embedding · Entity Embedding

1 Introduction

Short texts have become a prevalent form of information on the Internet. Snippets and news headlines on traditional media platforms, or Weibo, status updates, and personal profiles in emerging social media are examples of short texts. Content analysis is an important tool for many NLP tasks including recommending system [2], user interest profiling [4] and topic detection [5].

Topic models are widely used to extra the information of contextual content. Traditional topic models such as pLSA [6] and LDA [1] are proposed to discover the latent topics of documents. In these models, the documents are represented as a multinomial distribution over topics. Also, the topics are represented as a multinomial distribution over words. Based on word co-occurrence patterns,

© The Author(s), under exclusive license to Springer Nature Switzerland AG 2023
M. Tanveer et al. (Eds.): ICONIP 2022, LNCS 13625, pp. 521–532, 2023.
https://doi.org/10.1007/978-3-031-30111-7_44

statistical techniques [11] are utilized to learn the document-topic distribution and the topic-word distribution. However, due to data sparsity on short texts, directly employing these traditional topic models based on word co-occurrence patterns has trouble in achieving accurate results on short texts.

A simple solution is to alleviate the sparsity problem by aggregating short texts into lengthy pseudo-documents and then applying a standard topic model. The aggregation is guided by auxiliary contextual information [13,14]. However, the auxiliary information is not always available or just too costly for employers. Another way to eliminate the drawback of short texts is to simplify the assumption that each short text only belongs to a single topic, known as Dirichlet Multinomial Mixture (DMM) model [17,19]. Due to the limited content of short texts, this assumption is reasonable to deal with the data sparsity problem of short texts. But the problem of minimal word co-occurrence information in short texts has not been solved yet. The third strategy is incorporating prior knowledge into topic models to guide the modelling process. When a human being interprets a piece of text, the understanding is not only based on its contextual content but also on its background knowledge. For example, GPU-DMM [7] extends DMM and incorporates the learned word embeddings trained from large text corpus through the generalized Pólya urn (GPU) model [8] in topic inferences.

The characteristic of existing works is that these models primarily rely on a single form of knowledge to represent the content of short texts, which is restricted and undesirable. Embedded knowledge is only composed of words in a specific aspect rather than a comprehensive part. When a human interprets a text, the understanding is based not only on a particular form of knowledge but also on multi-form knowledge.

We propose a novel topic model named MultiKE-DMM. MultiKE-DMM incorporates knowledge of entity embedding and pre-trained word embeddings and extends Dirichlet Multinomial Mixture through generalized Pólya urn model in topic inferences. Access to multi-form knowledge permits the creation of an intelligent topic modelling algorithm that considers semantic relatedness and ontological relationships between words, offering improved performance. MultiKE-DMM can enhance more coherent topics on four real-world datasets and achieve better classification accuracy than existing state-of-the-art alternatives.

2   Related Work

Knowledge Enhanced Topic Models. Some previous works have contributed to knowledge-enhanced topic modelling. These models take advantage of different forms of knowledge to guide the modelling process. Researchers widely adopt knowledge bases and graphs. KGE-LDA [18] utilizes embedding entities in WordNet and relations into the Dirichlet multinomial topic model to capture the linguistic relations of words with better interpretation. WCM-LDA [16] not only models the relationship between words and topics but also utilizes concept and category knowledge of entities to model the semantic relation of entities and topics.

Unlike single-knowledge enhanced models, the proposed approach focuses on combining various knowledge forms to improve the quality of the topic generated by the topic model.

Dirichlet Multinomial Mixture Based Topic Models. DMM [19] model is a simple and effective topic-modelling method for short texts. It follows a reasonable assumption that each short text is sampled from only one topic, and all words in the short text are assigned to this topic. Various models take advantage of DMM's simplicity, and effectiveness and those models are improved based on DMM model. LF-DMM [9] is built on DMM and replaces the topic-word multinomial distribution with a two-component mixture of a Dirichlet multinomial component and a continuous word embedding component.

In terms of its simplicity and effectiveness, we develop our proposed model based on DMM as the name implies.

Generalized Pólya urn Guided Topic Models. The generalized Pólya urn (GPU) model is widely used in topic models to guide topics' semantic similarity during the inference process directly. GPU approach was proposed to topic modelling in [8] where GPU concentrates words with high co-document frequency based on corpus-specific co-occurrence statistics. But in that model, researchers did not use any prior knowledge in their model. Considering the adverse effect of using prior domain knowledge in the topic model, MDK-LDA [3] incorporates GPU to deal with this problem by increasing the counts of rare words in the knowledge sets.

3   MultiKE-DMM

MutliKE-DMM model incorporates generalized Pólya urn (GPU) model into Dirichlet Multinomial Mixture (DMM) model with multi-knowledge weighted embedding to enhance the model's ability to mine more coherent topics. The multi-knowledge embedding of words is combined with two forms of knowledge embedding learned from external information, which helps alleviate the data sparsity problem in short texts. During the inference process, highly related words with similar multi-knowledge backgrounds boost the Gibbs Sampling algorithm and promote similar words assigned to the same topic by utilising GPU model, which permits the extraction of highly coherent topics. The architecture of MultiKE-DMM is shown in Fig. 1. In the following part, we describe MultiKE-DMM in detail.

3.1   Dirichlet Mixture Multinomial Model

Dirichlet Mixture Multinomial model is a novel approach to infer the topics of short texts with a reasonable assumption that each document is sampled by only one topic and all words in the document are assigned to this topic.

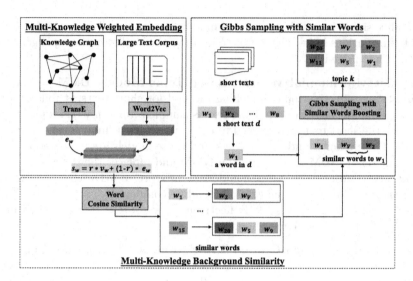

Fig. 1. MultiKE-DMM Overview

Given a short text corpus of D documents, with a vocabulary W, and K pre-defined latent topics, a document d associated with a special topic k has n_d words, which is represented as $\{w_{d,1}, w_{d,2}, ..., w_{d,n_d}\}$.

DMM chooses Dirichlet distribution for document-topic distribution Θ and topic-word distribution Φ as prior distribution with parameter α and β.

The generative process for DMM is described as follows.

1. Sample a topic proportion $\Theta \sim Dirichlet(\alpha)$
2. For each topic $k \in \{1, ..., K\}$
   (a) Sample a topic-word distribution $\Phi_k \sim Dirichlet(\beta)$
3. For each document $d \in \{1, ..., D\}$
   (a) Sample a topic $z_d \sim Multinomial(\Theta)$
   (b) For each word $w \in \{w_{d,1}, w_{d,2}, ..., w_{d,n_d}\}$
       Sample a word $w \sim Multinomial(\Phi_{z_d})$

Following the approach in DMM, the Gibbs sampler is given as follows.

$$p(z_d = k | \mathbf{Z}_{\neg d}, \mathbf{D}) \propto$$
$$\frac{m_{k,\neg d} + \alpha}{D - 1 + K\alpha} \times \frac{\prod_{w \in W} \prod_{j=1}^{n_d^w} (n_{k,\neg d}^w + \beta + j - 1)}{\prod_{i=1}^{n_d} (n_{k,\neg d} + V\beta + i - 1)} \tag{1}$$

3.2 Multi-knowledge Weighted Embedding

The main task of the topic model on short texts is to infer the latent topic by statistical techniques based on higher-order word cooccurrence patterns. Due to the limited content of short texts, topic models on short texts suffer from the data sparsity problem and lead to performance degradation.

Effective knowledge representation methods are now feasible with the development of neural network language models such as Word2Vec and TransE. In terms of the purpose of these models, word embeddings can map corresponding values closer together in embedding space, although words do not frequently co-occur in documents. On the other hand, when a human being interprets a piece of short text, the understanding is not solely based on its content words but also on its background knowledge. Therefore, it is natural to incorporate auxiliary knowledge representations into the topic model, which helps deal with the limited word cooccurrence in short texts and strengthens the model's ability to infer more coherent topics.

Every form of knowledge background represents specific information rather than wide-ranging information. It is restricted and undesirable to understand the content of the document comprehensively. The more forms of knowledge we incorporate, the more comprehensive our understanding of the short text. Motivated by this purpose, we combine word embedding and entity embedding related words to form the multiple knowledge background embedding of words, which is named as Multi-Knowledge Weighted Embedding (MKWE). Given the word embedding v_w and entity embedding e_w related to word w, the MKWE s_w is shown in Eq. 2. This method holds that different forms of knowledge embedding contribute differently to the comprehensive understanding of the short text. The contribution weight is described by r, which affects the ability of the model to generate coherent topics.

$$s_w = r \times v_w + (1 - r) \times e_w \tag{2}$$

3.3   Incorporating MKWE by Generalized Pólya urn Model

In this part, we present how MultiKE-DMM improves the performance of the topic model on short texts when it is integrated with both multi-knowledge embedding and generalized Pólya urn model.

Due to the power-law characteristics of natural language, most words are rare and will not co-occur with most other words regardless of their semantic similarity [20]. To encode this characteristic and assign highly similar words to the same topic, we use the generalized Pólya urn model. The generalized Pólya urn model is that when a ball with a certain colour is sampled from the urn, two same-colour balls are put back along with several balls with a similar colour. When the additional balls with similar colours are added to the urn, it increases their proportions in the urn, which is called "promotion". In our case, when a word in a topic is sampled during the inference process, highly similar words are put back to the same topic, which promotes the probability that words with high relatedness share the same topic.

Inspired by the promotion process, we take GPU model and the multi-knowledge embedding of words together to achieve the improvement. Given the multi-knowledge embedding of words, we can score the multiple knowledge background relatedness between words by the cosine similarity function. The multiple knowledge background relatedness between word w_i and word w_j with their corresponding multi-knowledge embedding s_i, s_j can calculated by Eq. 3. If $score(w_i, w_j) \geqslant \sigma$, w_i and w_j can be viewed as the word pair with high multi-knowledge relatedness where a threshold σ is to find out words with high multi-knowledge background similarity.

$$score(w_i, w_j) = \frac{s_i^\top \cdot s_j}{||s_i|| \, ||s_j||} \tag{3}$$

Then we construct a boost matrix $\mathbb{B}$ where $\mathbb{B}_{w_i, w_j}$ denotes the promotion of w_j when working on word w_i during the inference process. In this way, the boost matrix $\mathbb{B}$ of word pairs can be obtained via Eq. 4, where the hyper-parameter of promotion for each word pair with higher multi-knowledge relatedness is denoted as μ.

$$\mathbb{B}_{w_i, w_j} = \begin{cases} 1, & if & w_i = w_j \\ \mu, & if & score(w_i, w_j) \geqslant \sigma \text{ and } w_i \neq w_j \\ 0, & if & score(w_i, w_j) < \sigma \end{cases} \tag{4}$$

3.4   Collapsed Gibbs Sampling with Similar Words

We find highly similar word pairs and construct a boots matrix, which promotes identical words assigned to the same topic during the inference process. We propose the Gibbs sampling with GPU model to infer the variables in detail. The GPU model is non-exchangeable, indicating that the joint probability of the words assigned to a topic is not invariant to the permutation of those words. The Gibbs sampler can be more complex because of the non-exchangeable of words. We apply the approach [8], which approximates the actual Gibbs sampling distribution by viewing each word as if it was the last, ignoring its implications for the following words and their topic assignments. The approximate Gibbs sampler with GPU model has the conditional distribution in Eq. 5.

$$p(z_d = k | \mathbf{Z}_{\neg d}, \mathbf{D}) \propto$$

$$\frac{m_{k, \neg d} + \alpha}{D - 1 + K\alpha} \times \frac{\prod_{w \in W} \prod_{j=1}^{n_d^w} (\tilde{n}_{k, \neg d}^w + \beta + j - 1)}{\prod_{i=1}^{n_d} (\tilde{n}_{k, \neg d} + V\beta + i - 1)} \tag{5}$$

where $\tilde{n}_{k, \neg d}^w$ is the number of word w in the topic k, $\tilde{n}_{k, \neg d}$ is the number of words assigned to topic k and $m_{k, \neg d}$ is the number of documents belong to topic k. $\neg d$ is that document d is excluded from the counting.

4   Experiment

To evaluate MultiKE-DMM, we conduct experiments on some real datasets to compare with several comparison models. The main tasks of topic models are to learn the document-topic distribution and topic-word distribution. So we choose PMI (Pointwise Mutual Information) and text classification accuracy as two metrics to evaluate the model correspondingly. For topic coherence evaluation, a higher coherence score of PMI indicates that the word assigned to a topic is more relevant to others. For text classification, a high classification accuracy means the topic assignment for each document is reasonable.

4.1   Experiment Setup

External Knowledge. For word embedding, we use 300-dimensional word embeddings from Google's Word2Vec. For Entity embedding, we crawl the entities and their relationships from the knowledge graph of WordNet and adopt TransE model to learn the embedding of entities. Then we link words to entities in WordNet via NLTK and find out entity embeddings related to words. If a word has both its related word embedding and entity embedding, its multi-knowledge embedding is combined with two different forms of knowledge embedding. If a word only has a single form of knowledge embedding, its multi-knowledge embedding is set as its corresponding owned knowledge embedding. Otherwise, a word has no form of knowledge embedding; the word is considered as having no knowledge background similarity with other words.

Datasets. For SearchSnippets [10], Snippets belongs to 8 domains, which are Business, Computers, Culture-Arts, Education-Science, Engineering, Politics-Society, Sports and Health respectively. GoogleNews dataset [19] is downloaded from the Google news site and crawled the titles and snippets of 11109 news articles belonging to 152 clusters. StackOverflow [15] is a public topic model dataset on Kaggle.com and we randomly select 16407 question titles from 20 different labels. The Tweet dataset [21] consists of 2472 tweets that are highly related to 89 queries. The relevance between tweets and queries is manually labelled in the 2011 and 2012 microblog tracks at TREC. The detail statistics information of the two datasets show in Table 1.

Table 1. Statistics on the four datasets. Label: the number of true labels; Doc: the total number of documents; Len: the average length of each document.

Dataset	Label	Doc	Len	Vocab
SearchSnippets	8	12340	10.72	5581
GoogleNews	152	11108	6.23	8110
StackOverflow	20	16407	4.88	2200
Tweet	89	2472	8.56	5098

Baselines. LDA [1] is a traditional probabilistic model that learns latent topics from documents and words, by using Dirichlet prior to regularize the topic distributions. DMM [19] proposes a reasonable assumption that each short text owns only a single topic and all words in the document are assigned to the same topic. BTM [17] mines the topics by directly modelling the generation of word co-occurrence patterns in the whole short texts corpus. In BTM, a biterm is an unordered word pair co-occurrence in a short context. GPU-DMM [7] uses the generalized Pólya urn model with word embedding to promote the co-occurrence of semantically related words under the same topic, where the similarity is calculated by word embedding with cosine distance.

Parameter Setting. For α and β, we set $\alpha = \frac{K}{50}$ and $\beta = 0.01$ on DMM, BTM, GPU-DMM and our proposed model. We use the settings with $\alpha = 0.05$ and $\beta = 0.01$ of LDA. For **the boost amount** μ, we set the boost amount as 0.3 of all the models. For **the threshold** ϵ, we set ϵ as 0.7 on SearchSnippets, GoogleNews and Tweet. For StackOverflow, we set ϵ as 0.8. For **the contribution weight** r, we set r as 0.5 on StackOverflow and Tweet. For SearchSnippets, we set r as 0.1. For GoogleNews, we set r as 0.9. For **iteration and round**, to make the comparison equivalent, we run Gibbs sampling process of each model in the experiment for 1000 times and report the average results over 5 round.

4.2   Evaluation Metrics

We use the Pointwise Mutual Information (PMI) Score to evaluate topic coherence. It is a measure of whether the word distribution in a topic is consistent in an external corpus. A higher PMI-Score indicates more coherent topics. For SearchSnippets with 8 labels and StackOverflow with 20 labels, we set K as $\{20, 40, 60, 80\}$. For GoogleNews with 152 labels and Tweet with 89 labels, we set K as $\{100, 150, 200, 250\}$. We set T as 5 for all models.

We choose classification accuracy as a metric for classification and a higher accuracy score demonstrates the learnt topics of documents are more representative. We split the original dataset into the training dataset and test dataset. The training dataset with 80% is used to train the classifier and the other dataset is set as the test dataset to evaluate the accuracy. A linear kernel Support Vector Machine (SVM) classifier in LIBLINEAR is used with the default parameter set on this paper.

4.3   Result Analysis

The average PMI Topic Coherence of all models on four datasets is plotted at Table 2, 3, 4, 5. BTM and DMM perform slightly better than LDA, which means topic models based on the simple assumption can improve topic interpretability in short texts, but the improvements are not significant. Knowledge-enhanced topic models, such as GPU-DMM and MultiKE-DMM, outperform BTM, DMM and LDA on all short text datasets because incorporating external knowledge into the topic models on short texts can help models have a more comprehensive understanding of short texts. GPU-DMM perform well on GoogleNews and Tweet datasets but not on SearchSnippets and StackOverflow, which means word embedding trained from the Google News corpus can help short text topic modelling in the general domain, but may not use for a specific domain such as the business domain in SearchSnippets and technical topic in StackOverflow. MultiKE-DMM performs competitively against GPU-DMM on SearchSnippets and has a slightly better performance than that of GPU-DMM on StackOverflow. This is mostly because MultiKE-DMM uses the added knowledge in WordNets that focuses on lexical relation of words and incorporates with word semantic correlations by pre-trained word embedding, which suggests the combination can have a more improved effect on the topic coherence on specific domain datasets.

Table 2. Topic Coherence on SearchSnippets

Model	LDA	BTM	DMM	GPU-DMM	MultiKE-DMM
K=20	1.3674	1.3535	1.3324	1.7374	**2.1254**
K=40	1.402	1.5377	1.4731	1.8578	**2.1807**
K=60	1.4701	1.5603	1.5198	1.9244	**2.1705**
K=80	1.421	1.588	1.5385	1.9658	**2.1462**

Table 3. Topic Coherence on StackOverflow

Model	LDA	BTM	DMM	GPU-DMM	MultiKE-DMM
K=20	1.5392	1.5057	1.5527	**1.9305**	1.894
K=40	1.4239	1.2496	1.5049	1.7591	**1.7745**
K=60	1.3867	1.4012	1.448	1.6742	**1.7233**
K=80	1.5392	1.3466	1.448	1.6522	**1.7745**

Table 4. Topic Coherence on GoogleNews

Model	LDA	BTM	DMM	GPU-DMM	MultiKE-DMM
K=100	1.3801	1.474	1.4626	1.5672	**1.5801**
K=150	1.2023	1.4696	1.4487	1.5722	**1.5763**
K=200	1.3226	1.4316	1.4249	1.5499	**1.5509**
K=250	1.3039	1.3999	1.3804	1.5239	**1.5252**

Table 6 plots the average classification accuracy of models on four datasets. Although the performance is highly dependent on the dataset, DMM-based models which incorporate the knowledge representation outperform others on Tweet and GoogleNews. It's because GoogleNews and Tweet are general datasets, and word embeddings used in our model are trained in general datasets. The accuracy of classification tasks has been significantly improved on Tweet and Google-News, which is in line with the expectations of MultiKE-DMM. GPU-DMM

Table 5. Topic Coherence on Tweet

Model	LDA	BTM	DMM	GPU-DMM	MultiKE-DMM
K=100	1.2536	1.3333	1.3832	**1.6217**	1.5996
K=150	1.2329	1.1032	1.3642	**1.5263**	1.5205
K=200	1.1973	1.2924	1.35	1.4473	**1.4673**
K=250	1.1561	1.2687	1.3523	1.4191	**1.4358**

and MultiKE-DMM exploit external knowledge in the topic modelling process on Searchsnippets and StackOverflow, but MultiKE-DMM has a better result. The possible explanations are as follows. On the one hand, this is because of the characteristics of the dataset. Different from Tweet and GoogleNews, which are general datasets, SearchSnippet and StackOverflow are datasets focusing on a specific domain. On the other hand, the external knowledge representation that the two models use may also affect the result. GPU-DMM emphasises the

Table 6. Classification Accuracy of all models on four datasets

Dataset	Model	K=20	K=40	K=60	K=80
SearchSnippets	LDA	0.5734	0.5064	0.4719	0.4467
	BTM	0.7221	0.6495	0.6576	0.6013
	DMM	**0.7693**	0.7003	0.6698	0.6547
	GPU-DMM	0.7381	0.6697	0.6174	0.6192
	MultiKE-DMM	0.7348	**0.7693**	**0.7227**	**0.6785**
StackOverflow	LDA	0.4714	0.5767	0.6028	0.6207
	BTM	0.5652	0.652	0.6674	0.6771
	DMM	0.5711	**0.7090**	**0.7339**	**0.7611**
	GPU-DMM	0.5462	0.6534	0.6923	0.7087
	MultiKE-DMM	**0.5737**	0.6669	0.6832	0.6962
Dataset	Model	K=100	K=150	K=200	K=250
GoogleNews	LDA	0.7469	0.8077	0.8320	0.8491
	BTM	0.7841	0.8416	0.8764	0.8848
	DMM	0.8196	0.8801	0.8933	0.9085
	GPU-DMM	**0.8410**	0.8827	**0.9032**	**0.9151**
	MultiKE-DMM	0.8359	**0.8841**	0.9008	0.9134
Tweet	LDA	0.8486	0.8814	0.8838	0.8854
	BTM	0.8130	0.8316	0.8571	0.8530
	DMM	0.8745	0.9057	0.9219	0.9117
	GPU-DMM	**0.9000**	**0.9271**	**0.9275**	**0.9283**
	MultiKE-DMM	0.8968	0.9227	0.9255	0.9328

semantic relation between words learned from general dataset. At the same time, MultiKE-DMM addedly focuses on conceptual-semantic and cross-words lexical relations, and the latter may have a more significant effect on the classification accuracy of Searchsnippets and StackOverflow. If MultiKE-DMM tries to use pre-trained word embedding on domain-specific datasets, we may further improve the performance. These observations validate that incorporating words multiple knowledge background relation is beneficial for short text topic modelling on general or domain-specific datasets.

5   Conclusion

We propose a novel model topic model for short texts named MultiKE-DMM in this paper. By incorporating the GPU model into DMM, MultiKE-DMM successfully overcomes the sparsity problem of short texts. MultiKE-DMM utilizes multiple forms of knowledge to form multi-knowledge weighted embedding of words and combines generalized Pólya urn model with Multinomial Mixture model to boost highly similar words under the same topic. In this way, MultiKE-DMM uses abundant knowledge background to generate more coherent topics and make classification more accurate. Though the classification accuracy is dependent on dataset [12], MultiKE-DMM gets improved performance on four real-world datasets, which validates that incorporating word multi-knowledge background relation is beneficial for short text topic modelling.

Acknowledgement. This research work has been supported by the National key R & D program project (No. 2019YFC1605504).

References

1. Blei, D.M., Ng, A.Y., Jordan, M.I.: Latent Dirichlet allocation. J. Mach. Learn. Res. **3**(Jan), 993–1022 (2003)
2. Cami, B.R., Hassanpour, H., Mashayekhi, H.: User preferences modeling using Dirichlet process mixture model for a content-based recommender system. Knowl.-Based Syst. **163**, 644–655 (2019)
3. Chen, Z., Mukherjee, A., Liu, B., Hsu, M., Castellanos, M., Ghosh, R.: Leveraging multi-domain prior knowledge in topic models. In: Twenty-Third International Joint Conference on Artificial Intelligence (2013)
4. Dhelim, S., Aung, N., Ning, H.: Mining user interest based on personality-aware hybrid filtering in social networks. Knowl.-Based Syst. **206**, 106227 (2020)
5. Garcia, K., Berton, L.: Topic detection and sentiment analysis in twitter content related to Covid-19 from brazil and the USA. Appl. Soft Comput. **101**, 107057 (2021)
6. Hofmann, T.: Probabilistic latent semantic indexing. In: Proceedings of the 22nd Annual International ACM SIGIR Conference on Research and Development in Information Retrieval, pp. 50–57 (1999)
7. Li, C., Wang, H., Zhang, Z., Sun, A., Ma, Z.: Topic modeling for short texts with auxiliary word embeddings. In: Proceedings of the 39th International ACM SIGIR Conference on Research and Development in Information Retrieval, pp. 165–174 (2016)

8. Mimno, D., Wallach, H., Talley, E., Leenders, M., McCallum, A.: Optimizing semantic coherence in topic models. In: Proceedings of the 2011 Conference on Empirical Methods in Natural Language Processing, pp. 262–272 (2011)

9. Nguyen, D.Q., Billingsley, R., Du, L., Johnson, M.: Improving topic models with latent feature word representations. Trans. Assoc. Comput. Linguist. **3**, 299–313 (2015)

10. Phan, X.H., Nguyen, M.L., Horiguchi, S.: Learning to classify short and sparse text & web with hidden topics from large-scale data collections. In: Proceedings of the 17th International Conference on World Wide Web, WWW 2008, Beijing, China, 21–25 April 2008, pp. 91–100. ACM (2008)

11. Porteous, I., Newman, D., Ihler, A., Asuncion, A., Smyth, P., Welling, M.: Fast collapsed Gibbs sampling for latent Dirichlet allocation. In: Proceedings of the 14th ACM SIGKDD International Conference on Knowledge Discovery and Data Mining, pp. 569–577 (2008)

12. Qiang, J., Qian, Z., Li, Y., Yuan, Y., Wu, X.: Short text topic modeling techniques, applications, and performance: a survey. IEEE Trans. Knowl. Data Eng. **34**(3), 1427–1445 (2020)

13. Rajani, N.F.N., McArdle, K., Baldridge, J.: Extracting topics based on authors, recipients and content in microblogs. In: Proceedings of the 37th International ACM SIGIR Conference on Research & Development in Information Retrieval, pp. 1171–1174 (2014)

14. Wang, Y., Liu, J., Qu, J., Huang, Y., Chen, J., Feng, X.: Hashtag graph based topic model for tweet mining. In: 2014 IEEE International Conference on Data Mining, pp. 1025–1030. IEEE (2014)

15. Xu, J., et al.: Short text clustering via convolutional neural networks. In: Proceedings of the 1st Workshop on Vector Space Modeling for Natural Language Processing, VS@NAACL-HLT 2015, Denver, Colorado, USA, 5 June 2015, pp. 62–69. The Association for Computational Linguistics (2015)

16. Xu, K., Qi, G., Huang, J., Wu, T.: Incorporating Wikipedia concepts and categories as prior knowledge into topic models. Intell. Data Anal. **21**(2), 443–461 (2017)

17. Yan, X., Guo, J., Lan, Y., Cheng, X.: A biterm topic model for short texts. In: Proceedings of the 22nd International Conference on World Wide Web, pp. 1445–1456 (2013)

18. Yao, L., et al.: Incorporating knowledge graph embeddings into topic modeling. In: Thirty-First AAAI Conference on Artificial Intelligence (2017)

19. Yin, J., Wang, J.: A Dirichlet multinomial mixture model-based approach for short text clustering. In: Proceedings of the 20th ACM SIGKDD International Conference on Knowledge Discovery and Data Mining, pp. 233–242 (2014)

20. Zipf, G.K.: Selected studies of the principle of relative frequency in language. In: Selected Studies of the Principle of Relative Frequency in Language. Harvard University Press (2013)

21. Zubiaga, A., Ji, H.: Harnessing web page directories for large-scale classification of tweets. In: Proceedings of the 22nd International Conference on World Wide Web, pp. 225–226 (2013)

Adaptive Early Classification of Time Series Using Deep Learning

Anshul Sharma[1]([✉]), Saurabh Kumar Singh[2], Abhinav Kumar[3],
Amit Kumar Singh[4], and Sanjay Kumar Singh[3]

[1] The LNM Institute of Information Technology, Jaipur, Rajasthan, India
`anshul@lnmiit.ac.in`
[2] Tezpur University, Tezpur, Assam, India
[3] Indian, Institute of Technology (BHU) Varanasi, Varanasi, India
{`abhinav.rs.cse17,sks.cse`}`@iitbhu.ac.in`
[4] National Institute of Technology Patna, Patna, India
`amit.singh@nitp.ac.in`

Abstract. Early Classification of Time Series (ECTS) is a process of predicting the class label of time series at the earliest without observing the complete sequence. Time Series data is a collection of data points over time, and a decision has been made based on a complete sequence. However, early decision based on partial information is beneficial in time-sensitive applications. ECTS is an emerging research area with multiple applications in various domains such as health and disease prediction in medicine, Quality and Process Monitoring in Industry, Drought and Crop monitoring in agriculture. In this paper, we propose an adaptive early classification model composed of two components. The first component is the base classifier, which has been designed as a hybrid model of Convolutional Neural Network and Recurrent Neural Network. The Second component is the decision policy designed for adaptive halting capabilities, which has been defined as a reinforcement learning agent to determine when to stop and make a prediction. We evaluated our model on publicly available different kinds of time-series datasets. The proposed method outperformed the state-of-the-art in terms of both accuracy and earliness.

Keywords: Time Series · Early Classification · Deep Learning · RNN · CNN · Reinforcement Learning

1 Introduction

Classification of time series is a prominent problem in temporal data analysis. Time series is the sequence of observations collected/measured over time. Many time series classification algorithms have emerged and offer cutting-edge solutions [1]. The primary aim of time series classification is to predict the class level of given time series. In the traditional classification approach, time series are classified when the complete sequence becomes available. However, early decisions

M. Tanveer et al. (Eds.): ICONIP 2022, LNCS 13625, pp. 533–542, 2023.
https://doi.org/10.1007/978-3-031-30111-7_45

based on partial information are highly beneficial in time-sensitive applications, e.g., detecting high-risk events like a market crisis or earthquakes in advance. Applications of early classification are found in many areas, including agriculture [2], industries [3,4], medicine [5], Security [6] and transportation [7].

ECTS aims to classify the time series as early as possible based on incomplete sequences while ensuring an acceptable level of accuracy [8,9]. The problem of ECTS has two conflicting objectives accuracy and earliness. If we have more observations, we are more reliable about the prediction and vice-versa. Thus, the earliness of prediction is often inversely related o the accuracy, and balancing the trade-off between these two makes this problem challenging.

Rodríguez Diez and Alonso Gonzalez [10] introduced the problem of ECTS first time, and since then, the problem of ECTS has gained popularity among the research community [2,9,11–14]. They used relative and region-based predicates on segmented intervals from the time series. Predicates such as increments, stays, and decrements were relative predicates, and always, sometimes and true-percentage were region-based predicates. They constructed classifiers using predicated features, each of them containing only one predicate. To overcome the problem of incomplete data, they used Adaboost to make the base classifiers, which could make predictions on incomplete data. The significant contribution of this work was to identify the importance of the ECTS. Xing et al. [11] formally defined the ECTS and developed 1-NN based early classifier that examines the closest neighbour relationship in the training set. They also pointed out that the most critical aspect of early classification is to make the balance between earliness and accuracy. In another study, Xing et al. [15] presented a shapelet-based method named EDSC. Shapelets are the subsequence of time series defined as class representatives. The proposed method identified the shapelets with unique features in time series and learned to discriminate between classes as early as possible. In Ref. [16], the authors presented a method named Reliable Early Classification (RelClass) which utilized quadratic discriminant analysis (QDA) for early classification. The method used a user-defined threshold to determine a reliability score which is estimated as the probability that the true class level of complete time series and predicted class of truncated will be the same or higher at each timestep. In [8], the authors proposed an early classification approach based on differentiating classes over time, named ECDIRE. They defined the reliability threshold as a probabilistic difference between the two most probable classes and used it as a decision criterion for making an early prediction.

In recent years neural networks have been extensively used to classify time series and have shown promising results [9,17]. Ref. [9] presented an early classification model that uses LSTM as a classifier and a reinforcement learning agent for decision policy. This model focuses on optimizing both objectives simultaneously, which are accuracy and earliness.

1.1   Motivation and Contributions

Time-sensitive applications are major motivations for early classification. These applications can be found in every field, e.g., medical diagnosis of diseases with distinct symptoms at different timestamps, human activity classification, industrial process monitoring, and electricity usage monitoring. Literature indicates that there are many applications of ECTS in data mining and machine learning [2–4,9,14].

A recent application of early classification could be the detection of COVID-19 infected persons [18] during the pandemic, which is very crucial to stop the spread of the Virus. These applications directly impact the lives of the mass population and possibly save lives. Here we propose an adaptive early classification approach for sequence data. Contributions to our work are:

– We design a novel ECTS method that handles the Early classification problem in two parts. The first part extracts information from the partially observed time series using a base classifier, and the second part makes use of a Reinforcement Learning (RL) Agent to determine whether enough information is captured to make a prediction or not.
– The base classifier has been designed as a hybrid model of RNN and CNN that can capture temporal as well as spatial dependency from incomplete time series.
– The proposed solution utilized the specialty of RNN, CNN, and RL for adaptive early classification by optimizing both accuracy and earliness.
– We evaluated our proposed RCRL (RNN, CNN and Reinforcement Learning) model using seven datasets taken from the UCR Time series classification archive [19]. Our method provided a good balance between accuracy and earliness and outperformed state-of-the-art alternatives in terms of these objectives.

2   Methodology

We propose an early classification model by utilizing the characteristics of RNN, CNN, and RL, named RCRL. The proposed model architecture is depicted in Fig. 1. The RCRL model is logically composed of two components: a base classifier and a decision policy. The proposed RCRL model discriminates between classes based on the extracted features from the RNN and CNN blocks; these features are then fed to a reinforcement learning agent, which learns an early classification policy based on the observed information. The proposed model captures temporal as well as spatial information from the time series. A recurrent neural network helps to capture temporal information, and CNN helps to capture spatial information. The proposed model combined these networks as a hybrid classifier similar to MLSTM-FCN [17]. Further, this base classifier is combined with a reinforcement learning agent to make an adaptive early-classifier [9].

2.1   RCRL Model

We propose an RCRL model that uses an RNN block and a CNN block to extract features from the input data at each timestamp. The extracted features are then passed through an RL Agent, which decides whether or not to halt and predict a

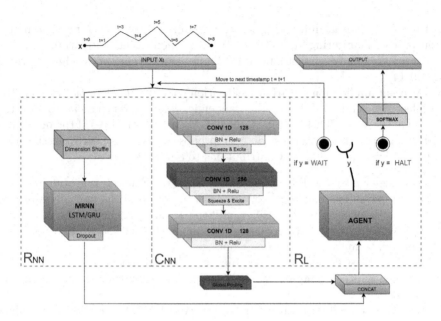

Fig. 1. RCRL-model

class label based on these input features. The agent solves a Partially-Observable Markov Decision Process (POMDP) in which the features extracted by the RNN-CNN block are received at each timestamp. Depending on the agent's learned policy, action is sampled, and based on the sampled action's quality, a reward is observed. The agent's job is to maximize its reward based on accuracy and earliness of prediction. The idea of decision policy is adopted from [9]. However, It has been learned with different kinds of features.

The agent's action a controls the halting of classifier proceedings, the agent can takes two actions $a = \{0, 1\}$, where $a = 1$ dictates the classifier to halt and produce a prediction and $a = 0$ dictates to continue observing more timestamps by classifier. We use ϵ-greedy action to minimize heavy exploitation by the agent. The value of ϵ decreases exponentially from 1 to 0. During training, random action is sampled with probability ϵ as shown in Eq. (1).

$$a_t = \begin{cases} a_t, & \text{with probability } 1 - \epsilon \\ \text{random action,} & \text{with probability } \epsilon \end{cases} \tag{1}$$

Current state H_t contains information about the extracted features by the classifier up to t timestamps. An action is sampled using the learned policy by the agent; the policy is learned using function approximation which maps state H_t to the probability of halting P_t.

The accuracy and earliness of the classification work against each other, while focusing too much on earliness, accuracy gets severely compromised. To address this problem of accuracy-earliness trade-off a hyperparameter λ has been introduced, which controls the earliness of the model by penalizing for late halting, this penalty addition to the loss function along with classification loss motivates the agent to classify early and also maintain the classification accuracy.

2.2 Network Architecture

The CNN block is worked as a feature extractor, it contains 3 temporal convolutional blocks, each block has a convolutional layer of kernel sizes 7, 5, and 3 and with several filters (128, 256, and 128). The next layer after each convolutional layer is Batch Normalization layer(momentum is 0.99 and epsilon is 0.001) followed by a ReLU activation function. At the end of the first two convolutional block a squeeze-and-excite [20] is added with a reduction ratio of 16. At the end of the third convolutional layer, a global pooling layer is added.

In addition to the CNN block, an RNN (LSTM [21] or GRU [22]) block is also introduced as a feature extractor. The RNN block contains a dimension shuffle layer, followed by an RNN layer and a dropout layer. The features extracted by these blocks are concatenated and then passed through an RL block, the RL agent learns a stochastic policy using a 1 layer fully-connected neural network to decide the halting timestamp at which the softmax function will be applied to produce the final class predictions.

2.3 Training

We trained the RNN and CNN block together first and then we used the trained model's parameters to train the reinforcement learning agent. The agent is trained separately on the extracted features from the trained RNN and CNN block, if the learned halting policy of the agent decides to halt at a certain timestamp the extracted features are passed through a softmax layer to produce the final prediction.

The model aims to maximize accuracy while training the classifiers, which are the RNN and CNN blocks. After that, we fixed the parameters of the trained model and introduced the RL agent into the model. While teaching the RL agent, its aim is to maximize the observed reward for classification. The reward function of the agent has a term introduced into it to penalize for earliness, the weightage of the penalty is controlled by hyperparameter λ. The higher value of λ yields a trained model with a high priority on earliness and low priority on the accuracy, while a lower value of λ yields a trained model with a high priority on accuracy and low priority on earliness.

3 Experimental Evaluations

The proposed RCRL has been evaluated on 7 datasets from the UCR archive [19], these datasets have also been used in the evaluation of prior related works on ECTS. We compared RCRL with state-of-the-art methods including, TEASER [23], ECTS [11], RelClass [24], EDSC [15] and ECDIRE [8] as shown in Table 1.

Table 1. Accuracy Comparison

Dataset	ACCURACY(%)					
	RCRL	TEASER	ECDIRE	RelClass	ECTS	EDSC
MoteStrain	65	89	80	58	88	78
ItalyPowerDemand	67	67	93	85	94	82
SonyAIBORobotSurface2	78	78	74	88	85	81
TwoLeadECG	80	77	81	72	73	88
SonyAIBORobotSurface1	85	85	83	79	69	80
ECG200	88	86	91	89	89	85
GunPoint	87	84	87	91	87	94
Dataset	EARLINESS(%)					
	RCRL	TEASER	ECDIRE	RelClass	ECTS	EDSC
MoteStrain	8.7	21	12.1	90.94	79.06	38.08
ItalyPowerDemand	19.63	22	70.16	35.92	79.33	67.08
SonyAIBORobotSurface2	15.9	21	17.66	70.86	54.54	35.51
TwoLeadECG	21	21	69.38	83.63	64.43	46.85
SonyAIBORobotSurface1	8.7	28	62.26	57.7	68.49	47.03
ECG200	9.7	20	90.1	68.81	60.11	23.24
GunPoint	16.8	23	32.37	71.33	46.92	45.58

The results of these methods have been taken from the source[1] for comparison, where these results have been recomputed by running the source codes of these papers.

3.1 Evaluation Metrics

The ECTS is a multi-objective problem involving two contradicting objectives accuracy and earliness. Accuracy Eq. (2) represents the percentage of true predictions out of total predictions. Earliness Eq. (3) presents the average predictive length of the time series. These metrics are formally defined as:

$$Accuracy = \frac{\text{Correct predictions}}{\text{Total predictions}} * 100 \tag{2}$$

$$Earliness = \frac{1}{N} \sum \frac{\text{Number of data points are used for making decision}}{\text{Total data points in the time series}} \tag{3}$$

where N is the number of time series in the dataset. A low value of Earliness represents that the model predicted the class at a very early stage whereas a higher value of earliness represents that the model predicted the class very late.

3.2 Dataset

The datasets used for evaluation are publicly available on UCR archive [19]. We used seven different datasets, four sensor type datasets *ItalyPowerDemand,*

---

[1] https://www2.informatik.hu-berlin.de/~schaefpa/teaser/.

SonyAIBORobotSurface1, *SonyAIBORobotSurface2* and *MoteStrain*, two ECG (Electrocardiogram) type datasets *ECG200* and *TwoLeadECG*, one motion type dataset *GunPoint* as shown in Table 1.

3.3   Dataset Preprocessing

Datasets are taken from UCR archive [19] which already provides a separate train and test set. We have not performed any pre-processing on time series. However, the proposed model accepts the fixed input length of time series. Therefore, to make the model adaptable to different length inputs, we used zero padding. For example, if the length of the complete time series is T and the model is trying to make a prediction at time-step t then $(T - t)$ data points are being padded with zero.

3.4   Results Analysis

The RCRL model has been evaluated on seven publically available datasets as shown in Table 1. The RCRL achieved good earliness comparable to other methods while maintaining acceptable accuracy. Table 2 presented the performance of RCRL by considering two kinds of RNN cells: LSTM and GRU. It is observed that both variants provided comparable results and none of them is superior. It depends on the sequence length, type of data the series has. The selection of λ helps to get desired earliness with acceptable accuracy.

3.5   Effect of Parameter λ

The RL agent maximizes its reward by selecting the optimal halting point for maximum accuracy. At some point during training, the agent learns that the halting criteria and receives maximum reward, probably at the completion of time series. To incorporate earliness into the RL agent, an additional term is being added to the loss function which promotes adaptive early halting based on the hyperparameter λ. The magnitude of λ determines the weightage of earliness for the RL agent, by increasing the λ. The effect of the λ parameter has been analyzed on accuracy and earliness for the "GunPoint" dataset, as demonstrated in Fig. 2. It can be seen that as the values of λ increases, the earliness improves and accuracy decreases. Moreover, the value of accuracy and earliness both increases with λ because of conflicts between these two objectives.

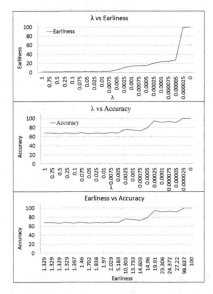

Fig. 2. Effect of hyperparameter λ.

Table 2. RCRL Accuracy

Dataset	RCRL(LSTM)			RCRL(GRU)		
	λ	Acc %	Ear %	λ	Acc %	Ear %
MoteStrain	0.0008	65	8.7	0.0008	80	10.99
SonyAIBORobotSurface1	0.0008	85	8.7	0.0008	85	9.5
TwoLeadECG	0.00001	80	21	0.0008	71	11.55
SonyAIBORobotSurface2	0.0008	78	15.9	0.0008	71	9.448
GunPoint	0.0008	87	16.1	0.0008	82.6	16.349
ItalyPowerDemand	0.02	67	19.63	0.0008	78.4	42.44
ECG200	0.0008	88	9.7	0.0008	86	10.84

3.6   Training Loss and Accuracy

The learning process of the RCRL has been analyzed on *GunPoint* dataset as shown in Fig. 3, First the classifier has been trained for 50 epochs as shown in Fig. 3(a) and 3(b). Figure 3(a) shows an epoch vs accuracy graph for train and test accuracy, while Fig. 3(b) shows an epoch vs test and train losses graph. Figure 3(c) and 3(d) represents the reinforcement learning agent's accuracy, earliness, and loss. As defined earlier the *Accuracy* and *Earliness* are contradictory terms, Fig. 3(c) demonstrates how the reinforcement learning agents balance the trade-off while training, and Fig. 3(d) shows an epoch vs loss graph with an additional term added to the loss function to penalise for late prediction.

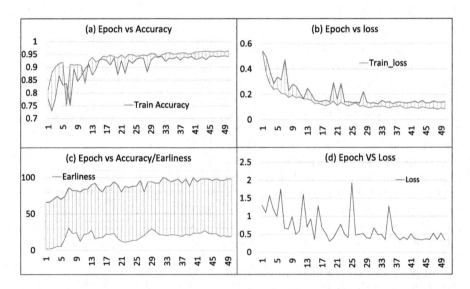

Fig. 3. (a) Epoch vs Accuracy of base classifier, (b) Epoch vs loss of base classifier, (c) Epoch vs Accuracy/Earliness of RL agent, (d) Epoch vs loss of RL agent.

4   Conclusion

We proposed the RCRL model for the early classification of time series. The RCRL utilizes the capabilities of Convolutional Neural Network, Recurrent Neural network, and Reinforcement Learning. The proposed RCRL extracts the features from incomplete time series using the proposed hybrid classifier and learns the decision policy with the help of a reinforcement learning agent. The RCRL model demonstrated the ability to balance the trade-off and give better performance than other alternative models. We compared RCRL with existing models and the results show that the RCRL model delivers state-of-the-art performance in terms of accuracy and earliness.

In future work, we will prepare a complex feature set to learn a more robust decision policy and optimize it for an application-specific problem.

References

1. Bagnall, A., Lines, J., Bostrom, A., Large, J., Keogh, E.: The great time series classification bake off: a review and experimental evaluation of recent algorithmic advances. Data Min. Knowl. Disc. **31**(3), 606–660 (2016)
2. Rußwurm, M., Tavenard, R., Lefèvre, S., Körner, M.: Early classification for agricultural monitoring from satellite time series. arXiv preprint arXiv:1908.10283 (2019)
3. Hatami, N., Chira, C.: Classifiers with a reject option for early time-series classification. In: 2013 IEEE Symposium on Computational Intelligence and Ensemble Learning (CIEL), pp. 9–16. IEEE (2013)
4. Nath, A.G., Sharma, A., Udmale, S.S., Singh, S.K.: An early classification approach for improving structural rotor fault diagnosis. IEEE Trans. Instrum. Measur. **70**, 1–13 (2021)
5. Ghalwash, M.F., Ramljak, D., Obradović, Z.: Patient-specific early classification of multivariate observations. Int. J. Data Min. Bioinform. **11**(4), 392 (2015)
6. Sharma, A., Singh, S.K.: A novel approach for early malware detection. Trans. Emerg. Telecommun. Technol. (2020)
7. Sharma, A., Singh, S.K., Udmale, S.S., Singh, A.K., Singh, R.: Early transportation mode detection using smartphone sensing data. IEEE Sens. J. **21**, 15651–15659 (2020)
8. Mori, U., Mendiburu, A., Keogh, E., Lozano, J.A.: Reliable early classification of time series based on discriminating the classes over time. Data Min. Knowl. Discov. **31**(1), 233–263 (2017)
9. Hartvigsen, T., Sen, C., Kong, X., Rundensteiner, E.: Adaptive-halting policy network for early classification. In: Proceedings of the 25th ACM SIGKDD International Conference on Knowledge Discovery & Data Mining, KDD 2019. ACM Press (2019)
10. Hetland, M.L.: A survey of recent methods for efficient retrieval of similar time sequences. In: Data Mining in Time Series Databases, pp. 23–42. World Scientific (2004)
11. Xing, Z., Pei, J., Philip, S.Y.: Early classification on time series. Knowl. Inf. Syst. **31**(1), 105–127 (2011)

12. Mori, U., Mendiburu, A., Miranda, I.M., Lozano, J.A.: Early classification of time series using multi-objective optimization techniques. Inf. Sci. **492**, 204–218 (2019)
13. Lv, J., Xuegang, H., Li, L., Li, P.: An effective confidence-based early classification of time series. IEEE Access **7**, 96113–96124 (2019)
14. He, G., Zhao, W., Xia, X., Peng, R., Wu, X.: An ensemble of shapelet-based classifiers on inter-class and intra-class imbalanced multivariate time series at the early stage. Soft. Comput. **23**(15), 6097–6114 (2018). https://doi.org/10.1007/s00500-018-3261-3
15. Xing, Z., Pei, J., Yu, P.S., Wang, K.: Extracting interpretable features for early classification on time series. In: Proceedings of the 2011 SIAM International Conference on Data Mining, pp. 247–258. SIAM (2011)
16. Anderson, H.S., Parrish, N., Tsukida, K., Gupta, M.R.: Reliable early classification of time series. In: 2012 IEEE International Conference on Acoustics, Speech and Signal Processing (ICASSP). IEEE (2012)
17. Karim, F., Majumdar, S., Darabi, H., Harford, S.: Multivariate LSTM-FCNs for time series classification. Neural Netw. **116**, 237–245 (2019)
18. Flores, C., Taramasco, C., Lagos, M.E., Rimassa, C., Figueroa, R.: A feature-based analysis for time-series classification of Covid-19 incidence in Chile: a case study. Appl. Sci. **11**(15) (2021)
19. Dau, H.A., et al.: The UCR time series archive. CoRR, abs/1810.07758 (2018)
20. Hu, J., Shen, L., Sun, G.: Squeeze-and-excitation networks. CoRR, abs/1709.01507 (2017)
21. Hochreiter, S., Schmidhuber, J.: Long short-term memory. Neural Comput. **9**(8), 1735–1780 (1997)
22. Chung, J., Gulcehre, C., Cho, K., Bengio, Y.: Empirical evaluation of gated recurrent neural networks on sequence modeling (2014)
23. Schäfer, P., Leser, U.: Teaser: early and accurate time series classification. Data Min. Knowl. Disc. **34**(5), 1336–1362 (2020)
24. Parrish, N., Anderson, H.S., Gupta, M.R., Hsiao, D.Y.: Classifying with confidence from incomplete information. J. Mach. Learn. Res. **14**(1), 3561–3589 (2013)

Introducing Multi-modality in Persuasive Task Oriented Virtual Sales Agent

Aritra Raut[1(✉)], Subrata Das[2], Abhisek Tiwari[2], Sriparna Saha[2], Anutosh Maitra[3], Roshni Ramnani[3], and Shubhashis Sengupta[3]

[1] Department of Computer Science, Ramakrishna Mission Vivekananda Educational and Research Institute, Belur, Howrah, India
aritraraut@gmail.com
[2] Department of Computer Science and Engineering, Indian Institute of Technology Patna, Patna, Bihar, India
[3] Accenture Labs, Bangalore, Karnataka, India

Abstract. In recent years, the usage of virtual assistants to complete tasks like service scheduling and online shopping has increased in both popularity and need. An end user's task goals are the main objectives of a task-oriented conversation agent, and those should be served effectively and successfully. Beside that, user satisfaction is one of the most important aspect that should be taken care of. Communication with multi-modal responses makes the conversation easier and more attractive. Responses through proper images can improve the quality of a task oriented conversation in terms of user satisfaction. Keeping these aspects in mind, we propose a framework which infuses multi-modality with an end-to-end persuasive task oriented dialogue generation module. Additionally, we create a personalised persuasive multi-modal dialogue (PPMD) corpus with slot, sentiment, and agent action annotation at turn level that contains multi-modal responses from both ends. The results and thorough analysis on this dataset show that the suggested multi-modal persuasive virtual assistant achieves better performance over traditional task-oriented frameworks in terms of user satisfaction.

Keywords: Persuasion · task-oriented dialogue generation · multi-modality

1 Introduction

Recent studies in natural language processing have concentrated on creating models for conversational agents, which have numerous uses in the business, sales, and healthcare sectors. Based on the nature of the purpose, conversational agents can be divided into two categories: task or goal-oriented virtual agents and chitchat agents. The former aims to help users complete tasks, whereas chit chat agents interact with users as a companion to satisfy communication needs and build a long-term connection.

M. Tanveer et al. (Eds.): ICONIP 2022, LNCS 13625, pp. 543–555, 2023.
https://doi.org/10.1007/978-3-031-30111-7_46

Conversational agents with a task-oriented approach have recently attracted attention in the field of natural language generation. These agents assist users with a variety of tasks, including hotel reservations, ticket bookings, product purchases, etc. There are several modules in a simple task-oriented dialogue generation setting, including a DST (dialogue state tracker) for extracting belief states, a module for searching a database query based on belief states, a policy learning module to determine the appropriate action against the context, and finally a NLG (Natural Language Generation) module to produce the response. Few attempts have been made to model these components independently [1,2]. The obvious drawback of the pipeline design is that errors propagating from cascaded components may impact following sub-tasks. To avoid this error, Yang et al. [3] proposed a fully end-to-end dialogue system, UBAR. This model collects belief states from the user's utterances, selects action, and generates responses on its own by learning to maximise the likelihood for next word prediction.

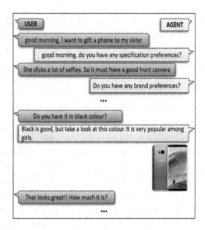

Fig. 1. An example of multi-modal persuasion

On the other hand, as instant messaging technology advanced in recent years, the medium of online communication also moved from pure text to a number of visual modalities (such as a picture, a gif animation, or a brief video). With the advancement in visual question answering and image captioning, the employment of many modalities in dialogue agents has demonstrated outstanding performance. An outstanding intelligent conversational agent should not only be able to communicate freely with plain text, but also be able to observe and share the real visual physical environment.

In some specific task domains like sales, the previous models would typically fall short of the task in situations where the goal is unavailable or the user is not satisfied. The likelihood of work completion in this situation increases if we can persuade while keeping the user's demands in mind. Not relying only on textual responses, if the agent can produce some images for products during persuasion it will be extremely effective. One example has been provided in Fig. 1. Additionally, it facilitates communication since a user may convey more information at once through visuals than if they solely use texts, which may require many passes.

To make the talks more enticing, in this paper we picked UBAR [3], adopted and made changes to it in a different context so that it may perform multi-modal persuasion as needed. Our main contributions are twofold:

1. We create a comprehensive, tailored, multi-modal dialogue corpus for the e-commerce industry that is annotated with semantic data on intent, slot, sentiment, user persona and dialogue act. This data collection includes multi-modal conversations that employ various persuasive techniques depending on the situation.

2. As far as we are aware, this is the first attempt to create a complete dialogue agent that can persuade the user with multi-modal responses in the event of a goal unavailability situation or user dissatisfaction.

2   Related Work

Personalized persuasive dialogue agents and the incorporation of multi-modality in dialogue agents are the key focus of our proposed work. Therefore, in the following section, we have provided summaries of the pertinent works.

Task Oriented Virtual Agents: In recent years, the application of pre-trained models, like GPT-2 [4], gained popularity. In the context of task-oriented dialogue, Budzianowski and Vulic (2019) [5] first highlighted the capacity to fine-tune all crucial information in plain text on GPT-2. Later with just a little tweaking, Yang et al. [3] were able to create an end-to-end task-oriented agent that surpassed all previous models.

Persuasive Virtual Agents: On the other hand, there had also been attempts to include persuasion in the NLG module. In some recent publications [6,7], researchers have concentrated on the DST module to perform persuasion in task-oriented conversation bots to successfully identify and react to dynamic user needs.

Multi-modal Dialogue Generation: Recent conversation system research has concentrated on adding several modalities, such as images, audio, and video, in order to build robust systems. The authors in [8] presented the task of Visual Dialog, which called for an AI agent to engage in meaningful conversation with people regarding visual content in natural conversational language. Apart from this, researchers in [10] designed a dialogue state tracker module which is capable of tackling dynamic user needs with multi-modal responses.

Table 1. Statistics of the existing datasets and our developed dialogue corpus (PPMD)

7 Dataset	Nature	Task	Multi-intent	Dynamic Goal	Task Unavalability	Persuasion	Personalization
ATIS [12]	Task-oriented	Flight booking	✓	×	×	×	×
MultiWoz [13]	Task-oriented	Service booking	✓	×	×	×	×
Persona-Chat [14]	Chit-Chat		×	×	×	×	✓
bAbi [15]	Task-oriented	restaurant reservation	✓	✓	×	×	×
Dear or not [16]	Task-oriented	Negotiation	×	✓	×	✓	×
MMD [17]	Chit-Chat	Fashion assistant	✓	✓	×	×	×
PFG [18]	Task-oriented + Chit-Chat	Donation appeal	×	✓	×	✓	✓
PPMD (our dataset)	Task-oriented + Chit-Chat	Electronics assistant (Phone, laptop and camera booking)	✓	✓	✓	✓	✓

3   Dataset

A variety of benchmark task-oriented corpora were examined, but we were unable to find a single dataset that was appropriate for the job. The properties of several existing discourse datasets are displayed in Table 1. Since persuasion is a key component of any sales agent, in the current work, we have created the PPMD (personalised persuasive multi-modal dialogue) corpus, which is a sizable dialogue database. This corpus is intended to inspire researchers to develop conversational persuasive conversational agents that are intelligent and persuasive. The establishment of this data collection aims to speed the design of conversational bots that can convince people to buy products if their expectations are not now feasible. The dataset includes a variety of discussions in which a salesperson is attempting to persuade a consumer to buy a product. Depending on the situation, different persuasion techniques are used throughout these interactions.

To include multi-modality in the dataset, instead of direct images we have included image links in the utterances. Such multi-modal utterances are frequently used during the section of the conversation where the user describes his or her requirements or where the agent is trying to convince the user with some persuasive appeal. For example, "Give me a phone like this - [link of the image of the phone]". At the end, the dataset contains 189 multi-modal conversations spread across all the domains.

PPMD: Data Creation and Annotation. Virtual assistants are widely used in commercial applications like online shopping. Thus, for our internal data production, we chose the duty of selling various technological devices. With the help of five mobile retailers, we extensively reviewed the assignment and produced 100 instances of dialogue conversations between sellers and buyers around the work of acquiring electronic items (Mobile, Tab and Laptop). The discussions that are produced involve the following three crucial elements: dynamic goal, goal unavailability, and personalised persuasion. The user intent, slot (BIO tag), user sentiment, user personality, persuasion strategy, and dialogue act of each statement in the interaction were also annotated.

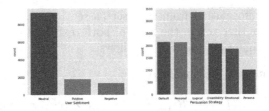

Fig. 2. (a) Sentiment and (b) persuasion strategy distribution across PPMD corpus

Role of Personalized Persuasive Strategy. Persuasion is a very individualised and dynamic process that is greatly influenced by the object of the persuasion and the persuadee's personality. Even the same persuasive technique can fail to convince the same person in two distinct situations. The suggested approach intends to harness both user personality and dialogue environment for convincing users in goal unavailability scenarios. It is motivated by the importance of customised and dynamic nature of persuasion task. We have examples of several techniques in Table 6 (See in Appendix). Distribution of emotion and persuasive tactics within the corpus are shown in Fig. 2.

For scaling up the conversational dataset in accordance with example conversations and a full guideline report, we hired five English linguists. We have utilized GSMArean's mobile database [19] for knowledge grounded conversation creation. We engaged five English linguists to scale up the conversational dataset in accordance with sample conversations and a comprehensive guideline report. For the establishment of knowledge-based conversations, we used the mobile database GSMAreana [19]. Each utterance in this dataset has been labelled with the appropriate intent, slot, user sentiment, dialogue act, and persuasion strategy. A considerable uniform annotation was found to exist when the kappa coefficient (k), which measures annotation agreement among annotators, was found to be 0.77. Table 2 contains statistics from the PPMD dataset. In Table 5 (See in Appendix), we have included reported metadata details like intent and slot listings.

Table 2. PPMD dataset statistics

Attribute	Value	Attribute	Value
Total no. of dialogues	1031	Avg. dialogue length	11.25
Total no. of multi-modal dialogues	189	No. of persuasion strategies	6
Total no. of utterances	11602	No. of samples in knowledge base	2697
No. of unique words	5937	No. of attributes	18

4 Problem Formulation

In order to better serve end users' goals in unavailability scenarios and reduce task failures, we developed a neural-based virtual assistant. The agent is capable of extracting information from the user passed texts as well as images and produces responses that too in the form of both texts and images. The agent's response (R_t) at time t is produced as follows:

1. We define context C_t at time step t as:

$$C_t = \{U_1, S_1, B_1, D_1, A_1, R_1, ..., U_{t-1}, S_{t-1}, B_{t-1} D_{t-1}, A_{t-1}, R_{t-1}, U_t\} \quad (1)$$

Where, $U_t, S_t, B_t, D_t, A_t, R_t$ stand for user utterance, sentiment, belief states, database query, agent action and agent response at t^{th} turn, respectively. U_t is defined as:

$$U_t = (T_t, Img_t) \quad (2)$$

Where, T_t stands for text information at turn t and Img_t stands for the image representation at turn t.

2. The proposed model first encodes (e) the information and generates one token $(R_t[j])$ at each time step depending upon encoded information and previously generated tokens. It can be expressed as follows:

$$R_t = \Pi_{j=1}^{j=n} d(e(C_t), R_t[1 : j - 1]) \quad (3)$$

where n is the number of words in the generated sequence (R_t) and $R_t[j]$ is j^{th} token of the generated sequence. This R_t can also be represented as a tuple of text (T_t) and image (Img_t) information.

5 Methodology

The work intends to create a neural-based framework for generating convincing multi-modal dialogues. In this section, we will first present a brief overview of the UBAR pipeline before going into more depth regarding our architectural (multi-USBAR) design.

5.1 UBAR Pipeline

The UBAR module's pipeline is fairly straightforward. Let's say, at turn $t = 0$, we get the very first user utterance, U_0. Following the receipt of the user utterance, UBAR creates the components as listed below.

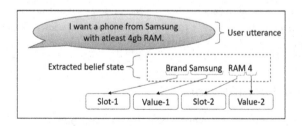

Fig. 3. An example of our belief state from texts

1. The model extracts the belief states B_0, based on U_0. Belief state at each turn is basically a set of decoupled slot-value pairs $\{slt_0, v_0, slt_1, v_1, ...slt_n, v_n\}$, where each pair (slt_t, v_t) consists of slot and value information extracted from the current utterance. One example of this belief state has been shown in Fig. 3.
2. After extracting the belief states, B_0, it performs the database query. This provides the number of database instances D_0 matching with the belief states, B_0.
3. Finally based on $[U_0, B_0, D_0]$, it generates agent action A_0, and the delexicalized response, R_0. Delexicalized response refers to the process through which the model creates unique placeholders in the responses for particular slots. For example, a brand name in the generated response is "$<value\_brand>$". Later these placeholders should be replaced by the respective values from the database query result.

This completes the first turn, and the flow keeps going all the way to the very end.

5.2   Multi-USBAR Pipeline

Our multi-USBAR model is basically a modification over the UBAR architecture in order to adopt this for multi-modal environment. The modified pipeline is as follows: Let's say, at turn $t = 0$ user passes his/her first utterance, U_0. This U_0 can contain both text and image. In our setting, instead of images user will provide the image link. From this point the flow of our model is as follows:

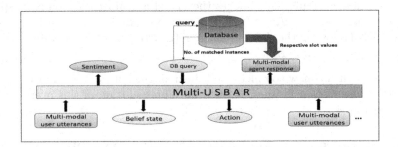

Fig. 4. Multi-USBAR (our proposed model) workflow

1. The model classifies the sentiment of the user's utterance (U_0). In essence, it generates a word (positive/negative/neutral) that expresses the user's sentiment (S_0).
2. The model next extracts the belief states B_0, based on $[U_0, S_0]$. Belief states from simple texts are as same as mentioned in Sect. 5.1. Additionally, our model has the ability to extract information from images. For this part, the

model will trace the image link from user utterances, then fetch the image from web following the link and will pass it through Google's application programming interface (API) in order to extract multiple information like the product (either it is phone or laptop or camera etc.), the brand and the colour of the product. These extracted information will be added to the belief state, B_0. One example of this belief state extraction pipeline from images has been shown in Fig. 5.

3. Next, based on the belief state, B_0, our model runs a database query to determine the number of instances (D_0) in the database that matches with B_0.

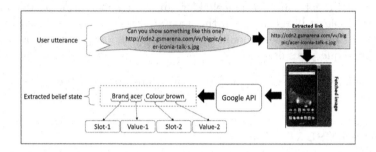

Fig. 5. The pipeline of belief state extraction from images

4. Finally based on $[U_0, S_0, B_0, D_0]$, the model chooses required agent action A_0 from the list of actions (including five different persuasion strategies) and generates delexicalized response R_0 (same as mentioned in Sect. 5.1). In addition, to reply with images the model is trained to produce a special placeholder "$<value\_link>$". Our database contains an image link of each product. Following the link, an image retrieved from the web will eventually replace this "$<value\_link>$".

The first turn is now complete, and the dialogue will continue in this manner until the conversation finishes. A pictorial representation of this overall flow is shown in Fig. 4.

6   Training Setup and Implementation Details

HuggingFace's Transformers [20] and DistilGPT2 [21], a distilled version of GPT-2, have been used to implement our model at the session level; this means the entire conversation has been supplied to the model, which then learns to produce or predict the next word depending on the current word. We used the cross-entropy function as our loss, AdamW as our optimizer, and the common greedy decoding technique with temperature of 0.7 and trained the system for 50 epochs. Through hyperparameter search of learning rate and batch size, we choose the model that performs the best on the validation set, and we then evaluate on the test set to obtain the results.

Table 3. Automatic evaluation results with different setups

Model	True bs	True act	BLEU-1	BLEU-2	BLEU-3	BLEU-4	Rouge-1	Rouge-L
UBAR	✓	✓	0.24	0.12	0.10	0.02	0.34	0.32
	✓	✗	0.21	0.10	0.08	0.01	0.32	0.30
	✗	✓	0.20	0.10	0.08	0.01	0.30	0.29
	✗	✗	0.21	0.10	0.06	0.01	0.28	0.27
multi-USBAR	✓	✓	0.23	0.12	0.10	0.02	0.34	0.32
	✓	✗	0.21	0.11	0.08	0.01	0.32	0.30
	✗	✓	0.20	0.10	0.06	0.01	0.30	0.28
	✗	✗	0.20	0.10	0.06	0.01	0.29	0.28

Table 4. Human evaluation results in an end-to-end setting

Model	Consistency	Repetitiveness	P.P Score	G.C Score	U.S Score
UBAR	0.67	0.35	2.31	0.87	2.11
multi-USBAR	0.74	0.27	2.87	0.94	2.41

7   Results

On a freshly created PPMD dataset, we trained the UBAR and our multi-USBAR module and measured their performances. On top of UBAR, in our multi-USBAR we implemented context-specific sentiment information, distinct placeholders for picture links, and an information extractor (only at the inference part) for pulling information from the images. To evaluate how closely the generated replies resemble the gold human responses, we have employed a few automated evaluation metrics like the BLEU (BiLingual Evaluation Understudy) [22] and Rouge scores [23]. In order to assess the qualitative performances of these models, we have also conducted human evaluations in terms of repetition, consistency, personalised persuasion, grammatical correctness, and user satisfaction. The human evaluation metrics are defined as follows:

- **Consistency:** This is defined as the number of slots fulfilled by the agent/number of slots asked by the user.
- **Repetitiveness:** It measures how much similar the generated responses are. We have defined it as the Jaccard Similarity between agent responses in a single conversation.
- **Personalized persuasion score (P.P Score):** These were graded according to the agent's perceived use of personalization techniques. On a scale of 1 to 5, the replies from the agents that were able to make use of the contextual information, received comparatively higher marks. The expected 2.5 points were awarded for a neutral response.
- **Grammatical Correctness Score (G.C Score):** It rates how well-written and grammatically sound the generated responses are. For each conversation, it is calculated as:

$$\text{Grammatical Correctness} = \frac{\text{number of grammatically correct responses}}{\text{total number of turns}}$$

Then, finally we have taken the mean over all the conversations in order to get the final Grammatical Correctness score (G.C Score).

– **User satisfaction score (U.S Score):** This metric is basically user given rating on the scale of 5 where the idea was to capture how soothing the generated responses are for an end user.

We have tested with several parameters and fine-tuned the models for 50 epochs each. We tested various combinations of generated and genuine **belief states(bs)** and **actions(act)** for each model, then evaluated the results using these automated evaluation criteria. The effectiveness of each model has also been evaluated in terms of the aforementioned human assessment measures, but only in an end-to-end environment (using generated belief states and actions in the context). Automatic evaluation results are shown in Table 3 and the human evaluation results are shown in Table 4.

We trained our multi-USBAR model while maintaining the current UBAR architecture, adding distinct placeholders for picture links only to aid the model in providing accurate image links to the user. Automatic assessment (see Table 3) demonstrates our model's ability to generate responses that are on par with UBAR in every different setting. On the other hand, we can easily observe the performance increase over UBAR when we look at the human assessment metrics at Table 4. Our model clearly produces responses that are more pleasing to the user, while maintaining superior consistency and reduced repetitiveness. One example of our model's conversation has been shown in Fig. 6 (in Appendix).

8   Error Analysis

We noticed the following two significant problems with the suggested model.

1. To learn the placeholders for the appropriate slots, the model is entirely dependent on the slot-value annotations of user utterances. However, there are a few slot values (such as processor, release date, etc.) that rarely occur at user utterances. Naturally, as a result, the model is unable to discover the appropriate placeholders for such slots.
2. We use the Google API to extract data from the photographs, but the API is not perfect, thus occasionally incorrect data is added to the belief state. This might steer a conversation into a wrong direction.

9   Conclusion and Future Work

This paper illustrates the design of a task-oriented dialogue agent capable of persuading via a variety of modalities. The agent decides a strategy (out of 5 available strategies) to respond, based on the context information. To the best of our knowledge, this is the first study on automated neural response generation for creating a persuasive conversational agent that can use several media. The model was constructed utilising a recently created dataset called PPMD (personalised persuasive multi-modal dialogue), and the addition of persuasion behaviour is making the model more effective in real-world scenarios, especially in the sales domain or when the agent needs to make some reservations. Results on the

PPMD dataset demonstrate how our Multi-USBAR really facilitates a more appealing and easier discourse while maintaining the consistency. Negotiation is another important aspect for our agent. In addition to persuasion, we will use negotiation in future developments since it can lead both the parties to a better agreement.

Appendix

Table 5. Intent, slot and dialogue act list of the PPMD dataset

Intent	greet, specification, inform, request, persuasion, thanks, preq, done
Slot	model, brand, battery, ram, p_camera, s_camera, radio, display_size, status, sim, gps, os, color, internal_ram, weight, released_year, released_month, price, phn_key, specifications, sp_done, features
Dialogue Act	greet, specification_request, specification_done, inform, request, result, recommend, persuasion, booking_request, close
Sentiment	positive, negative, neutral
Persuasion Strategy	Default, Credibility appeal, Logical appeal, Personal appeal, Emotional appeal, Persona based appeal

Table 6. Examples of different persuasion strategies

Persuasion strategy	Context (User)	Example
Credibility appeal	I trust only Nokia Brand. So please see something in it.	It is a Nokia brand, which ensures its outstanding quality. Many other brand phones with the same quantity do not perform equally well for a long time. You should buy this phone without a second thought.
Logical appeal	I do not like black color, please find a phone in silver color	You should buy this phone, it has lot of features such as a Radeon Pro 555X G2DDR5 (4 GB) graphic design with Intel Core i7 6 Core processor, 15.4 display size. Its rating is 4.1
Persona-based appeal	It is very costly, see something other.	Sure, but i still highly recommend this phone to you because of its special features particularly gorgeous titan black color.
Personal appeal	No, I do not like this phone. Its storage is less.	This is a great phone, it has received huge number of positive reviews across all countries.
Emotional appeal	Hi, I want to gift a phone to my girl friend on her birthday. She loves photography.	This phone will be perfect gift for a photographer, it has all the features and specifications which are necessary for a photographer. Your girlfriend will love this for sure.

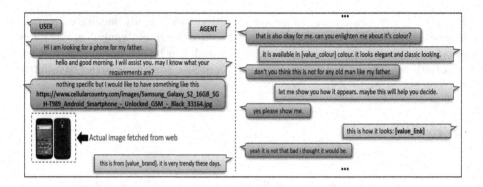

Fig. 6. An example of a generated conversation from multi-USBAR

References

1. Lei, W., Jin, X., Kan, M.Y., Ren, Z., He, X., Yin, D.: Sequicity: simplifying task-oriented dialogue systems with single sequence-to-sequence architectures. In: Proceedings of the 56th Annual Meeting of the Association for Computational Linguistics (Volume 1: Long Papers), Melbourne, Australia, pp. 1437–1447. Association for Computational Linguistics (2018). https://aclanthology.org/P18-1133
2. Liang, W., Tian, Y., Chen, C., Yu, Z.: MOSS: end-to-end dialog system framework with modular supervision. In: Proceedings of the AAAI Conference on Artificial Intelligence, vol. 34, no. 05, pp. 8327–8335 (2020). https://doi.org/10.1609/aaai.v34i05.6349
3. Yang, Y., Li, Y., Quan, X.: UBAR: towards fully end-to-end task-oriented dialog systems with GPT-2. In: AAAI (2021)
4. Radford, A., et al.: Language models are unsupervised multitask learners. OpenAI Blog **1**(8), 9 (2019)
5. Budzianowski, P., Vulić, I.: Hello, it's GPT-2-how can I help you? Towards the use of pretrained language models for task-oriented dialogue systems. arXiv preprint arXiv:1907.05774 (2019)
6. Tiwari, A., et al.: A dynamic goal adapted task oriented dialogue agent. PLoS ONE **16**(4), e0249030 (2021)
7. Tiwari, A., et al.: A persona aware persuasive dialogue policy for dynamic and co-operative goal setting. Expert Syst. Appl. **195**, 116303 (2022)
8. Das, A., et al.: Visual dialog. In: Proceedings of the IEEE Conference on Computer Vision and Pattern Recognition, pp. 326–335 (2017)
9. Guo, D., Wang, H., Wang, M.: Dual visual attention network for visual dialog. In: Proceedings of the Twenty-Eighth International Joint Conference on Artificial Intelligence, IJCAI 2019, Macao, China, 10–16 August 2019, pp. 4989–4995 (2019)
10. Tiwari, A., et al.: Multi-modal dialogue policy learning for dynamic and co-operative goal setting. In: 2021 International Joint Conference on Neural Networks (IJCNN), pp. 1–8. IEEE (2021)
11. Guo, D., Wang, H., Wang, S., Wangb, M.: Textual-visual reference-aware attention network for visual dialog. IEEE Trans. Image Process. **29**, 6655–6666 (2020)
12. Hemphill, C.T., Godfrey, J.J., Doddington, G.R.: The ATIS spoken language systems pilot corpus. In: Speech and Natural Language: Proceedings of a Workshop Held at Hidden Valley, Pennsylvania, 24–27 June 1990 (1990)

13. Budzianowski, P., et al.: MultiWOZ-a large-scale multi-domain wizard-of-oz dataset for task-oriented dialogue modelling. arXiv preprint arXiv:1810.00278 (2018)
14. Zhang, S., Dinan, E., Urbanek, J., Szlam, A., Kiela, D., Weston, J.: Personalizing dialogue agents: I have a dog, do you have pets too? In: Proceedings of the 56th Annual Meeting of the Association for Computational Linguistics (Volume 1: Long Papers), pp. 2204–2213 (2018)
15. Bordes, A., Boureau, Y.L., Weston, J.: Learning end-to-end goal-oriented dialog. arXiv preprint arXiv:1605.07683 (2016)
16. Lewis, M., Yarats, D., Dauphin, Y., Parikh, D., Batra, D.: Deal or no deal? End-to-end learning of negotiation dialogues. In: Proceedings of the 2017 Conference on Empirical Methods in Natural Language Processing, pp. 2443–2453 (2017)
17. Saha, A., Khapra, M., Sankaranarayanan, K.: Towards building large scale multimodal domain-aware conversation systems. In: Proceedings of the AAAI Conference on Artificial Intelligence, vol. 32 (2018)
18. Wang, X., et al.: Persuasion for good: towards a personalized persuasive dialogue system for social good. arXiv preprint arXiv:1906.06725 (2019)
19. Baichoo, A.: Kaggle GSMArean (2017). https://www.kaggle.com/arwinneil/gsmarena-phone-dataset
20. Wolf, T., et al.: HuggingFace's transformers: state-of-the-art natural language processing. arXiv preprint arXiv:1910.03771 (2019)
21. Sanh, V., Debut, L., Chaumond, J., Wolf, T.: DistilBERT, a distilled version of BERT: smaller, faster, cheaper and lighter. arXiv preprint arXiv:1910.01108 (2019)
22. Papineni, K., Roukos, S., Ward, T., Zhu, W.J.: Bleu: a method for automatic evaluation of machine translation. In: Proceedings of the 40th Annual Meeting of the Association for Computational Linguistics, pp. 311–318 (2002)
23. Lin, C.-Y.: ROUGE: a package for automatic evaluation of summaries. In: Text Summarization Branches Out (2004)

Low Dose CT Image Denoising Using Efficient Transformer with SimpleGate Mechanism

Lianjin Xiong[1], Wei Qiu[1], Ning Li[1], Yishi Li[1,2], and Yangsong Zhang[1(✉)]

[1] School of Computer Science and Technology, Laboratory for Brain Science and Medical Artificial Intelligence, Southwest University of Science and Technology, Mianyang 621010, China
zhangysacademy@gmail.com
[2] Department of Respiratory and Critical Care Medicine, The First Affiliated Hospital of Chongqing Medical University, Chongqing, China

Abstract. The widespread use of computed tomograph (CT) technology in clinic has caused more and more patients to worry that they will receive too much radiation during the scanning. The low-dose CT (LDCT) scanning is more likely to be accepted by the patients. But LDCT images can adversely affect doctors' diagnosis, owing to low quality of the images. Therefore, it is necessary to improve the diagnostic performance by denoising LDCT images. During the past few decades, the convolutional neural networks (CNNs) and Transformer models that achieve remarkable performance in natural image denoising provide new avenues for LDCT denoising. Although the existing methods have successfully achieved noise reduction, there is still large room for improvement in the denoising level. In this paper, we refer to the implementation of natural images denoising, and proposed a transformer-based U-shape network model to achieve denoising in LDCT images. In each transformer block, we used the depth-wise convolution, transposed self-attention mechanism, and SimpleGate to improve performance and speed up efficiency. Extensive experiments on the AAPM-Mayo clinic LDCT Grand Challenge dataset indicated that the proposed model yielded a competitive performance to the compared baseline denoising methods. In particular, good evaluation was achieved in noise suppression, structure preservation and lesion highlighting.

Keywords: Low dose CT · deep learning · denoising · transformer

1 Introduction

Computed tomography (CT) system, as noninvasive imaging equipment, has been widely used for medical diagnosis and treatment [16]. It works by collecting X-rays that pass through the body and then reconstructing each slice of organ tissues. But the radiation from X-rays causes some damage to cells and DNA in the body [1]. We can roughly assume that dose is positively related to risk and imaging quality, high dose means high risk and high imaging quality.

M. Tanveer et al. (Eds.): ICONIP 2022, LNCS 13625, pp. 556–566, 2023.
https://doi.org/10.1007/978-3-031-30111-7_47

The weaker the X-ray flux, the noisier a reconstructed CT image, which could influence the doctor's diagnosis. Therefore, many algorithms have been designed to alleviate this problem. In general, these algorithms can be divided into three categories, i.e., sinogram domain filtration, iterative reconstruction, and image post-processing.

Sinogram filtering techniques process the raw data before image reconstruction. The most famous and widely used algorithm is filtered backprojection (FBP) [18]. However, these methods often suffer spatial resolution loss or artifacts in the reconstructed images. Besides, it is very difficult for ordinary users to get raw data of commercial scanners. Over the past decade, researchers have worked to develop new iterative reconstruction (IR) algorithms in the field of LDCT. In general, these algorithms optimize an objective function, which includes a statistical noise model [19,25], priors information in the image domain and an image system model [8,12]. Although IR techniques obtained excellent results, it is impossible to apply it to practical applications because of the high computational cost. On the other hand, image post-processing is computationally efficient and directly perform on an image. Many methods have played a certain role in suppressing noise and artifacts, such as non-local means (NLM) method [15], dictionary-learning-based K-SVD method [7] and block-matching 3D (BM3D) algorithm [10,11].

In the past few years, deep learning technologies extensively applied in other fields have also attracted tremendous attention in the field of medical images, such as image registration [3,27], image segmentation [30], image classification [4] and LDCT denoising [5,13,21,23,26,28,32]. For example, Chen et al. [5] designed a residual encoder-decoder network for LDCT images, which greatly suppresses the image noise through the method based on CNN. Yang et al. [28] used WGAN network and perceptual loss to promise the reconstructed image quality. They greatly improved the smoothing problem by using perceptual loss, but obtained not very high values of the evaluation metrics PSNR and SSIM. Zhang et al. [31] designed CLEAR based on GAN network to achieve noise reduction very well. They used multi-level consistency loss and got a big improvement in the evaluation of PSNR and SSIM, but added multiple hyperparameters which brought difficulties to training and application. Liang et al. [13] used the soble operator to extract the edge information of the image, and then used convolution and dense connections to achieve denoising. And recently, Transformer has made great progress in the natural language processing (NLP) fields [22]. This also brings new research ideas to the computer vision (CV) fields. In particular, Vision Transformer (ViT) designed by Dosovitskiy et al. almost makes Transformer the main method in CV field [9]. Many computer vision tasks have reached the state-of-the-art performance through the use of Transformer architecture. However, the Transformer model has not been well applied in the field of LDCT denoising. For example, Zhang et al. proposed a dual-path TransCT network to predict the high quality images by fusing high-frequency and low-frequency features [32]. They successfully brought Transformer to the task of CT denoising, but their model did not outperform CNN by much in the evaluation metric PSNR. Wang et al. [24] proposed a convolution-free T2T vision Transformer to achieve a better noise reduction effect. These Transformer-based methods often require a lot of computational consumption when calculating self-attention. To alleviate the inherent problem and further improve denoising performance, we proposed a new network architecture, which was inspired by the facts that: (1) different methods for image denoising have complementary image prior models ability and can be incorporated to boost the performance [2]; (2) Restormer [29] and NAFNet [6] designed different network structures and achieved very good denoising performance. In the proposed model, we used a transformer block to combine depthwise convolution, transposed self-attention mechanism, and SimpleGate, and then plug it as the main building block into the UNet architecture. We termed the proposed model as SAGformer hereinafter. Extensive experiments on Mayo LDCT dataset demonstrate the superiority of our method over other baseline methods.

2 Method

In this study, our goal is to realize denoising of LDCT images by designing an efficient transformer block. In order to reduce the computational cost, we used the modification scheme of self-attention designed by Zamir et al. [29], and adopted the SimpleGate designed by Chen et al. [6] to replace the activation function. First, we introduced the overall structure of the proposed SAGformer (see Fig. 1). Then we presented the main components of the proposed transformer block: (a) Multi-Dconv Head Transposed Self-Attention (MDTSA) and (b) Dconv SimpleGate Feed-forward network (DSGFN).

Overall Structure. From the perspective of macrostructure, we built a standard UNet [20] structure network with four-level symmetric encoder-decoder. First, we applied a 3 × 3 convolution to learn low-dimensional features. In the encoder stage, downsampling was added to each layer to reduce the resolution and widen the channels of the image. With the deepening of the number of layers, the number of transformer blocks in each layer gradually increases. The resolution of the last level of encoder image is reduced by one-eighth, and the number of channels is eight times the original. On the other hand, in the decoder phase, upsampling between layers is used to restore information. In addition, we also used skip connections to assist the restoration of information and alleviate the problem of gradient disappearance during training. After concatenating the features of the encoder and the decoder, a 1 × 1 convolution is added to keep the number of channels unchanged, except the top one. Finally, the reconstructed features are added to the LDCT image to output normal dose CT (NDCT) image. Next, we describe the details of the Transformer block.

2.1 Multi-Dconv Head Transposed Self-attention

In the transformer network, one of the most important components is self-attention, which brings good results for vision tasks while incurring expensive

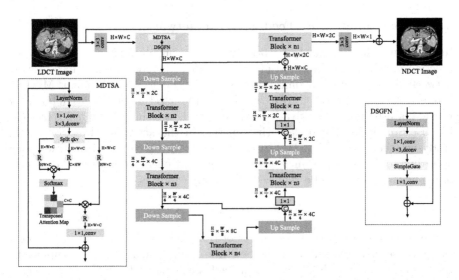

Fig. 1. Architecture of SAGformer for LDCT denoising. The macrostructure is a U-shaped network architecture, in which the transformer module is composed of MDTSA and DSGFN, which mainly realize denoising by performing attention calculation on the channel dimension. R in the figure stands for Reshape operation.

computational costs. The computing mechanism of self-attention (SA) determines that it has a large receptive field and can process global features, but this also leads to its insufficiency in local feature processing [22]. Therefore, we decided to introduce depthwise convolution to enrich local feature information. The reason for using depth-wise convolution is that it can reduce the amount of computation compared to common convolution. But before that, we added a 1×1 convolution to expand the channel to ensure that more features on the channel can be obtained from subsequent calculations. Considering that conventional self-attention cannot be used for such large-resolution image processing tasks, we take the approach proposed by Zamir et al. [29]: reshape the tensor before multiplying query (Q), key (K) and value (V); $Q \in \mathbb{R}^{HW \times C}$; $K \in \mathbb{R}^{C \times HW}$; $V \in \mathbb{R}^{HW \times C}$. After Q and K are multiplied, softmax operation is performed on the results to obtain the attention map of size $\mathbb{R}^{C \times C}$. The resulting attention map is multiplied by V to get the final attention feature. The advantage of transposing and reshaping before multiplication is that attention can be calculated based on channels. For this kind of large-resolution image, it is a very time-consuming and performance-consuming task to calculate attention map based on pixels. Overall, the MDTSA process is defined as:

$$\tilde{X} = W_\beta Attention(XW_1^Q W_3^Q, XW_1^K W_3^K, XW_1^V W_3^V) + X \tag{1}$$

where W_β is a learnable scaling parameter to control the size of the residual block, $W_1^{(\cdot)}$ is the 1×1 point-wise convolution and $W_3^{(\cdot)}$ is the 3×3 depth-wise convolution.

2.2 Dconv SimpleGate Feed-Forward Network

In the regular feed-forward network, two 1×1 convolutions and an activation function are included. In our study, we modified the conventional feed-forward network. After a 1×1 convolution, the same depth-wise convolution as MDTSA was introduced for local feature extraction. In the selection of activation function, we did not use nonlinear activation functions such as Gelu or Relu in this module, but use SimpleGate to replace the function of the activation function as the study by Chen et al. [6]. The feature map was directly divided into two parts according to the channel dimension, and then multiplied. In this way, the purpose of nonlinearity can also be achieved. The SimpleGate is defined as:

$$SimpleGate(X, Y) = X \cdot Y \tag{2}$$

where X and Y are two feature maps of the same size. $X \cdot Y$ can also achieve the purpose of nonlinearity of the activation function. And after SimpleGate, the number of channels of the feature will be reduced by 50%. This effect is difficult to achieve in common activation functions. In addition, we also add a 1×1 convolution to keep the number of channels unchanged.

2.3 Loss Function

We use the MSE loss function, which is widely used in the field of image restoration. MSE can well compare the pixel difference between the NDCT and model output, so that it can better guide the update of model parameters after gradient backward. It is defined as follows:

$$L_{mse} = \frac{1}{N} \sum_{i=1}^{N} \| D(x_i) - y_i \|^2 \tag{3}$$

3 Experiments and Results

3.1 Experimental Settings

Datasets. In this work, we used the pubicly released dataset from 2016 NIH-AAPM-Mayo LDCT Grand Challenge [17]. The dataset was obtained by performing normal-dose abdominal CT scans on 10 anonymous patients, and then simulating quarter-dose images by adding Poisson noise to the projection data. We used the images from 9 patients to train the model and the images from 1 patient to evaluate the performance of the model. We randomly extracted eight image patches of 64×64 from each 512×512 image during training.

Implementation Details. In our work, we used Pytorch 1.11.0 to build our model, the training was performed on a NVIDIA RTX 3090Ti GPU. During training, we set the number of the epoch to 400, and used the Adam optimizer to minimize our MSE loss with an initial learning rate of 1e−5. All 3 × 3 depth-wise convolutions were implemented with stride 1 and padding 1, and 1 × 1 convolutions ere implemented with stride 1 and padding 0. During the test, we no longer divided the images into patches, but directly input 512 × 512 images into the model.

Baseline Models. Three baseline methods, i.e., RED-CNN [5], EDCNN [13], CTformer [23] were used to compared with our method. These methods are deep learning methods that perform well on LDCT denoising. But since RED-CNN and EDCNN did not provide a trained model, we retrained with the same dataset. CTformer used the same dataset as ours, and the authors released their trained model. So, we directly used the trained model for testing.

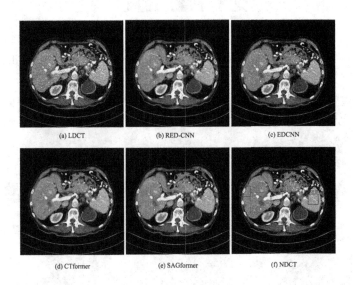

Fig. 2. Results from the different methods for comparison. The display window is [−160, 240] HU. (a) a LDCT image; (b) RED-CNN; (c) EDCNN; (d) CTformer; (e) SAGformer. (f) a NDCT image. The red rectangle is the several defined ROIs. (Color figure online)

Evaluation Metrics. We used three common metrics to evaluate the methods, i.e., peak signal to noise ratio (PSNR), structural similarity index measure (SSIM) and root mean square error (RMSE) [23]. The three metrics can be combined to evaluate the denoising level, PSNR is mainly to measure the reconstruction quality. SSIM is used to evaluate the structural similarity between images, which can reflect the visual quality to a certain extent. RMSE is used to reflect the difference between corresponding pixels [14]. Among them, PSNR and SSIM are positively correlated with the final image quality, and RMSE is negatively correlated with the image quality.

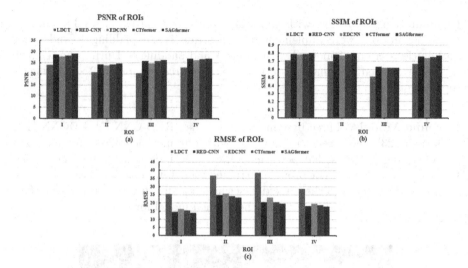

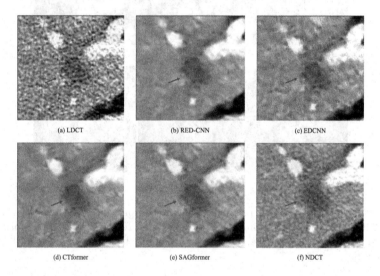

Fig. 3. The PSNR, SSIM and RMSE histogram of ROIs from Fig. 2 under different algorithms. (a) PSNR of ROIs; (b) SSIM of ROIs; (c) RMSE of ROIs.

Fig. 4. The amplified ROI images of different methods outputs in the blue rectangle marked from Fig. 2 (Color figure online).

3.2 Results

Quantitative Evaluation. First, we randomly selected four ROIs, as shown in the red rectangle of Fig. 2, to evaluate the reconstruction level of the local region. In order to make the results more intuitive, we display the values in the form of a histogram. As shown in Fig. 3, our results outperform other methods in PSNR, SSIM and RMSE. Second, we calculated the PSNR, SSIM and RMSE of each CT global image of patient L506, and then calculated the average values. Table 1 shows the overall quantitative results of all methods, and the best results are highlighted in bold. In terms of PSNR index comparison, SAGformer is 0.70 dB higher than the previous CNN-based method RED-CNN and 0.44 dB higher than the recent Transformer-based network CTformer. In the comparison of SSIM, our model is 0.005 higher than CTformer. Besides, the RMSE of the SAGformer is 0.45 lower than that of the CTformer.

Table 1. Quantitative results of different methods on L506

	PSNR↑	SSIM↑	RMSE↓
LDCT	29.2489	0.8759	14.2416
RED-CNN	32.8209	0.9092	9.3144
EDCNN	32.3036	0.9034	9.9121
CTformer	33.0793	0.9119	9.0726
SAGformer	**33.5281**	**0.9172**	**8.6172**

Qualitative Evaluation. Figure 2 shows the restoration effect of various models on a LDCT image. The abdominal CT slice was selected because the abdomen was the part with the most organs, so it can be seen clearly and intuitively that the noise reduction performance of our model was better than other models. Figure 4 shows the zoomed parts of the lesion position in the blue rectangle marked in Fig. 2. The lesion area pointed by the red arrow becomes more obvious after denoising, and the contour is also obvious.

3.3   Ablation Study

In this part, we verify the effectiveness of the introduced modules for LDCT denoising in the SAGformer through ablation experiments. The experimental results are shown in Table 2. We used multi-head transposed self-attention and feed-forward neural networks as benchmarks, which was termed as MTSA_FN. In the MTSA block, we used the normal 3×3 convolution to get QKV in self-attention, and in the FN network, we added the normal 3×3 convolution after the first 1×1 convolution. Although this model achieves better results than the CNN-based methods, the ordinary convolution operation results in a huge amount of parameters. Next, we replaced the 3×3 ordinary convolution with the depth-wise convolution, which resulted in the MDTSA_DFN model. The parameters of this model were reduced to 18.42 M, and the final PSNR, SSIM and RMSE are increased by 0.27, 0.004 and 0.27 respectively. Finally, we added SimpleGate to the feed-forward neural network module to form our final model (SAGformer), which once again improved PSNR, SSIM and RMSE. And the amount of parameters is reduced again because of the introduction of Simple-Gate.

Table 2. Quantitative Results of Ablation Experiments on L506

	Params	PSNR↑	SSIM↑	RMSE↓
MTSA_FN	92.79M	32.9941	0.9114	9.1574
MDTSA_DFN	18.42M	33.2641	0.9159	8.8809
SAGformer	**16.64M**	**33.5281**	**0.9172**	**8.6172**

4 Conclusion

In brief, we designed a novel Transformer-based UNet network to success-fully denoise low-dose CT images. Mainly through the modification of the self-attention calculation method, the tensor was reshaped and transposed to speed up the calculation speed. In addition, the use of depth-wise convolution and Sim-pleGate also greatly reduced the network parameters and improved the denoising performance. The experimental results show that the proposed model has great potential for structure preservation and lesion detection, and outperforms other models on both global and local regions after denoising. However, LDCT images still have some gaps compared to NDCT after denoising by our model, such as our images are over-smoothing. In addition, the very detailed textures have not been fully restored. In the future, we need to further optimize the SAGformer structure, and utilize the 3D CT image series to enhance the quality of low-dose CT images.

References

1. Brenner, D.J., Hall, E.J.: Computed tomographyan increasing source of radiation exposure. N. Engl. J. Med. **357**(22), 2277–2284 (2007)
2. Burger, H.C., Schuler, C., Harmeling, S.: Learning how to combine internal and external denoising methods. In: Weickert, J., Hein, M., Schiele, B. (eds.) GCPR 2013. LNCS, vol. 8142, pp. 121–130. Springer, Heidelberg (2013). https://doi.org/10.1007/978-3-642-40602-7_13
3. Cao, X., Yang, J., Gao, Y., Wang, Q., Shen, D.: Region-adaptive deformable regis-tration of CT/MRI pelvic images via learning-based image synthesis. IEEE Trans. Image Process. **27**(7), 3500–3512 (2018)
4. Chen, C.H., et al.: Computer-aided diagnosis of endobronchial ultrasound images using convolutional neural network. Comput. Methods Programs Biomed. **177**, 175–182 (2019)
5. Chen, H., et al.: Low-dose CT with a residual encoder-decoder convolutional neural network. IEEE Trans. Med. Imaging **36**(12), 2524–2535 (2017)
6. Chen, L., Chu, X., Zhang, X., Sun, J.: Simple baselines for image restoration. arXiv preprint arXiv:2204.04676 (2022)
7. Chen, Y., et al.: Improving abdomen tumor low-dose CT images using a fast dic-tionary learning based processing. Phys. Med. Biol. **58**(16), 5803 (2013)
8. De Man, B., Basu, S.: Distance-driven projection and backprojection in three dimensions. Phys. Med. Biol. **49**(11), 2463 (2004)

9. Dosovitskiy, A., et al.: An image is worth 16×16 words: transformers for image recognition at scale. arXiv preprint arXiv:2010.11929 (2020)
10. Feruglio, P.F., Vinegoni, C., Gros, J., Sbarbati, A., Weissleder, R.: Block matching 3D random noise filtering for absorption optical projection tomography. Phys. Med. Biol. **55**(18), 5401 (2010)
11. Kang, D., et al.: Image denoising of low-radiation dose coronary CT angiography by an adaptive block-matching 3D algorithm. In: Medical Imaging 2013: Image Processing, vol. 8669, pp. 671–676. SPIE (2013)
12. Lewitt, R.M.: Multidimensional digital image representations using generalized Kaiser-Bessel window functions. JOSA A **7**(10), 1834–1846 (1990)
13. Liang, T., Jin, Y., Li, Y., Wang, T.: EDCNN: edge enhancement-based densely connected network with compound loss for low-dose CT denoising. In: 2020 15th IEEE International Conference on Signal Processing (ICSP), vol. 1, pp. 193–198. IEEE (2020)
14. Luthra, A., Sulakhe, H., Mittal, T., Iyer, A., Yadav, S.: Eformer: edge enhancement based transformer for medical image denoising. arXiv preprint arXiv:2109.08044 (2021)
15. Ma, J., et al.: Low-dose computed tomography image restoration using previous normal-dose scan. Med. Phys. **38**(10), 5713–5731 (2011)
16. Mathews, J.P., Campbell, Q.P., Xu, H., Halleck, P.: A review of the application of X-ray computed tomography to the study of coal. Fuel **209**, 10–24 (2017)
17. McCollough, C.H., et al.: Low-dose CT for the detection and classification of metastatic liver lesions: results of the 2016 low dose CT grand challenge. Med. Phys. **44**(10), e339–e352 (2017)
18. Pan, X., Sidky, E.Y., Vannier, M.: Why do commercial CT scanners still employ traditional, filtered back-projection for image reconstruction? Inverse Prob. **25**(12), 123009 (2009)
19. Ramani, S., Fessler, J.A.: A splitting-based iterative algorithm for accelerated statistical X-ray CT reconstruction. IEEE Trans. Med. Imaging **31**(3), 677–688 (2011)
20. Ronneberger, O., Fischer, P., Brox, T.: U-Net: convolutional networks for biomedical image segmentation. In: Navab, N., Hornegger, J., Wells, W.M., Frangi, A.F. (eds.) MICCAI 2015. LNCS, vol. 9351, pp. 234–241. Springer, Cham (2015). https://doi.org/10.1007/978-3-319-24574-4_28
21. Shan, H., et al.: Competitive performance of a modularized deep neural network compared to commercial algorithms for low-dose CT image reconstruction. Nat. Mach. Intell. **1**(6), 269–276 (2019)
22. Vaswani, A., et al.: Attention is all you need. In: Advances in Neural Information Processing Systems, vol. 30 (2017)
23. Wang, D., Fan, F., Wu, Z., Liu, R., Wang, F., Yu, H.: CTformer: convolution-free token2token dilated vision transformer for low-dose CT denoising. arXiv preprint arXiv:2202.13517 (2022)
24. Wang, D., Wu, Z., Yu, H.: TED-Net: convolution-free T2T vision transformer-based encoder-decoder dilation network for low-dose CT denoising. In: Lian, C., Cao, X., Rekik, I., Xu, X., Yan, P. (eds.) MLMI 2021. LNCS, vol. 12966, pp. 416–425. Springer, Cham (2021). https://doi.org/10.1007/978-3-030-87589-3_43
25. Whiting, B.R., Massoumzadeh, P., Earl, O.A., O'Sullivan, J.A., Snyder, D.L., Williamson, J.F.: Properties of preprocessed sinogram data in X-ray computed tomography. Med. Phys. **33**(9), 3290–3303 (2006)
26. Wolterink, J.M., Leiner, T., Viergever, M.A., Išgum, I.: Generative adversarial networks for noise reduction in low-dose CT. IEEE Trans. Med. Imaging **36**(12), 2536–2545 (2017)

27. Wu, G., Kim, M., Wang, Q., Munsell, B.C., Shen, D.: Scalable high-performance image registration framework by unsupervised deep feature representations learning. IEEE Trans. Biomed. Eng. **63**(7), 1505–1516 (2015)
28. Yang, Q., et al.: Low-dose CT image denoising using a generative adversarial network with Wasserstein distance and perceptual loss. IEEE Trans. Med. Imaging **37**(6), 1348–1357 (2018)
29. Zamir, S.W., Arora, A., Khan, S., Hayat, M., Khan, F.S., Yang, M.H.: Restormer: efficient transformer for high-resolution image restoration. In: Proceedings of the IEEE/CVF Conference on Computer Vision and Pattern Recognition, pp. 5728–5739 (2022)
30. Zhang, W., et al.: Deep convolutional neural networks for multi-modality isointense infant brain image segmentation. Neuroimage **108**, 214–224 (2015)
31. Zhang, Y., et al.: Clear: comprehensive learning enabled adversarial reconstruction for subtle structure enhanced low-dose CT imaging. IEEE Trans. Med. Imaging **40**(11), 3089–3101 (2021)
32. Zhang, Z., Yu, L., Liang, X., Zhao, W., Xing, L.: TransCT: dual-path transformer for low dose computed tomography. In: de Bruijne, M., et al. (eds.) MICCAI 2021. LNCS, vol. 12906, pp. 55–64. Springer, Cham (2021). https://doi.org/10.1007/978-3-030-87231-1_6

iResSENet: An Accurate Convolutional Neural Network for Retinal Blood Vessel Segmentation

Proma Hossain Progga and Swakkhar Shatabda$^{(\boxtimes)}$

Department of Computer Science and Engineering, United International University, Plot-2, United City, Madani Avenue, Badda, Dhaka 1212, Bangladesh
pprogga192126@bscse.uiu.ac.bd, swakkhar@cse.uiu.ac.bd

Abstract. In this paper, we propose iResSENet, a novel deep learning based image segmentation model based on U-Net architecture. The proposed method enhances U-Net in three aspects. It replaces the encoder blocks with residual connections in addition to 1×1 convolutional layers and channel based attention. The proposed method was applied for segmentation task in retinal blood vessels. The experimental results shows that the proposed method is significantly superior compared to existing methods on several standard benchmark datasets.

Keywords: Channel based attention · 1×1 Convolution · Retinal Image Segmentation · Residual Networks

1 Introduction

Biomedical image processing has played a significant role in disease diagnosis and monitoring. Due to rapid technological advancements leading to high spatial and temporal resolution, multidisciplinary researchers have been working diligently to develop automated diagnosis systems. Segmentation is one of the most challenging tasks in biomedical domain. Biomedical image segmentation deals with partitioning an image into various regions that represent interesting anatomical objects in various types of 2D and 3D-medical images [1].

The retina is like a translator in a human eye that is responsible for vision. When light hits it, the retina transforms it into a signal the brain can interpret. The retinal artery and the vessels are the two primary blood sources for retina. Any variations or damages to the vascular morphology of the retinal blood vessels may lead to many diseases, including diabetic retinopathy, coronary heart disease, and atherosclerosis [2]. In diabetic retinopathy, the disease progression is imperceptible in the early stages and can only be detected by regular ophthalmic examinations which is a prolonged and labor-intensive task. Therefore, researchers have focused on developing automatic retinal vessel segmentation methods.

© The Author(s), under exclusive license to Springer Nature Switzerland AG 2023
M. Tanveer et al. (Eds.): ICONIP 2022, LNCS 13625, pp. 567–578, 2023.
https://doi.org/10.1007/978-3-031-30111-7_48

Over the past two decades, many different automatic retinal vessel segmentation methods have been developed by researchers, and they can be generally divided into two categories. One is image processing techniques, such as segmentation, pre-processing, and post-processing and another one is machine learning-based approaches, which primarily train a classifier to categorize pixels in the retina using the extracted vector features [1,2]. To extract retinal blood vessels from fundus images, a number of supervised and unsupervised methods have been suggested by researchers. Based on traditional machine learning, many approaches are applied before like thresholding, region-based, clustering, etc. [1,2].

Lately, deep learning-based methods have been used for retinal blood vessel segmentation and provide state-of-the-art performance. Since the breakthrough of Deep Convolutional Neural Networks (CNNs), several CNN based methods [3] have been proposed for retinal blood vessel segmentation. But a major drawback of the CNN architectures is that they require a huge amount of training data. However, CNNs have recently demonstrated significant promise in the segmentation of medical images [4]. For this, the majority of the credit goes to U-Net [5]. U-Net has the encoder and decoder network, that are connected by skip connections with feature upsampling to estimate both local and global image features. U-Net has demonstrated impressive results in segmenting medical images, even with a limited amount of labeled training data. However, it fails to segment thin vessels and pixels that are positioned at the edges of thicker vessels. Therefore, to enhance the extraction of the retinal vessels from the fundus images, several changes to the U-Net architecture have been accomplished such as AA Unet [6], Recurrent Residual Unet [7], H-DenseUNet [8], Swin-unet [9], etc. and the list goes on.

In this paper, we proposes a novel deep learning architecture based on U-Net. The encoder consists of four stages of convolutional blocks with channel based attention (SE) [10] in addition to 1×1 convolutional layers with residual connections (Res). SE block enhances the representational power of a network by enabling it to accomplish dynamic channel-wise feature recalibration. However, We have also applied GELU [11] as an activation function. The proposed architecture iResSENet achieves state-of-the-art performance in retinal blood vessel segmentation of fundus images on several standard benchmark datasets and significantly improves over the existing methods.

2 Related Work

Segmenting the vessels from retinal images has received a lot of attention over the past two decades. Earlier methods, such as matching filters, quadrature filters, and Gabor filters mainly depended on handcrafted features [2]. The latter approach, in particular, incorporates all vessel widths and orientations using a preset kernel bank. However, the handcrafted features are constrained by our ability to analytically model the segmentation technique. Since the breakthrough by convolutional neural networks (CNNs) on the ImageNet dataset, Deep learning has been extremely successful across a wide range of machine learning tasks.

In particular, the most successful deep architecture in this domain is U-net [5]. U-Net is the widely accepted architecture for biomedical segmentation, which consists of four encoder blocks and four decoder blocks that are connected via a bridge. By combining both features of expansive and contractive paths using depth concatenation, a good segmentation map can be predicted by the U-net architecture.

Since there aren't enough retinal images for network training and to improve the performance of a deep learning architecture, several researchers have explored the transfer learning techniques [2]. One of the extensions of the U-Net architecture for semantic medical image segmentation is called "CAR-UNet" [12]. Channel Attention Residual U- Net (CAR-UNet) considers the relationship between the feature channels, so a novel channel attention mechanism is introduced to strengthen the network's discriminative capability. Moreover, Sine-Net [13], is used to catch thin and thick vessel features using up-sampling and then down-sampling.

The potential successor MultiResUNet [14] of U-Net architecture has two incompatible sets of features, One is Res paths, which introduce some additional processing to make the two feature maps more homogeneous and another one is MultiRes blocks which are inspired by the Inception blocks and formulated a compact analogous structure, that was lightweight and demanded less memory. Contrarily, MultiResUNet seemed more dependable and durable in the experiments. MultiResUNet was robust in segmenting images with several perturbations and was rejectable to outliers. It was able to identify even the most modest boundaries. MultiResUNet was able to capture the finer features even while segmenting the majority class, where the U-Net tended to over-segment. Additionally, MultiResUNet's 3D adaption outperformed 3D U-Net in terms of performance.

However, Squeeze-and-Excitation Networks [10] shed some light on the inability of previous architectures to adequately model channel-wise feature dependencies. ConvNeXts [15], built entirely from standard ConvNet modules, compete favorably with Transformers in terms of accuracy and scalability, outperforming Swin Transformers on COCO detection and ADE20K segmentation while maintaining the simplicity and effectiveness of standard ConvNets. They achieved 87.8% ImageNet top-1 accuracy. In ConvNeXt, GELU [11] activation function is used. GELU (Gaussian Error Linear Unit) exceeded the accuracy of the ELU and ReLU consistently for the numerous datasets.

3   Proposed Architecture

The detailed architecture of our proposed method is illustrated in Fig. 1. Our proposed architecture is quite similar to U-net as it also has a contracting path (four encoder blocks) and an expansive path (four decoder blocks) with skip connections. In U-net, the contracting path follows the architecture of a convolutional network which consists of two 3×3 convolutions, each followed by a rectified linear unit (ReLU), and a 2×2 max pooling layer with a step size of 2 is used for downsampling.

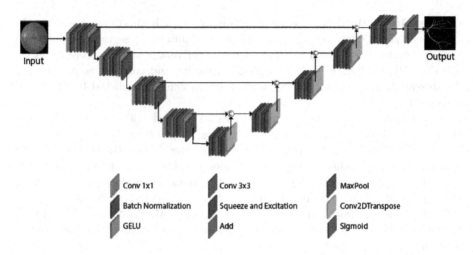

Fig. 1. The iResSENet architecture

3.1   Encoder Block

In the encoder section, a residual block with significant changes is used in place of the convolutional block. Generally, each residual block consists of stacked layers such as batch normalization (BN), ReLU, and weight layer (two 3 × 3 convolutional layers) [16]. Thus, three key enhancements have been made to the residual block in our suggested approach described in the following.

1×1 **Convolutional Layer.** A 1×1 convolutional layer is used instead of a 3×3 convolutional layer used in the residual block which enables it to be easily integrated into the current CNN pipelines. The most important use of 1×1 convolutions in our context is as dimension reduction modules to get rid of computational bottlenecks that would otherwise constrain the size of our networks. This enables us to expand our networks' depth and width without a significant performance penalty [17].

GELU. Instead of ReLU activation function, Gaussian Error Linear Unit, or GELU [11] is used in our proposed architecture. Due to GELU's better performance over ReLU, it is widely used as an activation function in both sector NLP and computer vision architectures. Moreover, Dropout regularization stochastically multiplies activations by 0 to make inactive while ReLU multiplies inputs with 0 or 1. However, GELU combines both functions by multiplying inputs by a value ranging from 0 to 1. Thus, GELU is a smoother variant of ReLU. Therefore, we used a single GELU activation in each block.

SE Residual Block. In a residual block, the output of one layer is added to another layer deeper in the block through a skip connection where ReLU activation is used, so that the summation of the input and the output of the residual block, should equal. Moreover, Skip connection between layers helps in better flow of gradient while backpropagation, which in turn helps the network to learn better representation. In general, with reference to the layer inputs, the residual block reformulates the layers as learning residual functions. However, our modified residual block with a 1×1 convolutional layer and GELU activation function also overcomes the problem of the degradation of a deeper network.

Moreover, Squeeze-and-Excitation recalibrates channel-wise feature responses through which it can learn to use global information to selectively emphasize informative features and suppress less useful ones. So, squeeze and excitation both act before summation with the identity branch in the residual block (Fig. 2). Therefore we added a squeeze and excitation block to the modified residual block.

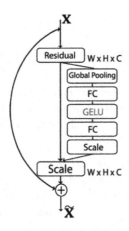

Fig. 2. SE-Residual

Furthermore, a 2×2 max pooling is used to reduce the dimensionality of the feature maps by taking the highest value from an identified region. This reduces the computational cost by decreasing the number of trainable parameters. In our model, trainable parameter: 7,530,973 and non- trainable parameter: 5,888.

However, The bridge completes the information flow by connecting the network of encoders and decoders. It consists of two 3×3 convolutions, where each convolution is followed by a GELU activation function.

3.2 Decoder Block

The decoder block starts with a 2×2 transpose convolution. The corresponding skip connection feature map from the encoder block is then concatenated with it. Then, two 3×3 convolutions are applied with a GELU activation function. Finally, it ends with a 1×1 convolution layer to map each feature vector to the desired number of classes with sigmoid activation. The network has 23 convolutional layers in total.

When compared to U-Net, the proposed architecture have multiple advantages. It does not increase the number of network parameters, so it has a significant impact on training and testing performance. Compared to 31,055,297 parameters of U-Net, iResSENet has only 7,536,861 parameters.

4   Experimental Analysis

For the experiments, we used Python programming language and network models implemented using the Keras framework with Tensorflow as the backend. All the experiments were conducted on a laptop with Intel core i5-8265U CPU (1.60 GHz) and 8 GB RAM.

4.1   Datasets

We have used four different datasets in our experiments. The details are given in the following.

Digital Retinal Images for Vessel Extraction (DRIVE) [18]: 40, This dataset is pre-divided into a training set of 20 images and a testing set of 20 images with a resolution of 565 × 584 pixels. Each image includes a circular field of view (FOV) mask (540 pixels).

Child Heart and Health Study (CHASEDB1) [19]: 28, Each image is 999 × 960 pixels. In this dataset, images were collected from both the left and right eyes of 14 school children. The dataset is partitioned into two sets where samples are selected randomly, of which the first 20 images are utilized for training, and the last 8 images are used for testing.

Structured Analysis of the Retina (STARE) [20]: 20, Each image resolution is 700 × 605 pixels. Due to the smaller number of samples, two approaches can be applied very often for training and testing on this dataset. First, training is sometimes performed with randomly selected samples from all 20 images. Another approach is the "leave-one-out" method, in which each image is tested, and training is conducted on the remaining 19 samples. Therefore, there is no overlap between training and testing samples. In this implementation, we used the "leave-one-out" approach for the STARE dataset.

High-Resolution Fundus (HRF) [21]: 45, Compared to the other three datasets, this dataset has a higher resolution which is 3,304 × 2,336 pixels. The HRF dataset consists of 15 healthy, 15 diabetic retinopathy, and 15 glaucomatous patient images. Therefore, we split the dataset into 22 training and 23 test images.

We have resized the DRIVE, CHASEDB1 and STARE to 512 × 512 pixels. We also crop a single image into 640 × 640 pixels, due to the high-resolution image in HRF dataset and the limited GPU capacity. In order to get more reasonable results, we augmented using Horizontal Flip, Vertical Flip, Transpose,

Random Rotation(90), Non-rigid transformations (Elastic Transform, Grid Distortion, Optical Distortion), Random Sized Crop(Allows to combine Random Crop and Random Scale). The OneOf technique is another intriguing aspect of this. Here, probabilities are allocated to the transformation specified in the OneOf block. The transformation with the greatest normalized value is chosen and applied to the image once these are normalized. This increases the effectiveness of implementing the appropriate transformations. Thus, from a single image, 13 images are generated including the original image. After augmentation, the training set instances were 260,260,247 and 286 for DRIVE, CHASEDB1, STARE and HRF datasets respectively.

4.2 Experimental Settings

To demonstrate the performance of iResSENet, we compare the outcomes of our retinal segmentation with the expert's provided ground truth. Therefore, Several performance metrics such as Accuracy (AC), Recall, Precision, F1-score, Jaccard similarity (JS), and Area Under the ROC Curve (AUC), are considered for the quantitative analysis of the experimental results. One of the most frequently used metrics for the medical image segmentation model is Area Under Curve. The AUC-ROC metric clearly helps determine and tell us about the capability of a model for evaluating the two classes. Higher AUC indicates a better model. Dice coefficient (DSC) and Intersection over Union (IoU) are two additional important matrices that are used in quantification. The amount of overlap between the predicted and ground truth is indicated by the IOU value, which ranges from 0 to 1.

The Dice similarity coefficient measures the spatial overlap between two segmentations. In the field of computer vision, the Dice coefficient is a widely known metric for determining how equivalent two images are. The DC is expressed as in Eq. 1 according to [22]. Here GT refers to the ground truth and SR refers the segmentation result.

$$DC = 2\frac{|GT \cap SR|}{|GT| + |SR|} \tag{1}$$

Later in 2016, it has also been adapted as a loss function known as Dice Loss [23]. Dice loss is defined as Eq. 2,

$$DL(y, \hat{p}) = 1 - \frac{2y\hat{p} + 1}{y + \hat{p} + 1} \tag{2}$$

Here, 1 is added in the numerator and denominator to ensure that the function is not undefined in edge case scenarios such as when $y = \hat{p} = 0$.

So, as the loss function of the network, we simply took the dice loss function and minimized it.

Moreover, we used GELU [11] as an activation function. Since the cumulative distribution of Gaussian distribution is often computed with the error function, hence Gaussian Error Linear Unit (GELU) is defined as Eq. 3,

$$GELU(x) = xP(X \le x) = x\Phi(x)$$
$$= \approx 0.5x(1 + tanh[\sqrt{2/\pi}(x + 044715x^3)]) \tag{3}$$

For all datasets, the training batch size is 2, and a total of 50 epochs are trained with a learning rate of 1×10^{-4} and utilized the Adam optimizer to optimize our network. Our implementation is built on TensorFlow.

4.3 Comparison with Existing Methods

We have compared the performances of our method with U-Net [5], DeepLabv3+ [24], MultiResUNet [14] and DR-VNet [25]. The results are reported in Table 1 (DRIVE and CHASE-DB) and Table 2 (STARE and HRF). Results are reported in terms of Dice Coefficient (DSC), IOU, Area Under Curve and Accuracy. In both of the tables, bold faced values indicate the best results achieved by any model.

From the tables, it can be observed that our proposed model outperforms the existing architecture in segmenting in terms of DSC and IOU. Figure 3 shows a comparison between U-Net and iResSENet for DRIVE, CHASE-DB, STARE and HRF datasets along with the ground truth. Also note that, the results reported in the tables shows comparative performance of iResSENet in terms of accuracy and area under curve. One interesting aspect of the resutls is that our proposed method works better in terms of DSC and IOU compared to the pixel based metrics like accuracy and AUC which do not reflect the performance of segmentation task in all situations.

Table 1. Results of Different Models on DRIVE and CHASEDB.

Methods	DRIVE				CHASE-DB			
	DSC	IOU	AUC	Accuracy	DSC	IOU	AUC	Accuracy
U-Net [5]	0.7074	0.5501	0.9255	0.9491	0.7292	0.5752	0.8948	0.9574
DeepLabv3+ [24]	0.6797	0.5180	0.9021	0.948	0.7707	0.6285	0.8747	0.9562
MultiResUNet [14]	0.7467	0.5992	0.8915	**0.9565**	0.7836	0.6467	0.8262	**0.9614**
DR-VNet [25]	0.6403	0.4743	0.9258	0.9405	0.6457	0.4818	**0.9157**	0.9487
iResSENet	**0.8603**	**0.7532**	**0.9259**	0.9496	**0.8716**	**0.7764**	0.8808	0.9568

4.4 Ablation Study

We have also performed ablation study to show the effectiveness of the proposed system. For ablation study, we have used U-Net as the baseline. Then we replaced the convolutional block with a residual block inspired by Deep Residual U-Net [26]. The core idea behind residual blocks, "skip connections", is what makes a neural network robust. The skip connections allow information to flow from the initial to the last layers. Moreover, The residual block will make network training easier. Moreover, we used max-pooling to extract low-level features.

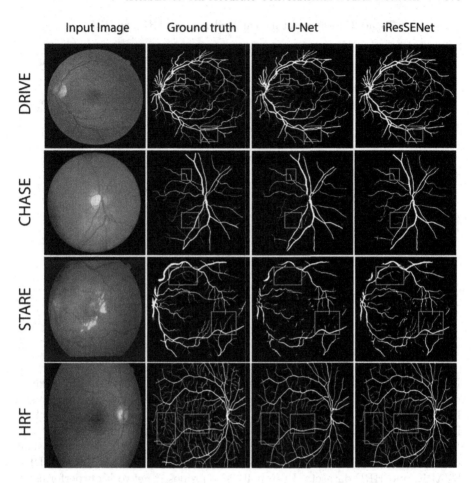

Fig. 3. Comparison between U-Net and iResSENet.

After that, we add the input to the residual block output after passing it through a squeeze-and-excitation (SE) block. As the SE block includes a simple gating mechanism that aims to improve the network's ability for representation by efficiently modeling channel-wise relationships. In addition, we change filter size for channel-wise pooling, and dimensionality reduction.

We have included these variants in the ablation experiments and employed two datasets: DRIVE and CHASE. The results in terms of DSC and IOU are reported in Table 3. The results reported in Table 3 shows the relative strength of each of the components.

Table 2. Results of Different Models on STARE and HRF.

Methods	STARE				HRF			
	DSC	IOU	AUC	Accuracy	DSC	IOU	AUC	Accuracy
U-Net [5]	0.6977	0.5369	0.8228	0.9502	0.6847	0.5250	0.8720	0.9556
DeepLabv3+ [24]	0.7234	0.5837	0.8045	0.9421	0.6216	0.4576	0.8189	0.9518
MultiResUNet [14]	0.7578	0.6111	0.8593	**0.9557**	0.7078	0.5547	0.8451	**0.9580**
DR-VNet [25]	0.6418	0.4763	0.8658	0.9409	0.6004	0.4354	0.8642	0.9475
iResSENet	**0.85018**	**0.7405**	**0.8734**	0.9464	**0.7882**	**0.6561**	**0.8801**	0.9531

Table 3. Ablation Study.

Models	DRIVE		CHASE	
	DSC	IOU	DSC	IOU
U-Net (baseline)	70.74	55.01	72.92	57.52
U-Net+Residual block	84.31	73.06	87.28	77.49
U-Net+SE+Residual block	83.17	71.35	85.80	75.21
iResSENet-SE	84.88	73.89	86.54	76.34
iResSENet	**86.03**	**75.32**	**87.16**	**77.64**

5 Conclusion

In this work, we have proposed the iResSENet for retinal blood vessel segmentation which considers the relationship between the feature channels. Thus, a lightweight gating mechanism SE block is introduced with residual block to improve the network's effectiveness. In order to ease training, we also used Inception blocks as inspiration for changing filter size. Our experiments show that iResSENet reaches state-of-the-art performance on DRIVE, CHASE DB1, STARE, and HRF datasets. Eventually, the proposed network outperforms U-Net with only a quarter of its parameters.

References

1. Chouhan, S.S., Kaul, A., Singh, U.P.: Soft computing approaches for image segmentation: a survey. Multimedia Tools Appl. **77**(21), 28483–28537 (2018). https://doi.org/10.1007/s11042-018-6005-6
2. Panda, N.R., Sahoo, A.K.: A detailed systematic review on retinal image segmentation methods. J. Digit. Imaging **35**(5), 1250–1270 (2022). https://doi.org/10.1007/s10278-022-00640-9
3. Badrinarayanan, V., Kendall, A., Cipolla, R.: SegNet: a deep convolutional encoder-decoder architecture for image segmentation. IEEE Trans. Pattern Anal. Mach. Intell. **39**(12), 2481–2495 (2017)
4. Anwar, S.M., Majid, M., Qayyum, A., Awais, M., Alnowami, M., Khan, M.K.: Medical image analysis using convolutional neural networks: a review. J. Med. Syst. **42**(11), 1–13 (2018). https://doi.org/10.1007/s10916-018-1088-1

5. Ronneberger, O., Fischer, P., Brox, T.: U-Net: convolutional networks for biomedical image segmentation. In: Navab, N., Hornegger, J., Wells, W.M., Frangi, A.F. (eds.) MICCAI 2015. LNCS, vol. 9351, pp. 234–241. Springer, Cham (2015). https://doi.org/10.1007/978-3-319-24574-4_28

6. Lv, Y., Ma, H., Li, J., Liu, S.: Attention guided U-Net with atrous convolution for accurate retinal vessels segmentation. IEEE Access **8**, 32826–32839 (2020)

7. Alom, M.Z., Hasan, M., Yakopcic, C., Taha, T.M., Asari, V.K.: Recurrent residual convolutional neural network based on U-Net (R2U-Net) for medical image segmentation. arXiv preprint arXiv:1802.06955 (2018)

8. Li, X., Chen, H., Qi, X., Dou, Q., Chi-Wing, F., Heng, P.-A.: H-DenseUNet: hybrid densely connected U-Net for liver and tumor segmentation from CT volumes. IEEE Trans. Med. Imaging **37**(12), 2663–2674 (2018)

9. Cao, H.: Swin-Unet: Unet-like pure transformer for medical image segmentation. arXiv preprint arXiv:2105.05537 (2021)

10. Hu, J., Shen, L., Sun, G.: Squeeze-and-excitation networks. In: Proceedings of the IEEE Conference on Computer Vision and Pattern Recognition, pp. 7132–7141 (2018)

11. Hendrycks, D., Gimpel, K.: Gaussian error linear units (GELUs). arXiv preprint arXiv:1606.08415 (2016)

12. Guo, C., Szemenyei, M., Hu, Y., Wang, W., Zhou, W., Yi, Y.: Channel attention residual U-Net for retinal vessel segmentation. In: ICASSP 2021–2021 IEEE International Conference on Acoustics, Speech and Signal Processing (ICASSP), pp. 1185–1189. IEEE (2021)

13. Atli, I., Gedik, O.S.: Sine-Net: a fully convolutional deep learning architecture for retinal blood vessel segmentation. Eng. Sci. Technol. Int. J. **24**(2), 271–283 (2021)

14. Ibtehaz, N., Rahman, M.S.: MultiResUNet: rethinking the U-Net architecture for multimodal biomedical image segmentation. Neural Netw. **121**, 74–87 (2020)

15. Liu, Z., Mao, H., Wu, C. Y., Feichtenhofer, C., Darrell, T., Xie, S.: A convnet for the 2020s. In: Proceedings of the IEEE/CVF Conference on Computer Vision and Pattern Recognition, pp. 11976–11986 (2022)

16. Khanna, A., Londhe, N.D., Gupta, S., Semwal, A.: A deep residual U-Net convolutional neural network for automated lung segmentation in computed tomography images. Biocybernetics Biomed. Eng. **40**(3), 1314–1327 (2020)

17. Szegedy, C., et al.: Going deeper with convolutions. In: Proceedings of the IEEE Conference on Computer Vision and Pattern Recognition, pp. 1–9 (2015)

18. Staal, J., Abràmoff, M.D., Niemeijer, M., Viergever, M.A., Van Ginneken, B.: Ridge-based vessel segmentation in color images of the retina. IEEE Trans. Med. Imaging **23**(4), 501–509 (2004)

19. Owen, C.G., et al.: Measuring retinal vessel tortuosity in 10-year-old children: validation of the computer-assisted image analysis of the retina (CAIAR) program. Invest. Ophthalmol. Vis. Sci. **50**(5), 2004–2010 (2009)

20. Hoover, A.D., Kouznetsova, V., Goldbaum, M.: Locating blood vessels in retinal images by piecewise threshold probing of a matched filter response. IEEE Trans. Med. Imaging **19**(3), 203–210 (2000)

21. Budai, A., Bock, R., Maier, A., Hornegger, J., Michelson, G.: Robust vessel segmentation in fundus images. Int. J. Biomed. Imaging **2013** (2013)

22. Dice, L.R.: Measures of the amount of ecologic association between species. Ecology **26**(3), 297–302 (1945)

23. Sudre, C.H., Li, W., Vercauteren, T., Ourselin, S., Jorge Cardoso, M.: Generalised dice overlap as a deep learning loss function for highly unbalanced segmentations.

In: Cardoso, M.J., et al. (eds.) DLMIA/ML-CDS -2017. LNCS, vol. 10553, pp. 240–248. Springer, Cham (2017). https://doi.org/10.1007/978-3-319-67558-9_28

24. Chen, L.C., Zhu, Y., Papandreou, G., Schroff, F., Adam, H.: Encoder-decoder with atrous separable convolution for semantic image segmentation. In: Proceedings of the European Conference on Computer Vision (ECCV), pp. 801–818 (2018)

25. Karaali, A., Dahyot, R., Sexton, D.J.: DR-VNet: retinal vessel segmentation via dense residual UNet. In: El Yacoubi, M., Granger, E., Yuen, P.C., Pal, U., Vincent, N. (eds.) Pattern Recognition and Artificial Intelligence. ICPRAI 2022. Lecture Notes in Computer Science, vol. 13363, pp 198–210. Springer, Cham (2022). https://doi.org/10.1007/978-3-031-09037-0_17

26. Zhang, Z., Liu, Q., Wang, Y.: Road extraction by deep residual U-Net. IEEE Geosci. Remote Sens. Lett. **15**(5), 749–753 (2018)

Evolutionary Action Selection
for Gradient-Based Policy Learning

Yan Ma[1], Tianxing Liu[1], Bingsheng Wei[1], Yi Liu[1], Kang Xu[1], and Wei Li[1,2(✉)]

[1] Academy for Engineering and Technology, Fudan University, Shanghai, China
fd_liwei@fudan.edu.cn
[2] Ji Hua Laboratory, Foshan, China

Abstract. Evolutionary Algorithms (EAs) and Deep Reinforcement Learning (DRL) have recently been integrated to take advantage of both methods for better exploration and exploitation. The evolutionary part of these hybrid methods maintains a population of policy networks. However, existing methods focus on optimizing the parameters of policy network, which is usually high-dimensional and tricky for EA. In this paper, we shift the target of evolution from high-dimensional parameter space to low-dimensional action space. We propose Evolutionary Action Selection-Twin Delayed Deep Deterministic Policy Gradient (EAS-TD3), a novel hybrid method of EA and DRL. In EAS, we focus on optimizing the action chosen by the policy network and attempt to obtain high-quality actions to promote policy learning through an evolutionary algorithm. We conduct several experiments on challenging continuous control tasks. The result shows that EAS-TD3 shows superior performance over other state-of-art methods.

Keywords: Evolutionary Algorithm · Reinforcement Learning · Policy Search

1 Introduction

Deep Reinforcement Learning (DRL) has achieved impressive performance in Go [28], Atari games [19], and continuous control tasks [10,15]. The purpose of DRL is to train an optimal or nearly-optimal policy network that maximizes the reward function or other user-provided reinforcement signal. Recently, Evolutionary Algorithms (EA) have been applied to search the parameter space of neural networks to train policies and showed competitive results as DRL [24,30]. Some work compares DRL and EA [16,25], highlighting their respective pros and cons. However, an emerging research direction tends to integrate them to promote each other for better policy search [23,27], in that they have several complementary properties and can be integrated to benefit from the best of both worlds. As one of the pioneers, [13] proposed Evolutionary Reinforcement Learning (ERL), which has shown promising results in continuous control tasks. ERL maintains a population containing n policy networks trained by EA and one policy network trained by DRL. The policy population provides a large number of samples

M. Tanveer et al. (Eds.): ICONIP 2022, LNCS 13625, pp. 579–590, 2023.
https://doi.org/10.1007/978-3-031-30111-7_49

(s, a, r, s') to the RL policy, and RL policy will be injected into the population periodically to replace the poorer individual. With the help of EA's global optimization ability, ERL explores the parameter space of the policy network to search for outstanding solutions. Most variants of the ERL [2,12,17,22] follow the parameter space of the policy network as the target of evolution. However, taking the policy network parameters as the evolution target may pose a potential problem. It is well known that EA is deficient in optimizing high-dimensional spaces. Although EA can be leveraged to train policy networks, it suffers from low sample efficiency since it is gradient-free. Corresponding to ERL methods, the contribution of the evolutionary part will be limited. In particular, we find that the evolutionary part may weaken the performance of the overall method at high dimensions of the parameter space. As illustrated in Fig. 1c (see more details in Sect. 4.1), the performance of ERL method decays with the increase of policy parameters. The problem in ERL methods is attributed to the use of EA to train the strategy network directly. In a nutshell, taking the parameter space of the policy network as the target of evolution will limit the contribution of EA to such hybrid methods. The motivation of this work is to transfer the evolutionary target from the high dimensional parameter space to a low dimensional space where EA is more proficient and avoids the problem mentioned earlier to serve the best of both worlds better.

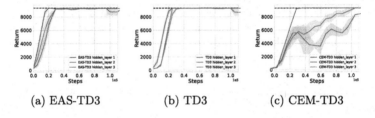

(a) EAS-TD3   (b) TD3   (c) CEM-TD3

Fig. 1. Comparative Experiment of EAS-TD3 (our work), CEM-TD3 (one of the sota methods in ERL variants), TD3 on Inverted Double Pendulum task from Gym [3] with different policy network parameters. The performance of CEM-TD3 decreases while that of EAS-TD3 remains stable.

Our goal is unambiguous, which aims to train an excellent policy to maximize the cumulative reward of the given task. A good policy $\mu(a|s)$ means it can choose a good action a according to the state s to maximize the cumulative reward. The quality of the action largely determines the quality of the policy. More importantly, the action space is relatively low dimensional and generally will not change for a given task. Thus, we intend to take the action space as the target of evolution. Concretely speaking, we leverage EA to optimize the action chosen by the RL policy and obtain better evolutionary actions. The policy network trained by RL collects samples in the environment. EA extracts actions from samples and evolves better evolutionary actions to promote the RL policy learning. It maintains the flow of information between the EA part and RL part.

Based on the above insight, we propose Evolutionary Action Selection (EAS). EAS utilizes actions selected by RL policy to form a population and utilizes Particle Swarm Optimization (PSO) [11] to evolve the action population from generation to generation. Finally, we obtain better evolutionary actions, which can promote the learning process through the evolutionary action gradient. We choose PSO for two reasons. Firstly, it has been widely used in many fields and proved effective in practice [32]. Secondly, it is easy to follow and will not add too much computational burden. In addition, PSO can be replaced by other evolutionary algorithms, such as genetic algorithm [18], cross-entropy method [7].

Our contributions are threefold: (1) We empirically demonstrate that taking the policy network parameter space as the evolutionary target may lead to performance degradation. (2) We transfer the target of evolution from high-dimensional parameter space to low-dimensional action space and propose a simple and effective mechanism to evolve action. (3) We apply EAS to TD3 [8] as EAS-TD3 and conduct a series of empirical studies on a benchmark suite of continuous control tasks to prove the feasibility and superiority of our approach.

2   Related Work

The idea of incorporating learning with evolution has been around for many years [1,9]. With the brilliance of reinforcement learning, recent literature [5,17] has begun to revisit the combination of the two to improve the performance of the overall approach.

As mentioned earlier, this paper is related to the recently proposed Evolutionary Reinforcement Learning (ERL) [13] framework. ERL combines Genetic Algorithm (GA) [18] with the off-policy DRL algorithm (DDPG) [4] and incorporates the two processes to run concurrently, formulating a framework. Specifically, ERL maintains a policy population trained by EA and a policy network trained by RL. By maintaining interactive information flow between EA and RL, the performance of the overall method is promoted. The ERL framework has triggered various variants, making the efficient combination of EA and RL an emerging research direction for both the EA and RL community. Collaborative Evolutionary Reinforcement Learning (CERL) [12] is the follow-up work of ERL. CERL attempts to train multiple policy networks with different hyperparameters to address the DRL's sensitivity to hyperparameters. Moreover, Proximal Distilled Evolutionary Reinforcement Learning (PDERL) [2] attempts to figure out the catastrophic forgetting of the neural network caused by the genetic operator used in ERL. CEM-RL [22] removes the single policy network trained by RL and instead allows half of the policy networks in the population to be trained directly by RL and the other half by cross entropy method (CEM) [7]. This approach magnifies the impact of gradient-based policy learning methods on the evolutionary population, which improves the sample efficiency of the ERL framework. AES-RL [14] proposes an efficient asynchronous method for integrating evolutionary and gradient-based policy search, which shortens the training time. QD-PG [21] introduces Quality-Diversity (QD) algorithm for RL to address the problem of deceptive reward.

Our approach focuses on integrating the advantages of EA and DRL to learn policies better. We shift the target of evolution from high-dimensional policy network parameter space to low-dimensional action space and empirically demonstrate that evolving the action is a favorable alternative to evolving the parameters of the policy network.

3   Methodology

3.1   Evolutionary Action Selection (EAS)

Taking the critic network Q_{μ_θ} of TD3 as the fitness evaluator, EAS follows the process of PSO and evolves the action a chosen by the current policy network $\mu_\theta(s)$. EAS's output is the evolutionary action a^e with a higher Q value than action a. The pseudocode of EAS is shown in Algorithm 1.

In EAS, we first add Gaussian noise to the action a and make it an action set $\mathbb{A}$ containing multiple noisy actions, which serves as the initial population. The fitness of each action is its Q value generated by the critic network Q_{μ_θ}. Secondly, we initialize the velocity vector, which determines the direction and step length when updating actions in the population. Then, we initialize the personal best action set $\mathcal{P} = (p_1^b...p_n^b...p_N^b)$, which records the best solution of each action in the population found so far. The best action in $\mathcal{P}$ is called the global best action g^b, which represents the best solution found so far. In each generation, action a_n^t in the population will be evaluated by fitness evaluator Q_{μ_θ} to obtain the fitness Q_n^t. Based on the magnitude of Q_n^t, we update the personal and global best action. Moreover, the velocity is updated by Eq. 2, which subsequently will be used to update actions and get the next generation of population. In Eq. 2, inertia weight ω describes previous velocity's influence on current velocity. Acceleration coefficients c_1 and c_2 represent the acceleration weights toward the personal best action and the global best action. r_1 and r_2 are random variables uniformly distributed in $[0, 1]$. Through several iterations, we can obtain the global best action g^b with the highest Q value searched so far and denote it as the evolutionary action a^e. The relationship of Q value between a^e and a is:

$$Q_{\mu_\theta}(s, a^e) \geq Q_{\mu_\theta}(s, a) \tag{1}$$

which reveals that EAS can increase the Q value of the action so that the action will have a higher expected reward. We claim the changing from a to a^e is the action evolution. The evolutionary action a^e is better than a and has a higher expected reward.

3.2   EAS-TD3 Framework

Figure 2 illustrates a diagram of EAS-TD3. At each timestep t, the policy observes s_t and outputs action a_t. Then, we receive a reward r_t and environment transitions to next state s_{t+1}. These four elements make up the sample

---

Algorithm 1. Evolutionary Action Selection

Input: State s, action a, critic network Q_{μ_θ}

PSO parameters: Inertia weight ω, acceleration coefficients c_1, c_2, random coefficients r_1, r_2

Output: Evolutionary action a^e

1: Extend the action a with Gaussian noise ϵ to form the initial action population $\mathbb{A} = (a_1...a_n...a_N)$, $a_n = a + \epsilon_n, \epsilon \in \mathcal{N}(0, \sigma)$, N is the number of actions

2: Initialize the velocity of action $\mathcal{V} = (v_1...v_n...v_N)$, $|v_n| = |a_n| = D$, $v_n \in [-v_{max}^D, v_{max}^D]$

3: Initialize personal best action set $\mathcal{P} = (p_1^b...p_n^b...p_N^b)$ and global best action g^b

4: **for** $t = 1$ **to** T **do**

5:     **for** $n = 1$ **to** N **do**

6:         $Q_n^t \leftarrow Q_{\mu_\theta}(s, a_n^t)$ ▷ Evaluate the action

7:         $p_n^b \leftarrow a_n^t$; **if** $Q_n^t > Q_{p_n^b}$ ▷ Update the personal best action

8:         $g^b \leftarrow p_n^b$; **if** $Q_{p_n^b} > Q_{g^b}$ ▷ Update the global best action

9:         $v_n^{t+1} = \omega * v_n^t + c_1 * r_1 * (p_n^b - a_n^t) + c_2 * r_2 * (g^b - a_n^t)$ (2)   ▷ Update the velocity

10:       $a_n^{t+1} = a_n^t + v_n^{t+1}$ (3)   ▷ Update the action

11:     **end for**

12: **end for**

13: Obtain the global best action g^b, representing evolutionary action a^e

---

(s_t, a_t, r_t, s_{t+1}), which will be stored into the replay buffer R. Then, EAS performs evolution on action a to obtain evolutionary action a^e, which will be stored into an archive $\mathcal{A}$. We draw the same mini-batches from $\mathcal{R}$ and $\mathcal{A}$ and update the current policy μ_θ with deterministic policy gradient [29] and evolutionary action gradient (will be described below). EAS adopts the way of delayed policy updates: one policy update for two Q function updates, which is the same as TD3.

Through EAS, we obtain evolutionary actions, storing their corresponding state-action pairs (s, a^e) into an archive $\mathcal{A}$. As mentioned in Sect. 3.1, the evolutionary action has a higher expected reward than the original action. Consequently, we intend to make the action space of RL policy similar to the space of evolutionary actions so that the evolutionary actions can contribute to the RL policy learning. Due to the idea given above, we construct a loss as below:

$$L_{evo}(\theta, \mathcal{A}) = \mathbb{E}_{(s_i, a_i^e) \sim \mathcal{A}} \left[\|\mu_\theta(s_i) - a_i^e\|^2 \right] \tag{4}$$

where s_i and a_i^e represent the state and evolutionary action sampled from $\mathcal{A}$ respectively, and θ represents the learning parameters in RL policy μ_θ. We call L_{evo} *evolutionary action gradient*. Furthermore, we weigh the constructed loss with an extra Q_{filter} as proposed in [20]:

$$Q_{filter} = \begin{cases} 1, & if \quad Q_{\mu_\theta}(s_i, a_i^e) > Q_{\mu_\theta}(s_i, \mu_\theta(s_i)), \\ 0, & else. \end{cases} \tag{5}$$

The purpose of Q_{filter} is to drop out $L_{evo}(\theta, \mathcal{A})$ when the action chosen by the current RL policy is superior to the evolutionary actions in archive $\mathcal{A}$. The reason is that if an evolutionary action generated long ago is sampled to update parameters, the action chosen by the current policy may be better than the previous evolutionary action. Thus, we need to filter out those outdated evolutionary actions. We periodically draw a batch of state-action pairs from $\mathcal{A}$ and utilize Eq. 6 to update parameters.

$$\nabla_\theta L_{Q_{filter}evo} = Q_{filter} \nabla_\theta L_{evo} \tag{6}$$

EAS promotes the evolution of actions, which can be used to guide strategy learning. The introduction of Q_{filter} effectively avoids a poor direction for learning when the action chosen by the current policy is superior to the previous evolutionary action in the archive. We refer to the dual integration with TD3 as Evolutionary Action Selection-Twin Delayed DDPG (EAS-TD3).

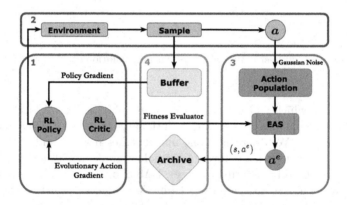

Fig. 2. A high-level view of EAS-TD3. RL policy interacts with the environment to generate the sample (s, a, r, s'), which will be stored in a replay buffer. Action a is added with Gaussian noise to form an evolutionary population. Then use EAS to update the action population to obtain the evolutionary action a^e. The state-action pair (s, a^e) will be stored in an archive. We draw batches from the replay buffer and archive to update the current RL policy with policy gradient and evolutionary action gradient.

4   Experiment

The main purpose of this section is to investigate the mechanism of EAS and the performance of EAS-TD3 compared to other evolutionary reinforcement learning methods.

4.1   A Toy Example

As mentioned in Sect. 1, ERL methods generally choose the high-dimensional parameter space as the target of evolution, which is tricky for EA to optimize and may lead to the collapse performance of the overall approach. To confirm this, we set up a toy example. Specifically, we increase the policy network's hidden layer and conduct experiments on the Inverted Double Pendulum from OpenAI gym. As a relatively easy continuous control task for most modern algorithms, Inverted Double Pendulum has an 11-dimensional state space and 1-dimensional action space.

Figure 1 shows the comparative performance of EAS-TD3, CEM-TD3, TD3. As shown in Fig. 1c, the performance of CEM-TD3 decays with increasing policy parameters. With the increase of hidden layers, the parameter space dimension that needs to be optimized increases dramatically. It makes it difficult for CEM to search for decent parameters of the policy network, which attenuates the performance of policy population and the contribution of the evolutionary part. Then, the stagnant policy population produces worthless samples to store in the replay buffer. The RL part may be drowning in these useless samples. In short, taking the policy network parameter space as the evolutionary target may lead to performance degradation. Correspondingly, the performance of EAS-TD3 is not affected as shown in Fig. 1a, since its evolutionary part optimizes the low-dimensional action space, which is rarely changed for a given task.

4.2   Experimental Setup

We select four continuous control locomotion tasks from OpenAI Gym simulated by MuJoCo [31]: HalfCheetah-v3, Walker2d-v3, Ant-v3, Humanoid-v3. We present the average reward and the associated standard deviation over ten runs. We test the learning policy for each run on ten evaluation episodes every 5000 steps. In all figures of learning curves, unless specified otherwise, the x-axis represents the number of steps performed in the environment, and the y-axis represents the mean return obtained by the policy.

By default, all the hyperparameters for TD3 are the same as those in the original paper [8]. Here, we only provide the unique hyper-parameters of the EAS-TD3. With regard to PSO, inertia weight ω is 1.2, acceleration coefficients c_1, c_2 are both 1.5, the number of iterations T is 10, v_{max} is 0.1, random coefficients r_1, r_2 are random numbers from 0 to 1. The size of archive $\mathcal{A}$ is 100,000 for all environments. The action population size is 10. Our method is compared against the official implementations for TD3 [26], CEM-TD3 [22], CERL [12], PDERL [2], ERL [13]. These baselines contain a series of studies on the combination of EA and RL. The population size of all baselines is set to 10.

4.3   Experiments on MuJoCo Environments

Figure 3 demonstrates the reward curves in MuJoCo. In general, EAS-TD3 performs consistently well across most MuJoCo environments, indicating that EAS plays a significant role in the process of policy learning.

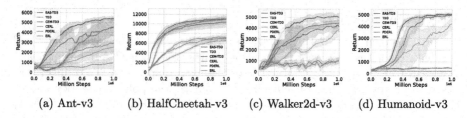

(a) Ant-v3	(b) HalfCheetah-v3	(c) Walker2d-v3	(d) Humanoid-v3

Fig. 3. The learning curves in MuJoCo environments. The shaded area represents mean ± standard deviation over the 10 runs.

Compared to ERL Methods. EAS-TD3 performs a considerable improvement on environments with high-dimensional state and action space like Humanoid, Ant, Walker2d. Why are ERL methods not performing as well as EAS-TD3 on these tasks? Note that the parameters of the policy network are associated with the state and action dimensions of the environment. For environments with higher dimensions, their policy networks also have more parameters, and more timesteps are needed for EA to search for sound policies. For example, a policy network in Humanoid-v3 task (state space 376 dims & action space 17 dims) consisting of a hidden layer with 400 and 300 nodes will have more than 270,000 parameters, which takes more timesteps to learn a decent policy for gradient-free evolutionary methods. Therefore, in high-dimensional tasks like Humanoid and Ant, the contribution of the evolutionary part in ERL methods will be weakened, resulting in the reduction of learning efficiency or even complete failure, which is consistent with the performance shown in the toy example. However, EAS-TD3 shifts the target of evolution from high-dimensional parameter space to low-dimensional action space to avoid the disaster of parameter dimension growth. The evolutionary part in EAS-TD3 focuses on evolving better actions and promoting policy learning through evolutionary action gradients. It is the main reason why EAS-TD3 performs better. Besides, CEM-TD3 performs better on HalfCheetah, probably because CEM can search for decent policies on these simple tasks. Moreover, we remark that the evolutionary part may even hinder the early learning of RL part in ERL methods. It is reflected in Ant and Walker2d tasks, where CEM-TD3, the best performer among the ERL methods, has a slightly slower learning efficiency than TD3. The reason may be attributed to the lackluster performance of the evolution population, which brings a vast amount of worthless experiences to the replay buffer, thus hindering the learning process of RL part. EAS-TD3 has no such concern since the evolution part does not directly train the policy network.

Compared to TD3. EAS-TD3 outperforms TD3 more or less in all environments. Compared with TD3, the increment of EAS-TD3 only lies in the influence of the evolutionary action gradient, which indicates the performance enhancement does come from the introduction of EAS. The evolutionary action can promote policy learning within different tasks. In HalfCheetah, EAS-TD3 achieves a minor improvement in experiments with multiple random seeds. Since

TD3 has already shown promising results in this environment, the evolutionary action can slightly enhance the sampling efficiency and performance. At the same time, EAS-TD3 performs well in high-dimensional complex tasks such as Ant and Humanoid, as the gradient of evolutionary action rapidly drives the policy's action space toward regions with higher expected reward and significantly improves learning speed and final performance. We will discuss more how EAS works in the following section.

4.4 Ablation Studies

The Size of Archive $\mathcal{A}$. The capacity of archive $\mathcal{A}$ is generally not significant in that the policy will continue to learn while evolutionary action selection is based on the policy at that time. Therefore, we should control the size of $\mathcal{A}$ to keep evolutionary actions fresh. Otherwise, the outdated evolutionary actions will not play a role in guiding policy learning. We select different archive sizes (10,000, 50,000, 100,000, 500,000) and perform

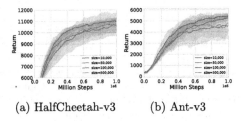

(a) HalfCheetah-v3   (b) Ant-v3

Fig. 4. Ablation study (mean ± standard deviation) on the size of archive $\mathcal{A}$. 100,000 seems to be a reasonable size.

an ablation study on the size of $\mathcal{A}$. As shown in Fig. 4, the archive size affects the performance. $\mathcal{A}$ stores state-action pairs (s, a^e) corresponding to evolutionary actions. As discussed earlier, since the evolutionary actions need to keep fresh, the archive size should not be too large. Otherwise, the archive $\mathcal{A}$ will store outdated actions, which may weaken the role of evolutionary action. In contrast, it should not be too small, or the diversity of actions in $\mathcal{A}$ may be poor. It will lead to the frequent sampling of limited evolutionary actions and hinder the exploration ability of policy. From Fig. 4, 100,000 seems to be a reasonable size for these tasks.

4.5 Evolutionary Action Evaluation

In this section, we will investigate the mechanism of EAS and mainly answer a question: how do evolutionary actions promote policy learning?

Figure 5 shows the first two-dimensional distributions of actions chosen by the policy and the corresponding evolutionary actions during the training process of Walker2d-v3. Each plot demonstrates the distribution of all actions in every 100,000 timesteps. The performance of the policy will gradually improve over timesteps. Therefore, the action distribution plot of each column in Fig. 5 indicates a higher expected reward than the previous column. From Fig. 5, we can see that evolutionary actions predict the action space with higher expected reward in advance and guide the current policy to move towards there, which may explain why EAS works. For example, within the 0 to 100,000 timesteps (corresponding to the two plots in the first column), the distribution of the

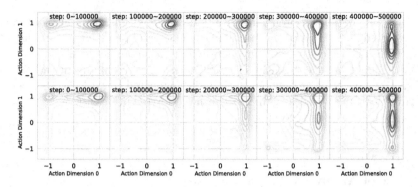

Fig. 5. The distribution of the first two action dimensions during the training process. The red contour represents the current policy's actions and the blue contour represents the corresponding evolutionary actions. The performance of the policy will gradually improve over timesteps. The action distribution plot of each column indicates a higher expected reward than the previous column. Evolutionary actions predict the action space with a higher expected reward in advance and guide the current policy to move towards there. (Color figure online)

evolutionary action exhibits a trend from the upper right downward. The downward trend subsequently becomes more pronounced. Correspondingly, the policy action distribution shows the same trend under the guidance of both the policy gradient and evolutionary action gradient. It indicates that the evolutionary action gradient promotes strategy learning in the same direction as the policy gradient, which significantly accelerates the progress of policy learning. From Fig. 5, we also discover that the space of evolutionary actions does not deviate too much from the current policy's action space, which is served as a gentle mentor and guides the current policy step by step. In conclusion, with the aid of evolutionary action gradient, we can make good use of evolutionary actions and finally play a role in the learning process.

5   Conclusion and Further Work

This paper proposes Evolutionary Action Selection (EAS) which applies evolution to the action chosen by the policy. We integrate EAS into TD3 as EAS-TD3, which is a novel hybrid method of EA and RL. Compared with other hybrid methods, our approach transforms the target of evolution from the high-dimensional parameter space to the low-dimensional action space. With the promotion of evolutionary actions, we train policies that substantially outperform those trained directly by TD3 and other compared methods. Extensive experiments and empirical analyses demonstrate that evolving action is a favorable alternative to evolving the parameters of the policy network. We believe that our work can trigger follow-up studies to address several interesting open questions. In future work, we can integrate the idea of Quality-Diversity (QD) [6], which evolves not only high-quality actions but also evolve various actions to deal with the environment with deceptive rewards.

Acknowledgments. This work was supported in part by Shanghai Municipal Science and Technology Major Project (No. 2021SHZDZX0103), in part by Ji Hua Laboratory (No. X190011TB190), in part by the Shanghai Engineering Research Center of AI and Robotics, and in part by the Engineering Research Center of AI and Robotics, Ministry of Education, China.

References

1. Ackley, D.: Interactions between learning and evolution. Artif. Life II **10**, 487–509 (1992)
2. Bodnar, C., Day, B., Lió, P.: Proximal distilled evolutionary reinforcement learning. In: Proceedings of the AAAI Conference on Artificial Intelligence, pp. 3283–3290 (2020)
3. Brockman, G., et al.: OpenAI gym. arXiv preprint arXiv:1606.01540 (2016)
4. Casas, N.: Deep deterministic policy gradient for urban traffic light control. arXiv preprint arXiv:1703.09035 (2017)
5. Colas, C., Sigaud, O., Oudeyer, P.Y.: GEP-PG: decoupling exploration and exploitation in deep reinforcement learning algorithms. In: International Conference on Machine Learning, pp. 1039–1048. PMLR (2018)
6. Cully, A., Clune, J., Tarapore, D., Mouret, J.B.: Robots that can adapt like animals. Nature **521**(7553), 503–507 (2015)
7. De Boer, P.T., Kroese, D.P., Mannor, S., Rubinstein, R.Y.: A tutorial on the cross-entropy method. Ann. Oper. Res. **134**(1), 19–67 (2005). https://doi.org/10.1007/s10479-005-5724-z
8. Fujimoto, S., Hoof, H., Meger, D.: Addressing function approximation error in actor-critic methods. In: International Conference on Machine Learning, pp. 1587–1596. PMLR (2018)
9. Grefenstette, J.J., Moriarty, D.E., Schultz, A.C.: Evolutionary algorithms for reinforcement learning. arXiv e-prints, p. arXiv-1106 (2011)
10. Haarnoja, T., et al.: Soft actor-critic algorithms and applications. arXiv preprint arXiv:1812.05905 (2018)
11. Kennedy, J., Eberhart, R.: Particle swarm optimization. In: Proceedings of ICNN'95-International Conference on Neural Networks, vol. 4, pp. 1942–1948. IEEE (1995)
12. Khadka, S., et al.: Collaborative evolutionary reinforcement learning. In: International Conference on Machine Learning, pp. 3341–3350. PMLR (2019)
13. Khadka, S., Tumer, K.: Evolution-guided policy gradient in reinforcement learning. In: Proceedings of the 32nd International Conference on Neural Information Processing Systems, pp. 1196–1208 (2018)
14. Lee, K., Lee, B.U., Shin, U., Kweon, I.S.: An efficient asynchronous method for integrating evolutionary and gradient-based policy search. arXiv preprint arXiv:2012.05417 (2020)
15. Lillicrap, T.P., et al.: Continuous control with deep reinforcement learning. arXiv preprint arXiv:1509.02971 (2015)
16. Majid, A.Y., Saaybi, S., van Rietbergen, T., Francois-Lavet, V., Prasad, R.V., Verhoeven, C.: Deep reinforcement learning versus evolution strategies: a comparative survey. arXiv preprint arXiv:2110.01411 (2021)
17. Marchesini, E., Corsi, D., Farinelli, A.: Genetic soft updates for policy evolution in deep reinforcement learning. In: International Conference on Learning Representations (2021)

18. Mitchell, M.: An Introduction to Genetic Algorithms. MIT Press, Cambridge (1998)

19. Mnih, V., et al.: Playing atari with deep reinforcement learning. arXiv preprint arXiv:1312.5602 (2013)

20. Nair, A., McGrew, B., Andrychowicz, M., Zaremba, W., Abbeel, P.: Overcoming exploration in reinforcement learning with demonstrations. In: 2018 IEEE International Conference on Robotics and Automation (ICRA), pp. 6292–6299. IEEE (2018)

21. Pierrot, T., et al.: Diversity policy gradient for sample efficient quality-diversity optimization. In: ICLR Workshop on Agent Learning in Open-Endedness (2022)

22. Pourchot, A., Sigaud, O.: CEM-RL: combining evolutionary and gradient-based methods for policy search. In: International Conference on Learning Representations (2019)

23. Qian, H., Yu, Y.: Derivative-free reinforcement learning: a review. arXiv preprint arXiv:2102.05710 (2021)

24. Salimans, T., Ho, J., Chen, X., Sidor, S., Sutskever, I.: Evolution strategies as a scalable alternative to reinforcement learning. arXiv preprint arXiv:1703.03864 (2017)

25. Schulman, J., Wolski, F., Dhariwal, P., Radford, A., Klimov, O.: Proximal policy optimization algorithms. arXiv preprint arXiv:1707.06347 (2017)

26. Fujimoto, S.: Open-source implementation for TD3. https://github.com/sfujim/TD3 (2018)

27. Sigaud, O.: Combining evolution and deep reinforcement learning for policy search: a survey. arXiv preprint arXiv:2203.14009 (2022)

28. Silver, D., et al.: Mastering the game of go with deep neural networks and tree search. Nature **529**(7587), 484–489 (2016)

29. Silver, D., Lever, G., Heess, N., Degris, T., Wierstra, D., Riedmiller, M.: Deterministic policy gradient algorithms. In: International Conference on Machine Learning, pp. 387–395. PMLR (2014)

30. Such, F.P., Madhavan, V., Conti, E., Lehman, J., Stanley, K.O., Clune, J.: Deep neuroevolution: genetic algorithms are a competitive alternative for training deep neural networks for reinforcement learning. arXiv preprint arXiv:1712.06567 (2017)

31. Todorov, E., Erez, T., Tassa, Y.: MuJoCo: a physics engine for model-based control. In: 2012 IEEE/RSJ International Conference on Intelligent Robots and Systems, pp. 5026–5033. IEEE (2012)

32. Wang, D., Tan, D., Liu, L.: Particle swarm optimization algorithm: an overview. Soft. Comput. **22**(2), 387–408 (2018)

Building Conversational Diagnosis Systems for Fine-Grained Diseases Using Few Annotated Data

Yiping Song[1], Wei Ju[2], Zhiliang Tian[1], Luchen Liu[2], Ming Zhang[2(✉)], and Zheng Xie[1]

[1] National University of Defense Technology, Changsha, China
{songyiping,tianzhiliang,xiezheng81}@nudt.edu.cn
[2] Peking University, Beijing, China
{juwei,liuluchen,mzhang_cs}@pku.edu.cn

Abstract. The conversational diagnosis system aims to interview patients and make the diagnosis just like doctors do. Most existing methods rely on medical dialog corpora collected from various medical forums. Compared to experts' annotated data, the dialog corpora are easily accessible, but lack medical knowledge and diagnostic decision logic. Thus, those systems can only handle coarse-grained diseases but achieve poor performance on fine-grained diseases. In this paper, we present a Reinforcement Learning (RL) framework that leverages a few annotated (from experts) and unannotated (from online forums) dialogs to make the diagnosis for fine-grained diseases. We summarize the doctor's diagnosis logic from the unannotated dialogs, then build a user simulator by annotated dialogs for RL training. In this way, a few annotated data are sufficient to support fine-grained disease diagnosis with the assistance of unannotated data. The experiments on eight fine-grained diseases show that our approach outperforms other competitive baselines.

Keywords: Diagnosis Systems · Fine-grained Diseases · Few Data

1 Introduction

The conversational diagnosis system [16,23,26], which is a typical task-oriented dialog system [22,27] in medical domain. It enables a smooth interaction between patients and computers via natural language and provides a reliable diagnosis for patients without too much time and expense. Existing conversational diagnosis systems usually employ reinforcement learning (RL) methods to build models for coarse-grained diseases using the dialog corpora from online forums. The training data from online forums are cheap and easy to access. Such dialogs contain medical common sense so the existing systems can do well on some easily diagnosed diseases, such as upper respiratory infection and infantile diarrhea. These diseases are from different clinical departments and can be easily diagnosed by some obvious symptoms, and we call them coarse-grained diseases. For example,

© The Author(s), under exclusive license to Springer Nature Switzerland AG 2023
M. Tanveer et al. (Eds.): ICONIP 2022, LNCS 13625, pp. 591–603, 2023.
https://doi.org/10.1007/978-3-031-30111-7_50

DX dataset [23, 26], the only available conversational diagnosis dataset, consists of four types of diseases from different clinical departments.

Compared with coarse-grained diseases, fine-grained diseases, where the diseases belong to the same clinical department, are of more importance in the real-world scenario. The reason is that people are more likely to seek help for a problem that cannot be solved with intuition. For fine-grained diseases, more training data is required because the differences between such diseases are so minor, and these diseases usually relate to fine-grained symptoms with various attributes. Moreover, diagnosing fine-grained diseases does not simply rely on the co-occurrence of several symptoms as the coarse-grained disease, but requires more medical knowledge and reliable decision logic from professional doctors. However, online dialogs, which may be written by laymen, cannot provide such professional and reliable information, so annotation from experts is vital for fine-grained diagnosis. Expert annotation in the medical domain is so expensive and time-consuming, and as a result, the insufficiency of annotated data becomes a bottleneck for building fine-grained disease diagnosis systems.

To remedy the lack of annotated data, we can exploit unannotated data. Particularly, we can extract and leverage the general diagnosis logic from the online dialogs, since the diagnosis logics on different diseases are similar and do not require exact medical knowledge. For example, when netizen A claims he has a fever, another netizen B may ask whether he or she has taken vaccines. No matter whether the final diagnosis is correct or not, this partial inquiry logic can be utilized to guide the policy of our diagnosis system. However, these online dialogs need to be used in a proper way because netizens are unprofessional.

In this paper, we propose to leverage a few expert annotated data and unannotated data from an online medical forum to train the fine-grained conversational diagnosis system. Our system is under the RL framework. For unannotated data, we extract the diagnostic logic from dialog sessions, and store them into the experience pool in the form of "state-action-next state-reward" quadruples. Even though the logic may not be accurate, it can be verified by the interaction under the RL framework. For annotated data, we build a user simulator to provide an environment for RL training. The simulator produces feedback to the system by imitating the annotated data, while the agent interacts with the simulator to learn an optimal policy under the guidance of the experience pool.

Our contributions are three-fold: (1) We are the first one to build an efficient conversational diagnosis system for fine-grained diseases which belong to the same clinical department. (2) We propose an RL-based framework that leverages a few annotated data and large-scaled unannotated data. (3) We extract diagnostic logic catering to the real doctors' diagnosis inference from unannotated data, and the logic is interpretable and proves to be useful in diagnosis.

2   Related Work

Task-Oriented Conversation Systems. The Task-oriented conversation system [11, 20] aims to accomplish a given task in a vertical domain. Existing works cover a limited number of domains including guiding, shopping, and education [5, 7, 8, 21, 27]. The approaches to building the task-oriented systems have two categories: the modular-based [23,24,26] and end-to-end trainable methods [3,25]. In this paper, we use the former one. Insufficiency of the corpus is a typical bottleneck for task-oriented conversation systems. Some works use domain adaptation [2,8], multi-task [6,15], few-shot learning [14,19,27] and transfer learning [1,4] technique. But they require multiple similar domains, but we cannot find similar ones for the medical domain. Another solution [17] is to use a self-play strategy to create new annotated data, but it requires crowdsourcing to correct the created dialogs, which is money-consuming. Li et al. [13] propose a probabilistic framework that can take advantage of expert knowledge, and can help to analyze fault cases. But this is not a purely data-driven method and highly relies on domain experts.

Diagnosis Systems. Automatic diagnosis systems are of great significance [9,18]. Tang et al. [20] use a symptom-checking list that allows the doctor to inquire about the symptoms and make the diagnosis, which can be regarded as a prototype of conversation systems. Kao et al. [11] build a system that takes the personal information of patients into consideration, but personal information is hard to collect due to the privacy policy. Wei et al. [23] are the first to formulate the inquiring process as a form of dialogue. Natural language is used for interaction, and many other actions such as greetings are allowed in the conversation to achieve a better user experience. They build the model under the RL framework, where the system works as the agent and a patient simulator works as the environment. Deep Q-network (DQN) is used to train the policy of the system. This is also the base model of our work. Xu et al. [26] propose to use two branches to predict a probability vector over all possible actions separately. One branch uses a DQN model like [23], and the other uses the statistical co-occurrence frequency of the symptoms and diseases. Then the two probability vectors are added together to make the final decision. Luo et al. [14] propose to learn the disease embedding by encoding the conversation histories in the training set and use the embedding to predict action at each conversational turn. The above models use the corpus from online forums to build the patient simulator and are only applicable to the diseases that can be diagnosed by common sense. While for fine-grained diseases, more medical knowledge and decision logic are required, so the corpus must be annotated by doctors instead of the netizens. Since the annotation from doctors is expensive, our work proposes an RL-based method for fine-grained diseases which only need a few annotated data.

3 Methodology

We first introduce the RL architecture, then explain how to use two different sources of data to train the dialog management module in two steps: diagnostic logic extraction and ensemble training. We describe the data flows in the end.

3.1 Reinforcement Learning (RL) Framework

Under the RL-based framework, the proposed conversational diagnosis system acts as the doctor, which is the agent in RL settings. A patient simulator acts as the patient, which is the environment in RL settings. The patient simulator first elaborates the illness condition, then the system continually interacts with the patient simulator to collect useful information (symptoms) to get a comprehensive understanding of his or her illness condition, then makes the final diagnosis at the end of the conversation. Following the settings of RL, other essential components are as follows.

State s. s_t records all behaviors of the system and patient until the current turn t. It contains all the symptoms with the attributes inquired by the system. **Action** a. a_t is agent's action at t-th turn. Each action has two types: the request action that inquires about the symptom, the inform action that makes a diagnosis. **Reward** r. The reward r is the immediate feedback given by the simulator after the system taking action a. The reward is +44 if the system makes the right diagnosis within the maximal turn, and the reward is +2 if the system hits a symptom or attribute in the user goal. Otherwise, the reward is -1 to encourage shorter conversation. **Policy**π. Policy π defines how the system acts in accordance with the current state s, noted as $\pi(a_t|s_t)$.

```
{
    "disease_tag": "缺血性心脏病(Ischaemic heart diseases)",
    "request_slots": {
        "disease": "UNK"
    },
    "goal": {
        "explicit_symptoms": {
            "心悸(palpitation)": {"frequency": "偶发(occasional)"},
            "出汗(sweatiness)": {"condition": "运动后(after exercise)"}
        },
        "implicit_symptoms": {
            "胸闷(chest distress)": {"performance": "加剧(aggravation)"},
            "发热(fever)": {"t/f": false},
            "呕吐(emesis)": {"occur": "数周前(a few weeks ago)"}
        }
    }
}
```

Fig. 1. An example of user goal.

Patient Simulator. The behaviors of the patient simulator during the interaction are based on the "user goal" (Fig. 1) extracted from the Electronic Health Record (EHR). It consists of 3 parts: the patient's disease, explicit symptoms, and implicit symptoms. The disease tag is the ground truth of the patient's disease, which is expected to be diagnosed by the system. The explicit symptoms are informed to the system at the beginning of the conversation. The implicit symptoms are other useful symptoms that the system needs to inquire about for diagnosis. Different from [23], we define several attributes for each symptom to better describe the patient's condition. The attributes are a binary variable (indicating whether the patient has this symptom), frequency (how frequent the symptom happens), duration (how long the symptom lasts), severity (how serious the symptom is), and condition (under which condition the symptom happens). These attributes are essential for fine-grained disease diagnosis. It is because the co-occurrence of symptoms and diseases is no longer sufficient to diagnose fine-grained diseases as the coarse-grained ones, while the differences lay in attributes. Notice that various attributes enlarge the action space for the system, making policy learning more challenging.

Diagnosis System. We use a modular-based framework to build the system, which consists of a Natural Language Understanding (NLU) module, a Dialog Management (DM) module and a Natural Language Generation (NLG) module. The NLU module takes the query from the patient as input and parses the user intent and other important information. In our scenario, the user's intent is to request a diagnosis, and other important information includes the patient's symptoms and corresponding attributes. Given the parsed information from NLU, the DM module determines the system's action. After that, the NLG module converts the action into the natural language to respond to the patient. The whole architecture is presented in Fig. 2. In this paper, NLU and NLG modules use the same template-based method in [23], and perform symptom and attribute normalization to eliminate synonyms. The difference lies in the DM module, where we use a few annotated dialogs and massive unannotated dialogs to train the model. We now describe the DM part in detail.

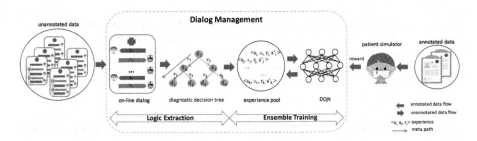

Fig. 2. The Dialog Management module of the proposed model where both annotated data (in orange arrows) and unannotated data (in blue arrows) are used. Unannotated data is converted into "experience" to feed into the experience pool of RL. (Color figure online)

3.2   Dialog Management

Dialog management, the core component of the system, decides how the system acts at each turn. To reduce the demand for annotated data for the fine-grained diagnosis system, we first extract diagnostic logic from unannotated data, then perform an ensemble training guided by the extracted logic in the RL framework. In this section, we first interpret these two steps separately, then illustrate how the unannotated and annotated data flow in the proposed framework.

Diagnostic Logic Extraction. In medicine, diagnostic logic is formally represented as diagnostic decision trees [12]. As shown in Fig. 3, the non-leaf node is the attribute of symptoms k enquired by the doctor, the edge is the value of the attribute v answered by the patient, and the leaf node is the final diagnosis result d. The doctor makes the decision according to the v on every node k until obtaining the result d. In practice, it is hard to build the diagnostic decision tree as it requires extensive medical expertise, but we can extract the "logical paths" of the decision tree from unannotated data, which is much cheaper and more efficient. In Fig. 3, we denote each path from the root to the leaf node as a logical path. In a conversation, the doctor's inquiry about the symptom's attribute corresponds to a node in the decision tree, and the patient's answer corresponds to an edge, so one conversation corresponds to one "logical path" in the tree. For the incomplete conversations without the final diagnosis (leaf-node), we treat the path from root to the last doctor's inquiry as a logical path.

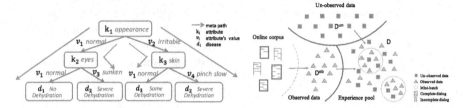

Fig. 3. The diagnostic decision tree presents the diagnosis logic of the doctor. The orange arrows point out the logical paths of diagnosis processes. (Color figure online)

Fig. 4. The experience pool consists of (1) "experience" extracted from unannotated data and (2) simulation records from the system-patient simulator.

Formally, we note the conversation as $c = \{q_1, r_1, \ldots, q_n, r_n\}$ where q_i is the doctor's utterance and r_i is the patient's utterance at the i-th turn. We reuse the NLU($\cdot$) module described in the former section to parse the symptom's attributes k and their value v for each utterance. For each conversation c, we can extract a "logical paths" $p = \{k_1, v_1, \ldots, k_n, v_n\}$ to represent the diagnostic logic.

Ensemble Training. To plug in the diagnostic logic extracted from the unannotated data for training, we employ RL using the experience replay strategy, where the experience pool contains the extracted diagnostic logic from both unannotated and annotated data (see Fig. 4).

Experience replay maintains an "experience pool", in which the "experience" records actions the model took during the exploration. At each training step, The model updates the parameters by minimizing the temporal-difference(TD) error on a mini-batch of "experience" sampled from the pool. The "experience" is formulated as a quadruple $\langle s, a, r, s' \rangle$, where s stands for the state, a stands for the action the model takes under s, r is the instant reward the model receives by taking a, and s' is the next state after taking a.

Since the logical paths obtained from unannotated data indicate how the doctor is supposed to act for diagnosis, we then propose to convert the logical paths into the format that the experience pool takes to substitute the random choices with more reliable ones. In the beginning stage of training, the exploration is almost randomly performed as the model has no medical knowledge about diagnosis, so the training of RL is unstable and may fall into the suboptimal point. We propose to add a warm start stage before training, where we feed the "experience" into the experience pool. In this way, so the initial exploration will be based on real data. Even though this kind of "experience" may not be totally correct, it is at least better than the random one. After the warm start, the model will further improve the policy based on the setting of a relatively reliable parameter, so the model converges more quickly.

Formally, one logical path is a series of symptoms and attributes in the doctor's inquiry order noted as $\{k_1, v_1, k_2, v_2, \cdots, k_n, v_n\}$. We separate former $2*n-1$ items as the state s and regard the last item as the action a, thus one path can be converted to one $\langle s, a \rangle$ pair. Then we obtain the instance reward r from the patient's answer in the next turn. For example, in the current turn, the doctor asks whether the patient has taken the vaccine recently. If the answer is yes, it means the doctor has asked about the right symptom that may help the diagnosis, so the reward is positive. If not, it means the doctor may ask about an irrelevant symptom, so the reward is negative. By taking action a, the state updates to s', so all the information can be noted as $\langle s, a, r, s' \rangle$. These quadruple can be inserted into the experience pool to guide the exploration direction.

To further improve the diagnosis efficiency, the symptom the systems ask about are desired to be more "distinguishable" for disease classification. Inspired by the term frequency-inverse document frequency (TF-IDF), which reflects the distinct degree of a word to the document, we propose an action frequency-inverse disease frequency (AF-IDF) to measure whether an action is distinguishable and worth for making diagnosis.

Specifically, we have a disease set $D = \{d_1, \cdots, d_n\}$, and each disease $d_i = \{c_i^1, \cdots, c_i^n\}$ consists of several dialogs c_i^j. For an action a extracted from d_i, AF-IDF is calculated as AF-IDF = AF $\cdot$ IDF, where

$$\mathrm{AF} = \frac{\mathrm{count}((a, c_i^j)|c_j^i \in d^i)}{|\{d_i\}|}, \quad \mathrm{IDF} = \log \frac{|D|}{\mathrm{count}((a, d_i)|\exists c_i^j \in d_i, a \in c_i^j)}. \quad (1)$$

We note AF-IDF as the confidence score δ of taking action a, and label the experience with it noted as $\langle s, a, r, s', \delta \rangle$. The higher the score δ is, the more distinguishable the action is. For actions from incomplete dialogs without a disease tag, we use the average score over all the diseases $\{d_i\}$ as its confidence score. We normalize the confidence score into $(0, 1)$ using the logistic function, multiply the normalized score δ_i' with the instant reward r_i, and then store the experience $\langle s, a, \delta' r, s' \rangle$ into the experience pool noted as D^{ob}. D^{ob} is the observed data as all the experiences are summarized from the real-world conversations.

As mentioned before, we adopt DQN using the experience replay and the target network. Q-learning updated at iteration i uses the objective function as,

$$\mathcal{L}_i(\theta_i) = E_{\langle s, a, r, s' \rangle \sim U(D_i)}[(r + \gamma \max_{a'} Q(s', a'; \theta_i^-) - Q(s, a; \theta_i))^2] \quad (2)$$

where $D_i = D_i^{un} \cup D^{ob}$, and $Q(s, a)$ is the function that estimates the maximum of cumulative future reward discounted by γ for taking action a at state s, $U(D_i)$ is the uniform distribution over the experience pool D_i, θ_i are the parameters of the Q-network at iteration i and θ_i^- are the parameters for the target network. The target network parameters θ_i^- are fixed and only updated with the Q-network parameters (θ) every C step. In Fig. 4, D consists of D^{un} from the simulation and D^{ob} from the unannotated data. D^{un} stands for the unobserved data, which is filled with the simulation results. D^{ob} contains the real dialog information that is closer to the real data distribution.

Bidirectional Data Flow. The model uses two kinds of data for training, as shown in the orange and blue arrows in Fig. 2. Annotated data is used as the supervised data to support the patient simulator, which is a straightforward way to guide the direction of gradient descent. Unannotated data is injected into the experience pool to affect the training direction indirectly. By doing so, the model is more likely to explore the space that has already been observed in the human-to-human conversation. The observed data helps the RL-based system to find a reliable warm start instead of a random one, which stabilizes the training process. The annotated data and unannotated data formulate a pair of collaborative learners and benefit the training simultaneously.

4   Experimental Settings

Datasets. We conduct the experiments on fine-grained diseases which are chosen from the *diseases of the circulatory system* category according to ICD-10[1]. The 8 diseases are: "Acute rheumatic fever", "Hypertensive diseases", "Ischaemic heart diseases", "Pulmonary heart disease and diseases of pulmonary circulation", "Other forms of heart disease", "Cerebrovascular diseases", "Diseases of arteries, arterioles, and capillaries", "Other and unspecified disorders of the circulatory system". We use 245 EHR annotated by qualified doctors as the annotated data and 10k unannotated dialogs from a Chinese inquiring system[2]. The dataset uses 121 different symptoms. Each symptom has 5 attributes, so the action space is 605. We do not use the dataset in [23] since the diseases in it belong to different disease categories in ICD-10, so a few simple symptoms such as diarrhea or cough are enough to distinguish functional dyspepsia and gastroenterology. In contrast, we use a fine-grained disease set coming from the "heart disease" category in ICD-10, which is more challenging and has practical significance.

Hyper-parameters. We use a two-layer multiple layer perception (MLP) DQN with the same hyper-parameters in [23]. The target network and experience replay strategy are used for policy training, as well as the ϵ-greedy exploration strategy where ϵ is set to 0.1. The maximal turn of the conversation is 22.

Competing Methods. DQN. The basic DQN-based method for the training of RL [23]. **DQN-A.** Integrate the action transition probability with DQN for joint

---

[1] https://icd.who.int/browse10/2010/en#/IX.

[2] zixun.haodf.com.

action prediction [26]. Here, we use a rule-based method for both NLU and NLG modules to make a fair comparison. **DQN-D.** Integrate the disease-symptom co-occurrence probability with DQN for joint action [26]. The experimental setting is the same as other baselines. **DQN-A-D.** Integrate both the action relation matrix and disease-symptom knowledge with DQN for joint action prediction [26]. **DQN-UN.** The proposed model uses the experience from unannotated data but does not uses AF-IDF to make the action more distinguishable. **DQN-UN-D.** The full model that uses both unannotated data and AF-IDF.

Evaluation Metrics. • *success.* The success rate of the correct disease prediction indicates the system has made the correct diagnosis within the maximal turn. • *wrong diagnosis.* The rate of the wrong disease prediction, indicating the system has made the wrong diagnosis within the maximal turn. • *reward.* The average reward over the whole conversation. • *turn.* The average number of conversation turn. Given the fact that RL is inherently unstable, we train all the above competing methods 5 times and run the testing of each model 5 times. All the results listed in the tables are the average results of 25 experiments.

Table 1. The overall performance of all competing methods in three categories. Our methods are listed in the third one. The "Avg" part shows the average results of 25 (5 pieces of training multiply 5 testings) experiments, while the "Best" part presents the best results among 25 experiments. A conversation is counted as "success" only when the system makes the diagnosis within the maximal turn.

	Avg				Best			
	success	wrong diagnosis	reward	turn	success	wrong diagnosis	reward	turn
Annotated								
DQN	0.285	0.586	−4.802	6.174	0.331	0.541	−1.826	6.343
Annotated + Knowledge								
DQN-A	0.259	0.664	−5.775	4.796	0.294	0.596	−3.634	5.141
DQN-D	0.247	0.632	−7.131	5.855	0.327	0.635	−1.102	4.368
DQN-A-D	0.256	0.681	−5.759	4.283	0.313	0.681	−1.641	4.283
Annotated + Unannotated								
DQN-UN	0.270	**0.531**	−6.299	7.213	0.305	0.560	−3.505	6.258
DQN-UN-D	**0.319**	0.580	**−2.173**	5.517	**0.361**	**0.518**	**0.358**	5.949

5   Experimental Results and Analyses

The upper bound of diagnosis accuracy is calculated by directly regarding all patient's symptoms as features for a classifier instead of collecting information via interaction. We use Support Vector Machine (SVM) and Logistic Regression (LR) to predict the disease tags. The maximal accuracy is about 0.37 (SVM:0.376, LR:0.370), showing that making the diagnosis in our setting is very challenging. Notice that both SVM and LR directly use all the useful symptoms for classification. According to an article, [10] in JAMA, a top medical journal, the correct diagnosis rate made by human experts in telemedicine is around 0.38, which is similar to our scenario (also focusing on fine-grained diseases). Both [10] and this paper uses online doctor-patient interactions without accessing medical examinations or personal records. Hence, 0.37 is a reasonable value.

- **Overall Performance** Table 1 presents the overall performance on all competing methods, and our two proposed methods achieve the highest success rates among all the competing methods. For the average performance among 25 experiments, DQN provides a borderline. Enhanced by external knowledge, DQN-A adds the action transition information to the policy network, which is expected to achieve better performance. However, DQN-A is worse than DQN in terms of all metrics. The problem may lie in the integrated way of action relation. DQN-A simply sums the action probabilities produced by action relation matrix and the predicted probabilities produced by DQN together, but these two distributions are probably in different scopes. Similarly, DQN-D introduces pre-defined symptom-disease knowledge into the framework. With similar integrated strategy, the results of DQN-D are even worse than DQN-A. The low performance of DQN-A-D is further evidence. Hence, these methods [26] are not robust.

Table 2. The standard deviation (std) of *success* of 5 times training.

	1	2	3	4	5	std
DQN	0.285	0.257	0.308	0.331	0.240	0.033
DQN-A-D	0.247	0.234	0.211	0.274	0.313	0.035
DQN-UN-D	0.296	0.288	0.338	0.314	0.361	**0.027**

Table 3. The warm start strategy on different methods. "*" indicates no rule-based warm start applied.

	1	2	3	4	5
DQN*	0.206	0.229	0.275	0.311	0.280
DQN	0.285	0.259	0.308	0.331	0.240
difference	+	+	+	+	−
DQN-UN-D*	0.302	0.301	0.253	0.234	0.373
DQN-UN-D	0.296	0.288	0.338	0.314	0.361
difference	−	−	+	+	−

As for our proposed methods, DQN-UN uses the unannotated data but not AF-IDF, and achieves a relatively good performance compared with the knowledge-based method. We also notice that DQN-UN performs worse than DQN, and the conversation turn is longer than other methods. The reason is that the actions that cannot be used to tell different diseases apart waste the interaction turn, so the model gains more negative rewards. DQN-UN-D has the best performance among all methods in terms of all evaluation metrics, indicating that the unannotated data indeed contains useful information as long as we use it in a proper way. In addition, the sum of *success* and *wrong diagnosis* can be regarded as the confidence score of making the diagnosis before the end of the conversation. DQN has 0.879 confidence score to make the diagnosis, while for DQN-UN, the confidence score drops to 0.801.

It shows if the exploration direction is effected by the wrong information, the system is less likely to make the diagnosis within the maximal turn. After the making the action more distinguishable, the system is equipped with reliable knowledge, and the confidence score of DQN-UN-D rises to 0.899.

For the best performance among 25 experiments of each method, DQN reaches the *success* of 0.33, and DQN-A is unsurprisingly low. Inconsistent with the average metrics, DQN-D has a better performance. The reason may be that when the predicted action distribution and knowledge-based distribution are in a similar scope, the overall performance will be much more satisfactory. While for DQN-A-D, it is difficult to find a balance between the three items, so the result is not so good. DQN-UN still has a poor performance, but DQN-UN-D gets 0.361 in terms of *success*, which is quite close to the performance of SVM and RL. Reaching almost the upper bound of *success* proves the potential of DQN-based policy network and the effectiveness of unannotated data.

- **Training Stability** RL is inherently unstable, and introducing unannotated data helps to maintain more stable training. We run all the approaches 5 times and record their success rate to estimate their stability as listed in Table 2. The lower the std score is, the lower the deviation between different experiments, and the more stable the approach is. Our model is much more stable than other DQN model. The std value of DQN is 0.033. When it comes to DQN-A-D, the std value arises 6% compared with DQN. In contrast, our full model DQN-UN-D drops 18%, which shows our method alleviates the instability of RL training.

- **Knowledge Summarization** From another perspective, the extraction of "experience" proposed in this paper can be regarded as a knowledge summarization process. In our model, we use the logical paths summarized from unannotated data to formulate the "experience" for the warm start, while conventional DQN designs rules to obtain "experience" for the warm start.

We compare the methods on whether to use the rule-based warm start strategy at the beginning or not. Table 3 shows that DQN benefits from the rule-based warm start in most cases (4/5), while our model DQN-UN-D is not (2/5). We believe it is because the knowledge from the warm start has been already included in the unannotated data, and our model does not rely on the manually-designed rules in the warm start stage. Hence, our model not only reduces the use of annotated data, but also waives the need for designing rules in the warm start.

6 Conclusion

Making the diagnosis of fine-grained diseases is of significance in practical, but it requires more complicated diagnosis logic and training data. To reduce the demand for annotated data in fine-grained diseases diagnosis, we propose an RL-based method to leverage both few annotated dialogs from experts and unannotated dialogs from online forums. The method extracts diagnostic logic from the unannotated dialogs, and build a user simulator using annotated dialogs. The experiments show that our model achieves good performance for 8 fine-grained "heart disease" diagnosis.

Acknowledgments. This paper is supported by National Natural Science Foundation of China (NSFC Grant No. 62106275, No.62106008 and No.62006004).

References

1. Adewumi, T., et al.: Småprat: DialoGPT for natural language generation of Swedish dialogue by transfer learning (2021)
2. Budzianowski, P., et al.: Sub-domain modelling for dialogue management with hierarchical reinforcement learning. In: SIGDIAL, pp. 86–92 (2017)
3. Dhingra, B., et al.: Towards end-to-end reinforcement learning of dialogue agents for information access. In: ACL, vol. 1, pp. 484–495 (2017)
4. Enayet, A., Sukthankar, G.: A transfer learning approach for dialogue act classification of GitHub issue comments (2020)
5. Glas, N., Prepin, K., Pelachaud, C.: Engagement driven topic selection for an information-giving agent. In: Workshop on the SPD (2015)
6. Golub, D., Huang, P., He, X., Deng, L.: Two-stage synthesis networks for transfer learning in machine comprehension. In: EMNLP, pp. 835–844 (2017)
7. Graesser, A.C., Chipman, P., Haynes, B.C., Olney, A.: AutoTutor: an intelligent tutoring system with mixed-initiative dialogue. IEEE Trans. Educ. **48**(4), 612–618 (2005)
8. Jaech, A., Heck, L.P., Ostendorf, M.: Domain adaptation of recurrent neural networks for natural language understanding. In: INTERSPEECH, pp. 690–694 (2016)
9. Jovanovic, M., Baez, M., Casati, F.: Chatbots as conversational healthcare services. IEEE Internet Comput. **PP**(99), 1 (2020)
10. Resneck, J.S., et al.: Choice, transparency, coordination, and quality among direct-to-consumer telemedicine websites and apps treating skin disease. JAMA Dermatol. **152**, 768–775 (2016)
11. Kao, H.C., Tang, K.F., Chang, E.Y.: Context-aware symptom checking for disease diagnosis using hierachical reinforcement learning. In: AAAI (2018)
12. Levine, A.C., et al.: Empirically derived dehydration scoring and decision tree models for children with diarrhea: assessment and internal validation in a prospective cohort study in Dhaka, Bangladesh. Global Health Sci. Pract. **3**(3), 405–418 (2015)
13. Li, T., Zhao, Y., Zhang, C., Luo, J., Zhang, X.: A knowledge-guided and data-driven method for building HVAC systems fault diagnosis. Build. Environ. **198**, 107850 (2021)
14. Luo, H., Li, S.W., Glass, J.: Prototypical q networks for automatic conversational diagnosis and few-shot new disease adaption. arXiv preprint arXiv:2005.11153 (2020)
15. Mo, K., Yang, Q., Fung, P.: Cross-domain dialogue policy transfer via simultaneous speech-act and slot alignment. arXiv preprint arXiv:1804.07691 (2018)
16. Moulya, S., Pragathi, T.R.: Mental health assist and diagnosis conversational interface using logistic regression model for emotion and sentiment analysis. In: Journal of Physics: Conference Series, no. 1, p. 012039 (2022)
17. Shah, P., et al.: Building a conversational agent overnight with dialogue self-play. arXiv preprint arXiv:1801.04871 (2018)

18. Sok, M., Svegl, E., Grabec, I.: A sensory-neural network for medical diagnosis. In: EAIS, pp. 1–6 (2017)
19. Song, Y., Liu, Z., Bi, W., Yan, R., Zhang, M.: Learning to customize model structures for few-shot dialogue generation tasks. In: ACL, pp. 5832–5841 (2020)
20. Tang, K.F., Kao, H.C., Chou, C.N., Chang, E.Y.: Inquire and diagnose: neural symptom checking ensemble using deep reinforcement learning. In: NeurIPS (2016)
21. Tran, V., Nguyen, L.: Adversarial domain adaptation for variational neural language generation in dialogue systems. In: COLING, pp. 1205–1217 (2018)
22. Tseng, B., et al.: Variational cross-domain natural language generation for spoken dialogue systems. In: SIGDIAL, pp. 338–343 (2018)
23. Wei, Z., et al.: Task-oriented dialogue system for automatic diagnosis. In: ACL, vol. 2, pp. 201–207 (2018)
24. Williams, J.D.: Web-style ranking and SLU combination for dialog state tracking. In: SIGDIA, pp. 282–291 (2014)
25. Williams, J.D., Zweig, G.: End-to-end LSTM-based dialog control optimized with supervised and reinforcement learning. arXiv preprint arXiv:1606.01269 (2016)
26. Xu, L., Zhou, Q., Gong, K., Liang, X., Tang, J., Lin, L.: End-to-end knowledge-routed relational dialogue system for automatic diagnosis. In: AAAI (2019)
27. Zhao, T., Eskenazi, M.: Zero-shot dialog generation with cross-domain latent actions. In: SIGDIAL, pp. 1–10 (2018)

Towards Improving EEG-Based Intent Recognition in Visual Search Tasks

Mansi Sharma$^{(\boxtimes)}$ ⓘ, Maurice Rekrut ⓘ, Jan Alexandersson ⓘ,
and Antonio Krüger ⓘ

Cognitive Assistants, German Research Center for Artificial Intelligence (DFKI),
Saarland Informatics Campus, Saarbrucken, Germany
{mansi.sharma,maurice.rekrut,jan.alexandersson,krueger}@dfki.de

Abstract. Accurate estimation of intentions is a prerequisite in a non-verbal human-machine collaborative search task. Electroencephalography (EEG) based intent recognition promises a convenient approach for recognizing explicit and implicit human intentions based on neural activity. In search tasks, implicit intent recognition can be applied to differentiate if a human is looking at a specific scene, i.e., Navigational Intent, or is trying to search a target to complete a task, i.e., Informational Intent. However, previous research studies do not offer any robust mechanism to precisely differentiate between the intents mentioned above. Additionally, these techniques fail to generalize over several participants. Thus, making these methods unfit for real-world applications. This paper presents an end-to-end intent classification pipeline that can achieve the highest mean accuracy of 97.89 ± 0.74 (%) for a subject-specific scenario. We also extend our pipeline to support cross-subject conditions by addressing inter and intra-subject variability. The generalized cross-subject model achieves the highest mean accuracy of 96.83 ± 0.53 (%), allowing our cross-subject pipeline to transfer learning from seen subjects to an unknown subject, thus minimizing the time and effort required to acquire subject-specific training sessions. The experimental results show that our intent recognition model significantly improves the classification accuracy compared to the state-of-the-art.

Keywords: Intent recognition · Brain-machine interaction · Machine learning

1 Introduction

Enabling machines to understand humans and their implicit or explicit intents accurately is a key objective of any efficient human-machine collaborative system [22]. For instance, one of the tasks of an assistive robotic arm in a noisy industrial scenario could be to hand over specific tools the human is searching for. If the robotic arm knows that the person is searching for something, it can assist with searching and handing over the desired object or tool. However, how can a robotic arm be sure if a human is searching for something or just scanning the

M. Tanveer et al. (Eds.): ICONIP 2022, LNCS 13625, pp. 604–615, 2023.
https://doi.org/10.1007/978-3-031-30111-7_51

environment? Object selection with eye-tracking has been studied in multiple literature [2,3] to identify implicit intents. The EEG analysis in Kang et al. [1] barely exceeds the chance level for distinguishing a search intent or a scene scan.

Another challenging task is to address the EEG variability. EEG signals are highly non-stationary and can differ a lot across days or even within the same day for the same user. Inter-subject variability refers to the differences in brain signals between multiple subjects, and intra-subject variability refers to the differences in brain activity for the same subject occurring in various repetitions of the same task [13]. Inter and Intra- subject variability is unavoidable due to the involved time-variant factors connected to the experimental recording setup and underlying psychological and neurophysiological parameters. Ideally, a real-world Brain-machine communication would need to be effective and efficient at all times, i.e., across sessions and participants, without re-calibration. To address these problems, we propose a combination of brain signals recorded via Electroencephalography (EEG) and eye-tracking in a simulated working environment to investigate correlations between EEG data and the eye-tracking data for two reasons (1) to create context from overt data by automatically labeling EEG data based on the eye-tracking input (2) to build models for intention recognition based on this labeled EEG data. We use eye-tracking information only to perform automatic labeling of the EEG data. Using eye-tracking information as an interaction modality is beyond the scope of this paper.

The main objective of this paper is to predict different human implicit intentions that occur during visual stimulus presentation, i.e., Navigational Intent (Free viewing) and Informational Intent (Target searching). This research collects the EEG data from various participants during a visual search task to identify brain state transitions between those intentions and classify users' implicit intentions using machine learning classification algorithms. We investigate a wide range of feature extraction methods and classification algorithms to provide the best setup for labeling Navigational and Informational Intent based on EEG activity with plausible accuracy. Our main contributions to this paper are as follows:

1. We design and develop an effective data acquisition paradigm and an end-to-end classification pipeline to categorize human intents using EEG signals.
2. We extensively evaluate our classification pipeline for single-subject to show a significant improvement compared to the existing state-of-the-art.
3. We extend the single-subject classification pipeline to enable the transfer of EEG-based learning to cross-subject scenarios for the first time.

The rest of the paper proceeds as follows. In Sect. 2, we present the related works concerning intent recognition and inter and intra-subject variability. Section 3 describes the data recording setup with recording devices and the overall recording procedure. Section 4 discusses the signal processing algorithm, including feature extraction techniques. Section 5 illustrates the model performance for a subject-wise and cross-subject scenario. Section 6 concludes the paper with some discussions and limitations.

2   Related Work

In the recent development of human-machine collaborative systems [15,22], intent recognition [14] plays a major role in making the collaboration much more efficient and successful. Recognition of human intentions using EEG signals offer strong research interest due to their quality of giving insights into the human mind and the ability to communicate or interact with external devices such as wheelchairs and intelligent robots [14,16]. In Slanzi et al., the authors propose a physiological-based analysis for predicting web users' click intention by combining EEG responses and pupil dilation [4]. Authors design ten questions for each website concerning finding certain information within the website. Participants follow a navigation path from the home page to the page where information is present. The authors chose a wide range of features like Hjorth parameters, Petrosian Fractal Dimension, Higuchi Fractal Dimension Hurst exponent, and statistical features to train the model. However, the performance of the classifiers is not satisfactory. This study achieved a maximum accuracy of 71.09% with logistic regression, which may not be sufficient for real-world scenarios.

Recent research shows that EEG-based intent recognition can understand the implicit intention, even when a human does not express his thoughts. For example, in Kang et al., authors develop advanced interactive web service engines which rely on identifying brain connectivity patterns related to the user's Navigational and Informational intentions through visual experiments based on static web images [1]. In this work, the authors analyze the differences in phase-locking value (PLV) to classify users' Navigational and Informational intentions. Authors use Support Vector Machines, Naïve Bayes, and Gaussian Mixture Model. However, accuracies mostly fall between 50% to 77% for all classifiers, which is not sufficient for real-world deployment where precise estimation of intents is of utmost importance to make the system robust.

Existing studies focus on subject-specific evaluation, which is not the best case for real-world settings where a generalized setup could save considerable training time and effort. Due to the complexity and high dimensionality of brain signals, intent recognition accuracy and signal interpretability heavily depend on feature vector representation in a sophisticated manner. Moreover, EEG signals reflect the fluctuations of the voltages from different cortical regions of the human brain over a time period [17]. It becomes necessary to effectively combine both spatial and temporal information to capture the uncertainties generated by inter, and intra-subject variability [13,17]. In Wei et al., authors use hierarchical clustering to explore the associations between EEG features and cognitive states to tackle inter and intra-subject variability within a large-scale dataset of EEG collected in a simulated driving task [18]. A subject transfer framework detects drowsiness, which reduces the calibration time by 90%. Still, some amount of training is needed. Other research studies that address these variabilities use a completely different EEG-based paradigm like Motor Imagery [19] or P300 speller [20] to reduce task-based calibration time. So far, the possibility of handling inter and intra-subject variability in improving intent recognition for visual search tasks is unexplored.

3   Data Acquisition Setup

Fifteen healthy subjects (age: 20 to 30 years) participated in the experiment without prior training or knowledge. Before the start of the experiment, participants were informed about the experiment process and asked to sign an informed consent form for the scientific use of the recorded data. The study was approved by the ethical review board of the Faculty of Mathematics and Computer Science at Saarland University[1]. The experiment was performed in a dim light room with minimum distractions from external noises, or electronic devices, where the voluntary participants are asked to sit in a comfortable chair to prevent unnecessary muscle movements to minimize noise and artifacts in the EEG signals which could unfold from mental stress, electrical interference, and other physiological motor activity [5,6]. The display resolution of the monitor was set to 1920 × 1080 pixels, the screen brightness is set to $300,00$ cd/m^2, the distance between the user and the screen is set to 60 cm, and the eyes of the user are about the same height as the center of the screen.

Recording Devices: EEG signals were recorded with a LivAmp 64 amplifier by Brain Products[2]. The sampling frequency was set to 500 Hz. The 10–20 international system of electrode placement was used to locate the electrodes [7]. Electrode impedances were kept below 25 kΩ throughout the duration, as it is a common practice for noise reduction in the EEG recordings [8]. Tobii pro fusion[3] is used to collect eye-tracking information, which is only used for automatically labeling the EEG data.

3.1   Experimental Procedure

The Experiment consists of 3 parts: (i) Navigational Intent or Free viewing, (ii) Target presentation, and (iii) Informational Intent or target searching. Figure 1 shows the experimental sequence. We designed the Experiment in Unity [9], where the industrial scenes are as close to the original working scene in an industrial context. The recording steps are as follows:

1. The participant glances over the input scene without knowing the target to get the overall overview of the scene.
2. The participant is shown a specific target tool as an image.
3. The participant searches for the shown target object in the input scene by looking around.
4. As soon as the participant finds the tool, the target object boundary appears with a red color which later changes to green color, ensuring that the participant found the correct tool.

The recorded dataset consists of 5 sessions for each subject recorded on the same day with short breaks in between sessions. Each session comprises 30 scenes

---

[1] https://erb.cs.uni-saarland.de/.

[2] https://brainvision.com/products/liveamp-64/.

[3] Tobii Pro AB. Tobii pro lab. 2014, https://www.tobiipro.com/siteassets/tobii-pro/user-manuals/Tobii-Pro-Lab-User-Manual/.

Fig. 1. Setup of the search task. The participant is shown the image on the left without a concrete target. Afterward (middle), the target is shown to the participant. The participant searches the target in the scene, and by gaze tracing, the object is highlighted and selected if the participant fixates on it for more than 2 s (right).

for both Navigational and Informational Intent. Each session consists of different images, resulting in a total of 150 in the unique input scenes resembling the industrial working conditions of manufacturing or production units.

4   Methods

This section presents different methods we used for EEG data signal processing, including pre-processing and feature extraction. We assemble the dataset for the individual subject using the following steps.

4.1   Data Preprocessing

Typically, EEG signals contain external noises and artifacts like muscle movement, eye blinks, etc., while recording [21]. Therefore, it is necessary to preprocess the recorded data before extracting meaningful information for further analysis. We preprocess the data in MATLAB[4] using functions from the EEGLAB toolbox [10]. Below are the preprocessing steps to clean the data:

1. **Filtering:** High-pass filtering at a cutoff frequency of 1 Hz is applied as recommended by [11] to remove low-frequency noise and low-frequency shifts before using the independent component analysis (IIR Filter, `pop-iirfilt` from EEGlab). A notch filter with a lower cutoff frequency of 48 Hz and an upper cutoff frequency of 52 Hz is applied to remove power line noise [6] which is followed by a low pass filtering done at a cutoff frequency of 40 Hz (IIR Filter).
2. **Artifact rejection:** We do electrode rejection using `pop_clean_rawdata` from EEGLAB as poor electrode-to-skin contact, broken recording device, and low signal quality hinder the quality of signals. Electrodes with a large

---

[4] MATLAB version 9.3.0.713579 (R2017b). 2017.

portion of noise are removed based on their standard deviation and channels, which poorly correlate with other channels. The rejection threshold for channel correlation is 0.8.

3. **Re-referencing:** All electrodes are re-referenced to a common average reference, as it minimizes uncorrelated signal and noise sources through averaging.

4. **Independent Component Analysis (ICA):** Since EEG data collected in a single channel is a composition of all neuron potentials in an area, the recordings between electrodes can be highly correlated [10]. We clean the data using Independent Component Analysis, which removes unwanted artifacts embedded in the data (muscle, eye blinks, or eye movements) without removing the affected data segments. We apply Second-Order Blind Identification (SOBI) algorithm as an ICA decomposition algorithm, following a subsequent automated IC_Label rejection (muscle, heart, and eye components with a 95% threshold).

5. **Channel interpolation:** The channels marked as bad are interpolated using spherical interpolation, `pop_interp`. The motivation behind channel interpolation is to avoid bias when calculating the average reference.

6. **Epoching:** We use preprocessed data to extract specific time windows from the continuous EEG signal, with reference to the stimulus onset from the preprocessed data. We took equal duration for Navigational and Informational Intent within each sample, as the feature extraction module expects the input to have the same dimensions. We also removed the period in the Informational part where the eyes are resting because the participant is only fixating on the object successfully located, see Sect. 3.1.

4.2   Feature Extraction

In this section, we present methods to assemble a feature vector using PyEEG [12] and Common Spatial Pattern (CSP) [21]. PyEEG is an open-source python module for EEG feature extraction [12]. We extract 15 features to generate a feature vector for further investigation. Table 1 shows the list of features extracted for each EEG channel. CSP extracts features from EEG data in a maximally discriminative manner. CSP's basic principle is applying a linear transformation to project the multi-channel EEG signal data to a lower-dimensional spatial subspace. The transformation results in the maximization of the variance of one class while minimizing the variance of other classes at the same time.

4.3   Classification Algorithms

Similar to past studies [1,21], we use Random Forest (RF) and Naïve Bayes classifiers (NB) to distinguish the EEG signals according to the users' implicit intention. Table 2 shows the hyper-parameters. We used default values for other parameters. RF uses bootstrap aggregation with multiple decision tree models. This strategy helps to improve predictive performance as compared to a single model. NB is a probabilistic machine learning model that uses algorithms based on the Bayes theorem. Each algorithm shares a common assumption, i.e., every pair of classified features is independent.

Table 1. List of extracted features from PyEEG

Feature name	Description
Power spectral intensity	distribution of signal power over frequency bands: delta, theta, alpha, beta, and gamma
Petrosian Fractal Dimension	ratio of number of self-similar pieces versus magnification factor
Hjorth mobility and complexity	mobility represents the proportion of the standard deviation of the power spectrum Complexity represents the change in frequency
Higuchi Fractal Dimension	computes fractal dimension of a time series directly in the time domain
Detrended Fluctuation Analysis	designed to investigate the long-range correlation in non-stationary series
Skewness	measure of asymmetry of an EEG signal
Kurtosis	used to determine if the EEG data has peaked or flat with respect to the normal distribution
Minimum, Maximum, and Standard deviation	measure of variability of an EEG signal

5   Experimental Evaluation

In this section, we present our results for subject-wise and cross-subject scenarios. We provide the best setup which is capable of generalizing across different participants. Kang et al. [1] are closely related to our work. We show their highest achieved accuracy as a baseline in all our box plots, depicted by a horizontal line. Since there exists no cross-subject evaluation on the search tasks, therefore, we cannot compare our cross-subject analysis.

5.1   Subject-Wise Analysis

We performed data assembly, training, and evaluation of the test set for each subject individually. To evaluate our classification pipeline, we use 80% of the data

Table 2. List of hyperparameters

Classifier	Hyperparameter
Random Forest	Number of trees, maximum depth, splitting criteria, maximum number of features, minimum samples to split in a node, and minimum number of data points in a leaf node
Naive Bayes	Gaussian distribution with variance smoothing parameter

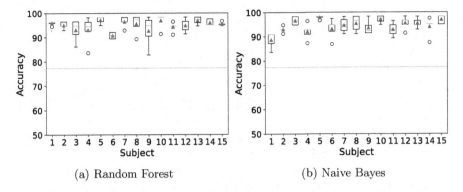

(a) Random Forest   (b) Naive Bayes

Fig. 2. Subject-wise accuracy using common spatial pattern with RF and NB classifiers. The Horizontal line shows the baseline. Triangle, orange line, and circular dots show the mean, median, and outlier values, respectively (Color figure online)

as the training set and 20% as the test set. The test set is assembled randomly at the start of the pipeline to keep it close to the online classification setup. We use hyperparameter optimization with grid search five-fold cross-validation, as it is a common practice for EEG classification [21]. Table 2 shows the parameters that influence the classification performance for both classifiers. The combination that yields the best classification accuracy is identified as the optimal meta-parameters for each subject. Finally, we use the test set to evaluate the performance of the trained classifier. The horizontal line shows the accuracy of the state-of-the-art. Results obtained for all subjects with the CSP feature extraction technique are shown as a box plot in Fig. 2. From the plot, it is evident that all subjects achieved admirable accuracy. For random forest, mean accuracy lies between 90.79% and 97.18%. The standard deviation falls in the range of 0.73 and 5.06. The highest mean intent recognition accuracy of 97.18% is attained by subject S_5. Moreover, other subjects, S_i, i $\in \{1, 2, 5, 7, 8, 10, 13, 14, 15\}$ achieve mean accuracy above 95%. As compared to Random Forest, Naive Bayes performs slightly worse, especially for subject S_1, where the mean accuracy is 88.73%. However, for subjects S_i, i $\in \{3, 6, 15\}$ Naive Bayes achieves better results with a mean accuracy of 96.55%, 93.16%, and 96.95% respectively. For Naive Bayes, the mean accuracy lies between 88.73% and 97.89%, and the standard deviation is between 0.74 and 3.57. Subject S_5 also attains the highest mean accuracy for the Naive Bayes classifier. Overall, for the CSP feature extraction technique, both the classifier perform similarly to mean accuracy. Figure 3 shows the results obtained from the assembled feature vector using the PyEEG toolbox with Random Forest and Naive Bayes classifiers. We use the same hyperparameters to compare different feature extraction techniques and classifiers, as shown in Table 2. Both the classifiers perform worse as compared to CSP. The mean accuracy lies between 81.07% and 93.93% with RF and 64.73% and 83.93% with NB. Figure 4 shows the confusion matrix for subject S_5 (which achieved the highest overall mean accuracy) using CSP and assembled feature vector with PyEEG. The diagonal

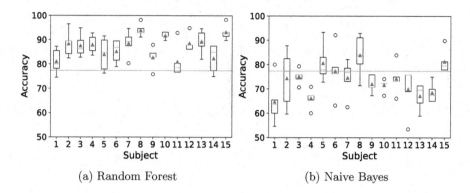

(a) Random Forest   (b) Naive Bayes

Fig. 3. Subject-wise accuracy using assembled feature vector (FV) using RF and NB classifiers. The Horizontal line shows the baseline. Triangle, orange line, and circular dots show the mean, median, and outlier values, respectively (Color figure online)

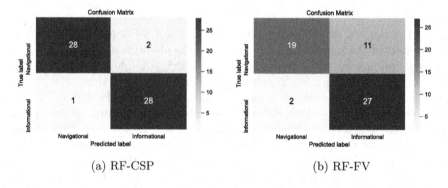

(a) RF-CSP   (b) RF-FV

Fig. 4. Confusion Matrix for subject S_5

elements show the number of correct classifications, while off-diagonal elements show misclassification. We have a balanced dataset between the two classes, with the highest number of correct predictions for CSP compared to PyEEG.

5.2 Cross-Subject Analysis

For the cross-subject case, we study two types of variability. Inter-subject: differences in brain activity across subjects. Intra-subject: differences in brain activity for the same subject occurring in multiple repetitions of the same task.

Inter-subject: Table 3 shows the result for inter-subject variability where each subject is taken as a test subject while the remaining subjects are in the training set. Thus, we acquire the test set from a different subject which is not a part of the training data. The performance evaluation is done using Random forest (RF) and Naive Bayes (NB) on the feature vector from CSP and assembled feature vector (FV) from the PyEEG toolbox. We compute the results using grid search five-fold cross-validation. We use the same parameters for hyperparameter optimization (shown in Table 2). Since we do not fix the random state of classifiers, we iterate the experiment 5 times to compute mean accuracy and standard deviation. The highest mean accuracy of 96.83% with RF-FV is achieved when S_8 is taken as the test set with S_i, $i \in \{1, 2, 3, 4, 5, 6, 7, 9, 10, 11, 12, 13, 14, 15\}$ as train set. Hyperparameter tuning plays a significant role in achieving optimal performance with an exhaustive and wide range of combinations. Overall, RF with assembled feature vectors performs best for all subjects.

Table 3. Inter-subject accuracy (%) for each subject as a test set with remaining subjects as train set

Classifier	P1	P2	P3	P4	P5	P6	P7	P8	P9	P10	P11	P12	P13	P14	P15	Mean
NB-FV	78.75	88.30	66.74	54.24	62.74	82.12	54.61	65.68	78.61	51.73	65.61	73.30	58.25	76.27	68.37	68.35
	± 0.14	± 0.0	± 0.14	± 0.29	± 0.40	± 0.63	± 0.15	± 0.17	± 0.17	± 0.01	± 1.01	± 0.16	± 0.14	± 0.01	± 0.01	± 0.09
RF-FV	85.95	93.05	83.88	89.57	90.63	93.61	83.75	**96.83**	87.71	88.06	77.70	75.86	79.57	79.80	**96.73**	86.84
	± 0.43	± 0.65	± 0.64	± 1.32	± 0.47	± 0.58	± 0.83	± 0.53	± 1.04	± 0.56	± 1.69	± 1.95	± 1.25	± 0.78	± 0.17	± 0.25
NB-CSP	50.00	65.25	77.78	71.09	65.41	77.13	61.15	67.55	63.89	65.62	72.66	73.86	71.74	65.15	52.31	66.70
	± 0.0	± 0.0	± 0.0	± 0.15	± 0.22	± 0.0	± 0.0	± 0.14	± 7.10	± 0.0	± 0.0	± 0.13	± 0.0	± 0.0	± 1.22	± 0.48
RF-CSP	50.00	53.05	68.47	71.09	53.29	63.19	54.46	59.57	61.74	69.86	65.90	68.97	66.16	61.82	50.41	61.18
	± 0.0	± 6.1	± 4.76	± 0.84	± 6.07	± 6.99	± 3.39	± 4.06	± 1.36	± 2.18	± 3.42	± 2.71	± 2.83	± 1.98	± 0.69	± 0.97

Intra-subject: For intra-subject estimation, we use one complete session from all the subjects and treat it as a test set while the remaining four sessions are in the training set. Since we do not fix the random state of the classifiers, we iterate 5 times and, thus, demonstrate the results in terms of the mean and standard deviation of classification accuracy. The same hyperparameters are tuned (shown in Table 2). Table 4 shows the mean accuracy, RF with assembled feature vector using PyEEG performs significantly better than other classifiers and feature extraction techniques. These results also align with the inter-subject analysis where RF-FV works best.

Table 4. Intra-subject mean accuracy (%)

Classifier	FV	CSP
RF	**92.46** ± 0.40	82.06 ± 2.05
NB	67.66 ± 0.44	71.88 ± 0.23

6   Conclusion and Discussion

This paper proposes a classification pipeline to classify users' intentions based on EEG data. The final prediction of the model is highly dependent on the methods used for data acquisition, preprocessing algorithms, computing features, and the choice of the classification algorithm. We evaluated our pipeline for subject-specific and cross-subject scenarios. In the case of the subject-specific analysis, our evaluation demonstrates that the CSP feature extraction method performs best for both Random Forest and Naive Bayes classifiers achieving a maximum mean accuracy of 97.18% and 97.89%, respectively. Our work is a significant improvement compared to state-of-the-art, which makes our pipeline applicable to a real-world setting. However, for PyEEG, our pipeline could only achieve a maximum mean accuracy of 93.93% and 83.93% for Random Forest and Naive Bayes classification algorithms, respectively. We also extend our pipeline to adapt to the cross-subject scenario by combining the subject-specific dataset. Our cross-subject model achieves the highest mean accuracy of 96.83% and 92.46% for inter and intra-subject variability, respectively. The implementation pipeline enables generalizing brain signals across different subjects, capable of reducing the necessity of exhaustive subject-specific training sessions and training processes with tedious calibration. We also recommend using PyEEG with a Random Forest classifier since it generalizes well over all the subjects while being comparable to other strategies for a subject-wise scenario. In the future, we would like to extend our implementation approach with a multi-modal intent recognition model for discovering users' intentions with complicated scenes. Additionally, it would be interesting to use data recorded on different days from the same subject and handle this type of variability.

Limitations: In this study, we claim that our intent recognition pipeline is generalized from trained subjects to the new unseen subject. However, we do not test this with subjects from diverse age groups or subjects with special conditions.

Acknowledgement. This work is partially funded by the German Ministry of Education and Research (BMBF) under project EXPECT (Grant Number: 01IW20003).

References

1. Kang, J.-S., et al.: Human implicit intent recognition based on the phase synchrony of EEG signals. Pattern Recognit. Lett. **66**, 144–152 (2015)
2. Jang, Y.-M., et al.: Human intention recognition based on eyeball movement pattern and pupil size variation. Neurocomputing **128**, 421–432 (2014)
3. Park, U., Mallipeddi, R., Lee, M.: Human implicit intent discrimination using EEG and eye movement. In: Loo, C.K., Yap, K.S., Wong, K.W., Teoh, A., Huang, K. (eds.) ICONIP 2014. LNCS, vol. 8834, pp. 11–18. Springer, Cham (2014). https://doi.org/10.1007/978-3-319-12637-1_2
4. Slanzi, G., Balazs, J.A., Velásquez, J.D.: Combining eye tracking, pupil dilation and EEG analysis for predicting web users click intention. Inf. Fusion **35**, 51–57 (2017)
5. Princy, R., Thamarai, P., Karthik, B.: Denoising EEG signal using wavelet transform. Int. J. Adv. Res. Comput. Eng. Technol. 4(3), 1070–1074 (2015)
6. Nottage, J.F., Horder, J.: State-of-the-art analysis of high-frequency (gamma range) electroencephalography in humans. Neuropsychobiology **72**(3–4), 219–228 (2015)
7. Sharbrough, F.C.G.E.: American Electroencephalographic Society guidelines for standard electrode position nomenclature. J. Clin. Neurophysiol. **8**, 200–202 (1991)

8. Ferree, T.C., et al.: Scalp electrode impedance, infection risk, and EEG data quality. Clin. Neurophysiol. **112**(3), 536–544 (2001)
9. Haas, J.K.: A history of the unity game engine (2014)
10. Delorme, A., Makeig, S.: EEGLAB: an open source toolbox for analysis of single-trial EEG dynamics including independent component analysis. J. Neurosci. Methods **134**(1), 9–21 (2004)
11. Klug, M., Gramann, K.: Identifying key factors for improving ICA-based decomposition of EEG data in mobile and stationary experiments. Eur. J. Neurosci. **54**(12), 8406–8420 (2021)
12. Bao, F.S., Liu, X., Zhang, C.: PyEEG: an open source python module for EEG/MEG feature extraction. Comput. Intell. Neurosci. **2011** (2011)
13. Saha, S., Baumert, M.: Intra-and inter-subject variability in EEG-based sensorimotor brain computer interface: a review. Front. Comput. Neurosci. **13**, 87 (2020)
14. Buerkle, A., et al.: EEG based arm movement intention recognition towards enhanced safety in symbiotic Human-Robot Collaboration. Robot. Comput.-Integr. Manuf. **70**, 102137 (2021)
15. Ajoudani, A., et al.: Progress and prospects of the human-robot collaboration. Auton. Rob. **42**(5), 957–975 (2018)
16. Mao, X., et al.: Progress in EEG-based brain robot interaction systems. Comput. Intell. Neurosci. **2017** (2017). https://doi.org/10.1155/2017/1742862
17. Betzel, R.F., et al.: The community structure of functional brain networks exhibits scale-specific patterns of inter-and intra-subject variability. Neuroimage **202**, 115990 (2019)
18. Wei, C.-S., et al.: A subject-transfer framework for obviating inter-and intra-subject variability in EEG-based drowsiness detection. NeuroImage **174**, 407–419 (2018)
19. Samek, W., Kawanabe, M., Müller, K.-R.: Divergence-based framework for common spatial patterns algorithms. IEEE Rev. Biomed. Eng. **7**, 50–72 (2013)
20. Lotte, F., Guan, C.: An efficient P300-based brain-computer interface with minimal calibration time. In: Assistive Machine Learning for People with Disabilities symposium (NIPS 2009 Symposium) (2009)
21. Rekrut, M., Sharma, M., Schmitt, M., Alexandersson, J., Krüger, A.: Decoding semantic categories from EEG activity in object-based decision tasks. In: 2020 8th International Winter Conference on Brain-Computer Interface (BCI), pp. 1–7 (2020).https://doi.org/10.1109/BCI48061.2020.9061628
22. Bengler, K., et al.: Interaction principles for cooperative human-machine systems, pp. 157–164 (2012)

RVFL Classifier Based Ensemble Deep Learning for Early Diagnosis of Alzheimer's Disease

Krishanu Maji, Rahul Sharma, Shradha Verma, and Tripti Goel[✉]

Biomedical and Imaging Lab, National Institute of Technology Silchar, Silchar, Assam, India

{krishanumaji_pg_21,rahul_rs,shradha_rs,triptigoel}@ece.nits.ac.in

Abstract. The prevalence of Alzheimer's Disease (AD) is increasing daily in elderly people, estimated to be 15 million by the year 2050. AD is an irreversible neurodegenerative disorder that may lead to the death of the affected person. Only early AD diagnosis at the Mild Cognitive Impairment (MCI) stage can help clinicians to convert MCI patients back to Cognitive Normal (CN) or slow down the progression of the disease. The advent of neuroimaging techniques like Magnetic Resonance Imaging (MRI) helps in observing the anatomical changes in the brain of MCI and AD patients with improved resolution. AD mainly affects the temporal lobe structure, hippocampus volume, and cerebral cortex, which are visible in MRI scans. In this paper, an ensemble of three planes of MRI is proposed using a deep learning model, and the extracted features are classified using Random Vector Functional Link (RVFL) neural networks. The experiments are done on the publicly available dataset, Alzheimer's Disease Neuroimaging Initiative (ADNI), to classify AD vs CN vs MCI. The performance of the proposed model is compared in terms of accuracy, specificity, sensitivity, and precision.

Keywords: Alzheimer's Disease · Ensemble Deep Learning · Machine learning · RVFL

1 Introduction

Alzheimer's Disease (AD) is an unpropitious chronic neural disorder distinguished by β amyloid plaques and neurofibrillary tangles, which are well-known neuropathological biomarkers, among other indications. The initial stage of AD involves symptoms like memory loss, loss of thinking and reasoning ability, and affecting day-to-day life. Age is one of the significant and noted threats in AD but age alone is not the only crucial cause of dementia [3]. The percentage of individuals suffering from AD dementia is 5.0% among the aged 65 to 74, 13.1% among the age group of 75 to 84, and 33.2% people above 85 years old. According to World Alzheimer's Report, 2021 75% of people with dementia are not diagnosed due to a lack of awareness, which is a major cause of late diagnosis [11]. Hitherto, AD has no remedy, early prognosis is the only method to decelerate this health crisis. As the disease progresses, AD affects most brain regions that can show

M. Tanveer et al. (Eds.): ICONIP 2022, LNCS 13625, pp. 616–626, 2023.
https://doi.org/10.1007/978-3-031-30111-7_52

symptoms like loss of reasoning ability, language problems, cognitive disability, and hampering social life. Nowadays, neuroimaging techniques are rampant for the prognosis of AD. Neuroimaging techniques are an imaging process that generates brain images to observe abnormalities. Magnetic Resonance Imaging (MRI) scan is a well-known imaging technique, among other techniques. MRI uses nuclear magnetic resonance property to record photographic illustrations of several brain regions. MRI manifests the damage to different regions in the brain due to cerebral atrophy [18]. MRI has a good contrasting ability through which differentiation between fat, muscles, water, and soft tissues can be seen effortlessly [8]. This approach illustrates the structural contortion of the region of interest (ROI).

A very large number of machine learning (ML) techniques like Support Vector Machine (SVM), k-nearest neighbor (kNN), and Naive Bayes are nowadays used for AD diagnosis [8,16]. Manual feature extraction and selection from the brain MRI scan is one major drawback of ML. Due to this human intervention in ML, it suffers from manual error, labor, and time. This drawback is overcome in deep learning (DL) architecture. Automatic feature selection in DL is one of the most advantageous and makes it a more advanced approach in today's working field. DL captures the finest feature discrimination in MRI scan images [7,17]. Deep Belief Networks (DBN) and Convolutional Neural networks (CNN) are the most popular networks that can build up a high-level diagnosis system by combining deep features implanted in MRI data sets. Once a deep neural network (DNN) is trained with an MRI dataset, it can easily prevail over ML. DL networks are conventionally trained by backward propagation methods to minimize loss factors, though this process is much more time consuming and needs lots of data set to train the network [10]. Randomized neural networks can overcome the backpropagation drawback. They are also fast to train. Among all the randomized networks, the random vector functional link (RVFL) is well known due to its performance in several domains like forecasting, classification, and regression. RVFL is a feedforward with a single hidden layer, and the weight and the bias in the hidden layer are taken randomly and kept fixed for the whole training period.

Ensemble Learning (EL) is more effective and resilient in learning techniques incorporating other learning architectures [9]. EL using multiple algorithms can give better predictive results than any other single algorithm [10]. Nowadays, researchers have started an ensemble of learning architectures to classify AD with a notable accuracy rate [13]. The combination of DNN and EL can make a more improved diagnosis system. It can also provide a better approximation than any other classifier. This paper uses the Ensembling of Deep Neural Networks followed by the RVFL model as a classifier for AD diagnosis. Here are some key contributions of this paper.

– We are proposing an ensemble-based DL architecture, including RVFL for classification, which will be more accurate and favorable for medical practitioners.

– In this model, we use ensemble of sagittal, coronal, and axial planes of MRI images to get a much clearer observation of the brain.
– In this study, we are doing an ensemble of MRI plane features to get a better amount of information, which can help the system to predict effortlessly.

2 Related Works

AD can be the primary reason for dementia in adults. It causes due to an imbalance in protein synthesis inside and outside the brain cell [6,12]. Hence, AD diagnosis is essential. This section discusses previous studies using MRI data for AD diagnoses.

In [17], authors have presented an approach of incorporating DL based model for feature extraction followed by a shallow network for performing classification. The authors incorporated a fuzzy-based activation function for the classification tasks. However, the model sensitivity obtained is low. Ahmed et al. [1] have proposed a DL-based learning framework for staging AD-based chosen ROI of MRI scan images. In this study, they have ensembled CNN and three view patches (TVPs) to analyze the contribution of all regions. This ensemble-based CNN achieved a notable rate of performance. The method demonstrated a higher accuracy rate in MRI-AD scan images and verified that the hippocampi, amygdalae, and insulae presented distinctive characteristics. However, selective ROI-based models do not offer enough information to diagnose AD using the structural MRI modality reliably.

Battineni et al. [4] have employed four individual learning models: Naive Bayes, kNN, Neural Network, and SVM. These four learning methods have been combined to create single EL or hybrid model. Hence, the approach has shown the merits of four techniques and achieved the highest prediction accuracy. Moreover, this hybrid model has provided the advantage of noise reduction and lower bias. However, the paper has not discussed the early detection of AD. Dongren et al. [21] have proposed an EL model for four-way AD classification. In that model, a new feature selection approach has been used to differentiate more informative features of 426 structural magnetic resonance imaging-based morphometric and three demographic features. This new method ensures that each binary classification can reach its highest accuracy. The method provides a framework for multi-way classification using hierarchical groups and precise feature selection. However, the methodology has illustrated limitations in terms of accuracy. Tan et al. [20] have proposed an instance transfer learning-based approach since the source and target data characteristics vary. Classification is performed using a combination of the optimal transferred and target domain training samples. However, the feature set used only two shape features and only thirty manually selected texture features which are very few for model learning.

Alam et al. [2] have suggested a novel method of linear discriminant analysis to extract the primary feature from dual-tree complex wavelet transform into a linear form and have used twin SVM (TWSVM) as a classifier. The model illustrates the advantage of TWSVM in terms of classification precision and processing speed. However, this methodology has high computational complexity.

Another effective methodology based on angle-based universum least squares twin support vector machine (AULSTWSVM) has been proposed by Richhariya and Tanveer [15]. The paper classifies data using the technique that minimizes the angle between the two hyperplanes. The quadratic loss is substituted by a linear loss, which lowers the computational cost and compared to other SVM techniques, AULSTWSVM has achieved decent accuracy. In contrast, angle based method required prior knowledge of data.

Several variants of RVFL are available and have been widely adopted by many researchers. Cheng et al. [5] have suggested another feature extraction approach that utilizes blocks-based residual networks. The technique has used diversified Ensemble Deep Random Vector based classifier to ensemble the features and perform classification. Similarly, in [19], authors have proposed several variants of RVFL, such as deep RVFL (dRVFL) and ensemble deep RVFL (edRVFL), for classification purposes. However, model parameter uncertainty is high, and the performance obtained was not outstanding for time series data.

This section motivates us to incorporate DL based ensemble approach followed by a shallow network as a classifier for the proposed work. The upcoming section provides details about the model architecture.

3 Methodology

3.1 Preprocessing

Statistical Parametric Mapping (SPM12) is used to preprocess all of the $3D$ T1-weighted sMRI scans taken from ADNI datasets in this research work. Voxel-based methodology (VBM) is a statistical neuro-scan analysis method to analyze local atrophies in complex brain architecture. Image normalization, image realignment, and image registration are preprocessing steps. A full explanation of the preprocessing methods is described in the following subsection and the pipeline is visualized in Fig. 1.

Image Normalisation. Data normalization is a crucial step that verifies each input parameter (in this case, voxel) has uniform data distribution. We subtracted the average intensity values and divided the deviation to achieve a zero-mean and one-standard-deviation distribution for the intensity values in all MRI scans.

Image Realignment. The main goal of image realignment is to remove artifacts created by the patient due to moving during the MRI scan. The least squares approach is used to realign, and six stiff transformations are performed spatially. Estimation and re-slicing are the two methods that are typically used for realignment procedures. All scan images are subjected to a 4th-degree B-spline without wrapping throughout the re-slicing process.

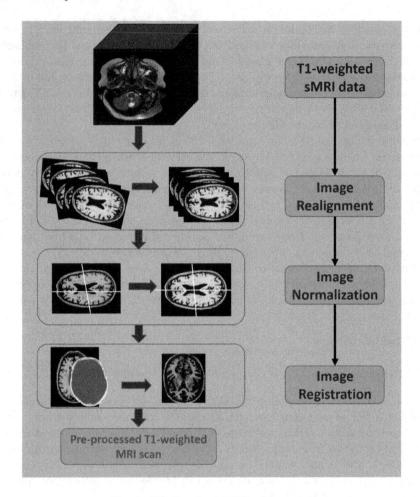

Fig. 1. Pipeline for MRI Preprocessing

Image Registration. MRI scans from any source might be used for the well-performing prognostic model for AD classification. Scanners have different scanning parameters such as slice count, slice thickness, flip angle, etc. The $3D$ sMRI scans must be registered on the reference picture using geometric adjustments to make one image fit into another. Slice-by-slice matching ensures that all scan images are identical to the reference picture. For registration, the recommended study operates with $4th$ degree B-spline interpolation, average separation of 4 mm $*$ 2 mm, and gaussian histogram smoothening of $7 * 7$. As an objective function for registration, normalized cross-correlation is utilized. Following registration, all scans' dimensions $256 * 256 * 166$ remained the same.

Slice Extraction. The entirety of the preprocessed samples are three-dimensional, and we aim to extract all sagittal, coronal, and axial plane slices from the MRI images. The grey level co-occurrence matrix method is used to choose the center slice. Slices from each data set were chosen based on atrophy to provide input to the DL model. After the registration process, the total dimensions of all processed images are $256 * 256 * 166$. Sagittal, coronal, and axial plane images are acquired from all registered structural MRI datasets, and slice extraction is executed using the MRIcro software. Five preferential and most relevant slices were extracted from the MRI scans for each of the three classes. All of the pre-processed images are $256 * 256$ in size.

3.2   Network Architecture

After the slice extraction procedure using grey level co-occurrence matrix, all the sagittal plane, axial plane, and coronal plane MRI scan images are taken. These extracted images also can be specified as the input data images are ensembled. As in this research work, we are using an ensemble process to enhance our model performance by using more than one classifier. By employing the ensembling approach, we can overcome the problem of underfitting and overfitting that can be shown due to the fewer or more feature selection processes. In this ensembling process, a Deep Neural Network (DNN) with fine-tuned ResNet-50 network is used for the feature extraction from every class of the scanned images.

Due to the rise in the prevalence of representational learning, randomized neural networks have also been used as a building block for deep neural networks. It takes less time to train a randomized neural network since it uses less processing power. In the literature, there are many multi-layer RVFL designs discussed. In the RVFL [14] network, the weights and bias of the hidden layer are randomly given and remain stable during the training process, making it a feed-forward neural network. Layer D is an output of the hidden layer's random non-linearly changed features, denoted as H, and direct linkages to the original layer are indicated as X, which is expressed as $D = [HX]$. The total number of input features to the output layer may be expressed as $x_o = x_{in} + n$ if x_{in} is an input feature and n is the number of neurons in the layer. The only parameter to be calculated in a hidden layer with fixed weights and biases is *beta*. To put it another way, the objective function is:

$$\min_{\beta} \parallel D\beta - O \parallel^2 + R \parallel \beta \parallel^2 \qquad (1)$$

where R resembles the regularization parameter, and O denotes the output target. For the case of $R = 0$, the Moore-Penrose pseudoinverse solution exists. β_s can be calculated using the Moore-Penrose pseudoinverse, and the expression for β_s can be emanated as $\beta_s = D^+ O$.

This investigation is conducted with the help of ensembling the MRI plane features extracted from fine-tuned ResNet-50 networks. As a result of the DNN model, training characteristics become more abstract, allowing for more flexibility in the effectiveness of model learning. Feature extracted are sent to RVFL classifiers. We passed the averaged 3 feature set to the RVFL classifier, which produced an output class that accurately predicted the final result. Deep CNN-based ensembling and a shallow classifier are used in this fashion as depicted in Fig. 2.

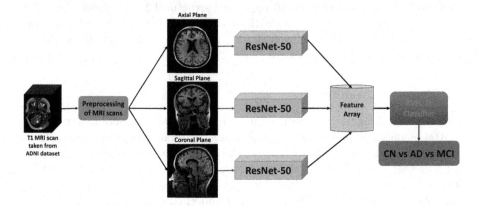

Fig. 2. Network Architecture for the proposed model

This section provides an overview of the proposed model and details the many steps that are involved. The proposed model is tested using the ADNI database, and the findings are discussed in the next section.

4   Result and Discussion

4.1   Performance Analogy Between Different MRI Plane

A comparison of sMRI efficacy in the different MRI planes has been tabulated in Table 1, which compares the three sMRI planes. The sagittal plane has proven to produce better categorization results than any other plane since prominent sites of the cerebral cortex are clearly visible in the sagittal plane. Also, the frontal lobe and corpus callosum are plainly evident in the sagittal plane. Individual ability to solve cognitive problems depends on the frontal lobe, thereby directly contributing to adopting the sagittal plane as the most relevant plane for AD diagnosis.

The proposed ensemble approach outperforms in terms of accuracy over individual planes. However, the coronal and axial planes show slightly better specificity and precision than the proposed approach.

4.2   Discussion

In this study, we applied deep learning methods to detect AD by extracting all levels of features. Specifically, we incorporated an ensemble approach to diagnosing AD with more robustness and generalization using different MRI planes.

Table 1. Performance analogy between different MRI planes (in %)

Performance	Axial only	Coronal only	Sagittal only	Ensemble Approach
Accuracy	90	84.44	91.33	**92.67**
Sensitivity	83.68	65	90.33	**92.33**
Specificity	93.78	**94.17**	91.83	92.83
Precision	**88.96**	84.78	84.69	86.56

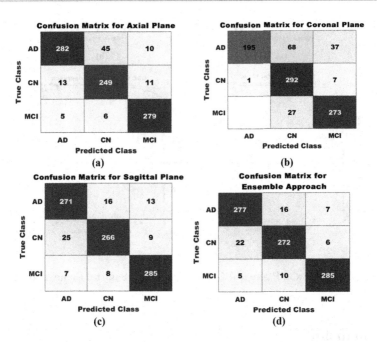

Fig. 3. Confusion Matrix for (a) Axial Plane, (b) Coronal Plane, (c) Sagittal Plane, and (d) Ensemble Approach

The proposed work relied on ADNI data sets and compared its findings over the different planes as input. Incorporating the aforementioned preprocessing approaches, a uniform dataset has been prepared, which allows deep models to learn features more effectively. Each plane visualizes different brain regions, thereby, a deep model extracts features from the entire brain. Ensembling features have been implemented, and classification has been made using RVFL as a classifier, providing a much more accurate and generalized performance in less time (Figs. 3 and 4).

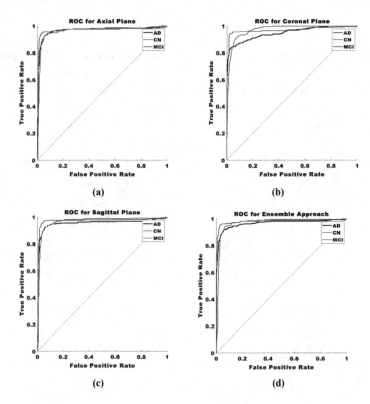

Fig. 4. ROC for (a) Axial Plane, (b) Coronal Plane, (c) Sagittal Plane, and (d) Ensemble Approach

5   Conclusion

Alzheimer's Disease is an incurable neurodegenerative disorder that damages brain cells and causes memory loss and cognitive decline. Only early diagnosis can help in reducing the progression of AD disease. This paper proposes the ensemble deep learning-based model for early AD diagnosis from MRI scans. Firstly, all the MRI scans taken from the ADNI dataset are preprocessed, which includes image realignment, normalization, and registration to make the model generalized for any type of scanner. After that, sagittal, axial, and coronal planes are extracted from the preprocessed image, and key slices are extracted based on the mutual information. These three planes are ensembled using ResNet-50 as the base model. The extracted features from the three planes are concatenated and fed to RVFL classifier for classification. The performance of the proposed ensemble network is compared with single plane slices to show the effectiveness of the proposed model. This model will assist clinicians in automatically diagnosing early-stage AD as that necessary treatment can be done to cure the patient.

Intending to improve upon the current findings, we want to look into the feasibility of integrating the different plane features based on the region of interest, in particular, to overcome the challenges in recognizing certain classes and a feature selection stage to lower the dimensionality of the features. Further work can be extended to include multi-modal data like positron emission tomography, diffusion tensor imaging, and quantitative susceptibility mapping in the ensemble deep learning so that more information can be extracted from different types of images to improve the diagnosis results.

References

1. Ahmed, S., Kim, B.C., Lee, K.H., Jung, H.Y., Initiative, A.D.N.: Ensemble of ROI-based convolutional neural network classifiers for staging the Alzheimer disease spectrum from magnetic resonance imaging. PLoS One **15**(12), e0242712 (2020)
2. Alam, S., Kwon, G.R., Kim, J.I., Park, C.S.: Twin SVM-based classification of Alzheimer's disease using complex dual-tree wavelet principal coefficients and LDA. J. Healthc. Eng. **2017**, 8750506 (2017)
3. An, N., Ding, H., Yang, J., Au, R., Ang, T.F.: Deep ensemble learning for Alzheimer's disease classification. J. Biomed. Inform. **105**, 103411 (2020)
4. Battineni, G., Chintalapudi, N., Amenta, F., Traini, E.: A comprehensive machine-learning model applied to magnetic resonance imaging (MRI) to predict Alzheimer's disease (ad) in older subjects. J. Clin. Med. **9**(7), 2146 (2020)
5. Cheng, W.X., Suganthan, P.N., Katuwal, R.: Time series classification using diversified ensemble deep random vector functional link and ResNet features. Appl. Soft Comput. 107826 (2021)
6. Cogswell, P.M., et al.: Associations of quantitative susceptibility mapping with Alzheimer's disease clinical and imaging markers. Neuroimage **224**, 117433 (2021)
7. Dwivedi, S., Goel, T., Sharma, R., Murugan, R.: Structural MRI based Alzheimer's disease prognosis using 3D convolutional neural network and support vector machine. In: 2021 Advanced Communication Technologies and Signal Processing (ACTS), pp. 1–4. IEEE (2021)
8. Dwivedi, S., Goel, T., Tanveer, M., Murugan, R., Sharma, R.: Multi-modal fusion based deep learning network for effective diagnosis of Alzheimers disease. IEEE MultiMed. **29**, 45–55 (2022)
9. Ganaie, M., Hu, M., Malik, A.K., Tanveer, M., Suganthan, P.N.: Ensemble deep learning: A review. Eng. Appl. Artif. Intell. **115**, 105151 (2022). Elsevier
10. Ganaie, M., Tanveer, M.: Ensemble deep random vector functional link network using privileged information for Alzheimer's disease diagnosis. IEEE/ACM Trans. Comput. Biol. Bioinform. (2022)
11. Gauthier, S., Rosa-Neto, P., Morais, J., Webster, C.: World Alzheimer report 2021: journey through the diagnosis of dementia. Alzheimer's Dis. Int. (2021)
12. Gong, N.J., Dibb, R., Bulk, M., van der Weerd, L., Liu, C.: Imaging beta amyloid aggregation and iron accumulation in Alzheimer's disease using quantitative susceptibility mapping MRI. Neuroimage **191**, 176–185 (2019)
13. Malik, A., Tanveer, M.: Graph embedded ensemble deep randomized network for diagnosis of Alzheimer's disease. IEEE/ACM Trans. Comput. Biol. Bioinform. (2022)
14. Pao, Y.H., Takefuji, Y.: Functional-link net computing: theory, system architecture, and functionalities. Computer **25**(5), 76–79 (1992)

15. Richhariya, B., Tanveer, M.: An efficient angle based universum least squares twin support vector machine for classification. ACM Trans. Internet Technol. **21**(3), 1–24 (2021)
16. Sharma, R., Goel, T., Murugan, R.: Prediction of Alzheimer's disease using machine learning algorithm. In: Das, K.N., Das, D., Ray, A.K., Suganthan, P.N. (eds.) Proceedings of the International Conference on Computational Intelligence and Sustainable Technologies. Algorithms for Intelligent Systems, pp. 11–17. Springer, Singapore (2022). https://doi.org/10.1007/978-981-16-6893-7_2
17. Sharma, R., Goel, T., Tanveer, M., Dwivedi, S., Murugan, R.: FAF-DRVFL: fuzzy activation function based deep random vector functional links network for early diagnosis of Alzheimer disease. Appl. Soft Comput. **106**, 107371 (2021)
18. Sharma, R., Goel, T., Tanveer, M., Murugan, R.: FDN-ADNet: fuzzy LS-TWSVM based deep learning network for prognosis of the Alzheimer's disease using the sagittal plane of MRI scans. Appl. Soft Comput. **115**, 108099 (2022)
19. Shi, Q., Katuwal, R., Suganthan, P.N., Tanveer, M.: Random vector functional link neural network based ensemble deep learning. Pattern Recogn. **117**, 107978 (2021)
20. Tan, X., et al.: Localized instance fusion of MRI data of Alzheimer's disease for classification based on instance transfer ensemble learning. Biomed. Eng. Online **17**(1), 1–17 (2018)
21. Yao, D., Calhoun, V.D., Fu, Z., Du, Y., Sui, J.: An ensemble learning system for a 4-way classification of Alzheimer's disease and mild cognitive impairment. J. Neurosci. Methods **302**, 75–81 (2018)

Anatomical Landmarks Localization for 3D Foot Point Clouds

Sheldon Fung[1], Xuequan Lu[1(✉)], Mantas Mykolaitis[2], Imran Razzak[3],
Gediminas Kostkevičius[2], and Domantas Ozerenskis[2]

[1] Deakin University, Geelong, Australia
xuequan.lu@deakin.edu.au
[2] Ortho Baltic, Kaunas, Lithuania
[3] University of New South Wales, Sydney, Australia

Abstract. 3D anatomical landmarks play an important role in health research. Their automated prediction/localization thus becomes a vital task. In this paper, we introduce a deformation method for 3D anatomical landmarks prediction. It utilizes a source model with anatomical landmarks which are annotated by clinicians and deforms this model non-rigidly to match the target model. Two constraints are presented in the optimization, which are responsible for alignment and smoothness, respectively. Experiments are performed on our dataset and the results demonstrate the robustness of our method and show that it yields better performance than the previous techniques in most cases.

Keywords: Anatomical landmarks prediction · Point cloud · Deformation

1 Introduction

As imaging technology becoming more advanced, 3D volumetric data is commonly used for a variety of fields, especially in the medical and manufacturing industries. Naturally, 3D anatomical landmarks are also becoming significant and prerequisites for those fields. Identifying landmarks is essential for body deformities diagnosis/treatment planning and professional sports equipment (e.g., sneakers) design. However, anatomical landmarks detection via visual assessment of the 3D model requires experienced specialists or clinicians and thus remains a difficult and costly task.

To tackle this problem, a variety of 3D anatomical landmarks detection techniques has been proposed in the previous studies. Some researchers took advantage of deep learning techniques. For example, Alison et al. [1] utilized an efficient deep learning method for locating the landmarks in computed tomography (CT) scan image data which involves regular pixels. Similar techniques are also adopted in other papers [2–4]. However, these methods are vulnerable to irregular 3D point cloud models. Some other methods resort to the template model

© The Author(s), under exclusive license to Springer Nature Switzerland AG 2023
M. Tanveer et al. (Eds.): ICONIP 2022, LNCS 13625, pp. 627–638, 2023.
https://doi.org/10.1007/978-3-031-30111-7_53

deformation strategy. For example, Fang et al. [5] introduced a technique utilizing volume morphing for 3D biological point landmark predictions. Whereas such methods often neglect the role of non-rigid deformation which will lead to smaller error when matching the source and the target model.

Given the above motivations, we introduce a novel 3D anatomical landmarks prediction approach in this paper. The core idea of our method is to deform a source model with known landmarks (manually annotated) to match the target model, then we can predict the target landmarks according to the transformed known landmarks. We achieve this by resorting to the deformation graph [6]. Concretely, inspired by the technique proposed in [6], we first randomly select vertices in a mesh model as nodes and build a deformation graph. Two constraints form the optimization energy function: aligning two models non-rigidly while keeping the smoothness of the template surface. Then, following Lu et al. [7], we can apply Singular Value Decomposition (SVD) to obtain the optimal rotation matrices and transformation vectors for the deformation equation. Finally, landmarks predictions can be made via deforming the known landmarks.

Extensive experiments on our private dataset which contains 211 3D foot models with 21 manually annotated anatomical landmarks by professional clinicians validate our method. We also compare it with Iterative Closest Point (ICP) (rigid alignment) and two other state-of-the-art non-rigid deformation approaches. Experimental results show that our method is robust and outperforms other methods in most cases, in terms of the root square error metric. Our main contributions are as follows:

- We propose a non-rigid deformation method to predict 3D anatomical landmarks on point clouds.
- We conduct experiments on a dataset that contains various shapes of foot models and compare the proposed method with other techniques.

The rest of the paper is organized as follows. Section 2 reviews previous research work on 3D landmarks prediction and non-rigid point registration methods. Section 3 illustrates our proposed approach. Detailed experimental results and analysis will be covered in Sect. 4. And Sect. 5 summarizes this work.

2 Related Work

In this section, we are first going to review some previous works on 3D landmark prediction. Second, we will cover some state-of-the-art works on non-rigid registration which were developed to align 3D scans of rigid objects that are distorted due to various factors.

3D Landmark Prediction Methods. Most current 3D anatomical landmark prediction methods rely on two categories of data: CT scan and magnetic resonance imaging (MRI) data. There are various prediction techniques. Some methods are based on deep learning techniques. Zheng et al. [2] proposed a two-step approach using a shallow network and a deep network. Similar techniques are also seen in other works. Alison et al. [1] presented an efficient deep learning method for predicting anatomical landmarks' location in CT scan data, which meets the variability across all landmarks classes. A slightly different approach was introduced by Amir et al. [8] which evaluates reinforcement learning (RL) strategies and uses RL agents to identify landmarks by interacting with an environment. With a similar idea, Florin et al. [9] also proposed a deep reinforcement learning (DRL) method to detect anatomical landmarks. More intriguingly, Ebner et al. [10] proposed a landmark localization algorithm for MRI using multiple random regression forests. Subburaj et al. [11] presented an anatomical landmark prediction method by segmenting the model surface into different regions based on surface curvature. Some other methods identify the landmarks by deforming the existing annotated model. For instance, Fang et al. [5] introduced a technique using volume morphing for 3D biological point landmark predictions. Baek et al. [12] developed an anatomically deformable model of the femur based on CT images in order to predict the bone landmarks.

Non-rigid Registration. As a widely studied and utilized technique, there are diverse solutions for non-rigid registration problems. We can roughly categorize them into two groups: optimization-based and statistics-based. For the former one, Sumner et al. [6] presented an algorithm that generates natural deformation according to the constraints provided by users. Similarly, Li et al. [13] developed a robust framework for the reconstruction of complex deforming shapes. It uses a smooth template that provides a coarse approximation of the scanned object which serves as a geometric and topological prior for the reconstruction process. Lu et al. [7,14] introduced an Expectation-Maximization (EM) algorithm for the non-rigid registration of human bodies as point cloud data. Yao et al. [15] proposed a non-rigid registration algorithm using Welsch's function [16] for both alignment error and regularization, addressing the drawback of slow convergence due to the high-accuracy solution in other methods [17]. From the statistics perspective, Myronenko et al. [18] presented Coherent Point Drift (CPD), a robust probabilistic multidimensional point set registration algorithm for both rigid and non-rigid transformation using Gaussian Mixture Model (GMM). The core mind of this method is to force the GMM to move coherently as a group to the target data points by maximizing the likelihood. As an extension of CPD, Dai et al. [19] proposed a fully automated pipeline to synthesize a shape-and-texture 3D morphable model. Unlike other methods, Wand et al. [20] introduced a novel topology-aware adaptive subspace deformation technique that no longer requires a model template.

3 Method

In this section, we introduce our approach in detail. As illustrated in Fig. 1, we first take both the annotated source model and the target model as input and align the former to the latter with rigid-CPD. Then, we deform the source model with a deformation graph to match the target model. Finally, we can predict the landmarks on the target model based on the known landmarks on the source model.

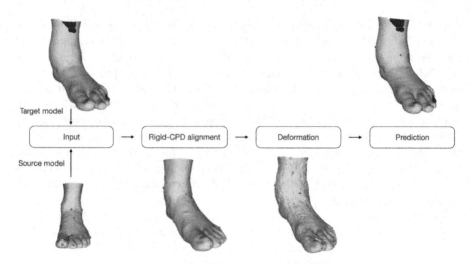

Fig. 1. Overview of our proposed method. Note that the red points represent the landmarks of the corresponding foot model. (Color figure online)

3.1 Deformation Formulation

The goal of our method is to achieve landmark prediction on the surface of 3D models (e.g. meshes or point clouds) using an existing 3D model with known ground-truth landmarks via non-rigid registration. Let $V_t = \{v_{t1}, v_{t2}, \cdots, v_{tn} \in \mathbb{R}^3\}$ be the target model and let $V_{sr} = \{V_s, R_l\}$ be the point set consisting of source model $V_s = \{v_1, v_2, \cdots, v_n \in \mathbb{R}^3\}$ with manually annotated landmarks $R_l = \{r_1, r_2, \cdots, r_n \in \mathbb{R}^3\}$. Note that here R_l is not necessary to be a subset of V_s. To ensure these two models are well aligned to avoid misleading results of closest point search in the following steps, we initialize V_{sr} and V_t with the global rigid point set registration in CPD [18]. To deform the source model V_s to match the target model V_t, we adopt the deformation graph following Sumner et al. [6]. Concretely, we randomly sample the source model V_s to construct the nodes $N = \{n_1, n_2, \cdots, n_j \in \mathbb{R}^3\}$. For each node $n_j \in N$, the affine transformation is specified by a rotation matrix $\mathbf{R} \in \mathbb{R}^{3\times3}$ and a translation vector $\mathbf{T} \in \mathbb{R}^{3\times1}$. Each vertex $v_n \in V_s$ is induced by the combined influence of the neighbouring nodes, which is defined as follows:

$$v_n' = \sum_{n_j \in \Bbbk(v_n)} \omega_j(v_n)[\mathbf{R}_j(v_n - n_j) + n_j + \mathbf{T}_j], \tag{1}$$

where v_n' is the transformed position of v_n, and $\Bbbk(v_n)$ denotes the neighbouring nodes of v_n. Note that there are different approaches for searching the close nodes, such as Euclidean distance and Breadth-first search (BFS). In our experiments, we adopt the former strategy. The weight between n_j and v_n is defined as:

$$\omega_j(v_n) = \mathbb{D}_{n_j}^{-1}/d, \tag{2}$$

where $\mathbb{D}_{nj}$ and d are respectively defined as follows:

$$d = \sum_{n_j \in \Bbbk(v_n)} \frac{1}{|v_n - n_j|}, \tag{3}$$

$$\mathbb{D}_{n_j} = \begin{cases} |v_n - n_j|, & (v_n \neq n_j) \\ \sum_{n_j \in \Bbbk(v_n)} |v_n - n_j|, & (v_n = n_j) \end{cases} \tag{4}$$

Then, we can obtain the predicted landmarks by applying the deformation on the known landmarks:

$$rp_n = \sum_{n_j \in \Bbbk(v_n)} \omega_j(r_n)[\mathbf{R}_j(r_n - n_j) + n_j + \mathbf{T}_j], \tag{5}$$

where rp_n is the landmark predicted by deforming $r_n \in R_l$ with the deformation graph.

3.2 Optimization

We now introduce the optimization strategy for the involved parameters $\mathbf{R}_j$ and $\mathbf{T}_j$ in the deformation equation. We adopt two constraints for the energy function: data term and smooth term.

Data Term. To non-rigidly align the template model to the target model, we design a non-rigid ICP method. For each vertex $v_n \in V_s$, we look for the closest vertex $c_k \in V_t$. And the alignment term minimizes the deviation between the corresponding transformed vertex v'_n and the closest vertex c_k. Following previous works [6,21], we solve it as a ℓ_2 minimization problem with the following objective equation:

$$E_{align} = \sum_{(v_n, c_k) \in C} \|v'_n - c_k\|_2^2, \tag{6}$$

where C is the correspondences (closest pairs) between V_s and V_t. Note that using $\ell_2 - norm$ for minimizing the point-wise distance can bring about a noisy surface on the transformed model due to the noisy data or incomplete structure induced by the camera scans. However, our goal is to force the template model to non-rigidly align the target model. Therefore, the smoothness of the model surface is not considered to be a top priority in practice.

Smooth Term. Each node serves as a localized leader point, influencing all other vertices in the local sense. Therefore, nodes should be resistant to the overlapping influence of each other. Specifically, considering a specific node $n_j \in N$ and one of its nearby node n_k, the influence I_j and I_k are induced by n_j itself and n_k, respectively. They can be computed as follows:

$$I_j = \mathbf{R}_j(n_j - n_j) + n_j + \mathbf{T}_j = (n_j + \mathbf{T}_j), \tag{7}$$

$$I_k = \mathbf{R}_k(n_j - n_k) + n_k + \mathbf{T}_k, \tag{8}$$

We hope that the position of n_j predicted by the affine transformation of n_k to match the actual position when only applying the transformation of n_j to itself. This is achieved by minimizing the ℓ_2 distance between every I_k and I_j for each node. In practice, the regularization term should account for each node in the deformation graph with all their neighboring nodes:

$$E_{smooth} = \sum_{n_j} \sum_{n_k \in \mathbb{k}(n_j)} \|I_j - I_k\|_2^2, \tag{9}$$

Finally, we define the total energy function E_{total} by combining the alignment function and the regularization function with a coefficient α:

$$E_{total} = \alpha E_{align} + E_{smooth} \tag{10}$$

Note that R_j in Eq. (1) should be restricted to be in SO(3) with the orthogonal constraint $R_j^T R_j = I, \forall j$.

3.3   Minimization

To efficiently solve the rotation matrix R as well as the translation vector T for each node transformation, we minimize E_{total} by following Lu et al. [7] and take advantage of the following scheme: with the fixing remaining nodes in the deformation graph, we can update an individual node instead at each time, through which not only the computational complexity can be enormously reduced, but also the non-positive growing trend of the energy function can be guaranteed.

We can obtain the following equation by taking the derivative of Eq. (10) with respect to the translation vector T_j of a specific node j and equating it to zero:

$$\mathbf{T}_j = \mu_v - \mathbf{R}_j \mu_y, \tag{11}$$

where $\mu_v, \mu_y \in \mathbb{R}^{3 \times 1}$ represent the rest parts of the equation that can be calculated. Next, we substitute Eq. (11) with Eq. (10), then organize it to obtain: $E = -tr(\mathbf{H}\mathbf{R}_j) + Z_j, \mathbf{H} \in R^{3 \times 3}$, where $tr(\cdot)$ and Z_j are the trace operator and a scalar, respectively. Following [22], we resort to Lemma 1 for achieving the closed-form solution of R_j which will maximize $-E$ (equivalent to minimizing E):

$$\mathbf{R}_{\hat{j}} = \mathbf{U}_{\hat{j}} \mathbf{C}_{\hat{j}} \mathbf{V}_{\hat{j}}^T, \tag{12}$$

where $\mathbf{C}_{\hat{j}} = diag(1, 1, det(\mathbf{U}_{\hat{j}} \mathbf{V}_{\hat{j}}^T))$. We apply SVD on H^T to obtain $\mathbf{U}_{\hat{j}} \mathbf{S}_{\hat{j}} \mathbf{V}_{\hat{j}}^T = svd(\mathbf{H}_{\hat{j}}^T)$. Finally, the translation vector T_j of node j can be easily calculated from Eq. (11).

4   Experimental Results

In this section, we first illustrate the details of the implementation used in our experiments. Then, we introduce the test data and error metric we use in our experiments. In the end, we show our experimental results and compare them with other methods.

4.1 Implementation Details

Our method is implemented in C++ using functions provided by libigl [23] and Eigen [24]. All the experiments are conducted on a computer equipped with a Quad-Core Intel Core i5 CPU (2.3 GHz) and 8 GB of memory. For the parameters in the experiments, we set the node number to be 500 and each vertex is set to be influenced by the 10 nearest nodes. The α in the objective function (see Eq. (10)) is set to 2000.

4.2 Database and Data Preprocessing

The database contains 1,511 real-world 3D-scan foot mesh models with 21 ground-truth landmarks which are marked by medical specialists. Figure 2 displays the foot models sampled from the database. Note that the huge diversity of the foot shapes: Fig. 2(b) is a foot model wearing socks during scanning; the foot model of Fig. 2(c) consists of an additional leg area that is not the region of interest; due to human error during the scanning, the model in Fig. 2(d) only contains the bottom part of the foot. Fortunately, cases like Fig. 2(c) and 2(d) take up only a negligible proportion in the database.

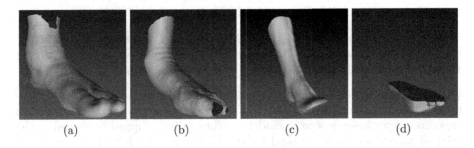

(a)   (b)   (c)   (d)

Fig. 2. Foot mesh model examples with various shapes from the database.

We randomly select approximately 20% of the database forming a test dataset for our experiments, which contains 211 foot models. The original foot model contains approximately 50,000 vertices including duplicated elements. For efficiency, we decimate the meshes with Meshlab [25] to 5,000 vertices per model.

Figure 3 displays different views of a single foot model after data preprocessing (containing 5,000 vertices) with manually annotated anatomical landmarks. Notice that, this specific foot model which originally contains 34,395 vertices does not belong to the test dataset and is chosen to be the source model throughout the experiments, as its shape is visually more universal.

4.3 Error Metric

To measure the performance of the landmark predictions, we compute the average of the square norm between the target landmarks and the predicted landmarks, which is defined as:

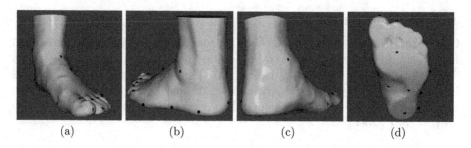

(a)   (b)   (c)   (d)

Fig. 3. Four different views of a foot model after data preprocessing along with the anatomical landmarks annotated by medical specialists.

$$Err_{avg} = \frac{1}{n}\sum_{i}^{n} \|Lp_i - Lt_i\|_2, \qquad (13)$$

where Lp_i and Lt_i are the i-th predicted landmark and the target landmark, respectively.

4.4   Comparison with Previous Methods

Source Model. The source foot model (see Sect. 4.2) is processed to different resolutions with Meshlab [25]. To be specific, four resolutions containing different amount of vertices (2k, 5k, 10k, original) are selected to perform the experiments.

Other Methods. In the experiments, three high-performing point set registration methods are chosen to be the competitors against our method. Note that experiments with some source model resolutions are skipped while using the original source codes of non-rigid CPD [18] and Fast-RNRR [15], in order to avoid the occurrence of certain issues of their codes (i.e. prompting errors).

Table 1. Error Comparisons using different methods in various point cloud resolution.

Resolution	2k		5k		10k		original (34k)	
Method	Avg.	Mid.	Avg.	Mid.	Avg.	Mid.	Avg.	Mid.
ICP	17.81	15.88	17.76	15.90	17.76	15.83	17.76	15.88
Non-rigid CPD	15.72	15.17	15.73	15.21	-	-	-	-
Fast-RNRR	-	-	-	-	-	-	17.92	15.11
Our method	**15.03**	**14.39**	**14.90**	**14.15**	**15.05**	**14.22**	**14.73**	**13.98**

In Table 1, we report the average (Avg.) and the median (Mid.) of the error results using the metric mentioned in Sect. 4.3. We can observe that our method overall outperforms other methods by a noticeable margin. Our method reaches the best performance when using the original source data, which outweighs the ICP and Fast-RNRR by 3.03 and 3.19 for average errors, respectively. Our method also outperforms other method when using different resolutions. The average errors for ICP are 2.78, 2.86 and 2.71 higher than our method when tested on 2k, 5k and 10k, respectively. Despite that Fast-RNRR saw a strong performance for 2k and 5k resolutions, reaching 15.72 and 15.73 on the average error, respectively. Our method still dominates by achieving 0.69 and 0.83 lower than Fast-RNRR, respectively. Another phenomenon we observed from the results of different resolutions is that the errors are negatively correlated with the resolutions of the target models.

4.5 Case Study

We randomly select three target models from the test set and compare the synthesized mesh models using Fast-RNRR, CPD, and our method in Fig. 4. CPD and Fast-RNRR have a common characteristic that they tend to preserve the original shape of the source model, which will probably lead to larger errors. We can observe that the model synthesized with our method can better fit the target model. Concretely, regarding the foot model with ID 43, we can notice the narrow footbridge and the small heel are reflected on the model synthesized with our method. In the other two cases, we can also observe how our method outperforms the other two methods in such detailed aspects (e.g., the narrow ankle area for the model with ID 711 and the forward-bending ankle for the model with ID 716). One minor drawback of our method is the coarse surface of the generated models. Nevertheless, the goal of our method focuses on accurate landmarks prediction, therefore this weakness is negligible.

Table 2. Corresponding errors of Fig. 4

model ID	43	711	716
CPD	10.95	9.35	**10.47**
Fast-RNRR	9.25	11.71	31.28
ours	**8.90**	**8.97**	10.82

Statistically, the effectiveness of our method is also demonstrated by the errors displayed in Table 2. The errors show that our method achieves top results for models with ID 43 and 711, reaching 8.90 and 8.97, respectively. Even though the model with ID 716 where CPD outperforms our method by 0.35, our approach still outperforms Fast-RNRR by a large margin (20.46), which illustrates the robustness of our method. Further, we also notice that the similarity between the source model and the target model is negatively correlated with the resulting error. Therefore, one limitation is, to minimize the prediction errors, we can choose the template model with a more similar shape to the target model when predicting the landmarks.

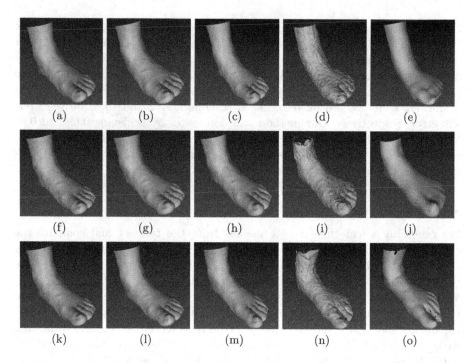

Fig. 4. (a), (f) and (k) are the source models. (e), (j) and (o) are the target models with ID 43, 711, and 716, respectively, which are randomly selected from the test set. (b), (g) and (l) are the models synthesized with Fast-RNRR. (c), (h), and (m) are the models synthesized with non-rigid CPD. (d), (i), and (n) are the models synthesized with our method.

5   Conclusion

We have presented a robust 3D anatomical landmark detection method for 3D foot models. It realizes the prediction of anatomical landmarks by deforming a given source model with annotated landmarks to a target model. After non-rigid deformation, the updated landmarks of the source model are naturally viewed as the landmarks of the target model. Experiments demonstrate that compared to previous point registration techniques, our method can predict landmarks on target models with smaller errors. In the future, it will be interesting to extend our framework to other types of data such as the human face and body data.

References

1. O'Neil, A.Q., et al.: Attaining human-level performance with atlas location auto-context for anatomical landmark detection in 3D CT data. In: Proceedings of the European Conference on Computer Vision (ECCV) Workshops (2018)

2. Zheng, Y., Liu, D., Georgescu, B., Nguyen, H., Comaniciu, D.: 3D deep learning for efficient and robust landmark detection in volumetric data. In: Navab, N., Hornegger, J., Wells, W.M., Frangi, A.F. (eds.) MICCAI 2015. LNCS, vol. 9349, pp. 565–572. Springer, Cham (2015). https://doi.org/10.1007/978-3-319-24553-9_69

3. Zhang, J., Liu, M., Shen, D.: Detecting anatomical landmarks from limited medical imaging data using two-stage task-oriented deep neural networks. IEEE Trans. Image Process. **26**(10), 4753–4764 (2017)

4. Bier, B., et al.: X-ray-transform invariant anatomical landmark detection for pelvic trauma surgery. In: Frangi, A.F., Schnabel, J.A., Davatzikos, C., Alberola-López, C., Fichtinger, G. (eds.) MICCAI 2018. LNCS, vol. 11073, pp. 55–63. Springer, Cham (2018). https://doi.org/10.1007/978-3-030-00937-3_7

5. Fang, S., Raghavan, R., Richtsmeier, J.T.: Volume morphing methods for landmark-based 3D image deformation. In: Medical Imaging 1996: Image Processing, vol. 2710. International Society for Optics and Photonics (1996)

6. Sumner, R.W., Schmid, J., Pauly, M.: Embedded deformation for shape manipulation. In: ACM SIGGRAPH 2007 Papers, p. 80-es (2007)

7. Lu, X., et al.: Unsupervised articulated skeleton extraction from point set sequences captured by a single depth camera. In: Proceedings of the AAAI Conference on Artificial Intelligence, vol. 32, no. 1 (2018)

8. Alansary, A., et al.: Evaluating reinforcement learning agents for anatomical landmark detection. Medical Image Anal. **53**, 156–164 (2019)

9. Ghesu, F.C., Georgescu, B., Grbic, S., Maier, A.K., Hornegger, J., Comaniciu, D.: Robust multi-scale anatomical landmark detection in incomplete 3D-CT Data. In: Descoteaux, M., Maier-Hein, L., Franz, A., Jannin, P., Collins, D.L., Duchesne, S. (eds.) MICCAI 2017. LNCS, vol. 10433, pp. 194–202. Springer, Cham (2017). https://doi.org/10.1007/978-3-319-66182-7_23

10. Ebner, T., Stern, D., Donner, R., Bischof, H., Urschler, M.: Towards automatic bone age estimation from MRI: localization of 3D anatomical landmarks. In: Golland, P., Hata, N., Barillot, C., Hornegger, J., Howe, R. (eds.) MICCAI 2014, Part II. LNCS, vol. 8674, pp. 421–428. Springer, Cham (2014). https://doi.org/10.1007/978-3-319-10470-6_53

11. Subburaj, K., Ravi, B., Agarwal, M.: Automated identification of anatomical landmarks on 3D bone models reconstructed from CT scan images. Comput. Med. Imaging Graph. **33**(5), 359–368 (2009)

12. Baek, S.-Y., et al.: Automated bone landmarks prediction on the femur using anatomical deformation technique. Comput.-Aided Design **45**(2), 505–510 (2013)

13. Li, H., et al.: Robust single-view geometry and motion reconstruction. ACM Trans. Graph. (ToG) **28**(5), 1–10 (2009)

14. Lu, X., et al.: 3D articulated skeleton extraction using a single consumer-grade depth camera. Comput. Vision Image Underst. **188**, 102792 (2019)

15. Yao, Y., et al.: Quasi-Newton solver for robust non-rigid registration. In: Proceedings of the IEEE/CVF Conference on Computer Vision and Pattern Recognition (2020)

16. Holland, P.W., Welsch, R.E.: Robust regression using iteratively reweighted least-squares. Commun. Stat.-Theory Methods **6**(9), 813–827 (1977)

17. Boyd, S., Parikh, N., Chu, E.: Distributed Optimization and Statistical Learning via the Alternating Direction Method of Multipliers. Now Publishers Inc. (2011)

18. Myronenko, A., Song, X.: Point set registration: coherent point drift. IEEE Trans. Pattern Anal. Mach. Intell. **32**(12), 2262–2275 (2010)

19. Dai, H., et al.: A 3D morphable model of craniofacial shape and texture variation. In: Proceedings of the IEEE International Conference on Computer Vision (2017)

20. Wand, M., et al.: Efficient reconstruction of nonrigid shape and motion from real-time 3D scanner data. ACM Trans. Graph. (TOG) **28**(2), 1–15 (2009)
21. Besl, P.J., McKay, N.D.: Method for registration of 3-D shapes. In: Sensor Fusion IV: Control Paradigms and Data Structures, vol. 1611. International Society for Optics and Photonics (1992)
22. Myronenko, A., Song, X.: On the closed-form solution of the rotation matrix arising in computer vision problems. arXiv preprint arXiv:0904.1613 (2009)
23. Jacobson, A., et al.: libigl: a simple C++ geometry processing library (2018)
24. Guennebaud, G., Jacob, B., et al.: Eigen v3 (2010)
25. Cignoni, P., et al.: MeshLab: an open-source mesh processing tool. In: Eurographics Italian Chapter Conference (2008)

Impact of the Composition of Feature Extraction and Class Sampling in Medicare Fraud Detection

Akrity Kumari[1], Narinder Singh Punn[1]([✉])(iD), Sanjay Kumar Sonbhadra[2](iD), and Sonali Agarwal[1](iD)

[1] Indian Institute of Information Technology Allahabad,
Jhalwa, Prayagraj, Uttar Pradesh, India
{mit2020087,pse2017002,sonali}@iiita.ac.in

[2] Department of CSE, ITER, Siksha 'O' Anusandhan, Bhubaneswar, Odisha, India
sanjaykumarsonbhadra@soa.ac.in

Abstract. With healthcare being critical aspect, health insurance has become an important scheme in minimizing medical expenses. Medicare is an example of such a healthcare insurance initiative in the United States. Following this, the healthcare industry has seen a significant increase in fraudulent activities owing to increased insurance, and fraud has become a significant contributor to rising medical care expenses, although its impact can be mitigated using fraud detection techniques. To detect fraud, machine learning techniques are used. The Centers for Medicaid and Medicare Services (CMS) of the United States federal government released "Medicare Part D" insurance claims is utilized in this study to develop fraud detection system. Employing machine learning algorithms on a class-imbalanced and high dimensional medicare dataset is a challenging task. To compact such challenges, the present work aims to perform feature extraction following data sampling, afterward applying various classification algorithms, to get better performance. Feature extraction is a dimensionality reduction approach that converts attributes into linear or non-linear combinations of the actual attributes, generating a smaller and more diversified set of attributes and thus reducing the dimensions. Data sampling is commonly used to address the class imbalance either by expanding the frequency of minority class or reducing the frequency of majority class to obtain approximately equal numbers of occurrences for both classes. The proposed approach is evaluated through standard performance metrics such as F-measure and AUC score. Thus, to detect fraud efficiently, this study applies autoencoder as a feature extraction technique, synthetic minority oversampling technique (SMOTE) as a data sampling technique, and various gradient boosted decision tree-based classifiers as a classification algorithm. The experimental results show the combination of autoencoders followed by SMOTE on the LightGBM (short for, Light Gradient Boosting Machine) classifier achieved best results.

Keywords: GBDTs · fraud · SMOTE · autoencoders · medicare · LightGBM

All authors contributed equally.

© The Author(s), under exclusive license to Springer Nature Switzerland AG 2023
M. Tanveer et al. (Eds.): ICONIP 2022, LNCS 13625, pp. 639–658, 2023.
https://doi.org/10.1007/978-3-031-30111-7_54

1   Introduction

Fraud is described as the misuse of a company's system that does not always result in direct legal consequences. Frauds are dynamic and have no patterns. Out of various categories of fraud, insurance fraud is one of the subtypes, which is committed frequently. Insurance fraud is defined as any action performed to obtain a false insurance claim. There are again various categories of insurance fraud, one of which is healthcare insurance fraud that is committed in the healthcare industry. Fraud in healthcare insurance could be committed through the claimant (insured person) or the provider (doctor).

The fraudulent activities by providers involve:

- Charging for more expensive services than were actually given,
- Providing and thereafter charging for non-medically necessary treatments,
- Scheduling additional visits for patients,
- Recommending patients to other doctors when they do not seek further treatment,
- Phantom billing i.e. demanding a fee for services that were not rendered,
- Ganging, i.e. demanding a fee for services provided to members of family or other people accompanying the patient who did not receive any treatment for themselves.

Whereas, fraud by claimants involve:

- Claims on behalf of members and/or dependents who are not eligible,
- Modifications to membership forms,
- Hiding pre-existing conditions,
- Other coverage not disclosed,
- Prescription medication fraud,
- Failure to disclose claims arising from work-related injuries.

Thus, in the health insurance program, both beneficiaries, as well as healthcare providers, send false insurance claims to insurance companies in order to benefit from reimbursements. Such fraudulent activities by individuals or groups impact the lives of many innocent people as the beneficiaries have to pay higher insurance premiums rate for the services being received. Thus, insurance fraud creates a serious issue, hence, governments, insurance companies, and other organizations make an effort to discourage these kinds of activities by identifying the maximum number of fraudsters. To minimize such fraudulent activities, fraud detection is needed. The traditional technique of fraud detection, which is still commonly employed today, is detecting fraud patterns that have previously been encountered. This approach is primarily focused on the application of pre-established business rules (basic or advanced) to past data, which is insufficient given a large number of insurance claims and the wide range of fraudulent patterns. This system also necessitates continuous supervision by the expert in order to keep the rules up to date.

In recent years, several advancements are introduced in healthcare domain [1] with machine learning and deep learning technologies [2–6]. Machine learning methods have been applied to various other fraud detection problems such as tax fraud detection, credit card fraud detection, and bankruptcy fraud detection, and thus, are also being applied to fraud detection in healthcare insurance claims. Models that are build on machine learning and artificial neural networks (ANN) have made it possible to automatically extract features and build patterns and hence detect fraudulent activities more effectively and efficiently. Medicare [7] is a government-run health insurance program that covers approximately 54.3 million people in the United States. It covers persons over the age of 65 as well as younger people with specified medical conditions and disabilities. In Medicare program, insurance companies receive a huge number of requests for payment for the services provided by healthcare providers to their patients. Such requests are called insurance claims and a part of these might be fraudulent.

Healthcare is one of the vulnerable areas for perpetrators of fraud and owing to the continuous increase in fraud, waste, and abuse (FWA) activities in medicare programs necessitate the need for a fraud detection system to prevent the possible fraudulent activities arising in the Medicare program. This paper proposes an efficient framework using machine learning techniques to detect Medicare fraud. Following are the techniques that have been applied in building the framework's architecture:

- Feature extraction: This work use autoencoders as a feature extraction technique to minimise the number of features in the dataset by generating new ones from old ones.
- Class imbalance: This work uses SMOTE for handling the imbalanced ratios of output classes since non-fraudulent class constitutes a significant part of the dataset.
- Classification: This work trains various implementations or improvements of gradient boosted decision tree classifiers. A comparative study is done between various gradient boosting algorithms like Catboost, XGBoost, AdaBoost, and LightGBM to get the best performing classifier on the medicare dataset.
- Analysis: The research work evaluates the performance of classifiers with F1-score and AUC metrics, and result shows that classifiers built with a combination of autoencoders and SMOTE attain better results.

The rest of the paper is divided into different sections where Sect. 2 describes background and related work behind the research work. Section 3 contains the architecture of the framework adopted to conduct the study. Section 4 describes the results and output of the algorithm. Finally, the concluding remarks are presented in Sect. 5.

2 Background and Related Work

There exists a body of research for the application of machine learning in the domain of anomaly detection. Thus, this work started over various publications in the area of anomaly detection and led to the study of fraud detection in the healthcare domain [6,8–12]. The previous publications on fraud detection in healthcare insurance claims and other related areas [13,14] lead us to the question of whether various recent GBDTs implementations, along with autoencoder's automatic feature extraction capability to address high dimensionality, followed by class sampling, is a suitable algorithm for Medicare fraud detection.

2.1   Handling Imbalanced Dataset with the Sampling Method

The primary challenge of applying machine learning models in fraud detection, especially for medicare data is the highly imbalanced distribution of two classes: normal and fraudulent providers. When there is an imbalanced distribution of classes in a dataset, such as when the negative class (majority class) has a large number of data points in comparison to the positive class (minority class), class imbalance occurs (minority class). They usually give incorrect results and can be misleading with too optimistic scores if accuracy measures are taken into account. One of the reasons for these failures is that minority class points are seen as outliers that contain no information and be inclined toward majority class.

In order to address class imbalance, different training strategies can be used such as resampling (oversampling and undersampling), membership probability thresholding, and cost-sensitive learning. One of the most popular methods for dealing with an imbalanced dataset is to resample the data. Undersampling and oversampling are the two most common strategies that comes under resampling. Majority of the studies concerning the fraud detection in medicare datasets have used resampling techniques (usually, undersampling technique by varying the sampling ratios) to overcome the imbalanced class problem [15–19]. These studies have come to a conclusion that undersampling (down-sampling) is more efficient than oversampling as adding new data samples results in overfitting and increases in training time of the classifier, so this work attempts to discover whether an oversampling technique is a good proposal for addressing class imbalance problem. Therefore, this work proposes a method based on the oversampling technique known as SMOTE where the artificial samples are created for the minority class. This technique helps to avoid the overfitting problem caused by random oversampling, which involves adding exact replicas of minority instances to the original dataset, whereas SMOTE employs a subset of data from the minority class as an example and then creates new synthetic identical instances.

SMOTE. This method synthetically increases the minority class by generating fresh examples of the minority class using specialized methods like the nearest neighbor and Euclidean distance. To generate fresh "synthesized" instances, the technique gives a set of simple rules. The created data is never an exact clone of one of its parents, despite the fact that each new synthetic data is built from its parents. There is no loss of essential information in SMOTE in contrast to undersampling.

To implement SMOTE, a library called imblearn is used that implements 85 variants of the SMOTE technique. Imbalanced-learn (also known as imblearn) is an open-source, MIT-licensed library that uses scikit-learn (also known as sklearn) and provides tools for dealing with imbalanced class categorization. It was first introduced by Chawla et al. [20].

2.2   Handling Heterogeneous Datasets with the Dimensionality Reduction Method

To achieve valuable characteristics and accurate outcomes, machine learning models tend to incorporate as many features as feasible at the beginning. However, as the number of characteristics increases, the model's performance begins to deteriorate. Curse of dimensionality is a term used to describe this problem, which can lead to overfitting. Dimensionality reduction is the process of obtaining a set of principal features that reduces the dimensionality of the feature space in consideration with the aim that lower dimension representation retains some meaningful characteristics of original instances of the dataset. Dimensionality reduction is commonly applied when the dataset contains a large number of features and the medicare dataset contains 1360 attributes after one-hot encoding of the categorical variables necessitating the need for a dimensionality reduction step.

The two primary approaches to dimensionality reduction are feature extraction and feature selection. Feature selection is the process of finding the subset of features from the original features. Feature extraction is the process of creating new features from the existing feature of higher dimensional space to lower feature subspace. It is used to compress the data. This research work uses the feature extraction technique.

Feature Extraction. Feature extraction is used to reduce the number of features in a dataset by creating a new set of features from the original set of features, afterwards the original set of features is removed. These new minimized sets of features should then be able to contain the maximum amount of information present in the actual features. In such a manner, a summarized version of the actual features can be created from a combination of the original set. Regardless of the difficulty with imbalanced datasets in Medicare, these also have a significant number of features that must be handled. A fraud detection system that is built using all features is usually not very efficient because the machine learning algorithms are impacted by insignificant or non-trivial features during the training process leading to overfitting. The reason for introducing the feature extraction are as follows:

- It produces better results than applying machine learning algorithms to original data, i.e. boosts the classification scores.
- It reduces the memory and computation load on the hardware resources.
- It allows for easier visualization of data.
- It provides a deeper understanding of the fundamental structure of the data.
- It also reduces overfitting by the classifier.

To address the aforementioned issues, this research employs a non-linear dimensionality reduction technique known as stacked autoencoder to generate robust and discriminative features for fraudulent instances, which will aid in the effective detection of fraudulent providers by grouping them into homogeneous clusters. Various alternative feature extraction strategies, such as early attempts to build on the projection method and involving mapping of input attributes in the original high-dimensional space to the new low-dimensional space with little information loss, have also been investigated. The two most well-known projection techniques are principal component analysis(PCA) and linear discriminant analysis (LDA). These techniques have been applied to anomaly detection in recent papers [21–24]. However, there are disadvantages associated with such projection techniques. The main drawback of the aforementioned approaches is that they perform linear projection among features while autoencoders can model complex, non-linear functions. Another drawback of these projection techniques is that most of these works tend to map data from high-dimensional to low-dimensional space by extracting features once, rather than stacking them to build deeper levels of representation gradually. Using artificial neural networks, autoencoders compress dimensionality by reducing reconstruction loss.As a result, it is simple to stack autoencoders by adding any number of hidden layers with the sequential API of the Python library. This gives the autoencoder the ability to extract meaningful features.

Autoencoders. Autoencoders are a special type of feedforward neural network in which input and output are the same. They compress the input into a lower-dimensional representation or code, that is used afterward to reconstruct the output. The code which is a condensed "summary" or "compression" of the input, is also known as the latent space representation. The representation obtained from the autoencoders has the following characteristics:

- They are data-specific which means they could only compress data that is identical to what the training was done on. This is in contrast to the MPEG-2 Audio Layer III (MP3) compression method, which only makes assumptions about "sound" in general, not specific sorts of sounds. Because the features it learns are face-specific, an autoencoder trained on photographs of faces would do a bad job compressing pictures of trees.
- They are lossy, which implies that when compared to the original inputs, the decompressed outputs will be degraded (similar to MP3 or JPEG compression). This is not to be confused with lossless arithmetic compression.
- They are automatically learned from data points, which is a useful property because the autoencoder makes it simple to train specialized instances of the algorithm to perform efficiently on a particular kind of input. It does not necessitate any new engineering, but it does necessitate appropriate training data.

An autoencoder is comprised of three parts: encoder, code, and decoder. The encoder compresses the input and generates the code, which the decoder subsequently uses to reconstruct the input. Building an autoencoder requires three components: an encoding function, a decoding function, and a distance function to calculate the amount of information loss between the compressed representation and the decompressed representation of the data (i.e. a "loss" function). The encoder and decoder are chosen to be parametric functions (generally, fully-connected feedforward neural networks, specifically the ANNs) that are differentiable with reference to the distance function, and allow the parameters of the encoding or decoding functions to be optimized using Stochastic Gradient Descent to minimize the reconstruction loss. Code is a single layer of an ANN with our desired dimensions. Before training the autoencoder, the number of nodes in the code layer (code size) and encoder-decoder layer is set as a hyperparameter. For the code generation, the input is first passed through the encoder, which is a fully-connected ANN. The output is subsequently generated solely using the code by the decoder, which has also a structure similar to ANN. The purpose is to get an output that is exactly the same as the input. The architecture of the decoder is usually identical to that of the encoder.

2.3 Choosing Among Various Classification Algorithms

Prokhorenkov et al. state that ensembles of gradient boosted decision trees (GBDT algorithms) are suitable for operating on heterogeneous datasets [25]. Heterogeneous data include features from a wide range of data types, i.e. from numerical to categorical features. Tabular datasets are frequently heterogeneous, and CMS's medicare claims data is an example of heterogeneous data. Khoshgoftaar et al. [25] show that CatBoost, LightGBM, and XGBoost, which are recent gradient boosted decision tree (GBDTs) implementations, are robust classifiers for highly imbalanced, insurance claims data. As a result of these findings, the current study examines the performance of four different types of GBDT algorithms (i.e. XGBoost, AdaBoost, CatBoost, and LightGBM) on Medicare claims data. This study mainly explores which GBDT improvement performs the best on the Medicare dataset. For all of the GBDTs classifiers, hyper-parameters are near to default values, allowing for a fair baseline comparison.

Gradient Boosted Decision Trees. Gradient boosting is a technique for improving the performance of a machine learning model by using an ensemble (i.e. combination) of weak learners. On each problem, the actual performance of boosting methods is clearly influenced by the input and the weak classifier. Decision trees, specifically Classification And Regression Tree (CART) trees, are usually the weak learners. A better prediction model is created by combining the output of several base learners. The class with the most votes from weak learners could be the final result of the classification task. For gradient boosting methods, weak learners work in a sequential order. Each model aims to reduce the mistake introduced by the previous model. Trees in boosting-based classifiers are weak learners, but by stacking multiple trees in a row, each concentrating on the preceding model's errors, boosting algorithms become a very efficient and accurate model. To determine the errors, a loss function is utilised. For example, mean squared error (MSE) can be used for regression tasks, while logarithmic loss (log loss) can be used for classification tasks. When a new tree is introduced to the ensemble, the current trees do not change.

The steps involved in the boosting process are:

1. Create a primary model with the input data,
2. Make predictions on the entire dataset,
3. Using the predictions and the actual values, calculate the error,
4. Give more weight to the wrong predictions,
5. and create a new model that tries to rectify errors from the previous model,
6. Make predictions on the whole dataset with the newly created model,
7. Create a number of models with each model aiming at rectifying the errors generated from the previous model,
8. Get the final model by weighting the mean of all the models.

The various boosting algorithms present in machine learning and used in this work are as follows:

- AdaBoost: AdaBoost (Adaptive Boosting) is a Machine Learning approach that is utilised as part of an Ensemble Method. It's quick, straightforward, and simple to programme. It does not have any tuning parameters. Decision trees with one level, or Decision trees with only one split, are the most popular algorithm used with AdaBoost. Decision Stumps is another name for these trees. This algorithm creates a model by giving all data points the same weight. It then gives points that are incorrectly categorised a higher weight. Then, in the following model, all of the points with greater weights are given more relevance. It will continue to train models until a smaller error is received [26].
- XGBoost: eXtreme Gradient Boosting, sometimes known as XGBoost, is a scalable machine learning approach based on tree boosting. It also employs a collection of weak decision trees. It's a linear model that uses parallel calculations to train trees. The following are the model's primary algorithmic implementation features:
  1. Sparse Aware implementation with automatic handling of missing data values.
  2. A block structure supports the parallelization of tree construction.
  3. Continued training to improve a model that has already been fitted using new data.
- LightGBM: Light Gradient Boosting Machine [27], is a decision tree-based gradient boosting architecture that improves model efficiency while reducing memory utilisation. It employs two innovative techniques: Gradient-based One Side Sampling (GOSS) and Exclusive Feature Bundling (EFB), which address the shortcomings of the histogram-based algorithm utilised in all GBDT frameworks. GOSS and EFB are the two strategies that make up the LightGBM algorithm, and they work together to make the model run smoothly and provide a competitive advantage over competing GBDT frameworks.

– CatBoost: Categorical Boosting [28], an open-source gradient boosting machine learning algorithm. Ordered Target Statistics and Ordered Boosting are two of the advances used. CatBoost is well-suited to machine learning tasks involving category, heterogeneous data as a Decision Tree-based method [29]. It produces good results without extensive data training and with a small amount of data.

Bauder et al. [6] conducted extensive studies on fraud detection utilizing both supervised and unsupervised learning techniques for fraud detection. Their experimentation was based on Medicare Part B provider data and applied methods for detecting outliers. They then merged a number of Medicare-related datasets. The combined medicare dataset was labeled with the LEIE data. Their work also takes into account the data imbalance problem, by using various data levels as well as algorithm level approaches. According to their research, the data level performed better In comparison to the algorithmic level.

Johnson and Khoshgoftaar [27] studied the performance of various deep learning algorithms on the Medicare fraud detection challenge [29,30]. The authors broaden the scope of sample procedures while altering the learners' classification thresholds. They showed that a hybrid strategy of random undersampling and random oversampling has an effect on the AUC of deep learning algorithms. It was found that GBDT algorithms show promising performance in the task of Medicare Fraud detection. For most experiments, they found CatBoost is the strongest performer when Random Undersampling was used for class balancing, the ratio was 1:1 for the minority to majority class. According to their research, the best performance was obtained for the non-aggregated Medicare Part B and Part D datasets.

From all the research work covered so far, it is observed that researchers have yet not compared the performance of CatBoost, LightGBM, XGBoost, and AdaBoost along with autoencoder's feature extraction capability on the work of identification of fraud in Medicare data. The Medicare data used in this study have various categorical features, for example, Drug Name, Provider State, and Specialty Description. Out of these, the Drug name feature has 1193 distinct values in the dataset. To represent this high cardinality feature in the building model, one-hot encoding is applied which increases the dimensionality of the dataset considerably. To address this curse of dimensionality, autoencoders are used to reduce dimensionality. By doing so, the inclusion of high cardinality categorical features like the drug name became easier during the training of models. Hence, this research work takes the advantage of the autoencoder's automatic feature extraction capability to contribute to the body of research area that these studies relate to. This research work compares four different types of GBDT algorithms, for the task of medicare fraud detection. We also address the class imbalance problem by the means of SMOTE technique, which is still unexplored in the medicare dataset.

3   Implementation of the Framework

The implementation of the proposed framework covers the following stages:

1. Data collection and preparation,
2. Feature selection and feature engineering,
3. Choosing the machine learning algorithm and training our model,
4. Evaluating our model

The structure of the framework is shown in Fig. 1. The techniques applied in building the framework consist of autoencoder as feature extractor, SMOTE for data sampling, and GBDTs as a classifier.

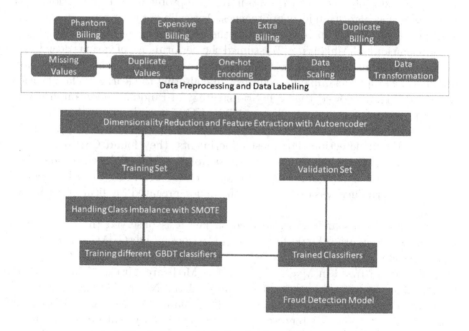

Fig. 1. Framework for the fraud detection.

3.1   Data Collection and Preparation

This paper uses the Medicare Part D insurance claims dataset [31]. CMS provides a number of publicly accessible files every year, and this dataset is one of them. These files are combinedly known as Medicare Provider Utilization and Payment Data: Physician and Other Supplier and Medicare Provider Utilization and Payment Data: Part D Prescriber (Part D). This data is present in character separated value (CSV) format, and there is one file for each year, this work uses the file for the year 2018. CMS prepares a document that specifies all the features or attributes of the Medicare Part D insurance claims data. Each row in the dataset describes a provider, mainly by the national provider identifier (NPI) of the provider, and secondarily by various features that provide information related to the name, demographics, and location (state and city) of the provider. This dataset contains information about the drug names they prescribe to their patients. It also contains records related to provider type indicating the nature of the provider's practice, such as ophthalmology, family practice, nursing, and so forth. For every drug prescription, the provider has submitted a drug cost claim to Medicare, and there is one record in the PUF file for the year. For each row, along with the drug name, there are certain aggregate statistics associated with it, namely the total number of unique Medicare Part D beneficiaries with at least one claim for the drug, and the aggregate drug cost paid for all associated claims. The features of the dataset used in the experiment are discussed in Table 1.

The second dataset used in this study is the List of Excluded Individuals and Entities (LEIE) data [32]. The LEIE dataset is updated by the Office of Inspector General (OIG) on a monthly basis. This file is also present in a CSV format. It contains information about the healthcare providers that are prohibited from sending claims to Medicare because they have previously broken Medicare's rules and regulations for submitting claims. In this study, the important attributes of the LEIE dataset are the NPI and the exclusion type. Initially, the part D dataset is unlabelled. This work derives a label for the Medicare Part D data from the LEIE data on the basis of the NPI attribute. If an NPI from the Medicare Part D data is present in the LEIE data, then all Medicare Part D data, having that NPI is labeled as fraudulent. To provide labels, the Part D dataset and LEIE dataset are merged using left join on NPI, and all the NAN records obtained after combining the datasets are labeled as non-fraudulent (class 0).

3.2   Data Preprocessing

All the required data pre-processing steps such as handling missing and duplicate values, data scaling, data transformation, and data filtering are performed on the dataset. Following are the two significant issues confronted in the preprocessing step of this study.

Handling Heterogenous Datasets. Since the medicare dataset is a heterogeneous dataset (i.e. consisting of both categorical and numerical features), the features like specialty_description, nppes_provider_state and drug_name are categorical in nature, which cannot be used with the classifiers in their raw form. Table 2 shows the counts of distinct values of categorical features in the dataset. So, the data need to be processed in order to convert the categorical features into numerical features to train the model with the GBDTs classifier. Hence, the categorical features were encoded by one-hot encoding. But performing this increased the dimensionality of data drastically. Autoencoder was then applied to reduce the dimension. The implementation uses Keras's deep learning

Table 1. Features used for the experiment.

Name	Description	Type
npi	Unique identification of provider, used for labeling	Numeric
nppes_provider_state	The state where the provider is located	Categorical
speciality_description	Medical provider's specialty (or practice)	Categorical
description_flag	A flag that indicates the source of the specialty_description	Categorical
drug_name	The name of the drug prescribed (or filled).	Categorical
bene_count	The total number of unique Medicare Part D beneficiaries with at least one claim for the drug	Numeric
total_claim_count	Number of Medicare Part D Claims, including refills	Numeric
total_30_day_fill_count	Number of standardized 30-day fills, including refills	Numeric
total_day_supply	Number of day's supply for all claims	Numeric
total_drug_cost	Aggregate cost paid for all claims.	Numeric

framework, which is a functional API of Python on the top of TensorFlow to build the autoencoder. An autoencoder was designed with the following architecture, two hidden layers in the encoder as well as the decoder. The hyperparameters set before training an autoencoder to get good results are as follows:

- Code size: It is defined as the number of nodes in the middle layer or code layer. More compression occurs when the size is smaller. Its value is set to 32 during experimentation.
- The number of layers: The depth of the autoencoder depends on the performance requirement. The architecture of the autoencoder in the experiment consists of 2 layers in the encoder as well as the decoder, without taking into consideration input and output.
- The number of nodes per layer: Since the layers are placed one after another, the autoencoder architecture in the implementation is a stacked autoencoder. Stacking autoencoders usually resemble a "sandwich". With each consecutive encoder layer, the number of nodes per layer drops and then increases in the decoder. The decoder is symmetric to the encoder in terms of the layer structure. The number of nodes is set to 100 in the first layer and to 50 nodes in the second layer. The activation function used is relu in the encoder layer and tanh in the decoder layer.

– Loss function: mean squared error (MSE) or binary cross-entropy can be used by the autoencoder. When the input values are in the range [0, 1], cross-entropy is used, otherwise, mean squared error is used. This work uses MSE as loss function and adam as optimizer.

To fit the model, the epoch value is set to 50 with early stopping to avoid overfitting, and the batch size is set to 128. The model is fitted to all non-fraud instances in the dataset to prevent the autoencoder from merely learning to replicate the inputs to the output, that is, without any meaningful representations being learned. In the implementation, three layers of autoencoders are stacked for building the final model.

Table 2. The cardinality of categorical features in dataset.

Features	Distinct Values
specialty_description	101
nppes_provider_state	59
description_flag	2
drug_name	1160

Handling Imbalanced Dataset. The medicare dataset is highly imbalanced in nature. The non-fraudulent class overwhelms the majority of the data. Fraudulent transactions comprise 341 instances or 0.055%, thus the dataset is highly imbalanced with respect to the majority to minority classes. It is necessary to deal with the class imbalance problem present in the medicare dataset. To deal with this problem, this research work applies a data sampling method called SMOTE. The minority class i.e. fraudulent class is oversampled by adding synthetic minority samples and making the minority to majority class proportion the same. By performing the class balancing step, class 1 would be learned as much as class 0. The SMOTE approach is used only on the training dataset to ensure that the classification algorithm fit the data adequately.

Choosing the Machine Learning Algorithm and Training Our Model. The latent space representation is obtained from the autoencoder's feature extraction technique ability. These extracted features are then used to train different classification models. In this work, various GBDTs implementations are trained on the extracted feature or latent space representation to obtain the best performing classifier with regard to F1-Score and AUC score with the aim to accurately predict fraud or non-fraud outcome for the data points of which class label is not known. For this research work following classification algorithms are used: XGBoost, CatBoost, AdaBoost and LightGBM. To provide a fair baseline comparison the hyperparameters for all the classifiers are set to default values.

Evaluating the Model. This study evaluates the impact of employing the data sampling technique (SMOTE) and feature extraction technique (autoencoders) on classifier performance. The implementation is divided into four sections, first is the use of SMOTE only, second is the use of feature extraction only, third is the use of both SMOTE and feature extraction, and the last one is a baseline (no SMOTE and no feature extraction). When used with the SMOTE preprocessing phase, the feature extraction preprocessing step is executed first in the experiment.

For all the learners stated above, all of them produced an accuracy score greater than 85%. But the problem with fraud detection is that it has a skewed distribution for the target class. Therefore the accuracy metric is always misleading. The purpose of such research is to see how successfully each fraud and non-fraud class is classified. The percentage of precision and recall (count of true positives, false positives, true negatives, and false negatives) are the important metrics to be considered. Precision shows the percentage of non-fraudulent classes labeled as a fraud, while recall shows the percentage of fraudulent classes classified as non-fraudulent, which is even more dangerous in the task of fraud detection. So, the primary metric for evaluation in this study is the F1-score, which is the harmonic mean of precision and recall and takes into account both metrics and is a more considerable indicator for datasets with a high-class imbalance ratio. Table 1 shows a comparison between the above-mentioned classification algorithms in terms of various performance metrics. It depicts the accuracy, precision, recall, F1-score, and AUC score for each learner. Comparing all the aforesaid algorithms, LightGBM produces better results concerning AUC and f1-score rate.

4   Experimental Results

4.1   Metrics for Evaluation of Classifier Performance

To evaluate the proposed systems, standard performance metrics are used to calculate the performance of the system (accuracy, precision, recall, f-score, and AUC score).

Accuracy. In a classification problem, the accuracy score is defined as the ratio of the number of correct predictions to the total number of instances.

$$\text{Accuracy Score} = \frac{\text{Number of correct predictions}}{\text{Total number of instances}} \tag{1}$$

But, this prediction score is unreliable for an unbalanced distribution of classes or skewed dataset because the training and evaluation as per this measure create a model that is likely to predict the non-fraud class (majority class) for all the test examples by increasing the percentage of True Negative and thus, the value rises to 99%. Hence, the confusion matrix is preferred for evaluating the model, which is a summary of correct and incorrect prediction values compared with the actual values of the input data, divided among classes as shown in Fig. 2, where TP (true positive) and TN (true negative) are correct predictions and FP (false positive) and FN (false negative) are wrong predictions. In terms of TP, TN, FP, and FN, accuracy is calculated as shown in Eq. 2

$$Accuracy = \frac{TP + TN}{TP + TN + FP + FN} \tag{2}$$

AUC-Score. The Receiver Operator Characteristic (ROC) curve is a statistic for evaluating binary classification issues. It's a probability curve that plots the TPR against the FPR at various threshold levels, allowing the signal to be distinguished from the noise. The AUC is a summary of the ROC curve that assesses the ability of a classifier to distinguish between classes. The AUC measures how successfully a model can distinguish between positive and negative classifications. The higher the AUC number, the better.

Precision. Precision indicates how many of the instances that were predicted positively by the model turned out to be actually positive. It is calculated as shown in Eq. 3.

$$Precision = \frac{TP}{TP + FP} \tag{3}$$

Recall. Recall indicates how many of the actual positive cases the model is able to correctly predict. It is computed using Eq. 4.

$$Recall = \frac{TP}{TP + FN} \tag{4}$$

F-Score. The harmonic mean or weighted average of Precision and Recall is the F-score. Both false positives and false negatives are taken into account in this score. It is calculated using Eq. 5.

$$F1 = \frac{2 * precision * recall}{precision + recall} \tag{5}$$

Prediction / Actual	Fraud	Not Fraud
Fraud	True Positive(TP)	False Negative(FN)
Not Fraud	False Positive(FP)	True Negative(TN)

Fig. 2. Confusion Matrix.

4.2   Results

The technique of t-distributed stochastic neighbour embedding (TSNE) [33] is used to visualise transaction data. TSNE is a statistical method for visualising high-dimensional data by assigning a two- or three-dimensional map to each datapoint. The method is a simplified form of Stochastic Neighbor Embedding, and it improves graphics by reducing the tendency for points to cluster in the map's centre. t-SNE surpasses existing approaches when it comes to creating a single map that exhibits structure at several sizes. This is especially important for high-dimensional data that is scattered across multiple low-dimensional manifolds but is related to one another, such as images of items from various classes taken from diverse angles. Figure 3 and Fig. 4 represent the scatter plot of the two dimensions before and after feature extraction from autoencoders, respectively.

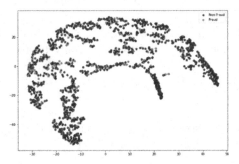

Fig. 3. Latent space representations before feature extraction using t-SNE projection.

Table 3 shows the values for precision, recall, F1-score, accuracy, and AUC values for each of the selected classifiers. Each row presents the outcomes of a unique and sequential combination of the first (feature extraction) and second (class balancing) preprocessing stages. The highest score for all the performance metrics and for each classifier is highlighted.

The combination of Autoencoder followed by SMOTE emerges to be the most effective. It has the highest F1-score (0.9714), the highest AUC (0.9950), the highest recall (1.0000), and the highest precision (0.9444) for the LightGBM classifier. Moreover, from the result, it is observed, that the baseline (no feature extractor and no class balancing) is performing the worst for all the mentioned classifiers. It yields the lowest recall, precision, F1-score, and AUC. The F1-score incorporates both precision and recall, as earlier mentioned. As a result, the F1-score is given greater significance than its individual parts, precision, and recall. Therefore, based on the F1-score and AUC score values from Table 3, the combinations of Autoencoders followed by SMOTE for LightGBM produces the best result.

The experiments involving only autoencoders are also giving some better results because the latent representation is robust toward the imbalanced class

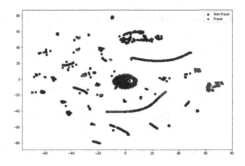

Fig. 4. Latent space representations after feature extraction using t-SNE projection.

due to the fact that the latent features extracted from the autoencoder have strong clustering power. The latent features allow the model to group the healthcare providers into clusters and make it easier to identify fraudulent behaviors, and this can be seen in Fig. 4 (fraudulent latent representations (i.e. red points) concentrate on the separate clusters of the latent space) [34].

Table 3. Scores for various classifier with Feature Extraction and SMOTE.

Classifier	Feature extraction	Data Sampling	Precision	Recall	F1-score	AUC	Accuracy
Catboost	none	none	0.0000	0.0000	0.0000	0.4999	0.9995
	Autoencoders	none	0.6582	0.6117	0.6341	0.7879	0.9282
	none	SMOTE	0.2096	0.5652	0.3058	0.7821	0.9988
	Autoencoders	SMOTE	0.8585	1.0000	0.9239	0.9860	0.9761
AdaBoost	none	none	0.0000	0.0000	0.0000	0.5000	0.9995
	Autoencoders	none	0.6129	0.4470	0.5170	0.7075	0.9150
	none	SMOTE	0.2631	0.5434	0.3546	0.8214	0.9981
	Autoencoders	SMOTE	0.6043	0.9882	0.7500	0.9392	0.9044
XGBoost	none	none	0.0000	0.0000	0.0000	0.5000	0.9995
	Autoencoders	none	0.7142	0.3529	0.4724	0.6684	0.9198
	none	SMOTE	0.0039	0.8378	0.0078	0.8425	0.8472
	Autoencoders	SMOTE	0.6439	1.0000	0.7834	0.9530	0.9197
LightGBM	none	none	0.0000	0.0000	0.0000	0.4994	0.9984
	Autoencoders	none	0.5955	0.6235	0.6091	0.7877	0.9186
	none	SMOTE	0.0692	0.8108	0.1276	0.9014	0.9920
	Autoencoders	SMOTE	0.9444	1.0000	0.9714	0.9950	0.9914

4.3 Comparison with Previous Literature

According to the publications that address the same domain, this study outperformed the other results mentioned in Table 4. It produces better result than Shamitha et al. [30] because their study use PCA for dimensionality reduction which fails to capture the non-linear correlations between features but the current work use Autoencoder which can extract the non-linearly correlations between mutiple features and also perform dimensionality reduction. Besides, this study produces better result than Hancock and Khoshgoftaar [27] because their work relies only on catBoost's internal mechanism for encoding categorical features while this study empolys autoencoder for dimensionality reduction in addition to catboost classifier.

Table 4. Comparative Performance Analysis with previous work.

Research article	Precision	Recall	F1-score	AUC	Accuracy
Shamitha et al. [30]	0.9700	0.7300	-	-	-
Hancock and Khoshgoftaar [27]	-	-	-	0.7250	-
This study	0.9444	1.0000	0.9714	0.9950	0.9914

5   Conclusion

Healthcare being an integral component of people's lives has increased the requirement of health insurance schemes over the past few years. But, increasing insurance programs have motivated fraudsters to accomplish fraudulent activities on such schemes for their monetary gain. In an attempt to increase transparency and lessen fraud, there is a requirement for an efficient fraud detection system for the health insurance claims. To address this, an efficient framework is designed which deals with efficient solutions to eliminate problems associated with highly imbalanced and heterogeneous data. With exhaustive experiments using combination of several techniques such as data preprocessing, dimensionality reduction, oversampling and classifiers. Several learners are trained and compared to find the most effective one in building the fraud detection model. Among the classifiers under consideration, LightGBM produced the best F1-score and AUC score when implemented with autoencoder followed by SMOTE technique. That is, applying feature extraction followed by data sampling outperformed the baseline architecture and produced better classification results. For further optimization, this work also performs L1-regularization and stacked various layers of the autoencoders, and the final goal of finding the best answer to the problem was fairly accomplished.

References

1. Punn, N.S., Agarwal, S.: Modality specific U-net variants for biomedical image segmentation: a survey. Artif. Intell. Rev. **55**, 1–45 (2022)
2. Nagabhushan, P., Sonbhadra, S.K., Punn, N.S., Agarwal, S.: Towards machine learning to machine wisdom: a potential quest. In: Srirama, S.N., Lin, J.C.-W., Bhatnagar, R., Agarwal, S., Reddy, P.K. (eds.) BDA 2021. LNCS, vol. 13147, pp. 261–275. Springer, Cham (2021). https://doi.org/10.1007/978-3-030-93620-4_19

3. Sudhanshu, Punn, N.S., Sonbhadra, S.K., Agarwal, S.: Recommending best course of treatment based on similarities of prognostic markers. In: Mantoro, T., Lee, M., Ayu, M.A., Wong, K.W., Hidayanto, A.N. (eds.) ICONIP 2021. LNCS, vol. 13109, pp. 393–404. Springer, Cham (2021). https://doi.org/10.1007/978-3-030-92270-2_34

4. Punn, N.S., Agarwal, S.: CHS-net: a deep learning approach for hierarchical segmentation of COVID-19 via CT images. Neural Process. Lett. **54**, 1–22 (2022)

5. Kaushik, D., Prasad, B.R., Sonbhadra, S.K., Agarwal, S.: Post-surgical survival forecasting of breast cancer patient: a novel approach. In: 2018 International Conference on Advances in Computing, Communications and Informatics (ICACCI), pp. 37–41. IEEE (2018)

6. Agarwal, S., Pandey, G.: SVM based context awareness using body area sensor network for pervasive healthcare monitoring. In: Proceedings of the First International Conference on Intelligent Interactive Technologies and Multimedia, pp. 271–278 (2010)

7. Medicare CMS (2022). https://www.cms.gov/Medicare/Medicare. Accessed 22 Dec 2021

8. Ketu, S., Agarwal, S.: Performance enhancement of distributed k-means clustering for big data analytics through in-memory computation. In: 2015 Eighth International Conference on Contemporary Computing (IC3), pp. 318–324. IEEE (2015)

9. Hancock, J., Khoshgoftaar, T.M.: Leveraging lightGBM for categorical big data. In: 2021 IEEE Seventh International Conference on Big Data Computing Service and Applications (BigDataService), pp. 149–154. IEEE (2021)

10. Tomar, D., Agarwal, S.: An effective weighted multi-class least squares twin support vector machine for imbalanced data classification. Int. J. Comput. Intell. Syst. **8**(4), 761–778 (2015)

11. Bauder, R., Khoshgoftaar, T.: Medicare fraud detection using random forest with class imbalanced big data. In: 2018 IEEE International Conference on information reuse and integration (IRI), pp. 80–87. IEEE (2018)

12. Tomar, D., Agarwal, S.: Predictive model for diabetic patients using hybrid twin support vector machine. In: Proceedings of the 5th International Conferences on Advances in Communication Network and Computing (CNC 2014), pp. 1–9 (2014)

13. Salekshahrezaee, Z., Leevy, J.L., Khoshgoftaar, T.M.: Feature extraction for class imbalance using a convolutional autoencoder and data sampling. In: 2021 IEEE 33rd International Conference on Tools with Artificial Intelligence (ICTAI), pp. 217–223. IEEE (2021)

14. Bouzgarne, I., Mohamed, Y., Bouattane, O., Mohamed, Q.: Composition of feature selection methods and oversampling techniques for banking fraud detection with artificial intelligence. Int. J. Eng. Trends Technol. **69**, 216–226 (2021). https://doi.org/10.14445/22315381/IJETT-V69I11P228

15. Bauder, R.A., Khoshgoftaar, T.M.: The detection of medicare fraud using machine learning methods with excluded provider labels. In: The Thirty-First International Flairs Conference (2018)

16. Liu, Q., Vasarhelyi, M.: Healthcare fraud detection: a survey and a clustering model incorporating geo-location information. In: 29th World Continuous Auditing and Reporting Symposium (29WCARS), Brisbane, Australia (2013)

17. Herland, M., Khoshgoftaar, T.M., Bauder, R.A.: Big data fraud detection using multiple medicare data sources. J. Big Data **5**(1), 1–21 (2018)

18. Johnson, J.M., Khoshgoftaar, T.M.: Medicare fraud detection using neural networks. J. Big Data **6**(1), 1–35 (2019). https://doi.org/10.1186/s40537-019-0225-0

19. Van Hulse, J., Khoshgoftaar, T.M., Napolitano, A.: Experimental perspectives on learning from imbalanced data. In: Proceedings of the 24th International Conference on Machine Learning, pp. 935–942 (2007)

20. Chawla, N.V., Bowyer, K.W., Hall, L.O., Kegelmeyer, W.P.: SMOTE: synthetic minority over-sampling technique. J. Artif. Intell. Res. **16**, 321–357 (2002)

21. Chen, Z., Yeo, C.K., Francis, B.S.L., Lau, C.T.: A MSPCA based intrusion detection algorithm tor detection of DDoS attack. In: 2015 IEEE/CIC International Conference on Communications in China, pp. 1–5. IEEE (2015)

22. Chen, Z., Yeo, C.K., Francis, B.S.L., Lau, C.T.: Combining mic feature selection and feature-based MSPCA for network traffic anomaly detection. In: 2016 Third International Conference on Digital Information Processing, Data Mining, and Wireless Communications, pp. 176–181. IEEE (2016)

23. Chen, Z., Yeo, C.K., Lee, B.S., Lau, C.T.: Detection of network anomalies using improved-MSPCA with sketches. Comput. Secur. **65**, 314–328 (2017)

24. Chen, Z., Yeo, C.K., Lee, B.S., Lau, C.T.: A novel anomaly detection system using feature-based MSPCA with sketch. In: 2017 26th Wireless and Optical Communication Conference (WOCC), pp. 1–6. IEEE (2017)

25. Hancock, J.T., Khoshgoftaar, T.M.: Gradient boosted decision tree algorithms for medicare fraud detection. SN Comput. Sci. **2**(4), 1–12 (2021)

26. Wu, P., Zhao, H.: Some analysis and research of the AdaBoost algorithm. In: Chen, R. (ed.) ICICIS 2011. CCIS, vol. 134, pp. 1–5. Springer, Heidelberg (2011). https://doi.org/10.1007/978-3-642-18129-0_1

27. Hancock, J., Khoshgoftaar, T.M.: Medicare fraud detection using CatBoost. In: 2020 IEEE 21st international conference on information reuse and integration for data science (IRI), pp. 97–103. IEEE (2020)

28. Dorogush, A.V., Ershov, V., Gulin, A.: CatBoost: gradient boosting with categorical features support. arXiv:1810.11363 (2018)

29. Hancock, J.T., Khoshgoftaar, T.M.: CatBoost for big data: an interdisciplinary review. J. Big Data **7**(1), 1–45 (2020)

30. Shamitha, S., Ilango, V.: A time-efficient model for detecting fraudulent health insurance claims using artificial neural networks. In: 2020 International Conference on System, Computation, Automation and Networking, pp. 1–6. IEEE (2020)

31. Medicare part d prescribers - by provider and drug (2018). https://data.cms.gov/provider-summary-by-type-of-service/medicare-part-d-prescribers/medicare-part-d-prescribers-by-provider-and-drug/data/2018. Accessed 25 Nov 2021

32. Leie downloadable databases (2022). https://oig.hhs.gov/exclusions/exclusions_list.asp. Accessed 25 Feb 2022

33. Van der Maaten, L., Hinton, G.: Visualizing data using t-SNE. J. Mach. Learn. Res. **9**(11) (2008)

34. Chen, Z., Yeo, C.K., Lee, B.S., Lau, C.T.: Autoencoder-based network anomaly detection. In: 2018 Wireless Telecommunications Symposium (WTS), pp. 1–5. IEEE (2018)

A Hybrid Feature Selection Approach for Data Clustering Based on Ant Colony Optimization

Rajesh Dwivedi[1]([envelope]), Aruna Tiwari[1], Neha Bharill[2], and Milind Ratnaparkhe[3]

[1] Indian Institute of Technology Indore, Indore 453552, India
anubhav.dwivedi8@gmail.com, artiwari@iiti.ac.in
[2] Ecole Centrale School of Engineering, Mahindra University,
Hyderabad 500043, India
neha.bharill@mahindrauniversity.edu.in
[3] ICAR-Indian Institute of Soybean Research, Indore 452001, India

Abstract. Machine learning, data mining, and pattern recognition all require feature selection when working with high-dimensional data. Feature selection helps in improving the prediction accuracy and significantly reduces the computation time. The problem is that many of the feature selection algorithms use a sequential search strategy to choose the most important features. This means that each time you add or remove a feature from the dataset, you get stuck in a local optimum. This paper proposes a hybrid feature selection technique based on ant colony optimization that randomly selects features and quantifies their quality using K-means clustering in terms of silhouette index and laplacian score. The proposed hybrid feature selection technique allows for random selection of features, which facilitates a better exploration of feature space and avoids the problem of being trapped in a local optimal solution, while also generating a global optimal solution. Furthermore experimental investigation shows that the proposed method outperforms the state-of-the-art method.

Keywords: Ant Colony Optimization · Jaccard index · K-means clustering · Laplacian Score · Silhouette Index

1 Introduction

Feature selection is widely used in various data mining and machine learning tasks such as classification, clustering, and regression to improve readability and interpretability. Due to the popularity of feature selection in such areas, so far, researchers have primarily attempted to analyze and explain feature selection tasks in supervised learning area, especially in classification, but in unsupervised learning area [15], especially clustering [8] it has not been explored extensively. A feature selection method generally consists of four main phases: selection, examination, terminating criterion, and validation. The first phase involves

© The Author(s), under exclusive license to Springer Nature Switzerland AG 2023
M. Tanveer et al. (Eds.): ICONIP 2022, LNCS 13625, pp. 659–670, 2023.
https://doi.org/10.1007/978-3-031-30111-7_55

selecting a feature subset using a predetermined search strategy, like sequential search, sequential floating search, or complete search. The second phase consists of examining the chosen feature subset by a specific criterion. After getting termination criteria, the third step selects the best performing subset from all possible subsets. The last step involves the validation of the chosen subset using the validation metrics.

The remaining portion of this paper is arranged as follows: Sect. 2 provides a literature review of the existing work. The proposed feature selection method based on Ant Colony Optimization is introduced in Sect. 3. Section 4 presents experimental results on various benchmark datasets. Finally, Sect. 5 presents the conclusion and future work.

2   Literature Review

This section presents the various techniques for features selection proposed by other researchers. Dash and Liu [2] developed a hybrid feature selection approach that calculates entropy from the similarity of data and uses a measure based on entropy to evaluate the features in the filter stage. The wrapper stage uses scatter separability criteria and k-means clustering to select the relevant feature subset. A drawback of this approach is its high computation cost. Later on, Hruschka et al. [7] proposed a hybrid feature selection approach that uses a Bayesian filter with k-means clustering to identify the relevant feature subset. They used a Bayesian network that uses Markov blanket property for the filter approach. A drawback of this approach is that they have only tested it for datasets having less than 30 features. By adopting the same idea proposed by Dash and Liu [2], Li et al. [9] proposed a new hybrid feature selection approach, in which they used Fuzzy Feature Evaluation Index (FFEI) with an exponential entropy index to evaluate the feature in the filter stage to increase the performance. They used a scatter separability criterion and Fuzzy C-Means algorithm for the wrapper stage. This approach also suffers from high computation costs.

In 2015 another feature selection approach suggested by Nahato et al. [10] uses rough set theory to identify the relevant features. They used rough indiscernibility relation to select the reducts and trained backpropagation neural networks using the selected reducts. This method was tested on statlog heart disease datasets, wisconsin breast cancer dataset, and hepatitis dataset taken from UCI machine learning library [1] and achieved an accuracy of 90.4%, 98.6%, and 97.3% with 6, 7, and 13 features, respectively. Later on, in 2016 Solorio et al. [14] proposed a hybrid feature selection technique that uses a laplacian score to rank the features and a modified Calinski Harabase index to measure a feature subset. They tested their approach on various benchmark datasets taken from the UCI Machine learning repository [1] and also on several synthetic datasets and thus achieve better results than approaches proposed by Dash and Liu [2] and Li et al. [9].

In 1990, Dorigo et al. [3] came up with the idea of an Ant Colony Optimization (ACO). This approach mimics the social behaviour of ants searching for food.

Initially, it was developed to solve the famous traveling salesman problem. Later it was applied to various complex optimization problems like feature selection [11]. An unsupervised ACO based feature selection was proposed by Tabakhi et al. [17] that uses Cosine similarity measures to measure the similarity between features. In this work, the number of artificial ants used was equal to the number of attributes in the dataset such that each ant was responsible for constructing a feature subset. The frequency of selected attributes on different subsets was used to update the pheromone value. Feature having low similarity, and high pheromone value was added to feature subset in every step till max iteration. They tested their approach on several UCI machine learning datasets [1] like wine and breast cancer datasets and got an average classification error of 19.8%. In 2017, Dhalia et al. [16] proposed another ACO based approach that uses a tandem run strategy to select the relevant feature subset. They used Cosine similarity measure to measure the similarity between features and support vector machine (SVM) for assigning the fitness to a feature and further SVM is used for classification. They tested their approach on Lung CT scan images to diagnose bronchitis and achieved 81.66% accuracy.

From the presented work, it can be inferred that hybrid methods are performing well in comparison of filter and wrapper methods. Also, ACO is used in various tasks for the feature selection and gives an increased accuracy; therefore, in the proposed work, a hybrid feature selection approach based on ACO is presented which uses silhouette index and laplacian score as a fitness measure and gives an increased clustering performance on various benchmark datasets.

3   Proposed Method

In this work, we proposed a novel hybrid feature selection technique based on ant colony optimization [4] (NHFS based ACO) that follows the tandem run strategy [5] to select the best feature subset.

The simulation model is expressed by a completely connected undirected graph G = (V, E) having a one-to-one mapping between vertices and features. Hence the number of vertices (v_n) equals the number of features (f_n). V denotes the set of vertices as $v_1, v_2, v_3....v_n$ and E denotes the set of edges $(e_1, e_2, e_3....e_n)$ joining any two vertices in the graph. In this model, the number of artificial ants (N_{ant}) is taken same as the number of features (f_n) to avoid being trapped in the local optimum, so $f_n = v_n = N_{ant}$. In ACO each artificial ant constructs a feature subset (F_i). The N denotes the set of all feature subsets created by ants and n_{max} indicates the maximum number of features possible in each subset, then $N_{ant} = N$ and $0 \leq n_{max} \leq n$.

To make the feature subsets, each ant starts from a vertex and creates a feature subset by traversing different vertices in between. Every feature is associated with a pheromone value (α) set to a constant initially. n_{tan} denotes the number of features selected by the tandem run strategy.

In this method, three steps are used to choose n_{max} features. In the first step, n feature subsets are made by picking n_{max} features at random. Then, we apply K-means clustering to these subsets and evaluate their efficacy in terms of silhouette index (SI) values. The leader subset is the one with the highest SI value ($g_{bestset}$). Algorithm 1 describes the working of first step. In the second step, n feature subsets are made in a different way, and the selection of n_{max} features is accomplished in three stages. Certain features are chosen randomly (n_{random}), while others ($n_{arbitary}$) are chosen based on their high pheromone and low laplacian scores. On the other hand, some features (n_{tan}) from the leader subset are chosen because they have a high pheromone score but a low laplacian score. Once again, these subsets are used in K-means clustering to evaluate how effective they are in terms of a SI value that is determined in the third step. The subset with the highest SI is called localbest ($l_{bestset}$), and if it is greater than globalbest, it becomes globalbest ($g_{bestset}$). Iterate the second and third steps until max_{iter}. After all iterations, the global bestset is the subset with the highest SI value. The working of second and third step are shown in Algorithm 2.

Algorithm 1

Input: Dataset, n_{max}

Output: Leader subset after first iteration ($g_{bestset}$)

 1: Create n feature subsets, each feature subset will have n_{max} features choosen randomly.
 2: These subsets are applied to K-means clustering and the efficacy of these subsets is evaluated in terms of SI value. Take the number of clusters equal to the number of class label as given in the dataset, if class label is not available then decide k randomly.
 3: Subset gives the best SI value which is considered as the leader subset and known as $g_{bestset}$.
 4: Return $g_{bestset}$.

3.1   Computation of Laplacian Score

Laplacian score [6] represents the local preserving power of a feature. It is used for feature ranking in many feature selection approaches. A good feature always has a low laplacian score value. For a given dataset with m instances, a similarity graph is constructed in the form of the weight matrix W of size $m * m$ such that $W = \{w_{11}, w_{12}, w_{13}, ... w_{ij}, ... w_{mm}\}$, where each edge connecting instances x_i to x_j represents similarity in form of weight w_{ij}. Laplacian matrix L is calculated as defined in Eq. (1).

$$L = D - W \qquad (1)$$

where D is the diagonal matrix and W is the weight matrix.

Algorithm 2

Input: Dataset, n_{max}, max_{iter}, $g_{bestset}$ after first iteration.
Output: Leader subset ($g_{bestset}$)
1: Select n_{max} features in n subsets using step 2.

$$n_{random} = n_{max} - n_{remain}$$
$$n_{remain} = n_{arbitary} + n_{tan}.$$

2: Select n_{random} features randomly and $n_{arbitary}$ number of features with high pheromone and max heuristic value.
3: Select n_{tan} features from leader subset with high pheromone and max heuristic value.
4: These subsets are applied to K-means clustering and the efficacy of these subsets is evaluated in terms of SI value. Take the number of clusters equal to the number of class label as given in the dataset, if class label is not available then decide k randomly.
5: Subset with maximum SI value is known as localbest.
6: Compare localbest and globalbest, if localbest if greater than globalbest make localbest as globalbest for further iteration.
7: Repeat step 1 to 6 till max_{iter}.
8: Return $g_{bestset}$.

Let's f_r is the rth feature in all m instances then
$f_r = (f_{r1}, f_{r2}, f_{r3}, f_{r4}, f_{r5}, ... f_{rm})^T$ where $r \in [1, n]$. Laplacian score of f_r is computed as defined in Eq. (2).

$$L_r = \tilde{f}_r^T L \tilde{f}_r / \tilde{f}_r^T D \tilde{f}_r \qquad (2)$$

$\tilde{f}_r$ denotes the f_r vector's deviation from the mean and computed as given in Eq. (3).

$$\tilde{f}_r = f_r - \left(\frac{f_r^T D 1}{1^T D 1} \right) \qquad (3)$$

where D is the diagonal matrix and $1 = [1, ..., 1]^T$. f_r^T is the transpose of f_r.
After getting laplacian score heuristic value (h_r) is computed using Eq. (4).

$$h_r = \frac{1}{L_r} \qquad (4)$$

3.2   Calculation of Pheromone

Step 1: Allocate initial pheromone (α) to all features as given in Eq. (5).

$$\alpha_{f_i} = \frac{1}{n} \qquad (5)$$

where $i \in [1, n]$.

Step 2: Each time a feature f_i is selected in a subset F_j, where $i \in [1, n]$ and $j \in [1, n]$ its pheromone update occurs in Eq. (6) and (7).

$$fitness_{f_i} = \frac{SI(F_j)}{n_{max}} \tag{6}$$

$$\alpha_{f_i}(t + 1) = \alpha_{f_i}(t) + fitness_{f_i} \tag{7}$$

where $\alpha_{f_i}(t)$ and $\alpha_{f_i}(t + 1)$ are pheromones value of a feature f_i at time t and t + 1, respectively.

4   Results and Discussion

4.1   Dataset Details

In the experimental study, we have used three benchmark datasets namely Ionosphere, Sonar and Vehicle silhouettes, which are collected from the UCI Machine Learning repository [1]. Preprocessing of these datasets is performed such as removal of missing values. After preprocessing, details of the datasets are presented in Table 1.

Table 1. Dataset Details

Dataset Name	No. of instances	No. of features	No. of classes
Ionosphere	351	33	2
Sonar	208	60	2
Vehicle silhouettes	813	18	3

4.2   Evaluation Measures

In this approach, two cluster measures and one visualizer are used to evaluate and visualize the clustering performance which are defined as follows:

Jaccard Index (JI). JI [12] is an external evaluation measure for any clustering approach. It evaluates clustering performance based on its similarity to the ground truth or Expert classification. Jaccard index value range from 0 to 1, where 0 represents no match between clustering and ground truth and 1 illustrates a perfect match.

Silhouette Index (SI). SI [13] is known as an internal evaluation measure for any clustering algorithm. It is based on the similarity of a data point within its cluster known as Cohesion and to its nearest cluster known as separation. It ranges from -1 to 1, and a high SI value represents well-clustered data points. The silhouette index is calculated by taking the average of all data point's silhouette coefficients.

Silhouette Visualizer. The silhouette visualizer visualizes which clusters are dense and which are not by displaying the silhouette coefficient for each sample on a per-cluster basis. It also shows how many clusters are achieving the average SI value.

4.3   Parameter Settings for NHFS Based ACO

In this work, experiments are carried out by taking different values of n_{max} for all datasets and for each dataset '50' independent runs of experiments were conducted, therefore $max_{iter} = 50$. Parameters settings for various variables is as follows:

$n_{arbitary} = 30\% of n_{max}$

$n_{tan} = 30\% of n_{max}$

In case of fraction value, consider its ceil value.

4.4   Experimental Analysis

Experiments are performed on the datasets listed in Table 1 and the NHFS based ACO approach is compared with a hybrid feature selection approach developed by Solorio et al. [14] because they used a similar strategy to obtain the best feature subset and measured the results in terms of jaccard index and silhouette index.

Experimental Findings on Ionosphere Dataset: The NHFS based ACO is applied on Ionosphere dataset for different values of n_{max} and results are presented in Table 2.

Table 2. Results on Ionosphere Dataset

Technique used	No. of feature selected	Jaccard Index	Silhouette Index
Solorio et al.	7	0.4376	0.5131
NHFS based ACO	1	**0.6132**	**0.7506**
NHFS based ACO	5	0.4486	0.5438
NHFS based ACO	7	0.4589	0.5681

It can be seen from Table 2 that NHFS based ACO is giving increased JI and SI values in comparison to Solorio et al. [14] approach when 1 feature is selected. Both the approaches also worked on 7 number of features, in spite of NHFS based ACO selected more relevant features and that is by performed better than Solorio et al. [14] approach. The silhouette visualizer obtained from Solorio et al. [14] approach shown in Fig. 1(a) showing that some data points having negative silhouette coefficient in blue colored cluster, which shows that those data points are wrongly clustered. On the other hand silhouette visualizer obtained from NHFS based ACO approach shown in Fig. 1(b) shows that all data points have positive silhouette coefficient value and gives better clustering.

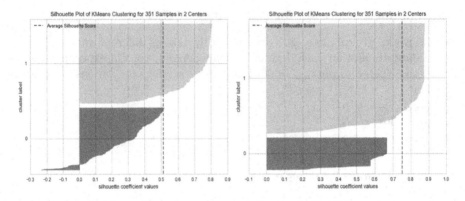

Fig. 1. (a) Silhouette visualizer for Ionosphere using Solorio et al. [14]. (b) Silhouette visualizer for Ionosphere using NHFS based ACO

Experimental Findings on Sonar Dataset: The NHFS based ACO is applied on Sonar dataset for different values of n_{max} and combined results are shown in Table 3.

Table 3. Results on Sonar Dataset

Technique used	No. of feature selected	Jaccard Index	Silhouette Index
Solorio et al.	1	0.3448	0.6304
NHFS based ACO	**1**	**0.4273**	**0.7501**
NHFS based ACO	3	0.4473	0.6319

It can be seen from Table 3 that NHFS based ACO is giving increased JI and SI value in comparison to Solorio et al. [14] approach when 1 feature is getting selected. silhouette visualizer obtained from Solorio et al. [14] and NHFS based ACO are also shown in Fig. 2(a) and Fig. 2(b) to visualize the clustering results.

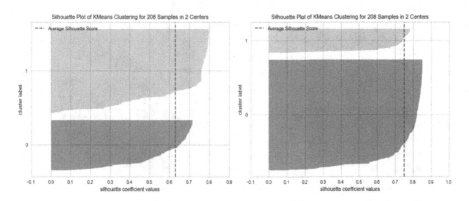

Fig. 2. (a) Silhouette visualizer for Sonar using Solorio et al. [14]. (b) Silhouette visualizer for Sonar using NHFS based ACO

Experimental Findings on Vehicle Silhouettes Dataset: The NHFS based ACO is applied on Vehicle silhouettes dataset for different values of n_{max} and combined results are presented in Table 4.

Table 4. Results on Vehicle silhouettes Dataset

Technique used	No. of feature selected	Jaccard Index	Silhouette Index
Solorio et al.	5	0.2935	0.5635
NHFS based ACO	5	0.3162	0.6603
NHFS based ACO	**4**	**0.3150**	**0.6650**
NHFS based ACO	3	0.3148	0.6562
NHFS based ACO	1	0.3319	0.6516

It can be seen from Table 4 that NHFS based ACO approach is giving increased JI and SI values in comparison to Solorio et al. [14] approach in all cases but the highest SI value when 4 features are taken. Both the approaches also worked on 5 number of features, in spite of NHFS based ACO selected more relevant features and that is by performed better than Solorio et al. [14] approach. Silhouette visualizer obtained from Solorio et al. [14] and NHFS based ACO approach with 4 features are also presented in Fig. 3(a) and Fig. 3(b) to observe the clustering results.

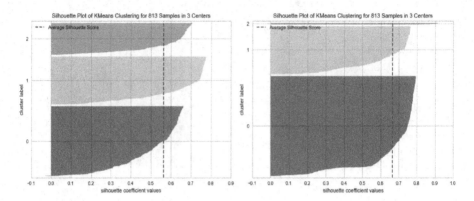

Fig. 3. (a) Silhouette visualizer for Vehicle silhouettes using Solorio et al. [14]. (b) Silhouette visualizer for Vehicle silhouettes using NHFS based ACO

4.5   Comparison

Comparison graphs presented in Fig. 4 showing the comparison between Solorio et al. [14] and NHFS based ACO in terms of silhouette index, shows that NHFS based ACO gives better SI value in all datasets. Whereas Fig. 5 showing the comparison between Solorio et al. [14] and NHFS based ACO in terms of jaccard index shows that NHFS based ACO gives better JI value for all datasets.

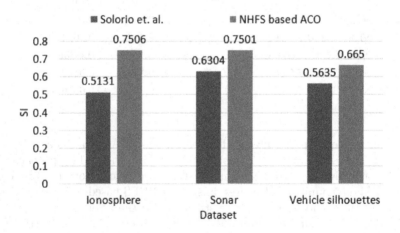

Fig. 4. Comparison between Solorio et al. [14] and NHFS based ACO in SI

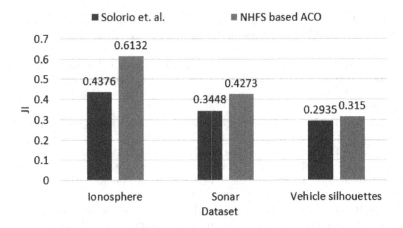

Fig. 5. Comparison between Solorio et al. [14] and NHFS based ACO in JI

5   Conclusion

In the proposed method a hybrid feature selection algorithm based on Ant Colony Optimization named as NHFS based ACO is presented, which removes redundant and irrelevant features that have a negative impact on model building and selects the more appropriate features from data having large number of features. The NHFS based ACO approach uses mixture of laplacian score as well as silhouette index to measure the relevancy of a feature rather than using laplacian score in the filter stages and then silhouette index in the wrapper stage, separately. It also uses tandem run strategy to select the most promising features from the leader subset, which improves the power of proposed approach. The proposed approach is tested on 3 benchmark datasets having a large number of features and achieved the better results than other state-of-the-art approach. The proposed method also worked well on Ionosphere dataset and clustered data points in such a way that all data points have a positive silhouette coefficient. The work on feature selection can be expanded by considering other bio-inspired feature selection algorithms as well by taking various other filter measures.

Acknowledgements. This research is funded by The Council of Scientific and Industrial Research (CSIR), Government of India under grant no. 22(0853)/20/EMR-II.

References

1. Blake, C.: UCI repository of machine learning databases (1998). http://www.ics.uci.edu/~mlearn/MLRepository.html
2. Dash, M., Liu, H.: Feature selection for clustering. In: Terano, T., Liu, H., Chen, A.L.P. (eds.) PAKDD 2000. LNCS (LNAI), vol. 1805, pp. 110–121. Springer, Heidelberg (2000). https://doi.org/10.1007/3-540-45571-X_13

3. Dorigo, M., Gambardella, L.M.: Ant colony system: a cooperative learning approach to the traveling salesman problem. IEEE Trans. Evol. Comput. **1**(1), 53–66 (1997)
4. Dwivedi, R., Kumar, R., Jangam, E., Kumar, V.: An ant colony optimization based feature selection for data classification. Int. J. Recent Technol. Eng. **7**, 35–40 (2019)
5. Franks, N.R., Richardson, T.: Teaching in tandem-running ants. Nature **439**(7073), 153 (2006)
6. He, X., Cai, D., Niyogi, P.: Laplacian score for feature selection. Adv. Neural Inf. Process. Syst. **18**, 507–514 (2005)
7. Hruschka, E.R., Covoes, T.F., Ebecken, N.F.: Feature selection for clustering problems: a hybrid algorithm that iterates between k-means and a Bayesian filter. In: Fifth International Conference on Hybrid Intelligent Systems (HIS 2005), pp. 6-pp. IEEE (2005)
8. Kumar, R., Dwivedi, R., Jangam, E.: Hybrid fuzzy C-means using bat optimization and maxi-min distance classifier. In: Singh, M., Gupta, P.K., Tyagi, V., Flusser, J., Ören, T., Kashyap, R. (eds.) ICACDS 2019. CCIS, vol. 1046, pp. 68–79. Springer, Singapore (2019). https://doi.org/10.1007/978-981-13-9942-8_7
9. Li, Y., Lu, B.L., Wu, Z.F.: A hybrid method of unsupervised feature selection based on ranking. In: 18th International Conference on Pattern Recognition (ICPR 2006), vol. 2, pp. 687–690. IEEE (2006)
10. Nahato, K.B., Harichandran, K.N., Arputharaj, K.: Knowledge mining from clinical datasets using rough sets and backpropagation neural network. Computat. Math. Methods Med. **2015** (2015). https://doi.org/10.1155/2015/460189
11. Nayar, N., Gautam, S., Singh, P., Mehta, G.: Ant colony optimization: a review of literature and application in feature selection. In: Smys, S., Balas, V.E., Kamel, K.A., Lafata, P. (eds.) Inventive Computation and Information Technologies. LNNS, vol. 173, pp. 285–297. Springer, Singapore (2021). https://doi.org/10.1007/978-981-33-4305-4_22
12. Real, R., Vargas, J.M.: The probabilistic basis of Jaccard's index of similarity. Syst. Biol. **45**(3), 380–385 (1996)
13. Rousseeuw, P.J.: Silhouettes: a graphical aid to the interpretation and validation of cluster analysis. J. Comput. Appl. Math. **20**, 53–65 (1987)
14. Solorio-Fernández, S., Carrasco-Ochoa, J.A., Martínez-Trinidad, J.F.: A new hybrid filter-wrapper feature selection method for clustering based on ranking. Neurocomputing **214**, 866–880 (2016)
15. Solorio-Fernández, S., Carrasco-Ochoa, J.A., Martínez-Trinidad, J.F.: A review of unsupervised feature selection methods. Artif. Intell. Rev. **53**(2), 907–948 (2020)
16. Sweetlin, J.D., Nehemiah, H.K., Kannan, A.: Feature selection using ant colony optimization with tandem-run recruitment to diagnose bronchitis from CT scan images. Comput. Methods Programs Biomed. **145**, 115–125 (2017)
17. Tabakhi, S., Moradi, P., Akhlaghian, F.: An unsupervised feature selection algorithm based on ant colony optimization. Eng. Appl. Artif. Intell. **32**, 112–123 (2014)

FaceMix: Transferring Local Regions for Data Augmentation in Face Recognition

Nikita Garaev[1][✉] , Evgeny Smirnov[1] , Vasiliy Galyuk[1] ,
and Evgeny Lukyanets[2]

[1] Speech Technology Center, Saint Petersburg, Russian Federation
garaevnick@gmail.com, {smirnov-e,galyuk}@speechpro.com
[2] ITMO University, Saint Petersburg, Russian Federation
lukyanets@speechpro.com

Abstract. Augmentation strategies for image recognition based on local image patches have gained widespread popularity. Their main idea is to replace or remove some local regions of the image. The advantage of these methods is that they change part of the image and force the network to pay attention to the less significant parts, which leads to a greater generalization capacity of the network. While these methods work good for image recognition, they do not perform as well for face recognition tasks. The purpose of this work is to create augmentation specialized for face recognition and devoid of the shortcomings of previous works. We present FaceMix: a flexible face-specific data augmentation technique that transfers a local area of an image to another image. The method has two operating modes: it can generate new images within a class, and it can generate images for a class, using face data from other classes, and these two modes also could be combined. FaceMix is helping to solve the problems of class imbalance and insufficient number of images per identity. A feature of this method is that the number of possible artificial images grows quadratically with the growth of real images. Experiments on face recognition benchmarks, such as CFP-FP, AgeDB, CALFW, CPLFW, XQLFW, SLLFW, RFW and MegaFace, demonstrate the effectiveness of the proposed method.

1 Introduction

In the last decade deep neural networks have made a huge leap forward and become the main family of algorithms for solving a variety of problems. Deep neural networks show excellent results in tasks of computer vision, such as: object detection [25], object recognition [36], semantic segmentation [30], head pose estimation [15], face recognition [4,17,19–22] and so on. Although neural networks show excellent performance, they need a lot of data for training. Along with the evolution of neural networks, the sizes of datasets also grow, but often the size of existing datasets is not enough for effective training. Often, due to the fact that the network contains a huge number of parameters, and the dataset for training is not large enough, overfitting occurs. To combat this problem, various regularization mechanisms are used, such as dropout, weight decay, early stopping and augmentation. Dropout turns off various neurons so that the network learns to find workarounds and turns out to be more robust. Weight decay

M. Tanveer et al. (Eds.): ICONIP 2022, LNCS 13625, pp. 671–682, 2023.
https://doi.org/10.1007/978-3-031-30111-7_56

restrains the weights by adding the L2 norm to the loss. Early stopping ends training before overfitting begins. Augmentation expands the original dataset with the help of, for example, geometric image transformations and as a result increases the variance of the data and strengthens the ability of the network to generalize and reduce overfitting. Augmentation can change the image slightly, like flipping, or very much, like GAN based algorithms. Also, augmentation can increase the size of the original set from several percent to several times. Therefore, augmentation is one of the most popular regularization methods, and many new approaches appear in this direction in all areas of computer vision.

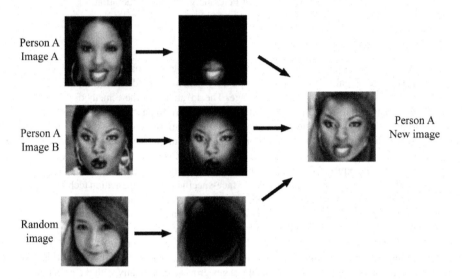

Fig. 1. General scheme of the proposed FaceMix augmentation method. The proposed method combines multiple images of one or more persons into a new image by selecting a piece on each image and joining the pieces together by overlaying.

The most popular for computer vision at the moment are the following augmentations: geometric transformations, color change, cropping, rotation. There are also more complex, but also popular augmentation algorithms, such as: noise injection, random erasure and blending of images, GAN-based algorithms and style transfer. And, although many common simple augmentations are well suited for face recognition tasks (flipping, color jittering), at the same time, a large number of augmentations do not work well enough or may not be applicable at all for face recognition. The face recognition task has its own characteristics, such as: all faces have strict patterns in the shape of the head, hair color, and others. This makes it difficult to apply standard geometry distortion or color change methods.

In this paper we propose the FaceMix augmentation algorithm, which involves the best properties of the patch-based algorithms for computer vision, such as applying patches of one image to another image (CutMix [29], RICAP [26]) or mixing two images in pixel space (MixUp [31]). Unlike above algorithms, our method was developed taking into account the specifics of face recognition. The general idea of the method is to combine several images of one or more people into a new image of the same person by selecting a part in each image and merging the parts together. Our algorithm has two operating modes:

- Background FaceMix - mixing a part of the image with a face and a background from a random image, producing a new image of the same person as a result. This can be especially useful if the dataset contains many classes with a small number of images. For each image in such class, you can create a large number of new images by replacing the background with a background from another random image, which as a result helps with the problem of long tails.
- Intra-class FaceMix - mixing face parts from several images of the same class, producing a new image of the same person as a result. This way we improve the diversity of the face images for each person in the dataset.

These operating modes could also be combined. This method can be applied online with little time and memory overhead. FaceMix shows performance improvements on benchmark datasets such as CFP-FP [18], AgeDB-30 [13], CALFW [35], CPLFW [34], XQLFW [12], SLLFW [33], RFW [28] and MegaFace [11].

2 Related Work

Augmentation is a popular way to increase dataset size and overcome overfitting. Currently, there are many methods of augmentation, differing in the amount of additional resources, implementation complexity, quality and number of synthesized images, etc. All augmentation methods can be divided into two groups: first one is based on simple image manipulations, and the second is based on neural networks. The first group includes such augmentations as random erasing, geometric transformations, occlusions, image distortion, color space transformations, and mixing images. Among the second group are methods based on GAN, style transfer and others. Next, we will describe the types of augmentations that are most related to our work.

2.1 Color Space Transformations

One of the most popular types of augmentation is color space manipulation. Examples include changing image contrast, manipulating color histograms, etc. Implementation of most of these methods can be found in [2]. Since each pixel of the image is encoded with a vector of three numbers for each of the RGB channels, we get a large matrix of vectors that can be changed in very different ways and therefore there are many different augmentation methods that work with color. The advantage of such methods is a small overhead in terms of resources, ease of development and a wide variety of methods. Among the shortcomings, it can be noted that not all methods that work with color are useful. For example, since often all the pixels of the image are involved in the transformation, the lightening of the dark part of the image will lead to the loss of information in the light part of the image due to excessive lightening. Also, it is difficult to apply color space transformations to the faces of images of people, as the faces become unnatural in color. Unlike classical methods of color manipulation, FaceMix can change the color scheme of an image by borrowing it from another image, for example, it can insert a face captured at night in daytime conditions and then the color space of the image will change, but the face will remain in its natural color.

2.2 Regional Dropout

There are such methods of augmentation, in which some part of the image is cut out [5]. These methods can be thought of as a modification of the dropout method that is applied to the input image. This allows the network not to overfit to some specific part of the image, but to learn more abstract features, which makes it possible to increase the overall generalization ability of the model. The disadvantage of these methods is that there is a possibility to cut out essential information, making correct classification impossible. FaceMix differs from this family of methods in that it fills an empty area in the image with another image, which allows not to lose information, but to provide the model with more diverse and useful training data.

2.3 Neural Data Augmentation

Another option is to use GAN [6] for augmentation [16]. In these approaches, new images are generated from the existing images in the dataset or directly from the distribution density function of the original dataset. The advantages of this approach are the diversity of the resulting images, the preservation of the original domain of the dataset and the ability to greatly expand the original set. But this approach requires a lot of time and computational resources for its work and most often cannot be performed online.

Another approach is to use neural networks [3,24,32] (usually with GAN architecture) to separate identity-related information from context-related information. In [3,24] the main idea is to create new image with the same identity but a new context from different (often more popular) class by GAN. As we mentioned earlier, the GAN architecture is usually slow and time consuming and often cannot be done online. In [32] authors proposed a network that used attention to find more identity-related data and then place it to background on image of another class. In this case, it also took a lot of time and resources to process, and as a result, we got combination of two images without any smoothing and color correction, which reduces photorealism and can affect the final accuracy of the neural network we are training.

Unlike these methods, our FaceMix algorithm does not require training a neural network and does not require a lot of time and computational resources, so it can work online.

2.4 Mixing Images

Another popular family of augmentation algorithms is image mixing. MixUp [31] and SamplePairing [10] are used to generate a new image by mixing two images pixel by pixel. There are many works based on combining several images by superimposing one image on another with rectangular pieces: Summers et al. [23] in their work offer different options for mixing two images using a different number of rectangles and oblique lines. Yun et al. [29] in their work propose the CutMix method, which cuts out a rectangular part of an image of one class and puts it on an image of another class, and also mixes their labels in proportion to the areas of the images, RICAP [26] offers a similar idea, but uses 4 images to generate a new one. There is also work [14] in which authors take a random crop from the target class and put it in the context of a more popular class. The advantages of these methods are simplicity of implementation, low consumption of time and computational resources. This allows these algorithms to work online. Disadvantage of these algorithms is that they do not perform well in the face recognition task. This is due to the fact that a person's face is very specific and any accidental cutting or pasting of rectangular areas can compromise the integrity of the face and reduce the similarity of the result to the real face. Our algorithm is designed to solve this problem, we use the fact that all faces are normalized by 5 landmarks, so we work with certain parts of the image (eyes, mouth, whole face) so that the final image looks like a face. We also use ellipses instead of rectangles and apply smoothing to make the blending of the two faces as seamless as possible.

2.5   Face Recognition Related Image Synthesis

Hu et al. [9] in their work propose an image synthesis method specific to the field of face recognition. The essence of the method is to transfer parts of one face (eyes, nose, mouth and others) in the form of rectangles to another face of the same person or another person. The advantage of this method is the speed and realism of the final images of faces relative to other methods such as CutMix. The disadvantage of this method is that transferred parts of the face are noticeable in the final image due to the different color and borders of the insertion, which is why the synthesized faces can be easily distinguished from the real ones. In our approach, we use smooth ellipses, which reduces the visibility of the insertion. The algorithm proposed by us, in contrast to previous works, can transfer not only parts of a face to another face, but also the whole face to the background from another image. We call background everything except the face, it can be hair, headwear, objects around and behind the face.

3   Proposed Method

In this article, we propose FaceMix augmentation algorithm, which synthesizes a new image based on two or three existing images by superimposing parts of the face (or the entire face) from one or more images onto the main image.

3.1   FaceMix

The basic idea of the algorithm is simple and is shown in Fig. 1. First, the images are selected (image selection method and their number will be described later). Let's look at the basic idea for two images without losing any generality. After selecting images, specific face parts are chosen to be cut from each of them. Next, a transparency mask is

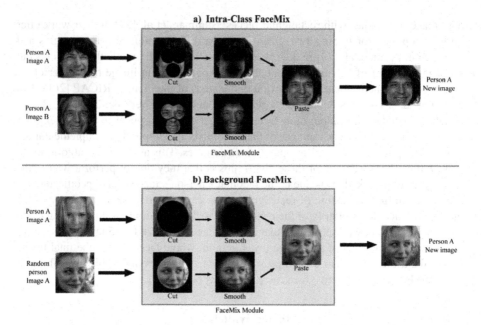

Fig. 2. The scheme of two FaceMix operating modes. All methods accept several images as input, then mix them using transparency masks and output one new image. a) Intra-Class FaceMix - increases the number of images within one class, using only the images of this class. b) Background FaceMix - increases the number of images in one class, using random images from the whole dataset.

created with the size and coordinates of each face part. After that, one image is superimposed on another using a transparency mask. The result is a new image that contains parts of both images and seamlessly blending them together. This is the general scheme of the algorithm, which differs for two specific operating modes. We will describe them below.

3.2 Intra-class FaceMix

The input to the algorithm is two random different images for person A (Fig. 2, a). Next, a part of the face for transfer is randomly selected, it can be: "eyes", "mouth" or "eyes and mouth". To properly perform Intra-Class FaceMix, the training dataset must contain images aligned by 5 landmarks (corners of the eyes, nose and corners of the mouth), so that all images in the dataset have eyes, nose and mouth in approximately the same locations. This property of the dataset allows us to know in advance where the parts of the face will be located, and therefore we can roughly outline the coordinates and sizes of the image areas for cutting in advance. Important note: any dataset that was obtained using a detector and aligned by 5 points (and this is the majority of datasets for face recognition) is suitable for the algorithm, since the location of the masks is based on the result of the alignment. After we have selected the parts of the face, we create a transparency mask corresponding to a specific area, the mask consists of circles around the eyes (if selected) and the mouth (if selected), then smoothing is applied to these circles for smoother blending. Next, the two images are combined using a transparency mask so that only the parts of the face we need are cut from the first image and inserted into the second image. As a result, we get a new image for person A.

The main purpose of this method is to increase the strength of the dataset. If there is more than one photo for a person in the dataset, then new photos can be generated to prevent overfitting. If for a certain person there is more than one photo with different types of the mouth area (mouth closed, mouth open, smiling, etc.), then you can create a new image for each existing image of this person, but with a modified mouth. The same examples can be given about different options for the eye area (closed, open, winking, etc.), as these two approaches can be combined, which as a result gives us a huge number of potential new images. Also, some datasets have a lot of variability in yaw face rotation angles. To eliminate this problem, we measured the rotation angle yaw for each face image in the dataset and impose a restriction that the rotation angles of the input images should differ no more than a certain value. It is also worth noting that combining images with different lighting parameters, facial expressions, etc. does not cause similar problems, but more often it helps, makes the image more diverse, keeping the face realistic, so we do not filter the images by these parameters, but only by the yaw angle of the face.

3.3   Background FaceMix

In this case, the input is an image for person A and a random image from a random person (Fig. 2, b). Next, a round transparency mask is created, the size of which is adjusted in such a way that as much of the face as possible and as little as everything else falls into it. After that, similarly to the previous method, the image with the face of person A and the random image of a random person are mixed pixel by pixel, applying a transparency mask to the first image. The result is a new image for person A, which contains the face of this person and various attributes from the second image, it can be hair, a hat, clothes, as well as a background, for example, a room or sky, rain or sun.

This method takes a face from the first image, and a background from the second, which is reflected in its name, and because of it, it can generate a very large number of images from one image simply by substituting different backgrounds. This way, you can create many images for each class, which can be useful for datasets with long tails. Modern datasets for face recognition, such as MS-Celeb-1M [7] and Glint360k [1], while being large, contain a lot of classes with a small number of images. This feature of face recognition datasets can lead to undesirable consequences. For example, in the process of training, images of a class with a large number of images will be fed to the network input much more often, and an image from a class with one image may only get into training a few times. Also, if there are many classes with one image, then the network can simply memorize the photo of each class and, as a result, will be overfitted. Using FaceMix, you can increase the number of images for all classes, including small classes, which will eliminate this unwanted property of datasets.

4   Experiments

In order to confirm the effectiveness of our method, we conducted experiments for each of the two algorithms: Intra-Class FaceMix and Background FaceMix, and their combination on popular face recognition benchmarks.

4.1   Implementation Details

Datasets. For training we used Glint360k-R [19] dataset with $353,658$ persons and $16,370,196$ images, which is a refined version of Glint360k [1] dataset. For testing, we used CFP-FP [18], AgeDB-30 [13], CALFW [35], CPLFW [34], XQLFW [12], SLLFW [33], RFW [28] and MegaFace (original [11] and cleaned [19] versions) datasets.

- CFP-FP [18] and CPLFW [34] datasets show how resistant the model is to face rotation.
- AgeDB-30 [13] and CALFW [35] datasets contain face images of different ages for the same person. They can show how resistant the model is to age differences.
- XQLFW [12] is a dataset to measure the robustness to image quality.
- SLLFW [33] is a dataset with a faces of similar-looking people.
- RFW [28] is a dataset to perform face recognition on different races.
- MegaFace [11] is a large-scale face recognition benchmark, containing more than a million images.

Training Settings. For small-scale experiments we used ResNet-34 [8] as an architecture. We used CosFace [27] with a margin of 0.4 and a scale of 64 as a loss function. For training with a large number of classes we used Prototype Memory [19] with embedding size $D = 256$, memory size $M = 100,000$ and refresh ratio $r = 0.2$. During training, a batch size of 512 was used. For the experiments we used composite mini-batch [22] with 384 images sampled by group-based iterate-and-shuffle [19] with $k = 4$, and 128 images sampled using a combination of Multi-Doppelganger Mining [19] and Hardness-aware example mining [19], with $h = 0.25$, 2 classes sampled at random and 30 classes sampled using doppelgangers. The initial learning rate was set to 0.1 and decreased 10 times at 100K, 200K, 250K, and 275K iterations, in total training was performed for 300K iterations.

To perform large-scale experiments, we used ResNet-100 [8] network, pre-trained on Glint-360k dataset, and fine-tuned for 200k more iterations with different augmentation options on Glint-360k-R dataset using batch size of 512, learning rate starting with 0.01 and decreased 10 times at 50k, 125k and 175k iterations. We used CosFace with $m = 0.4$ and $s = 64$, Prototype Memory with $D = 512$, $M = 200,000$ and $r=0.2$, and composite mini-batch with the same parameters, as for smaller models.

We have averaged the results, achieved by networks at five last snapshots (260k, 270k, 280k, 290k and 300k iterations for ResNet-34 nets, and 180k, 185k, 190k, 195k, 200k - for ResNet-100 nets), to get the final results.

Table 1. Comparison on small datasets. † denotes the switching off FaceMix for last 50k iterations

Method	CFP-FP	AgeDB	CALFW	CPLFW	XQLFW	SLLFW
Baseline	98.41	97.77	95.63	93.27	89.9	98.86
FaceMix-B	98.5	97.9	95.66	93.2	89.95	**99.01**
FaceMix-B†	98.45	97.87	**95.71**	93.19	90	98.97
FaceMix-I	98.4	97.85	95.66	**93.3**	**90.43**	98.79
FaceMix-I†	**98.53**	97.85	95.58	**93.3**	90.26	98.86
FaceMix-BI	98.47	97.9	95.62	92.95	89.72	98.91
FaceMix-BI†	98.51	**97.94**	95.61	92.91	89.99	98.94

Table 2. Comparison on RFW dataset. † denotes the switching off FaceMix for last 50k iterations

Method	African	Asian	Indian	Caucasian	Average
Baseline	91.92	92.94	94.82	97.01	93.96
FaceMix-B	92.22	93.13	95.07	97.2	94.11
FaceMix-B†	92.13	93.18	94.92	**97.38**	**94.2**
FaceMix-I	91.95	93.24	95.12	96.94	94.1
FaceMix-I†	91.95	93.21	94.96	97.02	94.14
FaceMix-BI	**92.52**	**93.25**	95.09	96.9	94.13
FaceMix-BI†	92.51	93.06	**95.2**	96.99	94.12

Table 3. Comparison with different augmentations on RFW dataset. R-100 network is used for all models. † denotes the switching off FaceMix for last 25k iterations

Method	African	Asian	Indian	Caucasian	Average
Baseline	96.13	97.7	98.19	98.99	97.28
MixUp	97.28	97.71	98.09	98.99	97.75
CutMix	97.32	97.62	98.12	99.01	97.89
FaceMix-BI	**97.77**	**98.13**	98.33	**99.18**	**98.26**
FaceMix-BI†	97.72	97.99	**98.34**	99.05	98.23

Table 4. Comparison with different augmentations on MegaFace dataset. R-100 network is used for all models. † denotes the switching off FaceMix for last 25k iterations

Method	MF-ID	MF-ID (R)
Baseline	81.77	98.33
MixUp	81.54	98.33
CutMix	81.79	98.32
FaceMix-BI	81.75	98.41
FaceMix-BI†	**81.91**	**98.45**

4.2   Results

Experiments with Background FaceMix, Intra-class FaceMix, and their Combination. To perform the experiments with the proposed FaceMix variants, we used ResNet-34. To evaluate Background FaceMix (FaceMix-B), we added a 10% chance that the FaceMix-B will be applied (Fig. 2, b). The radius of the circle for the mask is 45, and the center is at coordinates (56, 56). For the case of Intra-Class FaceMix (FaceMix-I), we added a 10% chance that the Intra-Class FaceMix will be applied to the image (Fig. 2, a). The second image is selected to have the yaw rotation angle difference with the face on the first image to be no more than $7°$ (we pre-calculated the yaw rotation angles for all faces in the dataset using [15]). The radius of the circles for the mask is 20, and their centers are located at coordinates (40, 50), (70, 50), (55, 90). To perform the experiment with a combination of the two FaceMix variants (FaceMix-BI), we used a 5% probability for each of them to be applied. We also tried to turn FaceMix off for the last 50k iterations to feed the network with only "real" images. The results in Tables 1 and 2 show that FaceMix variants improve the performance of face recognition models. They help to get more diverse, large and balanced datasets and mitigate the problems of long tails and insufficient numbers of images for some classes.

Comparison with Different Augmentation Variants. To compare proposed methods with the state-of-the-art augmentations, we also performed experiments with large networks and used MegaFace Identification benchmark. For this experiment we used ResNet-100, fine-tuned with 10% probability of either MixUp [31] or CutMix [29] augmentation. We performed mixing using samples only of the same class as it is impossible to "mix" the targets of different classes correctly using CosFace loss (because of margin subtraction operation). We used $\alpha = 0.2$ for MixUp and $\alpha = 1$ for CutMix. The baseline is a ResNet-100 without augmentation. We used FaceMix-BI with probability of 5% for each FaceMix variant in this experiment. The results are presented in Tables 3 and 4. FaceMix outperformed all other methods. This can be explained by the fact that MixUp and CutMix are ignorant to the structure of the human face and generate unrealistic samples.

5   Conclusion

Recently, neural networks have made a big step forward in solving problems in the field of machine vision and in face recognition in particular. With the development of neural networks, they need more data, and although datasets are also growing, they have various problems, such as a small number of images or long tails. To address these issues, we have presented a generic FaceMix augmentation method that helps to train better neural networks for a face recognition task. We also introduced two operating modes of the method: Background FaceMix, which helps in case of large number of identities, and Intra-Class FaceMix, which is useful for improving the diversity of face image variations in dataset. These methods are not difficult to implement, since they consist of the simplest transformations over images and add only insignificant computational costs. We have conducted experiments on popular face recognition benchmarks. FaceMix methods show their superiority in a large-scale MegaFace Identification benchmark, which confirms their effectiveness in improving the quality of face recognition models.

References

1. An, X., et al.: Partial FC: training 10 million identities on a single machine. arXiv preprint arXiv:2010.05222 (2020)
2. Buslaev, A., Iglovikov, V.I., Khvedchenya, E., Parinov, A., Druzhinin, M., Kalinin, A.A.: Albumentations: fast and flexible image augmentations. Information **11**(2), 125 (2020)
3. Conway, D., Simon, L., Lechervy, A., Jurie, F.: Training face verification models from generated face identity data. arXiv preprint arXiv:2108.00800 (2021)
4. Deng, J., Guo, J., Xue, N., Zafeiriou, S.: ArcFace: additive angular margin loss for deep face recognition. In: CVPR, pp. 4690–4699 (2019)
5. DeVries, T., Taylor, G.W.: Improved regularization of convolutional neural networks with cutout. arXiv preprint arXiv:1708.04552 (2017)
6. Goodfellow, I.J., et al.: Generative adversarial networks. arXiv preprint arXiv:1406.2661 (2014)
7. Guo, Y., Zhang, L., Hu, Y., He, X., Gao, J.: MS-Celeb-1M: a dataset and benchmark for large-scale face recognition. In: Leibe, B., Matas, J., Sebe, N., Welling, M. (eds.) ECCV 2016. LNCS, vol. 9907, pp. 87–102. Springer, Cham (2016). https://doi.org/10.1007/978-3-319-46487-9_6
8. He, K., Zhang, X., Ren, S., Sun, J.: Deep residual learning for image recognition. In: CVPR (2016)
9. Hu, G., Peng, X., Yang, Y., Hospedales, T.M., Verbeek, J.: Frankenstein: learning deep face representations using small data. IEEE Trans. Image Process. **27**(1), 293–303 (2017)
10. Inoue, H.: Data augmentation by pairing samples for images classification. arXiv preprint arXiv:1801.02929 (2018)
11. Kemelmacher-Shlizerman, I., Seitz, S.M., Miller, D., Brossard, E.: The MegaFace benchmark: 1 million faces for recognition at scale. In: CVPR, pp. 4873–4882 (2016)
12. Knoche, M., Hormann, S., Rigoll, G.: Cross-quality LFW: a database for analyzing cross-resolution image face recognition in unconstrained environments. In: FG (2021)
13. Moschoglou, S., Papaioannou, A., Sagonas, C., Deng, J., Kotsia, I., Zafeiriou, S.: AgeDB: the first manually collected, in-the-wild age database. In: CVPR Workshops (2017)
14. Park, S., Hong, Y., Heo, B., Yun, S., Choi, J.Y.: The majority can help the minority: context-rich minority oversampling for long-tailed classification. In: CVPR, pp. 6887–6896 (2022)
15. Ruiz, N., Chong, E., Rehg, J.M.: Fine-grained head pose estimation without keypoints. In: CVPR Workshops, pp. 2074–2083 (2018)
16. Sandfort, V., Yan, K., Pickhardt, P.J., Summers, R.M.: Data augmentation using generative adversarial networks (CycleGAN) to improve generalizability in CT segmentation tasks. Sci. Rep. **9**(1), 1–9 (2019)
17. Schroff, F., Kalenichenko, D., Philbin, J.: FaceNet: a unified embedding for face recognition and clustering. In: CVPR, pp. 815–823 (2015)
18. Sengupta, S., Chen, J.C., Castillo, C., Patel, V.M., Chellappa, R., Jacobs, D.W.: Frontal to profile face verification in the wild. In: WACV (2016)
19. Smirnov, E., Garaev, N., Galyuk, V., Lukyanets, E.: Prototype memory for large-scale face representation learning. IEEE Access **10**, 12031–12046 (2022)
20. Smirnov, E., Melnikov, A., Novoselov, S., Luckyanets, E., Lavrentyeva, G.: Doppelganger mining for face representation learning. In: ICCV Workshops, pp. 1916–1923 (2017)
21. Smirnov, E., Melnikov, A., Oleinik, A., Ivanova, E., Kalinovskiy, I., Luckyanets, E.: Hard example mining with auxiliary embeddings. In: CVPR Workshops, pp. 37–46 (2018)
22. Smirnov, E., et al.: Face representation learning using composite mini-batches. In: ICCV Workshops, pp. 551–559 (2019)

23. Summers, C., Dinneen, M.J.: Improved mixed-example data augmentation. In: WACV, pp. 1262–1270. IEEE (2019)
24. Sun, R., Masson, C., Hénaff, G., Thome, N., Cord, M.: Swapping semantic contents for mixing images. arXiv preprint arXiv:2205.10158 (2022)
25. Szegedy, C., Toshev, A., Erhan, D.: Deep neural networks for object detection. In: NeurIPS, vol. 26 (2013)
26. Takahashi, R., Matsubara, T., Uehara, K.: Data augmentation using random image cropping and patching for deep CNNs. IEEE Trans. Circuits Syst. Video Technol. **30**(9), 2917–2931 (2019)
27. Wang, H., et al.: CosFace: large margin cosine loss for deep face recognition. In: CVPR (2018)
28. Wang, M., Deng, W., Hu, J., Tao, X., Huang, Y.: Racial faces in the wild: reducing racial bias by information maximization adaptation network. In: ICCV (2019)
29. Yun, S., Han, D., Oh, S.J., Chun, S., Choe, J., Yoo, Y.: CutMix: regularization strategy to train strong classifiers with localizable features. In: ICCV, pp. 6023–6032 (2019)
30. Zhang, H., et al.: Context encoding for semantic segmentation. In: CVPR, pp. 7151–7160 (2018)
31. Zhang, H., Cisse, M., Dauphin, Y.N., Lopez-Paz, D.: Mixup: beyond empirical risk minimization. arXiv preprint arXiv:1710.09412 (2017)
32. Zhang, L., Huang, S., Liu, W.: Intra-class part swapping for fine-grained image classification. In: WACV, pp. 3208–3217 (2021)
33. Zhang, N., Deng, W.: Fine-grained LFW database. In: ICB (2016)
34. Zheng, T., Deng, W.: Cross-pose LFW: a database for studying cross-pose face recognition in unconstrained environments. Beijing University of Posts and Telecommunications, Technical report (2018)
35. Zheng, T., Deng, W., Hu, J.: Cross-age LFW: a database for studying cross-age face recognition in unconstrained environments. arXiv preprint arXiv:1708.08197 (2017)
36. Zheng, Y., Chen, Q., Zhang, Y.: Deep learning and its new progress in object and behavior recognition. J. Image Graph. **19**(2), 175–184 (2014)

Permissioned Blockchain-Based XGBoost for Multi Banks Fraud Detection

Septiviana Savitri Asrori[1], Lihua Wang[2], and Seiichi Ozawa[1(✉)]

[1] Graduate School of Engineering, Kobe University, Kobe, Japan
`213t265t@stu.kobe-u.ac.jp`, `ozawasei@kobe-u.ac.jp`
[2] National Institute of Information and Communications Technology, Koganei,
Tokyo, Japan
`lh-wang@nict.go.jp`

Abstract. Fraud detection is one of the financial institution problems which can utilize Machine Learning (ML). However, the fraud activity is hard to detect since the occurrence is relatively low compared to the actual transaction. Several banks can collaborate to gather more fraudulent transactions from their data. However, the collaboration can cause data leakage from each bank, where the customer data should be confidential. Decentralized ML is one of the approaches to tackle the privacy-preserving aspect. This work proposed a fully decentralized environment using a permissioned blockchain to detect multiple banks' fraud. The training process utilizes a continual eXtreme Gradient Boosting (XGBoost) model. We provided the architecture of blockchain implementation for multiple banks, where it is conducted as batch and streaming data processing. As we compared our approach with the centralized, individual, and federated GBDT models, it maintains a good prediction performance and fulfills the environment of a fully distributed system.

Keywords: Blockchain · XGBoost · Decentralized model

1 Introduction

Fraud activity usually happens among the bank transaction data. However, fraud detection usually comes with imbalanced data, where the proportion of fraud data is low compared to non-fraud data. This characteristic makes it difficult to find the pattern due to the lack of data. Therefore, some banks collaborate for ML training to gather more data and generate an aggregated models which can give better prediction results. In [1], the Secure Aggregation method is implemented for practical Privacy-Preserving Machine Learning (PPML).

The ML training process can generally be divided into centralized and decentralized. Centralized learning will collect raw data from the data owner and

This work was supported in part by Japan Society for the Promotion of Science (JSPS) KAKENHI under Grant JP20K11826, and in part by Japan Science and Technology Agency (JST) AIP Accelerated Program under Grant JPMJCR22U5.

perform the training on a central server. In comparison, decentralized training achieves the data owner's local training process and learns a shared model from a locally trained model. Decentralized training has various approaches, such as Federated Learning (FL) [4] and Swarm Learning (SL) [8].

FL [4] uses a local server for training and a central server for model aggregation. In [10], FL setting is implemented using the XGBoost algorithm. The efficient FL for Gradient-Based Decision Tree (GBDT) is introduced in [9]. However, FL needs the use of the central server to update the aggregated models. On the other hand, SL provides a fully decentralized model, where raw data and aggregated models are generated at the edge using the blockchain system. For the current implementation, SL only supports parametric ML algorithms, such as Neural Networks (NN).

Extending the use of blockchain in SL, our proposed method also utilized blockchain-based machine learning. Blockchain aims to achieve and maintain integrity in distributed systems [3]. That allows the blockchain to keep its privacy-preserving and put all the peers into equal positions. Thus, the utilization of blockchain in federated learning has its opportunities and challenges, as described in [6]. There are two types of blockchain: permissionless (also called public) and permissioned (also called private). Permissioned blockchains limit the access to the authorized nodes, and each node already trusts each other [11]. In [5], permissioned blockchain is also used for financial institution data by generating a Quorum blockchain network.

This paper's main contribution is to develop a fully decentralized XGBoost model in a permissioned blockchain network. Each data owner will train the local model and update the global model stored in the blockchain. The global model consists of tree representation, hyperparameter, and configuration for the XGBoost model. Since it's written in the blockchain network, each data owner can have an equal right to access the global model, and we can eliminate the use of the central server. To the best of our knowledge, this is the first implementation of XGBoost on a permissioned blockchain network that only utilized the shared model without involving the raw data in the blockchain network.

The organization of this paper explains as follows. Section 2 covers the problem statement. Then, Sect. 3 describes the related works on XGBoost, Blockchain-based Machine Learning, and the FL GBDT application. Section 4 represents the architecture of permissioned blockchain used for multiple banks. It also explains the overall proposed method. Experiments and Conclusions are described Sect. 5 and Sect. 6 respectively.

2   Problem Statement

In this section, we adopt the concern from real-life applications. We focus on fraud detection for transaction data of bank customers with several problems and limitations during the training process. First, the number of fraud data is small in the overall transaction data, which leads to an imbalanced data problem. Due to the increasing number of fraud data, several banks can collaborate and generate aggregated models, which expects to give a better model. However, it leads to the second problem: each data owner must secure their dataset and keep it confidential. Although each bank already trusts the other, sharing the raw dataset among banks is prohibited since it violates the customer's privacy. Thus, we need to provide a system where data owners can still conduct their training process locally. Still, the collaborated banks can share the resulting aggregated models among the data owners to increase the prediction performance.

3 Related Works

3.1 XGBoost

XGBoost [2] is an implementation of the Gradient Boosted Decision Tree (GBDT) algorithm that is claimed as an efficient and scalable method. It is based on the function approximation of a loss function and utilizes regularization. For a dataset with n samples and m features $D = \{(\mathbf{x}_i, y_i)\}(|D| = n, \mathbf{x}_i \in \mathbb{R}^m), y_i \in \mathbb{R}$ with $\hat{y}_i^{(t-1)}$ as the prediction calculated from previous trees, suppose that the t-th decision tree f_t is constructed to minimize the objective function $\mathcal{L}^t$ as

$$\mathcal{L}^t = \sum_{i=1}^{n} l(y_i, \hat{y}_i^{(t-1)} + f_t(x_i)) + \Omega f_t. \tag{1}$$

The objective function consists of the training loss $\sum_{i=1}^{n} l(y_i, \hat{y}_i^{(t-1)} + f_t(x_i))$, which measures how well the model fits into the training data, and regularization Ωf_t, which determines the complexity of the trees. In this work, we utilized XGBoost[1] due to the ability to control over-fitting and capability for continual learning. XGBoost allows continual learning by providing the current model as the parameter for the following training process.

Table 1. Information of Datasets

Notation	Definition
n	Number of DataOwners
BN	Blockchain Network
$BN.transact(addr, data)$	Add transaction of $data$ in BN from $addr$
D	Overall Training Set
$\{D^{(1)}, \ldots, D^{(n)}\}$	Training Set for $\{DataOwner_{1,\ldots,n}\}$
$\{D^j_{chunk_i}\}$	j-th Data Chunk from $DataOwner_i$
$\{(\mathbf{x}_1^{(i)}, y_1^{(i)}), \ldots, (\mathbf{x}_k^{(i)}, y_k^{(i)})\}$	Data training on each $DataOwner_i$
M^i_{global}	i-th Global Model
M^i_{local}	i-th Local Model from $DataOwner_i$
$M^j_{local_i}$	j-th Local Model from $DataOwner_i$
$|Dataset|$	Number of element in $Dataset$

---

[1] https://github.com/dmlc/xgboost.

3.2 Federated Learning GBDT

Maintaining the privacy between data owners is also done in FL-XGBoost [10] and eFL-Boost [9], it has S as the central server and n number of *DataOwner*. The *DataOwners* are denoted as $U = \{u_1, u_2, u_3, \ldots, u_n\}$. $u_{tr} \in U$ has a dataset that refers to the *DataOwner* that consecutively trains the model. This work introduces FL-XGBoost-G (FL-XGBoost with the absolute average gradient of the loss function applied). Here the selection of u_{tr} depends on the absolute average gradient of loss function g.

As the improvement of FL-XGBoost, eFL-Boost allows the federated GBDT, which minimizes accuracy loss, communication costs, and information leakage. In FL GBDT approaches, the central server is still utilized to control the aggregation mechanism among data owners. Later in Sect. 5, since we have a similar setting with these approaches, we compare our proposed method with FL-XGBoost-G and eFL-Boost.

3.3 Evaluation Metric: F1 Score

F1 score is an evaluation metric used to measure the performance of a model, especially for the binary classification model. It's a harmonic mean from a combination of precision and recall. Let True Positive (TP) denote the actual and predicted value is True, False Positive (FP) denote the actual value is False, but the predicted value is True, False Negative (FN) denote the actual value is True, but the predicted value is False. Given the equation as

$$Precision = \frac{TP}{(TP + FP)}, \tag{2}$$

$$Recall = \frac{TP}{(TP + FN)}, \tag{3}$$

$$F1score = 2 \times \frac{Precision \times Recall}{(Precision + Recall)}. \tag{4}$$

Equation (4) is used for our proposed method, and this evaluation will explain further in the next section.

4 Proposed Method

4.1 Development Architecture

We need to develop the blockchain network and all related technology to enable the blockchain system. Figure 1a explains the details of blockchain architecture. We use the Ethereum[2], an open-source blockchain that enables both permissionless and permissioned blockchain for developers. For the experiment, the permissioned blockchain is generated using software called Ganache[3], which allows the quickstart of a personal Ethereum blockchain.

---

[2] https://ethereum.org/en/.
[3] https://trufflesuite.com/ganache/.

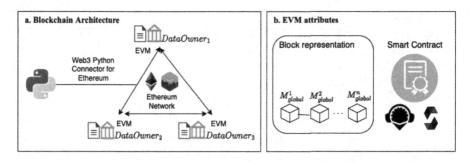

Fig. 1. Development Architecture for the Blockchain System

In the Ethereum network, the network member is called an Ethereum Virtual Machine (EVM). In our case, EVM represents each *DataOwner* who participates in the collaborative training process. Each *DataOwner* has its dataset in its local machine. To communicate with each EVM in the Ethereum network, we can utilize Web3[4], a Python3 [7] library connector to the Ethereum network. This architecture allows us to connect with the Ethereum network and do some operations inside the blockchain, such as creating a transaction and deploying the smart contract.

A smart contract is a fundamental rule that governs all activities inside the blockchain network. Solidity[5] is a high-level language for implementing smart contracts. Solidity can be compiled on Remix[6]. As illustrated in Fig. 1b, each EVM has a copy of the block representation and smart contract. When any EVM tries to make a new transaction or update the smart contract, the changes are recorded among other EVMs connected to the blockchain system. This paradigm of blockchain allows global model sharing in a fully distributed system.

4.2   Blockchain-Based XGBoost Training

To achieve both the privacy aspect and the sufficient training data, we proposed the permissioned blockchain-based application on the continual XGBoost algorithm. Table 1 explains all the notations used in the following proposed method. As we already defined the problem statement, our proposed method provides two different treatments depending on the arrival time of the data. We divided the scenario into *batch processing* and *stream processing*. The batch data is collected as bulk data $D^{(i)}$ for $DataOwner_i$ in a specific time frame. This approach can simplify the training process since we have all the data at the beginning of the training process. While the stream data $D^{(i)}$ consists of a small chunk of data $D^{j}_{chunk_i}$ that arrives from time to time.

---

[4] https://web3py.readthedocs.io/en/stable/.

[5] https://docs.soliditylang.org/en/v0.8.11/.

[6] https://remix.ethereum.org/.

Algorithm 1. Batch Data Training

Input: Training set D from n $DataOwners$

Output: M_{global}^n aggregated models

1: BN for n $DataOwners$

2: Training set $D = \{D^{(1)}, \ldots, D^{(n)}\}$

3: where $D^{(i)} = \{(\mathbf{x}_1^{(i)}, y_1^{(i)}), \ldots, (\mathbf{x}_k^{(i)}, y_k^{(i)})\}$

4: $M_{global}^0 = 0$

5: **for** $i = 1, 2, \ldots, n$ **do**

6:     $M_{local}^i \leftarrow \mathsf{train}(D^{(i)})$ from $DataOwner_i$ as continuation of M_{global}^{i-1}

7:     $M_{global}^i \leftarrow M_{local}^i$

8:     $BN.transact(DataOwner_i, M_{global}^i)$

9: **end for**

Since we proposed a collaborative training process, note that M_{local} refers to the local model trained at the edge, and M_{global} refers to the global model shared among data owners in the blockchain network. The information shared in a blockchain network is M_{global} and evaluation properties. There is no raw data sharing among the $DataOwners$.

Batch Data Training. As described in Algorithm 1, all the $DataOwners$ join the BN. Then, for the first training process, the M_{global}^1 refers to the M_{local}^1 and $DataOwner_1$ transacts M_{global}^1 to BN. Next, the M_{global}^2 is the continuation from the M_{global}^1 and M_{local}^2, make transaction to BN and so on. Using this method, each data owner has an equal position to affect the global model.

Stream Data Training. Describe in Algorithm 2, after joining the BN, sequentially $DataOwner_i$ transact the $M_{local_i}^j$ into the BN. The shared $M_{local_i}^j$ is saved locally by each $DataOwner_k$ to calculate prediction result for $M_{local_i}^j$ using $D_{chunk-test_k}^j$. The $EvalSet_k$ consists of $F1score$ as calculated in Eq. (4) and $|D_{chunk-test_k}^j|$.

Furthermore, we define Eq. (5):

$$SharedF1(M) = \frac{\sum_{k=1}^n w_k F1score_k(M)}{\sum_{k=1}^n |D_{chunk-test_k}^j|}, \tag{5}$$

where w_k refers to the proportion of the chunk data in overall data, i.e. $|D_{chunk-test_k}^j| / \sum_{k=1}^n |D_{chunk-test_k}^j|$. The equation gives the weighted F1 score from $M_{local_i}^j$ called as $SharedF1(M_{local_i}^j)$. This aggregation calculation is due to the model evaluation's fairness yet keeps the privacy of the raw data from each data owner. Once $SharedF1$ is calculated from all $M_{local_i}^j$, the $DataOwner$ with the highest $SharedF1$ is selected as $DataOwner_{best}$. The selection of $SharedF1$ in streaming training will eliminate the training set that does not increase the aggregated models performance. The $DataOwner_{best}$ then makes the transaction in BN by updating the M_{global}^j. Although it can be an infinite process in a

---

Algorithm 2. Stream Data Training for the j-th chunk

    Input: Training set D from n $DataOwners$
    Output: $M_{global}^{j\_last}$ aggregated models
1: BN for n $DataOwners$
2: Training set $D = \{D^{(1)}, \ldots, D^{(n)}\}$
3: where $D^{(i)} = \{(\mathbf{x}_1^{(i)}, y_1^{(i)}), \ldots, (\mathbf{x}_k^{(i)}, y_k^{(i)})\}$, and k for each $DataOwner_i$ may differ
4: $M_{global}^0 = 0$
5: **while** ∞ **do**
6:     $D_{chunk_i}^j = \{(\mathbf{x}_1^{(i)}, y_1^{(i)}), \ldots, (\mathbf{x}_l^{(i)}, y_l^{(i)})\}$ split into $D_{chunk-train_i}^j$, $D_{chunk-test_i}^j$
7:     **for** $i = 1, 2, \ldots, n$ **do**
8:         $M_{local_i}^j \leftarrow$ train$(D_{chunk-train_i}^j)$ from $DataOwner_i$ as continuation of M_{global}^{j-1}
9:         BN.$transact(DataOwner_i, M_{local_i}^j)$
10:         **for** $k = 1, 2, \ldots, n$ **do**
11:             $EvalSet_k \leftarrow$ test$(M_{local_i}^j, D_{chunk-test_k}^j)$
12:             BN.$transact(DataOwner_k, EvalSet_k)$
13:         **end for**
14:         $SharedF1(M_{local_i}^j) \leftarrow$ Calculate Equation (5)
15:     **end for**
16:     $DataOwner_{best} \leftarrow \underset{i \in \{1, \ldots, n\}}{\arg\max}\ (SharedF1[M_{local_i}^j])$
17:     $M_{global}^j \leftarrow$ train$(D_{chunk_{best}}^j)$ from $DataOwner_{best}$ as continuation of M_{global}^{j-1}
18:     BN.$transact(DataOwner_{best}, M_{global}^j)$
19: **end while**

---

real-life setting, in this experiment, we assume the final global model as $M_{global}^{j\_last}$, where j refers to the number of iterations until we proceed with all the training sets D. The calculation of $SharedF1$ is only proposed in Stream Data Training.

5 Experiment

5.1 Experiment Setting

As illustrated in Fig. 2, the experiment is divided into four mechanisms. Figure 2a represents centralized training (**Central**), where all the datasets and training processes are centralized in a central server. Figure 2b illustrates when each data owner performs an individual training (**Idv**) process relying on their dataset. Then, Fig. 2c is conducted in a federated setting, **FLXGB-G** [10] and **eFL-B** [9]. Finally, Fig. 2d is our proposed method where the training process is fully decentralized under a blockchain setting. We proposed **Batch** for Algorithm 1 Batch Data Training, and **Stream** for Algorithm 2 Stream Data Training.

For the experiment, we use a public dataset for fraud detection: **Credit**[7]. It consists of 284805 rows of data with 30 features. The proportion of fraud data is 0.0017 (492 out of 284805). In addition, we utilized the hyperparameters of XGBoost, such as max_depth, learning_rate, subsample, colsample_bytree, colsample_bylevel, min_child_weight, gamma, reg_lambda, and n_estimators.

---

[7] https://www.kaggle.com/mlg-ulb/creditcardfraud.

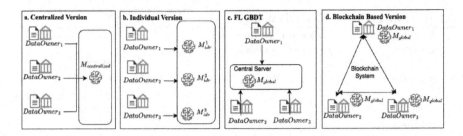

Fig. 2. Experiment setting from different perspective

Table 2. Experiment I Result

F1 score					
Central	Idv	FLXGB-G [10]	eFL-B [9]	Batch	Stream
0.853	0.818 ± 0.016	0.844 ± 0.00	0.850 ± 0.00	**0.861 ± 0.007**	0.850

5.2 Experiment Result

We provide three experiments to investigate the proposed method's performance. First, Experiment I compares the result for all mechanisms described in Fig. 2. We split the dataset into a 70:30 training and testing set. Although the testing data for each implementation are the same, the **Central** mechanism trains the data all at once, while others divide the data into n *DataOwners*. For the proposed method, we assume the number of data owners $n = 3$. The experiment result is shown in Table 2.

Experiment II investigates the prediction performance when the *Data Owners* increase. Here, we provide the experiment result for $n = 3$, $n = 5$ and $n = 10$. In this experimental setting, we assume each data owner has the same number of data. Table 3 shows the Experiment II result.

In addition, Experiment III is conducted to investigate the performance when the amount of data is imbalanced among data owners. This setting represents the real-life implementation where each *DataOwner* can provide a different number of data. We follow the proportion conducted in [9], which are 1:1:1, 8:1:1, and 6:2:2. The 8:1:1 shows that one data owner has 80% of the dataset, while the other only has 10% each. Table 4 shows the Experiment III result.

5.3 Experiment Analysis

We can see from the experiment result that both **Batch** and **Stream** methods can maintain a good performance compared to other mechanisms. Furthermore, experiments I show that the proposed method can outperform the **Idv** since **Idv** only relies on each *DataOwner* dataset. In contrast, our proposed method continually utilizes all the datasets during the training process.

Table 3. Experiment II Result

DataOwner	**FLXGB-G** [10]	**eFL-B** [9]	**Batch**	**Stream**
$n = 3$	0.799 ± 0.00	0.805 ± 0.00	$\mathbf{0.810 \pm 0.01}$	0.801
$n = 5$	0.797 ± 0.00	0.832 ± 0.00	$\mathbf{0.836 \pm 0.02}$	0.833
$n = 10$	0.825 ± 0.00	$\mathbf{0.843 \pm 0.00}$	0.842 ± 0.00	0.838

Table 4. Experiment III Result

Data Proportion	Idv	**FLXGB-G** [10]	**eFL-B** [9]	**Batch**	**Stream**
1:1:1	0.818 ± 0.016	0.844 ± 0.00	0.850 ± 0.00	$\mathbf{0.861 \pm 0.07}$	0.850 ± 0.00
8:1:1	0.849 ± 0.02	0.850 ± 0.00	$\mathbf{0.856 \pm 0.00}$	0.854 ± 0.02	0.849 ± 0.00
6:2:2	0.830 ± 0.00	0.843 ± 0.00	0.848 ± 0.00	$\mathbf{0.850 \pm 0.00}$	0.844 ± 0.00

Experiment II shows how adding data owners can improve prediction performance. For all algorithms compared, increasing the number of *DataOwners* can improve the prediction performance. **Batch** method shows slightly better performance than other methods. For Experiment III, the performance remains stable between proportions. Still, the blockchain-based method can maintain the prediction performance in a fully distributed manner. We need to mention that what we aim for in this proposed method is a secured system between *DataOwners*, so the prediction performance is not our primary goal.

From security analysis, the proposed blockchain method is secured from the attacker. As permissioned blockchain needs authentication of each member inside the network, it's hard for the attacker to enter the network. In the worst scenario, when any of the blockchain nodes is hacked by the attacker, the information stored in the blockchain network only consists of a shared global model (tree representation and parameter), evaluation result for shared F1 score calculation. Therefore, the attacker doesn't have access to the raw data of each data owner and is only left with the abstract tree representation.

6 Conclusion

This work introduces a fully distributed continual XGBoost approach utilizing the permissioned Ethereum blockchain network. The proposed method is designated for a secured training process among data owners, where each member is trustable and would like to generate the shared global model. However, each party still needs to secure the raw data due to privacy-preserving. To maintain data privacy, our blockchain system manages to store only the tree representation and evaluation attributes so that all the raw datasets are secured in the data owner's local machines. The proposed method provided the training process for both batch and stream depending on the characteristic of the dataset.

Our proposed method is still limited to XGBoost implementation in a blockchain setting. Investigating another algorithm, such as neural network-based or tree-based algorithms in a blockchain network, can be considered future work. In addition, exploring another aggregation mechanism to improve the global model is also an exciting topic. While we propose the Shared F1-score, we can extend it to another evaluation method later. Then, the additional dataset for the experiment can help measure the performance of the proposed method.

References

1. Bonawitz, K., et al.: Practical secure aggregation for privacy-preserving machine learning. In: proceedings of the 2017 ACM SIGSAC Conference on Computer and Communications Security, pp. 1175–1191 (2017)
2. Chen, T., Guestrin, C.: XGBoost: a scalable tree boosting system. In: Proceedings of the 22nd ACM SIGKDD International Conference on Knowledge Discovery and Data Mining, pp. 785–794 (2016)
3. Drescher, D.: Blockchain Basics: A Non-technical Introduction in 25 Steps, 1st edn. Apress, USA (2017)
4. Konecný, J., McMahan, H.B., Ramage, D., Richtárik, P.: Federated optimization: distributed machine learning for on-device intelligence. arXiv preprint arXiv:1610.02527 (2016)
5. Mazzoni, M., Corradi, A., Di Nicola, V.: Performance evaluation of permissioned blockchains for financial applications: the consensys quorum case study. Blockchain: Res. Appl. **3**(1), 100026 (2022)
6. Nguyen, D.C., et al.: Federated learning meets blockchain in edge computing: Opportunities and challenges. IEEE Internet Things J. **8**, 12806–12825 (2021)
7. Van Rossum, G., Drake Jr., F.L.: Python reference manual. Centrum voor Wiskunde en Informatica Amsterdam (1995)
8. Warnat-Herresthal, S., et al.: Swarm learning for decentralized and confidential clinical machine learning. Nature **594**(7862), 265–270 (2021)
9. Yamamoto, F., Ozawa, S., Wang, L.: EFL-boost: efficient federated learning for gradient boosting decision trees. IEEE Access **10**, 43954–43963 (2022)
10. Yamamoto, F., Wang, L., Ozawa, S.: New approaches to federated XGBoost learning for privacy-preserving data analysis. In: Yang, H., Pasupa, K., Leung, A.C.-S., Kwok, J.T., Chan, J.H., King, I. (eds.) ICONIP 2020. LNCS, vol. 12533, pp. 558–569. Springer, Cham (2020). https://doi.org/10.1007/978-3-030-63833-7_47
11. Yang, R., et al.: Public and private blockchain in construction business process and information integration. Autom. Constr. **118** (2020). https://doi.org/10.1016/j.autcon.2020.103276, https://www.sciencedirect.com/science/article/pii/S0926580520301886

Rethinking Image Inpainting with Attention Feature Fusion

Shuyi Qu[1], Kaizhu Huang[2(✉)], Qiufeng Wang[1(✉)], and Bin Dong[3]

[1] Xi'an Jiaotong-Liverpool University, Suzhou 215213, China
{shuyi.qu,qiufeng.wang}@xjtlu.edu.cn
[2] Duke Kunshan University, Kunshan 215316, China
kaizhu.huang@dukekunshan.edu.cn
[3] Ricoh Software Research Center, Beijing 100044, China
bin.dong@cn.ricoh.com

Abstract. Recent image inpainting models have archived significant progress through learning from large-scale data. However, restoring images under complicated scenarios (e.g. large masks or complex textures) remains challenging. We argue that the inadequate learning of global structure and local texture could lead to the artifacts and blur of current models. Inspired by feature fusion methods, we utilize Attention Feature Fusion (AFF) to better aggregate the different levels of features within our inpainting model from two perspectives. 1) We insert AFF through skip connections to pass long-distance textures to late semantics; 2) Our modified multi-dilated blocks with AFF residual could fuse features in different receptive fields. Both strategies aim to strengthen the texture and structure aggregation and reduce the inconsistency of semantics during learning. We show quantitatively and qualitatively that our approach outperforms current methods on benchmark datasets.

Keywords: Image inpainting · GAN · Feature fusion

1 Introduction

Image inpainting targets refilling an image's masked area without visual inconsistency. It has been a fundamental task in computer vision due to its wide applications in object removal, image editing/stitching, manipulation. Before the extensive adoption of deep learning, traditional approaches harness low-level image features that could handle images with simple geometric structures while failing in complex scenarios in the real world. Modern deep generative models are trained on large-scale datasets in an adversarial training fashion. Therefore, they better understand high-level image statistics and can generate plausible patches on complicated cases.

Recently, deep generative inpainting models are usually built in a two-stage manner. On the first stage, they either generate coarse output [25] or other auxiliary guidance (e.g. edge [17], semantic map [13], smoothed image [19]). These models borrow previously learned information in the second stage and generate the final refined outputs. Two-stage methods can separate the learning of global structure and local texture. Thus, these methods advance the single encoder-decoder architectures to more robust and powerful feature representative.

© The Author(s), under exclusive license to Springer Nature Switzerland AG 2023
M. Tanveer et al. (Eds.): ICONIP 2022, LNCS 13625, pp. 693–704, 2023.
https://doi.org/10.1007/978-3-031-30111-7_58

Apart from two-stage approaches, another group of methods tries to model the structure and texture within feature space. Some representative work build structure & texture branches [15], AOT block [27] and multi-scale attention module [23] on vanilla encoder-decoder features. Instead of progressively learning the structure and texture of different stages or pyramids, these works build novel modules to enhance the learned features with more accurate structure and texture.

However, these two categories of methods either introduce external information or complex modules which consume more computation. This paper investigates the efficient representative enhancement from a feature fusion perspective. The Attention Feature Fusion (AFF) [5] has recently shown advance in fusing different levels of features with inconsistent semantics. Motivated by this observation, we build our model on top of the existing two-stage method and adopt the AFF block to better model the structure and texture information in two parts. First, we build Skip Connection Attention Fusion (SCAF) which uses AFF to fuse the texture information from the encoder with semantics from the decoder. The aggregation is then passed to the decoder through skip connections. Second, to further strengthen the vanilla dilated convolutions architecture which is vital to capture the global structure and multi-scale objects on inpainting, we build Dilated Convolution Residual (DCR), which could better fuse the features with multi-receptive fields. With the help of SCAF and DCR, our model can better understand global and local information with an image and could alleviate the blur and artifacts of the current model. Our contribution can be concluded as:

- To alleviate the blurry results and artifacts caused by the lack of understanding of image global and local information, we propose a feature fusion-inspired method;
- We apply Attention Feature Fusion to better model the vanilla two-stage encoder-decoder features through Skip Connection Attention Fusion and Dilated Convolution Residual;
- The quantitative and qualitative results on Places2 and Celeba-HQ have shown that our model outperforms other state-of-the-arts.

2   Related Works

As one of the fundamental tasks of computer vision, the image inpainting technique refers to a model that synthesis pixels of the corrupt area in an incomplete image, with the capacity to restore the image with plausible textures which visually align with the rest of the pixels. The modern image inpainting can be categorized into two parts: traditional methods and deep learning methods. We discuss these two categories and the related approaches in the following part.

2.1   Traditional Image Inpainting

Before deep learning was widely used, the traditional image inpainting methods fell into diffusion-based and patch-based methods. The diffusion-based image inpainting approaches are based on the variation method and the Partial Differential Equation (PDE), which propagate the features from boundary regions to the inside regions in the isophotes direction in an iterative manner [2]. These methods are suitable for restoring small cracks or scratches. For the image with significant structure change, they often construct blurry edges or artifacts due to the lack of overall understanding of the semantics within the image. Meanwhile, these methods always need an iterative process to diffuse the pixels gradually, which is time-consuming.

The patch-based methods are also called exemplar-based approaches. These methods are derived from texture synthesis [6]. During the inpainting process, the surrounding patches of holes are regarded as a reference; these holes are filled in with similar pixels within the reference patches. Criminisi *et al.* proposed a priority mechanism to select the best surrounding patches, and these patches are used to fill the holes [4]. To reduce the high computation cost of the best patch searching, Patchmatch [1] proposed a fast randomized algorithm that applied Nearest Neighbour Fields (NNF). This approach uses an iterative searching mechanism for finding the best-matching patch. Patch-based methods assume that the holes have similar content to the unbroken area. They fill the holes by searching for the best candidate patches from the known region. Patch-based approaches have shown to be more efficient and accurate than previous diffusion-based ones, especially on an image with structure replication. However, if the targeted region has an independent texture from the rest regions, these methods struggle to generate good results.

2.2   Deep Learning-Based Image Inpainting

The emergence of deep learning and adversarial training has pushed forward image inpainting&outpainting to a great extent [9,18,22]. The utilization of deep generative networks has the advantage of learning more global context from a distance than traditional methods. Context encoder [18] is the first GAN-based work with an encoder-decoder model. However, the context encoder cannot handle images with fine details and is limited to square box masks. Iizuka [10] proposed using a discriminator with global and local branches to handle the high-resolution inpainting. The contextual attention [24] is introduced for filling the hole using the most similar patches from the background. Liu *et al.* [14] proposed a novel convolution operation called partial convolutional layer for generating plausible contents under free-form masks. Yu *et al.* [25] improved partial convolution with a learnable mask-update mechanism called gated convolution, and they adopted a coarse-to-fine architecture to synthesize high-quality results. To deal with the inefficiency of learning image texture and structure with a unified network, EdgeConnect [17] uses an edge completion model prior to the inpainting networks. In [15], they encodes the shallow layer features via their structure and texture branches, and they aggregate the output features into the decoder after equalization. PEN-Net [26] built a feature pyramid to obtain the enriched features through a non-local module. HiFill [23] applied the multi-scale attention module to learn the contexts from multi-level features. Unlike the above methods, we focus on building a model with enriched features to model images' texture and structure better.

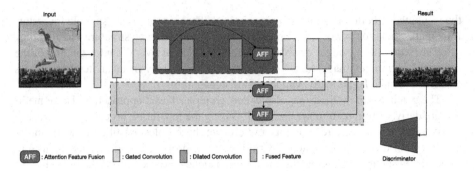

Fig. 1. Architecture of our proposed method. For simplicity, we only show the generator with single stage. The proposed Skip Connection Attention Fusion (SCAF) and Dilated Convolution Residual (DCR) is shown in the figure, and we apply Attention Feature Fusion (AFF) [5] within these two modules.

3   Method

This section introduces our model design and the components in detail. Section 3.1 includes the overall architecture of our model. In Sect. 3.2, we first revisit the attention feature fusion layer and then highlight how we integrate this mechanism with our model from two perspectives. The optimization objectives are elaborated in Sect. 3.3.

3.1   Architecture

In a modern inpainting task, given a color image I multiplied with a binary mask m, the masked results can be represented as $I' = I \odot m$. This image I' is concatenated with the mask as the final input $x = concat(I', m)$. A typical inpainting model uses the generative adversarial network (GAN) that consists of a generator $f_{\theta G}(\cdot)$ and a discriminator network $g_{\theta D}(\cdot)$. The final inpainted image is denoted as $I_o = f_{\theta G}(x)$. During training, each pair of inputs consists of one image from the selected dataset and one randomly generated mask.

The overall model is shown in Fig. 1. We build our algorithms on top of the baseline model from [25]. This baseline is a coarse-to-refine model with an attention branch and dilation branch in the refining stage. In the first stage, they generate a coarse result that mainly focuses on the global structure within the image. This first stage output is then fed into the next stage for refining the details within the global completion. Both stages applied the U-Net architecture constructed by stacking multiple fully convolutional layers. The convolutional layers used in the model are gated-convs, which have better results on free-from inpainting. The main problem of this architecture is the insufficient learning of global context and local details. We propose to address this problem by fusing features on different levels. According to the learning process of an encoder-decoder model, the low-level features from the encoder contain a more delicate texture, and the high-level features from the decoder have rich semantic structure. As shown in Fig. 1, we fuse the low-level feature to its corresponding high-level feature through skip-connection. We use the Attention Feature Fusion layer to guarantee this fusion without inconsistency. Meanwhile, dilated convolutions can bring more semantics. We further enhance the semantics aggregation by applying the Attention Feature Fusion in the multi-dilated residual learning, see Fig. 1.

More details about Attention Feature Fusion and the integration are introduced in the following part.

3.2 Feature Fusion Inpainting

In this section, we first revisit the attention feature fusion technique initially designed for image classification and semantic segmentation [5]. In the following part, we elaborate on how this feature fusion technique can benefit image inpainting and how we integrate it into our model from two perspectives.

Attention Feature Fusion. Feature fusion is a vital strategy to increase further the representation power of deep convolutional networks (DCNs). It has been an important research field and widely adopted into many popular models. Previous fusion strategies commonly exploit simple linear operations such as addition or concatenation. These strategies lack non-linear aggregation of features and are hence not optimal for representative learning. Inspired by the global channel attention mechanism [8] and [16], the proposed attention feature fusion framework is based on a novel Multi-scale channel attention mechanism (MS-CAM) [5]. Given two different features X and Y to be fused. The fusion process can be depicted as:

$$F = M(X + Y) \otimes X + (1 - M(X + Y)) \otimes Y, \tag{1}$$

where M is the MS-CAM, the element-wise addition is processed on the two features, $\otimes$ denotes the element-wise multiplication. The MS-CAM combines the global and local features within the attention mechanism. It is constructed with point-wise convolutions (PWConv) by varying the spatial pooling size along the channel dimension. The architecture of MS-CAM is shown in Fig. 2. There are two branches within the module: the global channel context $g(X) \in \mathbb{R}^{C \times 1 \times 1}$ and the local channel context $L(X) \in \mathbb{R}^{C \times H \times W}$. The global context is inspired by [8] and

$$G(X) = \mathcal{B}(PWConv_2^g(ReLU(\mathcal{B}(PWConv_1^g(GAP(X)))))), \tag{2}$$

where GAP is the global average pooling, $\mathcal{B}$ is the Batch Normalization (BN) [11]. The $PWConv$ is used for exploiting less parameters. For the local context, is can be depicted as:

$$L(X) = \mathcal{B}(PWConv_2^l(ReLU(\mathcal{B}(PWConv_1^l(X))))). \tag{3}$$

The output context has the same shape as the input feature, so the local information is well-preserved. For the input feature X, the feature after the attention computation $X' \in \mathbb{R}^{C \times H \times W}$ can be represented as:

$$X' = X \otimes W(X) = X \otimes Sigmoid(L(X) \oplus g(X)), \tag{4}$$

where $W \in \mathbb{R}^{C \times H \times W}$ is the computed weights and $\oplus$ is the broadcasting addition.

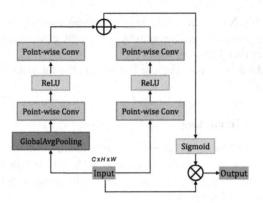

Fig. 2. The MS-CAM module [5].

Skip Connection Attention Fusion. Skip connection is widely used in many encoder-decoder models, such as semantic segmentation and object detection. This technique helps pass long-range information from encoder to decoder [20]. Features from the encoder are rich in low-level texture, and as the network goes deep, these textures are attenuated by the successive convolutions and poolings. Meanwhile, the learned features from the decoder have rich high-level semantic context. Using skip connections could help obtain a more robust feature with both low-level textures and high-level semantics and thus benefit the further tasks. The vanilla skip connection was proved that have a limited effect on the inpainting task as in [25]. The reason is mainly that a large part of masked regions is involved in the low-level features, and it is hard to extract helpful texture information (e.g., details, edges, colors) through the simple fusion operation, e.g., concatenation. More experimental results can be found in Sect. 4.3. We propose to use AFF as our fusion strategy for its advantage in integrating features with long distances. As shown in Fig. 1, a feature from the encoder and its corresponding feature in the decoder are fed into the AFF module as two distinct inputs. We follow the fusion process mentioned above and generate the enhanced feature map with attention computed on both low-level and high-level features. Then, we concatenate this feature with the high-level input feature as the final decoder feature. Ultimately, we archive to compensate for the original decoder features with fine textures from the encoder.

Dilated Convolution Residual. At the same time, the dilated convolution is also a vital component of the image inpainting model. At first, Iizuka [10] proposed to use dilated convolution in image inpainting to increase the receptive field. Many following works find that more sophisticated dilated architectures could generate better results [25,27]. In [25], they adopted a series of dilated convolution blocks with increasing dilation rates, which is as the same framework in semantic segmentation [3]. This dilation architecture could better model the multi-scale objects in the image and help generate a good semantic initial for the late decoder. We argue that for image inpainting, a more dedicated treatment is needed for a better fusion of these features with different receptive fields. Thus, we use AFF as a residual for the dilated block series. We found this simple addition help to improve the understanding of image global semantics and hence generates results with visually meaningful patches. Initially, the dilation blocks with dilation rates 2, 4, 8, and 16 are in a sequential arrangement. Inspired by residual learning [7], we use AFF to fuse these different receptive field features. The dilated convolution residual is implemented using the input of dilation blocks as identity mapping and the output of the last dilated convolution layer as learned residual. Finally, the two features are fused through the attention feature fusion module.

3.3 Training

We introduce the training details in this section. Our model is trained in the GAN fashion, including a generator and a discriminator. The generator is a commonly used encoder-decoder model described above, and we apply the patch discriminator [25] for stable and fast training. The objective function includes reconstruction loss and adversarial loss. The total loss function is:

$$\mathcal{L} = \mathcal{L}_{adv}(G, D) + \alpha \mathcal{L}_{re}(G), \tag{5}$$

where α is the hyperparameter for weight among different losses, we adopted $\alpha = 1$ in our experiments.

Reconstruction Loss. For fast convergence and accuracy, we choose the $\mathcal{L}_1$ distance to guarantee that the reconstructed image is as similar as possible to the ground truth. The $\mathcal{L}_{re}$ can be calculated with input I and the generated output I':

$$\mathcal{L}_{re}(G) = ||I' - I||_1. \tag{6}$$

Adversarial Loss. To further ensure the generated results have fine-grained details and semantically meaningful structure, we follow [25] to use SN-PatchGAN loss. Also, this loss helps to stabilize the training of the discriminator. The overall adversarial loss can be formulated as:

$$\mathcal{L}_G = -\mathbb{E}_{z \sim p_z, y \sim p_{data}} D\left(G\left(z\right), y\right), \tag{7}$$

and

$$\mathcal{L}_D = \mathbb{E}_{x \sim p_{data(x)}} ReLU\left(\mathbb{1} - D\left(x\right)\right) + \mathbb{E}_{z \sim p_{z(z)}} ReLU\left(\mathbb{1} + D\left(G\left(z\right)\right)\right), \tag{8}$$

where G and D denote the generator the discriminator network, respectively, and z denotes the input.

4   Experiments

We evaluate our proposed approach on Places2 [28] and CelebA-HQ faces [12]. During training, the images are with 256×256, and we use a single NVIDIA RTX 2080Ti

GPU to train our model with a batchsize of 8. For Places2, we adopt random-crop before sending the data into the model, and for CelebA-HQ, we downsample the data to 256 × 256 for training. We utilize the same arbitrary-shape mask generation process as in [25]. The whole model is built on TensorFlow v1.15, CUDNN v7.6.5, CUDA v10.2. We apply Adam optimizer with a learning rate of 0.0001 for both generator and discriminator. In Sect. 4.1 and Sect. 4.2, we show the qualitative and quantitative results compared with other SOTA methods that include CA [24], EC [17], Gated [25] and HiFill [23]. In the last part, we introduce the ablation study on our model, elaborating the importance of our proposed skip connection attention fusion and dilated convolution residual.

4.1 Qualitative Results

(a) Masked   (b) CA [17]   (c) EC [26]   (d) Gated [25]   (e) HiFill [23]   (f) Ours

Fig. 3. Qualitative comparisons on Places2 validation dataset. Images are in 512 × 512. Best viewed by zooming-in.

In Fig. 3, we show the visual comparison results. We obtain the results on Places2 dataset with a resolution of 512 × 512. The input of masked images is shown in Fig. 3(a). We choose masks with arbitrary shapes and show the results of object removal, people deletion, and free-form inpainting. The results show that CA tends to generate artifacts with irregular patches. EC could generate sound structure due to its edge completion network, but the blur results indicate that it struggled with filling in high-frequency details. For Gated, noisy patches still exist, i.e., the dissonant grass patches on the third row. The results of HiFill have shown that it could generate good textures, while these details are usually duplicated. We argue this phenomenon is because their model lacks modeling the image structure, and it is hard to organize the patterns well. In contrast, our model could learn the enriched features with image texture and structure through attention feature fusion. The last column's results show that our model could fill the hole with plausible content without blur and artifacts.

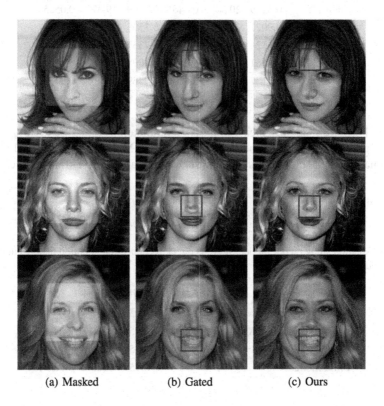

(a) Masked   (b) Gated   (c) Ours

Fig. 4. Qualitative results on CelebA-HQ val set (with image resolution 256 × 256).

We also show our results on CelabA-HQ faces in Fig. 4. We compare our method with Gated [25], showing that our methods could synthesize plausible content with meaningful human facial contour and fine details (e.g. nose in the second row and teeth in the last row).

4.2 Quantitative Results

The quantitative comparison is conducted on Places2 dataset with free-form and center square masks. For a fair comparison, we choose images with 256 × 256 from Places2 validation set. These images are first cropped into 512×512 randomly and then resized into 256×256. We obtain the free-from masks from the commonly-used mask

Table 1. Quantitative results on the Places2 validation dataset. Images are in 256 × 256. For the free-from mask, we adopt the settings from [14]. Up-arrow (↑) indicates higher score is better, down-arrow (↓) indicates lower is better.

Mask type	free-form mask			center square mask		
Metrics	SSIM↑	PSNR↑	L1↓	SSIM↑	PSNR↑	L1↓
CA [25]	0.8299	20.54	0.0461	0.7399	19.47	0.0505
EC [17]	0.9021	24.05	0.0382	**0.7505**	**20.53**	0.0546
Gated [25]	0.9124	24.23	0.0270	0.7357	19.17	0.0519
HiFill [23]	0.8561	21.86	0.0409	0.7008	19.06	0.0553
ours	**0.9137**	**24.36**	**0.0270**	0.7469	20.10	**0.0476**

set from [14]. We randomly pick 100 masks for validation from 20% to 30% mask to image ratio. We use standard metrics that include SSIM [21], PSNR, and $L1$. Table 1 shows that our method outperforms other state-of-the-art on most of the metrics.

4.3 Ablation Study

Table 2. Ablation study on our model. SCAF is our skip connection attention fusion, and DCR is our dilated convolution residual.

Model	SCAF	DCR	SSIM↑	PSNR↑	L1↓
Skip-connection w/o AFF			0.9531	27.35	0.0151
Skip-connection w/ AFF	✓		0.9562	27.65	0.0149
Dilated conv residual		✓	0.9575	27.78	0.0145
Final model	✓	✓	**0.9596**	**28.12**	**0.0140**

To evaluate our skip connection attention fusion (SCAF) and dilated convolution residual (DCR), we have done experiments and show the effectiveness of these two components. In these experiments, we choose a subset of images from Places2 validation set in the resolution 256 × 256. Random masks from [14] are used. From Table 2, we could find that both SCAF and DCR modules are essential for our model. The complete model with both modules reaches favorable results.

5   Conclusion

This paper proposes a feature fusion-based method for high-quality image inpainting. Current methods cannot model the image texture and structure appropriately, resulting in blur and artifacts in the synthesized results. We leverage the attention feature fusion technique to address this issue by passing the shallow details to deep semantics and building a modified multi-dilated residual block. In this way, our model could obtain features with rich global and local information. Extensive experiments on the benchmark datasets have shown the outperformance of our approach to the other state-of-the-art methods.

Acknowledgements. The work was funded by National Natural Science Foundation of China under no. 61876154 and no. 61876155; and Jiangsu Science and Technology Programme (Natural Science Foundation of Jiangsu Province) under no. BE2020006-4.

References

1. Barnes, C., Shechtman, E., Finkelstein, A., Goldman, D.B.: PatchMatch: a randomized correspondence algorithm for structural image editing. ACM Trans. Graph. (ToG) **28**(3), 24 (2009)
2. Bertalmio, M., Sapiro, G., Caselles, V., Ballester, C.: Image inpainting. In: Proceedings of the 27th Annual Conference on Computer Graphics and Interactive Techniques, pp. 417–424 (2000)
3. Chen, L.C., Papandreou, G., Schroff, F., Adam, H.: Rethinking atrous convolution for semantic image segmentation. arXiv preprint arXiv:1706.05587 (2017)
4. Criminisi, A., Pérez, P., Toyama, K.: Region filling and object removal by exemplar-based image inpainting. IEEE Trans. Image Process. **13**(9), 1200–1212 (2004)
5. Dai, Y., Gieseke, F., Oehmcke, S., Wu, Y., Barnard, K.: Attentional feature fusion. In: Proceedings of the IEEE/CVF Winter Conference on Applications of Computer Vision, pp. 3560–3569 (2021)
6. Efros, A.A., Leung, T.K.: Texture synthesis by non-parametric sampling. In: Proceedings of the IEEE International Conference on Computer Vision, vol. 2, pp. 1033–1038. IEEE (1999)
7. He, K., Zhang, X., Ren, S., Sun, J.: Deep residual learning for image recognition. In: Proceedings of the IEEE/CVF Conference on Computer Vision and Pattern Recognition, pp. 770–778 (2016)
8. Hu, J., Shen, L., Sun, G.: Squeeze-and-excitation networks. In: Proceedings of the IEEE/CVF Conference on Computer Vision and Pattern Recognition, pp. 7132–7141 (2018)
9. Huang, K., Hussain, A., Wang, Q.F., Zhang, R.: Deep Learning: Fundamentals, Theory and Applications, vol. 2. Springer, Cham (2019). https://doi.org/10.1007/978-3-030-06073-2
10. Iizuka, S., Simo-Serra, E., Ishikawa, H.: Globally and locally consistent image completion. ACM Trans. Graph. (ToG) **36**(4), 1–14 (2017)
11. Ioffe, S., Szegedy, C.: Batch normalization: accelerating deep network training by reducing internal covariate shift. In: International Conference on Machine Learning, pp. 448–456. PMLR (2015)
12. Karras, T., Aila, T., Laine, S., Lehtinen, J.: Progressive growing of GANs for improved quality, stability, and variation. In: 2018 Proceedings of International Conference on Learning Representations (ICLR) (2018). https://iclr.cc/Conferences/2018. International Conference on Learning Representations, ICLR; Conference date: 30-04-2018 Through 03-05-2018
13. Liao, L., Xiao, J., Wang, Z., Lin, C.-W., Satoh, S.: Guidance and evaluation: semantic-aware image inpainting for mixed scenes. In: Vedaldi, A., Bischof, H., Brox, T., Frahm, J.-M. (eds.) ECCV 2020. LNCS, vol. 12372, pp. 683–700. Springer, Cham (2020). https://doi.org/10.1007/978-3-030-58583-9_41
14. Liu, G., Reda, F.A., Shih, K.J., Wang, T.C., Tao, A., Catanzaro, B.: Image inpainting for irregular holes using partial convolutions. In: Proceedings of the European Conference on Computer Vision (ECCV), pp. 85–100 (2018)

15. Liu, H., Jiang, B., Song, Y., Huang, W., Yang, C.: Rethinking image inpainting via a mutual encoder-decoder with feature equalizations. In: Vedaldi, A., Bischof, H., Brox, T., Frahm, J.-M. (eds.) ECCV 2020. LNCS, vol. 12347, pp. 725–741. Springer, Cham (2020). https://doi.org/10.1007/978-3-030-58536-5_43

16. Liu, W., Rabinovich, A., Berg, A.C.: ParseNet: looking wider to see better. arXiv preprint arXiv:1506.04579 (2015)

17. Nazeri, K., Ng, E., Joseph, T., Qureshi, F., Ebrahimi, M.: EdgeConnect: structure guided image inpainting using edge prediction. In: The IEEE International Conference on Computer Vision (ICCV) Workshops (2019)

18. Pathak, D., Krahenbuhl, P., Donahue, J., Darrell, T., Efros, A.A.: Context encoders: feature learning by inpainting. In: Proceedings of the IEEE/CVF Conference on Computer Vision and Pattern Recognition, pp. 2536–2544 (2016)

19. Ren, Y., Yu, X., Zhang, R., Li, T.H., Liu, S., Li, G.: StructureFlow: image inpainting via structure-aware appearance flow. In: Proceedings of the IEEE International Conference on Computer Vision, pp. 181–190 (2019)

20. Ronneberger, O., Fischer, P., Brox, T.: U-net: convolutional networks for biomedical image segmentation. In: Navab, N., Hornegger, J., Wells, W.M., Frangi, A.F. (eds.) MICCAI 2015. LNCS, vol. 9351, pp. 234–241. Springer, Cham (2015). https://doi.org/10.1007/978-3-319-24574-4_28

21. Wang, Z., Bovik, A.C., Sheikh, H.R., Simoncelli, E.P.: Image quality assessment: from error visibility to structural similarity. IEEE Trans. Image Process. **13**(4), 600–612 (2004)

22. Yao, K., Gao, P., Yang, X., Sun, J., Zhang, R., Huang, K.: Outpainting by queries. In: Avidan, S., Brostow, G., Cissé, M., Farinella, G.M., Hassner, T. (eds.) ECCV 2022. LNCS, vol. 13683, pp. 153–169. Springer, Cham (2022). https://doi.org/10.1007/978-3-031-20050-2_10

23. Yi, Z., Tang, Q., Azizi, S., Jang, D., Xu, Z.: Contextual residual aggregation for ultra high-resolution image inpainting. In: Proceedings of the IEEE/CVF Conference on Computer Vision and Pattern Recognition (2020)

24. Yu, J., Lin, Z., Yang, J., Shen, X., Lu, X., Huang, T.S.: Generative image inpainting with contextual attention. In: Proceedings of the IEEE/CVF Conference on Computer Vision and Pattern Recognition, pp. 5505–5514 (2018)

25. Yu, J., Lin, Z., Yang, J., Shen, X., Lu, X., Huang, T.S.: Free-form image inpainting with gated convolution. In: Proceedings of the IEEE International Conference on Computer Vision, pp. 4471–4480 (2019)

26. Zeng, Y., Fu, J., Chao, H., Guo, B.: Learning pyramid-context encoder network for high-quality image inpainting. In: Proceedings of the IEEE/CVF Conference on Computer Vision and Pattern Recognition, pp. 1486–1494 (2019)

27. Zeng, Y., Fu, J., Chao, H., Guo, B.: Aggregated contextual transformations for high-resolution image inpainting. IEEE Trans. Vis. Comput. Graph. (2022)

28. Zhou, B., Lapedriza, A., Khosla, A., Oliva, A., Torralba, A.: Places: a 10 million image database for scene recognition. IEEE Trans. Pattern Anal. Mach. Intell. **40**, 1452–1464 (2017)

Towards Accurate Alignment and Sufficient Context in Scene Text Recognition

Yijie Hu[1], Bin Dong[2], Qiufeng Wang[1(✉)], Lei Ding[2], Xiaobo Jin[1], and Kaizhu Huang[3]

[1] Xi'an Jiaotong-Liverpool University, Suzhou, China
`Yijie.Hu20@student.xjtlu.edu.cn`, {`Qiufeng.Wang,Xiaobo.Jin`}`@xjtlu.edu.cn`
[2] Ricoh Software Research Center (Beijing) Co., Ltd., Beijing, China
{`Bin.Dong,Lei.Ding`}`@cn.ricoh.com`
[3] Duke Kunshan University, Kunshan, China
`kaizhu.huang@dukekunshan.edu.cn`

Abstract. Encoder-decoder framework has recently become cutting-edge in scene text recognition (STR), where most decoder networks consist of two parts: an attention model to align visual features from the encoder for each character, and a linear or LSTM-based model to predict label sequence. However, it is difficult for these attention models to obtain accurate alignment, and linear or LSTM model usually captures limited context. To emphasize the role of character feature alignment, we separate the attention alignment module from the decoder network in this work, forming an Encoder-Alignment-Decoder framework. Under this framework, we propose a deformable attention based model to accurately align visual features of each character. In this alignment model, we explicitly learn the spatial coordinate information of each character from the input reading order sequence and optimize it with learnable sampled offsets in the attention block to obtain accurate aligned features. To address the lack of context, we explore transformer-based decoder to capture global context by multi-head attention, where a mask matrix is integrated to keep attention weights focused on the relevant context during the decoding. Extensive experiments demonstrate the effectiveness of the Encoder-Alignment-Decoder framework in STR, achieving better performance than other language free methods with significant improvement on most benchmark STR datasets, and obtain the state-of-the-art performance on several datasets by integrating a language model.

Keywords: Scene text recognition · Character alignment · Deformable attention · Mask attention

1 Introduction

Scene text recognition (STR), which aims to read the text content in images, has attracted much attention. By regarding STR as a sequence-to-sequence task, the encoder-decoder framework is widely adopted and has made great progress

M. Tanveer et al. (Eds.): ICONIP 2022, LNCS 13625, pp. 705–717, 2023.
https://doi.org/10.1007/978-3-031-30111-7_59

aarrett   jarrett

(a)   (b)

Fig. 1. (a) Parallel attention alignment results of the character "J", where most of the alignment positions (red block) are outside of the character, leading to the misclassification of "j" − > "a". (b) Deformable attention alignment results, where both sampled points (blue) and reference points (yellow) are located on the character. (Color figure online)

[6,18,23,24,32], where the encoder extracts visual features from input images, and the decoder predicts the label sequence with an attention mechanism [2] aiming to align the visual features of each character.

Although these methods have made promising progress, they tend to fail when images have complex background clutter and severe distortion. We argue that these methods generally suffer from the following two issues: **(1) Misalignment issue**, note that the attention module does not obtain accurate alignment weights on the feature map for each character, resulting in the character label prediction failed in the decoder [24,28,33], e.g., Fig. 1(a); **(2) Insufficient context issue**, the decoder does not capture sufficient relevant context in the decoder, causing recognition failure [3,17,24,28,31].

In this work, we separate the attention alignment module from the decoder to emphasize the importance of character feature alignment, forming an Encoder-Alignment-Decoder framework, as is shown in Fig. 2. In this framework, the encoder extracts multi-scale feature maps from input images. In the alignment module, we propose to adopt the deformable attention mechanism [34] for accurate feature alignment. This mechanism only focuses on some sampled points per character (e.g., Fig. 1(b)). Consequently, this is a sparse attention method. Moreover, the spatial position of each point is explicitly optimized by a reference point plus its sampling offsets in this framework, which facilitates accurate alignment. For the decoder network, although the vanilla transformer decoder is able to capture the global context via the attention mechanism, it cannot focus on a specific relevant context. We argue that this is unreasonable because the extracted visual features may contain some irrelevant contexts, such as the noisy background in scene text images. To overcome this issue, we integrate an attention mask matrix in the transformer decoder to focus the attention weights on more relevant visual contexts.

2   Related Work

2.1   Scene Text Recognition

Attention-based frameworks have been widely adopted recently in STR [6,18, 24,28], where the attention mechanism replaces CTC strategy to make the alignment learnable. The pioneering work ASTER [24] successfully integrated the Bahdanau attention mechanism into the recognition process of the LSTM decoder, while this attention alignment suffered from the accumulation of historical errors. Wang et al. [28] proposed a convolutional alignment module (CAM), which obtains a 2D attention map to locate the features of each character, and utilized GRU [7] as a decoder to predict each character. Though relatively high performance can be achieved, the efficiency of these methods is usually limited because the recursive decoding process of LSTM or GRU only captures the historical visual contexts at each time step. In order to fully explore global visual information and achieve a faster decoding process, Lyu *et al.* [18] proposed a relation module to capture the dependencies between visual information, and utilized a two-stage decoder to predict all characters in parallel. Recently, many works have proposed semantic reasoning methods to adequately capture linguistic contextual information.

2.2 Deformable Attention

Although transformer [26] has been applied to computer vision tasks owing to its remarkable global and local pairwise relation modelling ability, high computational cost and slow convergence make it hard to apply to some scenarios, such as object detection [5]. Thus, deformbale attention [34] is proposed to address the complexity of the transformer, which improves the convergence of transformer by selecting a small number of keys for each query on the CNN backbone. Though a small amount of feature information is processed in the attention process, it is explicitly optimized in the framework, thereby improving the representational ability of the model.

3 Methodology

We propose an encoder-alignment-decoder framework for scene text recognition, which consists of three components: an encoder network, a deformable attention alignment module (DAAM), and a mask transformer decoder, as shown in Fig. 2. For an input image I, the encoder network aims to extract multi-scale 2D feature maps F. Then, DAAM aligns each character indexed by reading order on the extracted feature map to obtain a character feature sequence G with a maximum length of N, which is finally input to the transformer decoder to predict all character labels parallel lines. In the Transformer decoder, we integrate a mask matrix in the attention mechanism to capture the relevant context guided by the positional encoding during decoding.

3.1 Encoder Network

Given an input image I of $H \times W$, we first use ResNet-50 [9] as the backbone to extract feature maps according to the works [8,24,28]. Then, the last 2D feature

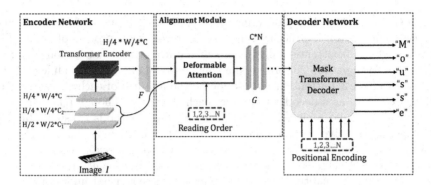

Fig. 2. The overall structure of the proposed method.

map is fed to the transformer encoder unit [26] to capture the global visual context. In summary, the overall process can be formulated as:

$$F = T(R(I)) \in \mathbf{R}^{\frac{H}{4} \times \frac{W}{4} \times C}, \tag{1}$$

where $R(\cdot)$ is ResNet [9], $T(\cdot)$ is Transformer encoder [26], C is the number of channels of the feature map.

3.2 Deformable Attention Alignment Module

In this work, we introduce a new alignment module to improve alignment accuracy by learning sample points, called Deformable Attention Alignment Module (DAAM), which is inspired by deformable attention in object detection [34]. The structure of DAAM is shown in Fig. 3. Following the parallel attention [33], the query comes from a reading order O whose value is a list of character indices $[1, 2, ..., N]$, where N is the maximum length. Based on the embedding $E(O)$, we generate reference points and sampling offsets and attention weights by the following three linear project functions, respectively:

$$P = \text{Sigmoid}(W_r E(O)) \in \mathbf{R}^{2 \times N}, \tag{2}$$

$$\Delta P = W_o E(O) \in \mathbf{R}^{2 \times N \times S \times L \times K}, \tag{3}$$

$$A = \text{Softmax}(W_a E(O)) \in \mathbf{R}^{N \times S \times L \times K}, \tag{4}$$

where W_r, W_o, W_a are three trainable weights. For each character, we generate a reference point $P(p_x, p_y)$ with normalized coordinates, where $(0, 0)$ and $(1, 1)$ represent the top-left and bottom-right image corners, respectively. In the sampling offsets, we exploit the multi-head [26] structure and multi-scale features [34] to learn more powerful representations, where S and L denote the number of heads and scales, respectively, and K denotes the number of sampling offsets for each attention head and scale feature map, thus we sample $R = S \times L \times K$ offsets for each reference point. Similarly, the scalar attention weight A_{nslk} is normalized to satisfy the condition $\sum_{l=1}^{L} \sum_{k=1}^{K} A_{nslk} = 1$ for the nth character in the sth attention head.

Once we have the reference points and their sampling offsets, we can construct their sampling points for each character and then combine the attention weights to generate the final aligned character features g_n from the feature maps by

$$g_n = \text{DeformAttn}(O_n, \{F^l\}_{l=1}^{L}) \tag{5}$$

$$= \sum_{s=1}^{S} W_s \left[\sum_{l=1}^{L} \sum_{k=1}^{K} A_{nslk} W_s' F^l \left(\phi_l \left(P_n \right) + \Delta P_{nslk} \right) \right], \tag{6}$$

where s, l, k index the attention head, feature scale level and sampling offset, respectively. $W_s' \in \mathbf{R}^{C_s \times C}$ and $W_s \in \mathbf{R}^{C \times C_s}$ are possible Weights of trained s^{th} attention heads, where $C_s = C/S$. P_n represents the reference point of the nth character, ΔP_{nslk} and A_{nslk} represent the kth offset and attention weight, respectively In s^{th} head and l^{th} scale feature maps, for n^{th} characters. The multi-scale feature map is represented as $\{F^l\}_{l=1}^{L}$, where $F^l \in \mathbf{R}^{C \times H_l \times W_l}$. $\phi_l \left(P_n \right)$ is to rescale the normalized coordinate P_n to the lth scale feature map, and then add the sampling offset Δp_{nslk} generates a sample point. Since $\phi_l \left(P_n \right) + \Delta P_{nslk}$ is a decimal, bilinear interploation is applied when calculating $F^l(\cdot)$ interpolation. Compared to other alignment methods [28,33], the number of sampling points per character and in each attention head and feature map is usually much smaller than the image size ($K = 16$), which helps to obtain more related aligned features (e.g., less interference from noisy background features, as shown in Fig. 1).

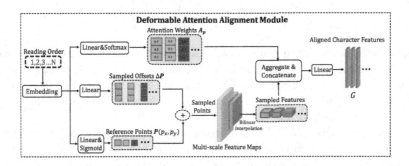

Fig. 3. Illustration of the deformable attention alignment module.

3.3   Mask Transformer Decoder

After obtaining aligned character features, existing works use linear classifiers [8,33] or LSTM-based models [6,24] to predict character labels. However, they usually only capture restricted context during decoding (e.g. in LSTM only local history information). To transfer this problem, we employ a Transformer decoder that captures the global context through a multi-head attention mechanism. Furthermore, we find that the attention mechanism in vanilla Transformer tends to assign credit to the whole context under a balanced distribution (e.g., Fig. 5b), resulting in some irrelevant capture from noisy backgrounds Context. To overcome this problem, we propose to integrate a mask attention in the transformer decoder. After inserting the mask matrix in the attention mechanism, the attention score becomes the focus to focus on more relevant contexts (e.g., Fig. 5a). Current works focus on modeling semantic context(e.g., ABINet [8] and SRN [33]), we argue that visual context (e.g., shape context) also exists in scene text recognition. Inspired by this, we directly input the aligned features instead of embedding character labels into the converter decoder.

Figure 4 shows an overall structure of the proposed mask transformer decoder, where the query Q is a predefined positional encoding to represent the character order information. The key K and value V are the visual features $G = \{g_1, g_2, g_3...g_N\}$ from the DAAM, then visual features are directly fed into the multi-head attention block:

$$Q_h^{'} = W_h^Q Q, K_h^{'} = W_h^K K, V_h^{'} = W_h^V V, \tag{7}$$

where W_h^Q, W_h^K, $W_h^V \in \mathbf{R}^{C \times C}$ are the learnable parameter matrices of the hth head in multi-head attention process, $Q \in \mathbf{R}^{N \times C}$ is positional encoding and $K, V \in \mathbf{R}^{N \times C}$ are visual features. To make the attention focus more relevant visual contexts, we integrate a mask matrix M in the attention score calculation:

$$M_{ij} = \begin{cases} 0, & i \neq j \\ -\infty, & i = j \end{cases} \tag{8}$$

$$S_h = \mathrm{softmax} \left(\frac{Q_h^{'} K_h^{'\top}}{\sqrt{C}} + M \right) V_h^{'}, \tag{9}$$

where S is the output of the multi-head attention. After the feed forward network (i.e., 'Add&Normalize', 'MLP' and 'Add&Normalize' in Fig. 4), final output of this layer of the transformer decoder, which will input to the next layer as the query matrix Q while the K and V keep the same as those in the first layer. This process can be repeated T times to generate a T-layer transformer decoder. The mask matrix $M \in \mathbf{R}^{N \times N}$ is a diagonal matrix, where the diagonal elements are minus infinite to force diagonal weights to zero in the softmax function, then the whole attention score is re-assigned. As the value matrix V represents the character-level features in the sequence, higher attention score means more important to this character in the contexts.

3.4   Loss Function

The model is trained by the end-to-end manner to minimize the cross-entropy loss function as

$$L = -\frac{1}{N} \sum_{n=1}^{N} \log p(y_n \mid I, \theta), \tag{10}$$

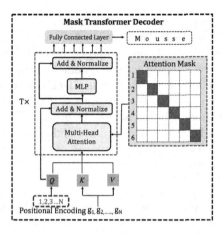

Fig. 4. The structure of the proposed mask decoder.

where I is the input image, θ represents all trainable parameters in the network, $\{y_n\}_{n=1}^N$ are the ground truth labels of a character sequence, and N is the maximum length of the sequence.

4 Experiments

4.1 Datasets and Implementation Details

Datasets. Model is trained on two synthetic datasets, Synth90K [11] and Synth-Text [12]. We evaluate the performance of the proposed method on six scene text benchmarks, including three regular datasets (i.e., IIIT5K-words (IIIT5k) [19], ICDAR2013 (IC13) [14] and Street View Text (SVT) [27]), and three irregular datasets (i.e., ICDAR2015 (IC15) [13], Street View Text-Perspective (SVTP) [20] and CUTE80 (CUTE) [22]). The recognition covers 37 character classes including a–z, 0–9, and an end-of-sequence symbol 'EOS'.

Model Training. Data augmentations such as random rotation, color jittering and perspective distortion are randomly performed on the input image during the training process. Adam optimizer with initial learning rate $1e^{-4}$ is used to optimize the model, and the learning rate is decayed to $1e^{-5}$ after 8 epochs. The batch size is set as 384 in all experiments.

4.2 Experimental Results

Different Alignment Modules. To show the effectiveness of the proposed DAAM, we compare the DAAM with two popular alignment methods (i.e., parallel attention [8,33] and CAM [28]) with the same encoder network (ResNet-50 and two-layer transformer-encoder), and the results are shown in Table 1. In the DAAM, we adopt multi-scale features (i.e., $L = 3$) in this experiment. To remove the interaction of the decoder network, we simply adopt a linear classifier as the decoder network to obtain each character label on the aligned character features directly. As is shown in Table 1, our proposed method outperforms other two methods at least 0.5% on average accuracy, demonstrating its effectiveness. For the CAM and parallel attention, they either lack of ordering information, or lack spatial information. As the DAAM aims to learn both spatial and order information of each character in the alignment process, it is worth noticing that the performance of the DAAM is significantly higher on irregular datasets such as IC15, SVTP and CUTE, where the majority of samples are curved texts and have complicated backgrounds.

Table 1. Experimental results (%) of comparing different alignment modules with the same encoder network and decoder network (linear classifier).

Method	IIIT5K	IC13	SVT	IC15	SVTP	CUTE	Avg
Parallel Attention	93.7	94.2	89.6	80.6	82.3	85.1	88.8
CAM	**94.5**	93.8	89.6	80.3	81.6	86.5	88.9
DAAM	93.9	**95.2**	**89.8**	**81.2**	**83.4**	**87.8**	**89.4**

Different Decoders. To show the effectiveness of the proposed mask transformer decoder, we compare different decoders as shown in Table 2. From Table 2, we can see that the linear classifier performs the worst because it does not capture any contexts during the decoding. By considering the historical contexts, the LSTM-decoder improves the average accuracy by 0.6%. Vanilla transformer decoder continually boosts average accuracy by 0.3%. Finally, after integrating the mask attention, the proposed decoder boosts the average accuracy significantly from 89.4% to 90.7%, achieving the best performance on the majority of benchmark datasets, verifying its effectiveness of the mask attention.

Table 2. Experimental results (%) of comparing different decoders with the same encoder network and DAAM.

Method	IIIT5K	IC13	SVT	IC15	SVTP	CUTE	Avg
Linear Classifier	93.9	95.2	89.8	81.2	83.4	87.8	89.4
LSTM-based	94.9	95.0	89.5	82.6	82.5	**89.2**	90.0
Vanilla Decoder	94.9	95.7	89.3	82.9	85.0	87.8	90.3
Mask Decoder	**95.1**	**95.8**	**90.3**	**83.4**	**85.3**	88.9	**90.7**

4.3   Ablation Study

Multi-scale Features VS Single-Scale Features. Multi-scale feature maps have been widely adopted in previous STR methods [17,28] to improve the recognition accuracy. Thus, we conduct an experiment to compare the recognition accuracy of single-scale feature maps and multi-scale feature maps in the DAAM. As is shown in the Table 3, it is not surprising that multi-scale features achieve higher recognition accuracy than single-scale features on most of benchmark datasets, and obtain an improvement of 0.4% on the average accuracy. The reason is that multi-scale feature maps allow DAAM to look over multiple sampling points from different resolutions, leading more accurate alignment of character features. On the other hand, the single-scale feature map obtains fewer computational cost due to using fewer sampled points, but higher accuracy than other alignment methods in Table 1.

Table 3. Experimental results (%) of comparing multi-scale and single-scale features in the DAAM.

Method	IIIT5K	IC13	SVT	IC15	SVTP	CUTE	Avg
Single-scale	**94.1**	94.2	88.4	80.8	**83.4**	86.1	89.0
Multi-scale	93.9	**95.2**	**89.8**	**81.2**	**83.4**	**87.8**	**89.4**

Visualization of Attention Mask. In order to show the effectiveness of the proposed mask attention intuitively, we visualize the attention score of mask Transformer decoder (Fig. 5(a)) and vanilla Transformer decoder (Fig. 5(b)), which is calculated by the softmax in Eq. (10). We can see that the original attention score is a very balanced distribution to all characters in the Fig. 5(b), and it becomes more concentrated (Fig. 5(a), deeper color means higher score) after the integration of mask attention. This comparison demonstrates that our proposed mask transformer decoder can capture more relevant contexts than the vanilla decoder.

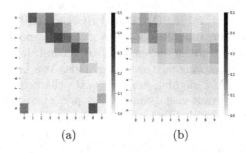

(a)   (b)

Fig. 5. Visualization of the comparison of the attention scores.

4.4   Comparison with State-of-the-Art Methods

We compare the proposed model with previous state-of-the-art methods on six benchmark datasets, and the results are shown in Table 4. Following the

Table 4. Comparisons of scene text recognition accuracy (%) with previous methods. **Bold** represents the best performance and underline represents the second best result.

	Method	IIIT5K	IC13	SVT	IC15	SVTP	CUTE
Lan-Free	Shi *et al.* [23] (CTC)	81.2	89.6	82.7	-	-	-
	Xie *et al.* [30] (ACE)	82.3	89.7	82.6	68.9	70.1	82.6
	Liao *et al.* [16] (FCN)	91.9	91.5	86.4	-	-	-
	Lyu *et al.* [18] (Parallel)	94.0	92.7	90.1	76.3	82.3	86.8
	Fang *et al.* [8] (ABINet, w/o LM)	94.6	94.9	90.4	<u>81.7</u>	84.2	86.5
	Atienza*et al.* [1] (ViTSTR)	88.4	93.2	87.7	78.5	81.8	81.3
	Yang *et.al* [31]	94.7	93.2	88.9	79.5	80.9	85.4
	Lu *et.al* [17] (MASTER)	<u>95.0</u>	**95.3**	<u>90.6</u>	79.4	<u>84.5</u>	<u>87.5</u>
	Ours	**96.0**	<u>95.0</u>	**91.7**	**83.8**	**85.3**	**89.2**
Lan-Aware	Shi *et al.* [24] (ASTER)	93.4	91.8	89.5	76.1	78.5	79.5
	Li *et al.* [15] (SAR)	91.5	91.0	84.5	69.2	76.4	83.3
	Wang *et al.* [28] (DAN)	94.3	93.9	89.2	74.5	80.0	84.4
	Qiao *et al.* [21] (SEED)	93.8	92.8	89.6	80.0	81.4	83.6
	Yu *et al.* [33] (SRN)	94.8	95.5	91.5	82.7	85.1	87.8
	Fang *et al.* [8] (ABINet, with LM)	96.2	**97.4**	93.5	**86.0**	89.3	89.2
	Wang *et al.* [29] (VisionLan)	95.8	95.7	91.7	83.7	86.0	88.5
	He *et al.* [10] (S-GTR)	95.8	96.8	94.1	84.6	87.9	<u>92.3</u>
	Tang *et al.* [25] (VST)	96.1	96.4	93.8	85.4	89.1	**94.8**
	Ours+LM	**96.4**	<u>97.3</u>	**94.4**	<u>85.7</u>	**89.9**	91.0

work [29], we roughly divide the methods into two categories: (1) language free (**Lan-Free**) methods without involving any linguistic information for recognition, (2) language-aware (**Lan-Aware**) methods using implicit linguistic information (e.g., using historical decoded results [24]) or adopting external language model (e.g., FastText [4] in SEED [21] and BCN in ABINet [8]) or training a visual semantic reasoning module (e.g., SRN [33] and VisionLan [29]) to provide robust context information for recognition. In general, language free methods perform worse than those language aware methods. As a Lan-Free method, with the benefits from DAAM aligning more accurate character features and mask attention capturing relevant visual contexts, its performance is even competitive to those Lan-Aware methods. To be specific, our method obtains much better performance than all language free methods on majority of datasets, especially on those irregular datasets (i.e., IC15, SVTP and CUTE), even superior to some language aware methods on several datasets. For example, we obtain the accuracy 89.2% on the dataset of CUTE, which is higher than the recent works using explicit linguistic information (e.g., SRN [33] and VisionLan [29]).

Furthermore, we conduct experiments on integrating a language model (Bidirectional Cloze Network [8]) to provide explicit linguistic information for final prediction. From Table 4, we can see that the integration of BCN language model obtains 0.4 %, 2.3%, 2.7%, 1.9% and 4.6% improvement on III5k, IC13, SVT, IC15 and SVTP datasets, respectively, demonstrating the complementary of language model in our framework. Compared to other lan-aware methods, our model obtains the highest performance on III5K, SVT and SVTP datasets, and the increases are 0.3%, 0.3% and 0.6%, respectively.

5 Conclusions and Future Work

In this work, we separate the alignment module from the decoder network to emphasize the importance of character feature alignment, forming an Encoder-Alignment-Decoder framework for scene text recognition. Under this framework, we propose a Deformable Attention Alignment Module (DAAM) to obtain more accurate aligned character features. In the decoder network, we propose a mask transformer decoder to capture the relevant visual context by integrating a label matrix in the attention mechanism. The proposed method significantly outperforms other language-free methods, and obtains the state-of-the-art performance on several datasets by simply integrating a language model. In the future, we will investigate the possibility of integrating high-level language contexts in decoder networks.

Acknowledgements. The work was funded by National Natural Science Foundation of China under no.61876154 and no.61876155; and Jiangsu Science and Technology Programme under no. BE2020006-4, and "Qing Lan Project" in Jiangsu universities.

References

1. Atienza, R.: Vision transformer for fast and efficient scene text recognition. In: Lladós, J., Lopresti, D., Uchida, S. (eds.) ICDAR 2021. LNCS, vol. 12821, pp. 319–334. Springer, Cham (2021)
2. Bahdanau, D., Cho, K., Bengio, Y.: Neural machine translation by jointly learning to align and translate. In: ICML (2014)
3. Bhunia, A.K., Sain, A., Kumar, A., Ghose, S., Chowdhury, P.N., Song, Y.Z.: Joint visual semantic reasoning: multi-stage decoder for text recognition. In: ICCV, pp. 14940–14949 (2021)
4. Bojanowski, P., Grave, E., Joulin, A., Mikolov, T.: Enriching word vectors with subword information. In: ACL, vol. 5, pp. 135–146 (2017)
5. Carion, N., Massa, F., Synnaeve, G., Usunier, N., Kirillov, A., Zagoruyko, S.: End-to-end object detection with transformers. In: Vedaldi, A., Bischof, H., Brox, T., Frahm, J.-M. (eds.) ECCV 2020. LNCS, vol. 12346, pp. 213–229. Springer, Cham (2020). https://doi.org/10.1007/978-3-030-58452-8_13
6. Cheng, Z., Bai, F., Xu, Y., Zheng, G., Pu, S., Zhou, S.: Focusing attention: towards accurate text recognition in natural images. In: CVPR, pp. 5076–5084 (2017)
7. Cho, K., et al.: Learning phrase representations using rnn encoder-decoder for statistical machine translation. arXiv preprint arXiv:1406.1078 (2014)
8. Fang, S., Xie, H., Wang, Y., Mao, Z., Zhang, Y.: Read like humans: Autonomous, bidirectional and iterative language modeling for scene text recognition. In: CVPR. pp. 7098–7107 (2021)

9. He, K., Zhang, X., Ren, S., Sun, J.: Deep residual learning for image recognition. In: CVPR, pp. 770–778 (2016)
10. He, Y., et al.: Visual semantics allow for textual reasoning better in scene text recognition. In: AAAI, pp. 888–896 (2022)
11. Jaderberg, M., Simonyan, K., Vedaldi, A., Zisserman, A.: Synthetic data and artificial neural networks for natural scene text recognition. In: NIPS (2014)
12. Jaderberg, M., Simonyan, K., Vedaldi, A., Zisserman, A.: Reading text in the wild with convolutional neural networks. In: CVPR, pp. 1–20 (2016)
13. Karatzas, D., et al.: ICDAR 2015 competition on robust reading. In: ICDAR, pp. 1156–1160. IEEE (2015)
14. Karatzas, D., et al.: ICDAR 2013 robust reading competition. In: ICDAR, pp. 1484–1493. IEEE (2013)
15. Li, H., Wang, P., Shen, C., Zhang, G.: Show, attend and read: a simple and strong baseline for irregular text recognition. In: AAAI, pp. 8610–8617 (2019)
16. Liao, M., et al.: Scene text recognition from two-dimensional perspective. In: AAAI, pp. 8714–8721 (2019)
17. Lu, N., et al.: MASTER: multi-aspect non-local network for scene text recognition. Pattern Recogn. **117**, 107980 (2021)
18. Lyu, P., Yang, Z., Leng, X., Wu, X., Li, R., Shen, X.: 2D attentional irregular scene text recognizer. arXiv preprint arXiv:1906.05708 (2019)
19. Mishra, A., Alahari, K., Jawahar, C.: Scene text recognition using higher order language priors. In: BMVC. BMVA (2012)
20. Phan, T.Q., Shivakumara, P., Tian, S., Tan, C.L.: Recognizing text with perspective distortion in natural scenes. In: ICCV, pp. 569–576 (2013)
21. Qiao, Z., Zhou, Y., Yang, D., Zhou, Y., Wang, W.: SEED: semantics enhanced encoder-decoder framework for scene text recognition. In: CVPR, pp. 13528–13537 (2020)
22. Risnumawan, A., Shivakumara, P., Chan, C.S., Tan, C.L.: A robust arbitrary text detection system for natural scene images. Expert Syst. Appl. **41**(18), 8027–8048 (2014)
23. Shi, B., Bai, X., Yao, C.: An end-to-end trainable neural network for image-based sequence recognition and its application to scene text recognition. IEEE Trans. Pattern Anal. Mach. Intell. **39**(11), 2298–2304 (2016)
24. Shi, B., Yang, M., Wang, X., Lyu, P., Yao, C., Bai, X.: ASTER: an attentional scene text recognizer with flexible rectification. IEEE Trans. Pattern Anal. Mach. Intell. **41**(9), 2035–2048 (2018)
25. Tang, X., Lai, Y., Liu, Y., Fu, Y., Fang, R.: Visual-semantic transformer for scene text recognition. In: AAAI (2022)
26. Vaswani, A., et al.: Attention is all you need. In: NIPS, pp. 5998–6008 (2017)
27. Wang, K., Babenko, B., Belongie, S.: End-to-end scene text recognition. In: ICCV, pp. 1457–1464. IEEE (2011)
28. Wang, T., et al.: Decoupled attention network for text recognition. In: AAAI, pp. 12216–12224 (2020)
29. Wang, Y., Xie, H., Fang, S., Wang, J., Zhu, S., Zhang, Y.: From two to one: a new scene text recognizer with visual language modeling network. In: ICCV, pp. 14194–14203 (2021)
30. Xie, Z., Huang, Y., Zhu, Y., Jin, L., Liu, Y., Xie, L.: Aggregation cross-entropy for sequence recognition. In: CVPR, pp. 6538–6547 (2019)
31. Yang, L., Wang, P., Li, H., Li, Z., Zhang, Y.: A holistic representation guided attention network for scene text recognition. Neurocomputing **414**, 67–75 (2020)

32. Yang, X., He, D., Zhou, Z., Kifer, D., Giles, C.L.: Learning to read irregular text with attention mechanisms. In: IJCAI, vol. 1, p. 3 (2017)
33. Yu, D., et al.: Towards accurate scene text recognition with semantic reasoning networks. In: CVPR, pp. 12113–12122 (2020)
34. Zhu, X., Su, W., Lu, L., Li, B., Wang, X., Dai, J.: Deformable DETR: deformable transformers for end-to-end object detection. In: ICLR (2020)

Author Index

M. Tanveer et al. (Eds.): ICONIP 2022, LNCS 13625, pp. 719–722, 2023.
https://doi.org/10.1007/978-3-031-30111-7